APPLIED MULTIVARIATE ANALYSIS

Using Bayesian and Frequentist Methods of Inference

APPLIED MULTIVARIATE ANALYSIS

Using Bayesian and Frequentist Methods of Inference

Second Edition

S. JAMES PRESS

University of California
Riverside, California

DOVER PUBLICATIONS, INC.
Mineola, New York

Bibliographical Note

This Dover edition, first published in 2005, is an unabridged republication of the second edition published by Robert E. Krieger Publishing Company, Malabar, Florida, 1982. (The first edition was published by Holt, Rinehart and Winston, Inc., New York, 1972). The Dover edition includes a newly added solutions manual that begins on page 601. An errata list has been added on page 671.

Library of Congress Cataloging-in-Publication Data

Press, S. James.
Applied multivariate analysis : using Bayesian and frequentist methods of inference / S. James Press.
p. cm.
Originally published: 2nd ed. Malabar, Fla. : R.E. Krieger Pub. Co., 1982. With new solutions manual.
Includes bibliographical references and index.
ISBN 0-486-44236-5 (pbk.)
1. Multivariate analysis. 2. Bayesian statistical decision theory. 3. Inference. I. Title.

QA278.P77 2005
519.5'35–dc22

2005043260

Manufactured in the United States of America
Dover Publications, Inc., 31 East 2nd Street, Mineola, N.Y. 11501

To Grace

for her patience, help, and understanding.

PREFACE

to the First Edition

The material in this book was developed for use as a text in a course at the graduate level in Applied Multivariate Analysis. Since there is more material than might normally be covered in such a course, topics can be selected for study to suit various purposes, and the book may serve various research purposes as well. The format is so designed that by omitting certain of the sections and certain of the proofs of the theorems (such as those found in Complements at the ends of chapters), shorter and less mathematically oriented courses will result.

The book has developed from notes used in a one-quarter course given jointly in the Graduate School of Business and the Statistics Department at the University of Chicago. The course has been taken mainly by students in the social sciences and statistics, and therefore, the examples reflect those disciplines. In that course, most chapters are covered in part (many sections and proofs are deleted). Some of the omitted sections are covered in a Special Topics course. Exercises are provided at the end of each chapter.

The book may be of interest not only to professional statisticians, but also to teachers and research workers in those areas of the behavioral and social sciences where multivariate statistics is heavily applied. The methodology is of course also applicable to the biological and physical sciences although text examples have not been aimed at those targets.

The book is divided into two parts. Part I deals with foundations of multivariate analysis and Part II deals with models and applications. Statisticians may prefer to focus more attention on Part I, whereas researchers and students, whose primary interest lies in the applications, may prefer to concentrate on Part II, referring to Part I only as is required for an understanding of the theory underlying a particular model.

It has been assumed that the reader is already familiar with the techniques of elementary univariate statistics such as those which would be studied in a first-year course. Moreover, although operations with matrices are summarized in Chapter 2, it is assumed in a one-quarter course that the reader already has some familiarity with that subject, so that time spent on this chapter can be minimal. In longer courses where little familiarity of matrices is assumed, the subject of matrix calculus might be examined in some detail.

Multivariate analysis is treated using both sampling theory and Bayesian techniques. Moreover, the Bayesian approach is explicated in context so that knowledge of this approach is not a prerequisite for the book.

A large appendix is included to provide summary information on computer programs available for implementing the various multivariate techniques discussed in the text. Most tables required are provided in a second appendix and an extensive bibliography is provided to direct the reader not only to original sources, but also, to practical applications which have been made of the various multivariate models.

In every book, a line must be drawn somewhere as to what topics are to be included within the space available. In this book, topics of multivariate analysis which are important and are sometimes mentioned, but which have been excluded on this basis from detailed treatment, include multivariate time series and spectral analysis, simultaneous equation systems involving more than one dependent variable in each equation, path analysis, multivariate "errors-in-the-variables" models, multivariate random coefficient models, many constructs which arise in multivariate experimental design including sample survey models, nonparametric multivariate analysis, and many problems involving discrete multivariate distributions (including multidimensional contingency tables and Markov processes).

The author's interest in multivariate analysis was inspired by Ingram Olkin, whose enthusiasm for the subject is contagious and whose ideas are tacitly in evidence, especially in Chapter 2. He was also gracious enough to provide many critical comments on the manuscript, although I claim full responsibility for any remaining errors.

Arnold Zellner helped finance this work with National Science Foundation grants GS–1350 and GS–2347 and also helped clarify several Bayesian problems which arose. The National Science Foundation helped bring the work to completion with NSF grant GP–17592.

Jacques Drèze, Thomas Ferguson, Seymour Geisser, Divakar Sharma, Henri Theil, and Robert Winkler provided helpful comments on various sections of the manuscript. Discussions with D. V. Lindley, M. Stone, and G. Tiao were helpful in putting various Bayesian concepts into prospective.

My colleagues at the University of Chicago (in particular Harry Roberts) have stimulated some of the ideas in this work, as have many of my students by their questions during and after class.

Much detailed bibliographical and compilation work was carried out by John Bildersee, Bruce Cleland, and Richard Winborne.

Ilene Haniotis and her associates devoted considerable personal time and effort to the seemingly endless task of typing and proofing various versions of the manuscript, summer vacations notwithstanding.

I am indebted, also, to the Literary Executor of the late Sir Ronald A. Fisher, F.R.S., to Dr. Frank Yates, F.R.S., and to Oliver & Boyd, Edinburgh, for permission to reprint the Cumulative Student's Distribution Table from their book *Statistical Tables for Biological, Agricultural* and *Medical Research.*

Finally, the Graduate School of Business of the University of Chicago provided a large part of the time, funds, and encouragement that made the writing of the manuscript possible.

London
August 1971

S. James Press

PREFACE TO SECOND EDITION

Applied Multivariate Analysis was written about ten years ago. This second edition attempts to bring the reader up to date with the current frontiers of the field. We have added approximately 100 new references to recent developments. In some instances whole new sections have been added to explicate important areas that are either new, or which were not previously discussed (zonal polynomials, hypergeometric functions of matrix argument, generalized distributions, estimation of factor scores, factor selection by MAICE criterion, transformations of inverted Wishart and matrix T-variates, expanded treatment of Bayesian inference and subjective assessment procedures, generalization of multivariate symmetric stable distributions, log-linear models, generalized inverses). Fifty new exercises have been added in this edition (and many exercises have been modified in various ways) to aid the student, and to challenge his understanding of the concepts presented. Some are simple and mechanical, intending to provide drill type learning; others will require thought, and even research.

A second edition, of course, affords the author opportunity to correct errors, clarify exposition, amplify important ideas, delete the less important concepts, and provide new insights that only time and experience can provide. From using the book repeatedly throughout the decade, the author has benefited from many suggestions, comments, questions, and ideas contributed by students, colleagues, interested readers who have written in, or telephoned, and by coworkers at research institutions and government and industrial organizations with which the author has been affiliated. This second edition reflects these suggestions that have resulted in important modifications.

During the decade since the first edition was written, computer software packages for carrying out various types of multivariate analyses have become abundantly available. Moreover, Bayesian procedures have been implemented with carefully documented computer programs; there are now even programs for helping the researcher to elicit (subjective) prior distributions. For these reasons, the appendix devoted to listings of computer programs for multivariate models has more than doubled in size to reflect these changes. The multivariate programs included in the major statistical software packages, such as BMDP, SPSS, and SAS, are covered, as well as many model-specific packages, such as Multivariance, and LISREL. A new section devoted entirely to Bayesian computer programs has also been added.

A major topic of multivariate analysis is the analysis of multivariate discrete data, as summarized in contingency tables. This field has seen substantial development in recent years in terms of log-linear modeling, and the author has contributed to this development. Because the literature has become so extensive, however, I decided not to discourse on such a large topic; instead, there is a brief new section included in the Second Edition to summarize the basic ideas and to direct the reader to the appropriate books and articles that explain the subject in detail. In my own teaching of the subject I draw upon and refer students to other sources.

I have used the book successfully as a text both for a one quarter course at the graduate level where the students were either Statistics graduate students, or well prepared graduate students from other fields, and for a two quarter course at the graduate level for graduate students in Statistics. In the one quarter course, emphasis is on the models in Part II; in the two quarter sequence, I am able to cover most of Chapters 1–7 during the first quarter, and most of Chapters 8, 9, 10, 13, and some selected topics from the remaining chapters during the second quarter. A third quarter is really needed to cover the remaining material. Students are taught both Bayesian and frequentist procedures and points of view.

To the casual reader who has seen the first edition, the second edition may not look too different. On closer inspection, however, it will be found that the differences are substantial.

I am grateful to Mirza Ali, Ashis Sen Gupta, and Hassan Zahedi for their assistance in proofreading, and to Hassan Zahedi for his help in revising Appendix A. Mrs. Peggy Franklin provided almost flawless typing of all mathematical symbols and revised text. The responsibility for any errors that remain is entirely mine. Finally, I am grateful to the University of California for its encouragement and financial support.

S. James Press

Riverside, California
March, 1981

CONTENTS

APPENDIX B

APPENDIX C

NOTATION

In this book, vectors and matrices will appear in boldface type. Vectors will usually be denoted by lower case letters, and matrices which are not vectors will generally be denoted by upper case letters. Primes are generally used to denote transposed vectors or matrices, and unprimed vectors should be assumed to be column vectors, unless specified otherwise. Thus, if **a** denotes a column vector, **a**′ will denote a row vector; **A** might be used to denote a matrix.

Script $\mathcal{L}$ will be used to denote probability law, so that, for example, $\mathcal{L}(\mathbf{x}) = N(\boldsymbol{\theta},\boldsymbol{\Sigma})$ should be interpreted as: The probability law of the random vector $\mathbf{x}$ is multivariate Normal with mean vector $\boldsymbol{\theta}$ and covariance matrix $\boldsymbol{\Sigma}$.

APPLIED MULTIVARIATE ANALYSIS

Using Bayesian and Frequentist
Methods of Inference

1

INTRODUCTION

1.1 DEFINITIONS

Multivariate analysis is that branch of statistics that is devoted to the study of random variables which are correlated with one another. If two random variables are correlated, knowledge of the behavior of one provides some knowledge about the behavior of the other. To specify the probability of occurrence of many events of practical interest, it is usually necessary to make probabilistic assertions about several correlated random variables simultaneously. For example, suppose the success of some social welfare program is being evaluated. Since program success is generally assessed in terms of several different but correlated measures, the problem must be handled by multivariate analysis.

The term "applied multivariate analysis," as used in this book, refers to the motivation for using the models developed. Thus, the models presented in Part II will be seen to be useful for studying phenomena which arise in real problems.

Bayesian statistics is a term applied to the body of inferential techniques that uses Bayes' theorem to combine observational data with personalistic or subjective beliefs. Multivariate Bayesian techniques were integrated into the text wherever it was deemed appropriate.

The essence of applied multivariate analysis involves the motivation to solve problems and arrive at numerical answers, or to generate strong degrees of belief about natural phenomena, or to provide results which can be used as the basis for decision making. Often an hypothesis is proposed, multivariate data is collected and examined to test the hypothesis, and if the hypothesis is rejected by the data, it is modified accordingly, and new data is gathered to test the new hypothesis. There is also the notion that, inherently, the problem of interest involves repeated observations on several correlated variables. Thus, the field not only embodies a collection of tools, techniques, and methods of thinking which may be brought to bear upon problems involving the treatment and interpretation of many correlated random variables simultaneously, but also it involves the translation of the multidimensional techniques and models into numerical decision strategies or findings.

1.2 ORIENTATION OF TEXT

Until recently, application of results from multivariate analysis to real data and their integration with judgmental information required considerable effort, long hours, much hand computation, and an artistic flair for choosing significance levels. Two fairly new developments that have drastically altered this situation are the large scale availability of computers and the accelerated growth of the use of the Bayesian approach in statistics.

The development that may have had the greatest impact was the advent of high-speed large-storage-capacity digital computers. Although such computers have been around since the post World War Two era, it has only been in recent years that computer software development permitted large numbers of people to communicate with these machines without considerable advance preparation, that libraries of prepackaged programs were made available to users who did not want to develop their own computer routines to solve problems many other people had already solved, and that computer speed and storage capacity were sufficiently great to solve realistic problems that typically demand examination in many dimensions simultaneously.

The other major development that has greatly affected the applications of multivariate analysis is the rapid development of multivariate Bayesian results. That is, during roughly the same time period that computers were becoming impressively useful and accessible, the multivariate Bayesian approach was being successfully brought to bear on problems that had been difficult to treat from other points of view.

The above, and other developments (such as the steadily growing mathematical sophistication of the social sciences) helped multivariate analysis to make a strong impact in the applied fields. Moreover, these applied fields have continued to press their demands that the gap be bridged between theory (mathematical statistics) and application.

For such reasons, this book provides discussions of the interpretation of the main results, sometimes presenting proofs in context or in Complements at the end of the chapter, and sometimes referring the more technically inclined reader to proofs in the literature. References are also given for additional applications of the various multivariate models discussed. As a result the bibliography, although certainly not exhaustive, was designed to be sufficiently broad in scope and deep in concept to serve the interests of a wide variety of readers.

For the subject "applied multivariate analysis" not to be a contradiction in terms, it is necessary that the reader become familiar with some of the computer programs available for applying multivariate procedures. The general characteristics of computer routines associated with the models discussed in this book are described in an appendix and are sometimes referred to in the text. The computer routines will prove most helpful to those users who fully understand the assumptions underlying the model for which the routine was developed.

Exercises are provided at the end of each chapter. Although some straightforward computational exercises are sometimes included for practice in the use of certain relations, most problems are not of this type; rather, they are intended to gauge the reader's overall grasp of the sense of the material in the chapter. That is, attention is often focused on questions such as: When should a particular kind of model be applied, what are the underlying assumptions, what other models might be used, why is one model better than another for a given problem, and so on?

1.3 GENESIS OF MULTIVARIATE MODELS

The models of multivariate analysis described in this book arose in real problems in many different disciplines. Correlation was used by F. Galton in the second half of the nineteenth century. Factor analysis was introduced in education and psychology by C. Spearman to explain human intelligence. R. A. Fisher introduced many notions of multivariate analysis (including that of intraclass correlation) to explain phenomena in genetics. Regression models were first well formulated by Gauss for multidimensional applications in astronomy. Many of the variants of the regression model were investigated and developed for applications in economics. Experimental design models were developed for use in agriculture. Latent structure models and multidimensional scaling methods were developed in education, psychology, and sociology, while control models have been pressed forward by chemists, economists, and engineers. Advances in multivariate stable distribution theory have been stimulated by problems in finance, and classification and discrimination models are finding increasing application in marketing, after having served well for many years in anthropology and taxonomy.

In summary, the models to be described have been drawn from many of the scientific disciplines, especially from biometrics, econometrics, psychometrics, sociometrics, education, and the subfields of business. It is expected that by exposing models drawn from diverse disciplines, the models will find new application in fields different from those in which they originally arose.

1.4 SAMPLING THEORY VERSUS BAYESIAN APPROACH

This book does not take a dogmatic position on the sampling theory versus the Bayesian approach toward solving problems. There is no claim that there is a right and a wrong way. Rather, it is believed that cogent arguments can be made for both approaches to inference and decision making, and each may involve some subjective or technical difficulties. In the sampling theory approach the analyst must use his prior beliefs relative to the trade-off between sample size and Type I and Type II errors. Moreover, he obtains confidence intervals applicable only to averages taken over many samples, rather than to the sample at hand. The Bayesian needs no Type I and II errors, but he does need to assess prior distributions, which may not always be an easy task. In many situations, in which the sampling theory approach is used for making inferences based upon a given sample, results are marginal in the sense that, for example, a statistic might be significant at the five percent level but not at the one percent level, and the analyst is not really strongly attached to either significance level. In such circumstances, calculating the entire posterior distribution with respect to a diffuse prior will often put the analyst in a much better position to make a decision about which he will feel confident.

It sometimes happens that a problem which is extremely difficult to analyze from a sampling theory viewpoint becomes simpler to study from a Bayesian viewpoint. Such is the case, for example in both multivariate regression and in Normal classification problems, for certain prior densities. In multivariate regression, a likelihood ratio approach to testing the coefficients for significance leads to an extremely complicated distribution which has been approximated in several ways. However, use of the Bayes approach (with diffuse priors) for testing requires the relatively simple application of the Student t-distribution (it is more complicated for natural conjugate priors). Similarly, there is not yet general agreement on the finite sample solution to the problem of classifying a vector into one of two multivariate Normal populations with unequal and unknown parameters. But a Bayesian will establish his basis for decision when he computes the posterior odds, which he accomplishes by inserting the observed data into the ratio of two Student t-densities. In the case of principal components and canonical correlations, the Bayesian result is complicated (it is expressed in terms of zonal polynomials), while the sampling theory result is relatively simple for the usual types of inferences desired. In the case of the square of the multiple correlation coefficient both approaches yield complicated results. More generally, no matter which model is involved, and no matter which approach is simpler, if prior information is easily assessable and if proach provides a formalism for combining judgemental information with observational data. If prior information is difficult to assess, Bayesian procedures based upon "vague" priors are sometimes used. Such procedures are sometimes equivalent to sampling theory procedures, but not always. In the latter case, a philosophical choice must be made.

The rationale for presenting multivariate analysis from both sampling theory and Bayesian viewpoints is to provide the reader with a broader base from which real problems may be studied.

1.5 SCOPE

The book is divided into two parts. Part I deals with foundations. As such, it provides a convenient summary of the main theoretical results that are needed in applied multivariate analysis. Since the study of many variables simultaneously is most efficiently and expeditiously handled in a vector and matrix context, Chapter 2 is devoted to a review of the pertinent matrix algebra and calculus required.

Chapter 3 begins with a brief discussion of multivariate distributions in general, and continues with a more detailed treatment of the multivariate Normal distribution in particular. In this book attention is directed largely to the class of continuous distributions that has densities. Thus, there will never be any need for measure theoretic arguments (exceptions for sets of Lebesgue measure zero are always to be understood, but will usually be omitted).

The multivariate Normal distribution is the principal one used for drawing inferences. In univariate analysis, the Normal distribution is fundamental largely because of its role as the limiting distribution for sums of independent and identically distributed random variables with finite second moments (central limit theorem). The multivariate analogue of the central limit theorem is the basis for the importance of the multivariate Normal distribution. This theorem, along with other asymptotic multidimensional results, is discussed in Chapter 4.

Chapters 5 and 6 treat the Wishart distribution (the distribution of sample variances and covariances), various multivariate versions of the

Student t-distribution, and other major multivariate distributions (such as Hotelling's T^2-distribution) which will be needed in making inferences in multivariate models. The subject of stable distributions, which has become of interest in the study of securities markets, income variation, portfolio analysis, and other areas of finance and economics, is discussed from a multivariate standpoint in Chapter 6.

Chapter 7 presents a brief treatment of elementary multivariate statistics, including estimation and testing using the multivariate Normal and Wishart distributions. Some attention is given to the fundamental work of Stein on the inadmissibility, in higher dimensions, of the sampling theory estimator of the mean vector in a multivariate Normal distribution.

Part II treats various models that might be used singly or sequentially to analyze multivariate data. It begins with Chapter 8, which focuses on linear models. Results are provided for univariate regression models under a wide variety of underlying assumptions likely to be encountered in applications (such as unequal variances and serially correlated disturbances). Both multivariate regression and generalized multivariate regression models (different regressor matrices for each dependent variable) are discussed. One and two way layouts for the analysis of variance are presented (univariate and multivariate), as are the multivariate analysis of covariance and multivariate multiple comparison techniques for obtaining simultaneous confidence intervals for regression parameters. Both the Bayesian and sampling theory methods of analyzing linear models are exposed.

Study of the variance within observed data is simplified by the principal components model, discussed in Chapter 9. In this model, data interpretation is simplified by representing the data in a rotated coordinate system.

Study of the correlation structure underlying the observed data variables is facilitated with the factor analysis and latent structure analysis models discussed in Chapter 10. The factor analysis model attempts to discover a few elemental factors that may have generated the data. The latent structure analysis model attempts to categorize the data, after the fact.

Chapter 11 treats the canonical correlations model, which is an attempt to correlate two or more groups of variates rather than pairs of random variables.

Chapter 12 discusses the problem of allocating scarce resources when the variables follow multivariate stable distributions (stable portfolio analysis).

In Chapter 13, Bayesian and sampling theory procedures are given for classifying data vectors (attribute profiles) into predesignated populations.

Chapter 14 discusses some adaptive multivariate control problems in the context of controlling a multivariate regression output.

The final chapter of the book (Chapter 15) is devoted to the new and important topic of structuring of multivariate populations. One area discussed involves the optimal grouping of multivariate observations into clusters. Another related problem is multidimensional scaling, or the locating of points in a multidimensional space based upon preference, or ordered data. These subjects are currently undergoing extensive investigation. Moreover, although many interesting results have already been obtained, rigorous statistical underpinning is still lacking.

Appendix A provides reference material on computer programs which will implement the models discussed in the text. Information is given, such as limitations on the input and output variables, location of the program, and references for additional program details. Several alternative programs (with different properties) are given for some of the models so that a program can be selected which best fits the problem at hand. Appendix B contains a variety of numerical tables useful for work in applied multivariate analysis, while Appendix C lists all references in the book by author.

part I

Foundations

2

MATRIX THEORY USEFUL IN MULTIVARIATE ANALYSIS

An efficient construct for studying many variables simultaneously is that of vectors and matrices. This chapter contains only a certain subcollection of results that appears to bear directly upon the requirements of multivariate analysis. Most of the elementary results are assumed to be known and are only included here for convenience. In general, these elementary results are presented without proof, although references are often provided where additional information and applications can be found. For the less elementary relationships, either proofs are given, or indication is given as to how to establish the proof. For the most part, results presented will be restricted to those required later in the book.

2.1 NOTATION AND DEFINITIONS

A rectangular array of real or complex numbers is called a matrix. Unless otherwise specified it will always be assumed that the numbers are real. Let $\mathbf{A}: p \times n$ denote the matrix $\mathbf{A}$ with p rows and n columns and write it as

$$\mathbf{A} = \begin{pmatrix} a_{11} & \cdots & a_{1n} \\ \cdot & & \cdot \\ \cdot & & \cdot \\ \cdot & & \cdot \\ a_{p1} & \cdots & a_{pn} \end{pmatrix},$$

or, $\mathbf{A} = (a_{ij})$, $i = 1,\ldots,p$; $j = 1,\ldots,n$. The entries in the array are called the elements of $\mathbf{A}$. If $p = n$, $\mathbf{A}$ is called *square* and of *order* p. If $n = 1$, $\mathbf{A}$ is called a column vector (and if $p = 1$, $\mathbf{A}$ is called a row vector). If $p = n = 1$, $\mathbf{A}$ is called a *scalar* (see Example (2.3.2)).

If $a_{ij} = 0$ for all i, j, **A** is called the *zero matrix*. If $n = p$, and if $a_{ij} = 1$ for $i = j$, and $a_{ij} = 0$ for $i \neq j$, **A** is called the *identity matrix* of order p and is usually denoted by **I**, or $\mathbf{I}_p$. The elements a_{ii} of a square matrix **A** are called its *diagonal elements*. Let **A**′ denote the matrix obtained from **A** by interchanging its rows and columns. **A**′ is called the *transposed matrix*. If in $\mathbf{A}: p \times n$, $n = 1$, so that **A** is a column vector, **A**′ is a row vector.

To motivate the matrix algebra and calculus to be discussed, the notion of a matrix of covariances will be used. Let $X_1, \ldots, X_p$ denote scalar random variables. Then the covariance between X_i and X_j is defined as

$$\sigma_{ij} \equiv \text{cov}\,(X_i, X_j) = E(X_i - EX_i)(X_j - EX_j),$$

where E denotes the expectation, so that EX_i is the mean of X_i. If $i = j$, σ_{ii} is called the *variance*. Let $\mathbf{\Sigma} = (\sigma_{ij})$ denote the $p \times p$ matrix of covariances. $\mathbf{\Sigma}$ is called the covariance matrix of the vector of random variables $\mathbf{X} \equiv (X_1, \ldots, X_p)'$. It is also sometimes expressed as $\mathbf{\Sigma} = \text{var}\,\mathbf{X} = E(\mathbf{X} - E\mathbf{X})(\mathbf{X} - E\mathbf{X})'$. This notion will be discussed formally in more detail in Section 3.2.2. The correlation between X_i and X_j is defined as

$$\rho_{ij} = \frac{\sigma_{ij}}{(\sigma_{ii}\sigma_{jj})^{1/2}}.$$

When $\rho_{ij} = 0$, X_i and X_j are said to be uncorrelated.

2.2 SPECIALLY STRUCTURED MATRICES (PATTERN MATRICES)

Pattern matrices arise often in multivariate analysis. If they can be justified, they are often useful for reducing the parameter space in a problem to a manageable size. Fewer parameters means the available observations will yield better estimators and more powerful tests since the number of observations per parameter is greater. For these reasons, various pattern matrices and their properties will be presented in this chapter.

A *diagonal matrix* of order p is a $p \times p$ matrix whose off-diagonal elements are zero. It will often be helpful to use the notation

$$\mathbf{D}_\lambda = \begin{pmatrix} \lambda_1 & & & 0 \\ & \cdot & & \\ & & \cdot & \\ 0 & & & \lambda_p \end{pmatrix}, \tag{2.2.1}$$

or $\mathbf{D}_\lambda = \text{diag}\,(\lambda_1, \ldots, \lambda_p)$, to denote a diagonal matrix of order p whose diagonal elements are λ_k, $k = 1, \ldots, p$. If random variables are uncorrelated, their covariance matrix is diagonal, since in this case, the covariances are all zero, although the variances are positive. A diagonal covariance matrix whose diagonal elements are equal is called a *scalar* covariance matrix.

A is a *symmetric matrix* if $\mathbf{A} = \mathbf{A}'$. Of course this is a property of square matrices only. It is clear from the definition that all covariance matrices are symmetric.

Example (2.2.1): The following matrices are symmetric

$$\begin{pmatrix} 2 & 1 \\ 1 & 3 \end{pmatrix}, \quad \begin{pmatrix} 4 & -9 & 2 \\ -9 & 1 & -1 \\ 2 & -1 & 7 \end{pmatrix}, \quad \begin{pmatrix} 1 & 0 & 0 \\ 0 & 1 & 0 \\ 0 & 0 & 1 \end{pmatrix}, \quad \begin{pmatrix} 2 & 3 & 3 \\ 3 & -7 & 3 \\ 3 & 3 & 0 \end{pmatrix}.$$

An *orthogonal matrix* $\mathbf{\Gamma} = (\gamma_{ij})$, $i, j = 1,\ldots,p$ is a square matrix of order p in which

(i) $\displaystyle\sum_{j=1}^{p} \gamma_{ij}\gamma_{kj} = 0, \quad i \neq k, \quad$ or $\quad \displaystyle\sum_{i=1}^{p} \gamma_{ij}\gamma_{ik} = 0, \quad j \neq k;$

(ii) $\displaystyle\sum_{j=1}^{p} \gamma_{ij}^2 = 1, \quad$ or $\quad \displaystyle\sum_{i=1}^{p} \gamma_{ij}^2 = 1; \quad i, j = 1,\ldots,p.$

That is, (i) asserts that the rows of $\mathbf{\Gamma}$ are all orthogonal, as are the columns, and (ii) asserts that each row vector (column vector) has length unity. Equivalently, $\mathbf{\Gamma}$ is orthogonal if and only if

$$\mathbf{\Gamma\Gamma}' = \mathbf{\Gamma}'\mathbf{\Gamma} = \mathbf{I}. \tag{2.2.2}$$

Example (2.2.2): Consider the matrix

$$\mathbf{\Gamma} = \begin{pmatrix} \frac{1}{\sqrt{p}} & \frac{1}{\sqrt{1\cdot 2}} & \frac{1}{\sqrt{2\cdot 3}} & \cdots & \frac{1}{\sqrt{(p-1)p}} \\ \frac{1}{\sqrt{p}} & -\frac{1}{\sqrt{1\cdot 2}} & \frac{1}{\sqrt{2\cdot 3}} & \cdots & \vdots \\ \vdots & 0 & -\frac{2}{\sqrt{2\cdot 3}} & \cdots & \vdots \\ \vdots & \vdots & 0 & \ddots & \vdots \\ \vdots & \vdots & \vdots & & \vdots \\ \frac{1}{\sqrt{p}} & 0 & \cdots & & -\frac{(p-1)}{\sqrt{(p-1)p}} \end{pmatrix}. \tag{2.2.3}$$

The $p \times p$ matrix given in (2.2.3) is not only orthogonal, it also has the property that all the columns sum to zero except the first (and the first column has common elements). Such an orthogonal matrix is called a *Helmert* matrix (see Helmert (1876)).

One common use of orthogonal matrices is for rotating coordinate systems to explain data in terms of uncorrelated or orthogonal variables, as in the principal components and factor analysis models of Chapters 9 and 10. A commonly used two-dimensional rotation of coordinates is given by the orthogonal matrix

$$\mathbf{\Gamma} = \begin{pmatrix} \cos\theta & \sin\theta \\ \sin\theta & -\cos\theta \end{pmatrix}.$$

A covariance matrix whose off-diagonal elements are common and whose diagonal elements are common is said to have the *intraclass covariance matrix* pattern. It may be written in the form [see also (2.4.7)].

$$\underset{(p \times p)}{\mathbf{A}} = \begin{pmatrix} a & & & \\ & \cdot & b & \\ & & \ddots & \\ & b & \cdot & \\ & & & a \end{pmatrix}, \qquad b \geq -\frac{a}{p-1}, \tag{2.2.4}$$

where the notation implies that all off-diagonal elements have the value b. This terminology dates back to R. A. Fisher (1925) who was concerned with the correlation of factors in individuals within a family or class. These matriçes have since been used in many problems in multivariate analysis. See, for instance, Geisser (1963), Press (1967a), Srivastava (1965), Votaw (1948), and Wilks (1936). In some applications, such as in stable portfolio analysis (Chapter 12), matrices with the structure in (2.2.4) are used for expressing this pattern of association even though the matrices are not actually covariance matrices. It follows from Sect. 2.11.1 and page 65 that the inequality constraint in (2.2.4) represents a requirement of positive semidefiniteness for $\mathbf{A}$. Such a condition might not be required if $\mathbf{A}$ is not a covariance matrix.

A *circular matrix* is one which has the structure

$$\mathbf{A} = \begin{pmatrix} a_1 & a_2 & \cdots & a_p \\ a_p & a_1 & \cdots & a_{p-1} \\ \cdot & \cdot & & \cdot \\ \cdot & \cdot & & \cdot \\ \cdot & \cdot & & \cdot \\ a_2 & a_3 & \cdots & a_1 \end{pmatrix}. \tag{2.2.5}$$

If **A** were a covariance matrix, the additional structure of symmetry would be imposed upon (2.2.5), further constraining the parameters. Such circular covariance matrices are relevant in applications involving cyclical stationary time series, or in problems such as signal detection in which points are ordered on a circle (see, for instance, Press, 1964, p. 27; Mirsky, 1955, p. 36; and Olkin and Press, 1969). It has also been pointed out to the author by I. Olkin that many animals exhibit a naturally occurring structural symmetry, such as the starfish and octopus. Study of the nervous or circulatory systems in such animals should involve pattern matrices. For an extensive mathematical discussion of circular matrices and applications, see Davis, 1979.

The elements just to the right (left) of those along the diagonal of a square matrix are called the elements of the *super* (*sub*) *diagonal*, while those along the diagonal are often called the elements of the *principal diagonal*.

A *Jacobi matrix* is a square matrix in which the elements *not* on the principal, super, or sub diagonals are zero. Thus, **W** is a Jacobi matrix if it can be written in the form

$$\mathbf{W} = \begin{pmatrix} w_{11} & w_{12} & & \\ & & \cdot & 0 \\ w_{21} & \cdot & & \cdot \\ & \cdot & \cdot & \cdot \\ & \cdot & \cdot & w_{p-1,p} \\ 0 & \cdot & & \\ & & w_{p,p-1} & w_{pp} \end{pmatrix}. \qquad (2.2.6)$$

Such matrices arise naturally in time series problems in which correlations between observations at different times decay exponentially with increasing time differences (Example 8.3.3) and in distributed lag problems (Example 8.3.5).

A Green's matrix is a square symmetric matrix which has the structure

$$\mathbf{G} = \begin{pmatrix} u_1v_1 & u_1v_2 & \cdots & u_1v_p \\ u_1v_2 & u_2v_2 & \cdots & u_2v_p \\ \cdot & \cdot & & \cdot \\ \cdot & \cdot & & \cdot \\ \cdot & \cdot & & \cdot \\ u_1v_p & u_2v_p & \cdots & u_pv_p \end{pmatrix}, \qquad (2.2.7)$$

or $\mathbf{G} = (g_{ij})$, where $g_{ij} = u_iv_j$, $i \leq j$, and $g_{ij} = u_jv_i$, $i \geq j$; $i, j = 1,\ldots,p$. This specially structured matrix is useful for expressing certain types of patterns of association among random variables [see for instance, Example (6.5.8)]. The matrix may also be generated from the Green's function kernel (see Courant and Hilbert, 1953).

Toeplitz matrices arise in time series applications and are directly associated with stationary series; that is, series in which the covariance of the series' values at any two times depends only on the time difference. Such matrices have the characteristic form

$$\mathbf{K} = \begin{pmatrix} C_0 & C_1 & \cdots & C_{p-1} & C_p \\ C_{-1} & C_0 & \cdots & C_{p-2} & C_{p-1} \\ \cdot & \cdot & & \cdot & \cdot \\ \cdot & \cdot & & \cdot & \cdot \\ \cdot & \cdot & & \cdot & \cdot \\ C_{-p+1} & C_{-p+2} & \cdots & C_0 & C_1 \\ C_{-p} & C_{-p+1} & \cdots & C_{-1} & C_0 \end{pmatrix}, \tag{2.2.8}$$

in which all diagonals have common elements. In applications it is often the case that $C_{-j} = C_j$, so that $\mathbf{K} = \mathbf{K}'$. A special case that arises in exponential smoothing applications (used in business forecasting, for example) is the one in which $C_{-j} = C_j$, $C_0 = 1$, $C_j = \rho^j$, $j = 1, \ldots, p$, $0 < |\rho| < 1$. An extended discussion of the properties and applications of these matrices is given in Grenander and Szegö (1958). A specific application in regression is given in Example (8.3.3).

A special case of the symmetric Toeplitz matrix which arises frequently enough to be studied separately is the *tri-diagonal* matrix, in which all elements in (2.2.8) are zero except those on the super, principal, and sub diagonals; thus, $\mathbf{A}$ is a tri-diagonal matrix if $\mathbf{A}$ has the structure

$$\mathbf{A} = \begin{pmatrix} a & b & & & \\ & & \cdot & 0 & \\ b & \cdot & \cdot & & \\ & \cdot & \cdot & \cdot & \\ & & \cdot & \cdot & b \\ & 0 & \cdot & & \\ & & & b & a \end{pmatrix}. \tag{2.2.9}$$

Covariance matrices with tri-diagonal structure arise, for example, in distributed lag models in regression [see Chapter 8, Example (8.3.5)].

As another example, suppose there are identical signal receivers located at p equidistant points on a circle. The signal received at any point is correlated with that received at any adjacent point, but beyond adjacent points, the correlation is zero. In such a case, $\mathbf{A}$ in (2.2.9) could be taken as the covariance matrix of the p signals received.

2.3 MATRIX ALGEBRA

2.3.1 Addition

Addition and subtraction of two matrices $\mathbf{A} = (a_{ij}): p \times n$ and $\mathbf{B} = (b_{ij}): p \times n$ is defined by $\mathbf{A} \pm \mathbf{B} = \mathbf{C} = (c_{ij})$, where $c_{ij} = a_{ij} \pm b_{ij}$, $i = 1,\ldots,p$, $j = 1,\ldots,n$.

Example (2.3.1):

$$\begin{pmatrix} 3 & 2 & 4 \\ 7 & -1 & 0 \\ 8 & -2 & 3 \end{pmatrix} + \begin{pmatrix} -1 & 2 & 6 \\ 0 & 0 & 1 \\ 3 & -2 & 2 \end{pmatrix} = \begin{pmatrix} 2 & 4 & 10 \\ 7 & -1 & 1 \\ 11 & -4 & 5 \end{pmatrix}.$$

2.3.2 Multiplication

Now let $\mathbf{E}: p \times n$, $\mathbf{F}: n \times p$ be two matrices. $\mathbf{E}$ and $\mathbf{F}$ are said to be *conformable* if the number of columns of the first is equal to the number of rows of the second. If two matrices are conformable, they may be multiplied. Let $\mathbf{G}: p \times p$ be the *product matrix* so that $\mathbf{G} = \mathbf{EF}$. Then if $\mathbf{G} = (g_{ij})$, $\mathbf{E} = (e_{ij})$, $\mathbf{F} = (f_{ij})$, $\mathbf{G}$ is defined by

$$g_{ij} = \sum_{\alpha=1}^{n} e_{i\alpha} f_{\alpha j}, \qquad i, j = 1,\ldots,p. \tag{2.3.1}$$

Example (2.3.2): If

$$\mathbf{A} = \begin{pmatrix} 1 & 2 \\ 2 & 3 \\ 3 & 4 \end{pmatrix}, \qquad \mathbf{B} = \begin{pmatrix} 2 & 1 & 4 \\ 0 & 3 & 1 \end{pmatrix},$$

$$\mathbf{AB} = \begin{pmatrix} 1 \times 2 + 2 \times 0 & 1 \times 1 + 2 \times 3 & 1 \times 4 + 2 \times 1 \\ 2 \times 2 + 3 \times 0 & 2 \times 1 + 3 \times 3 & 2 \times 4 + 3 \times 1 \\ 3 \times 2 + 4 \times 0 & 3 \times 1 + 4 \times 3 & 3 \times 4 + 4 \times 1 \end{pmatrix}.$$

Recall (it is easy to check) that even for square matrices, it is *not* generally true that $\mathbf{AB} = \mathbf{BA}$. Moreover, in general, $\mathbf{BA}$ may not even exist when $\mathbf{AB}$ exists, due to lack of conformability.

A square matrix of order 1 is called a *scalar*. It is just a number. Multiplication of a scalar by a matrix multiplies each element of the matrix by the scalar. Thus if b is a scalar and if $\mathbf{A} = (A_{ij})$, $b\mathbf{A} = (ba_{ij})$. For example,

$$5\begin{pmatrix} 2 & 3 \\ 1 & 4 \end{pmatrix} = \begin{pmatrix} 10 & 15 \\ 5 & 20 \end{pmatrix}.$$

Algebra of matrices includes the operations

$$
\begin{aligned}
(\mathbf{A}+\mathbf{B})' &= \mathbf{A}'+\mathbf{B}', \\
(\mathbf{AB})' &= \mathbf{B}'\mathbf{A}', && \text{if } \mathbf{A} \text{ and } \mathbf{B} \text{ are conformable,} \\
(\mathbf{A}')' &= \mathbf{A}, \\
\mathbf{A}(\mathbf{B}+\mathbf{C}) &= \mathbf{AB}+\mathbf{AC}, \\
\mathbf{AI} &= \mathbf{A}, && \text{for all } \mathbf{A}.
\end{aligned}
$$

In matrix algebra, the analogue of taking the reciprocal of a number is the operation of matrix inversion. This will be defined in terms of the notion of determinants.

2.4 DETERMINANTS

Definition: A square matrix $\mathbf{A} = (a_{ij})$, of order p, has associated with it a scalar, real valued function of its elements called its *determinant*, denoted by $|\mathbf{A}|$ (see, e.g. Mirsky, 1956, page 6).

Example (2.4.1): For a 2×2 matrix,

$$\begin{vmatrix} a_{11} & a_{12} \\ a_{21} & a_{22} \end{vmatrix} = a_{11}a_{22} - a_{12}a_{21},$$

so that

$$\begin{vmatrix} 3 & 2 \\ 4 & 7 \end{vmatrix} = (3)(7) - (2)(4) = 13.$$

It is often necessary to carry out simple operations on matrices and groups of matrices. For this reason it is convenient to have available a storehouse of elementary relationships that exist among determinants. The determinantal properties provided below are demonstrated in many elementary books on matrix algebra (see for instance, Mirsky, (1955)).

2.4.1 Elementary Properties of Determinants

(1) The value of a determinant vanishes, $|\mathbf{A}| = 0$, if all elements of any row or column are zero; also, if all elements of any row (column) are identical with, or are multiples of, the corresponding elements of any other row (column).

(2) The value of a determinant is unchanged if rows and columns are interchanged ($|\mathbf{A}| = |\mathbf{A}'|$), or if a linear combination of any number of rows (columns) is added to any row (column).

(3) If two rows or columns are interchanged, the determinant is multiplied by minus one.

(4) If all elements of a row (column) are multiplied by a constant factor, the determinant is multiplied by the same factor. Thus, if

$$\mathbf{B} = \begin{pmatrix} \alpha a_{11} & \alpha a_{12} & \cdots & \alpha a_{1p} \\ a_{21} & a_{22} & \cdots & a_{2p} \\ \cdot & \cdot & & \cdot \\ \cdot & \cdot & & \cdot \\ \cdot & \cdot & & \cdot \\ a_{p1} & a_{p2} & \cdots & a_{pp} \end{pmatrix},$$

where α is some scalar and $\mathbf{A} = (a_{ij})$, $|\mathbf{B}| = \alpha|\mathbf{A}|$.

(5) If every element in any row (column) is the sum of two quantities, the determinant may be expanded as the sum of two determinants of the same order. Thus, if

$$\mathbf{B} = \begin{pmatrix} a_{11} + b_{11} & a_{12} + b_{12} & \cdots & a_{1p} + b_{1p} \\ a_{21} & a_{22} & \cdots & a_{2p} \\ \cdot & \cdot & & \cdot \\ \cdot & \cdot & & \cdot \\ \cdot & \cdot & & \cdot \\ a_{p1} & a_{p2} & \cdots & a_{pp} \end{pmatrix},$$

$$|\mathbf{B}| = \begin{vmatrix} a_{11} & \cdots & a_{1p} \\ a_{21} & \cdots & a_{2p} \\ \cdot & & \cdot \\ \cdot & & \cdot \\ \cdot & & \cdot \\ a_{p1} & \cdots & a_{pp} \end{vmatrix} + \begin{vmatrix} b_{11} & \cdots & b_{1p} \\ a_{21} & \cdots & a_{2p} \\ \cdot & & \cdot \\ \cdot & & \cdot \\ \cdot & & \cdot \\ a_{p1} & \cdots & a_{pp} \end{vmatrix}.$$

2.4.2 Minors and Cofactors

The *minor* of the element a_{ij} of $\mathbf{A} = (a_{ij})$ is the value of the determinant obtained from $\mathbf{A}$ by deleting the ith row and the jth column.

The *cofactor* of a_{ij}, denoted by A_{ij}, is given by $(-1)^{i+j}$ times the minor of a_{ij}.

The determinant of a matrix of order p is sometimes obtained by means of the following, called the *Laplace development*.

$$\begin{aligned} |\mathbf{A}| &= \sum_{j=1}^{p} a_{ij}A_{ij}, \text{ for any fixed } i = 1,\ldots,p, \\ &= \sum_{i=1}^{p} a_{ij}A_{ij}, \text{ for any fixed } j = 1,\ldots,p. \end{aligned}$$

If $|\mathbf{A}| = 0$, $\mathbf{A}$ is called *singular.*

Example (2.4.2): Suppose

$$\mathbf{A} = \begin{pmatrix} 3 & 5 & 2 \\ -1 & 1 & 0 \\ 0 & -3 & 2 \end{pmatrix}.$$

Then, the Laplace development of $|\mathbf{A}|$ by elements of the first row gives

$$\begin{aligned} |\mathbf{A}| &= 3\begin{vmatrix} 1 & 0 \\ -3 & 2 \end{vmatrix} - 5\begin{vmatrix} -1 & 0 \\ 0 & 2 \end{vmatrix} + 2\begin{vmatrix} -1 & 1 \\ 0 & -3 \end{vmatrix} \\ &= 3[1 \times 2 - 0(-3)] - 5[(-1) \times 2 - 0 \times 0] \\ &\quad + 2[(-1) \times (-3) - 1 \times 0] \\ &= 22. \end{aligned}$$

Example (2.4.3): Consider the matrix

$$\mathbf{B} = 2\mathbf{A} = \begin{pmatrix} 2 \times 3 & 2 \times 5 & 2 \times 2 \\ 2 \times (-1) & 2 \times 1 & 2 \times 0 \\ 2 \times 0 & 2 \times (-3) & 2 \times 2 \end{pmatrix} = \begin{pmatrix} 6 & 10 & 4 \\ -2 & 2 & 0 \\ 0 & -6 & 4 \end{pmatrix}.$$

It follows from Property (4) that

$$|\mathbf{B}| = 2 \times 2 \times 2|A| = 2^3|\mathbf{A}| = 8|\mathbf{A}|.$$

More generally, if $\mathbf{A}: p \times p$ and $\mathbf{B} = \alpha\mathbf{A}$, where α is any scalar, $|\mathbf{B}| = \alpha^p|\mathbf{A}|$.

2.4.3 Special Properties of Determinants

(1) Let $\mathbf{A}_1, \ldots, \mathbf{A}_k$ all be $p \times p$ (of order p). Then

$$|\mathbf{A}_1\mathbf{A}_2 \cdots \mathbf{A}_k| = |\mathbf{A}_1| \cdot |\mathbf{A}_2| \cdots |\mathbf{A}_k| \tag{2.4.1}$$

(for a simple proof see, for instance, Mirsky, 1955, p. 12).

If $\mathbf{\Sigma}: p \times p$ is the covariance matrix of a random p-vector $\mathbf{X}: p \times 1$, and $\mathbf{Y} = \mathbf{AX}$, where $\mathbf{A}$ is $p \times p$, the covariance matrix of $\mathbf{Y}$ is $\mathbf{A\Sigma A}'$. Then $|\mathbf{A\Sigma A}'| = |\mathbf{A}|^2|\mathbf{\Sigma}|$. For instance, see (3.2.8) for such transformations.

(2) If $\mathbf{A}: p \times n$ and $\mathbf{B}: n \times p$, then

$$|\mathbf{I}_p + \mathbf{AB}| = |\mathbf{I}_n + \mathbf{BA}|, \tag{2.4.2}$$

a result which has been attributed to Sylvester. It will be seen in (2.8.1) that (2.4.2) is equivalent to $\mathbf{AB}$ and $\mathbf{BA}$ having the same characteristic equation. Thus, if $n = 1$, $\mathbf{A}$ is a column vector, $\mathbf{a}$, and if $\mathbf{B}$ is a row vector, $\mathbf{b}'$, then

$$|\mathbf{I}_p + \mathbf{AB}| = |\mathbf{I}_p + \mathbf{ab}'| = 1 + \mathbf{b}'\mathbf{a}. \tag{2.4.3}$$

This result will prove useful, for example, in developing a Bayes estimator of the mean vector in a multivariate Normal distribution [see (7.1.9)].

2.4.4 Determinants of Pattern Matrices

If $\mathbf{\Gamma}: p \times p$ is *orthogonal* [defined in (2.2.2)],

$$|\mathbf{\Gamma}| = \pm 1. \tag{2.4.4}$$

Since $\mathbf{\Gamma\Gamma}' = \mathbf{I}$, $|\mathbf{\Gamma\Gamma}'| = |\mathbf{\Gamma}|^2 = 1$, from which (2.4.4) follows.

If $\mathbf{D}_\lambda = \text{diag}(\lambda_1, \ldots, \lambda_p)$,

$$|\mathbf{D}_\lambda| = \prod_{j=1}^{p} \lambda_j. \tag{2.4.5}$$

This relation follows from the definition of determinant.

If $\mathbf{A}: p \times p$ is the *intraclass correlation* matrix [see (2.2.4)],

$$\mathbf{A} = \begin{pmatrix} a & & & \\ & \cdot & b & \\ & & \cdot & \\ & b & \cdot & \\ & & & a \end{pmatrix}, \qquad b \geq -\frac{a}{p-1}, \tag{2.4.6}$$

$$|\mathbf{A}| = (a-b)^{p-1}[a + b(p-1)]. \tag{2.4.7}$$

This result follows from (2.8.2) and (2.8.3) below.

More generally, if $\mathbf{G}: pk \times pk$ has the structure of the *matrix intraclass correlation* matrix, so that

$$\mathbf{G} = \begin{pmatrix} \mathbf{A} & \mathbf{B} & \cdots & \mathbf{B} \\ \mathbf{B} & \mathbf{A} & \cdots & \mathbf{B} \\ \cdot & \cdot & & \cdot \\ \cdot & \cdot & & \cdot \\ \cdot & \cdot & & \cdot \\ \mathbf{B} & \mathbf{B} & \cdots & \mathbf{A} \end{pmatrix}, \tag{2.4.8}$$

where $\mathbf{A}$ and $\mathbf{B}$ are each square matrices of order k [it will be seen in (2.6.1) that $\mathbf{G}$ is a "partitioned" matrix],

$$|\mathbf{G}| = |\mathbf{A} - \mathbf{B}|^{p-1}|\mathbf{A} + \mathbf{B}(p-1)|. \tag{2.4.9}$$

This result is demonstrated in Complement 2.1. An application of this matrix in econometrics is referenced following (2.8.6). Other properties and applications of these matrices are given in Arnold (1973, 1976, 1979a, 1979b); and Press (Feb. 1979, Sept. 1978, 1980c, and 1981).

For the *circular* matrix defined in (2.2.5),

$$|\mathbf{A}| = \prod_{k=1}^{p} \left[\sum_{j=1}^{p} a_j \exp\left\{ \frac{2\pi i}{p}(j-1)(k-1) \right\} \right]. \tag{2.4.10}$$

This result follows from (2.8.2) and (2.8.4) below. Note that the circular matrix defined in (2.2.5) with scalar elements can also be generalized, by letting the elements in (2.2.5) be square matrices.

The determinant of the tri-diagonal matrix $\mathbf{A}: p \times p$ defined in (2.2.9) is given by

$$|\mathbf{A}| = \prod_{k=1}^{p} \left(a + 2b \cos \frac{k\pi}{p+1} \right). \tag{2.4.11}$$

(2.4.11) will be obtained from (2.8.2) and (2.8.5).

2.5 MATRIX INVERSION

Definition: Let $\mathbf{A} = (a_{ij})$ denote a square matrix of order p. If $|\mathbf{A}| \neq 0$, $\mathbf{A}$ has associated with it an inverse matrix denoted by $\mathbf{A}^{-1} = (a^{ij}): p \times p$. The inverse matrix is defined by

$$a^{ij} = \frac{A_{ji}}{|\mathbf{A}|}, \qquad |\mathbf{A}| \neq 0, \tag{2.5.1}$$

where A_{ij} is the cofactor of a_{ij}, and (A_{ji}) denotes the transpose of the cofactored matrix. Then

$$\mathbf{A}\mathbf{A}^{-1} = \mathbf{A}^{-1}\mathbf{A} = \mathbf{I}, \qquad (\mathbf{A}^{-1})' = (\mathbf{A}')^{-1}. \tag{2.5.2}$$

If $\mathbf{A}$ is singular, $\mathbf{A}^{-1}$ does not exist.

Example (2.5.1): Suppose

$$\mathbf{A} = (a_{ij}) = \begin{pmatrix} 3 & 2 \\ -1 & 4 \end{pmatrix}.$$

The cofactors are

$$A_{11} = 4, \qquad A_{12} = 1, \qquad A_{21} = -2, \qquad A_{22} = 3,$$

and the matrix of cofactors is

$$\begin{pmatrix} 4 & 1 \\ -2 & 3 \end{pmatrix}.$$

Therefore, the transpose of the cofactored matrix is

$$\begin{pmatrix} 4 & -2 \\ 1 & 3 \end{pmatrix}.$$

Since $|\mathbf{A}| = 14$,

$$\mathbf{A}^{-1} = \frac{1}{14}\begin{pmatrix} 4 & -2 \\ 1 & 3 \end{pmatrix}.$$

2.5.1 Elementary Properties of Matrix Inverses

(1) Let $\mathbf{A}$ and $\mathbf{B}$ both be of order p and assume $|\mathbf{A}| \neq 0$ and $|\mathbf{B}| \neq 0$. Then

$$(\mathbf{AB})^{-1} = \mathbf{B}^{-1}\mathbf{A}^{-1}. \tag{2.5.3}$$

This result follows from the fact that $(\mathbf{AB})(\mathbf{B}^{-1}\mathbf{A}^{-1}) = \mathbf{I}$.

(2) If $\mathbf{A}$ is nonsingular,

$$|\mathbf{A}^{-1}| = |\mathbf{A}|^{-1}.$$

This useful property may be seen to hold because from $\mathbf{A}^{-1}\mathbf{A} = \mathbf{I}$, it follows that $|\mathbf{A}^{-1}\mathbf{A}| = |\mathbf{A}^{-1}| \cdot |\mathbf{A}| = 1$.

(3) If $\mathbf{D}_\lambda = \text{diag}(\lambda_1, \ldots, \lambda_p)$,

$$\mathbf{D}_\lambda^{-1} = \text{diag}\left(\frac{1}{\lambda_1}, \ldots, \frac{1}{\lambda_p}\right). \tag{2.5.4}$$

(2.5.4) follows immediately from (2.5.1).

(4) If $\mathbf{\Gamma}$ is an orthogonal matrix, $\mathbf{\Gamma}^{-1} = \mathbf{\Gamma}'$. It is easy to see this by premultiplying by $\mathbf{\Gamma}^{-1}$ in $\mathbf{\Gamma\Gamma}' = \mathbf{I}$.

(5) *Binomial Inverse Theorem (Woodbury, 1950):* Assume $\mathbf{A}: p \times p$, $\mathbf{U}: p \times q$, $\mathbf{B}: q \times q$, $\mathbf{V}: q \times p$. Then, if $\mathbf{A}$ and $\mathbf{B}$ are nonsingular,

$$(\mathbf{A} + \mathbf{UBV})^{-1} = \mathbf{A}^{-1} - \mathbf{A}^{-1}\mathbf{UB}(\mathbf{B} + \mathbf{BVA}^{-1}\mathbf{UB})^{-1}\mathbf{BVA}^{-1}. \tag{2.5.5}$$

This well-known theorem is often useful for expressing matrix inverses in alternative ways. It is proved in Complement 2.2. One often arising case occurs for $\mathbf{A} = \mathbf{B} = \mathbf{I}$, and this results in

$$(\mathbf{I} + \mathbf{UV})^{-1} = \mathbf{I} - \mathbf{U}(\mathbf{I} + \mathbf{VU})^{-1}\mathbf{V}. \tag{2.5.5$'$}$$

As a special case of this theorem, take $\mathbf{B} = \mathbf{I}$, $q = 1$. Then $\mathbf{U}$ is a column vector, which is denoted by $\mathbf{u}$.[1] Similarly, $\mathbf{V}$ is a row vector, denoted by $\mathbf{v}'$. The theorem then implies that

(6)
$$(\mathbf{A} + \mathbf{uv}')^{-1} = \mathbf{A}^{-1} - \frac{\mathbf{A}^{-1}\mathbf{uv}'\mathbf{A}^{-1}}{1 + \mathbf{v}'\mathbf{A}^{-1}\mathbf{u}}. \tag{2.5.6}$$

If $\mathbf{A} = \mathbf{I}$ in (2.5.6),

$$(\mathbf{I} + \mathbf{uv}')^{-1} = \mathbf{I} - \frac{\mathbf{uv}'}{1 + \mathbf{v}'\mathbf{u}}. \tag{2.5.7}$$

(7) Consider the case of the intraclass correlation matrix defined in (2.2.4). $\mathbf{A}$ may be written

$$\mathbf{A} = (a - b)\mathbf{I} + b\mathbf{ee}', \tag{2.5.8}$$

where $\mathbf{e}$ is a column vector of ones. Hence, if $\mathbf{A}$ is of order p, applying (2.5.6) gives

$$\mathbf{A}^{-1} = \frac{\mathbf{I}}{a - b} - \frac{b\mathbf{ee}'}{(a - b)[a + (p - 1)b]}, \tag{2.5.9}$$

which is again a matrix with intraclass correlation structure.

[1] Unless otherwise specified, vectors will be denoted by lowercase letters, and should be assumed to be columns unless they are primed.

(8) Let $\mathbf{K}$ denote the Toeplitz matrix, defined in (2.2.8), for the case in which $\mathbf{K} = \mathbf{K}'$, $C_0 = 1$, $C_j = \rho^j$, $j = 1, \ldots, p$, $0 < |\rho| < 1$. The inverse is given by

$$\mathbf{K}^{-1} = \frac{1}{1-\rho^2}\begin{pmatrix} 1 & -\rho & & & 0 \\ -\rho & (1+\rho^2) & \ddots & & \\ & \ddots & \ddots & \ddots & \\ & & \ddots & (1+\rho^2) & -\rho \\ 0 & & & -\rho & 1 \end{pmatrix}, \tag{2.5.10}$$

which is a Jacobi matrix [see (2.2.6)]. This result may be checked by showing that $\mathbf{KK}^{-1} = \mathbf{I}$, for $\mathbf{K}$ defined in (2.2.8). Note that if ρ is small, $\mathbf{K}^{-1}$ is approximately a tri-diagonal matrix, a fact sometimes useful in numerical work.

(9) Let $\mathbf{G}$ denote the Green's matrix defined in (2.2.7). Then, $\mathbf{G}^{-1}$ is a symmetric Jacobi matrix. Analogously, the inverse of a symmetric Jacobi matrix is a Green's matrix. A proof of these assertions may be found in Karlin (1968).

2.5.2 Generalized Inverses

Let $\mathbf{A}$ be of order p and assume $|\mathbf{A}| = 0$. Hence, $\mathbf{A}^{-1}$ does not exist. However, it is possible to define a generalization of the usual inverse, $\mathbf{A}^*$, as a matrix that exists even when $\mathbf{A}^{-1}$ does not, that equals $\mathbf{A}^{-1}$ when $\mathbf{A}^{-1}$ exists, and that satisfies the property

$$\mathbf{AA}^*\mathbf{A} = \mathbf{A}. \tag{2.5.11}$$

$\mathbf{A}^*$ is called a generalized inverse of $\mathbf{A}$. This matrix is not necessarily unique although it is possible, by imposing additional constraints, to define it uniquely.

Moore (1935) and Penrose (1955) defined a particular generalized inverse (often called the pseudoinverse) as a matrix $\mathbf{A}^+$ satisfying

$$\begin{aligned} &\text{(a)}\quad \mathbf{AA}^+\mathbf{A} = \mathbf{A}, &&\text{(c)}\quad \mathbf{A}^+\mathbf{AA}^+ = \mathbf{A}^+, \\ &\text{(b)}\quad (\mathbf{AA}^+)' = \mathbf{AA}^+, &&\text{(d)}\quad (\mathbf{A}^+\mathbf{A})' = \mathbf{A}^+\mathbf{A}. \end{aligned} \tag{2.5.12}$$

Such a matrix not only always exists, but it is unique. A method of constructing $\mathbf{A}^+$ is discussed in 2.12.5 in terms of matrix factorizations.

Other generalized inverses have been defined so that, in addition to satisfying (2.5.11), they possess particular properties of special interest in certain kinds of problems. For additional background see, for instance, Rao (1965).

Generalized inverses can be very useful in the regression model (see Chapter 8), where inverses arise naturally. For example, computer solutions for models involving inverses can be simplified if generalized inverses are used in place of ordinary inverses.

2.6 PARTITIONED MATRICES

Define

$$\mathbf{A}: 3 \times 4 = \left(\begin{array}{c|ccc} a_{11} & a_{12} & a_{13} & a_{14} \\ a_{21} & a_{22} & a_{23} & a_{24} \\ \hline a_{31} & a_{32} & a_{33} & a_{34} \end{array}\right), \tag{2.6.1}$$

and insert horizontal and vertical lines as shown in (2.6.1). The division of the elements of $\mathbf{A}$ as shown represents a *partitioning* of the matrix. Define the matrices

$$\mathbf{A}_{11} = \begin{pmatrix} a_{11} \\ a_{21} \end{pmatrix}, \qquad \mathbf{A}_{12} = \begin{pmatrix} a_{12} & a_{13} & a_{14} \\ a_{22} & a_{23} & a_{24} \end{pmatrix},$$

$$A_{21} = (a_{31}), \qquad \mathbf{A}_{22} = (a_{32} \quad a_{33} \quad a_{34}).$$

Then $\mathbf{A}$ may be rewritten as

$$\mathbf{A} = \begin{pmatrix} \mathbf{A}_{11} & \mathbf{A}_{12} \\ A_{21} & \mathbf{A}_{22} \end{pmatrix},$$

and $\mathbf{A}_{ij}$ are *submatrices* of $\mathbf{A}$. Of course other divisions of $\mathbf{A}$ are possible and all such divisions define possible partitions and submatrices. The horizontal and vertical lines will usually be omitted.

Partitioned vectors and matrices are of interest, for example, when for inference purposes, the marginal distribution of a subvector of random variables is required. Moreover, it is often much easier to obtain determinants and inverses of submatrices than of the full matrix.

2.6.1 Inverse of a Partitioned Matrix

Now assume $\mathbf{A}: (p + n) \times (p + n)$ is an arbitrary matrix with $|\mathbf{A}| \neq 0$. Partition $\mathbf{A}$ as

$$\mathbf{A} = \begin{pmatrix} \mathbf{A}_{11} & \mathbf{A}_{12} \\ \mathbf{A}_{21} & \mathbf{A}_{22} \end{pmatrix}, \tag{2.6.2}$$

where $\mathbf{A}_{11} : p \times p$, $\mathbf{A}_{12} : p \times n$, $\mathbf{A}_{21} : n \times p$, $\mathbf{A}_{22} : n \times n$. Let

$$\mathbf{B} = \mathbf{A}^{-1} = \begin{pmatrix} \mathbf{B}_{11} & \mathbf{B}_{12} \\ \mathbf{B}_{21} & \mathbf{B}_{22} \end{pmatrix},$$

with $\mathbf{B}_{11} : p \times p$, $\mathbf{B}_{12} : p \times n$, $\mathbf{B}_{21} : n \times p$, $\mathbf{B}_{22} : n \times n$. It is often useful to be able to express the elements of $\mathbf{B}$ in terms of the elements of $\mathbf{A}$. Multiplication of two partitioned matrices is carried out just as it is for unpartitioned matrices (as long as the submatrices are conformable). Multiplying $\mathbf{A}$ and $\mathbf{B}$, setting the result equal to $\mathbf{I}$, and solving, gives the relations

$$\mathbf{B}_{11} = (\mathbf{A}_{11} - \mathbf{A}_{12}\mathbf{A}_{22}^{-1}\mathbf{A}_{21})^{-1} \equiv \mathbf{A}_{11\cdot 2}^{-1}, \tag{2.6.3}$$

$$\mathbf{B}_{22} = (\mathbf{A}_{22} - \mathbf{A}_{21}\mathbf{A}_{11}^{-1}\mathbf{A}_{12})^{-1} \equiv \mathbf{A}_{22\cdot 1}^{-1}, \tag{2.6.4}$$

$$\mathbf{B}_{12} = -\mathbf{A}_{11}^{-1}\mathbf{A}_{12}\mathbf{A}_{22\cdot 1}^{-1}, \qquad \mathbf{B}_{21} = -\mathbf{A}_{22}^{-1}\mathbf{A}_{21}\mathbf{A}_{11\cdot 2}^{-1}. \tag{2.6.5}$$

2.6.2 Determinant of a Partitioned Matrix

Consider the special matrices

$$\begin{pmatrix} \mathbf{A}_{11} & \mathbf{0} \\ \mathbf{A}_{21} & \mathbf{A}_{22} \end{pmatrix}, \quad \text{and} \quad \begin{pmatrix} \mathbf{A}_{11} & \mathbf{A}_{12} \\ \mathbf{0} & \mathbf{A}_{22} \end{pmatrix}.$$

Using the Laplace developments, it is easy to see that

$$\begin{vmatrix} \mathbf{A}_{11} & \mathbf{0} \\ \mathbf{A}_{21} & \mathbf{A}_{22} \end{vmatrix} = \begin{vmatrix} \mathbf{A}_{11} & \mathbf{A}_{12} \\ \mathbf{0} & \mathbf{A}_{22} \end{vmatrix} = |\mathbf{A}_{11}| \cdot |\mathbf{A}_{22}|. \tag{2.6.6}$$

The determinant of a general partitioned matrix is given in the following.

Theorem (2.6.1): Suppose $\mathbf{A}$ is partitioned as in (2.6.2). Then

(i) if $|\mathbf{A}_{22}| \neq 0$,

$$|\mathbf{A}| = |\mathbf{A}_{22}| \cdot |\mathbf{A}_{11} - \mathbf{A}_{12}\mathbf{A}_{22}^{-1}\mathbf{A}_{21}| \equiv |\mathbf{A}_{22}| \cdot |\mathbf{A}_{11\cdot 2}|; \tag{2.6.7}$$

(ii) if $|\mathbf{A}_{11}| \neq 0$,

$$|\mathbf{A}| = |\mathbf{A}_{11}| \cdot |\mathbf{A}_{22} - \mathbf{A}_{21}\mathbf{A}_{11}^{-1}\mathbf{A}_{12}| \equiv |\mathbf{A}_{11}| \cdot |\mathbf{A}_{22\cdot 1}|. \tag{2.6.8}$$

To prove (i) of this theorem use the extension of the properties in Section 2.4.1 to entire rows and columns; multiply the second matrix column of $\mathbf{A}$ in (2.6.2) by $\mathbf{A}_{22}^{-1}\mathbf{A}_{21}$; and subtract the result from the first matrix column. The resulting matrix has the same determinant as $\mathbf{A}$ and is obtained from (2.6.6). Part (ii) is completely analogous.

It is simple to check (by multiplication) that if $\mathbf{A}$ is defined as in (2.6.2) with $\mathbf{A}_{12} = \mathbf{0}$ and $\mathbf{A}_{21} = \mathbf{0}$, and if $|\mathbf{A}_{11}| \neq 0$ and $|\mathbf{A}_{22}| \neq 0$,

$$\mathbf{A}^{-1} = \begin{pmatrix} \mathbf{A}_{11} & \mathbf{0} \\ \mathbf{0} & \mathbf{A}_{22} \end{pmatrix}^{-1} = \begin{pmatrix} \mathbf{A}_{11}^{-1} & \mathbf{0} \\ \mathbf{0} & \mathbf{A}_{22}^{-1} \end{pmatrix}. \tag{2.6.9}$$

2.6.3 Transpose of a Partitioned Matrix

Another property of partitioned matrices that is often required concerns the transpose. Let $\mathbf{A}$ be partitioned as in (2.6.2). Then

$$\mathbf{A}' = \begin{pmatrix} \mathbf{A}_{11}' & \mathbf{A}_{21}' \\ \mathbf{A}_{12}' & \mathbf{A}_{22}' \end{pmatrix}. \tag{2.6.10}$$

For example, suppose $\mathbf{A}_{11}$ is a scalar α, $\mathbf{A}_{12}$ is a row vector $\mathbf{a}'$, $\mathbf{A}_{21}$ is a column vector $\mathbf{b}$, and $\mathbf{A}_{22}$ is a square matrix $\mathbf{C}$. Then since

$$\mathbf{A} = \begin{pmatrix} \alpha & \mathbf{a}' \\ \mathbf{b} & \mathbf{C} \end{pmatrix}, \quad \mathbf{A}' = \begin{pmatrix} \alpha & \mathbf{b}' \\ \mathbf{a} & \mathbf{C}' \end{pmatrix}.$$

2.7 RANK

Definition: Let $\mathbf{A}: p \times n$ be given, and consider the collection of determinants of all possible square submatrices which can be formed by partitioning or deletion of certain rows and/or columns of $\mathbf{A}$ (including $|\mathbf{A}|$ itself). The *rank of* $\mathbf{A}$ is the largest order of the submatrices with nonvanishing determinants.

Example (2.7.1): Let

$$\mathbf{A}: 3 \times 4 = \begin{pmatrix} -1 & 0 & 2 & 1 \\ 0 & 1 & 1 & -1 \\ 2 & 0 & -4 & -2 \end{pmatrix}.$$

First of all the rank of $\mathbf{A}$ cannot exceed 3 since the largest determinant that can be formed from $\mathbf{A}$ has order 3. Examination of the four submatrices of order 3 shows that they all have determinants which vanish. However, since there is at least one (there are many) submatrix of order 2 which has a nonvanishing determinant, $\mathbf{A}$ has rank 2.

2.7.1 Elementary Properties

In the following useful properties of rank, the notation $r(\mathbf{A})$ denotes the rank of a matrix $\mathbf{A}$. These elementary properties are proved in most books on matrix algebra and are reproduced here for reference. It will be important in the multivariate regression model to establish that the data matrix has sufficiently high rank [see (8.4.4)].

(1) Let $\mathbf{A}$ and $\mathbf{B}$ be two conformable matrices. Then

$$r(\mathbf{AB}) \leq \min\,[r(\mathbf{A}),r(\mathbf{B})]. \tag{2.7.1}$$

(2) If $\mathbf{A}: p \times n$ and $\mathbf{B}: p \times n$,

$$r(\mathbf{A} + \mathbf{B}) \leq r(\mathbf{A}) + r(\mathbf{B}). \tag{2.7.2}$$

(3) If $\mathbf{A}: p \times n$,

$$r(\mathbf{A}) \leq \min\,(p,n). \tag{2.7.3}$$

(4) If $\mathbf{P}: m \times m$, $\mathbf{Q}: n \times n$, $|\mathbf{P}| \neq 0$, and $|\mathbf{Q}| \neq 0$, then, for $\mathbf{A}: m \times n$,

$$r(\mathbf{PAQ}) = r(\mathbf{A}). \tag{2.7.4}$$

(5)

$$r(\mathbf{A}) = r(\mathbf{A}'). \tag{2.7.5}$$

(6) If $\mathbf{D} = \text{diag}\,(\lambda_1,\ldots,\lambda_p)$,

$$r(\mathbf{D}) = \text{number of } \lambda_j\text{'s} \neq 0. \tag{2.7.6}$$

Example: Suppose $\mathbf{x}$ is a $p \times 1$ column vector and $\mathbf{A}$ and $\mathbf{B}$ are $p \times p$ nonsingular matrices. Then, by (2.7.4),

$$r(\mathbf{Axx'B}) = r(\mathbf{xx'}),$$

and by (2.7.1) and (2.7.5),

$$r(\mathbf{Axx'B}) \leq \min\,[r(\mathbf{x}),r(\mathbf{x}')] = 1 \qquad \text{if } \mathbf{x} \not\equiv \mathbf{0};$$

that is, for $\mathbf{x} \not\equiv \mathbf{0}$, $r(\mathbf{Axx'B}) = 1$, and $r(\mathbf{Axx'B}) = 0$ only if $\mathbf{x} \equiv \mathbf{0}$.

2.8 LATENT ROOTS AND VECTORS

2.8.1 Latent Roots

Let $\mathbf{A}$ denote a square matrix of order p. The *characteristic equation* of $\mathbf{A}$ is given by

$$|\mathbf{A} - \lambda\mathbf{I}_p| = 0, \tag{2.8.1}$$

where λ is a scalar. Equation (2.8.1) represents an algebraic equation of degree p in λ,

$$\lambda^p + b_{p-1}\lambda^{p-1} + \cdots + b_1\lambda + b_0 = 0,$$

the p solutions of which are called the latent roots of $\mathbf{A}$. Other names for the λ_j's include proper values, eigenvalues, and characteristic values. It will be seen in Chapter 9 that the latent roots of a covariance matrix are the variances of the associated principal components.

In general, latent roots are complex numbers (occurring in complex conjugate pairs when the roots are not real), although if $\mathbf{A} = \mathbf{A}'$, the latent roots are all real.

If $(\lambda_1, \ldots, \lambda_p)$ denote the latent roots of $\mathbf{A}: p \times p$,

$$|\mathbf{A}| = \prod_{j=1}^{p} \lambda_j. \tag{2.8.2}$$

Thus, if $\mathbf{A}$ is of order p, $r(\mathbf{A}) < p$ if and only if at least one latent root is zero. The rank of a symmetric matrix, $\mathbf{A}$, is the number of nonzero latent roots it possesses. This can be shown using (2.12.1) and (2.7.6).

The proof of (2.8.2) follows from the fact that the characteristic equation of $\mathbf{A}$ yields a pth-degree polynomial (which is factorable in terms of the p latent roots). That is, since

$$|\mathbf{A} - \lambda\mathbf{I}| = \prod_{j=1}^{p} (\lambda_j - \lambda),$$

setting $\lambda = 0$ gives (2.8.2).

Latent Roots of Pattern Matrices

If $\mathbf{\Gamma}: p \times p$ is *orthogonal*, its latent roots lie on the unit circle (in the complex plane), that is, the roots all have modulus unity; reciprocals of the latent roots are also latent roots. These assertions are proven, for example, in Mirsky (1955), pp. 225–226.

If $\mathbf{A}$ is the pth-order *intraclass correlation* matrix given in (2.2.4), its latent roots are given by

$$\lambda_1 = a + (p-1)b, \qquad \lambda_2 = \cdots = \lambda_p = a - b. \tag{2.8.3}$$

This result is obtained easily by writing $\mathbf{A}$ as in (2.5.8) and using (2.4.3) in the characteristic equation.

If $\mathbf{A}$ is the *circular* matrix given in (2.2.5), the latent roots are given by (see, for instance, Press, 1964)

$$\lambda_k = \sum_{j=1}^{p} a_j \exp\left[\frac{2\pi i}{p}(j-1)(k-1)\right], \qquad k = 1, \ldots, p. \tag{2.8.4}$$

Equation (8.3.20) gives the result for the case of $\mathbf{A} = \mathbf{A}'$.

If $\mathbf{A}$ is the *tri-diagonal* matrix defined in (2.2.9), its latent roots are given by (Rutherford, 1952)

$$\lambda_k = a + 2b \cos \frac{k\pi}{p+1}, \qquad k = 1, \ldots, p. \tag{2.8.5}$$

If $\mathbf{G}$ is the special *matrix intraclass correlation* matrix defined in (2.4.8), where $\mathbf{A}$ and $\mathbf{B}$ each have intraclass correlation structure as well, that is,

$$\underset{(k\times k)}{\mathbf{A}} = \begin{pmatrix} a_1 & & & \\ & \cdot & b_1 & \\ & & \cdot & \\ & b_1 & \cdot & \\ & & & a_1 \end{pmatrix}, \qquad \underset{(k\times k)}{\mathbf{B}} = \begin{pmatrix} a_2 & & & \\ & \cdot & b_2 & \\ & & \cdot & \\ & b_2 & \cdot & \\ & & & a_2 \end{pmatrix},$$

the latent roots of $\mathbf{G}: pk \times pk$ are given by

$$\begin{aligned} \lambda_1 &= a_1 + (k-1)b_1 + (p-1)[a_2 + (k-1)b_2], \\ \lambda_2 &= \cdots = \lambda_k = (a_1 - b_1) + (p-1)(a_2 - b_2), \\ \lambda_{k+1} &= \cdots = \lambda_{p+k-1} = (a_1 - a_2) + (k-1)(b_1 - b_2), \\ \lambda_{p+k} &= \cdots = \lambda_{pk} = (a_1 - a_2) + (b_2 - b_1). \end{aligned} \tag{2.8.6}$$

A proof of this assertion is given in Complement 2.3.

An application of the matrix intraclass correlation matrix in the problem of combining cross section data with time series data may be found in Wallace and Hussain (1969), p. 58.

2.8.2 Latent Vectors

Let λ denote a scalar, and $\mathbf{A}$ any $p \times p$ matrix. If $\mathbf{x}$ is a $p \times 1$ vector not identically zero satisfying

$$(\mathbf{A} - \lambda\mathbf{I})\mathbf{x} = \mathbf{0}, \tag{2.8.7}$$

then $\mathbf{x}$ is called a latent vector of $\mathbf{A}$ associated with λ. Note that the solution of (2.8.7) is not unique although it is possible to solve this system of equations for the ratios of the components of $\mathbf{x}$. However, by normalizing the latent vectors so that $\mathbf{x}'\mathbf{x} = 1$, the ambiguity in $\mathbf{x}$ is removed up to a change in sign.

Latent vectors associated with the distinct latent roots of a symmetric matrix are orthogonal. If a latent root has multiplicity m, it is always possible to find m mutually orthogonal latent vectors of unit length that are also orthogonal to the latent vectors corresponding to the distinct latent roots (see, for instance, Mirsky, 1961, p. 304).

Latent Vectors of Pattern Matrices

The *intraclass correlation* matrix $\mathbf{A}: p \times p$ of (2.2.4) has the property, useful for statistical inference, that its latent vectors do not depend upon the elements of $\mathbf{A}$. Any p mutually orthogonal vectors of unit length, the first of which has components that are all identical, will suffice. For example, the columns of the Helmert matrix in (2.2.3) are a set of latent vectors for **A**.

The *circular* matrix **A** of (2.2.5) also has latent vectors not dependent upon the elements of **A** (the intraclass correlation matrix is a special case of the circular matrix). The kth normalized latent vector corresponding to the kth latent root given in (2.8.4) is

$$\frac{1}{p^{1/2}}\left(1, \exp\left[\frac{2\pi i(k-1)}{p}\right], \exp\left[\frac{4\pi i(k-1)}{p}\right], \ldots, \exp\left[\frac{2\pi i(k-1)(p-1)}{p}\right]\right)', \tag{2.8.8}$$

for $k = 1,\ldots,p$ [see also (8.3.21) for the result when $\mathbf{A} = \mathbf{A}'$].

Finally, if **A** is the *tri-diagonal* matrix defined in (2.2.9), the kth latent vector corresponding to the kth latent root in (2.8.5) is given by

$$\left(\sin\frac{k\pi}{p+1}, \sin\frac{2k\pi}{p+1}, \ldots, \sin\frac{pk\pi}{p+1}\right)', \tag{2.8.9}$$

where $k = 1,\ldots,p$. Note that these latent vectors are also independent of the elements of the **A** matrix.

The assertions above about the latent vectors of pattern matrices are easily checked by imposing the condition that every latent vector $\mathbf{x}$ must satisfy; that is, $\mathbf{Ax} = \lambda\mathbf{x}$.

2.9 TRACE OF A MATRIX

Definition: The sum of the diagonal elements of a square matrix is called its *trace*. Thus, if $\mathbf{A} = (a_{ij})$, $i, j = 1,\ldots,p$,

$$\text{tr } \mathbf{A} = \sum_{j=1}^{p} a_{jj}. \tag{2.9.1}$$

This operation on a matrix will prove to be of great use, for example, in helping to define the probability distributions of multivariate analysis.

2.9.1 Properties

(1) Assume $\mathbf{A}: p \times n$, $\mathbf{B}: n \times p$. Then

$$\text{tr }(\mathbf{AB}) = \text{tr }(\mathbf{BA}). \tag{2.9.2}$$

For example, if $\mathbf{x}$ is a $p \times 1$ vector,

$$\text{tr}(\mathbf{xx}') = \text{tr}(\mathbf{x}'\mathbf{x}) = \mathbf{x}'\mathbf{x}.$$

This result is found by noting that the ijth element of **AB** is $\Sigma_1^n a_{i\alpha} b_{\alpha j}$, so that

$$\operatorname{tr}(\mathbf{AB}) = \sum_{i=1}^{p} \sum_{\alpha=1}^{n} a_{i\alpha} b_{\alpha i}.$$

Moreover, the ijth element of **BA** is $\Sigma_1^p b_{i\alpha} a_{\alpha j}$, so that

$$\operatorname{tr}(\mathbf{BA}) = \sum_{i=1}^{n} \sum_{\alpha=1}^{p} b_{i\alpha} a_{\alpha i}.$$

Interchanging subscripts shows the equality of the two.

(2) The trace of a square matrix is equal to the sum of its latent roots [see for instance, Mirsky (1955), p. 198].

(3) If $\mathbf{A}: p \times n$, $\mathbf{B}: p \times n$, then

$$\operatorname{tr}(\mathbf{A} + \mathbf{B}) = \operatorname{tr}(\mathbf{A}) + \operatorname{tr}(\mathbf{B}). \tag{2.9.3}$$

This follows from the fact that

$$\operatorname{tr}(\mathbf{A} + \mathbf{B}) = \sum_i (a_{ii} + b_{ii}) = \sum_i a_{ii} + \sum_i b_{ii} = \operatorname{tr}\mathbf{A} + \operatorname{tr}\mathbf{B}.$$

(4) If α is a scalar and if $\mathbf{A}: p \times p$,

$$\operatorname{tr}(\alpha\mathbf{A}) = \alpha \operatorname{tr}\mathbf{A}. \tag{2.9.4}$$

The proof is that

$$\operatorname{tr}(\alpha\mathbf{A}) = \sum_i \alpha a_{ii} = \alpha \sum_i a_{ii} = \alpha \operatorname{tr}\mathbf{A}.$$

(5) Let $\mathbf{S}: p \times p$, $|\mathbf{S}| \neq 0$, $\mathbf{A}: p \times p$. Then

$$\operatorname{tr}(\mathbf{S}^{-1}\mathbf{AS}) = \operatorname{tr}\mathbf{A}. \tag{2.9.5}$$

This follows immediately from (2.9.2). The transformation $\mathbf{S}^{-1}\mathbf{AS}$ is often called a *similarity transformation*. (In Section 15.2.3 it will be seen that these transformations arise in the multidimensional scaling model.)

(6) Let $r(\mathbf{A}) = 1$ for $\mathbf{A}: p \times p$. Then, if $|\mathbf{A} - \lambda\mathbf{I}| = 0$, it follows that $\lambda = \operatorname{tr}(\mathbf{A})$. For example, if $\mathbf{A} = \mathbf{xx}'$, where $\mathbf{x}$ is a $p \times 1$ vector, $\lambda = \mathbf{x}'\mathbf{x}$; that is,

$$|\mathbf{xx}' - \lambda\mathbf{I}| = 0 \Rightarrow \lambda = \mathbf{x}'\mathbf{x}. \tag{2.9.6}$$

Since the rank of a matrix is the number of its nonvanishing latent roots [see the remark following (2.8.2)], if $r(\mathbf{A}) = 1$, **A** has but one nonzero latent root. From Property (2), the root must be given by $\operatorname{tr}\mathbf{A}$.

(7) Let $\mathbf{A}: p \times n$, $\mathbf{B}: n \times p$, for $\mathbf{A} = (a_{ij})$, $\mathbf{B} = (b_{ij})$. Then

$$\operatorname{tr}(\mathbf{AB}) = \sum_{i=1}^{p} \sum_{j=1}^{n} a_{ij} b_{ji}. \tag{2.9.7}$$

This was proved following (2.9.2). If $\mathbf{A}$ and $\mathbf{B}$ are symmetric matrices,

$$\operatorname{tr}(\mathbf{AB}) = \sum_{i=1}^{n} \sum_{j=1}^{n} a_{ij} b_{ij}.$$

More generally,

$$\operatorname{tr}(\mathbf{ABC}) = \sum_{\alpha} \sum_{\beta} \sum_{\gamma} a_{\alpha\beta} b_{\beta\gamma} c_{\gamma\alpha}. \tag{2.9.8}$$

These results follow by definition of a matrix product and of a trace.

(8) If A is a scalar, $\operatorname{tr}(A) = A$.
As an example, if $\mathbf{x}: p \times 1$, $\mathbf{A}: p \times p$, $\mathbf{x}'\mathbf{A}\mathbf{x}$ is a scalar. Hence, $\operatorname{tr} \mathbf{A}\mathbf{x}\mathbf{x}' = \mathbf{x}'\mathbf{A}\mathbf{x}$.

2.10 DIRECT PRODUCT

Definition: Let $\mathbf{A} = (a_{ij}) : p_1 \times p_2$ and $\mathbf{B} = (b_{kl}) : n_1 \times n_2$. Then the $p_1 n_1 \times p_2 n_2$ matrix $\mathbf{C} = \mathbf{A} \otimes \mathbf{B}$ is called the direct product of $\mathbf{A}$ and $\mathbf{B}$ and is given by

$$\mathbf{C} = \mathbf{A} \otimes \mathbf{B} = \begin{pmatrix} a_{11}\mathbf{B} & a_{12}\mathbf{B} & \cdots & a_{1p_2}\mathbf{B} \\ \cdot & \cdot & & \cdot \\ \cdot & \cdot & & \cdot \\ \cdot & \cdot & & \cdot \\ a_{p_1 1}\mathbf{B} & a_{p_1 2}\mathbf{B} & \cdots & a_{p_1 p_2}\mathbf{B} \end{pmatrix}. \tag{2.10.1}$$

Direct products are very useful, for example, in the multivariate regression model [see (8.4.6)].

2.10.1 Properties

(1) If $\mathbf{A}$, $\mathbf{B}$, $\mathbf{C}$, $\mathbf{F}$ are such that the products are defined,

$$(\mathbf{A} \otimes \mathbf{B})(\mathbf{C} \otimes \mathbf{F}) = \mathbf{AC} \otimes \mathbf{BF}. \tag{2.10.2}$$

This property may be checked directly from the definition.

(2) Let $\mathbf{A}: p \times p$, and $\mathbf{B}: n \times n$. Then it may be verified by multiplication that if $\mathbf{A}$ and $\mathbf{B}$ are nonsingular,

$$(\mathbf{A} \otimes \mathbf{B})^{-1} = \mathbf{A}^{-1} \otimes \mathbf{B}^{-1}. \tag{2.10.3}$$

To prove (2.10.3), multiply $(\mathbf{A} \otimes \mathbf{B})(\mathbf{A}^{-1} \otimes \mathbf{B}^{-1})$ using (2.10.2).

(3) Suppose the latent roots of $\mathbf{A}: p \times p$ are $(\lambda_1, \ldots, \lambda_p)$, and those of $\mathbf{B}: n \times n$ are $(\theta_1, \ldots, \theta_n)$. Then

$$|A \otimes B| = \left(\prod_{j=1}^{p} \lambda_j \right)^n \left(\prod_{k=1}^{n} \theta_k \right)^p . \tag{2.10.4}$$

This follows from (2.8.2) and Property (5) below.

(4) For $\mathbf{A}: p \times p$, and $\mathbf{B}: n \times n$,

$$|\mathbf{A} \otimes \mathbf{B}| = |\mathbf{A}|^n \cdot |\mathbf{B}|^p. \tag{2.10.5}$$

Use (2.8.2) and (2.10.4) for proof.

(5) Using the same notation as in Property (3), the latent roots of $\mathbf{A} \otimes \mathbf{B}$ are $(\lambda_j \theta_k)$, $j = 1, \ldots, p$, $k = 1, \ldots, n$.

For proof, see for instance, Anderson, 1958.

(6)
$$(\mathbf{A} \otimes \mathbf{B})' = \mathbf{A}' \otimes \mathbf{B}'. \tag{2.10.6}$$

This may be checked from the definition.

(7)
$$\text{tr} \, (\mathbf{A} \otimes \mathbf{B}) = (\text{tr} \, \mathbf{A}) \cdot (\text{tr} \, \mathbf{B}). \tag{2.10.7}$$

(8) For α a scalar,

$$\alpha \otimes \mathbf{A} = \mathbf{A} \otimes \alpha. \tag{2.10.8}$$

Property (7) is proved by noting that $\text{tr} \, \mathbf{A} \otimes \mathbf{B} = a_{11} \, \text{tr} \, \mathbf{B} + a_{22} \, \text{tr} \, \mathbf{B} + \cdots + a_{pp} \, \text{tr} \, \mathbf{B}$. Property (8) follows by inspection of the defining matrix.

2.11 QUADRATIC FORMS AND DEFINITENESS

Definition: Let $\mathbf{x}: p \times 1$ denote a p vector, and $\mathbf{A} = (a_{ij})$ denote a $p \times p$ symmetric matrix. Then

$$Q = \mathbf{x}'\mathbf{A}\mathbf{x} = \sum_{i=1}^{p} \sum_{j=1}^{p} x_i x_j a_{ij} \tag{2.11.1}$$

is called a *quadratic form* in the variables $x_1, \ldots, x_p$. It shall always be assumed that the square matrix $\mathbf{A}$ of a quadratic form is symmetric, since if it is not, it can always be made so by defining

$$\mathbf{B} = \frac{\mathbf{A} + \mathbf{A}'}{2},$$

and considering $\mathbf{x}'\mathbf{B}\mathbf{x}$.

The density function of a multivariate Normal distribution will involve a quadratic form in the variables [see (3.3.1)].

Example (2.11.1): If $p = 2$ and $\mathbf{x}' = (x_1, x_2)$,

$$Q = a_{11}x_1^2 + 2a_{12}x_1x_2 + a_{22}x_2^2.$$

2.11.1 Positive Definiteness

If $Q \equiv \mathbf{x}'\mathbf{A}\mathbf{x} > 0$ (<0) for all vectors $\mathbf{x}$, $\mathbf{x}$ not identically zero, Q is called a *positive* (*negative*) *definite quadratic form* and the matrix $\mathbf{A}$ is called positive (negative) definite; it is often written $\mathbf{A} > 0$ $(\mathbf{A} < 0)$. If $Q \geq 0$ for all $\mathbf{x} \not\equiv \mathbf{0}$, Q is called a positive semidefinite quadratic form, and $\mathbf{A}$ is called a positive semidefinite matrix; this latter fact is denoted by $\mathbf{A} \geq 0$. If none of these conditions holds, Q is called an indefinite quadratic form, and $\mathbf{A}$ is called an indefinite matrix.

A positive definite matrix can also be defined strictly in terms of the properties of matrices. Consider all submatrices of $\mathbf{A}: p \times p$, $\mathbf{A} = \mathbf{A}'$, whose principal diagonals are part of the principal diagonal of $\mathbf{A}$. The determinants of these *principal submatrices* are called the *principal minors* of $\mathbf{A}$. The definition of positive definiteness above shows that $\mathbf{A} > 0$ if and only if $|\mathbf{A}| > 0$ and all principal minors of $\mathbf{A}$ are positive. We will use the notation $\mathbf{A} > \mathbf{B}$ to mean that $\mathbf{A} - \mathbf{B} > 0$.

Example (2.11.2): An important special case is a covariance matrix $\mathbf{\Sigma}$. That is, $\mathbf{\Sigma} = \mathbf{\Sigma}'$ and $\mathbf{\Sigma} \geq 0$ for every covariance matrix. If $|\mathbf{\Sigma}| = 0$, which occurs, for example, when random variables are perfectly correlated, there exist vectors $\mathbf{x} \not\equiv \mathbf{0}$ for which $\mathbf{x}'\mathbf{\Sigma}\mathbf{x} = 0$. Thus, if $\mathbf{\Sigma} = \begin{pmatrix} 4 & 2 \\ 2 & 1 \end{pmatrix}$, $\mathbf{x}'\mathbf{\Sigma}\mathbf{x} = 0$ if $\mathbf{x} = (1, -2)'$.

2.11.2 Properties of Positive Definite Matrices

The following properties are proven, for example, in Beckenbach and Bellman (1961).

(1) If $\mathbf{A} > 0$, $\mathbf{A}^{-1} > 0$. (2.11.2)

(2) Assume $\mathbf{A}: n \times n$, $\mathbf{B}: p \times n$, $p \leq n$, $\mathbf{A} > 0$, and rank $(\mathbf{B}) = r$. Then

$$\mathbf{BAB}' > 0 \quad \text{if } r = p, \qquad \text{and} \qquad \mathbf{BAB}' \geq 0 \quad \text{if } r < p. \tag{2.11.3}$$

(3) If $\mathbf{A} > 0$, and $\mathbf{B} > 0$, and both are of order p,

$$|\mathbf{A}|^{1/p} + |\mathbf{B}|^{1/p} \leq |\mathbf{A} + \mathbf{B}|^{1/p}, \tag{2.11.4}$$

with equality only if $\mathbf{A}$ is proportional to $\mathbf{B}$.

(4) If $\mathbf{A} > 0$, $\mathbf{B} > 0$, and $\mathbf{A} - \mathbf{B} > 0$, then $|\mathbf{A}| > |\mathbf{B}|$. (2.11.5)

(5) If $\mathbf{A} > 0$, there exists at least one matrix $\mathbf{B}$ such that $\mathbf{A} = \mathbf{BB}'$. Conversely, if for some $\mathbf{B}$, $\mathbf{A} = \mathbf{BB}'$, then $\mathbf{A} \geq 0$.

(6) *Hadamard's Inequality (1893):* Let $\mathbf{A}: p \times p = (a_{ij})$, and assume $\mathbf{A} > 0$. Then

$$|\mathbf{A}| \leq \prod_{1}^{p} a_{jj}. \tag{2.11.6}$$

(7) If $\mathbf{A} > 0$, there exists a nonsingular matrix $\mathbf{M}$ such that

$$\mathbf{MAM}' = \mathbf{I}. \tag{2.11.7}$$

(8) If $\mathbf{A} > 0$, all its latent roots are positive, and conversely.

(9) If $\mathbf{A} > 0$, $\mathbf{B} > 0$, and $\mathbf{A} - \mathbf{B} > 0$, then

$$\mathbf{B}^{-1} - \mathbf{A}^{-1} > 0. \tag{2.11.8}$$

(10) Let $\lambda_j(\mathbf{A})$, $\lambda_j(\mathbf{B})$ denote the jth latent roots of $\mathbf{A}$ and $\mathbf{B}$, respectively, where $\mathbf{A} = \mathbf{A}'$ and $\mathbf{B} = \mathbf{B}'$. Then, if $\mathbf{A} - \mathbf{B} \geq 0$, $\lambda_j(\mathbf{A}) \geq \lambda_j(\mathbf{B})$, for all j. For proof, see, for instance, Beckenbach and Bellman, p. 73.

(11) If $\mathbf{A} > 0$, then there exists a lower triangular matrix $\mathbf{T} = (t_{ij})$, $(t_{ij} = 0, i < j)$, such that $\mathbf{A} = \mathbf{TT}'$. Moreover, $\mathbf{T}$ is unique except for signs. For proof, see Graybill, 1969, p. 189.

2.11.3 Extrema for Quadratic Forms

If $\mathbf{A} = \mathbf{A}'$ and λ_j's denote the latent roots of $\mathbf{A}$, $j = 1, \ldots, p$,

$$\min_{1 \leq j \leq p} (\lambda_j) \leq \frac{\mathbf{x}'\mathbf{A}\mathbf{x}}{\mathbf{x}'\mathbf{x}} \leq \max_{1 \leq j \leq p} (\lambda_j). \tag{2.11.9}$$

This ratio of quadratic forms occurs, for example, as the test statistic (von Neumann ratio) in testing the disturbances in a regression for serial correlation (see Press, 1969b). Use of (2.12.1) easily establishes this result.

2.12 MATRIX FACTORIZATIONS

In multivariate analysis, many of the techniques that have been developed to facilitate the interpretation of large quantities of correlated data involve transforming the matrices of data into new matrices whose structures are simpler to interpret than those of the original data matrix. In this connection, matrix factorizations are fundamental. For example, it will be seen in Chapter 10 that the factor analysis model involves a factorization of the covariance matrix of the underlying data matrix. It will be shown below that by means of matrix factorizations, it is possible to define inverses for matrices that are singular, and to define such things as powers and square roots of matrices. Some basic factorizations are given in the sections below. Unless otherwise noted, proofs can be found in elementary texts on matrix algebra such as Bellman (1960) and Mirsky (1955).

2.12.1 Factorization of Symmetric Matrices

If $\mathbf{A} = \mathbf{A}'$, there exists an orthogonal matrix $\mathbf{\Gamma}$ such that

$$\mathbf{A} = \mathbf{\Gamma} \mathbf{D}_\lambda \mathbf{\Gamma}', \tag{2.12.1}$$

where $\mathbf{D}_\lambda = \text{diag}(\lambda_1, \ldots, \lambda_p)$ and the λ_j's are the latent roots of $\mathbf{A}$ (some of which may be zero and some of which may be multiple roots). If the elements of the first row of $\mathbf{\Gamma}$ have the same sign, and if the latent roots are distinct, the representation, with the elements of $\mathbf{D}$ arranged so that $\lambda_1 > \lambda_2 > \cdots > \lambda_p$, is unique. In any case, the columns of $\mathbf{\Gamma}$ are an orthonormal set of latent vectors of $\mathbf{A}$. For proof of the main result, see for instance, Mirsky (1955), p. 304.

2.12.2 Simultaneous Diagonalization

Let $\mathbf{A} > 0$, $\mathbf{B} \geq 0$. Then there exists a matrix $\mathbf{V}$, $|\mathbf{V}| \neq 0$, such that

$$\mathbf{VAV}' = \mathbf{I}, \qquad \text{and} \qquad \mathbf{VBV}' = \mathbf{D}_\lambda = \text{diag}(\lambda_1, \ldots, \lambda_p), \tag{2.12.2}$$

where the λ_j's are the roots of $|\mathbf{B} - \lambda \mathbf{A}| = 0$.

2.12.3 Nonuniqueness of Factorizations

Vinograd's Theorem (1950): For $\mathbf{A} : p \times n$ and $\mathbf{B} : p \times n$, assume $\mathbf{AA}' = \mathbf{BB}'$. Then there exists an orthogonal matrix $\mathbf{\Delta} : n \times n$ such that

$$\mathbf{A} = \mathbf{B}\mathbf{\Delta}. \tag{2.12.3}$$

That is, two different factorizations of a positive definite matrix are related by a rotation (see Section 10.6 for an application in factor analysis).

2.12.4 Orthogonal Factorization

For $\mathbf{A} : p \times n$, $p \leq n$, there exist orthogonal matrices $\mathbf{\Gamma} : p \times p$ and $\mathbf{\Delta} : n \times n$, such that

$$\mathbf{A} = \mathbf{\Gamma}(\mathbf{D}_\lambda \; \mathbf{0})\mathbf{\Delta}', \tag{2.12.4}$$

where $\mathbf{D}_\lambda = \text{diag}(\lambda_1, \ldots, \lambda_p)$ and $|\mathbf{AA}' - \lambda^2 \mathbf{I}| = 0$. Note that if $n = p$ and $\mathbf{A} = \mathbf{A}'$, by taking $\Delta = \Gamma$, this property reduces to (2.12.1). If γ_j denotes the jth column of Γ, $(\mathbf{AA}')\gamma_j = \lambda_j^2 \gamma_j$, $j = 1, \ldots, p$; $\gamma_j' \gamma_j = 1$. If δ_j denotes the jth column of Δ, $(\mathbf{A}'\mathbf{A})\delta_j = \bar{\lambda}_j^2 \delta_j$, $j = 1, \ldots, n$; $\delta_j'\delta_j = 1$, $|\mathbf{A}'\mathbf{A} - \bar{\lambda}^2 \mathbf{I}_n| = 0$. Note that $(\bar{\lambda}_1^2, \ldots, \bar{\lambda}_n^2) \equiv (\lambda_1^2, \ldots, \lambda_p^2, 0, \ldots, 0)$. λ_j may be taken as the positive root of λ_j^2.

2.12.5 Pseudoinverse

Suppose $\mathbf{A}: p \times n$, $p \leq n$. Decompose $\mathbf{A}$ using (2.12.4) above. Then replace the nonzero elements of $\mathbf{D}_\lambda$ by their reciprocals and call the resulting matrix $\mathbf{D}_\lambda{}^*$. The pseudoinverse of $\mathbf{A}$ is given by

$$\mathbf{A}^+ = \mathbf{\Delta}(\mathbf{D}_\lambda{}^* \ \mathbf{0})'\mathbf{\Gamma}'. \tag{2.12.5}$$

Note that $\mathbf{\Delta}$ and $\mathbf{\Gamma}'$ are obtained as in (2.12.4). Note also that $(\mathbf{A}^+)' = (\mathbf{A}')^+$, and $\mathbf{A}^+ = (\mathbf{A}'\mathbf{A})^+\mathbf{A}'$. Moreover, when the inverse exists, it is equal to the pseudoinverse. (See also Sect. 2.5.2).

Example (2.12.1): Suppose $\mathbf{A} = \begin{pmatrix} 4 & 2 \\ 2 & 1 \end{pmatrix}$, so that $|\mathbf{A}| = 0$. Compute the pseudoinverse of $\mathbf{A}$. First find the latent roots and vectors.

$$|\mathbf{A} - \lambda\mathbf{I}| = 0 \qquad \text{implies} \qquad \lambda_1 = 5,\ \lambda_2 = 0.$$

The matrix of latent vectors, $\mathbf{\Gamma}$, is computed from $\mathbf{A}\mathbf{x}_k = \lambda_k\mathbf{x}_k$, $\mathbf{x}_k{}'\mathbf{x}_k = 1$, $k = 1, 2$. One $\mathbf{\Gamma}$ satisfying these equations is easily found to be

$$\mathbf{\Gamma} = \begin{pmatrix} \frac{2}{\sqrt{5}} & -\frac{1}{\sqrt{5}} \\ \frac{1}{\sqrt{5}} & \frac{2}{\sqrt{5}} \end{pmatrix} = \frac{1}{\sqrt{5}}\begin{pmatrix} 2 & -1 \\ 1 & 2 \end{pmatrix}.$$

Another $\mathbf{\Gamma}$ satisfying the required equations is

$$\mathbf{\Gamma} = \begin{pmatrix} \frac{2}{\sqrt{5}} & \frac{1}{\sqrt{5}} \\ \frac{1}{\sqrt{5}} & -\frac{2}{\sqrt{5}} \end{pmatrix} = \frac{1}{\sqrt{5}}\begin{pmatrix} 2 & 1 \\ 1 & -2 \end{pmatrix}.$$

As mentioned above, $\mathbf{\Gamma}$ is unique only up to a sign change. Adopt the first solution for $\mathbf{\Gamma}$.

Since $\mathbf{D}_\lambda = \text{diag}\,(5,0)$, $\mathbf{D}_\lambda{}^* = \text{diag}\,(\frac{1}{5},0)$. Then since $\mathbf{A} = \mathbf{\Gamma}\mathbf{D}_\lambda\mathbf{\Gamma}'$,

$$\mathbf{A}^+ = \mathbf{\Gamma}\mathbf{D}_\lambda{}^*\mathbf{\Gamma}' = \begin{pmatrix} \frac{4}{25} & \frac{2}{25} \\ \frac{2}{25} & \frac{1}{25} \end{pmatrix} = \frac{1}{25}\,\mathbf{A}.$$

It is easy to check that $\mathbf{A}^+$ satisfies $\mathbf{A}\mathbf{A}^+\mathbf{A} = \mathbf{A}$, in addition to the other properties in (2.5.12).

Example (2.12.2): Suppose $\mathbf{A} \equiv \mathbf{a}$: $n \times 1$. Find its pseudoinverse.

Factoring $\mathbf{b} \equiv \mathbf{a}'$, $\mathbf{b}$: $1 \times n$ according to (2.12.4) gives $\mathbf{\Gamma} = 1$, $\mathbf{D}_\lambda = (\mathbf{b}\mathbf{b}')^{1/2}$, and

$$\mathbf{b} = (\sqrt{\mathbf{bb}'}, 0, \ldots, 0)\Delta',$$

where

$$\Delta = \left(\frac{\mathbf{a}}{\sqrt{\mathbf{a}'\mathbf{a}}}, H \right),$$

and H is chosen arbitrarily except for the constraints that $H'H = \mathbf{I}$, and $H'\mathbf{a} = \mathbf{0}$. Then, $\Delta\Delta' = \mathbf{I}$. Substituting into (2.12.5) gives

$$\mathbf{b}^+ = (\mathbf{a}')^+ = (\mathbf{a}^+)' = (\mathbf{a}'\mathbf{a})^{-1}\mathbf{a}'.$$

Note that this result could also have been obtained directly by substituting in the result of Section 2.12.5 that

$$\mathbf{A}^+ = (\mathbf{A}'\mathbf{A})^+ \mathbf{A}',$$

and noting that in this case $(\mathbf{A}'\mathbf{A})^+ = (\mathbf{a}'\mathbf{a})^+ = (\mathbf{a}'\mathbf{a})^{-1}$.

Example (2.12.3): Suppose $\mathbf{A} = \mathbf{B} \otimes \mathbf{C}$. It is straightforward to check that

$$\mathbf{A}^+ = \mathbf{B}^+ \otimes \mathbf{C}^+.$$

Example (2.12.4): Suppose $\mathbf{A} = \begin{pmatrix} 1 & 0 & 1 \\ -1 & 1 & 0 \end{pmatrix}$. Find the pseudo-inverse.

The latent roots of $\mathbf{AA}'$ are 3, 1. Take the latent vectors of $\mathbf{AA}'$ as $1/\sqrt{2}(1,-1)'$, and $1/\sqrt{2}(1,1)'$, respectively. The latent roots of $\mathbf{A}'\mathbf{A}$ are (3, 1, 0) and its latent vectors are taken as $(2/\sqrt{6}, -1/\sqrt{6}, 1/\sqrt{6},)$ $(0, 1/\sqrt{2}, 1/\sqrt{2})$, $(1/\sqrt{3}, 1/\sqrt{3}, -1/\sqrt{3})$, respectively. Substituting into (2.12.4) and (2.12.5) gives

$$\mathbf{A}^+ = 1/3 \begin{pmatrix} 1 & -1 \\ 1 & 2 \\ 2 & 1 \end{pmatrix}.$$

Remark: Suppose $\mathbf{A}: p \times n$, and $\mathbf{g}: p \times 1$ are known, and $\mathbf{x}: n \times 1$ is unknown. Then, if $\mathbf{Ax} = \mathbf{g}$,

$$\mathbf{x} = \mathbf{A}^+\mathbf{g} + (\mathbf{I} - \mathbf{A}^+\mathbf{A})\mathbf{h},$$

where $\mathbf{h}$ denotes any arbitrary vector (see Rao, 1965). It may also be found that if $\mathbf{AXB} = \mathbf{C}$, with $(\mathbf{A}, \mathbf{B}, \mathbf{C})$ known, $\mathbf{X} = \mathbf{A}^+\mathbf{CB}^+ + \mathbf{H} - \mathbf{A}^+\mathbf{AHBB}^+$, where $\mathbf{X}$ denotes an unknown matrix, all products are defined, and $\mathbf{H}$ is any conformable matrix.

2.12.6 Powers of Symmetric Matrices

Assume $\mathbf{A} = \mathbf{A}'$ so that from (2.12.1), $\mathbf{A} = \mathbf{\Gamma D}_\lambda\mathbf{\Gamma}'$. Then

$$\mathbf{A}^2 = \mathbf{A} \cdot \mathbf{A} = (\mathbf{\Gamma D}_\lambda\mathbf{\Gamma}')(\mathbf{\Gamma D}_\lambda\mathbf{\Gamma}') = \mathbf{\Gamma D}_\lambda{}^2\mathbf{\Gamma}' = \mathbf{\Gamma D}_{\lambda^2}\mathbf{\Gamma}'.$$

Since $\mathbf{A}^n = \mathbf{A}^{n-1} \cdot \mathbf{A}$, it is clear that for any positive integer n,

$$\mathbf{A}^n = \mathbf{\Gamma D}_\lambda{}^n\mathbf{\Gamma}'. \tag{2.12.6}$$

Since $\mathbf{A}^{-1} = (\mathbf{\Gamma D}_\lambda \mathbf{\Gamma}')^{-1} = (\mathbf{\Gamma}')^{-1}\mathbf{D}_\lambda{}^{-1}\mathbf{\Gamma}^{-1} = \mathbf{\Gamma D}_\lambda{}^{-1}\mathbf{\Gamma}'$, and $\mathbf{A}^{-n} = (\mathbf{A}^{-1})^n$, $\mathbf{A}^n = \mathbf{\Gamma D}_\lambda{}^n\mathbf{\Gamma}'$ for any integer n whether positive or negative.

If $\mathbf{A} > 0$, this definition may be extended to the case in which n is any rational number.

2.12.7 Simultaneous Orthogonal Diagonalization

If $\mathbf{A} = \mathbf{A}'$, $\mathbf{B} = \mathbf{B}'$, and $\mathbf{AB} = \mathbf{BA}$, there exists an orthogonal matrix $\mathbf{P}: p \times p$ such that

$$\begin{aligned} \mathbf{P}'\mathbf{AP} &= \operatorname{diag}(\alpha_1, \ldots, \alpha_p), \\ \mathbf{P}'\mathbf{BP} &= \operatorname{diag}(\beta_1, \ldots, \beta_p), \end{aligned} \tag{2.12.7}$$

where the α_j, β_j are the latent roots of $\mathbf{A}$, $\mathbf{B}$, respectively. Moreover, the simultaneous diagonalization is not possible unless $\mathbf{AB} = \mathbf{BA}$ (see Bellman, (1960, p. 56). Since in applied multivariate analysis matrices do not usually commute, simultaneous diagonalization is not usually possible. (For an exception, see Complement 2.3.)

2.13 IDEMPOTENT MATRICES

Definition: A square symmetric matrix $\mathbf{A}: p \times p$ is said to be *idempotent*[2] if $\mathbf{A}^2 = \mathbf{A}$. If $\mathbf{y} = \mathbf{Ax}$, and $\mathbf{A}: p \times p$ is idempotent, $\mathbf{y}$ is a projection of $\mathbf{x}$. Such matrices occur as an integral part of multivariate regression and the analysis of variance to be discussed later [see for instance the remark following (8.4.10); also, Exercise 2.3]. An extension of the idempotency notion is tripotency. A symmetric matrix $\mathbf{A}$ is tripotent if $\mathbf{A}^3 = \mathbf{A}$ [see Luther (1965)].

Example (2.13.1): It is straightforward to check that

$$\mathbf{A} = \frac{1}{3}\begin{pmatrix} 2 & -1 & 1 & 0 \\ -1 & 1 & 0 & -1 \\ 1 & 0 & 1 & -1 \\ 0 & -1 & -1 & 2 \end{pmatrix}$$

is an idempotent matrix.

2.13.1 Properties

(1) If $\mathbf{A}$ is idempotent, its latent roots are zeros or ones.
Use (2.12.6) for $n = 1, 2$ to prove this result.

(2) If $\mathbf{A}$ is idempotent and $\mathbf{A} \geq 0$,

[2] In the sequel, all idempotent matrices will be assumed to be symmetric.

$$r(\mathbf{A}) = \text{tr}\ (\mathbf{A}).$$

From Property (1) and the fact that tr (**A**) is the sum of its latent roots, tr (**A**) is the number of its nonzero latent roots, which is its rank.

(3) If **A** is idempotent, $(\mathbf{I} - \mathbf{A})$ is also idempotent.
For proof, merely square $(\mathbf{I} - \mathbf{A})$.

(4) If **A** is idempotent and $\mathbf{A} = \mathbf{A}'$, then $\mathbf{A} \geq 0$.
Let **x** denote any conformable vector. Then $\mathbf{x}'\mathbf{A}\mathbf{x} = \mathbf{x}'\mathbf{A}^2\mathbf{x} = \mathbf{y}'\mathbf{y}$, where $\mathbf{y} = \mathbf{A}\mathbf{x}$. But $\mathbf{y}'\mathbf{y} \geq 0$.

(5) If **A** is idempotent and $\mathbf{A} \geq 0$, there exists an orthogonal matrix $\mathbf{\Gamma}$ such that

$$\mathbf{\Gamma}\mathbf{A}\mathbf{\Gamma}' = \begin{pmatrix} \mathbf{I}_r & \mathbf{0} \\ \mathbf{0} & \mathbf{0} \end{pmatrix}, \tag{2.13.1}$$

where $r = r(\mathbf{A})$. From (2.12.1), **A** can be diagonalized by $\mathbf{\Gamma}$. But from Property (1) above, the latent roots of **A** are zeros or ones, only (and r is the number of nonzero latent roots).

2.14 DIFFERENTIATION AND INTEGRATION

To make inferences and decisions on the basis of data, it is often necessary to maximize likelihood functions (or minimize error sums of squares or cost functions) of random vectors and matrices. To find these extreme values for functions of vectors and matrices it is convenient to have available a calculus which permits differentiation and integration operations to be carried out on matrices. See Dwyer and Macphail (1948), Dwyer (1967), and Nel (1980), for additional results. It will be assumed throughout that all required derivatives exist.

2.14.1 Derivative of a Scalar Function of a Matrix

The derivative of a scalar function f of a matrix $\mathbf{X} = (x_{ij}) : p \times n$ is defined as

$$\frac{d}{d\mathbf{X}} f(\mathbf{X}) = \left(\frac{\partial f(\mathbf{X})}{\partial x_{ij}} \right), \qquad \begin{array}{l} i = 1, \ldots, p, \\ j = 1, \ldots, n. \end{array} \tag{2.14.1}$$

Example (2.14.1): Suppose $\mathbf{x} : p \times 1$, and $f(\mathbf{x}) = \mathbf{x}'\mathbf{x}$. Then, by (2.14.1),

$$\frac{d}{d\mathbf{x}} f(\mathbf{x}) = \begin{pmatrix} \dfrac{\partial}{\partial x_1} \mathbf{x}'\mathbf{x} \\ \cdot \\ \cdot \\ \cdot \\ \dfrac{\partial}{\partial x_p} \mathbf{x}'\mathbf{x} \end{pmatrix} = 2\mathbf{x}.$$

2.14.2 Properties of the Derivative of a Scalar Function of a Matrix

(1) For general $\mathbf{X} \neq \mathbf{X}'$,

$$\frac{d}{d\mathbf{X}}|\mathbf{X}| = |\mathbf{X}|(\mathbf{X}^{-1})', \qquad \text{for } \mathbf{X}: p \times p,\ |\mathbf{X}| \neq 0. \quad (2.14.2)$$

For $\mathbf{X} = \mathbf{X}'$,

$$\frac{d}{d\mathbf{X}}|\mathbf{X}| = 2|\mathbf{X}|\mathbf{X}^{-1} - \text{diag}(|\mathbf{X}|\mathbf{X}^{-1}), \quad (2.14.2)$$

for $\mathbf{X}: p \times p$, $|\mathbf{X}| \neq 0$.

The first case is easy to prove using the Laplace development of a determinant.

The second case may be proven using the method of implicit differentiation. Thus, if $\mathbf{X} = (x_{ij})$, and $\mathbf{X} = \mathbf{X}'$, and if we take

$$x_{ij} = f_{ij}(y_1, \ldots, y_n),$$

for some arbitrary n, then

$$\frac{d|\mathbf{X}|}{dy_k} = \sum_{i=1}^{p} \sum_{j=1}^{p} \frac{\partial |\mathbf{X}|}{\partial x_{ij}} \frac{\partial f_{ij}}{\partial y_k}.$$

Next note that $d|\mathbf{X}|/\partial x_{ij} = X_{ij}$ = cofactor of x_{ij}, and take $y_k = x_{\alpha\beta}$, $f_{ij} = x_{ij} = x_{ji}$. Then consider the cases of $\alpha = \beta$, and $\alpha \neq \beta$, separately. Note that in this case, since $\alpha, \beta = 1, \ldots, p$, $k = 1, \ldots, p^2$.

(2) $$\frac{d}{d\mathbf{X}}|\mathbf{X}|^{\alpha} = \alpha|\mathbf{X}|^{\alpha-1}\frac{d}{d\mathbf{X}}|\mathbf{X}|, \qquad \text{for } \mathbf{X}: p \times p. \quad (2.14.3)$$

This result follows from the univariate calculus.

(3) *Product rule:* $$\frac{d}{d\mathbf{X}}[f(\mathbf{X})g(\mathbf{X})] = f(\mathbf{X})\frac{dg(\mathbf{X})}{d\mathbf{X}} + \frac{df(\mathbf{X})}{d\mathbf{X}}g(\mathbf{X}), \quad (2.14.4)$$

for scalar functions $f(\cdot)$ and $g(\cdot)$.

This result follows from the univariate calculus.

For general $\mathbf{A}, \mathbf{X}$,

(4) $$\frac{d}{d\mathbf{X}}\text{tr}(\mathbf{A}'\mathbf{X}) = \mathbf{A}, \qquad \text{for } \mathbf{A}': p \times q,\ \mathbf{X}: q \times p, \quad (2.14.5)$$

and

(5) $$\frac{d}{d\mathbf{X}}\text{tr}(\mathbf{A}\mathbf{X}') = \mathbf{A}, \qquad \text{for } \mathbf{A}: p \times q,\ \mathbf{X}: p \times q. \quad (2.14.6)$$

The above two properties are proved by expressing $\mathbf{A}'\mathbf{X}$ or

$\mathbf{AX}'$ as a product in component form, and differentiating. If $\mathbf{A} = \mathbf{A}'$, and $\mathbf{X} = \mathbf{X}'$, $p = q$,

$$\frac{d}{d\mathbf{X}} \operatorname{tr}(\mathbf{AX}) = 2\mathbf{A} - \operatorname{diag}(\mathbf{A}). \tag{2.14.6$'$}$$

(6) For general $\mathbf{X}, \mathbf{A}$,

$$\frac{d}{d\mathbf{X}} \operatorname{tr}(\mathbf{X}^{-1}\mathbf{A}) = -\mathbf{X}^{-1}\mathbf{A}\mathbf{X}^{-1}, \qquad \text{for } |\mathbf{X}| \neq 0. \tag{2.14.7}$$

This result may be demonstrated by direct differentiation.

(7) For general $\mathbf{X}$, if $\mathbf{X}: p \times p$, $|\mathbf{X}| \neq 0$, $\mathbf{a}: p \times 1$,

$$\frac{d}{d\mathbf{X}} (\mathbf{a}'\mathbf{X}^{-1}\mathbf{a}) = -\mathbf{X}^{-1}\mathbf{a}\mathbf{a}'\mathbf{X}^{-1}. \tag{2.14.8}$$

To see this, express $\mathbf{a}'\mathbf{X}^{-1}\mathbf{a}$ as $\operatorname{tr}(\mathbf{X}^{-1}\mathbf{a}\mathbf{a}')$ and apply Property (6).

(8) If $\mathbf{x}: p \times 1$, $\mathbf{A}: p \times p$,

$$\frac{d}{d\mathbf{x}} (\mathbf{x}'\mathbf{A}\mathbf{x}) = 2\mathbf{A}\mathbf{x}. \tag{2.14.9}$$

This result follows immediately by direct substitution.

(9) If $\mathbf{x}: p \times 1$, $\mathbf{a}: p \times 1$,

$$\frac{d}{d\mathbf{x}} (\mathbf{a}'\mathbf{x}) = \mathbf{a}. \tag{2.14.10}$$

This result is immediate from the definition.

(10) If $\mathbf{X}: p \times p$ and $\mathbf{X} = \mathbf{X}'$,

$$\frac{d}{d\mathbf{X}} \operatorname{tr}(\mathbf{X}^2) = 4\mathbf{X} - 2\operatorname{diag}(\mathbf{X}). \tag{2.14.11}$$

Merely apply the definition.

2.14.3 Derivative of a Matrix with Respect to an Element

Suppose $\mathbf{X} = (x_{ij})$, $i = 1, \ldots, p$; $j = 1, \ldots, n$. By definition, if $\mathbf{Y} = (y_{ij})$, $i = 1, \ldots, p$, $j = 1, \ldots, n$, then for any (i,j),

$$\frac{d\mathbf{Y}}{dx_{ij}} = \left(\frac{\partial y_{\alpha\beta}}{\partial x_{ij}}\right), \qquad \alpha = 1, \ldots, p; \qquad \beta = 1, \ldots, n. \tag{2.14.12}$$

Define $\mathbf{U}_{ij}$ to be the matrix with a one in the (i,j)th place and zeros elsewhere. Then, for $\mathbf{X}: p \times p$ general, and $|\mathbf{X}| \neq 0$,

$$\frac{d}{dx_{ij}} \mathbf{X}^{-1} = -\mathbf{X}^{-1}\mathbf{U}_{ij}\mathbf{X}^{-1}. \tag{2.14.13}$$

2.14.4 Derivative of a Vector with Respect to a Vector—Hessians

Let $f(\mathbf{x})$ be a scalar function of the $p \times 1$ vector $\mathbf{x}$. Then $df(\mathbf{x})/d\mathbf{x}$ is a $p \times 1$ vector. The second derivative is obtained by differentiating each element of $df(\mathbf{x})/d\mathbf{x}'$ with respect to the column vector $\mathbf{x}$. Thus,

$$\frac{d^2 f(\mathbf{x})}{d\mathbf{x}d\mathbf{x}'} = \frac{d}{d\mathbf{x}} \frac{df(\mathbf{x})}{d\mathbf{x}'}.$$

Example (2.14.2): Suppose $f(\mathbf{x}) = \mathbf{x}'\mathbf{x}$. Then if $\mathbf{x} = (x_i), i = 1, \ldots, p$.

$$\frac{d^2 f(\mathbf{x})}{d\mathbf{x}d\mathbf{x}'} = \frac{d}{d\mathbf{x}}\left[\frac{d}{d\mathbf{x}'}(\mathbf{x}'\mathbf{x})\right] = \frac{d}{d\mathbf{x}}(2\mathbf{x}').$$

But

$$\frac{d}{d\mathbf{x}}(\mathbf{x}') = \left[\frac{d}{d\mathbf{x}}(x_1), \ldots, \frac{d}{d\mathbf{x}}(x_p)\right]$$

$$= \begin{pmatrix} 1 & 0 & \cdots & 0 \\ 0 & 1 & \cdots & 0 \\ \cdot & \cdot & & \cdot \\ \cdot & \cdot & & \cdot \\ \cdot & \cdot & & \cdot \\ 0 & 0 & \cdots & 1 \end{pmatrix} = \mathbf{I}_p.$$

Hence, for $\mathbf{x}: p \times 1$, $d\mathbf{x}/d\mathbf{x}' = \mathbf{I}_p$ and

$$\frac{d^2}{d\mathbf{x}d\mathbf{x}'}(\mathbf{x}'\mathbf{x}) = 2\mathbf{I}. \tag{2.14.14}$$

An immediate generalization of this result is the following. If $\mathbf{A}: p \times p$ and $\mathbf{x}: p \times 1$,

$$\frac{d^2}{d\mathbf{x}d\mathbf{x}'}(\mathbf{x}'\mathbf{A}\mathbf{x}) = \mathbf{A}. \tag{2.14.15}$$

The Hessian matrix $\mathbf{H} \equiv (h_{ij})$ of a scalar function $f(\mathbf{x})$, of a $p \times 1$ vector $\mathbf{x}$ is defined as the symmetric matrix

$$\mathbf{H} = \frac{d^2 f(\mathbf{x})}{d\mathbf{x}d\mathbf{x}'}; \tag{2.14.16}$$

that is,

$$h_{ij} = \frac{d^2 f(\mathbf{x})}{dx_i dx_j}. \tag{2.14.17}$$

The Hessian matrix is useful for examining whether a function has an extreme value.

Suppose $\mathbf{x}: p \times 1$ and that at $\mathbf{x} = \mathbf{x}_0$, $df(\mathbf{x})/d\mathbf{x} = \mathbf{0}$. Thus, $\mathbf{x}_0$ is a stationary point of $f(\mathbf{x})$. If in addition, however, $\mathbf{H} > 0$ for all $\mathbf{x}$, $\mathbf{x}_0$ corresponds to a global minimum of $f(\mathbf{x})$. Alternatively, if $\mathbf{H} < 0$ for all $\mathbf{x}$, $\mathbf{x}_0$ corresponds to a global maximum of $f(\mathbf{x})$. If $\mathbf{H} > 0(\mathbf{H} < 0)$ at least in a neighborhood of the point $\mathbf{x}_0$, $\mathbf{x}_0$ corresponds to a local minimum (maximum). Although, in general, $\mathbf{H}$ is neither positive definite nor negative definite over the entire range of $\mathbf{x}$, in applied multivariate analysis, the objective function often involves Hessians which are of fixed sign over all $\mathbf{x}$.

Functions $f(\mathbf{x})$ for which $\mathbf{H} > 0$ are called *convex;* if $\mathbf{H} < 0$ they are called *concave.* More generally, $f(\mathbf{x})$ might not be sufficiently smooth for $\mathbf{H}$ to exist. In general, $f(\mathbf{x})$ is convex if for any two points $\mathbf{x}_1$ and $\mathbf{x}_2$ and for any λ, $0 < \lambda < 1$, $f[\lambda\mathbf{x}_1 + (1 - \lambda)\mathbf{x}_2] \leq \lambda f(\mathbf{x}_1) + (1 - \lambda)f(\mathbf{x}_2)$. If the inequality is reversed, $f(\mathbf{x})$ is concave. The definitions in terms of $\mathbf{H}$ are implied by the more general definitions when $f(\mathbf{x})$ is twice differentiable.

2.14.5 Integration

Suppose $f(\mathbf{X})$ is a scalar function of the matrix $\mathbf{X}$. Then

$$\int_R f(\mathbf{X})d\mathbf{X}$$

is defined to be the iterated integral of $f(\mathbf{X})$ for each element of $\mathbf{X}$ separately, over a region R in the space defined by the simplex bounding the ranges of the elements of $\mathbf{X}$. This notion will be needed to define expectations of random vectors and matrices [see (3.2.1) and Section 5.1.5].

Example (2.14.3): Suppose $\mathbf{x}: p \times 1$, $f(\mathbf{x}) = \mathbf{x}'\mathbf{x}$, and $R = \{\mathbf{x}: \mathbf{x}'\mathbf{x} = 1\}$. Then

$$\int_R f(\mathbf{x})d\mathbf{x} = \int_R \mathbf{x}'\mathbf{x}d\mathbf{x} = \int \cdots \int_R \left(\sum_1^p x_i^2\right) dx_1 \cdots dx_p.$$

It is necessary to complete such integrations in order to evaluate probabilities in multidimensional space.

2.15 JACOBIANS OF MATRIX TRANSFORMATIONS

Definition: Suppose $\mathbf{X}$, $\mathbf{Y}$ are matrices which have the same number of distinct elements, r. Then, if $\mathbf{X} = f(\mathbf{Y})$, the Jacobian of the transformation is defined as

$$J(\mathbf{Y} \to \mathbf{X}) = ||\mathbf{A}||, \qquad \mathbf{A} = \left(\frac{\partial y_i}{\partial x_j}\right), \qquad i, j = 1, \ldots, r,$$

where $||\mathbf{A}||$ means the absolute value of $|\mathbf{A}|$, and $(x_1, \ldots, x_r)$ and $(y_1, \ldots, y_r)$

denote the distinct values of $\mathbf{X}$ and $\mathbf{Y}$, respectively.

Jacobians are often used in multivariate analysis to evaluate the densities of functions of random vectors or matrices [see (3.1.13)] and to evaluate marginal distributions by integration over unwanted subvectors or matrices (see proof of Corollary 5.1.3).

Example (2.15.1): Suppose

$$\mathbf{Y} = \begin{pmatrix} a & b \\ b & c \end{pmatrix},$$

where a, b, c are scalars. Since $\mathbf{Y}: 2 \times 2$ is symmetric, it has only three distinct elements. Hence, take $(y_1, y_2, y_3) \equiv (a, b, c)$.

Using the above definition, it is possible to derive the Jacobians for many complicated transformations such as those given below. Details and proofs of these results are given in Deemer and Olkin (1951) and in Olkin (1953).

2.15.1 Some Useful Jacobians

The following Jacobians are required frequently in multivariate analysis. Absolute values have been omitted for simplicity, but are to be understood.

(1) If $\mathbf{y}: p \times 1$, $\mathbf{x}: p \times 1$, $\mathbf{A}: p \times p$, and $\mathbf{y} = \mathbf{A}\mathbf{x}$,

$$J(\mathbf{y} \to \mathbf{x}) = |\mathbf{A}|. \tag{2.15.1}$$

This is the Jacobian of a linear transformation of a vector.

(2) If $\mathbf{Y}: p \times q$, $\mathbf{A}: p \times p$, $\mathbf{X}: p \times q$, and $\mathbf{Y} = \mathbf{A}\mathbf{X}$,

$$J(\mathbf{Y} \to \mathbf{X}) = |\mathbf{A}|^q. \tag{2.15.2}$$

Here, the q columns of $\mathbf{Y}$ are linear transformations of the q columns of $\mathbf{X}$; so (2.15.1) is repeated q times.

(3) If $\mathbf{Y}: p \times q$, $\mathbf{X}: p \times q$, $\mathbf{B}: q \times q$, and $\mathbf{Y} = \mathbf{X}\mathbf{B}$,

$$J(\mathbf{Y} \to \mathbf{X}) = |\mathbf{B}|^p. \tag{2.15.3}$$

This result is analogous to that of (2.15.2), except the transformations are applied to the p rows.

(4) If $\mathbf{Y}: p \times q$, $\mathbf{A}: p \times p$, $\mathbf{B}: q \times q$, $\mathbf{X}: p \times q$, and $\mathbf{Y} = \mathbf{A}\mathbf{X}\mathbf{B}$,

$$J(\mathbf{Y} \to \mathbf{X}) = |\mathbf{A}|^q|\mathbf{B}|^p. \tag{2.15.4}$$

This result combines (2.15.2) and (2.15.3).

(5) If $\mathbf{Y}: p \times p$, $\mathbf{X}: p \times p$, $\mathbf{X} = \mathbf{X}'$, and $\mathbf{Y} = \mathbf{A}\mathbf{X}\mathbf{A}'$,

$$J(\mathbf{Y} \to \mathbf{X}) = |\mathbf{A}|^{p+1}, \qquad |\mathbf{A}| \neq 0. \tag{2.15.5}$$

This result is useful for transformations on covariance matrices.

(6) If $\mathbf{Y}: p \times q$, $\mathbf{X}: p \times q$, a is a scalar, and $\mathbf{Y} = a\mathbf{X}$,

$$J(\mathbf{Y} \to \mathbf{X}) = a^{pq}. \qquad (2.15.6)$$

This transformation corresponds to a scale change of units of all the elements of $\mathbf{Y}$.

(7) If $\mathbf{Y}: p \times p$, $\mathbf{X}: p \times p$, $\mathbf{Y} = \mathbf{Y}'$, $\mathbf{X} = \mathbf{X}'$, a is a scalar, and $\mathbf{Y} = a\mathbf{X}$,

$$J(\mathbf{Y} \to \mathbf{X}) = a^{p(p+1)/2}. \qquad (2.15.7)$$

The result is analogous to (2.15.6).

(8) If[3] $|\mathbf{A}| \neq 0$, $d\mathbf{A}^{-1} = -\mathbf{A}^{-1}(d\mathbf{A})\mathbf{A}^{-1}$.
If $\mathbf{X} = \mathbf{A}^{-1}$,

$$J(\mathbf{A} \to \mathbf{X}) = |\mathbf{X}|^{-(p+1)}, \qquad (2.15.8)$$

where $\mathbf{X} = \mathbf{X}'$, $\mathbf{X}: p \times p$.

This result follows from (2.15.5) since $J(\mathbf{X} \to \mathbf{A}) = J(d\mathbf{X} \to d\mathbf{A}) = J(d\mathbf{A}^{-1} \to d\mathbf{A})$, and $J(\mathbf{A} \to \mathbf{X}) = [J(\mathbf{X} \to \mathbf{A})]^{-1}$.

Property (8) is often useful for checking the second-order conditions required for extreme values of a function. For example, in the multivariate regression model, use of this property shows that the method of differentiation of the log-likelihood function actually gives rise to a maximum of the likelihood function (see Section 8.4.2).

(9) If $f(\mathbf{X}) \equiv \operatorname{tr}(\mathbf{A}\mathbf{X}'\Sigma^{-1}\mathbf{X})$, $\mathbf{A} = \mathbf{A}'$, $\mathbf{A}: q \times q$, $\mathbf{X}: p \times q$, $\Sigma > 0$, $\Sigma: p \times p$,

$$\frac{\partial f(\mathbf{x})}{\partial \mathbf{X}} = 2\Sigma^{-1}\mathbf{X}\mathbf{A}.$$

REFERENCES

Anderson (1958).
Arnold (1973, 1976, 1979a, 1979b).
Balestra (1976).
Bellman (1960).
Beckenbach and Bellman (1961).
Courant and Hilbert (1953).
Davis (1979).
Deemer and Olkin (1951).
Dwyer (1967).
Dwyer and Macphail (1948).
Fisher (1925).
Gantmacher (1959).
Geisser (1963).
Graybill (1969).
Grenander and Szego (1958).
Hadamard (1893).
Helmert (1876)
Karlin (1968).
Kaufman (1969).
Luther (1965).
Mirsky (1955).
Moore (1935).
Nel (1980).
Olkin (1953).
Olkin and Press (1969).
Penrose (1955).
Press (1967a, 1969b, 1978, 1979, 1980c, 1981).
Rao (1965).

[3] If $\mathbf{A}: p \times n$, $d\mathbf{A}$ denotes the $p \times n$ matrix of increments of the elements of $\mathbf{A}$.

Rutherford (1952).
Srivastava (1965).
Tracy (1969).
Vinograd (1950).
Votaw (1948).
Wallace and Hussain (1969).
Wilks (1936).
Woodbury (1950).

COMPLEMENT 2.1 DETERMINANT OF THE MATRIX INTRACLASS COVARIANCE MATRIX

Let **A** and **B** be square matrices of order k, and define [see (2.4.8)] the $pk \times pk$ matrix

$$\mathbf{G}_p = \begin{pmatrix} \mathbf{A} & \mathbf{B} & \cdots & \mathbf{B} \\ \mathbf{B} & \mathbf{A} & \cdots & \mathbf{B} \\ \cdot & \cdot & & \cdot \\ \cdot & \cdot & & \cdot \\ \cdot & \cdot & & \cdot \\ \mathbf{B} & \mathbf{B} & \cdots & \mathbf{A} \end{pmatrix}.$$

Then

$$|\mathbf{G}_p| = |\mathbf{A} - \mathbf{B}|^{p-1}|\mathbf{A} + \mathbf{B}(p-1)|.$$

Proof: Subtract the second matrix column of $\mathbf{G}_p$ (that is, columns $k+1$, $k+2,\ldots,2k$ of $\mathbf{G}_p$) from its first matrix column without changing the value of $|\mathbf{G}_p|$. Then

$$|\mathbf{G}_p| = \begin{vmatrix} \mathbf{A}-\mathbf{B} & \mathbf{B} & \cdots & \mathbf{B} \\ \mathbf{B}-\mathbf{A} & \mathbf{A} & \cdots & \mathbf{B} \\ \mathbf{0} & \mathbf{B} & \cdots & \mathbf{B} \\ \cdot & \cdot & & \cdot \\ \cdot & \cdot & & \cdot \\ \cdot & \cdot & & \cdot \\ \mathbf{0} & \mathbf{B} & \cdots & \mathbf{A} \end{vmatrix}.$$

Now add the first matrix row to the second matrix row, again leaving $|\mathbf{G}_p|$ unchanged. This gives

$$|\mathbf{G}_p| = \begin{vmatrix} \mathbf{A}-\mathbf{B} & \mathbf{B} & \mathbf{B} & \cdots & \mathbf{B} \\ \mathbf{0} & \mathbf{A}+\mathbf{B} & 2\mathbf{B} & \cdots & 2\mathbf{B} \\ \hline \mathbf{0} & \mathbf{B} & \mathbf{A} & \cdots & \mathbf{B} \\ \cdot & \cdot & \cdot & & \cdot \\ \cdot & \cdot & \cdot & & \cdot \\ \cdot & \cdot & \cdot & & \cdot \\ \mathbf{0} & \mathbf{B} & \mathbf{B} & \cdots & \mathbf{A} \end{vmatrix}.$$

Hence, by (2.6.6),

$$|\mathbf{G}_p| = |\mathbf{A} - \mathbf{B}| \cdot \begin{vmatrix} \mathbf{A} + \mathbf{B} & 2\mathbf{B} & \cdots & 2\mathbf{B} \\ \mathbf{B} & \mathbf{A} & \cdots & \mathbf{B} \\ \cdot & \cdot & & \cdot \\ \cdot & \cdot & & \cdot \\ \cdot & \cdot & & \cdot \\ \mathbf{B} & \mathbf{B} & \cdots & \mathbf{A} \end{vmatrix},$$

and the matrix intraclass covariance matrix structure is maintained in the submatrix within the heavy bars, $\mathbf{G}_{p-2}$. Carrying out exactly this procedure a total of $(p - 2)$ times gives

$$|\mathbf{G}_p| = |\mathbf{A} - \mathbf{B}|^{p-2} \cdot \begin{vmatrix} \mathbf{A} + (p-2)\mathbf{B} & (p-1)\mathbf{B} \\ \mathbf{B} & \mathbf{A} \end{vmatrix}.$$

Subtracting the first matrix column from the second, and then applying (2.6.7) gives the result.

COMPLEMENT 2.2 BINOMIAL INVERSE THEOREM [EQUATION (2.5.5)]

$$(\mathbf{A} + \mathbf{UBV})^{-1} = \mathbf{A}^{-1} - \mathbf{A}^{-1}\mathbf{UB}(\mathbf{B} + \mathbf{BVA}^{-1}\mathbf{UB})^{-1}\mathbf{BVA}^{-1}.$$

Proof: Multiply $(\mathbf{A} + \mathbf{UBV})$ by the right-hand side of the equation in the theorem. The identity matrix should result.

$$\begin{aligned} &(\mathbf{A} + \mathbf{UBV})[\mathbf{A}^{-1} - \mathbf{A}^{-1}\mathbf{UB}(\mathbf{B} + \mathbf{BVA}^{-1}\mathbf{UB})^{-1}\mathbf{BVA}^{-1}] \\ &\qquad = \mathbf{I} + \mathbf{UB}[\mathbf{B}^{-1} - (\mathbf{I} + \mathbf{VA}^{-1}\mathbf{UB})(\mathbf{B} + \mathbf{BVA}^{-1}\mathbf{UB})^{-1}]\mathbf{BVA}^{-1} \\ &\qquad = \mathbf{I} + \mathbf{UB}[\mathbf{B}^{-1} - \mathbf{B}^{-1}(\mathbf{B} + \mathbf{BVA}^{-1}\mathbf{UB})(\mathbf{B} + \mathbf{BVA}^{-1}\mathbf{UB})^{-1}]\mathbf{BVA}^{-1} \\ &\qquad = \mathbf{I}. \end{aligned}$$

COMPLEMENT 2.3 LATENT ROOTS OF THE MATRIX INTRACLASS COVARIANCE MATRIX

Define the special matrix intraclass covariance matrix

$$\underset{(pk \times pk)}{\mathbf{G}} = \begin{pmatrix} \mathbf{A} & & \\ & \cdot \ \mathbf{B} & \\ & \cdot & \\ \mathbf{B} & \cdot & \\ & & \mathbf{A} \end{pmatrix},$$

where

$$\underset{(k\times k)}{\mathbf{A}} = \begin{pmatrix} a_1 & & & \\ & \cdot & b_1 & \\ & & \cdot & \\ b_1 & \cdot & & \\ & & & a_1 \end{pmatrix}, \qquad \underset{(k\times k)}{\mathbf{B}} = \begin{pmatrix} a_2 & & & \\ & \cdot & b_2 & \\ & & \cdot & \\ b_2 & \cdot & & \\ & & & a_2 \end{pmatrix}.$$

The problem is to find the latent roots of $\mathbf{G}$.

Let $\mathbf{\Gamma}: k \times k$ denote any orthogonal matrix whose first row has equal elements. Then $\mathbf{\Gamma}$ simultaneously diagonalizes $\mathbf{A}$ and $\mathbf{B}$; that is, $\mathbf{A}$ and $\mathbf{B}$ may be written as $\mathbf{A} = \mathbf{\Gamma D}_1\mathbf{\Gamma}'$, $\mathbf{B} = \mathbf{\Gamma D}_2\mathbf{\Gamma}'$, where $\mathbf{D}_1 \equiv \text{diag}\,(\alpha_1,\beta_1,\ldots,\beta_1)$, $\mathbf{D}_2 \equiv \text{diag}\,(\alpha_2,\beta_2,\ldots,\beta_2)$, $\alpha_j = a_j + (k-1)b_j$, $\beta_j = (a_j - b_j)$, and $j = 1, 2$. This result was shown in (2.8.3). It is easy to see that $\mathbf{G}$ may be written as $\mathbf{G} = \mathbf{UDU}'$, where $\mathbf{U} \equiv \text{diag}\,(\mathbf{\Gamma},\ldots,\mathbf{\Gamma})$, so that $\mathbf{UU}' = \mathbf{I}$ is orthogonal, and

$$\mathbf{D} = \begin{pmatrix} \mathbf{D}_1 & & & \\ & \cdot & \mathbf{D}_2 & \\ & & \cdot & \\ \mathbf{D}_2 & \cdot & & \\ & & & \mathbf{D}_1 \end{pmatrix}.$$

Let λ denote a latent root of $\mathbf{G}$. Then $|\mathbf{G} - \lambda\mathbf{I}| = |\mathbf{UDU}' - \lambda\mathbf{I}| = |\mathbf{D} - \lambda\mathbf{I}| = 0$. But if $\mathbf{D}_3 \equiv \mathbf{D}_1 - \lambda\mathbf{I}_k$, it is required that

$$\begin{vmatrix} \mathbf{D}_3 & & & \\ & \cdot & \mathbf{D}_2 & \\ & & \cdot & \\ \mathbf{D}_2 & \cdot & & \\ & & & \mathbf{D}_3 \end{vmatrix} = 0.$$

But the determinant of a matrix intraclass covariance matrix is given in (2.4.9). Hence,

$$|\mathbf{D}_3 - \mathbf{D}_2|^{p-1}|\mathbf{D}_3 + (p-1)\mathbf{D}_2| = 0,$$

or

$$|\mathbf{D}_1 - \mathbf{D}_2 - \lambda\mathbf{I}_k|^{p-1}|\mathbf{D}_1 + (p-1)\mathbf{D}_2 - \lambda\mathbf{I}_k| = 0,$$

from which the result in (2.8.6) follows directly.

EXERCISES

2.1 Let $\mathbf{U}_j: k \times 1$ denote a vector of k scores on a battery of tests given to a group of children in city j. It is believed that

$$\operatorname{var}(\mathbf{U}_j) = \mathbf{A} = 9\begin{pmatrix} 1 & & & \\ & \cdot & \frac{1}{2} & \\ & & \cdot & \\ \frac{1}{2} & \cdot & & \\ & & & 1 \end{pmatrix}, \qquad j = 1,\ldots,p,\ p = 3,$$

and the matrix of covariances between $\mathbf{U}_i$ and $\mathbf{U}_j$, $i \neq j$, is

$$\mathbf{B} = 4\begin{pmatrix} 1 & & & \\ & \cdot & \frac{1}{3} & \\ & & \cdot & \\ \frac{1}{3} & \cdot & & \\ & & & 1 \end{pmatrix}.$$

Define

$$\mathbf{\Sigma} = \operatorname{var}\begin{pmatrix} \mathbf{U}_1 \\ \cdot \\ \cdot \\ \cdot \\ \mathbf{U}_p \end{pmatrix} = \begin{pmatrix} \mathbf{A} & \mathbf{B} & \cdots & \mathbf{B} \\ \mathbf{B} & \mathbf{A} & \cdots & \mathbf{B} \\ \cdot & \cdot & & \cdot \\ \cdot & \cdot & & \cdot \\ \cdot & \cdot & & \cdot \\ \mathbf{B} & \mathbf{B} & \cdots & \mathbf{A} \end{pmatrix}.$$

Evaluate $|\mathbf{\Sigma}|$. Interpret the pattern within $\mathbf{\Sigma}$ in the context of this problem.

2.2 Let x_t denote sales of a company during period t, and suppose the year is divided into four periods. Let the matrix of variances and covariances of x_i and x_j have the structure

$$\mathbf{\Sigma} = \operatorname{var}\begin{pmatrix} x_1 \\ x_2 \\ x_3 \\ x_4 \end{pmatrix} = 2\begin{pmatrix} 1 & \frac{1}{5} & \frac{1}{10} & \frac{1}{5} \\ \frac{1}{5} & 1 & \frac{1}{5} & \frac{1}{10} \\ \frac{1}{10} & \frac{1}{5} & 1 & \frac{1}{5} \\ \frac{1}{5} & \frac{1}{10} & \frac{1}{5} & 1 \end{pmatrix}.$$

Thus, sales during successive periods have greater correlation than sales two periods apart. Noting that $\mathbf{\Sigma}$ is both circular and symmetric, evaluate

(a) $\mathbf{\Sigma}^{-1}$,
(b) $\mathbf{\Sigma}^2$.

What advantages might accrue from being able to establish that $\mathbf{\Sigma}$ follows such a circular pattern?

2.3 Suppose $\mathbf{y}: n \times 1$ is a vector of response times of n individuals in a learning experiment, $\mathbf{u}: n \times 1$ and $\mathbf{v}: n \times 1$ are vectors of scores of the n individuals on each of two types of tests, and $\mathbf{y} = (\mathbf{u}\ \mathbf{v})\boldsymbol{\beta} + \mathbf{e}$, where $\boldsymbol{\beta}' = (\beta_1\ \beta_2)$ is a 1×2 vector of coefficients, and $\mathbf{e}$ is an $n \times 1$ vector of errors in the relationship. Let $\mathbf{Z} \equiv (\mathbf{u}\ \mathbf{v})$. Find the latent roots of $\mathbf{A} \equiv \mathbf{Z}(\mathbf{Z}'\mathbf{Z})^{-1}\mathbf{Z}'$. [*Hint:* Calculate $\mathbf{A}^2$.]

2.4 Define

$$\mathbf{A} = 4\begin{pmatrix} 1 & \frac{1}{2} & \frac{1}{4} \\ \frac{1}{2} & 1 & \frac{1}{3} \\ \frac{1}{4} & \frac{1}{3} & 1 \end{pmatrix}.$$

Find the latent roots and all sets of normalized latent vectors of **A**. Is **A** a covariance matrix? Is **A** singular? What are the latent roots of $\mathbf{A}^2$?

2.5 Let X_1, X_2, X_3 denote changes in the incidence of crime from 1967 to 1968, for three types of crime. Suppose there is so much data available that the matrix of variances and covariances of (X_1,X_2,X_3) is known to be given approximately by

$$\mathbf{\Sigma} = \operatorname{var}\begin{pmatrix} X_1 \\ X_2 \\ X_3 \end{pmatrix} = \begin{pmatrix} 2 & -1 & 1 \\ -1 & 1 & 0 \\ 1 & 0 & 3 \end{pmatrix}.$$

(a) Find $\mathbf{\Sigma}^{-1}$.
(b) Show by numerical computation that $\mathbf{\Sigma}^{-1}$ is positive definite.
(c) Interpret the negative signs in $\mathbf{\Sigma}$.

2.6 Find the pseudoinverse of

$$\mathbf{\Sigma} = \begin{pmatrix} 3 & 4 & -5 \\ 8 & 7 & -2 \\ 2 & -1 & 8 \end{pmatrix}.$$

Is $\mathbf{\Sigma}$ positive definite? What is the advantage of computing the pseudoinverse instead of any other generalized inverse?

2.7 Let x_j denote the time required to perform job j, $j = 1,\ldots,5$. Suppose the cost of a project is given by

$$C(\mathbf{x}) = \mathbf{x}'\mathbf{A}\mathbf{x},$$

where $\mathbf{x}' = (x_1,\ldots,x_5)$, and

$$\mathbf{A} = \begin{pmatrix} 2 & 1 & 0 & 0 & 0 \\ 1 & 2 & 1 & 0 & 0 \\ 0 & 1 & 2 & 1 & 0 \\ 0 & 0 & 1 & 2 & 1 \\ 0 & 0 & 0 & 1 & 2 \end{pmatrix}.$$

Assuming the units are defined so that $\Sigma_1^5 x_j^2 = 1$, find the lowest and highest possible costs of the project.

2.8 Let $\mathbf{x}: p \times 1$ denote a vector of positive control variables associated with the administration of a hospital. That is, $\mathbf{x}$ can be set as desired to achieve a particular effect.

Let $f(\mathbf{x})$ denote an objective or criterion function whose behavior is to be examined. Let

$$f(\mathbf{x}) = \mathbf{x}' \, \overline{\mathbf{log}} \, \mathbf{x} - \mathbf{x}'\mathbf{e},$$

where $\overline{\mathbf{log}}\,\mathbf{x}$ is defined as the $p \times 1$ vector whose ith component is $\log x_i$, and $\mathbf{e}$ is a $p \times 1$ vector of ones.

(a) Show that $\dfrac{df(\mathbf{x})}{d\mathbf{x}} = \overline{\mathbf{log}}\,\mathbf{x}$.

(b) Find the stationary points of $f(\mathbf{x})$.

(c) Find the Hessian matrix of $f(\mathbf{x})$.

(d) Determine whether any of the stationary points found in (b) are extreme points, and explain your reasoning.

2.9 Suppose **A** and **B** are defined as in Exercise 2.1. Evaluate or determine the following:

(a) $\mathbf{A} \otimes \mathbf{B}$,

(b) $|\mathbf{A} \otimes \mathbf{B}|$,

(c) $\operatorname{tr}(\mathbf{A} \otimes \mathbf{B})$,

(d) $(\mathbf{A} \otimes \mathbf{B})^{-1}$,

(e) the latent roots of $\mathbf{A} \otimes \mathbf{B}$,

(f) Is $\mathbf{A} \otimes \mathbf{B}$ a covariance matrix? Does it have special structure?

2.10 In Exercise 2.1, find the latent roots of $\mathbf{\Sigma}$. In Chapter 9 it will be seen that these latent roots will correspond to the variances of the principal components under an intraclass covariance assumption.

2.11 Define $f(\mathbf{Z}) = E[\operatorname{tr}(\mathbf{\Sigma} - \mathbf{Z})^2]$, for $\mathbf{\Sigma} : p \times p$, $\mathbf{Z} : p \times p$, $\mathbf{\Sigma} = \mathbf{\Sigma}'$, $\mathbf{Z} = \mathbf{Z}'$, and the expectation is taken with respect to the distribution of $\mathbf{\Sigma}$. Show that $f(\mathbf{Z})$ is minimized for $\mathbf{Z} = E(\mathbf{\Sigma})$. It will be seen in (7.1.11) that **Z** is the Bayes estimator under a diffuse prior when $f(\mathbf{Z})$ is the loss function of interest. [*Hint:* See Section 2.14.1.]

2.12 Suppose $\mathbf{V} : p \times p$ is positive definite and **V** is partitioned as

$$\mathbf{V} = \begin{pmatrix} \mathbf{V}_{11} & \mathbf{V}_{12} \\ \mathbf{V}_{21} & \mathbf{V}_{22} \end{pmatrix},$$

where $\mathbf{V}_{11} : q \times q$ and $\mathbf{V}_{11}$ is positive definite. Define

$$f \equiv \int |\mathbf{V}|^{(n-p-1)/2} d\mathbf{V}_{12},$$

where $n > p + 1$, and the integration extends over all possible values of the elements of $\mathbf{V}_{12}$. Using (2.6.7), (2.15.2), and (2.15.3), show that

$$f = C|\mathbf{V}_{11}|^{(n-q-1)/2}|\mathbf{V}_{22}|^{(n-(p-q)-1)/2},$$

where C is a numerical constant depending upon n, p, and q only. It will be seen in Corollary (5.1.3) that this result may be used to show that when **V** is a random matrix following a Wishart distribution (5.1.1) with block diagonal scale matrix, $\mathbf{V}_{11}$ and $\mathbf{V}_{22}$ are independently distributed. [*Remark:* A *block diagonal* matrix is one whose principal submatrices have possibly nonzero elements, but all other elements are zero.]

2.13 Write the following equations in matrix form:

$$3x + 2y - z = 1,$$
$$7x - y + 2z = 0,$$
$$-2x + 3y - 5z = 6.$$

2.14 Given the vectors

$$\mathbf{a} = (3,1,2)', \qquad \text{and} \qquad \mathbf{b} = (-1,7,4)',$$

find:
(a) $\mathbf{b}'\mathbf{b}$, (b) $\mathbf{a}'\mathbf{b}$, (c) $\mathbf{a}\mathbf{a}'$,
(d) $5\mathbf{a} + 2\mathbf{b}$, (e) $\mathbf{a}'\mathbf{H}\mathbf{b}$,
where

$$\mathbf{H} = \operatorname{diag}(2,-1,7).$$

2.15 Find the ranks of the following matrices:

(a) $\begin{pmatrix} 3 & 4 \\ 2 & -2 \end{pmatrix}$, (b) $\begin{pmatrix} 3 & 1 & -3 \\ 5 & 2 & 0 \\ 1 & 3 & 1 \end{pmatrix}$,

(c) $\begin{pmatrix} 3 & 2 & 1 & 2 \\ 4 & -1 & 8 & 5 \\ 6 & 3 & -8 & 2 \\ -4 & 6 & 2 & -3 \end{pmatrix}$

2.16 Let

$$\mathbf{A} = \begin{pmatrix} 4 & 1 \\ 1 & 2 \end{pmatrix}.$$

Find the latent roots and latent vectors of:
(a) $4\mathbf{A}$, (b) $\mathbf{A} + 5\mathbf{I}$, (c) $\mathbf{A}^{-1}$,
(d) $\mathbf{A}^2$, (e) $\mathbf{A}^{1/2}$.

2.17 Write each of the following quadratic forms in terms of a symmetric matrix;
(a) $2x_1^2 + 4x_1x_2 + x_2^2$,
(b) $3x_1^2 - 2x_1x_2 + 4x_2^2$.
Are they positive definite?

3

CONTINUOUS MULTIVARIATE DISTRIBUTIONS:

THE NORMAL DISTRIBUTION, BAYESIAN INFERENCE

This chapter is concerned with continuous multivariate distributions. They are introduced in terms of the matrix notation required for working with random vectors. Multivariate distributions derived from Bayes' theorem are called posterior distributions. After the multivariate Normal distribution is introduced, posterior distributions based upon the multivariate Normal distribution are developed. The Bayesian notions of diffuse and natural conjugate prior density are discussed and predictive distributions are introduced. These ideas will be used in later chapters. Capital letters will usually denote random variables, random vectors, or random matrices, while lower case letters will usually be reserved for observed values of random variables, or vectors. Chapters 3–6 will discuss a wide variety of multivariate distributions. Additional multivariate continuous distributions and their properties are discussed in Johnson and Kotz. 1972.

3.1 MULTIVARIATE DISTRIBUTIONS AND BAYESIAN INFERENCE

3.1.1 Cumulative Distribution Function (cdf)

Let X, Y be two random variables that are defined jointly. That is, X and Y may have a joint probability distribution with joint cdf:

$$F(x,y) \equiv P\{X \leq x, Y \leq y\}.$$

More generally, $\mathbf{X}' = (X_1,\ldots,X_p)$ may be a vector of random variables that are jointly distributed with cdf

$$F(\mathbf{x}) \equiv F(x_1,\ldots,x_p) \equiv P\{X_1 \le x_1,\ldots,X_p \le x_p\}.$$

Every multivariate cdf $F(\mathbf{x})$ satisfies the following properties:

(1) $F(\mathbf{x})$ is monotone nondecreasing in each component of $\mathbf{x}$,

(2) $$0 \le F(\mathbf{x}) \le 1,$$

(3) $$\begin{aligned} F(-\infty,x_2,\ldots,x_p) &= F(x_1,-\infty,x_3,\ldots,x_p) \\ &= \cdots = F(x_1,\ldots,x_{p-1},-\infty) = 0, \end{aligned}$$

(4) $$F(+\infty,\ldots,+\infty) = 1,$$

(5) The probability of any p-dimensional rectangle is nonnegative; for example, if $p = 2$, and the random variables are X, Y,

$$\begin{aligned} P\ \{\text{rectangle}\} &= P\{x_1 \le X \le x_2,\ y_1 \le Y \le y_2\} \\ &= F(x_2,y_2) - F(x_2,y_1) - F(x_1,y_2) + F(x_1,y_1) \\ &\ge 0, \qquad \text{for } x_1 < x_2,\ y_1 < y_2. \end{aligned}$$

All of the above properties have analogues in univariate distributions except (5), and there exist functions that are not cdf's, that satisfy the first four properties without satisfying the fifth.[1]

In the sequel, emphasis will be on continuous distributions; that is, those for which $F(\mathbf{x})$ is continuous. Moreover, it will be assumed throughout that the cdf of any continuous distribution is expressible as the integral of some function $f(\mathbf{x})$ called its density; that is,

$$F(\mathbf{x}) = \int_{-\infty}^{x_p} \cdots \int_{-\infty}^{x_1} f(\mathbf{x})\, d\mathbf{x}. \tag{3.1.1}$$

3.1.2 Density

Suppose $F(\mathbf{x})$ is continuous; then, from (3.1.1) there is a *joint density function* for $\mathbf{X}$,

$$f(\mathbf{x}) \equiv f(x_1,\ldots,x_p) = \frac{\partial^p F(\mathbf{x})}{\partial x_1 \cdots \partial x_p}. \tag{3.1.2}$$

Strictly speaking, there are some exceptional sets of values of x for which (3.1.2) does not hold. However, in the interest of simplicity all such (measure theoretic) considerations will be ignored in the sequel.

Analogous to the univariate case, there is a relationship for the probability of any event (or set of values in p-dimensional space) in terms of the joint density. For $\mathbf{X}: p \times 1$,

$$P\ \{\mathbf{X} \text{ lies in the region } R\} = \int \underset{R}{\cdots} \int f(\mathbf{x})\, d\mathbf{x}, \tag{3.1.3}$$

for any region R.

[1] Suppose $F(x_1,x_2) = 0$, if $x_1 \le 0$ or $x_2 \le 0$, or $x_1 + x_2 \le 1$, and $F(x_1,x_2) = 1$ everywhere else in the (x_1,x_2) plane. $F(x_1,x_2)$ is not continuous. $F(x_1,x_2)$ satisfies Properties (1)–(4), which would be sufficient for a cdf in one dimension. However, since $F(1,1) - F(1,\frac{1}{2}) - F(\frac{1}{2},1) + F(\frac{1}{2},\frac{1}{2}) = -1$, Property (5) is not satisfied; so $F(x_1,x_2)$ cannot be a bivariate cdf.

3.1.3 Marginal Distributions

In multivariate data analysis, it is typical to start out with a vector of many components and then find subsequently that only a particular subvector is of interest. In such a case, the marginal distribution of the subvector is important for inferential purposes.

Let $\mathbf{X}' = (\mathbf{Y}',\mathbf{Z}')$, where $\mathbf{Y}$ and $\mathbf{Z}$ are subvectors of $\mathbf{X}: p \times 1$. [For example, $\mathbf{Y}' \equiv (X_1,X_2)$, $\mathbf{Z}' \equiv (X_3,\ldots,X_p)$.] Then, if $g(\mathbf{y})$, $h(\mathbf{z})$ denote the densities of $\mathbf{Y}$, $\mathbf{Z}$, and if $f(\mathbf{x}) = f(\mathbf{y},\mathbf{z})$ denotes the density of $\mathbf{X}$,

$$g(\mathbf{y}) = \int_{-\infty}^{\infty} \cdots \int_{-\infty}^{\infty} f(\mathbf{y},\mathbf{z})\, d\mathbf{z}, \tag{3.1.4}$$

and

$$h(\mathbf{z}) = \int_{-\infty}^{\infty} \cdots \int_{-\infty}^{\infty} f(\mathbf{y},\mathbf{z})\, d\mathbf{y}, \tag{3.1.5}$$

where all the integrations are taken over $(-\infty, \infty)$. $g(\mathbf{y})$ and $h(\mathbf{z})$ are called the marginal densities of $\mathbf{Y}$ and $\mathbf{Z}$.

3.1.4 Conditional Distributions

It is often the case that a group of random variables is studied while a second group is held fixed. That is, the conditional distribution is of interest.

Let A and B be two events that can occur in a two-dimensional space. Then by definition, the conditional probability of B, given A has occurred, is given by

$$P(B|A) = \frac{P(AB)}{P(A)},$$

provided $P(A) \neq 0$. If A is the event that a random variable X falls in the interval $a \leq X \leq b$, and B the event that a random variable Y falls in the interval $c \leq Y \leq d$,

$$P\{c \leq Y \leq d | a \leq X \leq b\} = \frac{P\{a \leq X \leq b,\, c \leq Y \leq d\}}{P\{a \leq X \leq b\}},$$

and by (3.1.3),

$$P\{c \leq Y \leq d | a \leq X \leq b\} = \frac{\int_c^d \int_a^b f(x,y)\, dx\, dy}{\int_a^b g(x)\, dx}, \tag{3.1.6}$$

where $f(x,y)$ is the joint density of X, Y, and $g(x)$ is the marginal density of X.

The conditional density of Y given $X = x$ is defined as

$$h(y|x) = \frac{f(x,y)}{g(x)}. \tag{3.1.7}$$

Hence,

$$P\{c \leq Y \leq d | X = x\} = \int_c^d h(y|x)\, dy. \tag{3.1.8}$$

Generalizing to p dimensions, let $\mathbf{X}' = (X_1,\ldots,X_p)$, $\mathbf{Y}' = (X_1,\ldots,X_k)$, $\mathbf{Z}' = (X_{k+1},\ldots,X_p)$ with lower case letters used to denote observed values. The conditional density of $\mathbf{Y}$ given $\mathbf{Z}$ is given by

$$g(\mathbf{y}|\mathbf{z}) = \frac{f(\mathbf{y},\mathbf{z})}{h(\mathbf{z})} = \frac{f(\mathbf{x})}{h(\mathbf{z})}, \tag{3.1.9}$$

where $f(\mathbf{x})$ denotes the density of the random vector $\mathbf{X}$, and $h(\mathbf{z})$ denotes the marginal density of the random vector $\mathbf{Z}$.

3.1.5 Independence

Two random vectors, $\mathbf{Y}$, $\mathbf{Z}$, are said to be *independent* if any of the following relations hold.

$$f(\mathbf{y},\mathbf{z}) = g(\mathbf{y})h(\mathbf{z}), \tag{3.1.10}$$

or

$$F(\mathbf{y},\mathbf{z}) = G(\mathbf{y})H(\mathbf{z}), \tag{3.1.11}$$

or

$$p(\mathbf{y}|\mathbf{z}) = g(\mathbf{y}), \tag{3.1.12}$$

where $f(\mathbf{y},\mathbf{z})$, $g(\mathbf{y})$, $h(\mathbf{z})$ are the densities of $\mathbf{X} \equiv (\mathbf{Y},\mathbf{Z})$, $\mathbf{Y}$, $\mathbf{Z}$, respectively; F, G, H are the respective cdf's; and $p(\mathbf{y}|\mathbf{z})$ is the conditional density of $\mathbf{Y}|\mathbf{Z}$.

3.1.6 Transformations

It is often useful to reexamine a problem in terms of a transformed set of random variables. If such a transformation is made, the next theorem shows how the density is affected.

Theorem (3.1.1): Let $Y_k = f_k(X_1,\ldots,X_p)$, $k = 1,\ldots,p$, denote a one-to-one differentiable transformation to $\mathbf{Y} \equiv (Y_1,\ldots,Y_p)'$, from $\mathbf{X} = (X_1,\ldots,X_p)'$, with $X_k = g_k(\mathbf{Y})$, $k = 1,\ldots,p$, denoting the inverse transformation. If $p(\mathbf{x}) \equiv p(x_1,\ldots,x_p)$ denotes the density of $\mathbf{X}$, the density of $\mathbf{Y}$ is given by

$$q(\mathbf{y}) = p[g_1(\mathbf{y}),g_2(\mathbf{y}),\ldots,g_p(\mathbf{y})] \cdot J(\mathbf{X} \rightarrow \mathbf{Y}), \tag{3.1.13}$$

where $J(\mathbf{X} \rightarrow \mathbf{Y})$ denotes the Jacobian of the transformation, as defined in Section 2.15.

Proof: See, for instance, Cramér (1946).

Example(3.1.1): Suppose $\mathbf{X}: 2 \times 1$ is given by $\mathbf{X} = (Y,Z)'$, where Y and Z are scalar random variables, and the density of $\mathbf{X}$ is given by

$$f(\mathbf{x}) \equiv f(y,z) = \frac{1}{2\pi\sigma^2} \exp\left\{-\frac{1}{2\sigma^2}[(y-\theta_y)^2 + (z-\theta_z)^2]\right\},$$

for $-\infty < y < \infty$, and $-\infty < z < \infty$. It will be seen in Section 3.3 that X follows a bivariate Normal distribution with independent components.

Next suppose that $Y = \log U$, and $Z = \log V$. What is the joint distribution of (U,V)?

Using Theorem (3.1.1), substitute the inverse transformation into the old density to get the new density. Thus, if $g(u,v)$ is the density of (U,V),

$$g(u,v) = f[y(u,v),z(u,v)] \cdot J[(Y,Z) \rightarrow (U,V)],$$

or

$$g(u,v) = \frac{1}{2\pi\sigma^2} \exp\left\{-\frac{1}{2\sigma^2}[(\log u - \theta_y)^2 + (\log v - \theta_z)^2]\right\} J[(Y,Z) \rightarrow (U,V)].$$

To find the Jacobian, form the table whose entries are the partial derivatives

	U	V
Y	$\frac{1}{U}$	0
Z	0	$\frac{1}{V}$.

Noting that U and V are positive random variables, it follows that

$$J[(Y,Z) \rightarrow (U,V)] = \frac{1}{UV}.$$

Hence, the transformed density is given by

$$g(u,v) = \frac{1}{2\pi\sigma^2 uv} \exp\left\{-\frac{1}{2\sigma^2}[(\log u - \theta_y)^2 + (\log v - \theta_z)^2]\right\},$$

for $0 < u < \infty$, and $0 < v < \infty$.

Note that since the joint density factors into the product of a function of u and a function of v, by (3.1.10), U and V must be independent.

The transformed variables (U,V) are said to have independent logarithmic Normal distributions since their logarithms are Normally distributed and independent. This definition is applicable even when the "logged" variables have a joint Normal distribution in which the variables are not independent, as they are in this example. There are many applications, in regression for example, in which $\log u$ and $\log v$ are linearly related, whereas u and v are not. To make inferences about the relationship, it is necessary to use the log Normal distribution.

[*Remark:* If $Y_k = f_k(\mathbf{X})$, $k = 1, \ldots, p$ denotes a many-to-one differentiable transformation, the density of **Y** may be obtained by applying (3.1.13) to each solution of the inverse transformation separately and then summing the transformed densities for each solution.]

3.1.7 Bayes Theorem, Prior and Posterior Distributions

Let **X**, $\boldsymbol{\Theta}$ denote p-dimensional and k-dimensional random vectors that are jointly distributed with the conditional density of **X** given $\boldsymbol{\Theta}$ denoted by $f(\mathbf{x} \mid \theta)$. Denote the marginal density of Θ by $g(\theta)$. We think of data being generated by observing **X**, for some fixed, unobservable Θ. We would like to make inferences about Θ which take into account both our preconceptions (prior beliefs) about Θ, and also our observations of $(\mathbf{X} \mid \Theta)$ that indirectly relate to it. Bayes Theorem provides a formal mechanism for accomplishing this. In Bayesian terminology, $g(\theta)$ is called the "a priori" or prior density of Θ since it is the density of Θ prior to observing the data.

Theorem (3.1.2) (Bayes): Let $g(\boldsymbol{\theta})$ denote the prior density of $\boldsymbol{\Theta}$, and $f(\mathbf{x}|\boldsymbol{\theta})$ the density of **X** for given $\boldsymbol{\theta}$. Then, the density of $\boldsymbol{\Theta}$ given $\mathbf{X} = \mathbf{x}$ is given by

$$h(\boldsymbol{\theta}|\mathbf{x}) = \frac{f(\mathbf{x}|\boldsymbol{\theta})g(\boldsymbol{\theta})}{\int_{-\infty}^{\infty} f(\mathbf{x}|\boldsymbol{\theta})g(\boldsymbol{\theta})d\boldsymbol{\theta}}. \tag{3.1.14}$$

[*Remarks:* (1) $h(\boldsymbol{\theta}|\mathbf{x})$ is often referred to as the posterior density since it is the density of $\boldsymbol{\Theta}$ subsequent to observing the data. (2) $f(\mathbf{x}|\boldsymbol{\theta})$ or any constant multiple of it is often referred to as the *likelihood function* (**x** may denote a vector of observations) if the function is viewed as depending upon the parameters, with **x** fixed and known.]

Proof: The theorem is proven easily by applying (3.1.7).

Since the integral in (3.1.14) depends on **x** only, and since **x** is assumed fixed and known, in (3.1.14) the denominator behaves merely as a proportionality constant. Therefore, the relation will often be written

$$h(\boldsymbol{\theta}|\mathbf{x}) \propto f(\mathbf{x}|\boldsymbol{\theta})g(\boldsymbol{\theta}), \tag{3.1.15}$$

and the notation means that $h(\boldsymbol{\theta}|\mathbf{x})$ is proportional to the product of the likelihood function and the prior density. Bayesian methods have undergone extensive development since 1950. The interested reader should see: Aykac et al., 1977; Box and Tiao, 1973; de Finetti, Volumes 1 and 2, 1974, 1975; DeGroot, 1970; Fienberg and Zellner, 1975.

Prior Distributions

A prior distribution for $\Theta : k \times 1$ possesses all the usual properties of distributions of observable random variables, except Θ is not observable. The distribution is determined on the basis of degrees of belief about Θ. For this reason, such distributions are often called *subjective* probability distributions. They are also often referred to as *personal* distributions since they are determined on the basis of the degrees of belief about Θ of a given person. That is, the prior distribution for Θ may reflect *your* beliefs about Θ. *My* prior distribution would most likely differ from yours. So your prior distribution is personal. Using your prior distribution you will develop your posterior distribution from which you will draw inferences about Θ that represent your belief about Θ, after having seen data bearing upon Θ.

For example, suppose $k = 1$, and Θ is one or zero according to whether or not it rains tomorrow. Suppose you strongly believe it will rain tomorrow, based upon your study of the frequency of rain on that day in the past, based upon your having consulted five meteorologists, and based upon your own evaluation of the rain clouds that appear to be forming. This type of approach to assessing prior probabilities is called *introspection.* You are not certain it will rain tomorrow but you would be willing to give someone 9:1 odds that it will rain tomorrow. Then, your probability mass function for Θ is

$$p(\theta) = (.9)^{\theta}(.1)^{1-\theta}, \qquad \theta = 0,1.$$

That is, $p(1) = P(\text{rain}) = 0.9$. The same kind of thinking can be used to develop densities for continuous prior distributions.

In some situations, such as those where public policy decisions must be made, we don't wish to adopt a particular individual's prior distribution for fear of introducing what might be called an undesirable bias. In those situations we often adopt "vague" prior distributions that reflect minimal prior information, and so minimize subjective bias. Such prior distributions are discussed in Sect. 3.6.1.

Prior probabilities may often be found by using the *reference lottery* approach (see Schlaifer, 1969, p. 336). As an illustration, suppose we wish to determine your subjective probability for the event A, that it will rain tomorrow (or for the event there is a pool of underground oil in a given location). We give you a choice of engaging in either of two lotteries (games). In Lottery (1) you will receive X dollars with probability $P(A)$, and you will receive Y dollars with probability $[1 - P(A)]$; X might be \$1,000 and Y might be zero. In Lottery (2) you will receive X if A occurs, and you will receive Y if A does not occur. To begin with, take $P(A) = 0.5$; that is, in Lottery (1) there is a 50:50 chance you will receive X, or Y. Suppose that under these circumstances you would prefer to play Lottery (2). This means that you believe the chance of A occurring is greater than 0.5. Now suppose we repeat the situation, but with $P(A) = 0.75$. That is, you are asked whether you would prefer to engage in Lottery (1) or in Lottery (2), but in Lottery (1) you receive X with probability ¾ and receive Y with probability ¼. Now which lottery would you prefer to play? Suppose that now you choose Lottery (1). This decision implies you believe that $.5 < P(A) < .75$. We now repeat the question, with $P(A) = .6$ (or any other intermediate value); and by continuing this process we come as close as we desire to *your* true, subjective probability for event A. This approach is useful in this form for discrete prior distributions; but it must be modified for continuous prior distributions. (See below).

Suppose $\Theta : 1 \times 1$ is continuous. We might then ask you to determine a point $\theta(.5)$ so that it is equally likely, in your view, that Θ is less than $\theta(.5)$ as it is that Θ is greater than $\theta(.5)$. The point $\theta(.5)$ is the median of your prior distribution. Now suppose you are told that $\Theta < \theta(.5)$, some specific number. Then you are asked to find a number $\theta(.25)$ such that it is equally likely, in your view, that $\Theta < \theta(.25)$ and $\Theta > \theta(.25)$. This point is the 25% fractile of your prior distribution. By repeating this process for many fractiles, we are assessing many points of the graph of the *cdf* of your prior distribution for Θ. We may then connect these points in a smoothed way to find *your* prior *cdf* for Θ. The more points we assess, the closer will be the determination of the distribution.

A prior distribution for $\Theta \equiv (\theta_i)$ is sometimes assumed to be *exchangeable*, in that the distribution is invariant under changes in the ordering of the θ_i's (see de Finetti, 1974, 1975; and Lindley and Novick, 1981). In problems in which such an assumption is reasonable, the number of parameters of the prior distribution that must be assessed is thereby reduced to a small number.

3.1.8 Multivariate Characteristic Functions

The characteristic function of a random variable (or vector) is an alternative representation to its density, or cdf. In fact, there are many continuous distributions of interest in applied multivariate analysis whose densities are too complicated to consider (for example, many are expressible only as infinite series). In particular, the *stable* family of distributions which is of great interest in the areas of finance and economics is best represented in terms of characteristic functions. This family of distributions, whose densities are generally expressible only as infinite series, is discussed in Chapter 6.

Recall from univariate analysis that the characteristic function of a continuous scalar random variable X is defined as

$$\phi_X(t) = Ee^{itX} = \int_{-\infty}^{\infty} e^{itx}f(x)dx, \tag{3.1.16}$$

where $f(x)$ is the density function of X, and $i = \sqrt{-1}$. It is clear from (3.1.16) that $\phi_X(t)$ always exists for all X. The properties given below are often useful.

3.1.9 Properties of Characteristic Functions

(1) $\phi_X(0) = 1$.
Set $t = 0$ in (3.1.16).

(2) $|\phi_X(t)| \leq 1$.
This follows from (3.1.16) using the fact that $|e^{itx}| = 1$.

(3) $\phi_X(-t) = \overline{\phi_X(t)}$, where the bar denotes complex conjugate.
Take complex conjugates in (3.1.16).

(4) $\phi_X(t)$ is uniformly continuous on $(-\infty, \infty)$.
For proof see, for instance, Lukacs (1960).

(5) If X and Y are independent with characteristic functions $\phi_X(t)$ and $\phi_Y(t)$, the characteristic function of $(X + Y)$ is the product $\phi_X(t)\phi_Y(t)$.
This follows from the fact that under independence,

$$\phi_{X+Y}(t) = E[\exp(itX)\exp(itY)] = E[\exp(itX)]E[\exp(itY)].$$

(6) Two continuous distribution functions $F_1(x)$ and $F_2(x)$ are identical if and only if their characteristic functions are identical. Thus, specifying the characteristic function of a random variable is completely equivalent to specifying its density or cdf.
For proof see, for instance, Lukacs (1960).

(7) If $\phi(\mathbf{t})$ is absolutely integrable for $\mathbf{t}: p \times 1$, the distribution has a density given by

$$f(\mathbf{x}) = \frac{1}{(2\pi)^p} \int \cdots \int e^{-i\mathbf{t}'\mathbf{x}}\phi(\mathbf{t})d\mathbf{t},$$

for $\mathbf{x}: p \times 1$. This result is often called the inversion formula.

A characteristic function is, in general, a complex variable (although its argument is a real variable) whose complete variation takes place within the unit circle in the complex plane.

The notion of characteristic function may easily be extended to vector random variables. Let $\mathbf{Y}: p \times 1$ denote a random vector with multivariate density $f(\mathbf{y})$. Then the characteristic function of $\mathbf{Y}$ is defined as

$$\phi_{\mathbf{Y}}(\mathbf{t}) \equiv \phi_{\mathbf{Y}}(t_1,\ldots,t_p) \equiv Ee^{i\mathbf{t}'\mathbf{Y}}$$
$$= E \exp\left\{i \sum_{j=1}^{p} t_j Y_j\right\} = \int e^{i\mathbf{t}'\mathbf{y}} f(\mathbf{y})\, d\mathbf{y}, \tag{3.1.17}$$

where $\mathbf{Y}' \equiv (Y_1,\ldots,Y_p)$, $\mathbf{y}' \equiv (y_1,\ldots,y_p)$, $\mathbf{t}' \equiv (t_1,\ldots,t_p)$, and the integration is assumed to extend over all possible values of all of the components of $\mathbf{y}$.

Analogously, the definition extends to matrix random variables. Let $\mathbf{V} \equiv (v_{ij})$, $\mathbf{V}: p \times q$ be a matrix of random variables and let $\mathbf{T}: q \times p$ be a matrix of characteristic function variables: $\mathbf{T} \equiv (t_{ij})$, $i = 1,\ldots,q$, $j = 1,\ldots,p$. Then the characteristic function of $\mathbf{V}$ is defined as

$$\phi_{\mathbf{V}}(\mathbf{T}) = E \exp\,[i\, tr(\mathbf{VT})] = \int \exp\,[i\, tr(\mathbf{VT})] f(\mathbf{V})\, d\mathbf{V}$$
$$= \int \exp\left[i \sum_j \sum_k v_{jk} t_{kj}\right] f(\mathbf{V})\, d\mathbf{V}$$

where $f(\mathbf{V})$ denotes the density of $\mathbf{V}$. In general, unless otherwise indicated, the integration will always be assumed to extend over all possible values of all variables.

The first six properties of (univariate) characteristic functions given above generalize immediately, in the obvious way, to multivariate characteristic functions. For example, if $\mathbf{X}$ is a vector, $\phi_{\mathbf{X}}(0, \ldots, 0) = 1$. Note also that if $\phi_{X_1,X_2}(t_1, t_2) = \phi_{X_1}(t_1)\phi_{X_2}(t_2)$, X_1 and X_2 are independent, and conversely. This follows from the definition and use of (3.1.10) or (3.1.11).

3.2 MOMENTS OF MULTIVARIATE DISTRIBUTIONS

3.2.1 Mean

Let $\mathbf{X}: p \times 1$ denote a column vector of random components X_i, $i = 1, \ldots, p$, which have joint density function $f(\mathbf{x}) \equiv f(x_1, \ldots, x_p)$. When it exists, the *expectation of a vector* $\mathbf{X}$ is defined as

$$E\mathbf{X} = \begin{pmatrix} EX_1 \\ \cdot \\ \cdot \\ \cdot \\ EX_p \end{pmatrix}. \tag{3.2.1}$$

Analogously, if $\mathbf{V}: p \times n$, $E\mathbf{V} = (Ev_{ij})$, where $\mathbf{V} \equiv (v_{ij})$.

3.2.2 Second-Order Moments

The *covariance* between two scalar random variables Y and Z with finite second moments is defined as

$$\operatorname{cov}(Y,Z) = E[(Y - EY)(Z - EZ)]. \tag{3.2.2}$$

It is a quantity that may be positive, negative, or zero.

The covariances of all of the components of the $\mathbf{X}$ vector are given simultaneously in the definition of the *covariance matrix* of the vector $\mathbf{X}$:

$$\mathbf{\Sigma} \equiv (\sigma_{ij}) = E[(\mathbf{X} - E\mathbf{X})(\mathbf{X} - E\mathbf{X})'], \tag{3.2.3}$$

for $i, j = 1,\ldots,p$. A typical element of $\mathbf{\Sigma}$ is

$$\sigma_{ij} = E(X_i - EX_i)(X_j - EX_j), \qquad i, j = 1,\ldots,p.$$

If $i = j$, the element is located along the diagonal of $\mathbf{\Sigma}$ and is called the *variance* of X_i.

$$\operatorname{var} X_i = E(X_i - EX_i)^2. \tag{3.2.4}$$

If $i \neq j$, σ_{ij} is the covariance of X_i and X_j.

The *correlation coefficient* between two scalar random variables Y and Z with finite second moments is defined as

$$\rho \equiv \operatorname{corr}(Y,Z) \equiv \frac{\operatorname{cov}(Y,Z)}{[(\operatorname{var} Y)(\operatorname{var} Z)]^{1/2}}. \tag{3.2.5}$$

It is a measure of closeness of association of Y and Z, rather than of cause and effect. In general, $-1 \leq \rho \leq 1$, although in some cases, ρ is restricted to an even smaller interval.

For example, suppose $\mathbf{\Sigma}: p \times p$ has the intraclass correlation structure

$$\mathbf{\Sigma} = \sigma^2 \begin{pmatrix} 1 & & & \\ & \cdot & \rho & \\ & & \cdot & \\ & \rho & \cdot & \\ & & & 1 \end{pmatrix},$$

so that $\mathbf{\Sigma} = \sigma^2[(1 - \rho)\mathbf{I} + \rho\mathbf{e}\mathbf{e}']$, where $\mathbf{e}: p \times 1$ is a vector of ones. If $\mathbf{\Sigma}$ is to be a positive definite covariance matrix, one necessary (but not sufficient) condition is that $|\mathbf{\Sigma}| > 0$. But for $\rho < 1$, since from eqn. (2.4.7), $|\mathbf{\Sigma}| = \sigma^{2p}(1 - \rho)^{p-1}[1 + (p - 1)\rho]$, it is necessary to have $\rho > -(p - 1)^{-1}$.

In general, if ρ denotes the correlation between two scalar random variables Y, Z, $\rho > 0$ implies that on the average, Y tends to increase when Z increases and if $\rho < 0$, Y tends to decrease when Z increases. Of course, $\rho = 0$ implies there is no association (although there might be dependence), and in this case, Y and Z are said to be *uncorrelated*.

A *correlation matrix* is a matrix of correlation coefficients; the random vector $\mathbf{X}$ has an associated correlation matrix $\mathbf{R} \equiv (\rho_{ij})$, $i, j = 1,\ldots,p$. This matrix is useful for studying all associations among the components of a vector of random variables simultaneously. The correlation matrix is computed as a matter of course in many models used in multivariate data analysis since the **R** matrix often provides rapid insight into many unsuspected relationships. The diagonal elements, ρ_{ii}, of a correlation matrix must all be unity. The off-diagonal elements

$$\rho_{ij} \equiv \text{corr}\,(X_i,X_j) = \frac{\text{cov}\,(X_i,X_j)}{[(\text{var}\,X_i)(\text{var}\,X_j)]^{1/2}}, \qquad i \neq j,$$

must always satisfy the inequality $-1 \leq \rho_{ij} \leq +1$, for $i, j = 1,\ldots,p$.

The notion of correlation represents a standardization of covariance because the correlation coefficient is a dimensionless quantity that is independent of the units of measurement. In fact, the correlation coefficient is invariant under any linear transformation. That is, if $Y^* = aY + b$, and $Z^* = cZ + d$, it is trivial to prove that the correlation between the starred and the unstarred random variables is the same; that is, $\text{corr}\,(Y^*,Z^*) = \text{corr}\,(Y,Z)$, for any constants a, b, c, d, $a \neq 0$, $c \neq 0$, a and c of the same sign (see also Lemma 11.2.1).

From the related definitions of correlation and covariance above, it is clear that in matrix form they must also be related. If **R** and $\mathbf{\Sigma}$ are, respectively, the correlation and covariance matrices of the random vector **X**, the relationship is given by

$$\mathbf{\Sigma} = \mathbf{D}^{1/2}\mathbf{R}\mathbf{D}^{1/2}, \tag{3.2.6}$$

where $\mathbf{D} \equiv \text{diag}\,(\sigma_{11},\ldots,\sigma_{pp})$ and σ_{jj} are the diagonal elements of $\mathbf{\Sigma}$. Equivalently, if $\sigma_j^2 = \sigma_{jj}$, $j = 1,\ldots,p$,

$$\mathbf{\Sigma} = \mathbf{D}_\sigma\mathbf{R}\mathbf{D}_\sigma, \tag{3.2.7}$$

where $\mathbf{D}_\sigma \equiv \text{diag}\,(\sigma_1,\ldots,\sigma_p)$.

From (3.2.6) and Property (2) of Section 2.11, it is clear that all correlation matrices are symmetric and positive semidefinite.

Example (3.2.1): Suppose $\text{var}\,X_1 = \text{var}\,X_2 = \cdots = \text{var}\,X_p = 1$, and $\text{cov}\,(X_i,X_j) = \frac{1}{2}$, for all $i, j = 1,\ldots,p$. Then $\mathbf{\Sigma} = \mathbf{R}$, and

$$\mathbf{R} = \begin{pmatrix} 1 & & & \\ & \cdot & \frac{1}{2} & \\ & & \cdot & \\ & \frac{1}{2} & & \cdot \\ & & & 1 \end{pmatrix},$$

where all off-diagonal elements are equal to $\frac{1}{2}$. This is a special intraclass correlation matrix with properties outlined in Chapter 2. It is easy to check that **R** is positive definite. However, if

$$\mathbf{R} = \begin{pmatrix} 1 & & & \\ & \cdot & -\frac{1}{2} & \\ & & \cdot & \\ -\frac{1}{2} & & & \cdot \\ & & & 1 \end{pmatrix},$$

and if $\mathbf{R}: p \times p$, then for $p \geq 4$, $\mathbf{R}$ is not a correlation matrix since in that case, $\mathbf{R}$ is not positive definite.

Example (3.2.2): Suppose the correlation matrix of $\mathbf{X}: 4 \times 1$ is given by

$$\mathbf{R} = \begin{pmatrix} 1 & .9 & .1 & .1 \\ .9 & 1 & .01 & .01 \\ .1 & .01 & 1 & .2 \\ .1 & .01 & .2 & 1 \end{pmatrix}.$$

Note that as expected for all correlation matrices, $\mathbf{R} = \mathbf{R}'$, $\mathbf{R} > 0$, and the diagonal elements are all unity. Moreover, it is clear that if $\mathbf{X} = (X_i)$, $i = 1, \ldots, 4$, X_1 and X_4 have little correlation, as do X_2 and X_4, and X_2 and X_3. X_1 and X_2 are highly correlated, while all other pairs of variables are only slightly correlated with one another.

3.2.3 Linear Transformations

Let $\mathbf{Y} \equiv \mathbf{AX} + \mathbf{b}$ denote a linear transformation of the random vector $\mathbf{X}: p \times 1$ with $\mathbf{A}: k \times p$ and $\mathbf{b}: k \times 1$, so that $\mathbf{Y}$ is a $k \times 1$ vector, $k \leq p$. Let $\mathbf{\Sigma_y}$ and $\mathbf{\Sigma_x}$ denote the covariance matrices of $\mathbf{Y}$ and $\mathbf{X}$, respectively. Then

$$\mathbf{\Sigma_y} = \mathbf{A\Sigma_x A}'. \tag{3.2.8}$$

This relation is very useful in applied work since it is often necessary or desirable to make linear transformations of data matrices. The result may be proved as follows.

$$\begin{aligned} \text{var}\ (\mathbf{Y}) &= E(\mathbf{Y} - E\mathbf{Y})(\mathbf{Y} - E\mathbf{Y})' \\ &= E[(\mathbf{AX} + \mathbf{b}) - (\mathbf{A}E\mathbf{X} + \mathbf{b})][(\mathbf{AX} + \mathbf{b}) - (\mathbf{A}E\mathbf{X} + \mathbf{b})]' \\ &= \mathbf{A}E(\mathbf{X} - E\mathbf{X})(\mathbf{X} - E\mathbf{X})'\mathbf{A}', \end{aligned}$$

which gives the result.

Now suppose $k = 1$ so that $\mathbf{A}$ is a row vector. Then Σ_y is a scalar and var $(Y) = \Sigma_y$. But a variance must be nonnegative. Since $\mathbf{A\Sigma_x A}' \geq 0$ for all $\mathbf{A}$, $\mathbf{\Sigma_x}$ must be positive semidefinite. Since $\mathbf{\Sigma_x}$ is arbitrary, the result must hold for all covariance matrices.

3.2.4 Higher-Order Moments

The kth-order moment of a component X_i of $\mathbf{X}: p \times 1$ is given by

$$\mu_{ik} \equiv E(X_i^k) = \int x_i^k f(\mathbf{x}) d\mathbf{x}, \tag{3.2.9}$$

$k = 0, 1, 2, \ldots, i = 1, \ldots, p$, where $f(\mathbf{x})$ is the density of the (continuous) random vector $\mathbf{X}$. Similarly, the mixed moments are defined by

$$E[X_1^{k_1} X_2^{k_2} \cdots X_p^{k_p}] = \int x_1^{k_1} \cdots x_p^{k_p} f(\mathbf{x}) d\mathbf{x}, \tag{3.2.10}$$

for $k_j = 0, 1, 2, \ldots, j = 1, \ldots, p$.

3.2.5 Moment of a Quadratic Form

Suppose that $\mathbf{X}: p \times 1$. Then, if $E(\mathbf{X}) = \boldsymbol{\theta}$, and $\text{Var}(\mathbf{X}) = \boldsymbol{\Sigma}$, and $\mathbf{A}: p \times p$ is symmetric and not dependent upon $\mathbf{X}$,

$$E(\mathbf{X}'\mathbf{A}\mathbf{X}) = \boldsymbol{\theta}'\mathbf{A}\boldsymbol{\theta} + \text{tr}(\boldsymbol{\Sigma}\mathbf{A}). \tag{3.2.11}$$

This result follows from the fact that

$$E(\mathbf{X}'\mathbf{A}\mathbf{X}) = E \text{ tr}(\mathbf{X}\mathbf{X}'\mathbf{A}) = \text{tr}(E\mathbf{X}\mathbf{X}')\mathbf{A}, \qquad \text{and} \qquad \boldsymbol{\Sigma} = E\mathbf{X}\mathbf{X}' - \boldsymbol{\theta}\boldsymbol{\theta}'.$$

3.3 THE MULTIVARIATE NORMAL DISTRIBUTION

The most important and fundamental distribution of applied multivariate analysis is the multivariate Normal distribution. Its major role is due largely to the fact that standardized sums of independent data vectors following arbitrary multivariate distributions often tend, in large samples, to follow the multivariate Normal distribution (see Chapter 4). This result is a direct generalization of the univariate central limit theorem to higher dimensions. The central role played by the Normal distribution is also attributable in no small way to the fact that results can often be obtained for the Normal distribution and not for other distributions that, for many reasons, are not as tractable as the Normal.

3.3.1 General Density

Let $\mathbf{X}: p \times 1$ be a random vector with density function $f(\mathbf{x})$. $\mathbf{X}$ is said to follow a nonsingular multivariate (p-variate) Normal distribution with mean vector $\boldsymbol{\theta}: p \times 1$ and covariance matrix $\boldsymbol{\Sigma}: p \times p$ if

$$f(\mathbf{x}) = \frac{1}{(2\pi)^{p/2}|\boldsymbol{\Sigma}|^{1/2}} \exp\left[-\tfrac{1}{2}(\mathbf{x} - \boldsymbol{\theta})'\boldsymbol{\Sigma}^{-1}(\mathbf{x} - \boldsymbol{\theta})\right], \tag{3.3.1}$$

for $\boldsymbol{\Sigma} > 0$. If $|\boldsymbol{\Sigma}| = 0$, the distribution of $\mathbf{x}$ is called *singular* or *degenerate* Normal and the density does not exist. All of the probability mass of the distribution then lies in a linear subspace of p-dimensional space (thus, if $p = 2$, all the mass lies on the line $x_2 = ax_1 + b$, for some $a \neq 0$, and some b, or in a point) and probability statements must be made with reference to the subspace. The probability mass of singular but non-Normal multivariate distributions will generally lie in a nonlinear subspace of p-dimensional space.

The notation

$$\mathcal{L}(\mathbf{X}) = N(\boldsymbol{\theta},\boldsymbol{\Sigma}) \tag{3.3.2}$$

will often be used to denote the fact that the density of $\mathbf{X}$ is given by (3.3.1). In Chapter 6 it will be shown that $N(\boldsymbol{\theta},\boldsymbol{\Sigma})$ is a multivariate stable distribution. Hence, instead of using (3.3.1), it will be seen that it could be defined as the distribution for which every linear combination of its components follows a univariate normal distribution. This definition is not as convenient for the sequel as is (3.3.1), so it will not be used.

Example (3.3.1): Consider the univariate general linear model (regression and the analysis of variance) in which N independent observations $y_1,\ldots,y_N$ on a dependent variable are combined in the N-vector $\mathbf{y} = (y_1,\ldots,y_N)'$ and expressed as (see Section 8.2.1 for an elaboration of this model)

$$\mathbf{y} = \mathbf{X}\boldsymbol{\beta} + \mathbf{u},$$

where $\mathbf{u}: N \times 1$ is a vector of disturbances, $\boldsymbol{\beta}: k \times 1$ is a vector of k unknown coefficients, and $\mathbf{X}: N \times k$ is a regressor matrix of fixed independent variables. Suppose $\mathcal{L}(\mathbf{u}) = N(\mathbf{0},\sigma^2\mathbf{I})$, and $|\mathbf{X}'\mathbf{X}| = 0$. The case of $|\mathbf{X}'\mathbf{X}| = 0$ is called *multicollinearity* and occurs, for example, when there are more independent variables than observations. Let $\hat{\boldsymbol{\beta}}$ denote the maximum likelihood estimator of $\boldsymbol{\beta}$. Then $\hat{\boldsymbol{\beta}}$ follows a singular multivariate Normal distribution with all mass distributed in a subspace of dimension no greater than the number of available observations. Moreover, under the same assumption, the same type of result is obtained in the multivariate regression model (in which there is more than one dependent variable).

3.3.2 Bivariate Normal Distribution

Let $\mathbf{X}: 2 \times 1$ be a bivariate random vector with $\mathcal{L}(\mathbf{X}) = N(\boldsymbol{\theta},\boldsymbol{\Sigma})$ and $\boldsymbol{\Sigma} > 0$. Let $\boldsymbol{\theta} = (\theta_i)$, and $\boldsymbol{\Sigma} = (\sigma_{ij})$, $i, j = 1, 2$. For simplicity, take $\sigma_{11} = \sigma_1^2$, $\sigma_{22} = \sigma_2^2$, and $\sigma_{12} = \rho\sigma_1\sigma_2$, where ρ is the correlation coefficient between X_1 and X_2. By expanding (3.3.1) for $p = 2$, it is easily found that the bivariate density is given by

$$f(\mathbf{x}) \equiv f(x_1,x_2) = \frac{1}{2\pi\sigma_1\sigma_2\sqrt{1-\rho^2}}$$
$$\exp\left\{-\frac{1}{2(1-\rho^2)}\left[\left(\frac{x_1-\theta_1}{\sigma_1}\right)^2 - 2\rho\left(\frac{x_1-\theta_1}{\sigma_1}\right)\left(\frac{x_2-\theta_2}{\sigma_2}\right) + \left(\frac{x_2-\theta_2}{\sigma_2}\right)^2\right]\right\}. \tag{3.3.3}$$

Here,

$$\mathbf{\Sigma} = \begin{pmatrix} \sigma_1^2 & \rho\sigma_1\sigma_2 \\ \rho\sigma_1\sigma_2 & \sigma_2^2 \end{pmatrix}, \quad \mathbf{\Sigma}^{-1} = \begin{pmatrix} \dfrac{1}{\sigma_1^2(1-\rho^2)} & -\dfrac{\rho}{\sigma_1\sigma_2(1-\rho^2)} \\ -\dfrac{\rho}{\sigma_1\sigma_2(1-\rho^2)} & \dfrac{1}{\sigma_2^2(1-\rho^2)} \end{pmatrix}. \tag{3.3.4}$$

The expression within brackets in (3.3.3) controls the variation of $f(\mathbf{x})$. That is, if the bracketed expression is constant, $f(\mathbf{x})$ is constant, and conversely. Reference to (3.3.3) shows that the loci of points in (x_1,x_2) space along which $f(\mathbf{x})$ is constant are concentric ellipses centered at (θ_1,θ_2) with major and minor axes tilted at an angle with respect to the (x_1,x_2) axes, and the angle depends upon the values of (ρ,σ_1,σ_2). If $\rho = 0$, the ellipses are parallel to the coordinate axes. Such curves along which the density is constant are called the *density contours* of the distribution. A typical case is shown in Figure 3.3.1. Thus the density contours of a bivariate Normal distribution are ellipses (and more generally, those of a multivariate Normal distribution are hyperellipses). The contours are often a useful device for studying the behavior of a distribution in two or three dimensions. It will be seen in Chapter 6 that the *characteristic function contours* are also ellipses and hyperellipses (for the multivariate Normal distribution with zero mean).

The values of (ρ,σ_1,σ_2) scale the shape of the contours in Figure 3.3.1. Note that if $t_j = (x_j - \theta_j)/\sigma_j$, $j = 1, 2$, and t_2 is graphed as a function of t_1, the curves will still be ellipses; however, they will be centered at the new origin and the semimajor axis will be along the 45° line through the origin if $\rho > 0$, and along the 135° line if $\rho < 0$. For $\rho > 0$, the semimajor and semiminor axes are proportional to $(1 + \rho)^{1/2}$ and $(1 - \rho)^{1/2}$ respectively, with the reverse holding for $\rho < 0$. These assertions follow

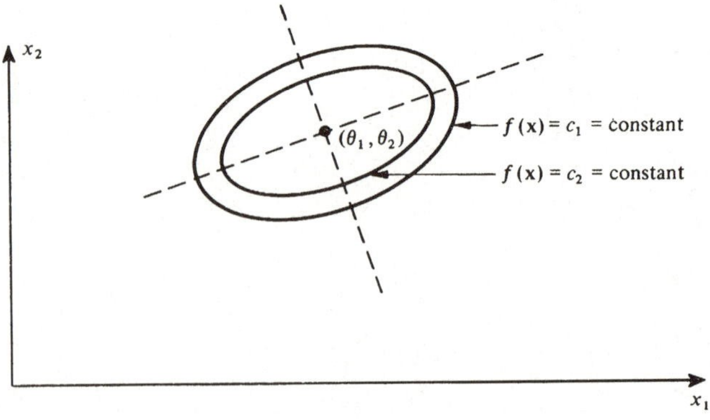

Figure 3.3.1. Density Contours of a Bivariate Normal Distribution

from the fact that any second degree equation of the form $At_1^2 + Bt_1t_2 + At_2^2 + g = 0$ can always be transformed to $A_0v_1^2 + C_0v_2^2 + g = 0$, where $A_0 = A + B/2$, and $C_0 = A - B/2$, by rotating the coordinate axes through a 45° angle. It should be noted that integrating the bivariate density in (3.3.3) over arbitrary portions of the (x_1,x_2) plane is not generally simple [see Exercise 3.3(d)]. Moreover, it is even harder in higher dimensions.

3.3.3 Independence

Let $\mathcal{L}(\mathbf{X}) = N(\boldsymbol{\theta},\boldsymbol{\Sigma})$ and $\mathbf{X}: 2 \times 1$. Then, if $\rho = 0$ in (3.3.4), X_1 and X_2 are not only uncorrelated, they are also independent. This is easily seen by substituting $\rho = 0$ in (3.3.3) and observing that $f(x_1,x_2)$ reduces to the product of a function of x_1 and a function of x_2. Of course the converse is also true; that is, if X_1 and X_2 are independent, they are also uncorrelated; in this direction, the result holds for all bivariate distributions (whereas in the other direction, lack of correlation does not generally imply independence, although it does for the Normal distribution).

3.3.4 Standardization

If $\mathcal{L}(\mathbf{X}) = N(\boldsymbol{\theta},\boldsymbol{\Sigma})$, the distribution may be standardized by the transformation $\mathbf{Y} = \boldsymbol{\Sigma}^{-1/2}(\mathbf{X} - \boldsymbol{\theta})$; that is, $\mathcal{L}(\mathbf{Y}) = N(\mathbf{0},\mathbf{I})$.

This is easily seen by substituting in Theorem (3.1.1) above. Since $J(\mathbf{X} \to \mathbf{Y}) = |\boldsymbol{\Sigma}|^{1/2}$, the density of $\mathbf{Y}$ is

$$g(\mathbf{y}) = \frac{1}{(2\pi)^{p/2}} \exp\left[-\tfrac{1}{2}\mathbf{y}'\mathbf{y}\right].$$

3.4 MARGINAL NORMAL DISTRIBUTIONS

Suppose $\mathcal{L}(\mathbf{X}) = N(\boldsymbol{\theta},\boldsymbol{\Sigma})$. Then the marginal distribution of any subvector is also Normally distributed.

Specifically, let

$$\mathbf{X} = \begin{pmatrix} \mathbf{Y} \\ \mathbf{Z} \end{pmatrix}, \qquad \boldsymbol{\theta} = \begin{pmatrix} \boldsymbol{\theta}_{\mathbf{Y}} \\ \boldsymbol{\theta}_{\mathbf{Z}} \end{pmatrix}, \qquad \boldsymbol{\Sigma} = \begin{pmatrix} \boldsymbol{\Sigma}_{11} & \boldsymbol{\Sigma}_{12} \\ \boldsymbol{\Sigma}_{21} & \boldsymbol{\Sigma}_{22} \end{pmatrix}, \tag{3.4.1}$$

where $\mathbf{Y}: q \times 1$, $\mathbf{Z}: r \times 1$, $\boldsymbol{\theta}_{\mathbf{Y}}: q \times 1$, $\boldsymbol{\theta}_{\mathbf{Z}} = r \times 1$, $\boldsymbol{\Sigma}_{11}: q \times q$, $\boldsymbol{\Sigma}_{22}: r \times r$ $\boldsymbol{\Sigma}_{12}: q \times r$, and $q + r = p$.

Then, if $\boldsymbol{\Sigma}_{11} > 0$, $\boldsymbol{\Sigma}_{22} > 0$, $\mathbf{Y}$ and $\mathbf{Z}$ have proper densities and

$$\mathcal{L}(\mathbf{Y}) = N(\boldsymbol{\theta}_{\mathbf{Y}},\boldsymbol{\Sigma}_{11}) \tag{3.4.2}$$

and

$$\mathcal{L}(\mathbf{Z}) = N(\boldsymbol{\theta}_{\mathbf{Z}},\boldsymbol{\Sigma}_{22}). \tag{3.4.3}$$

These relations are found by substituting (3.4.1) into (3.3.1) and integrating out the unwanted subvector using the normalizing constant of a multivariate Normal cdf. The method is demonstrated below for the marginal distribution of **Y**. The proof for **Z** is analogous.

Define the quadratic form

$$Q(\mathbf{y},\mathbf{z}) = [(\mathbf{y} - \boldsymbol{\theta}_Y),(\mathbf{z} - \boldsymbol{\theta}_Z)]' \begin{pmatrix} \boldsymbol{\Sigma}^{11} & \boldsymbol{\Sigma}^{12} \\ \boldsymbol{\Sigma}^{21} & \boldsymbol{\Sigma}^{22} \end{pmatrix} \begin{bmatrix} (\mathbf{y} - \boldsymbol{\theta}_Y) \\ (\mathbf{z} - \boldsymbol{\theta}_Z) \end{bmatrix},$$

where

$$\boldsymbol{\Sigma}^{-1} \equiv \begin{pmatrix} \boldsymbol{\Sigma}^{11} & \boldsymbol{\Sigma}^{12} \\ \boldsymbol{\Sigma}^{21} & \boldsymbol{\Sigma}^{22} \end{pmatrix}.$$

Then, from (3.3.1) and (3.4.1),

$$f(\mathbf{x}) \equiv f(\mathbf{y},\mathbf{z}) = \frac{1}{(2\pi)^{p/2}|\boldsymbol{\Sigma}|^{1/2}} \exp\left[-\tfrac{1}{2}Q(\mathbf{y},\mathbf{z})\right].$$

But

$$Q(\mathbf{y},\mathbf{z}) = (\mathbf{y} - \boldsymbol{\theta}_Y)'\boldsymbol{\Sigma}^{11}(\mathbf{y} - \boldsymbol{\theta}_Y) + (\mathbf{z} - \boldsymbol{\theta}_Z)'\boldsymbol{\Sigma}^{22}(\mathbf{z} - \boldsymbol{\theta}_Z) + 2(\mathbf{y} - \boldsymbol{\theta}_Y)'\boldsymbol{\Sigma}^{12}(\mathbf{z} - \boldsymbol{\theta}_Z).$$

Hence, if $g(\mathbf{y})$ denotes the marginal density of **Y**,

$$\begin{aligned} g(\mathbf{y}) &= \int f(\mathbf{y},\mathbf{z})d\mathbf{z} \\ &= \frac{H}{(2\pi)^{p/2}|\boldsymbol{\Sigma}|^{1/2}} \exp\left\{-\tfrac{1}{2}[(\mathbf{y} - \boldsymbol{\theta}_Y)'\boldsymbol{\Sigma}^{11}(\mathbf{y} - \boldsymbol{\theta}_Y) + \boldsymbol{\theta}_Z'\boldsymbol{\Sigma}^{22}\boldsymbol{\theta}_Z - 2(\mathbf{y} - \boldsymbol{\theta}_Y)'\boldsymbol{\Sigma}^{12}\boldsymbol{\theta}_Z]\right\}, \end{aligned}$$

where

$$H = \int \exp\left\{-\tfrac{1}{2}[\mathbf{z}'\boldsymbol{\Sigma}^{22}\mathbf{z} - 2(\boldsymbol{\theta}_Z'\boldsymbol{\Sigma}^{22} + \boldsymbol{\theta}_Y'\boldsymbol{\Sigma}^{12} - \mathbf{y}'\boldsymbol{\Sigma}^{12})\mathbf{z}]\right\} d\mathbf{z}.$$

To evaluate H, complete the square on **z**. Accordingly, define

$$\mathbf{a} \equiv \boldsymbol{\theta}_Z - (\boldsymbol{\Sigma}^{22})^{-1}\boldsymbol{\Sigma}^{21}(\mathbf{y} - \boldsymbol{\theta}_Y).$$

Then

$$\begin{aligned} H &= \int \exp\left\{-\tfrac{1}{2}[(\mathbf{z} - \mathbf{a})'\boldsymbol{\Sigma}^{22}(\mathbf{z} - \mathbf{a}) - \mathbf{a}'\boldsymbol{\Sigma}^{22}\mathbf{a}]\right\}d\mathbf{z} \\ &= (2\pi)^{r/2}|\boldsymbol{\Sigma}^{22}|^{-1/2} \exp\left(\frac{\mathbf{a}'\boldsymbol{\Sigma}^{22}\mathbf{a}}{2}\right). \end{aligned}$$

Now combine terms using (2.6.3)–(2.6.5) to get

$$g(\mathbf{y}) = \frac{|\boldsymbol{\Sigma}^{22}|^{-1/2} \exp\left\{-\tfrac{1}{2}(\mathbf{y} - \boldsymbol{\theta}_Y)'\boldsymbol{\Sigma}_{11}^{-1}(\mathbf{y} - \boldsymbol{\theta}_Y)\right\}}{(2\pi)^{q/2}|\boldsymbol{\Sigma}|^{1/2}}.$$

Use of (2.6.4) and (2.6.8) gives (3.4.2).

3.5 CONDITIONAL NORMAL DISTRIBUTIONS

Theorem (3.5.1): For $\mathbf{X}: p \times 1$, let $\mathcal{L}(\mathbf{X}) = N(\boldsymbol{\theta},\boldsymbol{\Sigma})$, and let $\mathbf{X}$ be partitioned as in (3.4.1). Then the conditional distribution of $\mathbf{Y}$ given $(\mathbf{Z} = \mathbf{z})$ is also Normal with mean vector a linear function of $\mathbf{z}$, and covariance matrix independent of $\mathbf{z}$; that is, the conditional distribution of $\mathbf{Y}$, given that $\mathbf{Z} = \mathbf{z}$, is

$$\mathcal{L}(\mathbf{Y}|\mathbf{Z} = \mathbf{z}) = N[\boldsymbol{\theta}_Y + \boldsymbol{\Sigma}_{12}\boldsymbol{\Sigma}_{22}^{-1}(\mathbf{z} - \boldsymbol{\theta}_Z), \boldsymbol{\Sigma}_{11\cdot 2}], \tag{3.5.1}$$

where $\boldsymbol{\Sigma}_{11\cdot 2} \equiv \boldsymbol{\Sigma}_{11} - \boldsymbol{\Sigma}_{12}\boldsymbol{\Sigma}_{22}^{-1}\boldsymbol{\Sigma}_{21}$.

This relation can be demonstrated by applying the definition of conditional density given in (3.1.9), and the expression for marginal density given in (3.4.3), and then combining terms.

Example (3.5.1): Let $\mathcal{L}(\mathbf{X}) = N(\boldsymbol{\theta},\boldsymbol{\Sigma})$, where $\mathbf{X}$ is a bivariate vector. The density of $\mathbf{X} = (X_1,X_2)'$ is given by (3.3.3). The conditional density of X_1 given $X_2 = x_2$, $g(x_1|x_2)$, is found as follows. From (3.5.1), $\mathcal{L}(X_1|X_2 = x_2) = N[\theta_1 + \Sigma_{12}\Sigma_{22}^{-1}(x_2 - \theta_2),\ \Sigma_{11\cdot 2}]$, where $\boldsymbol{\theta}' = (\theta_1,\theta_2)$. Use of (3.3.4) shows that $\Sigma_{12}\Sigma_{22}^{-1} = (\rho\sigma_1\sigma_2)(1/\sigma_2^2) = \rho(\sigma_1/\sigma_2)$, and

$$\Sigma_{11\cdot 2} = \sigma_1^2 - (\rho\sigma_1\sigma_2)\left(\frac{1}{\sigma_2^2}\right)(\rho\sigma_1\sigma_2) = \sigma_1^2(1 - \rho^2).$$

Thus,

$$g(x_1|x_2) = \frac{1}{\sqrt{2\pi}\,\sigma_1(1 - \rho^2)^{1/2}} \exp\left\{-\frac{1}{2\sigma_1^2(1 - \rho^2)}\left[x_1 - \theta_1 - \frac{\rho\sigma_1}{\sigma_2}(x_2 - \theta_2)\right]^2\right\}.$$

In particular, $E(X_1|X_2 = x_2) = \theta_1 + (\rho\sigma_1/\sigma_2)(x_2 - \theta_2)$. Since for $\rho > 0$ $E(X_1|X_2 = x_2)$ increases as x_2 increases (with an opposite effect for $\rho < 0$), there is a "tendency" for x_1 to increase as x_2 increases when $\rho > 0$, and a "tendency" for x_1 to decrease as x_2 increases when $\rho < 0$.

3.6 PRIOR DISTRIBUTIONS ASSOCIATED WITH THE NORMAL DISTRIBUTION

Bayesian analyses involving the multivariate Normal distribution require that a choice be made for the prior distributions of the mean vector and covariance matrix parameters. If information is available of either the rationalistic (subjective) or empiricistic (historic, or data based) type, then this information should be used to formulate *informative prior* distributions on the parameters. Often, however, such information is difficult to assess, so procedures are sought for assigning noninformative prior distributions; that is, those that reflect a high degree of ignorance or vagueness about the parameter values.

The problem of assessing a prior distribution on the parameters which reflects only a modest degree of ignorance is sometimes solved by assessing a particular member of the family of "natural conjugate" prior distributions.

These families of distributions, as they relate to the multivariate Normal distribution, are discussed below.

3.6.1 Vague Prior Distributions

Suppose $\mathcal{L}(\mathbf{X}|\boldsymbol{\theta},\boldsymbol{\Sigma}) = N(\boldsymbol{\theta},\boldsymbol{\Sigma})$. The problem of the subjectivist is to assess a prior distribution for $(\boldsymbol{\theta},\boldsymbol{\Sigma})$. First consider θ_j, a component of $\boldsymbol{\theta}$; $-\infty < \theta_j < \infty, j = 1,\ldots,p$.

One way to express vagueness is to use the notion that all values of a random variable, which may lie anywhere on the real line, are equally likely. Thus, it is assumed that the distribution of the random variable is uniform on the interval (a,b), where a is "large" in magnitude and negative, and b is "large" in magnitude and positive. In fact by (3.1.15), since only the *product* of the prior density and the conditional density of the observed random variable (likelihood function) is of importance in determining the posterior distribution, it is only necessary that the interval (a,b) extend over the region in which the likelihood function is appreciably different from zero. Reasoning in this manner is to employ the *theory of stable estimation* (see Savage, 1962).

For example, when the likelihood function is based upon the Normal distribution, a uniform prior distribution defined on a range of plus or minus three standard deviations from the mean of the Normal distribution will be sufficient. Outside of this range, the prior probabilities of the parameter are not important since they get multiplied by the tails of the Normal density function, and therefore do not significantly affect inferences based upon the posterior distribution.

Using the theory of stable estimation provides a posterior density function defined on a finite range. Therefore, to compute posterior probabilities it is necessary to integrate the posterior density over a finite range. Often, however, it is quite cumbersome mathematically to work with such proper integrals when in fact improper integrals (those with infinite limits) would be much easier to evaluate and study. For this reason, proper uniform prior densities are often approximated by *improper* uniform densities; that is, functions that are constant on the entire real line, and therefore cannot be proper densities (since they do not integrate to unity). The approximation usually does not significantly affect the posterior distribution (upon which inferences will be based) and the attendant computations are simplified. When, to express vague beliefs, the prior probability mass for a random variable θ_j is spread out (diffused) uniformly over the entire real line, the resulting density is written

$$p(\theta_j) \propto \text{constant}, \qquad -\infty < \theta_j < \infty,$$

and the distribution is said to be *diffuse*. It is important to note that this assessment approach is only of interest when, in spite of the fact that $p(\theta_j)$ is not proper, the resulting posterior distribution *is* proper. The use of improper priors in a decision theoretic framework will be discussed in terms of *admissibility* of procedures in Chapter 7. A diffuse prior density is sometimes referred to as a *vague*, or *noninformative* prior since "very little" information is actually contributed to the analysis by the subjective prior density (see Complement 3.1).

If $\boldsymbol{\theta}$ is a vector with components that range over the real line and that may be assumed to be distributed independently, a diffuse prior density for the vector of random variables is again one given by $p(\boldsymbol{\theta}) \propto$ constant. Thus, if $\mathcal{L}(\mathbf{X}|\boldsymbol{\theta},\boldsymbol{\Sigma}) = N(\boldsymbol{\theta},\boldsymbol{\Sigma})$, to be diffuse on $\boldsymbol{\theta}: p \times 1$, assuming its components are independently distributed, assume its density is given by

$$p(\boldsymbol{\theta}) \propto \text{constant}, \qquad -\infty < \theta_j < \infty, \qquad j = 1,\ldots,p. \tag{3.6.1}$$

Now consider the assessment of a vague prior density for a variable σ^2, $0 < \sigma^2 < \infty$. By taking the logarithm, the problem of assessing a vague prior density for the new variable $\log \sigma^2$, where $-\infty < \log \sigma^2 < \infty$, is reduced to the same problem as the one of assessing a vague prior density for the random mean vector, $\boldsymbol{\theta}$. That is, to be diffuse on $\log \sigma^2$, take

$$p(\log \sigma^2) \propto \text{constant}, \qquad 0 < \sigma^2 < \infty.$$

Transforming variables back again by (3.1.13) gives the vague prior density

$$p(\sigma^2) \propto \frac{1}{\sigma^2}, \qquad 0 < \sigma^2 < \infty, \tag{3.6.2}$$

or

$$p(\sigma) \propto \frac{1}{\sigma}, \qquad 0 < \sigma < \infty, \tag{3.6.3}$$

depending upon which parameterization is preferable. It should be noted that this result is the same as that which will be arrived at using Jeffreys' principle of invariance (see Section 3.6.2). More generally, if $\mathcal{L}(\mathbf{X}|\mathbf{D}) = N(\mathbf{0},\mathbf{D})$, where $\mathbf{D} = \text{diag}\,(\sigma_1{}^2,\ldots,\sigma_p{}^2)$, and if the σ_j's are independent, a vague prior density for the p elements is given by

$$p(\mathbf{D}) \equiv p(\sigma_1^2,\ldots,\sigma_p^2) = \prod_{j=1}^{p} p(\sigma_j^2) \propto \frac{1}{\sigma_1^2 \ldots \sigma_p^2},$$

or

$$p(\mathbf{D}) \propto \frac{1}{|\mathbf{D}|}, \qquad \mathbf{D} > 0. \tag{3.6.4}$$

The notion of vague priors is now developed for a full covariance matrix, $\mathbf{\Sigma}$. First note that since $\mathbf{\Sigma} = \mathbf{\Sigma}'$, there are p distinct elements in the first row, only $(p - 1)$ distinct elements remaining in the second row, only $(p - 2)$ distinct elements remaining in the third row, and so on. Hence, by addition over the rows, there is a total of $p(p + 1)/2$ distinct elements in the matrix, or $(p + 1)/2$ groups of p distinct elements, and the result for each group of p elements is given by (3.6.4). A multivariate generalization of (3.6.2) and (3.6.4) is in terms of the *generalized variance* $|\mathbf{\Sigma}|$. Vagueness about the distribution of $\mathbf{\Sigma}$ is represented by asserting that $\mathbf{\Sigma}$ follows the (vague) prior density

$$p(\mathbf{\Sigma}) \propto \frac{1}{|\mathbf{\Sigma}|^{(p+1)/2}}, \qquad \mathbf{\Sigma} > 0. \tag{3.6.5}$$

In the sequel, the terms "diffuse prior" or "vague prior" will be used in the sense of that in (3.6.1) and (3.6.5).

The prior in (3.6.5) was first proposed by Jeffreys, 1961, for $p = 1, 2$. His argument for its use was based upon invariance [see Example (3.6.2) for a generalization of this prior to higher dimensions]. When $p = 1$, the same prior arises as the limiting case of a natural conjugate prior. However, for $p > 1$, the limiting argument breaks down [see the note following (3.6.18)]. Geisser and Cornfield (1963) and Geisser (1965a) invoke the prior in (3.6.5) and show that its use reproduces many of the sampling theory results of multivariate analysis. The use of this prior has been discussed by Dempster (1963) (see Exercise 3.11), and Fraser (1968) shows that this same prior arises as a structural probability distribution. From the fiducial probability point of view, Villegas (1969) gave essentially the argument summarized below.

Let $\mathbf{x}_1,\ldots,\mathbf{x}_p$ denote independent p-vector observations following the law $N(\mathbf{0},\mathbf{\Sigma})$ and let $\mathbf{X} \equiv (\mathbf{x}_1,\ldots,\mathbf{x}_p)$. The likelihood function is

$$\begin{aligned} L(\mathbf{x}_1,\ldots,\mathbf{x}_p|\mathbf{\Sigma}^{-1}) &\propto \frac{1}{|\mathbf{\Sigma}|^{p/2}} \exp\left\{-\tfrac{1}{2}\sum_{j=1}^{p} \mathbf{x}_j'\mathbf{\Sigma}^{-1}\mathbf{x}_j\right\} \\ &\propto \frac{1}{|\mathbf{\Sigma}|^{p/2}} \exp\left\{-\tfrac{1}{2} \operatorname{tr} \mathbf{XX}'\mathbf{\Sigma}^{-1}\right\}. \end{aligned}$$

Let $\mathbf{y}_j = \mathbf{\Sigma}^{-\frac{1}{2}}\mathbf{x}_j$, $j = 1,\ldots, p$ and let $\mathbf{Y} \equiv (\mathbf{y}_1,\ldots,\mathbf{y}_p) = \mathbf{\Sigma}^{-\frac{1}{2}}\mathbf{X}$. Then, from

3.3.4, the $\mathbf{y}_j$'s follow the law $N(\mathbf{0},\mathbf{I})$, so that if $\mathbf{W} = \mathbf{Y}'\mathbf{Y} = \mathbf{X}'\mathbf{\Sigma}^{-1}\mathbf{X}$, $\mathbf{W}$ follows the Wishart distribution (see 5.1), with density proportional to

$$|\mathbf{W}|^{-\frac{1}{2}} \exp \{-\tfrac{1}{2} \operatorname{tr} \mathbf{W}\}, \qquad \mathbf{W} > 0.$$

Since the distribution of $\mathbf{W}$ does not depend upon $\mathbf{\Sigma}^{-1}$, assume the posterior distribution of $\mathbf{W} = \mathbf{X}'\mathbf{\Sigma}^{-1}\mathbf{X}$ (treating $\mathbf{\Sigma}^{-1}$ as unknown and $\mathbf{X}$ as fixed and given) is the same Wishart distribution. That is, assume

$$p(\mathbf{X}'\mathbf{\Sigma}^{-1}\mathbf{X}|\mathbf{X}) \propto |\mathbf{X}'\mathbf{\Sigma}^{-1}\mathbf{X}|^{-\frac{1}{2}} \exp \{-\tfrac{1}{2} \operatorname{tr} \mathbf{X}'\mathbf{\Sigma}^{-1}\mathbf{X}\}.$$

Transforming from $\mathbf{W}$ to $\mathbf{\Sigma}^{-1}$ [using (2.15.5)] gives (since $\mathbf{X}$ is just a constant),

$$p(\mathbf{\Sigma}^{-1}|\mathbf{X}) \propto |\mathbf{\Sigma}^{-1}|^{-\frac{1}{2}} \exp \{-\tfrac{1}{2} \operatorname{tr} \mathbf{X}\mathbf{X}'\mathbf{\Sigma}^{-1}\}.$$

Since by Bayes' theorem $p(\mathbf{\Sigma}^{-1}|\mathbf{X}) \propto L(\mathbf{x}_1,\ldots,\mathbf{x}_n|\mathbf{\Sigma}^{-1})p(\mathbf{\Sigma}^{-1})$, [see (3.1.15)], the required prior density corresponding to the case of minimum prior information is the one given in (3.6.5).

3.6.2 Invariance

The prior density given in (3.6.1) was arrived at based upon a vague belief, or theory of stable estimation, approach. Another approach, due to Jeffreys (1961), is based upon the notion that probability statements made about the observable random variables should remain invariant under changes in the parameterization of the problem. To satisfy this criterion, Jeffreys shows that a random parameter vector $\boldsymbol{\delta}$ must possess a prior density $p(\boldsymbol{\delta})$ of the form

$$p(\boldsymbol{\delta}) \propto |\mathbf{J}|^{1/2}, \tag{3.6.6}$$

where $\mathbf{J}$ is the *information matrix* associated with the likelihood function; that is, if $\boldsymbol{\delta}$ is the $p \times 1$ vector of parameters (δ_i), $p(\mathbf{x}|\boldsymbol{\delta})$ denotes the likelihood function for an observation $\mathbf{x}$, and if

$$J_{ij} \equiv -E\,\frac{\partial^2 \log p(\mathbf{x}|\boldsymbol{\delta})}{\partial \delta_i \partial \delta_j}, \qquad i, j = 1,\ldots,p,$$

then $\mathbf{J} = (J_{ij})$ is the $p \times p$ information matrix. Upon examining the result (3.6.6), Jeffreys finds that such a prior density is not necessarily reasonable when there is more than one parameter in the problem (such as in the Normal distribution with both parameters unknown). To correct for such cases, he recommends that the criterion (3.6.6) be applied to one scalar parameter at a time. Then, although invariance properties still apply, but to a restricted set of problems, most objections to such a prior density are avoided. In any case, however, because the Jeffreys invariant prior depends upon the form of the likelihood function instead of solely upon how one thinks about the parameters, this prior must be somewhat suspect.

Example (3.6.1): Suppose X is a scalar and $\mathcal{L}(X|\theta,\sigma^2) = N(\theta,\sigma^2)$. The density is

$$p(x|\theta,\sigma^2) \propto \frac{1}{\sigma} \exp\left[-\frac{1}{2\sigma^2}(x-\theta)^2\right], \qquad -\infty < x < \infty,$$

and the log density is

$$L \equiv \log p(x|\theta,\sigma^2) = -\log\sigma - \frac{1}{2\sigma^2}(x-\theta)^2 + c,$$

for some constant, c. It is easy to check that if $\mathbf{J} = (J_{ij})$, $i = 1, 2$, where

$$J_{11} = -E\frac{\partial^2 L}{\partial\theta^2}, \qquad J_{22} = -E\frac{\partial^2 L}{\partial(\sigma^2)^2}, \qquad J_{12} = -E\frac{\partial^2 L}{\partial\theta\partial\sigma^2},$$

$$J_{21} = -E\frac{\partial^2 L}{\partial\sigma^2\partial\theta},$$

then

$$J_{11} = \frac{1}{\sigma^2}, \qquad J_{22} = \frac{1}{2\sigma^4}, \qquad J_{12} = J_{21} = 0.$$

Use of (3.6.6) directly without treating the parameters separately gives $p(\theta,\sigma^2) \propto 1/\sigma^3$.

Now suppose that σ^2 is a known constant although θ is unknown. Then J is a scalar and $J = J_{11} =$ constant. So applying (3.6.6) gives for the invariant prior density,

$$p(\theta) \propto \text{constant}.$$

Next suppose that θ is a known constant while σ^2 is free to vary. Then $J = J_{22} = 1/2\sigma^4$, and by (3.6.6) for known θ,

$$p(\sigma^2) \propto \frac{1}{\sigma^2}.$$

Combining terms for the parameter by parameter case, assuming θ and

σ^2 are independent, gives $p(\theta,\sigma^2) \propto \sigma^{-2}$. Hence, Jeffreys' invariance argument applied to one parameter at a time gives the same result as does the diffuseness argument (the limiting case of the theory of stable estimation).

Example (3.6.2): Suppose $\mathbf{X}$: $p \times 1$ and $\mathcal{L}(\mathbf{X}|\boldsymbol{\theta},\boldsymbol{\Lambda}) = N(\boldsymbol{\theta},\boldsymbol{\Lambda}^{-1})$, $\boldsymbol{\Lambda} > 0$, with $\boldsymbol{\Lambda} \equiv \boldsymbol{\Sigma}^{-1}$. The density is $p(\mathbf{x}|\boldsymbol{\theta},\boldsymbol{\Lambda}) \propto |\boldsymbol{\Lambda}|^{\frac{1}{2}} \exp[-\frac{1}{2}(\mathbf{x} - \boldsymbol{\theta})'\boldsymbol{\Lambda}(\mathbf{x} - \boldsymbol{\theta})]$, and the log density is

$$L \equiv \log p(\mathbf{x}|\boldsymbol{\theta},\boldsymbol{\Lambda}) = \tfrac{1}{2} \log |\boldsymbol{\Lambda}| + c - \tfrac{1}{2}(\mathbf{x} - \boldsymbol{\theta})'\boldsymbol{\Lambda}(\mathbf{x} - \boldsymbol{\theta}),$$

for some constant c. In this problem we will develop the Jeffreys invariant prior by applying the procedure to one matrix parameter at a time.

First assume $\boldsymbol{\Lambda}$ is constant. Then,

$$\frac{\partial L}{\partial \boldsymbol{\theta}} = -\tfrac{1}{2}\{2\boldsymbol{\Lambda}\boldsymbol{\theta} - 2\boldsymbol{\Lambda}\mathbf{x}\},$$

and

$$-E\left(\frac{\partial^2 L}{\partial \boldsymbol{\theta} \partial \boldsymbol{\theta}'}\right) = -E(-\boldsymbol{\Lambda}) = \boldsymbol{\Lambda} = \text{constant}.$$

Hence, the information matrix is constant, and by (3.6.6), the invariant prior for $\boldsymbol{\theta}$ is

$$p(\boldsymbol{\theta}) \propto \text{constant}.$$

Now assume $\boldsymbol{\theta}$ is constant. Then, if L is written in the more convenient form

$$L = \tfrac{1}{2} \log |\boldsymbol{\Lambda}| + c - \tfrac{1}{2} \operatorname{tr} (\mathbf{x} - \boldsymbol{\theta})(\mathbf{x} - \boldsymbol{\theta})'\boldsymbol{\Lambda},$$

use of (2.14.2) and (2.14.5) gives

$$\frac{\partial L}{\partial \boldsymbol{\Lambda}} = \boldsymbol{\Lambda}^{-1} - \frac{1}{2} \operatorname{diag}(\boldsymbol{\Lambda}^{-1}) - \frac{1}{2}(\mathbf{x} - \boldsymbol{\theta})(\mathbf{x} - \boldsymbol{\theta})'.$$

From (2.15.8) and the definition of a Jacobian, if $\boldsymbol{\Lambda} \equiv (\lambda_{ij})$,

$$\frac{\partial^2 L}{\partial \lambda_{ij} \partial \lambda_{kl}} \propto |\boldsymbol{\Lambda}|^{-(p+1)}.$$

Taking expectations with respect to $\mathbf{x}$ leaves the result unchanged; that is, if $\mathbf{J}$ denotes the information matrix,

$$|\mathbf{J}| \propto |\boldsymbol{\Lambda}|^{-(p+1)}.$$

Hence, the invariant prior for $\boldsymbol{\Lambda}$ is, by (3.6.6),

$$p(\mathbf{\Lambda}) \propto \frac{1}{|\mathbf{\Lambda}|^{(p+1)/2}}.$$

Since the Jacobian from $\mathbf{\Lambda}$ to $\mathbf{\Sigma}$ is $|\mathbf{\Sigma}|^{-(p+1)}$, the implied invariant prior density for $\mathbf{\Sigma}$ is

$$p(\mathbf{\Sigma}) \propto \frac{1}{|\mathbf{\Sigma}|^{(p+1)/2}},$$

the same prior as in (3.6.5).

Finally, if $\boldsymbol{\theta}$ and $\mathbf{\Sigma}$ are a priori independent, the invariant prior implied by the Jeffreys argument is

$$p(\boldsymbol{\theta},\mathbf{\Sigma}) \propto \frac{1}{|\mathbf{\Sigma}|^{(p+1)/2}}.$$

3.6.3 Natural Conjugate Prior Distributions

If prior information is easily assessable, it can be brought to bear in a problem by means of an *informative prior* density. One vehicle for introducing informative prior information is the natural conjugate distribution.

The notion of a natural conjugate prior distribution is due to Barnard (1954), but the idea was refined and detailed by Raiffa and Schlaifer (1961). The basic idea is to write the density or likelihood function for the observable random variables and then interchange the roles of the observable random variables and the parameters, assuming the latter to be random and the former to be fixed and known. Modifying the proportionality constant appropriately so that the new "density" integrates to unity and letting the fixed parameters be arbitrary (but not dependent upon current sample data) to enrich the family of distributions provides a density that in this sense is "conjugate" to the original. Results involving natural conjugate priors are useful in Bayesian analyses involving distributions with sufficient statistics (under very general conditions, such distributions constitute the exponential family). The advantages of using a natural conjugate distribution to express prior beliefs are: (1) the posterior distribution has the same form as the prior (see Theorem 3.6.1), (2) the posterior distribution is generally tractable mathematically, and (3) the natural conjugate distribution is posterior to a vague prior [see the Remark following (3.6.20)].

Example (3.6.3): For illustration, assume $\mathbf{X}: p \times 1$ follows the law $\mathcal{L}(\mathbf{X}|\boldsymbol{\theta}) = N(\boldsymbol{\theta},\mathbf{A}_0)$, where $\mathbf{A}_0$ is a fixed, known, nonsingular $p \times p$ matrix. The density of this one observation is given by

$$p(\mathbf{x}|\boldsymbol{\theta},\mathbf{A}_0) \propto \exp\left[-\tfrac{1}{2}(\mathbf{x}-\boldsymbol{\theta})'\mathbf{A}_0^{-1}(\mathbf{x}-\boldsymbol{\theta})\right].$$

Interchanging the roles of $\mathbf{x}$ and $\boldsymbol{\theta}$ gives the prior density (just rewriting)

$$p(\boldsymbol{\theta}) \propto \exp\left[-\tfrac{1}{2}(\boldsymbol{\theta} - \mathbf{x})'\mathbf{A}_0^{-1}(\boldsymbol{\theta} - \mathbf{x})\right].$$

This expression, when viewed as a density in $\boldsymbol{\theta}$, has the form of a Normal density with mean vector $\mathbf{x}$ and covariance matrix $\mathbf{A}_0$. Thus, the family of prior densities may be enriched by adopting an arbitrary mean vector $\mathbf{a}$, and an arbitrary covariance matrix $\mathbf{A}$. Then the family of natural conjugate prior densities is of the form

$$p(\boldsymbol{\theta}) \propto \exp\left[-\tfrac{1}{2}(\boldsymbol{\theta} - \mathbf{a})'\mathbf{A}^{-1}(\boldsymbol{\theta} - \mathbf{a})\right],$$

or

$$\mathcal{L}(\boldsymbol{\theta}) = N(\mathbf{a},\mathbf{A}).$$

3.6.4 Natural Conjugate Priors for the Normal Distribution

Now consider the general case of $\mathbf{X}: p \times 1$ with $\mathcal{L}(\mathbf{X}|\boldsymbol{\theta},\boldsymbol{\Sigma}) = N(\boldsymbol{\theta},\boldsymbol{\Sigma})$, and $\boldsymbol{\Sigma}$ is unknown. Suppose there are n independent and identically distributed observations $\mathbf{X}_1,\ldots,\mathbf{X}_n$. The problem is to find natural conjugate prior densities for $\boldsymbol{\theta}$ and $\boldsymbol{\Sigma}$. This may be accomplished since there are sufficient statistics. The joint density of the observations is given by

$$p(\mathbf{x}_1,\ldots,\mathbf{x}_n|\boldsymbol{\theta},\boldsymbol{\Sigma}) \propto \frac{1}{|\boldsymbol{\Sigma}|^{n/2}} \exp\left\{-\tfrac{1}{2}\sum_{i=1}^{n}\left[(\mathbf{x}_i - \boldsymbol{\theta})'\boldsymbol{\Sigma}^{-1}(\mathbf{x}_i - \boldsymbol{\theta})\right]\right\}.$$

It is sometimes convenient to reparameterize in terms of $(\boldsymbol{\theta},\boldsymbol{\Lambda}) \equiv (\boldsymbol{\theta},\boldsymbol{\Sigma}^{-1})$. $\boldsymbol{\Lambda}$ is called the *precision matrix* (since the smaller the variance, the greater the precision). The density becomes

$$p(\mathbf{x}_1,\ldots,\mathbf{x}_n|\boldsymbol{\theta},\boldsymbol{\Lambda}) \propto |\boldsymbol{\Lambda}|^{n/2} \exp\left\{-\tfrac{1}{2}\sum_{i=1}^{n}\left[(\mathbf{x}_i - \boldsymbol{\theta})'\boldsymbol{\Lambda}(\mathbf{x}_i - \boldsymbol{\theta})\right]\right\}.$$

Expanding the exponent gives

$$p(\mathbf{x}_1,\ldots,\mathbf{x}_n|\boldsymbol{\theta},\boldsymbol{\Lambda}) \propto |\boldsymbol{\Lambda}|^{n/2} \exp\left\{-\tfrac{1}{2}\sum_{1}^{n}\left[\boldsymbol{\theta}'\boldsymbol{\Lambda}\boldsymbol{\theta} - 2\boldsymbol{\theta}'\boldsymbol{\Lambda}\mathbf{x}_i + \mathbf{x}_i'\boldsymbol{\Lambda}\mathbf{x}_i\right]\right\},$$

or

$$p(\mathbf{x}_1,\ldots,\mathbf{x}_n|\boldsymbol{\theta},\boldsymbol{\Lambda}) \propto |\boldsymbol{\Lambda}|^{n/2} \exp\left\{-\frac{n}{2}\left[\boldsymbol{\theta}'\boldsymbol{\Lambda}\boldsymbol{\theta} - 2\boldsymbol{\theta}'\boldsymbol{\Lambda}\bar{\mathbf{x}} + \frac{1}{n}\sum_{1}^{n}\mathbf{x}_i'\boldsymbol{\Lambda}\mathbf{x}_i\right]\right\},$$

where $\bar{\mathbf{x}} \equiv n^{-1}\Sigma_1^n\mathbf{x}_i$ is the usual expression for the mean of a sample. Completion of the square on $\boldsymbol{\theta}$ gives

$$p(\mathbf{x}_1,\ldots,\mathbf{x}_n|\boldsymbol{\theta},\boldsymbol{\Lambda}) \propto |\boldsymbol{\Lambda}|^{n/2}$$

$$\exp\left\{-\frac{n}{2}\left[(\boldsymbol{\theta} - \bar{\mathbf{x}})'\boldsymbol{\Lambda}(\boldsymbol{\theta} - \bar{\mathbf{x}}) + \frac{1}{n}\sum_{1}^{n}\mathbf{x}_i'\boldsymbol{\Lambda}\mathbf{x}_i - \bar{\mathbf{x}}'\boldsymbol{\Lambda}\bar{\mathbf{x}}\right]\right\}.$$

Now rewrite this expression as the product of two factors:

$$p(\mathbf{x}_1,\ldots,\mathbf{x}_n|\boldsymbol{\theta},\boldsymbol{\Lambda}) \propto \left[|\boldsymbol{\Lambda}|^{1/2}\exp\left\{-\frac{n}{2}(\boldsymbol{\theta}-\bar{\mathbf{x}})'\boldsymbol{\Lambda}(\boldsymbol{\theta}-\bar{\mathbf{x}})\right\}\right] \cdot \left[|\boldsymbol{\Lambda}|^{(n-1)/2}\exp\left\{-\tfrac{1}{2}\left(\sum_1^n \mathbf{x}_i'\boldsymbol{\Lambda}\mathbf{x}_i - n\bar{\mathbf{x}}'\boldsymbol{\Lambda}\bar{\mathbf{x}}\right)\right\}\right]. \quad (3.6.7)$$

Note that only the first factor contains $\boldsymbol{\theta}$, so that taking advantage of Example (3.6.2) suggests a family of conditional natural conjugate prior distributions for $\boldsymbol{\theta}$ in the form

$$\mathcal{L}(\boldsymbol{\theta}|\boldsymbol{\Lambda}) = N(\mathbf{a},(\boldsymbol{\Lambda}b)^{-1}), \quad (3.6.8)$$

where $\mathbf{a}$ is an arbitrary p-vector obtained by "enrichment" from $\bar{\mathbf{x}}$, and b is an arbitrary positive scalar obtained by enrichment from n. However, what is really sought is a joint prior density $p(\boldsymbol{\theta},\boldsymbol{\Lambda})$. Since

$$p(\boldsymbol{\theta},\boldsymbol{\Lambda}) = p(\boldsymbol{\theta}|\boldsymbol{\Lambda})p(\boldsymbol{\Lambda}), \quad (3.6.9)$$

it is only necessary to establish $p(\boldsymbol{\Lambda})$ from the second bracketed expression in (3.6.7). Since the trace of a scalar equals the scalar, that expression may be rewritten as

$$|\boldsymbol{\Lambda}|^{(n-1)/2}\exp\left\{-\tfrac{1}{2}\operatorname{tr}\left[\sum_1^n \mathbf{x}_i'\boldsymbol{\Lambda}\mathbf{x}_i - n\bar{\mathbf{x}}'\boldsymbol{\Lambda}\bar{\mathbf{x}}\right]\right\}.$$

Recalling (2.9.2), the cyclical property of the trace gives

$$|\boldsymbol{\Lambda}|^{(n-1)/2}\exp\left\{-\tfrac{1}{2}\operatorname{tr}\boldsymbol{\Lambda}\left(\sum_1^n \mathbf{x}_i\mathbf{x}_i' - n\bar{\mathbf{x}}\bar{\mathbf{x}}'\right)\right\},$$

or

$$|\boldsymbol{\Lambda}|^{(n-1)/2}\exp\{-\tfrac{1}{2}\operatorname{tr}\boldsymbol{\Lambda}\mathbf{V}\}, \quad (3.6.10)$$

where $\mathbf{V} \equiv \Sigma_1^n(\mathbf{x}_i - \bar{\mathbf{x}})(\mathbf{x}_i - \bar{\mathbf{x}})'$ is the unnormed sample covariance matrix, a known observable quantity.

In Chapter 5, it will be seen that if $\mathbf{V}: p \times p$ is nonsingular with density

$$p(\mathbf{V}|\mathbf{G}) \propto \frac{|\mathbf{V}|^{(m-p-1)/2}}{|\mathbf{G}|^{m/2}}\exp[-\tfrac{1}{2}\operatorname{tr}\mathbf{V}\mathbf{G}^{-1}], \qquad \mathbf{V} > 0,\ \mathbf{G} > 0, \quad m \geq p, \quad (3.6.11)$$

$\mathbf{V}$ will be said to have a Wishart distribution with scale matrix $\mathbf{G}$, dimension p, and m degrees of freedom; it will be written $\mathcal{L}(\mathbf{V}) = W(\mathbf{G},p,m)$. Noting that the expression in (3.6.10) is of exactly this form after n and $\mathbf{V}$ are "enriched" to arbitrary values, the marginal natural conjugate prior distribution for $\boldsymbol{\Lambda}$ becomes

$$\mathcal{L}(\boldsymbol{\Lambda}) = W(\mathbf{G},p,m), \quad (3.6.12)$$

where $\mathbf{G}$ and m are such that $\mathbf{G} > 0$ and $m \geq p$, but otherwise, they are arbitrary.

Properties of the Wishart distribution are fundamental to multivariate analysis and are considered in detail in Chapter 5.

Substituting from (3.6.8) and (3.6.12) into (3.6.9) shows that the joint natural conjugate prior distribution for $\boldsymbol{\theta}$ and $\boldsymbol{\Lambda}$ is given by the *Normal-Wishart distribution*

$$p(\boldsymbol{\theta},\boldsymbol{\Lambda}) \propto |\boldsymbol{\Lambda}|^{(m-p)/2} \exp\{-\tfrac{1}{2}[(\boldsymbol{\theta} - \mathbf{a})'\boldsymbol{\Lambda}(\boldsymbol{\theta} - \mathbf{a})b + \operatorname{tr}\boldsymbol{\Lambda}\mathbf{G}^{-1}]\}. \quad (3.6.13)$$

To use the natural conjugate prior density in future analyses it is only necessary to assess values for m, $\mathbf{a}$, b and $\mathbf{G}$ (see Section 3.8) based upon subjective feelings or historical evidence.

To obtain a natural conjugate prior for $(\boldsymbol{\theta},\boldsymbol{\Sigma})$ from $(\boldsymbol{\theta},\boldsymbol{\Lambda})$ it is only necessary to transform variates in (3.6.12). The result will be a "Normal-inverted Wishart" distribution. (The inverted Wishart distribution is discussed in Chapter 5.) Alternatively, the natural conjugate prior density for $(\boldsymbol{\theta},\boldsymbol{\Sigma})$ can be obtained just as that for $(\boldsymbol{\theta},\boldsymbol{\Lambda})$ was obtained. Returning to the likelihood function in (3.6.7) and parameterizing it in terms of $(\boldsymbol{\theta},\boldsymbol{\Sigma})$ by changing variables gives

$$p(\mathbf{x}_1,\ldots,\mathbf{x}_n|\boldsymbol{\theta},\boldsymbol{\Sigma}) \propto \left[|\boldsymbol{\Sigma}|^{-1/2} \exp\left\{-\frac{n}{2}(\boldsymbol{\theta} - \bar{\mathbf{x}})'\boldsymbol{\Sigma}^{-1}(\boldsymbol{\theta} - \bar{\mathbf{x}})\right\}\right]$$
$$\cdot\left[|\boldsymbol{\Sigma}^{-1}|^{(n-1)/2} \exp\left\{-\tfrac{1}{2}\left(\sum_1^n \mathbf{x}_i'\boldsymbol{\Sigma}^{-1}\mathbf{x}_i - n\bar{\mathbf{x}}'\boldsymbol{\Sigma}^{-1}\bar{\mathbf{x}}\right)\right\}\right]. \quad (3.6.14)$$

Again use the conditional prior $\mathcal{L}(\boldsymbol{\theta}|\boldsymbol{\Sigma}) = N(\mathbf{a},\boldsymbol{\Sigma})$. However, instead of (3.6.9) use

$$p(\boldsymbol{\theta},\boldsymbol{\Sigma}) = p(\boldsymbol{\theta}|\boldsymbol{\Sigma})p(\boldsymbol{\Sigma}). \quad (3.6.15)$$

The second factor in (3.6.14) may be written as the analogue of (3.6.10),

$$|\boldsymbol{\Sigma}^{-1}|^{(n-1)/2} \exp\{-\tfrac{1}{2}\operatorname{tr}\boldsymbol{\Sigma}^{-1}\mathbf{V}\}. \quad (3.6.16)$$

In (3.6.11), let $\mathbf{W} = \mathbf{V}^{-1}$; use of (3.1.13) and Property (8) of 2.15 yields the inverted Wishart density (see Section 5.2)

$$p(\mathbf{W}|\mathbf{G}) \propto |\mathbf{W}^{-1}|^{(m+p+1)/2}|\mathbf{G}^{-1}|^{m/2} \exp(-\tfrac{1}{2}\operatorname{tr}\mathbf{W}^{-1}\mathbf{G}^{-1}). \quad (3.6.17)$$

More generally, $\mathbf{W}$ follows the inverted Wishart distribution and is written

$$\mathcal{L}(\mathbf{W}) = W^{-1}(\mathbf{H},p,\nu),$$

if the density of $\mathbf{W}$ may be expressed as

$$p(\mathbf{W}|\mathbf{H}) \propto |\mathbf{W}^{-1}|^{\nu/2} \exp\{-\tfrac{1}{2}\operatorname{tr}\mathbf{W}^{-1}\mathbf{H}\}, \qquad \mathbf{W} > 0,\ \mathbf{H} > 0,\ \nu > 2p.$$

This definition represents an extension to the multivariate case of the inverted gamma distribution discussed by Raiffa and Schlaifer (1961).

Comparison of (3.6.16) and (3.6.17) shows that after enrichment, $\mathbf{\Sigma}$ follows the inverted Wishart distribution. Hence, to be consistent with (3.6.12) take as the marginal natural conjugate prior distribution for $\mathbf{\Sigma}$,

$$\mathcal{L}(\mathbf{\Sigma}) = W^{-1}(\mathbf{H},p,\nu), \qquad \mathbf{H} > 0, \mathbf{\Sigma} > 0, \nu \equiv m + p + 1, \quad (3.6.18)$$

where $m > p - 1$ for a proper distribution.

[*Note:* It may be seen from (3.6.18) and the defining density that in the limit as $\mathbf{H} \to \mathbf{0}$ and $m \to p - 1$, the resulting improper density takes the form

$$p(\Sigma) \propto \frac{1}{|\Sigma|^{(m+p+1)/2}} \to \frac{1}{|\Sigma|^p}.$$

Therefore, for $p = 1$, the result is the same as that for the invariant density of (3.6.5). However, for $p > 1$, the results diverge.]

The joint natural conjugate prior, the *Normal-inverted Wishart density* is obtained from (3.6.8), (3.6.15), and (3.6.18), for $\nu > 2p$, as

$$p(\boldsymbol{\theta},\mathbf{\Sigma}) \propto |\mathbf{\Sigma}^{-1}|^{(\nu+1)/2} \exp\{-\tfrac{1}{2}[(\boldsymbol{\theta} - \mathbf{a})'\mathbf{\Sigma}^{-1}(\boldsymbol{\theta} - \mathbf{a})b + \operatorname{tr} \mathbf{\Sigma}^{-1}\mathbf{H}]\}. \quad (3.6.19)$$

Some discussion of applications of natural conjugate prior densities associated with the Normal distribution is given in Ando and Kaufman (1965). Moreover, it will be seen in Chapter 8 that in multivariate Normal regression problems, use of the natural conjugate prior distribution may impose constraints on the parameters which may not be desired (Rothenberg's Problem, 1963).

A useful property of natural conjugate prior densities is given in the following theorem.

Theorem (3.6.1): Suppose the distribution of $\mathbf{X}$ for given $\boldsymbol{\theta}$ possesses sufficient statistics for $\boldsymbol{\theta}$. Let $p(\boldsymbol{\theta})$ denote a natural conjugate density for $\boldsymbol{\theta}$, and $\mathbf{x}_1,\ldots,\mathbf{x}_n$ denote a sample of observations. Then, if $p(\boldsymbol{\theta}|\mathbf{x}_1,\ldots,\mathbf{x}_n)$ denotes the posterior density for $\boldsymbol{\theta}$, $p(\boldsymbol{\theta}|\mathbf{x}_1,\ldots,\mathbf{x}_n)$ is in the same family as $p(\boldsymbol{\theta})$.

The proof of this theorem is straightforward and is based upon the fact that $\Sigma_1^n \mathbf{x}_j$ is sufficient for the exponential family. It is given in Raiffa and Schlaifer, p. 48. An implication of the theorem is that Normal priors generate Normal posteriors; Normal-Wishart priors generate Normal-Wishart posteriors, and so on. Note that the family of sampling distributions of $\mathbf{X}|\theta$ we are considering in theorem (3.6.1) is essentially the exponential family.

As an example, examine the case of $\mathbf{X}: p \times 1$, $\mathcal{L}(\mathbf{X}|\boldsymbol{\theta},\mathbf{\Sigma}) = N(\boldsymbol{\theta},\mathbf{\Sigma})$, where a natural conjugate prior density for $(\boldsymbol{\theta},\mathbf{\Sigma})$ was shown to be given by the Normal-inverted Wishart density in (3.6.19). Hence, from (3.6.14) and (3.6.19), the posterior density of $(\boldsymbol{\theta},\boldsymbol{\Sigma})$ given the sample is given by $p(\boldsymbol{\theta},\boldsymbol{\Sigma}|\mathbf{x}_1,\ldots,\mathbf{x}_n) \propto p(\mathbf{x}_1,\ldots,\mathbf{x}_n|\boldsymbol{\theta},\boldsymbol{\Sigma})p(\boldsymbol{\theta},\boldsymbol{\Sigma})$, or

$$p(\boldsymbol{\theta},\boldsymbol{\Sigma}|\mathbf{x}_1,\ldots,\mathbf{x}_n) \propto \frac{1}{|\boldsymbol{\Sigma}|^{n/2}} \exp\{-\tfrac{1}{2}\operatorname{tr}[n(\boldsymbol{\theta}-\bar{\mathbf{x}})(\boldsymbol{\theta}-\bar{\mathbf{x}})' + \mathbf{V}]\boldsymbol{\Sigma}^{-1}\}$$

$$\cdot \frac{1}{|\boldsymbol{\Sigma}|^{(\nu+1)/2}} \exp\{-\tfrac{1}{2}\operatorname{tr}[b(\boldsymbol{\theta}-\mathbf{a})(\boldsymbol{\theta}-\mathbf{a})' + \mathbf{H}]\boldsymbol{\Sigma}^{-1}\}.$$

Combining terms gives

$$p(\boldsymbol{\theta},\boldsymbol{\Sigma}|\mathbf{x}_1,\ldots,\mathbf{x}_n) \propto |\boldsymbol{\Sigma}^{-1}|^{(n+\nu+1)/2} \cdot \exp\left\{-\tfrac{1}{2}\left[(n+b)\left(\boldsymbol{\theta}-\frac{n\bar{\mathbf{x}}+\mathbf{a}b}{n+b}\right)'\boldsymbol{\Sigma}^{-1}\left(\boldsymbol{\theta}-\frac{n\bar{\mathbf{x}}+\mathbf{a}b}{n+b}\right) + \operatorname{tr}(\mathbf{V}+\mathbf{H})\boldsymbol{\Sigma}^{-1} + \frac{nb}{n+b}(\bar{\mathbf{x}}-\mathbf{a})'\boldsymbol{\Sigma}^{-1}(\bar{\mathbf{x}}-\mathbf{a})\right]\right\}. \quad (3.6.20)$$

Comparison of (3.6.20) with (3.6.19) shows that the prior and posterior are both in the Normal-inverted Wishart family of distributions.

[*Remark:* If, in a distribution with sufficient statistics, the posterior distribution with respect to a diffuse prior is used as the prior in a two stage process, the second stage prior is natural conjugate.]

3.6.5 Minimum Information Priors

In trying to express vagueness about the parameters of a distribution, attempts have been made to draw upon the results of information theory (see Shannon, 1948). It is useful to define the *information in a continuous distribution* with density f as

$$I(f) = \int f(\mathbf{x}) \log f(\mathbf{x})d\mathbf{x}, \quad (3.6.21)$$

where the integration extends over all possible values of the variable (vector) $\mathbf{x}$. If there are no constraints on $f(\cdot)$ other than that it be a bona fide density, it is straightforward to solve for that f which will minimize the information in this distribution. $I(f)$ may be minimized (see Complement 3.1 for details) subject to $\int f(\mathbf{x})d\mathbf{x} = 1$ by the Calculus of Variations. The result is

$$f(\mathbf{x}) \propto \text{constant}.$$

That is, the distribution that contains minimum information is the uniform, the same result found earlier in considering diffuse priors. Thus, the notion of information conforms to the idea of diffuse probability mass.

It is interesting to note that if it is required that f have positive probability mass on the entire real line, and not only be a bona fide density, but also, that it have finite first and second moments, the Normal distribution will be least informative, rather than the uniform.

3.7 PREDICTIVE DISTRIBUTIONS

Let $\mathbf{X}: p \times 1$ denote a random vector, and $\mathbf{x}_1, \ldots, \mathbf{x}_N$, independent observations of $\mathbf{X}$. Denote the likelihood function of the observations for parameter vector $\boldsymbol{\theta}$ by $f(\mathbf{x}_1, \ldots, \mathbf{x}_N | \boldsymbol{\theta})$. Suppose $p(\boldsymbol{\theta})$ is a prior density for $\boldsymbol{\theta}$ and it is of interest to study the behavior of a future observation $\mathbf{x}_{N+1}$ with density $g(\mathbf{x}_{N+1} | \boldsymbol{\theta})$. The predictive density of $\mathbf{x}_{N+1}$ given all the previous observations is defined to be

$$h(\mathbf{x}_{N+1} | \mathbf{x}_1, \ldots, \mathbf{x}_N) = \int g(\mathbf{x}_{N+1} | \boldsymbol{\theta}) q(\boldsymbol{\theta} | \mathbf{x}_1, \ldots, \mathbf{x}_N) d\boldsymbol{\theta}, \tag{3.7.1}$$

where

$$q(\boldsymbol{\theta} | \mathbf{x}_1, \ldots, \mathbf{x}_N) \propto p(\boldsymbol{\theta}) f(\mathbf{x}_1, \ldots, \mathbf{x}_N | \boldsymbol{\theta})$$

denotes the posterior density of $\boldsymbol{\theta}$. Since the predictive distribution is parameter free, it is useful for making inferences about the magnitude of future observations (see Chapters 8 and 14). For example, if a linear relationship is established for sales of a company over time, based upon past data, it is possible to develop the predictive distribution for sales at some future time. See Exercise 3.9. Thus the predictive distribution is useful for forecasting, and for comparing alternative models based upon their goodness of predictability of future observations.

Credibility Intervals: If $1 - \alpha = P\{a \leq \theta \leq b | \text{sample}\}$, where P denotes the posterior probability for θ given the sample data, the interval (a,b) is called a $(1 - \alpha)$ credibility interval. Credibility intervals (regions, in higher dimensions) will coincide with confidence intervals for the case in which the prior is improper and the sampling density $f(\mathbf{x} | \boldsymbol{\theta})$ is a function of $(\mathbf{x} - \boldsymbol{\theta})$.

3.8 ASSESSMENT OF MULTIVARIATE PRIOR PARAMETERS

Making inferences about the parameters of multivariate distributions by Bayesian techniques requires, of course, the assessment of prior distributions. Such assessments may be carried out by means of one of the general procedures discussed in Section 3.6. However, if use of an informative prior procedure is adopted (such as the natural conjugate) there are practical difficulties associated with assigning values to many parameters simultaneously, a problem which is relatively minimal in the univariate case.

For example, suppose $\mathcal{L}(\mathbf{X}|\boldsymbol{\theta}) = N(\boldsymbol{\theta},\mathbf{I})$. A natural conjugate prior approach dictates a prior with density

$$\mathcal{L}(\boldsymbol{\theta}) = N(\boldsymbol{\phi},\mathbf{A}), \qquad \mathbf{A} > 0,$$

where $\boldsymbol{\phi}: p \times 1$ and $\mathbf{A}: p \times p$ are assumed to be known a priori. However, to assess $\boldsymbol{\phi}$ means assessing p distinct numbers, a task which need not be easy, especially if p is large. Often, a simple solution to the problem of assessment (in the absence of information to the contrary) is to use a patterned matrix or vector for the preassigned elements—reducing the dimension of the assessment problem. In this example, if it is not inappropriate, one might take $\boldsymbol{\phi} \equiv c\mathbf{e}$, where c is a scalar and $\mathbf{e}$ is a vector of ones. Now it is only necessary to assess a value for c to start the process of posterior inference. Analogously, to assess the $(p + 1)$ parameters of a p-variate Dirichlet distribution (see Section 6.3.2), it is sometimes appropriate to take all parameters equal, a priori, thus reducing the number of parameters which must be assessed, to one. This distribution will be seen to arise as the natural conjugate prior for a multinomial sampling distribution (see also Exercise 6.8).

Assessing an entire covariance matrix for a p-vector, such as assessing $\mathbf{A}$ in the above problem, would require, in general, the assessment of $p(p + 1)/2$ separate numbers. Thus, for a problem of dimension $p = 10$, 55 separate assessments would normally be required. However if, for example, an intraclass correlation structure [see Equation (2.2.4)] can be assumed, all variances are the same and all covariances are the same and only two parameters need to be assessed to start the process. Other patterns such as tri-diagonal and circular could of course be used, according to the constraints permitted by the problem at hand.

A method for assessing the center of prior distributions for the covariance matrix and correlation coefficient in a bivariate normal distribution was developed by Gakhale and Press, 1979. They assessed the correlation coefficient by eliciting Kendall's concordance probability. This procedure, while useful in the bivariate case, has not yet been extended to the general multivariate case. In the bivariate case, however, only one point assessment is required.

In another approach to assessment, Kadane et al., 1980, propose inferring hyperparameters through elicitation of observables in a predictive distribution. Many points need to be assessed to establish regression relationships.

In summary, if prior information is difficult to assess, a diffuse prior density approach (one not requiring the assessment of any parameters) could be taken. If *some* prior information is available, a natural conjugate prior approach might be taken with pattern matrices or vectors used to simplify the assessment problem. If additional information is available, a full natural conjugate approach might be taken (with general, unpatterned prior parameters, each assessed individually). Finally, in some situations, a complete prior density might be assessed, thereby avoiding any arbitrariness associated with using the natural conjugate prior density.

For a discussion of assessment of parameters of multivariate prior distributions in the context of sample survey design, see Ericson (1965).

REFERENCES

Anderson (1958).
Andrews et al. (1975).
Ando and Kaufman (1965).
Aykac et al. (1977).
Barnard (1954).
Box and Cox (1964).
Box and Tiao (1973).
Cramer (1946).
de Finetti (1970).
DeGroot (1970).
Dempster (1963).
Ericson (1965).
Fienberg and Zellner (1975).
Fraser (1968).
Geisser (1965a)
Geisser and Cornfield (1963)
Gokhale and Press (1979).
Houle (1973).
Jeffreys (1961).
Johnsen and Kotz (1972).
Kadane et al. (1980).
Lindley (1972).
Lindley and Novick (1981).
Lukacs (1960).
Owen (1962).
Raiffa and Schlaifer (1961).
Rothenberg (1963).
Savage (1962).
Schlaifer (1969).
Shannon (1948).
Stein (1956).
Villegas (1969).
Zellner (1980).

COMPLEMENT 3.1 INFORMATION IN A DISTRIBUTION WITH DENSITY $f(x)$

It is shown below that the probability distribution associated with minimum information is the uniform (see Section 3.6.5).

Define the information in a distribution with density $f(x)$ and range (x_0,x_1) as

$$I\{f\} = \int_{x_0}^{x_1} F(x,f,f')\,dx,$$

where $f \equiv f(x)$ is a function of x, $f'(x) \equiv df(x)/dx$, $f(x_0)$, and $f(x_1)$ are assumed fixed, and $F \equiv f \log f$. Let

$$H\{f\} = \int_{x_0}^{x_1} G(x,f,f')\,dx = c = \text{constant}$$

denote a subsidiary condition on f, for some function G. Then it is shown in the Calculus of Variations that a necessary condition for f to yield a minimum of I is that

$$\frac{d}{dx}\left[\frac{\partial F}{\partial f'} + \lambda \frac{\partial G}{\partial f'}\right] - \left[\frac{\partial F}{\partial f} + \lambda \frac{\partial G}{\partial f}\right] = 0. \qquad \text{(C3.1.1)}$$

Since in (C3.1.1), $F \equiv f \log f$, and to satisfy $\int_{x_0}^{x_1} f\, dx = 1$, it is necessary to take $G \equiv f$, substitution into (C3.1.1) gives

$$1 + \log f + \lambda = 0.$$

Solving for f gives

$$f(x) = e^{-(\lambda+1)}.$$

Using the subsidiary condition shows that

$$f(x) = \frac{1}{x_1 - x_0},$$

a uniform distribution on (x_0, x_1).

EXERCISES

3.1 Let x_1 denote the change in the number of users of the hallucinogenic drug LSD in Chicago from 1968 to 1969; x_2 denotes the change in the number of users of heroin; x_3 denotes the change in the number of users of marijuana; and x_4 denotes the change in the number of users of opium. Suppose the covariance matrix of $\mathbf{X} = (X_i)$, $i = 1, 2, 3, 4$ is given by

$$\boldsymbol{\Sigma} = \begin{pmatrix} 9 & .6 & 9 & 1.2 \\ .6 & 4 & .1 & 4 \\ 9 & .1 & 25 & .2 \\ 1.2 & 4 & .2 & 16 \end{pmatrix}.$$

Find the correlation matrix of $\mathbf{X}$ and interpret it in terms of the relationships among the components of $\mathbf{X}$.

3.2 Let $\mathbf{X}: 5 \times 1$ denote the vector of demand for 5 separate items in a retail business, and $\mathcal{L}(\mathbf{X}|\boldsymbol{\theta},\boldsymbol{\Sigma}) = N(\boldsymbol{\theta},\boldsymbol{\Sigma})$. Suppose that for $\boldsymbol{\Sigma}$ fixed, $\boldsymbol{\theta}$ follows the natural conjugate prior distribution

$$\mathcal{L}(\boldsymbol{\theta}|\boldsymbol{\Sigma}) = N(2\mathbf{e},\boldsymbol{\Sigma}),$$

where $\mathbf{e}' = (1,1,1,1,1)$, and suppose $\boldsymbol{\Sigma}$ follows the natural conjugate prior distribution

$$\mathcal{L}(\boldsymbol{\Sigma}) = W^{-1}(\mathbf{H},5,10),$$

where

$$\mathbf{H} = \begin{pmatrix} 5 & 1 & 1 & 1 & 1 \\ 1 & 5 & 1 & 1 & 1 \\ 1 & 1 & 5 & 1 & 1 \\ 1 & 1 & 1 & 5 & 1 \\ 1 & 1 & 1 & 1 & 5 \end{pmatrix}.$$

Suppose that independent observations of $\mathbf{X}$ taken at 11 random times yield the summary statistics for the sample mean vector and unnormed sample covariance matrix

$$\bar{\mathbf{x}} = \begin{pmatrix} 2.1 \\ 3 \\ 1.5 \\ 1 \\ 2 \end{pmatrix} \qquad \mathbf{V} = \begin{pmatrix} 9 & 1.2 & .9 & 4.2 & 7.2 \\ 1.2 & 4 & .2 & .4 & 2.4 \\ .9 & .2 & 1 & 1 & 3.6 \\ 4.2 & .4 & 1 & 4 & 4.8 \\ 7.2 & 2.4 & 3.6 & 4.8 & 16 \end{pmatrix}.$$

Find the joint posterior density for $\boldsymbol{\theta}$ and $\boldsymbol{\Sigma}$ given the sample data, and express the density in its simplest numerical form, wherever possible. (Combine all matrix products, invert any matrices, evaluate any determinants, and so on.)

3.3 Let X denote the change in the number of international confrontations into which a nation entered from the quarter century preceding World War II to the quarter century following World War II; let Y denote the change in the gold reserves of a nation over the same two time periods. Suppose

$$\mathcal{L}(X,Y) = N(\boldsymbol{\theta},\boldsymbol{\Sigma}),$$

where

$$\boldsymbol{\theta} = \begin{pmatrix} 5 \\ 2 \end{pmatrix}, \qquad \boldsymbol{\Sigma} = \begin{pmatrix} 9 & 3 \\ 3 & 4 \end{pmatrix}.$$

Find:

(a) $E(Y|X)$, $E(X|Y)$,

(b) var $(Y|X)$, var $(X|Y)$,

(c) $P\{X \leq 6\}$,

(d) $P\{4 \leq X \leq 6,\ 1 \leq Y \leq 3\}$.

[*Hint:* For part (d), see, for example, the tables in Owen, 1962.]

3.4 The scores on a battery of three tests are given by $\mathbf{z}' = (z_1,z_2,z_3)$. Suppose 100 people are tested and the results are summarized from observations $\mathbf{z}_1,\ldots,\mathbf{z}_{100}$ by

$$\bar{\mathbf{z}} = \begin{pmatrix} 62 \\ 38 \\ 47 \end{pmatrix}, \qquad \mathbf{V} = \begin{pmatrix} 13 & 5 & 3 \\ 5 & 12 & 6 \\ 3 & 6 & 10 \end{pmatrix},$$

where $\bar{\mathbf{z}}$ and $\mathbf{V}$ are the sample mean vector and unnormed sample covariance matrix. Suppose $\mathcal{L}(\mathbf{Z}|\boldsymbol{\theta},\boldsymbol{\Sigma}) = N(\boldsymbol{\theta},\boldsymbol{\Sigma})$. Assume a diffuse prior for $\boldsymbol{\theta}$ and $\boldsymbol{\Sigma}$ and find, assuming $\boldsymbol{\Sigma} > 0$,

(a) the predictive distribution for a new observation $\mathbf{y}' = (y_1,y_2,y_3)$,

(b) $P\{65 \leq y_1 \leq 70|\mathbf{z}_1,\mathbf{z}_2,\ldots,\mathbf{z}_{100}\}$.

3.5 Let $\mathbf{X}: p \times 1$ denote a vector of random variables representing various observable quantities related to learning rate, and suppose for $\mathbf{\Sigma} > 0$, $\mathcal{L}(\mathbf{X}|\boldsymbol{\theta},\mathbf{\Sigma}) = N(\boldsymbol{\theta},\mathbf{\Sigma})$. It is found that the first q variables are of interest as a group, so $\mathbf{X}$ is partitioned according to

$$\mathbf{X} = \begin{pmatrix} \mathbf{Y} \\ \mathbf{Z} \end{pmatrix}, \qquad \mathbf{Y}: q \times 1.$$

Find the distribution of $\mathbf{AY} + \mathbf{BZ}$, where $\mathbf{A}: q \times q$, and $\mathbf{B}: q \times p - q$, $2q < p$.

[*Hint:* Let

$$\mathbf{\Sigma} = \begin{pmatrix} \mathbf{\Sigma}_{11} & \mathbf{\Sigma}_{12} \\ \mathbf{\Sigma}_{21} & \mathbf{\Sigma}_{22} \end{pmatrix}, \qquad \boldsymbol{\theta} = \begin{pmatrix} \boldsymbol{\theta}_1 \\ \boldsymbol{\theta}_2 \end{pmatrix}.$$

3.6 Suppose $\mathcal{L}(\mathbf{X}|\boldsymbol{\theta}) = N(\boldsymbol{\theta},\mathbf{\Sigma}_0)$, where $\mathbf{x}' = (x_1,x_2)$, x_1 denotes change in personal income, x_2 denotes change in personal consumption, and

$$\mathbf{\Sigma}_0 = \begin{pmatrix} 4 & 3 \\ 3 & 9 \end{pmatrix}.$$

Suppose there is one observation $\mathbf{x} = (5,3)'$. Test the hypothesis (at the 5 percent level) that $\boldsymbol{\theta} = \boldsymbol{\theta}_0 \equiv (4,4)'$ against the alternative that $\boldsymbol{\theta} \neq \boldsymbol{\theta}_0$, by computing the statistic $\mathbf{y}'\mathbf{y}$, where $\mathbf{y} \equiv \mathbf{\Sigma}_0^{-1/2}(\mathbf{x} - \boldsymbol{\theta}_0)$.

3.7 It is of interest to study sentiment of people toward racially integrated neighborhoods. Let X_1 denote time an individual has lived in a given neighborhood; X_2 denotes his feelings toward an integrated neighborhood (measured on a 100 point scale from -50 to $+50$, with the extremes implying least or greatest satisfaction with integration); X_3 denotes number of years of education of the individual. Many observations are taken in a given neighborhood so the means, variances, and covariances may be assumed to be known. Suppose

$$\mathbf{X}' \equiv (X_1,X_2,X_3),$$

and

$$\mathcal{L}(\mathbf{X}) = N(\boldsymbol{\theta},\mathbf{\Sigma}),$$

where

$$\boldsymbol{\theta} = \begin{pmatrix} 3 \\ 10 \\ 8 \end{pmatrix}, \qquad \mathbf{\Sigma} = \begin{pmatrix} 1 & 3 & 1 \\ 3 & 16 & 2 \\ 1 & 2 & 4 \end{pmatrix}.$$

Find the distribution of:
(a) X_1, for given X_2 and X_3,
(b) (X_1,X_2) for given X_3,
(c) X_3, for given X_1 and X_2.

3.8 Suppose the joint density of X and Y is given by

$$f(x,y) = \frac{1}{55\pi} \exp\left\{\left(-\frac{1}{1512}\right)[36x^2 - 24xy + 25y^2 - 120x - 2y + 121]\right\}.$$

Find the means, variances, and correlation for X and Y.

3.9 In an experiment, r_0 successes were obtained out of n_0 binomial trials, with probability of success p on a single trial. Suppose, a priori, that p follows a beta distribution [see (6.3.1); this is the natural conjugate prior distribution].

(a) Find the posterior distribution of p given the sample.

(b) Find the predictive density of the number of successes, r, in a new sequence of n binomial trials, given the results of the earlier experiment. This distribution is called the beta-binomial.

3.10 (a) Show that if X and Y are independent random variables, X and Y are also uncorrelated.

(b) Show that if X and Y follow a bivariate Normal distribution and X and Y are uncorrelated, then X and Y are also independent.

(c) Show that if X and Y are dichotomous random variables (that is, X and Y can each assume only one of two possible values such as zero or one) following a joint distribution, and X and Y are uncorrelated, then X and Y must also be independent. This result will be of interest in the latent class model of latent structure analysis (see Chapter 10).

(d) Show by example that if X and Y follow a joint distribution, and Y given X follows a Normal distribution, X and Y need not follow a bivariate Normal distribution. This result is of interest, for instance, in the regression model of Chapter 8.

3.11 In Dempster, 1963, an invariance argument was used to cast doubt upon sampling theory methods of making inferences about the covariance matrix in a large class of distributions. Discuss the implications of this argument in terms of the prior in (3.6.5).

3.12 Show that for $x > 0$,

$$\int_x^{\infty} e^{-t^2/2} dt \leq \frac{1}{x} e^{-x^2/2}.$$

3.13 Find the distribution functions corresponding to the characteristic functions:

$$\cos t, \cos^2 t, \frac{1}{4} + \left(\frac{1}{8}\right)e^{it} + \left(\frac{5}{8}\right)e^{-2it}.$$

[Use the inversion formula.]

3.14 Explain how you would test a set of data for multivariate normality. (Hint: See Andrews et al., 1973].

4

MULTIVARIATE LARGE SAMPLE DISTRIBUTIONS AND APPROXIMATIONS

4.1 INTRODUCTION

It is all too often the case that statistics of interest for inference purposes have complicated or, perhaps unknown, distributions. For example, the exact distributions required for inference in the principal components, canonical correlations and classification models are quite complicated. For this reason, it is useful to be able to approximate (at least in large samples) functions of vectors and matrices and their distributions. In this chapter, methods are presented for approximating matrices, for approximating the distributions of random matrices in large samples (central limit theorems), and for approximating posterior distributions in large samples.

Matrices may be approximated by expanding them in series and neglecting high-order terms. To this end it is necessary to define the notion of convergence of a sequence or series of matrices.

4.2 CONVERGENCE OF A SEQUENCE OF MATRICES

Let $\mathbf{A}_1$, $\mathbf{A}_2, \ldots$, or $\{\mathbf{A}_N\}$, $N = 1, 2, \ldots$ denote a sequence of $p \times q$ matrices. A sequence of matrices $\{\mathbf{A}_N\}$, $N = 1, 2, \ldots$ is said to *converge* to a matrix $\mathbf{A}$ if each element of $\mathbf{A}_N$ converges, as $N \to \infty$, to the corresponding element of $\mathbf{A}$. Otherwise, the sequence *diverges*.

Example (4.2.1): Let $\mathbf{X}: p \times 1$ denote a vector of scores obtained by an individual on p different types of tests. All scores have been centered so that $E(\mathbf{X}) = 0$, and it is known that var $(\mathbf{X}) = \mathbf{\Sigma}$. N individuals are given the battery of tests independently, so that if $\mathbf{X}_j$ denotes the values of $\mathbf{X}$ for the jth individual, the sample mean is given by

$$\bar{\mathbf{X}} = \frac{1}{N} \sum_{j=1}^{N} \mathbf{X}_j.$$

Then

$$\text{var } (\bar{\mathbf{X}}) = E\bar{\mathbf{X}}\bar{\mathbf{X}}' = \frac{1}{N^2} E\left(\sum_{i=1}^{N} \mathbf{X}_i\right)\left(\sum_{j=1}^{N} \mathbf{X}_j'\right) = \frac{1}{N^2} \sum_{i=1}^{N} \sum_{j=1}^{N} E\mathbf{X}_i\mathbf{X}_j'.$$

But by independence of the $\mathbf{X}_i$'s, $E\mathbf{X}_i\mathbf{X}_j' = \mathbf{0}$ if $i \neq j$, and $E\mathbf{X}_i\mathbf{X}_j' = \mathbf{\Sigma}$ if $i = j$. Hence,

$$\text{var } (\bar{\mathbf{X}}) = \frac{1}{N} \mathbf{\Sigma}.$$

Let $\mathbf{\Sigma} = (\sigma_{ij})$, and $\mathbf{H}_N = (h_{ij}[N]) = \text{var } (\bar{\mathbf{X}})$. Then, for all i, j, $h_{ij}[N] = \sigma_{ij}/N$, so that $\lim_{N \to \infty} h_{ij}[N] = 0$. Hence, the sequence of covariance matrices $\{\mathbf{H}_N\}$, $N = 1, 2, \ldots$ converges to the zero matrix. That is, the larger the sample size, the smaller the variances and covariances of the elements of the sample mean vector.

Theorem (4.2.1): Let $\lambda_1, \ldots, \lambda_p$ denote the latent roots of $\mathbf{A}: p \times p$, and assume they are distinct. Then the sequence of $p \times p$ matrix functions of a matrix $\mathbf{A}$, $\{\mathbf{f}_N(\mathbf{A})\}$, $N = 1, 2, \ldots$ converges to a limit matrix $\mathbf{f}(\mathbf{A})$, as $N \to \infty$, if and only if

(i) $$\lim_{N \to \infty} [f_N(\lambda_j)] = f(\lambda_j), \qquad \text{for all } j = 1, 2, \ldots, p,$$

and

(ii) $$\lim_{N \to \infty} \left[\frac{d^m f_N(\lambda_j)}{d\lambda_j^m}\right] = \frac{d^m f(\lambda_j)}{d\lambda_j^m}, \qquad \text{for all } j = 1, \ldots, p,$$

and $m = 1, \ldots, p$.

Proof: See Gantmacher (1959), Vol. I, p. 111.

Example (4.2.2): Suppose $\mathbf{f}_N(\mathbf{A}) = \mathbf{A}^N$, and suppose all latent roots, λ_j, of $\mathbf{A}$ are distinct, and of magnitude less than one. Then, for all $j = 1, \ldots, p$,

$$f_N(\lambda_j) = \lambda_j^N, \qquad \text{and} \qquad \lim_{N \to \infty} f_N(\lambda_j) = 0.$$

Also, all derivatives of $f_N(\lambda_j) \to 0$ as $N \to \infty$. Therefore, the sequence converges, and the theorem requires that $\mathbf{f}(\mathbf{A}) = \mathbf{0}$. In fact [see also Mirsky (1955), p. 328], a necessary and sufficient condition for the relation $\mathbf{A}^N \to \mathbf{0}$ to hold is that all of the latent roots of **A** should have magnitude less than one.

As a special case, suppose **A** is the 2×2 covariance matrix

$$\mathbf{A} = \frac{1}{2}\begin{bmatrix} 1 & \frac{2}{3} \\ \frac{2}{3} & 1 \end{bmatrix}.$$

The latent roots of **A** are easily found to be $\frac{5}{6}$ and $\frac{1}{6}$, so that the conditions of this example are satisfied. Hence, the sequence $\mathbf{A}, \mathbf{A}^2, \mathbf{A}^3, \ldots$ converges to the zero matrix.

4.3 CONVERGENCE OF A SERIES OF MATRICES

It is often possible to expand matrix functions of a matrix in convergent infinite series, as one would expand scalar functions of a scalar. The next theorem provides conditions under which this is possible.

Theorem (4.3.1): If the scalar function, $f(\lambda)$, of the scalar λ, can be expanded in a power series in the circle of radius r centered at λ_0, $|\lambda - \lambda_0| < r$, with the series given by

$$f(\lambda) = \sum_{j=0}^{\infty} \alpha_j (\lambda - \lambda_0)^j,$$

then this expansion remains valid when λ and λ_0 are replaced by matrices $\mathbf{A}$ and $\mathbf{A}_0$ such that the latent roots of $(\mathbf{A} - \mathbf{A}_0)$ lie within the circle of convergence.

Proof: See Gantmacher (1959), Vol. I, p. 113.

Example (4.3.1): Suppose $\mathbf{f}(\mathbf{A}) = (\mathbf{I} - \mathbf{A})^{-1}$ and it is desired to find a matrix series expansion for $\mathbf{f}(\mathbf{A})$ so that $\mathbf{f}(\mathbf{A})$ might be approximated by the first few terms. First examine

$$f(\lambda) = (1 - \lambda)^{-1} = \frac{1}{1 - \lambda}.$$

Since for $|\lambda| < 1$,

$$\frac{1}{1 - \lambda} = \sum_{j=0}^{\infty} \lambda^j,$$

if all the latent roots of **A** lie within the unit circle, $\mathbf{f}(\mathbf{A})$ has the convergent series expansion

$$(\mathbf{I} - \mathbf{A})^{-1} = \sum_{j=0}^{\infty} \mathbf{A}^j, \tag{4.3.1}$$

where $\mathbf{A}^0$ equals the identity matrix. If at least one latent root of $\mathbf{A}$ does not lie strictly within the unit circle, the statement (4.3.1) cannot be made. Thus, if $\mathbf{A}$ is idempotent (and $\mathbf{A}$ is not identically zero), the series diverges. However, the numerical illustration used in Example (4.2.2) would yield a convergent series (since its latent roots, $\frac{1}{6}$, $\frac{5}{6}$, both lie inside the unit circle).

4.4 MULTIVARIATE CENTRAL LIMIT THEOREMS

The univariate notion of appropriately normalized sums of random variables following a Normal distribution in sufficiently large samples is generalized below to the multivariate case. This result can be extremely useful for drawing approximate inferences when large samples of data are available, but the data vectors have intractable distributions.

Theorem (4.4.1): Let the p-component vectors $\mathbf{X}_1, \mathbf{X}_2, \ldots, \mathbf{X}_n$ be independent and identically distributed with means $\mathbf{E}\mathbf{X}_j = \boldsymbol{\theta}$, and covariance matrices $\mathbf{E}[(\mathbf{X}_j - \boldsymbol{\theta})(\mathbf{X}_j - \boldsymbol{\theta})'] = \boldsymbol{\Sigma}$, for all $j = 1, 2, \ldots, n$. Then, if $\boldsymbol{\Sigma} > 0$,

$$\lim_{n \to \infty} \mathfrak{L}\left[\frac{\sum_{j=1}^{n} (\mathbf{X}_j - \boldsymbol{\theta})}{\sqrt{n}}\right] = N(\mathbf{0}, \boldsymbol{\Sigma}). \tag{4.4.1}$$

Proof: See, for instance, Cramér (1946).

Remark: Note that the $\mathbf{X}_j$ vectors in Theorem (4.4.1) are required to be independent and identically distributed. For extensions of the multivariate central limit theorem to dependent and non-identically distributed random vectors, see Dvoretzky, 1977.

Example (4.4.1): In a survey of prevalence of illicit drug users, subjects are classified according to their responses as: nonusers, narcotics users only, users of stimulants only, users of depressants only, users of hallucinogens only, and users of more than one type of illicit drug. The responses of the subjects may be denoted by 6×1 vectors with a one in the place corresponding to the "use" category, and zeros elsewhere. Thus, if the jth respondent uses narcotic drugs only, his response vector $\mathbf{X}_j$ is observed as $\mathbf{x}_j$, and is given by

$$\mathbf{x}_j = \begin{pmatrix} x_{j1} \\ \cdot \\ \cdot \\ \cdot \\ x_{j6} \end{pmatrix} = \begin{pmatrix} 0 \\ 1 \\ 0 \\ 0 \\ 0 \\ 0 \end{pmatrix}.$$

Suppose that for every subject j and every response category k, $P\{X_{jk} = 1\} = \theta_k$, $k = 1,\ldots,6$; $\Sigma_1^6\theta_k = 1$. Let $\boldsymbol{\theta} = (\theta_1,\ldots,\theta_6)' = E(\mathbf{X}_j)$, and $\boldsymbol{\Sigma} = (\sigma_{ij}) = \text{var}(\mathbf{X}_j)$. Since the responses follow a *multinomial* distribution, it is well known (see for instance Wilks, 1962, p. 139) that

$$\sigma_{ij} = \begin{cases} \theta_i(1-\theta_i), & i = j; \\ -\theta_i\theta_j, & i \neq j. \end{cases}$$

Suppose n subjects are interviewed independently in the survey. Then, if n is large, Theorem (4.4.1) asserts it is approximately true that

$$\mathcal{L}\left[\frac{1}{\sqrt{n}}\sum_1^n (\mathbf{X}_j - \boldsymbol{\theta})\right] = N(\mathbf{0},\boldsymbol{\Sigma}),$$

or, if $\bar{\mathbf{X}} = n^{-1}\Sigma_1^n\mathbf{X}_j$ denotes the sample mean vector, it is approximately true that

$$\mathcal{L}(\bar{\mathbf{X}}) = N\left(\boldsymbol{\theta}, \frac{1}{n}\boldsymbol{\Sigma}\right). \tag{4.4.2}$$

The components of $\bar{\mathbf{X}}$ denote the sample fraction of subjects in each of the drug use categories. Thus, in large samples, (4.4.2) may be used to make inferences about θ from $\bar{\mathbf{X}}$. The one dimensional marginal densities are binomial.

4.5 ASYMPTOTIC DISTRIBUTION OF A FUNCTION OF A RANDOM MATRIX

The following two theorems assert that under certain conditions, a function of a random vector or matrix follows a Normal distribution in infinitely large samples. Since the regularity conditions required of the function are not severe, these results are approximately applicable to a wide variety of problems involving finite, but large samples.

Theorem (4.5.1): Let $\mathbf{X}_n$ be a p-component random vector and $\boldsymbol{\theta}$ a fixed vector. Assume[2] that

[1] The probability mass function of a multinomial distribution involving n trials, for obtaining r_1 category 1 responses, r_2 category 2 responses, $\ldots, r_6$ category 6 responses is

$$\frac{n!}{r_1!\cdots r_6!}\theta_1^{r_1}\cdots\theta_6^{r_6}, \qquad \sum_1^6 r_i = n, \qquad \sum_1^6 \theta_i = 1.$$

[2] $\operatorname{plim}_{n\to\infty} \mathbf{X}_n = \boldsymbol{\theta}$ if, for given $\epsilon > 0$, $\lim_{n\to\infty} P\{|X_n - \theta| > \epsilon\} = 0$, for each component of $(\mathbf{X}_n - \boldsymbol{\theta})$, separately. Note that if $\operatorname{plim}_{n\to\infty} \mathbf{X}_n = \boldsymbol{\theta}$, $\operatorname{plim}_{n\to\infty} f(\mathbf{X}_n) = f(\boldsymbol{\theta})$ for all continuous functions f. This definition extends to matrices as well.

For example, suppose for $\mathbf{X}_n : p \times q$, $\mathbf{X}_n = (\mathbf{Y}_n,\mathbf{Z}_n)$, $\operatorname{plim}_{n\to\infty} \mathbf{X}_n = (\mathbf{A},\mathbf{B})$, and $\mathbf{Y}_n$ and $\mathbf{Z}_n$ are conformable. Then, if $f(\mathbf{X}_n) \equiv \mathbf{Y}_n \cdot \mathbf{Z}_n$, $\operatorname{plim}_{n\to\infty} (\mathbf{Y}_n \cdot \mathbf{Z}_n) = \mathbf{A} \cdot \mathbf{B}$.

(i)
$$\operatorname*{plim}_{n\to\infty} \mathbf{X}_n = \boldsymbol{\theta}, \tag{4.5.1}$$

(ii)
$$\lim_{n\to\infty} \mathcal{L}[\sqrt{n}\,(\mathbf{X}_n - \boldsymbol{\theta})] = N(\mathbf{0},\boldsymbol{\Sigma}). \tag{4.5.2}$$

Note that (4.5.1) is the definition that $\mathbf{X}_n$ is a multivariate consistent estimator of $\boldsymbol{\theta}$.

Let $w = f(\mathbf{x})$ be a scalar function of a vector $\mathbf{x}$ with first and second derivatives existing in a neighborhood of $\mathbf{x} = \boldsymbol{\theta}$.

Let $\left.\dfrac{\partial f(\mathbf{x})}{\partial x_i}\right|_{\mathbf{x}=\boldsymbol{\theta}}$ be the ith component of the vector $\boldsymbol{\phi}$. Then

$$\lim_{n\to\infty} \mathcal{L}\{\sqrt{n}\,[f(\mathbf{X}_n) - f(\boldsymbol{\theta})]\} = N(0,\boldsymbol{\phi}'\boldsymbol{\Sigma}\boldsymbol{\phi}). \tag{4.5.3}$$

Proof: See Anderson (1958), p. 76.

A variant of this theorem is a generalization of a theorem due to Cramér (1946), p. 366, and essentially proved by him.

Theorem (4.5.2): Let $\boldsymbol{\Omega} = (\omega_{ij})$ denote a $p \times q$ random matrix whose columns (or subcollections of columns) are sample central moments (based on a sample of size n), and which are functionally independent of one another (distinct). Let $f(\Omega)$ be a scalar function of Ω which is continuous, with continuous derivatives of the first and second order with respect to the elements of Ω, and not an explicit function of the sample size. Then

$$\lim_{n\to\infty} \mathcal{L}\{\sqrt{n}[f(\Omega) - f(E\Omega)]\} = N(0,\sigma^2), \tag{4.5.4}$$

where

$$\sigma^2 = n \sum_{i=1}^{p}\sum_{j=1}^{q}\sum_{k=1}^{p}\sum_{l=1}^{q} \frac{\partial f(E\boldsymbol{\Omega})}{\partial \omega_{ij}} \frac{\partial f(E\boldsymbol{\Omega})}{\partial \omega_{kl}} \operatorname{cov}(\omega_{ij},\omega_{kl}),$$

and $[\partial f(E\boldsymbol{\Omega})]/\partial\omega_{ij}$ denotes a derivative evaluated at $\boldsymbol{\Omega} = E\boldsymbol{\Omega}$.

Remark: If, for example, Ω denotes a symmetric matirx, the elements of Ω are not distinct, so the theorem is not applicable.

Example (4.5.1): Let $\mathbf{X}_j : p \times 1$ denote a vector of differences, for an individual j, between his scores on a battery of tests, and the average scores of a control group of "normals." This battery of p tests is given, independently, to n patients in a mental hospital to see if the tests may be used to identify psychotics from normals. Some of the components of $\mathbf{X}_j$ are positive and some are negative, depending upon the nature of the mental illness. Let

$$\bar{\mathbf{X}} = \frac{1}{n}\sum_{1}^{n} \mathbf{X}_j.$$

Since the sign of a component of $\bar{\mathbf{X}}$ is not really of interest, but only the magnitude of the deviation from the normal value, it may be of interest to study the behavior of the group scores by means of $\bar{\mathbf{X}}'\bar{\mathbf{X}}$. Let $E(\mathbf{X}_j) = \boldsymbol{\theta}$, and var $(\mathbf{X}_j) = v^2\mathbf{I}_p$, for all $j = 1,\ldots,n$, where $\boldsymbol{\theta}: p \times 1$ is unknown, and v is some unknown scalar. Except for these first two moments, the distribution of $\mathbf{X}_j$ is otherwise assumed to be unknown, or at best, difficult to work with analytically. However, if n is large, $\bar{\mathbf{X}}'\bar{\mathbf{X}}$ may be studied by means of its asymptotic distribution, available from (4.5.4).

Define the p-vector $\boldsymbol{\Omega} = \bar{\mathbf{X}} = (\omega_i)$, $i = 1,\ldots,p$, and let $f(\boldsymbol{\Omega}) = \bar{\mathbf{X}}'\bar{\mathbf{X}}$. Then, from (4.5.4),

$$\lim_{n\to\infty} \mathfrak{L}\{\sqrt{n}\,[\bar{\mathbf{X}}'\bar{\mathbf{X}} - \boldsymbol{\theta}'\boldsymbol{\theta}]\} = N(0,\sigma^2),$$

where

$$\sigma^2 = n \sum_{i=1}^{p}\sum_{k=1}^{p} \left(\frac{\partial f}{\partial \omega_i}\right)\left(\frac{\partial f}{\partial \omega_k}\right) \operatorname{cov}(\omega_i,\omega_k)\Big|_{\boldsymbol{\Omega}=\boldsymbol{\theta}}.$$

Since cov $(\omega_i,\omega_k) =$ cov $(\bar{X}_i,\bar{X}_k) = 0$ unless $i = k$, and equals v^2/n when $i = k$,

$$\sigma^2 = v^2 \sum_{i=1}^{p} \left(\frac{\partial f}{\partial \omega_i}\right)\Big|_{\boldsymbol{\Omega}=\boldsymbol{\theta}}.$$

But

$$\frac{\partial f}{\partial \boldsymbol{\Omega}}\Big|_{\boldsymbol{\Omega}=\boldsymbol{\theta}} = \frac{\partial}{\partial \bar{\mathbf{X}}}(\bar{\mathbf{X}}'\bar{\mathbf{X}})\Big|_{\boldsymbol{\Omega}=\boldsymbol{\theta}} = 2\bar{\mathbf{X}}\Big|_{\bar{\mathbf{X}}=\boldsymbol{\theta}} = 2\boldsymbol{\theta}.$$

Hence,

$$\sigma^2 = v^2 \sum_{i=1}^{p} (2\theta_i)^2 = 4v^2\boldsymbol{\theta}'\boldsymbol{\theta}.$$

So the final approximation is that for large n,

$$\mathfrak{L}(\bar{\mathbf{X}}'\bar{\mathbf{X}}) \cong N\left[\boldsymbol{\theta}'\boldsymbol{\theta}, \frac{4v^2\boldsymbol{\theta}'\boldsymbol{\theta}}{n}\right].$$

It is sometimes useful in studying multivariate summary sampling statistics to be able to expand densities of sums of independent and identically distributed random vectors in terms of standardized multivariate Normal densities and their derivatives. In the univariate case, Edgeworth gave such an expansion, and it was asymptotic in that successive terms were inversely proportional to successive powers of the square root of sample size. Sobel (1963) gave such an Edgeworth expansion for the multivariate case.

4.6 ASYMPTOTIC POSTERIOR DISTRIBUTIONS

In this section it is asserted that in large samples, under a wide variety of circumstances, a Bayesian should "select that decision which maximizes his utility at the maximum likelihood value of the parameter." That is, "in large samples, the exact form of the prior distribution is irrelevant and maximum likelihood estimation is optimum" (see Lindley, 1961, p. 456). In fact, the typical asymptotic multivariate posterior distribution is multivariate Normal with mean vector equal to the vector of maximum likelihood estimators, and covariance matrix equal to the inverse of a matrix closely related to the Fisher Information matrix. This result was obtained by Le Cam (1956), and was discussed informally by Lindley (1965); see also Walker (1969).

Theorem (4.6.1): Let $p(\boldsymbol{\theta}|\mathbf{x}_1,\mathbf{x}_2,\ldots,\mathbf{x}_n)$ denote the posterior density of a parameter vector $\boldsymbol{\theta}: k \times 1$, given a sample of observations $\mathbf{x}_1,\ldots,\mathbf{x}_n$ from the density $p(\mathbf{x}|\boldsymbol{\theta})$, where $\mathbf{x}_j: p \times 1$. Let $\hat{\boldsymbol{\theta}}: k \times 1$ denote the maximum likelihood estimator of $\boldsymbol{\theta}$. Then, under regularity conditions discussed below, and ignoring improbable sets,

$$\lim_{n\to\infty} \mathcal{L}\{\mathbf{A}^{-1/2}(\boldsymbol{\theta} - \hat{\boldsymbol{\theta}})|\mathbf{x}_1,\ldots,\mathbf{x}_n\} = N(\mathbf{0},\mathbf{I}_k), \tag{4.6.1}$$

where $\mathbf{A}^{-1}$ is the matrix with typical element

$$-\left.\frac{\partial^2 \log p(\mathbf{x}_1,\mathbf{x}_2,\ldots,\mathbf{x}_n|\boldsymbol{\theta})}{\partial\theta_i\partial\theta_j}\right|_{\boldsymbol{\theta}=\hat{\boldsymbol{\theta}}}$$

and $\boldsymbol{\theta} = (\theta_i)$, $i = 1,\ldots,k$.

Most of the regularity conditions required for Theorem (4.6.1) to hold are somewhat technical although they are usually satisfied in applications. The conditions that sometimes need to be checked are as follows.

(1) The prior density $p(\boldsymbol{\theta})$ must be positive and continuous at $\boldsymbol{\theta}_0$, any interior point of the parameter space (see *note* below).
(2) $p(\mathbf{x}|\boldsymbol{\theta})$ must have continuous first- and second-order derivatives with respect to the components of $\boldsymbol{\theta}$, in a neighborhood of $\boldsymbol{\theta}_0$.
(3) All elements of $\mathbf{A}^{-1}$ must be bounded in $\boldsymbol{\theta}$ by some function of $\mathbf{x}_1,\ldots,\mathbf{x}_n$.
(4) The usual conditions for differentiating under an integral sign which are required for existence of the information matrix must be satisfied.
(5) $\mathbf{A}$ must be nonsingular.

[*Note:* If $\boldsymbol{\theta}_0$ were the true value of $\boldsymbol{\theta}$ and if $p(\boldsymbol{\theta}_0)$ were zero, no sample evidence with any size sample could modify the prior belief that $p(\boldsymbol{\theta}_0) = 0$.]

Example (4.6.1): Consider the drug survey setting of Example (4.4.1).

The likelihood function for $\boldsymbol{\theta}: 6 \times 1$ based upon the responses of n subjects is

$$p(\mathbf{x}_1,\ldots,\mathbf{x}_n|\boldsymbol{\theta}) = \prod_{i=1}^{n} p(\mathbf{x}_i|\boldsymbol{\theta}),$$

where

$$p(\mathbf{x}_i|\boldsymbol{\theta}) = \prod_{j=1}^{6} \theta_j^{x_{ij}}, \qquad \sum_1^6 \theta_j = 1,$$

and $\mathbf{x}_i$ is a 6×1 vector whose components x_{ij} may be zeros or ones, with $\sum_{j=1}^{6} x_{ij} = 1$. Define $\boldsymbol{\theta}^*$ as the vector consisting of the first five components of $\boldsymbol{\theta}$.

Suppose a priori that $\mathcal{L}(\boldsymbol{\theta}^*) = N(\boldsymbol{\theta}_0,\boldsymbol{\Sigma}_0)$, where

$$\underset{(5\times 1)}{\boldsymbol{\theta}_0} = \frac{1}{6}\begin{pmatrix} 1 \\ \cdot \\ \cdot \\ \cdot \\ 1 \end{pmatrix}, \qquad \underset{(5\times 5)}{\boldsymbol{\Sigma}_0} = \begin{pmatrix} 1 & & & & \\ & \cdot & & \frac{1}{2} & \\ & & \cdot & & \\ & \frac{1}{2} & & \cdot & \\ & & & & 1 \end{pmatrix}.$$

Then the posterior distribution of $\boldsymbol{\theta}^*$ given the observations $\mathbf{x}_1,\ldots,\mathbf{x}_n$ is

$$p(\boldsymbol{\theta}^*|\mathbf{x}_1,\ldots,\mathbf{x}_n) \propto \exp\left[-\tfrac{1}{2}(\boldsymbol{\theta}^* - \boldsymbol{\theta}_0)'\boldsymbol{\Sigma}_0^{-1}(\boldsymbol{\theta}^* - \boldsymbol{\theta}_0)\right]\theta_1^{\Sigma_1^n x_{i1}} \cdots \theta_6^{\Sigma_1^n x_{i6}},$$

where $\Sigma_1^6\theta_j = 1$. The maximum likelihood estimator of the true proportions of people in each of the drug use categories is easily found from the likelihood function above to be given by

$$\hat{\boldsymbol{\theta}} = \frac{1}{n}\sum_{i=1}^{n} \mathbf{x}_i = \bar{\mathbf{x}}.$$

Suppose n is large. To make inferences about $\boldsymbol{\theta}$ it is useful to study the asymptotic posterior distribution. Using Theorem (4.6.1) gives the approximate relation

$$\mathcal{L}[(\boldsymbol{\theta}^* - \bar{\mathbf{x}}^*)|\mathbf{x}_1,\ldots,\mathbf{x}_n] \cong N(\mathbf{0},\mathbf{A}),$$

where $\mathbf{A}$ is given in (4.6.1), and $\bar{\mathbf{x}}^*$ denotes the first five components of $\bar{\mathbf{x}}$. Since

$$\log p(\mathbf{x}_1,\ldots,\mathbf{x}_n|\boldsymbol{\theta}) = n\sum_{k=1}^{5} \bar{x}_k \log \theta_k + n\bar{x}_6 \log\left(1 - \sum_{k=1}^{5} \theta_k\right),$$

for $j = 1,\ldots,5$, and

$$\bar{x}_k \equiv \frac{1}{n}\sum_{j=1}^{n} x_{kj},$$

$$\frac{\partial \log p}{\partial \theta_j} = n\left(\frac{\bar{x}_j}{\theta_j} - \frac{\bar{x}_6}{\theta_6}\right),$$

and

$$\left.\frac{\partial^2 \log p}{\partial \theta_j \partial \theta_k}\right|_{\theta=\hat{\theta}} = \begin{cases} -n\left(\dfrac{1}{\bar{x}_j} + \dfrac{1}{\bar{x}_6}\right), & j = k; \\ -\dfrac{n}{\bar{x}_6}, & j \neq k. \end{cases}$$

That is,

$$\mathbf{A}^{-1} = n \begin{pmatrix} \left[\dfrac{1}{\bar{x}_1} + \dfrac{1}{\bar{x}_6}\right] & & \dfrac{1}{\bar{x}_6} \\ & \cdot & \\ & \cdot & \\ \dfrac{1}{\bar{x}_6} & & \left[\dfrac{1}{\bar{x}_5} + \dfrac{1}{\bar{x}_6}\right] \end{pmatrix} \equiv n\mathbf{\Phi}^{-1}.$$

Hence, for n large,

$$\mathcal{L}[(\boldsymbol{\theta}^* - \bar{\mathbf{x}}^*)|\mathbf{x}_1, \ldots, \mathbf{x}_n] \cong N\left(\mathbf{0}, \frac{1}{n}\mathbf{\Phi}\right),$$

where $\mathbf{\Phi}$ remains bounded as $n \to \infty$. Note again that the parameters of the prior distribution, $\boldsymbol{\theta}_0$, $\mathbf{\Sigma}_0$ do not enter the asymptotic posterior distribution.

4.7 Law of Large Numbers (LLN)

Suppose $\mathbf{X}_1, \ldots, \mathbf{X}_n, \ldots$ are mutually independent and identically distributed p-vectors, with mean $\boldsymbol{\theta}$. Then,

$$\operatorname*{plim}_{n\to\infty} \frac{1}{n} \sum_{j=1}^{n} \mathbf{X}_j = \boldsymbol{\theta}.$$

Proof: Apply the (weak) LLN (univariate) to each component separately.

REFERENCES

Anderson (1958).
Cramér (1946)
Dvoretzky (1977).
Gantmacher (1959).
Le Cam (1956).
Lindley (1961).
Lindley (1965).
Mirsky (1955).
Sobel (1963).
Walker (1969).
Wilks (1962).

EXERCISES

4.1 Let $\mathbf{X}: p \times 1$ denote the numbers of sales for each of p items of a door-to-door salesman, and let $\mathbf{X}_1, \ldots, \mathbf{X}_N$ denote the independent sales vectors of N salesman (selling the same items) on a given day. Suppose $\mathcal{L}(\mathbf{X}_j) = N(\boldsymbol{\theta}, \boldsymbol{\Sigma}_0)$, where $\boldsymbol{\theta}$ is unknown and

$$\boldsymbol{\Sigma}_0 = 5 \begin{pmatrix} 1 & & & \\ & \cdot & .7 & \\ & & \cdot & \\ .7 & & \cdot & \\ & & & 1 \end{pmatrix}.$$

Adopt a diffuse prior for $\boldsymbol{\theta}$ and find the asymptotic posterior distribution for $\boldsymbol{\theta}$ based upon the N observations. Show that the result is the same as that obtained by the use of Theorem (4.6.1).

4.2 Let $\mathbf{X}: p \times 1$ denote the scores of unemployed males 18–60 years of age, in a ghetto area, on a battery of p tests of their job skills and abilities. Let $\mathbf{X}_j$, $j = 1, \ldots, N$ denote the vector of scores for the jth subject, and $\bar{\mathbf{X}}$, the scores for the sample mean vector for N independent observations. Let $E(\mathbf{X}_j) = \boldsymbol{\theta}$, and $\operatorname{var}(\mathbf{X}_j) = \boldsymbol{\Sigma}$, $j = 1, \ldots, N$. Let $f(\bar{\mathbf{X}})$ denote a scalar function which is used as a general measure of the job skill level of the group taking the battery of tests. Suppose

$$f(\bar{\mathbf{X}}) = \mathbf{a}' \, \overline{\mathbf{log}} \, (\bar{\mathbf{X}}) + b,$$

where b is a known scalar, $\mathbf{a}$ is a known $p \times 1$ vector, and $\overline{\mathbf{log}} \, (\bar{\mathbf{X}})$ is a $p \times 1$ vector whose ith component is $\log (\bar{X}_i)$ (see Exercise 2.8). Find the asymptotic distribution of $f(\bar{\mathbf{X}})$ so that inferences can be drawn about the population from which the group was drawn.

4.3 Referring to Example (4.6.1), find
(a) the prior distribution of θ_6;
(b) $P\{\theta_6 \leq 4\}$;
(c) the asymptotic marginal posterior distribution of θ_6 given the n observations.

4.4 Suppose $X_1, \ldots, X_N$ are independent observations all following the law $N(\theta, 1)$. Adopt a natural conjugate prior distribution for θ and find the asymptotic posterior distribution for θ given the sample data.

4.5 Let $\mathbf{X}: p \times 1$ follow the law $N(\boldsymbol{\theta}, \mathbf{I})$. $\mathbf{X}_1, \ldots, \mathbf{X}_N$ are independent observations of $\mathbf{X}$. Suppose the prior density of $\boldsymbol{\theta} \equiv (\theta_i)$ is given by $p(\boldsymbol{\theta}) = 1$, for $0 \leq \theta_i \leq 1$ for every $i = 1, \ldots, p$, and $p(\boldsymbol{\theta}) = 0$ otherwise. Discuss the asymptotic posterior distribution of $\boldsymbol{\theta}$ given the sample.

4.6 Suppose $\mathbf{W}: p \times p$ and $\mathbf{W} = \mathbf{W}'$. Show that for all $\alpha \neq 0$,

$$|\mathbf{I} + \mathbf{W}|^{\alpha} = \exp\left[\alpha(\text{tr } \mathbf{W} - \tfrac{1}{2} \text{tr } \mathbf{W}^2 + \tfrac{1}{3} \text{tr } \mathbf{W}^3 - \cdots)\right],$$

if the latent roots of $\mathbf{W}$ lie within the unit circle. Moreover, if the latent roots are sufficiently small,

$$|\mathbf{I} + \mathbf{W}|^{\alpha} \cong \exp(\alpha \text{ tr } \mathbf{W}).$$

This result will be used to prove Theorem (8.6.4).

4.7 Suppose $\mathbf{X}: p \times 1$ is random and all third order moments of the components of $\mathbf{X}$ exist. Let $f(\mathbf{X})$ denote a scalar function of $\mathbf{X}$. The Taylor series expansion is given by

$$f(\mathbf{X}) = f(\boldsymbol{\theta}) + (\mathbf{X} - \boldsymbol{\theta})'\mathbf{q} + \ldots,$$

where:

$$\mathbf{q} \equiv \left.\frac{\partial f(\mathbf{X})}{\partial \mathbf{X}}\right|_{\boldsymbol{\theta}}.$$

Find the next term in the expansion in terms of vectors and matrices.

4.8 Suppose $\mathbf{V}: p \times p$ and $\mathcal{L}(\mathbf{V}) = W(\Sigma, p, n)$, $p \leq n$, $\Sigma > 0$. Find the asymptotic distribution of $|\mathbf{V}|$.

4.9 Let (X, Y) denote scalar random variables following the law $\mathcal{L}(X, Y) = N(\mathbf{0}, \Sigma)$,

$$\Sigma = \begin{pmatrix} \sigma_{11} & \sigma_{12} \\ \sigma_{21} & \sigma_{22} \end{pmatrix}.$$

Let $(X_1, Y_1), \ldots, (X_N, Y_N)$ denote independent bivariate observations. Evaluate:

(a) $\displaystyle \operatorname*{plim}_{N\to\infty} \frac{1}{N} \sum_1^N X_i Y_i,$

(b) $\displaystyle \operatorname*{plim}_{N\to\infty} \frac{1}{N} \sum_1^N X_i^3,$

(c) $\displaystyle \operatorname*{plim}_{N\to\infty} \frac{1}{N} \sum_1^N Y_i^4.$

4.10 $\mathbf{y} = (y_i): p \times 1$ follows a multinomial distribution with parameters $\boldsymbol{\theta} = (\theta_i): p \times 1$, and $n = \Sigma_i^p y_i$. We want to estimate $\gamma = \boldsymbol{\theta}'\boldsymbol{\theta}$.

(a) Show that $\hat{\gamma}$, the MLE of γ is $\hat{\gamma} = \hat{\boldsymbol{\theta}}'\hat{\boldsymbol{\theta}}$, where $\hat{\theta}_i = y_i/n$, $\hat{\boldsymbol{\theta}} = (\hat{\theta}_i)$;

(b) Show that: $E\hat{\gamma} \doteq \gamma + (1 - \gamma)/n$. *Hint*: Expand $\hat{\gamma}$ above γ in Taylor series.

(c) Show that:

$$\lim_{n\to\infty} \mathcal{L}\{\sqrt{n}(\hat{\gamma} - \gamma)\} = N(0, \sigma^2),$$

for $\sigma^2 \equiv 4(\Sigma_1^p \theta_i^3 - \gamma^2)$.

5

THE WISHART AND RELATED DISTRIBUTIONS

The Wishart distribution ranks next to the Normal distribution in order of importance and usefulness in multivariate analysis. It is a multivariate generalization of the univariate gamma distribution and as such, plays a fundamental role in inference. For example, estimation of factor loadings by maximum likelihood in the factor analysis model depends upon the Wishart distribution and its properties (Chapter 10); inferences about the variances or covariances in a multivariate Normal distribution are based upon the Wishart distribution (Chapter 7); and Bayesian approaches to multivariate analysis often involve the Wishart distribution, either as the underlying distribution for the sample variances and covariances [see Theorem (5.1.2)], or as the natural conjugate prior distribution for the precision matrix in Normal multivariate regression models (Chapter 8).

Some of the properties of the Wishart distribution which are useful in applications are developed below. Related distributions such as the inverted Wishart (the natural conjugate prior distribution for the covariance matrix in a Normal distribution), and the noncentral Wishart distribution, useful for inference in simultaneous equation system problems that arise in econometrics (see Section 5.3.2) are also discussed.

5.1 THE WISHART DISTRIBUTION (WISHART, 1928)

5.1.1 Density

Let $\mathbf{V}: p \times p$ be symmetric and positive definite. The random matrix $\mathbf{V}$ is said to follow the nonsingular p-dimensional Wishart distribution with scale matrix $\mathbf{\Sigma}$, and n degrees of freedom, $p \leq n$, if the joint distribution of the distinct elements of $\mathbf{V}$ is continuous with density function

$$p(\mathbf{V}) = \frac{c|\mathbf{V}|^{(n-p-1)/2}}{|\mathbf{\Sigma}|^{n/2}} \exp\left(-\tfrac{1}{2} \operatorname{tr} \mathbf{\Sigma}^{-1}\mathbf{V}\right), \qquad \mathbf{V} > 0,\ \mathbf{\Sigma} > 0, \quad (5.1.1)$$

and $p(\mathbf{V}) = 0$, otherwise, where c is a numerical constant defined as

$$c = \left[2^{np/2}\pi^{p(p-1)/4} \prod_{j=1}^{p} \Gamma\left(\frac{n+1-j}{2}\right)\right]^{-1}.$$

If $n < p$, the distribution is singular and there is no density. Thus, if $\mathbf{V} \equiv (v_{ij})$, and $\mathbf{\Sigma}^{-1} \equiv (\sigma^{ij})$, for $\mathbf{V} > 0$,

$$p(\mathbf{V}) \propto \frac{|\mathbf{V}|^{(n-p-1)/2}}{|\mathbf{\Sigma}|^{n/2}} \exp\left(-\tfrac{1}{2} \sum_{i=1}^{p} \sum_{j=1}^{p} v_{ij}\sigma^{ij}\right).$$

This relationship will be expressed as

$$\mathcal{L}(\mathbf{V}) = W(\mathbf{\Sigma},p,n). \qquad (5.1.2)$$

Multivariate Gamma Function

Define the p-dimensional gamma function

$$\Gamma_p(\theta) = \int_{\mathbf{X}>0} |\mathbf{X}|^{\theta-(p+1)/2} e^{-\operatorname{tr}(\mathbf{X})} d\mathbf{X},$$

where $\mathbf{X}$ denotes a positive definite symmetric matrix of order p. Write $\mathbf{X} = \mathbf{TT}'$, where $\mathbf{T}$ denotes a lower triangular matrix (see Sec. 2.11.2). Then, transforming from $\mathbf{X} \to \mathbf{T}$, with

$$J(\mathbf{X} \to \mathbf{T}) = 2^p t_{11}^p \cdots t_{pp}, \qquad \mathbf{T} = (t_{ij}),$$

and integrating, we find

$$\Gamma_p(\theta) = \pi^{p(p-1)/4} \prod_{1}^{p} \Gamma\left(\theta - \frac{j-1}{2}\right).$$

That is, for c in eqn. (5.1.1),

$$c^{-1} = 2^{\frac{np}{2}} \Gamma_p(n/2).$$

Example (5.1.1): Suppose $\mathcal{L}(\mathbf{V}) = W(\mathbf{I},2,n)$. Then $\mathbf{V}: 2 \times 2$ has density

$$p(\mathbf{V}) \propto |\mathbf{V}|^{(n-3)/2} \exp(-\tfrac{1}{2} \operatorname{tr} \mathbf{V}), \qquad \mathbf{V} > 0.$$

Let $\mathbf{V} = (v_{ij})$, $i, j = 1, 2$. The density then becomes (recall that $\mathbf{V} = \mathbf{V}'$)

$$p(\mathbf{V}) \equiv p(v_{11}, v_{12}, v_{22}) \propto (v_{11}v_{22} - v_{12}^2)^{(n-3)/2} \exp[-\tfrac{1}{2}(v_{11} + v_{22})].$$

Theorem (5.1.1): For $\mathbf{X}_\alpha : p \times 1$, let $\mathbf{X}_1, \ldots, \mathbf{X}_n$ be mutually independent; let $\mathcal{L}(\mathbf{X}_\alpha) = N(\mathbf{0}, \boldsymbol{\Sigma})$, $\boldsymbol{\Sigma} > 0$, for $\alpha = 1, \ldots, n$; and let $\mathbf{X} \equiv (\mathbf{X}_1, \ldots, \mathbf{X}_n)$, $\mathbf{X} : p \times n$, $p \leq n$. Define $\mathbf{V} = \mathbf{X}\mathbf{X}'$. Then $\mathbf{V} > 0$ and $\mathcal{L}(\mathbf{V}) = W(\boldsymbol{\Sigma}, p, n)$.

Proof: See Wishart, 1928 who used a geometric argument. The result may also be shown by demonstrating that the characteristic function of $\mathbf{XX}'$ is the same as that of a Wishart distribution (see Cramér, 1946, p. 405).

Lemma (5.1.1): If the density of $\mathbf{X} : p \times n$ is a function of $\mathbf{XX}'$, $g(\mathbf{XX}')$, then the density of $\mathbf{V} = \mathbf{XX}'$ is

$$p(\mathbf{V}) = \frac{|\mathbf{V}|^{(n-p-1)/2} g(\mathbf{V}) \pi^{(p/2)(n-p/2+1/2)}}{\prod_{j=1}^{p} \Gamma\left[\frac{n+1-j}{2}\right]}. \tag{5.1.3}$$

The proof of this lemma is based upon finding the distribution of the latent roots of a matrix (see for instance Anderson, 1958, p. 319).

[*Remark:* Note that $\mathbf{V}$ defined in Theorem (5.1.1) is essentially the covariance matrix of the sample observation vectors (without the scaling factor). That is,

$$\mathbf{V} = \sum_{\alpha=1}^{n} \mathbf{X}_\alpha \mathbf{X}_\alpha'.$$

Hence, the Wishart distribution is the joint distribution of the sample variances and covariances. Estimation of the covariance matrix is discussed in Chapter 7. If $p = 1$ there are no covariances and $\mathbf{V}$ is just proportional to a sample variance

$$V = \sum_{\alpha=1}^{n} X_\alpha^2.$$

Since in the univariate case it is well known that the distribution of $\mathbf{V}$ is proportional to that of a chi-square variate, the Wishart distribution represents a generalization to p-dimensions of distributions proportional to chi-square distributions (that is, gamma distributions).]

Theorem (**5.1.2**)**:** For $\mathbf{X}_\alpha : p \times 1$, and $\mathbf{X}_1, \ldots, \mathbf{X}_N$, mutually independent, let $\mathcal{L}(\mathbf{X}_\alpha) = N(\boldsymbol{\theta}, \boldsymbol{\Sigma})$, $\boldsymbol{\Sigma} > 0$, for $\alpha = 1, \ldots, N$. Denote the vector of sample means by $\bar{\mathbf{X}} = \frac{1}{N} \sum_{\alpha=1}^{N} \mathbf{X}_\alpha$. Define

$$\mathbf{V} = \sum_{\alpha=1}^{N} (\mathbf{X}_\alpha - \bar{\mathbf{X}})(\mathbf{X}_\alpha - \bar{\mathbf{X}})' = \sum_{\alpha=1}^{N} \mathbf{X}_\alpha \mathbf{X}_\alpha' - N\bar{\mathbf{X}}\bar{\mathbf{X}}'. \tag{5.1.4}$$

Then, if $p + 1 \leq N$, $\mathbf{V} > 0$ and $\mathcal{L}(\mathbf{V}) = W(\boldsymbol{\Sigma}, p, N - 1 \equiv n)$.

Proof: See Complement 5.1.

Example (**5.1.2**)**:** Let x_α and y_α, $\alpha = 1, \ldots, 10$, denote changes in sales of two different items in the same department store in ten different weeks. Suppose the actual pairs of values are

$$(1,1),\ (2,1),\ (3,1),\ (3,2),\ (-1,3)$$
$$(1,2),\ (0,1),\ (2,2),\ (2,0),\ (1,2),$$

and suppose all pairs to be mutually independent, with $\mathbf{X}_\alpha' = (x_\alpha, y_\alpha)$, and $\mathcal{L}(\mathbf{X}_\alpha) = N(\boldsymbol{\theta}, \boldsymbol{\Sigma})$, $\alpha = 1, \ldots, N$. This example will consider the question of whether the underlying variances and covariances are "significantly large." In the notation of Theorem (5.1.2), $N = 10$, $p = 2$. Then,

$$\bar{\mathbf{X}} = \begin{pmatrix} \frac{1}{10} \sum_1^{10} x_\alpha \\ \frac{1}{10} \sum_1^{10} y_\alpha \end{pmatrix} = \begin{pmatrix} 1.4 \\ 1.5 \end{pmatrix}.$$

By definition,

$$\mathbf{X}_\alpha \mathbf{X}_\alpha' = \begin{pmatrix} 1 & 1 \\ 1 & 1 \end{pmatrix}, \begin{pmatrix} 4 & 2 \\ 2 & 1 \end{pmatrix}, \begin{pmatrix} 9 & 3 \\ 3 & 1 \end{pmatrix}, \begin{pmatrix} 9 & 6 \\ 6 & 4 \end{pmatrix}, \begin{pmatrix} 1 & -3 \\ -3 & 9 \end{pmatrix},$$
$$\begin{pmatrix} 1 & 2 \\ 2 & 4 \end{pmatrix}, \begin{pmatrix} 0 & 0 \\ 0 & 1 \end{pmatrix}, \begin{pmatrix} 4 & 4 \\ 4 & 4 \end{pmatrix}, \begin{pmatrix} 4 & 0 \\ 0 & 0 \end{pmatrix}, \begin{pmatrix} 1 & 2 \\ 2 & 4 \end{pmatrix},$$

for $\alpha = 1, 2, \ldots, 10$, respectively. Hence,

$$\sum_{j=1}^{10} \mathbf{X}_\alpha \mathbf{X}_\alpha' = \begin{pmatrix} 34 & 17 \\ 17 & 29 \end{pmatrix}.$$

Since $\mathbf{V} = \Sigma_1^N \mathbf{X}_\alpha \mathbf{X}_\alpha' - N\bar{\mathbf{X}}\bar{\mathbf{X}}'$,

$$\mathbf{V} = \begin{pmatrix} 34 & 17 \\ 17 & 29 \end{pmatrix} - \begin{pmatrix} 19.6 & 21.0 \\ 21.0 & 22.5 \end{pmatrix},$$

or

$$\mathbf{V} = \begin{pmatrix} 14.4 & -4 \\ -4 & 6.5 \end{pmatrix}.$$

Clearly, $\mathbf{V} > 0$. Theorem (5.1.2) asserts that a random matrix, $\mathbf{V}$ (before it is actually observed), follows the law $W(\mathbf{\Sigma},2,9)$. Let $\mathbf{V} = (v_{ij})$, $i, j = 1, 2$, and assume

$$\mathbf{\Sigma} = \begin{pmatrix} \frac{1}{2} & -\frac{1}{2} \\ -\frac{1}{2} & \frac{3}{4} \end{pmatrix},$$

so that

$$\mathbf{\Sigma}^{-1} = \begin{pmatrix} 6 & 4 \\ 4 & 4 \end{pmatrix}.$$

The probability of observing matrices with smaller elements than those of the sample matrix given above [using (5.1.1)] is given by

$$P\{v_{11} \leq 14.4,\ v_{12} \leq -4,\ v_{22} \leq 6.5\}$$
$$= c \iiint_G (v_{11}v_{22} - v_{12}^2)^3(8^{9/2}) \exp\left[-\tfrac{1}{2}(6v_{11} + 8v_{12} + 4v_{22})\right] d\mathbf{V},$$

where

$$c = [2^9 \sqrt{\pi}\, \Gamma(\tfrac{9}{2})\Gamma(4)]^{-1},$$

and

$$G \equiv \{\mathbf{V}: v_{11}v_{22} \geq v_{12}^2,\ 0 \leq v_{11} \leq 14.4,\ 0 \leq v_{22} \leq 6.5,\ -\infty < v_{12} \leq -4\}.$$

This integral may be evaluated numerically.

5.1.2 Marginals

In applications it is often important to find parsimonious relationships by eliminating variables that contribute little to an understanding of the scatter in the data. To make inferences about the remaining variables it is necessary to use their marginal distributions.

Diagonal Marginals

The marginal distributions of submatrices centered around the diagonal of a matrix following a Wishart distribution are also Wishart.

Theorem (5.1.3): Let $\mathbf{V}: p \times p$ follow the distribution $\mathcal{L}(\mathbf{V}) = W(\mathbf{\Sigma},p,n)$, $\mathbf{\Sigma} > 0$, and partition $\mathbf{V}$ and $\mathbf{\Sigma}$ as

$$\mathbf{V} = \begin{pmatrix} \mathbf{V}_{11} & \mathbf{V}_{12} \\ \mathbf{V}_{21} & \mathbf{V}_{22} \end{pmatrix}, \qquad \mathbf{\Sigma} = \begin{pmatrix} \mathbf{\Sigma}_{11} & \mathbf{\Sigma}_{12} \\ \mathbf{\Sigma}_{21} & \mathbf{\Sigma}_{22} \end{pmatrix}, \tag{5.1.5}$$

where $\mathbf{V}_{11} : q \times q$ and $\mathbf{\Sigma}_{11} : q \times q$. Then

$$\mathcal{L}(\mathbf{V}_{11}) = W(\mathbf{\Sigma}_{11}, q, n). \tag{5.1.6}$$

Proof: From Theorem (5.1.1) write $\mathbf{V} = \Sigma_1^n \mathbf{X}_\alpha \mathbf{X}_\alpha'$, where the $\mathbf{X}_\alpha$'s are independent, and $\mathcal{L}(\mathbf{X}_\alpha) = N(\mathbf{0}, \mathbf{\Sigma})$. Let $\mathbf{x}_\alpha : q \times 1$ be defined by

$$\mathbf{X}_\alpha = \begin{pmatrix} \mathbf{y}_\alpha \\ \mathbf{z}_\alpha \end{pmatrix}.$$

Then from Chapter 3, $\mathcal{L}(\mathbf{x}_\alpha) = N(\mathbf{0}, \mathbf{\Sigma}_{11})$, for all $\alpha = 1, \ldots, n$ and the $\mathbf{y}_1, \ldots, \mathbf{y}_n$ are independent. Hence, if $\mathbf{V}_{11} = \Sigma_1^n \mathbf{y}_\alpha \mathbf{y}_\alpha'$, by Theorem 5.1.1, $\mathcal{L}(\mathbf{V}_{11}) = W(\mathbf{\Sigma}_{11}, q, n)$.

Example (5.1.3): Suppose $q = 1$ so that V_{11} is a scalar. Then $\mathcal{L}(V_{11}) = W(\Sigma_{11}, 1, n)$. The density is

$$p(V_{11}) = \frac{c V_{11}^{(n-2)/2}}{\Sigma_{11}^{n/2}} \exp\left(-\frac{1}{2}\frac{V_{11}}{\Sigma_{11}}\right), \qquad V_{11} > 0.$$

In more conventional notation, let $V_{11} = v$, $\Sigma_{11} = \sigma^2$; then

$$p(v) \propto \frac{v^{n/2-1}}{\sigma^n} \exp\left(-\frac{v}{2\sigma^2}\right), \qquad v > 0.$$

This univariate density should be recognized as that of a gamma variate.

Corollary (5.1.3): If $\mathbf{\Sigma}_{12} = 0$, $\mathbf{V}_{11}$ and $\mathbf{V}_{22}$ are independently distributed.

Proof: This result follows immediately by integrating the density of $\mathbf{V}$ with respect to $\mathbf{V}_{12}$ for $\mathbf{\Sigma}_{12} = 0$, and noting the result factors[1] into the product of a function of $\mathbf{V}_{11}$ and a function of $\mathbf{V}_{22}$. See Exercise 2.12.

[*Remark:* It should be noted that since the ordering of components was arbitrary, all diagonal elements of a Wishart matrix have marginal gamma distributions, and all blocks of elements centered on the diagonal have marginal Wishart distributions.]

Off-Diagonal Marginals

The marginal distributions of the off-diagonal elements of a p-dimensional Wishart matrix are not generally known, although results for $p = 2$

[1] The trace term in the joint density factors immediately. The term in $|\mathbf{V}|$ is rewritten using (2.6.7) and variables are changed, using (2.15.2) and (2.15.3), to effect a simple integration.

are summarized, for instance, in Press (1967b). For $p = 2$, the marginal Wishart is the distribution of the sample covariance between two jointly distributed Normal variates. Inferences about the relationship are easily carried out by means of the sample correlation coefficient so that the distribution of the sample covariance is usually not needed.

In Bayesian approaches to multivariate analysis, the precision matrix $\mathbf{\Sigma}^{-1}$ is of interest. Since the marginals involve $\mathbf{\Sigma}_{11\cdot2}$ and $\mathbf{\Sigma}_{22\cdot1}$, the distributions of such quantities are required. The $\mathbf{\Sigma}_{11\cdot2}$ and $\mathbf{\Sigma}_{22\cdot1}$ matrices also arise in the canonical correlations model (Chapter 11).

Theorem (5.1.4): Let $\mathbf{V}: p \times p$ follow the distribution $\mathcal{L}(\mathbf{V}) = W(\mathbf{\Sigma},p,n)$, $\mathbf{\Sigma} > 0$, and partition $\mathbf{V}$ and $\mathbf{\Sigma}$ as in (5.1.5). Define for $\mathbf{V}_{22} > 0$, $\mathbf{\Sigma}_{22} > 0$,

$$\mathbf{V}_{11\cdot2} \equiv \mathbf{V}_{11} - \mathbf{V}_{12}\mathbf{V}_{22}^{-1}\mathbf{V}_{21}, \qquad \mathbf{\Sigma}_{11\cdot2} \equiv \mathbf{\Sigma}_{11} - \mathbf{\Sigma}_{12}\mathbf{\Sigma}_{22}^{-1}\mathbf{\Sigma}_{21}.$$

Then the marginal distribution of $\mathbf{V}_{11\cdot2}$ is given by

$$\mathcal{L}(\mathbf{V}_{11\cdot2}) = W(\mathbf{\Sigma}_{11\cdot2}, q, n - p + q),$$

where $\mathbf{V}_{11}: q \times q$, and $\mathbf{\Sigma}_{11}: q \times q$. Moreover, $\mathbf{V}_{11}$ and $\mathbf{V}_{11\cdot2}$ are independent.

Proof: Analogous to Complement 5.2.

Theorem (5.1.5): Let $\mathbf{V}: p \times p$ follow the distribution $\mathcal{L}(\mathbf{V}) = W(\Sigma, p, n)$, $p \leqslant n$, $\Sigma > 0$, and partition $\mathbf{V}$ and Σ as in (5.1.5). Assume $\Sigma_{22} > 0$, and define $\mathbf{H} = \Sigma^{-1}$, with $\mathbf{H}$ partitioned as Σ, as $\mathbf{H} = (\mathbf{H}_{ij})$, $i, j = 1, 2$.

(a) If $\underset{q \times (p-q)}{\mathbf{X}} \equiv \mathbf{V}_{12}\mathbf{V}_{22}^{-1}$, $\mathbf{V}_{22} > 0$,

$$\mathcal{L}(\mathbf{X} \mid \mathbf{V}_{22}) = N(\mathbf{H}_{11}^{-1}\mathbf{H}_{12}, \mathbf{H}_{11}^{-1} \otimes \mathbf{V}_{22}^{-1}); \tag{5.1.7}$$

(b) The density of $\mathbf{X}$, unconditional on $\mathbf{V}_{22}$, is matrix $\mathbf{T}$ (see eqn. (6.2.6)), and is given by

$$f(\mathbf{X}) \propto |\mathbf{H}_{11}^{-1} + (\mathbf{X} - \mathbf{H}_{11}^{-1}\mathbf{H}_{12})(\Sigma_{22})(\mathbf{X} - \mathbf{H}_{11}^{-1}\mathbf{H}_{12})'|^{(n+q)/2}. \tag{5.1.8}$$

Proof: See Kaufman (1969).

5.1.3 Reproductive Property

If the sum of two independent random variables in the same distributional family yields a random variable also in the family, the family is said to possess the reproductive property. For the Wishart family the reproductive property is useful for pooling data from several samples and then drawing inferences from the pooled data. The required result is given in the next theorem.

Theorem (5.1.6): If $\mathbf{V}_1 : p \times p$ and $\mathbf{V}_2 : p \times p$ are independent, and if $\mathcal{L}(\mathbf{V}_1) = W(\mathbf{\Sigma},p,m)$ and $\mathcal{L}(\mathbf{V}_2) = W(\mathbf{\Sigma},p,n)$, then $\mathcal{L}(\mathbf{V}_1 + \mathbf{V}_2) = W(\mathbf{\Sigma}, p, m + n)$.

Proof: By Theorem (5.1.1), $\mathbf{V}_1 = \Sigma_1^m \mathbf{X}_\alpha \mathbf{X}_\alpha'$, and $\mathbf{V}_2 = \Sigma_{m+1}^{m+n} \mathbf{X}_\alpha \mathbf{X}_\alpha'$, where the $\mathbf{X}_\alpha$'s are independent and $\mathcal{L}(\mathbf{X}_\alpha) = N(\mathbf{0},\mathbf{\Sigma})$, $\alpha = 1,\ldots,m + n$. Thus, $\mathbf{V}_1 + \mathbf{V}_2 = \Sigma_1^{m+n} \mathbf{X}_\alpha \mathbf{X}_\alpha'$.

Let $\mathbf{V}_1/m$ and $\mathbf{V}_2/n$ be two independent estimates of $\mathbf{\Sigma}$, the covariance matrix of a population, based upon m and n degrees of freedom, respectively. A better estimate of $\mathbf{\Sigma}$ is obtained by pooling the data and using as an estimate $(\mathbf{V}_1 + \mathbf{V}_2)/(m + n)$, and drawing inferences based upon the distribution of $\mathbf{V}_1 + \mathbf{V}_2$ (see Chapter 7).

5.1.4 Transformations of Wishart Matrices

It is often more convenient to work with symmetric transformations of Wishart matrices (such as orthogonal transformations) instead of the original matrices. The distribution of the transformed matrix is similarly affected. Such transformations are typically used in the principal components and factor analysis models (Chapters 9 and 10). The main result is given below.

Theorem (5.1.7): Let $\mathbf{V} : p \times p$ follow the law $\mathcal{L}(\mathbf{V}) = W(\mathbf{\Sigma},p,n)$. Suppose $\mathbf{A} : q \times p$ is a constant matrix with $q \leq p$. Then, if $\mathbf{Z} = \mathbf{AVA}'$, $\mathcal{L}(\mathbf{Z}) = W(\mathbf{A\Sigma A}',q,n)$.

Proof: Let $\mathbf{V} = \Sigma_1^n \mathbf{X}_\alpha \mathbf{X}_\alpha'$, where $\mathcal{L}(\mathbf{X}_\alpha) = N(\mathbf{0},\mathbf{\Sigma})$, and the $\mathbf{X}_\alpha$ are independent. Then if $\mathbf{Z} = \mathbf{AVA}'$,

$$\mathbf{Z} = \mathbf{A} \sum_1^n \mathbf{X}_\alpha \mathbf{X}_\alpha' \mathbf{A}' = \sum_1^n (\mathbf{AX}_\alpha)(\mathbf{AX}_\alpha)',$$

where the $\mathbf{AX}_\alpha$'s are independent and $\mathcal{L}(\mathbf{AX}_\alpha) = N(\mathbf{0},\mathbf{A\Sigma A}')$.

Example (5.1.4): Suppose $\mathbf{a}$ is a p-vector, $\mathbf{V} : p \times p$ is a random matrix following $\mathcal{L}(\mathbf{V}) = W(\mathbf{\Sigma},p,n)$, and

$$Z = \frac{\mathbf{a}'\mathbf{V}\mathbf{a}}{(\mathbf{a}'\mathbf{a})}.$$

In terms of Theorem (5.1.7), $\mathbf{A}$ is a $1 \times p$ row vector given by $\mathbf{A} = \mathbf{a}'/(\mathbf{a}'\mathbf{a})^{1/2}$. Then

$$\mathcal{L}(Z) = W\left(\frac{\mathbf{a}'\mathbf{\Sigma}\mathbf{a}}{\mathbf{a}'\mathbf{a}}, 1,n\right).$$

Note that the dimension of the distribution is one since Z is a scalar. From Example (5.1.3) it is clear that Z follows a gamma distribution.

As a special case, suppose $\mathbf{a}$ is a vector of ones. Then Z is proportional to the sum of all the variances and covariances.

5.1.5 Moments of a Wishart Distribution

To make inferences about sample variances and covariances, it is necessary to have the moments of the elements of a Wishart matrix. Suppose $\mathcal{L}(\mathbf{V}) = W(\mathbf{\Sigma},p,n)$ so that $\mathbf{V} = \Sigma_1^n \mathbf{X}_\alpha \mathbf{X}_\alpha'$ with the $\mathbf{X}_\alpha$'s independent, and $\mathcal{L}(\mathbf{X}_\alpha) = N(\mathbf{0},\mathbf{\Sigma})$, $\alpha = 1,\ldots,n$. Let $\mathbf{V} = (v_{ij})$, and $\mathbf{\Sigma} = (\sigma_{ij})$, i, $j = 1,\ldots,p$. Let $\mathbf{X}_\alpha = (x_{i\alpha})$, $i = 1,\ldots,p$; $\alpha = 1,\ldots,n$.

Theorem (5.1.8): The means, variances and covariances of the components of a Wishart matrix $\mathbf{V}$ are given by

$$\begin{aligned} E(v_{ij}) &= n\sigma_{ij}, \\ \text{var}\,(v_{ij}) &= n(\sigma_{ij}^2 + \sigma_{ii}\sigma_{jj}), \\ \text{cov}\,(v_{ij},v_{kl}) &= n(\sigma_{ik}\sigma_{jl} + \sigma_{il}\sigma_{jk}). \end{aligned}$$

Proof: $v_{ij} = \Sigma_{\alpha=1}^n x_{i\alpha}x_{j\alpha}$. Hence,

$$E(v_{ij}) = \sum_{\alpha=1}^{n} E(x_{i\alpha}x_{j\alpha}) = \sum_{\alpha=1}^{n} \sigma_{ij} = n\sigma_{ij}.$$

The variances, covariances, and higher-order moments are derived analogously.

[*Remark:* If $\mathbf{V}$ were a parameter assigned the prior distribution in (5.1.1), specifying $E(\mathbf{V})$ and n would be sufficient to define the complete distribution (which may not always be satisfactory since then, the variances and other moments are fixed as well).]

Example (5.1.5): The Washington State Apple Commission was interested in studying the sales effects of two different campaign themes for apples (see Henderson, Hind, and Brown, 1961). One advertising theme stressed the various uses of apples (fruit combination salads, baked apples, other dishes), while the other emphasized the healthful qualities of apples (builds strong bodies, dental benefits, and so on). To determine relative sales effectiveness of the two promotional themes (in addition to other objectives) two regressions were run on independent variables such as length of time of campaign, and location of campaign (city, and location within city).

Suppose that the covariance matrix for the disturbances of the two regression equations is given by (for simplicity, assume it is known; in Chapter 8, the covariance matrix will generally be unknown)

$$\mathbf{\Sigma} = \begin{pmatrix} 4 & 1 \\ 1 & 2 \end{pmatrix}.$$

That is, the disturbances pertaining to the first equation have variance 4, those pertaining to the second equation have variance 2, and the covariance of the errors between the two equations is unity. Assume that analysis of the regression yields an estimator $\mathbf{V} = (v_{ij})$ such that $\mathcal{L}(\mathbf{V}) = W(\Sigma, 2, 10)$. Then, using Theorem (5.1.8), it may be checked that

$$E(\mathbf{V}) = \begin{pmatrix} 40 & 10 \\ 10 & 20 \end{pmatrix},$$

and the covariance matrix of $\mathbf{v} \equiv (v_{11}, v_{12}, v_{22})'$ is

$$\text{var}(\mathbf{v}) = \begin{pmatrix} 320 & 80 & 20 \\ 80 & 90 & 40 \\ 20 & 40 & 80 \end{pmatrix}.$$

It is now possible to study the residuals in the two equation regression system (as represented by the vector $\mathbf{v}$) for significantly large values, by computing probabilities, as in Example (5.1.2).

5.1.6 Generalized Variance

Let $\mathbf{\Sigma}$ denote the covariance matrix of a p-variate distribution. Wilks called the scalar $|\mathbf{\Sigma}|$ the *generalized variance of the population.*[2] If $\mathbf{X}_1, \ldots, \mathbf{X}_N$ are independent and $\mathcal{L}(\mathbf{X}_\alpha) = N(\mathbf{0}, \mathbf{\Sigma})$, a moment estimator of $\mathbf{\Sigma}$ (adjusted for bias) is

$$\mathbf{S} = \frac{1}{N-1} \sum_{\alpha=1}^{N} (\mathbf{X}_\alpha - \bar{\mathbf{X}})(\mathbf{X}_\alpha - \bar{\mathbf{X}})',$$

where $\bar{\mathbf{X}}$ is the sample mean vector. Hence, the *sample generalized variance* is $|\mathbf{S}|$. This generalized notion of variance is introduced here since $(N-1)\mathbf{S}$ is what has been called $\mathbf{V}$ in this chapter.

In applications, $|\mathbf{\Sigma}|$ is sometimes used to rank distinct groups or populations in order of their dispersion or spread. Other competing measures proposed for the same purpose include $(\text{tr}\, \mathbf{\Sigma}^2)^{1/2}$ and the difference between the largest and smallest latent roots of $\mathbf{\Sigma}$. An example of the use of $|\mathbf{\Sigma}|$ is given below.

Example (5.1.6): A certain article, such as a transistor, produced by Company k, is characterized by a vector of p measurements, $\mathbf{X}_k : p \times 1$. Suppose a buyer is facing the choice of which of r companies that manufacture

[2] $|\mathbf{\Sigma}|$ reduces to the ordinary variance when $p = 1$ and is a scalar measure of internal scatter when $p \geq 1$.

the article should be selected to supply it. Suppose that all the companies produce a transistor with the same characteristics on the average so that for all $k = 1,\ldots,r$, $\mathcal{L}(\mathbf{X}_k) = N(\boldsymbol{\theta},\boldsymbol{\Sigma}_k)$. Although the same product is produced on the average, the companies can be distinguished on the basis of their associated covariance matrices. That is, some of the suppliers tend to have great variation in the quality of their output, while the average is constant. Thus, the buyer can rank the supplying companies by ranking the normal populations using the generalized variance. In an effort to minimize his chances of receiving a product that is sometimes of unacceptably poor quality, the buyer may decide to choose the supplier for which the generalized variance is minimum; that is, the rule is, choose that k for which $|\boldsymbol{\Sigma}_k| = |\boldsymbol{\Sigma}^*|$, where

$$|\boldsymbol{\Sigma}^*| = \min_{1\leq k\leq r} |\boldsymbol{\Sigma}_k|.$$

5.2 THE INVERTED WISHART DISTRIBUTION

The inverted Wishart distribution is the multivariate generalization of the univariate inverted gamma distribution (see, for instance, Raiffa and Schlaifer, 1961, p. 227). It is the distribution of the inverse of a random matrix following the Wishart distribution, and is the distribution which is natural conjugate prior for the covariance matrix in a Normal distribution.

5.2.1 Density

Let $\mathbf{U} : p \times p$ be a random matrix following the inverted Wishart distribution with positive definite scale matrix $\mathbf{G}$, and n d.f. Then for $2p < n$, the density of $\mathbf{U}$ is given by

$$p(\mathbf{U}) = \frac{c_0|\mathbf{G}|^{(n-p-1)/2}}{|\mathbf{U}|^{n/2}} \exp\left(-\tfrac{1}{2}\operatorname{tr}\mathbf{U}^{-1}\mathbf{G}\right), \qquad \mathbf{U} > 0, \qquad (5.2.1)$$

and $p(\mathbf{U}) = 0$ otherwise; the relationship will be denoted by $\mathcal{L}(\mathbf{U}) = W^{-1}(\mathbf{G},p,n)$; the constant c_0 is given by

$$c_0^{-1} = 2^{(n-p-1)p/2}\pi^{p(p-1)/4} \prod_{j=1}^{p} \Gamma\left(\frac{n-p-j}{2}\right).$$

Note that if $\mathcal{L}(\mathbf{V}) = W(\boldsymbol{\Sigma},p,n)$, it follows from the above with $\mathbf{U} \equiv \mathbf{V}^{-1}$, and (2.15.8), that $\mathcal{L}(\mathbf{V}^{-1}) = W^{-1}(\boldsymbol{\Sigma}^{-1}, p, n+p+1)$. It is easy to identify the parameter n since it is the power of $|\mathbf{U}|^{1/2}$ in (5.2.1). The density expression will sometimes be written as

$$p(\mathbf{U}) \propto |\mathbf{U}|^{-n/2} \exp\left(-\tfrac{1}{2} \operatorname{tr} \mathbf{U}^{-1}\mathbf{G}\right), \qquad \mathbf{U} > 0, \mathbf{G} > 0. \qquad (5.2.2)$$

Example (5.2.1): Suppose $\mathbf{X}_1,\ldots,\mathbf{X}_n$ are independent p-variate observations from $N(\mathbf{0},\mathbf{\Sigma})$. Perhaps $\mathbf{X}_\alpha$ is the αth vector of observations of price changes for p commodities during a postwar period compared with a pre-war period. To determine the *Bayesian predictive density* (see Section 3.7) for a new observation $\mathbf{X}_{n+1}$, given the previous n observations (forecasting), it is necessary to compute

$$p(\mathbf{X}_{n+1}|\mathbf{X}_1,\ldots,\mathbf{X}_n) = \int p(\mathbf{X}_{n+1}|\mathbf{\Sigma})p(\mathbf{\Sigma}|\mathbf{X}_1,\ldots,\mathbf{X}_n)d\mathbf{\Sigma},$$

where $p(\mathbf{\Sigma}|\mathbf{X}_1,\ldots,\mathbf{X}_n)$ denotes the posterior distribution of $\mathbf{\Sigma}$ given the observed data. But

$$p(\mathbf{\Sigma}|\mathbf{X}_1,\ldots,\mathbf{X}_n) \propto p(\mathbf{X}_1,\ldots,\mathbf{X}_n|\mathbf{\Sigma})p(\mathbf{\Sigma}),$$

where $p(\mathbf{\Sigma})$ is the prior density of $\mathbf{\Sigma}$. Since

$$p(\mathbf{X}_1,\ldots,\mathbf{X}_n|\mathbf{\Sigma}) \propto |\mathbf{\Sigma}|^{-n/2} \exp\left[-\tfrac{1}{2} \operatorname{tr} \mathbf{\Sigma}^{-1}\left(\sum_1^n \mathbf{X}_i\mathbf{X}_i'\right)\right],$$

it is clear that a natural conjugate prior distribution for $\mathbf{\Sigma}$ is given by $\mathcal{L}(\mathbf{\Sigma}) = W^{-1}(\mathbf{G},p,\nu)$, $\nu > 2p$, for some parameters $\mathbf{G}$ and ν (see Chapter 3 to review the method of finding natural conjugate prior distributions).

The inverted Wishart distribution was used as a prior distribution for the covariance matrix of the disturbances in a Bayesian analysis of multivariate regression by Tiao and Zellner (1964a).

5.2.2 Marginals of the Inverted Wishart Distribution

The submatrices centered along the diagonal of an inverted Wishart matrix also follow an inverted Wishart distribution. The diagonal elements in particular follow an inverted gamma distribution (Raiffa and Schlaifer, 1961, p. 227), with density given in (5.2.3).

Theorem (5.2.1): Suppose $\mathcal{L}(\mathbf{U}) = W^{-1}(\mathbf{G},p,n)$, and $\mathbf{U}$ and $\mathbf{G}$ are partitioned as

$$\mathbf{U} = \begin{pmatrix} \mathbf{U}_{11} & \mathbf{U}_{12} \\ \mathbf{U}_{21} & \mathbf{U}_{22} \end{pmatrix}, \qquad \mathbf{G} = \begin{pmatrix} \mathbf{G}_{11} & \mathbf{G}_{12} \\ \mathbf{G}_{21} & \mathbf{G}_{22} \end{pmatrix},$$

with $\mathbf{U}_{11}: q \times q$, $\mathbf{G}_{11}: q \times q$. Then, for $q < p$,

$$\mathcal{L}(\mathbf{U}_{11}) = W^{-1}(\mathbf{G}_{11}, q, n - 2p + 2q).$$

Proof: The proof of this theorem is given in Complement 5.2.

Example (5.2.2): Suppose $q = 1$ so that $U_{11} \equiv u$ is a scalar. Let $g \equiv G_{11}$. Theorem (5.2.1) asserts that

$$\mathcal{L}(u) = W^{-1}(g, 1, n - 2p + 2),$$

or, using (5.2.1),

$$p(u) = \frac{c_0 g^{(n-2p)/2}}{u^{(n-2p+2)/2}} e^{-g/2u}, \qquad u > 0.$$

Substituting for c_0 gives

$$p(u) = \frac{g^{(n-2p)/2} e^{-g/2u}}{2^{(n-2p)/2}\Gamma\left(\frac{n-2p}{2}\right) u^{(n-2p+2)/2}}, \qquad u > 0, \qquad (5.2.3)$$

which is the density of the univariate inverted gamma distribution.

5.2.3 Moments of the Inverted Wishart Distribution

When using the inverted Wishart distribution it is often necessary to evaluate moments of the elements of the random matrix. The first- and second-order moments follow (Kaufman, 1967).

Theorem (5.2.2): Let $\mathbf{U} = (u_{ij})$ be a $p \times p$ random matrix with $\mathbf{U} > 0$. Then, if $\mathcal{L}(\mathbf{U}) = W^{-1}(\mathbf{G}, p, n)$, its first- and second-order moments are given by

$$E(\mathbf{U}) = \frac{\mathbf{G}}{n - 2p - 2}, \qquad n - 2p > 2, \qquad (5.2.4)$$

$$\operatorname{var}(u_{ii}) = \frac{2g_{ii}^2}{(n - 2p - 2)^2(n - 2p - 4)}, \qquad n - 2p > 4, \qquad (5.2.5)$$

$$\operatorname{var}(u_{ij}) = \frac{\left[g_{ii}g_{jj} + \dfrac{n - 2p}{n - 2p - 2} g_{ij}^2\right]}{(n - 2p - 1)(n - 2p - 2)(n - 2p - 4)}, \qquad (5.2.6)$$

for $n - 2p > 4$, and $i \neq j$, and

$$\operatorname{cov}(u_{ij}, u_{kl}) = \frac{\left[\dfrac{2}{n - 2p - 2} g_{ij}g_{kl} + g_{ik}g_{jl} + g_{il}g_{kj}\right]}{(n - 2p - 1)(n - 2p - 2)(n - 2p - 4)} \qquad (5.2.7)$$

for $n - 2p > 4$ (for all i, j, k, l), and $\mathbf{G} \equiv (g_{ij})$.

[*Remark:* If $\mathbf{U}$ were a parameter assigned the prior distribution in (5.2.1), specifying $E(\mathbf{U})$ and n would be sufficient to define the complete distribution. See also the remark following Theorem 5.1.8.]

5.2.4 Transformations of Inverted Wishart Matrices

Theorem (5.2.3): Let $\mathbf{V}: p \times p$ follow the distribution

$$\mathcal{L}(\mathbf{V}) = W^{-1}(\mathbf{G}, p, n),\ 2p < n,\ \mathbf{G} > 0.$$

Let $\mathbf{W} = \mathbf{AVA}'$, where $\mathbf{A}$: $q \times p$, $q \leqslant p$. Then

$$\mathcal{L}(\mathbf{W}) = W^{-1}(\mathbf{AGA}', q, n - 2p + 2q),\ r(\mathbf{A}) = q.$$

Proof: First take $q = p$, $r(A) = p$. Then, since

$$\mathcal{L}(\mathbf{W}) = W^{-1}(\mathbf{AGA}', q, n - 2p + 2q),\ r(\mathbf{A}) = q.$$

for any $\mathbf{B}: p \times p$, $r(\mathbf{B}) = p$, and $\mathcal{L}(\mathbf{BV}^{-1}\mathbf{B}') = W(\mathbf{BG}^{-1}\mathbf{B}', p, n - p - 1)$, so that $\mathcal{L}[(\mathbf{BV}^{-1}\mathbf{B}')^{-1}] = W^{-1}[(\mathbf{BG}^{-1}\mathbf{B}')^{-1}, p, n]$, by Theorem (5.1.7). Now take $\mathbf{B}' = \mathbf{A}^{-1}$.

For the case of $q < p$, augment the $\mathbf{A}$ matrix with $(p - q)$ rows and apply the result just proven. Taking the $(q \times q)$ marginal proves the theorem.

5.3 THE NONCENTRAL WISHART DISTRIBUTION

The noncentral Wishart distribution is quite complicated in form in even the simplest cases. However, it has come into increasing use in econometrics to provide a base of interpretation for distributional problems that arise in the study of simultaneous equation systems (see, for instance, Kabe, 1964).

5.3.1 Density

Let $\mathbf{X}_1, \ldots, \mathbf{X}_N$ be independent p-variate vectors with $\mathcal{L}(\mathbf{X}_\alpha) = N(\boldsymbol{\theta}_\alpha, \boldsymbol{\Sigma})$; that is, each vector observation has a different mean vector. Define

$$\boldsymbol{\Theta} = \sum_{\alpha=1}^{N} (\boldsymbol{\theta}_\alpha - \bar{\boldsymbol{\theta}})(\boldsymbol{\theta}_\alpha - \bar{\boldsymbol{\theta}})', \qquad \bar{\boldsymbol{\theta}} \equiv \frac{1}{N} \sum_{\alpha=1}^{N} \boldsymbol{\theta}_\alpha,$$

and let $r = \text{rank}\,(\boldsymbol{\Theta})$. Define

$$\mathbf{V} = \sum_{\alpha=1}^{N} (\mathbf{X}_\alpha - \bar{\mathbf{X}})(\mathbf{X}_\alpha - \bar{\mathbf{X}})', \qquad \bar{\mathbf{X}} \equiv \frac{1}{N} \sum_{\alpha=1}^{N} \mathbf{X}_\alpha.$$

Suppose not all $\boldsymbol{\theta}_\alpha = \mathbf{0}$. Then, if $n \equiv N - 1$, and $p \leq n$, $\mathbf{V}$ will be said to follow the noncentral Wishart distribution $W_r'(\mathbf{\Sigma},p,n)$.

Clearly if all $\boldsymbol{\theta}_\alpha = \mathbf{0}$, $r = 0$, so that $\mathcal{L}(\mathbf{V}) = W(\mathbf{\Sigma},p,N)$. Moreover, if all $\boldsymbol{\theta}_\alpha = \boldsymbol{\theta}$, $\mathbf{\Theta} = \mathbf{0}$, with the result that for $n \equiv N - 1$,

$$\mathcal{L}(\mathbf{V}) = W(\mathbf{\Sigma},p,n).$$

That is, in both cases, $r = 0$, and there is a reduction to the central Wishart distribution (of course when $\boldsymbol{\theta}_\alpha = \mathbf{0}$, $\mathbf{V}$ should be defined as $\mathbf{V} = \Sigma_{\alpha=1}^N \mathbf{X}_\alpha \mathbf{X}_\alpha'$). Thus, $W_0'(\mathbf{\Sigma},p,N) = W(\mathbf{\Sigma},p,N)$.

When $r > 0$, things become more complicated. Results were reported by Anderson and Girshick (1944), by Anderson in his doctoral dissertation (1945), and by Anderson (1946). The main results are given below.

Let c be defined as in (5.1.1). Then, if $r = 1$, the density of $\mathbf{V}$ is given by

$$p(\mathbf{V}) = \left\{\frac{c|\mathbf{V}|^{(n-p-1)/2}}{|\mathbf{\Sigma}|^{n/2}} \exp\left(-\tfrac{1}{2} \operatorname{tr} \mathbf{\Sigma}^{-1}\mathbf{V}\right)\right\} \cdot \left\{\frac{\exp\left(-\frac{1}{2} \operatorname{tr} \mathbf{\Sigma}^{-1}\mathbf{\Theta}\right)}{[\operatorname{tr} \mathbf{V\Theta}]^{(n-2)/4}} I_{n/2-1}[(\operatorname{tr} \mathbf{V\Theta})^{1/2}]\right\}, \tag{5.3.1}$$

where $I_k(z)$ is defined to be a Bessel function of purely imaginary argument (see Watson, 1944, p. 79). Thus, the density of $\mathbf{V}$ is proportional to the product of a central Wishart density and a Bessel function.

For $r > 1$ the density gets even more complicated (for $r = 2$ it is a central Wishart density multiplied by an infinite series of Bessel functions). A compact form for the general case (arbitrary r) was obtained by Herz (1955) and is given by

$$p(\mathbf{V}) = \frac{\Gamma_r\left(\frac{n}{2}\right) \exp\left[-\frac{1}{2} \operatorname{tr} \mathbf{\Sigma}^{-1}(\mathbf{V} + \mathbf{\Theta})\right]|\mathbf{V}|^{(n-p-1)/2}}{\Gamma_p\left(\frac{n}{2}\right)|2\mathbf{\Sigma}|^{n/2}} \cdot A_{(n-r-1)/2}^{(r)}\{-\tfrac{1}{4}\mathbf{\Sigma}^{-1}\mathbf{V}\mathbf{\Sigma}^{-1}\mathbf{\Theta}\}, \qquad \mathbf{V} > 0, \tag{5.3.2}$$

where $A_\gamma^{(r)}$ is the Bessel function of matrix argument of order γ, and

$$\Gamma_r(\delta) = \pi^{r(r-1)/4} \prod_{j=1}^{r} \Gamma\left(\delta - \frac{j-1}{2}\right).$$

Bessel functions of matrix arguments are discussed in Herz (1955), and in the related work of James (1961) and of Constantine (1963). They are defined and expressed in terms of zonal polynomials in Sect. 6.6.

5.3.2 An Application of the Noncentral Wishart Distribution to Simultaneous Equations

It was pointed out by Anderson (1946), p. 410, that the W' distribution may be useful in problems of estimating stochastic structural linear relations which arise in econometrics.

In 1964 Kabe found the W' distribution useful for studying a three simultaneous linear structural equation system studied earlier by Basmann (1963) (in 1963, Kabe studied a similar two equation system). The problem is developed as follows.

Define the equation system

$$\begin{aligned} y_{t1} &= \beta_{12}y_{t2} + \beta_{13}y_{t3} + \gamma_{11}x_{t1} + \gamma_{12}x_{t2} + \gamma_{13}x_{t3} + e_{t1}, \\ y_{t1} &= \beta_{22}y_{t2} + \beta_{23}y_{t3} + \gamma_{21}x_{t1} + \gamma_{22}x_{t2} + \gamma_{23}x_{t3} + e_{t2}, \\ y_{t1} &= \beta_{32}y_{t2} + \beta_{33}y_{t3} + \gamma_{31}x_{t1} + \gamma_{32}x_{t2} + \gamma_{33}x_{t3} + e_{t3}, \end{aligned} \tag{5.3.3}$$

where $t = 1,\ldots,N$; the y_{t1}, y_{t2}, y_{t3} are dependent variables determined by the model (called endogenous variables); the x_t's are determined external to the model (called exogenous variables); the e_t's are errors or disturbances; and the β's and γ's are parameters to be estimated. The equations in (5.3.3) are called structural because they relate the endogenous variables to the exogenous variables through the basic theoretical structure of the phenomena being described. If the system of equations is "solved" for the endogenous variables (so that there is only one in each equation), the resulting system is no longer structural, but is called the "reduced form," and it may then often be subjected to the methods of multivariate regression (Chapter 8). If it is of interest to make inferences about the original structural parameters (the betas and gammas) rather than the reduced form parameters, it will be seen that the noncentral Wishart distribution can be helpful. The treatment of the structural system (5.3.3) by Basmann and Kabe is given below.

For simplicity, define the matrices

$$\underset{(3\times3)}{\mathbf{B}} = \begin{pmatrix} 1 & -\beta_{12} & -\beta_{13} \\ 1 & -\beta_{22} & -\beta_{23} \\ 1 & -\beta_{32} & -\beta_{33} \end{pmatrix}, \qquad \underset{(3\times3)}{\mathbf{\Gamma}} = \begin{pmatrix} \gamma_{11} & \gamma_{12} & \gamma_{13} \\ \gamma_{21} & \gamma_{22} & \gamma_{23} \\ \gamma_{31} & \gamma_{32} & \gamma_{33} \end{pmatrix},$$

and the vectors

$$\underset{(3\times1)}{\mathbf{y}_t} = \begin{pmatrix} y_{t1} \\ y_{t2} \\ y_{t3} \end{pmatrix}, \qquad \underset{(3\times1)}{\mathbf{x}_t} = \begin{pmatrix} x_{t1} \\ x_{t2} \\ x_{t3} \end{pmatrix}, \qquad \underset{(3\times1)}{\mathbf{e}_t} = \begin{pmatrix} e_{t1} \\ e_{t2} \\ e_{t3} \end{pmatrix}, \qquad t = 1,\ldots,N.$$

Now the equation system (5.3.3) may be rewritten more compactly as

$$\mathbf{B}\mathbf{y}_t = \mathbf{\Gamma}\mathbf{x}_t + \mathbf{e}_t, \qquad t = 1,\ldots,N. \tag{5.3.4}$$

Assuming $|\mathbf{B}| \neq 0$, solve for the endogenous vector $\mathbf{y}_t$, and write the system as

$$\mathbf{y}_t = \mathbf{\Pi}\mathbf{x}_t + \mathbf{v}_t, \qquad t = 1,\ldots,N, \tag{5.3.5}$$

where $\mathbf{\Pi}$ is 3×3, $\mathbf{\Pi} = \mathbf{B}^{-1}\mathbf{\Gamma}$, and $\mathbf{v}_t = \mathbf{B}^{-1}\mathbf{e}_t$. (5.3.5) is the "reduced form" of the model. Suppose $\mathbf{v}_t$ is Normally distributed.

Let $\mathbf{P} = (p_{ij})$, $i, j = 1, 2, 3$ be the least squares (maximum likelihood) estimator of $\mathbf{\Pi}$, which has the property that the nine variates follow the joint multivariate Normal distribution

$$\mathcal{L}(\mathbf{P}) = N(\mathbf{\Pi}, \mathbf{I}_3 \otimes \mathbf{I}_3),$$

where $\mathbf{I}_3$ is the identity matrix of order 3 (least-squares estimation is discussed in Chapter 8).

Suppose it is of interest to estimate the structural parameters of the first equation in (5.3.3), namely $(\beta_{12},\beta_{13},\gamma_{11},\gamma_{12},\gamma_{13})$. It may be checked that in general, all five parameters cannot be estimated (this is called an *identification problem*). However, by constraining the problem, the remaining coefficients become estimable. For example, suppose a priori that $\gamma_{11} = \gamma_{12} = \gamma_{13} = 0$. Then β_{12} and β_{13} become estimable; this is the case treated below.

Define $\mathbf{V} = \mathbf{P}\mathbf{P}'$, where $\mathbf{V} = (v_{ij})$ is a 3×3 positive definite matrix. Partition $\mathbf{V}$ as

$$\mathbf{V} = \left(\begin{array}{c|cc} v_{11} & v_{12} & v_{13} \\ \hline v_{21} & v_{22} & v_{23} \\ v_{31} & v_{32} & v_{33} \end{array}\right) = \begin{pmatrix} \mathbf{V}_{11} & \mathbf{V}_{12} \\ \mathbf{V}_{21} & \mathbf{V}_{22} \end{pmatrix}.$$

It was shown by Kabe (1964) that it is possible to represent the estimators[3] of β_{12} and β_{13} in the form

$$\begin{pmatrix} \hat{\beta}_{12} \\ \hat{\beta}_{13} \end{pmatrix} = \mathbf{V}_{22}^{-1}\mathbf{V}_{12}. \tag{5.3.6}$$

However, from Section 5.3.1, it is seen that

$$\mathcal{L}(\mathbf{V}) = W_r'(\mathbf{I}, 3, N-1),$$

where r denotes the rank of $\mathbf{\Pi}$. Thus, to obtain the distributions of $\hat{\beta}_{12}$ and $\hat{\beta}_{13}$, it is necessary to find the distribution of $\mathbf{V}_{22}^{-1}\mathbf{V}_{12}$. The final density was given by Basmann (1963), and then verified by Kabe (1964) by means of the properties of the W' distribution, using $r = 1$ and the

[3] There are several possible alternatives that have been proposed for estimating structural parameters. The one used by Basmann and Kabe involves minimizing the generalized variance (see Section 5.1.6) of the errors in the reduced form equations, subject to constraints on the parameters (see, for instance, Goldberger, 1964), for elaboration of this technique—called "least residual variance estimation."

assumption that all $\pi_{ij} = 0$ except π_{33}. For all $\hat{\beta}_{12}$ and $\hat{\beta}_{13}$, $-\infty < \hat{\beta}_{12}$, $\hat{\beta}_{13} < \infty$, the joint density is given by

$$f(\hat{\beta}_{12},\hat{\beta}_{13}) = \frac{e^{-\pi_{33}^2/2}}{\pi(1+\hat{\beta}_{12}^2+\hat{\beta}_{13}^2)^2}\,{}_1F_1\left\{2;\tfrac{3}{2};\frac{\pi_{33}^2(1+\hat{\beta}_{12}^2)}{2(1+\hat{\beta}_{12}^2+\hat{\beta}_{13}^2)}\right\},$$

where ${}_1F_1\{\cdot\}$ denotes the confluent hypergeometric function (see, for instance, Abromovitz and Stegun, 1964), and $\mathbf{\Pi} \equiv (\pi_{ij})$. Equivalently,

$$f(\hat{\beta}_{12},\hat{\beta}_{13}) = \frac{e^{-\pi_{33}^2/2}}{\pi}\sum_{\alpha=0}^{\infty}\frac{\Gamma(2+\alpha)\Gamma(\frac{3}{2})\pi_{33}^{2\alpha}(1+\hat{\beta}_{12}^2)^{\alpha}}{\Gamma(\frac{3}{2}+\alpha)2^{\alpha}\alpha!(1+\hat{\beta}_{12}^2+\hat{\beta}_{13}^2)^{\alpha+2}}. \tag{5.3.7}$$

The marginal density of $\hat{\beta}_{13}$ is given by

$$f(\hat{\beta}_{13}) = \frac{e^{-\pi_{33}^2/2}}{\pi}\sum_{\alpha_1=0}^{\infty}\sum_{\alpha_2=0}^{\infty}\frac{\Gamma(\frac{3}{2})\Gamma(\frac{3}{2}+\alpha_1)\Gamma(\frac{1}{2}+\alpha_2)\pi_{33}^{2(\alpha_1+\alpha_2)}}{\Gamma(\frac{3}{2}+\alpha_1+\alpha_2)2^{\alpha_1+\alpha_2}\alpha_1!\alpha_2!(1+\hat{\beta}_{13}^2)^{\alpha_1+3/2}}. \tag{5.3.8}$$

Kabe also treated the case of $r = 2$. Results obtained for both $r = 1$ and $r = 2$ for this three equation system are generally of the same type and form as those Kabe obtained in 1963 for the two equation case. Thus, it will not be surprising to find that the general noncentral Wishart distribution of matrix argument will be useful for making inferences about the structural parameters of more general systems of simultaneous equations.

As an illustration of the problem treated above consider the system

$$\begin{aligned} y_{t1} &= \beta_{12}y_{t2} + \beta_{13}y_{t3} + e_{t1}, \\ y_{t1} &= \beta_{21}y_{t2} + e_{t2}, \\ y_{t1} &= \beta_{32}y_{t2} + \beta_{33}y_{t3} + \gamma_{33}x_{t3} + e_{t3}, \end{aligned}$$

for $t = 1, \ldots, N$. Here, the rank of the reduced form system is one, since the corresponding Γ matrix has only one non-zero element. Moreover, the coefficients in the first equation are estimable, as in (5.3.6). The joint density of the estimators (under Normal errors, $\mathbf{e}_t$) is given in (5.3.7).

REFERENCES

Abromovitz and Stegun (1964).
Anderson (1945).
Anderson (1946).
Anderson (1958).
Anderson and Girshick (1944).
Basmann (1963).
Constantine (1963).
Goldberger (1964).
Henderson, Hind, and Brown (1961).
Herz (1955).
James (1961).
Kabe (1963).
Kabe (1964).
Kaufman (1967, 1969).
Press (1967b).
Raiffa and Schlaifer (1961).

Tiao and Zellner (1964a).
Tiao and Zellner (1964b).
Watson (1944).
Wishart (1928).

COMPLEMENT 5.1

Theorem (5.1.2): If $\mathbf{X}_1,\ldots,\mathbf{X}_N$ are mutually independent with $\mathcal{L}(\mathbf{X}_\alpha) = N(\boldsymbol{\theta},\boldsymbol{\Sigma})$, $\boldsymbol{\Sigma} > 0$, then $\mathcal{L}(\mathbf{V}) = W(\boldsymbol{\Sigma},p,n)$, $n = N - 1$.

Proof: See Section 5.1 for definitions.

It is sufficient to reduce the $\mathbf{V}$ statistic of (5.1.5) to the one given in Theorem (5.1.1). Define a set of p-vectors, $\mathbf{Y}_1, \mathbf{Y}_2,\ldots,\mathbf{Y}_n$, $n = N - 1$, with the properties

(1) $\mathbf{Y}_1,\ldots,\mathbf{Y}_n$ are mutually independent,
(2) $\mathcal{L}(\mathbf{Y}_\alpha) = N(\mathbf{0},\boldsymbol{\Sigma})$, $\alpha = 1,\ldots,n$,
(3) $\mathbf{V} = \Sigma_{\alpha=1}^n \mathbf{Y}_\alpha \mathbf{Y}_\alpha'$.

Transform the $\mathbf{X}_\alpha$ vectors using any orthogonal matrix whose last row elements all equal $N^{-1/2}$ [one such matrix is an N-dimensional permuted version of the Helmert matrix defined in (2.2.3)]. That is, let $\boldsymbol{\Gamma} = (\gamma_{ij})$, i, $j = 1,\ldots,N$ denote any $N \times N$ orthogonal matrix which has the property that $\gamma_{Nj} = N^{-1/2}$, $j = 1,\ldots,N$ (the elements of the last row are all equal).

Define

$$\mathbf{Y}_i = \sum_{j=1}^{N} \gamma_{ij}\mathbf{X}_j, \qquad i = 1,\ldots,N. \tag{C5.1.1}$$

Since $\mathcal{L}(\mathbf{X}_j) = N(\boldsymbol{\theta},\boldsymbol{\Sigma})$, all j, $\mathcal{L}(\gamma_{ij}\mathbf{X}_j) = N(\gamma_{ij}\boldsymbol{\theta},\gamma_{ij}^2\boldsymbol{\Sigma})$; summing over j gives

$$\mathcal{L}(\mathbf{Y}_i) = N(\boldsymbol{\phi}_i,a\boldsymbol{\Sigma}),$$

where $\boldsymbol{\phi}_i \equiv (\Sigma_{j=1}^N \gamma_{ij})\boldsymbol{\theta}$, and $a = \Sigma_{j=1}^N \gamma_{ij}^2$. By orthogonality of $\boldsymbol{\Gamma}$, $a = 1$. Moreover, since $N^{1/2}\gamma_{Nj} = 1$ by definition, $\boldsymbol{\phi}_i = \Sigma_{j=1}^N \gamma_{ij}\gamma_{Nj}N^{1/2}\boldsymbol{\theta}$. By orthogonality of $\boldsymbol{\Gamma}$, for $i \neq N$, $\Sigma_{j=1}^N \gamma_{ij}\gamma_{Nj} = 0$. Hence, $\boldsymbol{\phi}_i = \mathbf{0}$, $i = 1,\ldots,n$, which establishes Property 2.

For $i \neq k$, $i \neq N$, and $k \neq N$,

$$E\mathbf{Y}_i\mathbf{Y}_k' = E\left(\sum_{j=1}^{N} \gamma_{ij}\mathbf{X}_j\right)\left(\sum_{\alpha=1}^{N} \gamma_{k\alpha}\mathbf{X}_\alpha'\right) = \sum_{j=1}^{N}\sum_{\alpha=1}^{N} \gamma_{ij}\gamma_{k\alpha}E\mathbf{X}_j\mathbf{X}_\alpha'.$$

For $j = \alpha$, $\Sigma_{j=1}^N\Sigma_{\alpha=1}^N \gamma_{ij}\gamma_{k\alpha} = \Sigma_{j=1}^N \gamma_{ij}\gamma_{kj} = 0$. Hence,

$$EY_iY_k' = \Big(\sum_{j\neq\alpha}\sum \gamma_{ij}\gamma_{k\alpha}\Big)\,\theta\theta' = \Big(\sum_{j=1}^{N}\sum_{\alpha=1}^{N} \gamma_{ij}\gamma_{k\alpha} - \sum_{j=1}^{N}\gamma_{ij}\gamma_{kj}\Big)\,\theta\theta'$$

$$= \Big[N\Big(\sum_{j=1}^{N}\gamma_{ij}\gamma_{Nj}\Big)\Big(\sum_{\alpha=1}^{N}\gamma_{k\alpha}\gamma_{N\alpha}\Big) - \Big(\sum_{j=1}^{N}\gamma_{ij}\gamma_{kj}\Big)\Big]\,\theta\theta' = \mathbf{0},$$

since each term within parentheses vanishes by orthogonality. This establishes Property (1).

Finally, from (C5.1.1),

$$\sum_{i=1}^{N} Y_iY_i' = \sum_{i=1}^{N}\Big(\sum_{j=1}^{N}\gamma_{ij}X_j\Big)\Big(\sum_{\alpha=1}^{N}\gamma_{i\alpha}X_\alpha'\Big)$$

$$= \sum_{j=1}^{N}\sum_{\alpha=1}^{N}\Big(\sum_{i=1}^{N}\gamma_{ij}\gamma_{i\alpha}\Big)X_jX_\alpha'$$

$$= \sum_{j=1}^{N}\Big(\sum_{i=1}^{N}\gamma_{ij}^2\Big)X_jX_j' + \sum_{\substack{j=1\\ j\neq\alpha}}^{N}\sum_{\alpha=1}^{N}\Big(\sum_{i=1}^{N}\gamma_{ij}\gamma_{i\alpha}\Big)X_jX_\alpha'.$$

Since the terms in parentheses must be 1 and 0, respectively,

$$\sum_{i=1}^{N} Y_iY_i' = \sum_{i=1}^{N} X_iX_i'.$$

Next note from (C5.1.1) that $Y_N = \sqrt{N}\,\bar{X}$. Hence,

$$V = \sum_{\alpha=1}^{N} X_\alpha X_\alpha' - N\bar{X}\bar{X}' = \sum_{\alpha=1}^{N} Y_\alpha Y_\alpha' - Y_NY_N' = \sum_{\alpha=1}^{N-1} Y_\alpha Y_\alpha',$$

which completes the proof of the theorem.

COMPLEMENT 5.2

Theorem (5.2.1): If $\mathcal{L}(U) = W^{-1}(G,p,n)$, and U is partitioned as $U = (U_{ij})$, $i, j = 1, 2$, then $\mathcal{L}(U_{11}) = W^{-1}(G_{11}, q, n - 2p + 2q)$.

Proof: See Section 5.2 for definitions.

Use the results (2.5.5), and (2.6.3)–(2.6.6), for the inverse and determinant of a positive definite partitioned matrix in terms of submatrices:

$$U^{-1} = \begin{pmatrix} U_{11}^{-1} + U_{11}^{-1}U_{12}U_{22\cdot1}^{-1}U_{21}U_{11}^{-1} & -U_{11}^{-1}U_{12}U_{22\cdot1}^{-1} \\ -U_{22\cdot1}^{-1}U_{21}U_{11}^{-1} & U_{22\cdot1}^{-1} \end{pmatrix},$$

and

$$|\mathbf{U}| = |\mathbf{U}_{11}|\,|\mathbf{U}_{22\cdot 1}|,$$

where

$$\mathbf{U}_{22\cdot 1} = \mathbf{U}_{22} - \mathbf{U}_{21}\mathbf{U}_{11}^{-1}\mathbf{U}_{12}. \tag{C5.2.1}$$

$\mathbf{U}^{-1}$ may be written

$$\mathbf{U}^{-1} = \begin{pmatrix} \mathbf{U}_{11}^{-1} & \mathbf{0} \\ \mathbf{0} & \mathbf{0} \end{pmatrix} + \mathbf{W},$$

where

$$\mathbf{W} = \begin{pmatrix} \mathbf{U}_{11}^{-1}\mathbf{U}_{12}\mathbf{U}_{22\cdot 1}^{-1}\mathbf{U}_{21}\mathbf{U}_{11}^{-1} & -\mathbf{U}_{11}^{-1}\mathbf{U}_{12}\mathbf{U}_{22\cdot 1}^{-1} \\ -\mathbf{U}_{22\cdot 1}^{-1}\mathbf{U}_{21}\mathbf{U}_{11}^{-1} & \mathbf{U}_{22\cdot 1}^{-1} \end{pmatrix},$$

so that if $\mathbf{X} = \mathbf{U}_{11}^{-1}\mathbf{U}_{12}$, and $\mathbf{Y} = \mathbf{U}_{22\cdot 1}$,

$$\mathbf{W} = \begin{pmatrix} \mathbf{X}\mathbf{Y}^{-1}\mathbf{X}' & -\mathbf{X}\mathbf{Y}^{-1} \\ -\mathbf{Y}^{-1}\mathbf{X}' & \mathbf{Y}^{-1} \end{pmatrix}.$$

From (5.2.2) and (C5.2.1),

$$p(\mathbf{U}_{11},\mathbf{U}_{12},\mathbf{U}_{22}) \propto |\mathbf{U}_{11}|^{-n/2}|\mathbf{Y}|^{-n/2} \exp\left(-\tfrac{1}{2} \operatorname{tr} \mathbf{U}_{11}^{-1}\mathbf{G}_{11} - \tfrac{1}{2} \operatorname{tr} \mathbf{W}\mathbf{G}\right).$$

Next, establish that

$$J(\mathbf{U}_{12},\mathbf{U}_{22} \to \mathbf{X},\mathbf{Y}) = |\mathbf{U}_{11}|^{p-q}.$$

Arrange the variables in a table whose entries are partial derivatives, as shown.

	$\mathbf{U}_{12}$	$\mathbf{U}_{22}$
$\mathbf{X}$	$\mathbf{A}$	$\mathbf{0}$
$\mathbf{Y}$	—	$\mathbf{I}$

The 1-2 place in the table is zero because $\mathbf{X}$ does not depend upon $\mathbf{U}_{22}$. Also, an $\mathbf{I}$ appears in the 2-2 place because, as seen in (C5.2.1), $(\mathbf{U}_{22\cdot 1} - \mathbf{U}_{22})$ does not depend on $\mathbf{U}_{22}$. There is a dash in the 2-1 place because the value does not matter (since in the Jacobian it will be multiplied by zero). Thus, $J(\mathbf{X},\mathbf{Y} \to \mathbf{U}_{12},\mathbf{U}_{22}) = J(\mathbf{X} \to \mathbf{U}_{12}) \equiv |\mathbf{A}|$. But from (2.15.2), $J(\mathbf{X} \to \mathbf{U}_{12}) = |\mathbf{U}_{11}^{-1}|^{p-q}$. Therefore, $J(\mathbf{U}_{12},\mathbf{U}_{22} \to \mathbf{X},\mathbf{Y}) = [J(\mathbf{X} \to \mathbf{U}_{12})]^{-1}$. Since $p(\mathbf{U}_{11},\mathbf{X},\mathbf{Y}) = p(\mathbf{U}_{11},\mathbf{U}_{12},\mathbf{U}_{22})J(\mathbf{U}_{12},\mathbf{U}_{22} \to \mathbf{X},\mathbf{Y})$, integrating $p(\mathbf{U}_{11},\mathbf{X},\mathbf{Y})$ with respect to $\mathbf{X}$, $\mathbf{Y}$ gives the result.

[*Remark:* Note that since $p(\mathbf{U}_{11},\mathbf{X},\mathbf{Y})$ factors into the product of a function of $\mathbf{U}_{11}$ and a function of $(\mathbf{X},\mathbf{Y})$, $\mathbf{U}_{11}$ is independent of $\mathbf{U}_{22\cdot 1}$.]

EXERCISES

5.1 Let $p_j(t)$ denote the price of security j at time t, and let $x_j(t) \equiv \log p_j(t) - \log p_j(t-1)$. Suppose there are q securities in the same industry so that their price changes are correlated in the same way. Let $\mathbf{x}(t) \equiv [x_1(t),\ldots,x_q(t)]'$.

Suppose the prices at time zero are fixed and known, and N independent observations of $\mathbf{x}(t)$ are taken at times $t = 1,\ldots,N$ yielding the vectors $\mathbf{x}(1),\ldots,\mathbf{x}(N)$. Assume

$$\text{var}\,[\mathbf{x}(t)] \equiv \mathbf{\Sigma} = 4\begin{pmatrix} 1 & & & \\ & \cdot & & \rho \\ & & \cdot & \\ & \rho & & \cdot & \\ & & & & 1 \end{pmatrix}, \qquad \rho > 0,$$

and that approximately, $\mathcal{L}[\mathbf{x}(t)] = N(\mathbf{0},\mathbf{\Sigma})$, for all $t = 1,\ldots,N$. Express the density of $\mathbf{V} = \Sigma_{t=1}^{N}[\mathbf{x}(t) - \bar{\mathbf{x}}][\mathbf{x}(t) - \bar{\mathbf{x}}]'$ where $N\bar{\mathbf{x}} = \Sigma_{t=1}^{N}\mathbf{x}(t)$, as an explicit function of ρ and the elements of $\mathbf{V}$.

[*Hint:* Use the expressions for $|\mathbf{\Sigma}|$ and $\mathbf{\Sigma}^{-1}$ developed in Chapter 2 for the intraclass correlation matrix.]

5.2 In Exercise 5.1, find the marginal density of $\mathbf{V}_{11}$, the upper left 2×2 submatrix of $\mathbf{V}$. Express the density as an explicit function of ρ and the elements of $\mathbf{V}_{11}$.

5.3 In Exercises 5.1 and 5.2, find the density of $\mathbf{V}_{11\cdot 2}$, and express it as an explicit function of ρ (recall that $\mathbf{V}_{11\cdot 2} \equiv \mathbf{V}_{11} - \mathbf{V}_{12}\mathbf{V}_{22}^{-1}\mathbf{V}_{21}$), and the elements of $\mathbf{V}$.

5.4 Let $\mathbf{V}$ be defined as in Exercise 5.1, and let

$$\mathbf{W} \equiv \sum_{t=1}^{M} [\mathbf{y}(t) - \bar{\mathbf{y}}][\mathbf{y}(t) - \bar{\mathbf{y}}]',$$

where $\mathbf{y}(t)$ is also $q \times 1$; $\mathbf{y}(t)$ is based upon q securities in another industry; $M\bar{\mathbf{y}} = \Sigma_1^M \mathbf{y}(t)$; and $\mathbf{y}(1),\ldots,\mathbf{y}(M)$ are mutually independent and independent of the $\mathbf{x}(t)$'s and follow the same distribution for all t. Define

$$\mathbf{A} = \begin{pmatrix} 1 & & & & & \\ & 2 & & & & \\ & & \cdot & 0 & & \\ & & & \cdot & & \\ & 0 & & & \cdot & \\ & & & & 2 & \\ & & & & & 1 \end{pmatrix}, \qquad \mathbf{B} = \begin{pmatrix} 1 & & & & & \\ & 3 & & & & \\ & & \cdot & 0 & & \\ & & & \cdot & & \\ & 0 & & & \cdot & \\ & & & & 3 & \\ & & & & & 1 \end{pmatrix},$$

and let $\mathbf{V}^* = \mathbf{AVA}'$, and $\mathbf{W}^* = \mathbf{BWB}'$. Evaluate the density of $\mathbf{U} = (\mathbf{V}^* + \mathbf{W}^*)$ as an explicit function of ρ, and the elements of $\mathbf{V}$ and $\mathbf{W}$.

5.5 Let $\mathbf{\Sigma} : p \times p$ denote the covariance matrix of the $p \times 1$ vector $\mathbf{X} = (x_i)$, $i = 1,\ldots,p$, where x_i denotes the location in a coordinate system of the ith fire out of the first p fires recorded in a city on a given day (assume there are always at least p recorded fires). Suppose that based upon many observations, it is well known that

$$\mathbf{\Sigma} = \begin{pmatrix} 3 & 1 & & & \\ 1 & 3 & \cdot & 0 & \\ & \cdot & \cdot & & \\ & & \cdot & \cdot & \cdot \\ & & & \cdot & \cdot \\ & 0 & \cdot & 3 & 1 \\ & & & 1 & 3 \end{pmatrix},$$

the tri-diagonal matrix discussed in Chapter 2. Let $\mathbf{X}(1),\ldots,\mathbf{X}(N)$ denote independent observations of $\mathbf{X}$ and let $\mathbf{V} = (v_{ij})$, where

$$\mathbf{V} = \sum_1^N [\mathbf{X}(t) - \bar{\mathbf{X}}][\mathbf{X}(t) - \bar{\mathbf{X}}]', \qquad \text{and} \qquad N\bar{\mathbf{X}} = \sum_1^N \mathbf{X}(t).$$

If $\mathcal{L}[\mathbf{X}(t)] = N(\mathbf{\theta},\mathbf{\Sigma})$, and $N = 10$, find

(a) $E(v_{11})$,

(b) $\operatorname{var}(v_{11})$,

(c) $P\{v_{11} \le 40\}$,

(d) the population generalized variance.

5.6 For Example (5.2.1) involving price changes for commodities during a postwar period compared with a prewar period, find the predictive density for a new observation given n previous observations, using a natural conjugate prior distribution for $\mathbf{\Sigma}$.

5.7 Suppose $\mathbf{V} : p \times p$ is a random symmetric matrix with $\mathcal{L}(\mathbf{V}) = W(\mathbf{\Sigma},p,n)$. Calculate the value of $\mathbf{\Sigma}$ which maximizes the density of $\mathbf{V}$, and show that it yields a maximum.

5.8 For the three equation simultaneous equation system given in (5.3.3), the density of the structural coefficient estimator of β_{13} is given in (5.3.8). Discuss the moments of $\hat{\beta}_{13}$. What is the "standard error"? [*Hint:* See Section 6.2.1.]

5.9 Show that the characteristic function of the distribution of $\mathbf{V} : p \times p$, where $\mathcal{L}(\mathbf{V}) = W(\Sigma,p,n)$, $\Sigma > 0$, $p \le n$, is given by:

$$\phi(\mathbf{T}) = |\mathbf{I} - 2i\Sigma\mathbf{T}|^{-n/2}, \qquad \text{for } \mathbf{T} : p \times p, \text{ and } i = \sqrt{-1}.$$

5.10 Show that if $\mathcal{L}(\mathbf{V}) = W(\Sigma,p,n)$, $\Sigma > 0$, $p \le n$, and if $\mathbf{U} = \mathbf{V}^{-1}$, then $\mathcal{L}(\mathbf{U}) = W^{-1}(\Sigma^{-1},p,n+p+1)$.

5.11 Show that if $\mathcal{L}(\mathbf{V}_1) = W(\Sigma_1,p,n)$, $p \leq n$, $\Sigma_1 > 0$, and $\mathcal{L}(\mathbf{V}_2) = W(\Sigma_2,p,n)$, $\Sigma_2 > 0$, and $\mathbf{V}_1$ and $\mathbf{V}_2$ are independent, $\mathcal{L}(\mathbf{V}_1 + \mathbf{V}_2)$ is Wishart if and only if $\Sigma_1 = \Sigma_2$.

5.12 (a) Suppose $\mathcal{L}(\mathbf{V}) = W(\Sigma,p,n)$, $p \leq n$, $\Sigma > 0$. Find the kth moment of $|\mathbf{V}|$.

(b) Suppose $\mathcal{L}(\mathbf{V}) = W^{-1}(\Sigma,p,n)$, $p < 2n$, $\Sigma > 0$. Find the kth moment of $|\mathbf{V}|$.

6

OTHER CONTINUOUS MULTIVARIATE DISTRIBUTIONS

This chapter summarizes the properties and applications of the remaining major continuous distributions of interest in the models of multivariate analysis to be discussed in Part II. The Normal and Wishart distributions have already been discussed (see Chapters 3 and 5).

Hotelling's T^2-distribution (a multivariate extension of the square of a Student t-variate) is fundamental to classical hypothesis testing problems including tests of the equality of mean vectors of Normal populations with unknown variances and covariances. This distribution and its extension, the T_0^2-distribution which arises in multivariate analysis of variance (Chapter 8), are discussed in Section 6.1.

Section 6.2 outlines the distributional properties of various extensions to the multivariate case of the Student t-distribution. Such distributions arise, for example, as the principal distributions in Bayesian analyses of multivariate regression.

Beta and related distributions (and their multivariate analogues) are discussed in Section 6.3. Such distributions arise as natural conjugates to multinomial families of data; they arise in a fundamental way in classical tests for serial correlation in regression; and they are needed in regression and analysis of variance problems for classical tests of hypotheses about regression coefficients.

Section 6.4 discusses the multivariate logarithmic Normal distribution that is often useful in multivariate regression problems.

Section 6.5 introduces the univariate family of stable distributions and proposes a new class of multivariate stable distributions. The multivariate stable distributions will be used in Chapter 15 to structure an approach to risk minimization in portfolio analysis based upon a stable distribution hypothesis for the variation of prices of assets in an investment portfolio.

6.1 HOTELLING'S T^2-DISTRIBUTION

6.1.1 Density of T^2

Let $\mathbf{x}_1, \ldots, \mathbf{x}_N$ be mutually independent, p-variate vector (column) observations each of which follows the law $\mathcal{L}(\mathbf{x}_k) = N(\boldsymbol{\theta}, \boldsymbol{\Sigma})$, $k = 1, \ldots, N$, for $\boldsymbol{\Sigma} > 0$. Denote the sample mean vector and sample covariance matrix by $\bar{\mathbf{x}}$ and $\mathbf{S}$, respectively, where

$$\bar{\mathbf{x}} = \frac{1}{N} \sum_{1}^{N} \mathbf{x}_k, \qquad \mathbf{S} = \frac{1}{N-1} \sum_{1}^{N} (\mathbf{x}_k - \bar{\mathbf{x}})(\mathbf{x}_k - \bar{\mathbf{x}})'.$$

Suppose $\boldsymbol{\theta}_0 : p \times 1$ is a known vector. Then, Hotelling's T^2-statistic is defined as

$$T^2 = N(\bar{\mathbf{x}} - \boldsymbol{\theta}_0)' \mathbf{S}^{-1} (\bar{\mathbf{x}} - \boldsymbol{\theta}_0). \tag{6.1.1}$$

In the univariate case ($p = 1$), (6.1.1) would reduce to

$$T^2 = \frac{N(\bar{x} - \theta_0)^2}{S},$$

which is the square of a Student t-variate. The distribution of the T^2-statistic for $\boldsymbol{\theta} = \boldsymbol{\theta}_0$ known, was derived by Hotelling (1931). The distribution for arbitrary $\boldsymbol{\theta}$ was derived by Bose and Roy (1938), and by Hsu (1938). A compact proof for the general case was given by Bowker (1960). The distribution of T^2 is given in the following theorem.

Theorem (6.1.1): Let

$$U = \left(\frac{T^2}{N-1} \right) \left(\frac{N-p}{p} \right),$$

where T^2 is defined in (6.1.1). Then

$$\mathcal{L}(U) = F_{p,N-p}(\lambda), \tag{6.1.2}$$

where

$$\lambda = N(\boldsymbol{\theta} - \boldsymbol{\theta}_0)' \boldsymbol{\Sigma}^{-1} (\boldsymbol{\theta} - \boldsymbol{\theta}_0), \tag{6.1.3}$$

and $F_{p,N-p}(\lambda)$ denotes an F-distribution with p and $N - p$ degrees of freedom, and noncentrality parameter λ.

Recall that if $(X_1, \ldots, X_m;\ Y_1, \ldots Y_n)$ are all independent, and if $\mathcal{L}(X_j) = N(\theta_j, 1)$, $\mathcal{L}(Y_j) = N(0,1)$, and $\lambda = \Sigma_1^m \theta_j^2$,

$$\mathcal{L}\left(\frac{n}{m} \frac{\Sigma_1^m X_j^2}{\Sigma_1^n Y_j^2} \right) = F_{m,n}(\lambda).$$

Thus Theorem (6.1.1) asserts that T^2 is proportional to an F-variate. Note from (6.1.3) that if $\theta = \theta_0$, the distribution of U is a central F-distribution.

6.1.2 Invariance Property of T^2

It follows from (6.1.1) that T^2 is invariant with respect to linear transformations. Thus, if the units of the x_k's are changed, the value of T^2 is unchanged. To see this let

$$\mathbf{y}_k = \mathbf{A}\mathbf{x}_k + \mathbf{b}, \qquad k = 1,\ldots,N,$$

for some nonsingular $\mathbf{A}: p \times p$, $\mathbf{b}: p \times 1$. Then

$$\mathcal{L}(\mathbf{y}_k) = N(\mathbf{A}\boldsymbol{\theta} + \mathbf{b}, \mathbf{A}\boldsymbol{\Sigma}\mathbf{A}').$$

Hence, if Hotelling's T^2-statistic based upon the $\mathbf{y}_k$'s is denoted by $\bar{T}^2$,

$$\begin{aligned}\bar{T}^2 &= N[\bar{\mathbf{y}} - (\mathbf{A}\boldsymbol{\theta} + \mathbf{b})]'(\mathbf{A}\mathbf{S}\mathbf{A}')^{-1}[\bar{\mathbf{y}} - (\mathbf{A}\boldsymbol{\theta} + \mathbf{b})] \\ &= N[\mathbf{A}\bar{\mathbf{x}} + \mathbf{b} - \mathbf{A}\boldsymbol{\theta} - \mathbf{b}]'(\mathbf{A}')^{-1}\mathbf{S}^{-1}(\mathbf{A}^{-1})[\mathbf{A}\bar{\mathbf{x}} + \mathbf{b} - \mathbf{A}\boldsymbol{\theta} - \mathbf{b}] \\ &= N(\bar{\mathbf{x}} - \boldsymbol{\theta})'\mathbf{S}^{-1}(\bar{\mathbf{x}} - \boldsymbol{\theta}) = T^2.\end{aligned}$$

Example (6.1.1): Recall the data given in Example (5.1.2) for changes in sales of two different items in the same department store in ten different weeks. It was found there that $\bar{\mathbf{X}}' = (1.4,1.5)$, and

$$\mathbf{V} \equiv \sum_1^{10} (\mathbf{X}_\alpha - \bar{\mathbf{X}})(\mathbf{X}_\alpha - \bar{\mathbf{X}})' = \begin{pmatrix} 14.4 & -4 \\ -4 & 6.5 \end{pmatrix}.$$

Hence,

$$\mathbf{S} \equiv \frac{\mathbf{V}}{N-1} = \frac{1}{9}\mathbf{V} = \begin{pmatrix} 1.600 & -.444 \\ -.444 & .722 \end{pmatrix}.$$

Substituting into (6.1.1) gives for Hotellings' T^2-statistic

$$T^2 = 10(1.4 - \theta_1,\ 1.5 - \theta_2)\mathbf{S}^{-1}\begin{pmatrix} 1.4 - \theta_1 \\ 1.5 - \theta_2 \end{pmatrix},$$

where $\boldsymbol{\theta}' \equiv (\theta_1,\theta_2)$. Inverting $\mathbf{S}$ gives

$$\mathbf{S}^{-1} = \begin{pmatrix} .754 & .463 \\ .463 & 1.670 \end{pmatrix}.$$

If, in fact, $\mathcal{L}(\mathbf{X}) = N(\boldsymbol{\theta},\boldsymbol{\Sigma})$, where

$$\boldsymbol{\theta} = \begin{pmatrix} 2.0 \\ 1.0 \end{pmatrix}, \qquad \boldsymbol{\Sigma} = \begin{pmatrix} \frac{1}{2} & -\frac{1}{2} \\ -\frac{1}{2} & \frac{3}{4} \end{pmatrix},$$

$$T^2 = 10(-.6,.5)\begin{pmatrix} .753 & .463 \\ .463 & 1.670 \end{pmatrix}\begin{pmatrix} -.6 \\ .5 \end{pmatrix} = 4.111.$$

To determine whether or not $T^2 = 4.111$ is significant at a given significance level, use Theorem (6.1.1).

First compute U:

$$U = \left(\frac{4.111}{10-1}\right)\left(\frac{10-2}{2}\right) = 1.827.$$

Then

$$\mathcal{L}(U) = F_{2,8}(\lambda),$$

where, for $\boldsymbol{\theta}_0' = (2.0, 1.0) = \boldsymbol{\theta}'$, $\lambda = 0$. Hence,

$$\mathcal{L}(U) = F_{2,8}(0).$$

Referring to a table of fractiles of the central F-distribution shows that at the 5 percent level of significance (for example), $U = 4.46$. Thus, $U = 1.827$ (or equivalently, $T^2 = 4.111$) is not significant at the 5 percent level.

6.1.3 The T_0^2-Distribution

The T_0^2-distribution was developed by Hotelling during World War II as a measure of multivariate dispersion (see Hotelling, 1944) in connection with a quality control problem associated with air testing of sample bombsights. The T_0^2-statistic is a generalization of the T^2-statistic to many samples of data, and may be used as a generalization to the multivariate case of the statistics used in the univariate analysis of variance (ANOVA) to test hypotheses(see Chapter 8). That is, in the ANOVA, hypothesis testing for main effects is typically based upon a ratio of a sum of squares (quadratic form) under the hypothesis to a sum of squares of all the residuals. If $\mathbf{H}$ and $\mathbf{G}$ denote the multivariate analogues of these sums of squares (in the multivariate ANOVA), then T_0^2 is defined as

$$T_0^2 = \operatorname{tr}(\mathbf{H}\mathbf{G}^{-1}).$$

Thus, for example, it will be seen that by using this trace criterion, the T_0^2-distribution is useful for testing whether several multivariate normal populations have the same mean vectors, given that they have the same covariance matrices.

The details of the quality control problem analyzed by Hotelling using T_0^2 are discussed in Chapter 3 of Eisenhart, Hastay, and Wallis (1947). Some of the underlying theory is presented in Hotelling (1951).

Definition of T_0^2: Let $(\mathbf{y}_1, \ldots, \mathbf{y}_N)$ denote a sample of p-variate independent observations, each of which follows the law $N(\boldsymbol{\theta}, \boldsymbol{\Sigma})$, $\boldsymbol{\Sigma} > 0$. Let $\mathbf{G}$ denote the sample covariance matrix

$$\mathbf{G} = \frac{1}{N-1} \sum_{1}^{N} (\mathbf{y}_i - \bar{\mathbf{y}})(\mathbf{y}_i - \bar{\mathbf{y}})',$$

where $\bar{\mathbf{y}}$ is the sample mean $N^{-1}\Sigma_1^N \mathbf{y}_i$.

Let $(\mathbf{x}_1,\ldots,\mathbf{x}_M)$ denote a second sample of p-variate independent observations from the same distribution (the two samples are also assumed independent). Define, for $\bar{\mathbf{x}} = M^{-1}\Sigma_1^M \mathbf{x}_i$,

$$T_\alpha^2 = (\mathbf{x}_\alpha - \bar{\mathbf{x}})'\mathbf{G}^{-1}(\mathbf{x}_\alpha - \bar{\mathbf{x}}), \tag{6.1.4}$$

for $\alpha = 1,\ldots,M$. Then define T_0^2 as

$$T_0^2 = \sum_{\alpha=1}^{M} T_\alpha^2 = \operatorname{tr}(\mathbf{H}\mathbf{G}^{-1}), \tag{6.1.5}$$

where $\mathbf{H} \equiv \Sigma_1^M(\mathbf{x}_\alpha - \bar{\mathbf{x}})(\mathbf{x}_\alpha - \bar{\mathbf{x}})'$.

Distribution of T_0^2

The distribution of T_0^2 is quite complicated in general. For example, consider the case of $p = 2$. Define $n \equiv N - 1$, $m = M - 1$. Then the density of T_0^2 is given by

$$f(t) = \frac{c}{(1 + t^2/n)^{(n+1)/2}} \cdot \int_0^{t^4/(2n+t^2)^2} z^{(m-3)/2}(1-z)^{(n-1)/2}dz, \tag{6.1.6}$$

where

$$c = \frac{(m+n-2)!}{4n(m-2)!(n-2)!}.$$

The integral in (6.1.6) is the incomplete beta function, which has already been tabulated [Pearson (1934)]. The multiplier of the integral is just a constant multiple of the ordinary T^2-statistic discussed in 6.1.1. For approximations to the cdf and some additional results, see Hotelling (1951).

Ito (1962) has compared (for simple alternatives) the power functions of tests in the multivariate analysis of variance based upon T_0^2 (using the trace criterion), and based upon statistics derivable by the likelihood ratio criterion [see Chapter 8 and Wilks (1932)] and found them almost identical. However, for more general alternatives, whether T_0^2 or the likelihood ratio statistic should be used depends upon the alternative hypothesis.

The distribution of T_0^2 for general p was given by Constantine (1966) in terms of zonal polynomials (Sect. 6.6.2) and generalized Laguerre polynomials. He also gave the mean and variance.

6.2 STUDENT t-TYPE DISTRIBUTIONS

6.2.1 The Multivariate Student t-Distribution

Let $\mathcal{L}(X) = N(0,1)$, and let Y be independent of X with $\mathcal{L}(Y) = \chi^2(n)$, where $\chi^2(n)$ denotes a chi-square distribution with n degrees of freedom (d.f.). Student, in 1908, defined the statistic

$$t = \frac{X}{(Y/n)^{1/2}} \tag{6.2.1}$$

and showed that t has the density (for $n > 0$)

$$f_0(t) = \frac{c}{(n + t^2)^{(n+1)/2}}, \qquad -\infty < t < \infty, \tag{6.2.2}$$

where

$$c = \frac{n^{n/2}}{B\left(\frac{1}{2}, \frac{n}{2}\right)},$$

and B is the beta function. [A multivariate generalization of (6.2.1) is given in Theorem (6.2.2).] This is a density centered at $t = 0$ and is called the standardized *Student t-distribution.* A generalization of this distribution to one with arbitrary location and scale is the one with density.

$$f(t) = \frac{c\sigma^{-1}}{\left[n + \left(\frac{t - \theta}{\sigma}\right)^2\right]^{(n+1)/2}}. \tag{6.2.3}$$

This distribution is centered at $t = \theta$, and t is scaled by σ. Thus, if t_0 follows (6.2.2), and if t follows (6.2.3), $\text{var } t = \sigma^2 \text{ var } t_0$, when the variance exists ($n > 2$). The mean and variance of t are easily shown to be

$$E(t) = \theta, \qquad n > 1,$$

$$\text{var } (t) = \frac{n}{n-2}\sigma^2, \qquad n > 2.$$

One generalization of the univariate Student t-distribution to the multivariate case was carried out by Cornish (1954), and Dunnett and Sobel (1954) (other generalizations are possible but this one appears to be the most useful in applications). They defined the *multivariate Student* t-*density* for $\mathbf{t}: p \times 1$ as

$$g(\mathbf{t}) = \frac{C_p|\mathbf{\Sigma}|^{-1/2}}{[n + (\mathbf{t} - \boldsymbol{\theta})'\mathbf{\Sigma}^{-1}(\mathbf{t} - \boldsymbol{\theta})]^{(n+p)/2}}, \qquad -\infty < t_j < \infty, n > 0, \tag{6.2.4}$$

for $j = 1,\ldots,p$, where $\boldsymbol{\theta}: p \times 1$ and $\boldsymbol{\Sigma}: p \times p$, $\boldsymbol{\Sigma} > 0$, and where

$$C_p = \frac{n^{n/2}\Gamma\left(\frac{n+p}{2}\right)}{\pi^{p/2}\Gamma\left(\frac{n}{2}\right)}.$$

The mean and covariance matrix of t are easily shown to be

$$E(\mathbf{t}) = \boldsymbol{\theta}, \qquad n > 1,$$

$$\operatorname{var}(\mathbf{t}) = \frac{n}{n-2}\boldsymbol{\Sigma}, \qquad n > 2.$$

Cornish (1954) and Ando and Kaufman (1965) showed that extensions of (6.2.1) to the general multivariate case (to the multivariate Student t-distribution) are also valid.

Marginals

If in the density (6.2.4), $\mathbf{t}$, $\boldsymbol{\theta}$, and $\boldsymbol{\Sigma}$ are partitioned so that $\mathbf{t} \equiv (\mathbf{t}_1', \mathbf{t}_2')'$, $\boldsymbol{\theta} \equiv (\boldsymbol{\theta}_1', \boldsymbol{\theta}_2')'$, and $\boldsymbol{\Sigma} \equiv (\boldsymbol{\Sigma}_{ij})$, $i, j = 1, 2$, where $\mathbf{t}_1 : p_1 \times 1$, and $\boldsymbol{\Sigma}_{11} : p_1 \times p_1$, the marginal density of $\mathbf{t}_1$ is given by

$$h(\mathbf{t}_1) = \frac{C_{p_1}|\boldsymbol{\Sigma}_{11}|^{-1/2}}{[n + (\mathbf{t}_1 - \boldsymbol{\theta}_1)'\boldsymbol{\Sigma}_{11}^{-1}(\mathbf{t}_1 - \boldsymbol{\theta}_1)]^{(n+p_1)/2}}, \qquad \boldsymbol{\Sigma}_{11} > 0,\ n > 0. \tag{6.2.4$'$}$$

Proof: Expand the quadratic form in (6.2.4), complete the square in $\mathbf{t}_2$, and integrate with respect to $\mathbf{t}_2$.

A useful property of the multivariate Student t-distribution is that a linear transformation of a multivariate Student t-variate also follows a multivariate Student t-distribution. Specifically:

Theorem (6.2.1): Let $\mathbf{t}$ follow the law with density given in (6.2.4). Then, if $\mathbf{y} = \mathbf{A}\mathbf{t} + \mathbf{b}$, where $\mathbf{A}: q \times p$, $\mathbf{b}: q \times 1$, $q \leq p$, the density of $\mathbf{y}$ is given by

$$h(\mathbf{y}) = \frac{C_q|\boldsymbol{\Phi}|^{-1/2}}{[n + (\mathbf{y} - \mathbf{A}\boldsymbol{\theta} - \mathbf{b})'\boldsymbol{\Phi}^{-1}(\mathbf{y} - \mathbf{A}\boldsymbol{\theta} - \mathbf{b})]^{(n+q)/2}},$$

where $\boldsymbol{\Phi} = \mathbf{A}\boldsymbol{\Sigma}\mathbf{A}'$.

This theorem is easily proven for $q = p$, by substituting in the theorem on transformations [Theorem (3.1.1)] and noting that the Jacobian of this linear transformation is $|\mathbf{A}^{-1}|$. For $q < p$, $(p - q)$ row vectors may always be added to A so that the resulting augmented matrix is non-singular, and then the same proof may be applied.

The multivariate Student t-distribution is the principal distribution encountered in Bayesian analyses of multivariate regression problems (see Chapter 8), using diffuse or natural conjugate prior distributions. Fractile points are tabled in Dunnett and Sobel (1954) and Dunnett (1955).

If $\mathbf{t} = (t_1,\ldots,t_p)'$ follows a multivariate t-distribution, (t_i/t_j) follows the *t-ratio distribution* for $i \neq j$. This distribution is required, for example, for posterior inferences on structural parameters in econometric systems of equations (see Press, 1969a). Also, in a Normal regression model (see Chapter 8) with a diffuse prior for the unknown parameters, the slope coefficients will jointly follow a posterior multivariate t-distribution. Therefore, the ratio of any two slope coefficients will follow a posterior t-ratio distribution.

6.2.2 The Inverted Multivariate Student t-Distribution

A p-variate vector $\mathbf{t}$ is said to follow the inverted multivariate Student t-distribution (see Raiffa and Schlaifer, 1961, p. 259) if its density is expressible as

$$p(\mathbf{t}) = k|\mathbf{\Sigma}|^{-1/2}[n - (\mathbf{t} - \boldsymbol{\theta})'\mathbf{\Sigma}^{-1}(\mathbf{t} - \boldsymbol{\theta})]^{(n-2)/2}, \qquad (6.2.5)$$

for $(\mathbf{t} - \boldsymbol{\theta})'\mathbf{\Sigma}^{-1}(\mathbf{t} - \boldsymbol{\theta}) \leq n$, $\mathbf{\Sigma} > 0$, and $n > 0$, where

$$k = \frac{\left(\dfrac{n + p}{2} - 1\right)!}{n^{[(n+p)/2]-1}\pi^{p/2}\left(\dfrac{n}{2} - 1\right)!}.$$

This distribution is fundamental to preposterior analysis associated with the Normal distribution (see Ando and Kaufman, 1965).

6.2.3 The Matrix T-Distribution

A random matrix $\mathbf{T} : p \times q$ is said to follow the matrix T-distribution if its density is expressible as

$$p(\mathbf{T}) = \frac{[k(m,q,p)]^{-1}}{|\mathbf{P}|^{(m-q)/2}|\mathbf{Q}|^{p/2}} \cdot \frac{1}{|\mathbf{P}^{-1} + \mathbf{T}\mathbf{Q}^{-1}\mathbf{T}'|^{m/2}}, \qquad (6.2.6)$$

where $\mathbf{P} : p \times p$, $\mathbf{P} > 0$, $\mathbf{Q} : q \times q$, $\mathbf{Q} > 0$, $m > p + q - 1$, and

$$k(m,p,q) = k(m,q,p) = \frac{\pi^{pq/2}\Gamma_q\left(\dfrac{m - p}{2}\right)}{\Gamma_q\left(\dfrac{m}{2}\right)},$$

where

$$\Gamma_p(\lambda) \equiv \pi^{p(p-1)/4}\Gamma(\lambda)\Gamma(\lambda - \tfrac{1}{2}) \cdots \Gamma\left(\lambda - \frac{p}{2} + \frac{1}{2}\right).$$

Note that $\Gamma_p(\lambda)$ denotes the multivariate gamma function (see Sect. 5.1).

An alternative form of the same density is given by

$$p(\mathbf{T}) = \frac{|\mathbf{Q}|^{(m-p)/2}|\mathbf{P}|^{q/2}}{k(m,p,q)} \cdot \frac{1}{|\mathbf{Q} + \mathbf{T}'\mathbf{P}\mathbf{T}|^{m/2}}. \tag{6.2.7}$$

This density appears to have been given first by Kshirsagar in 1960. It has also been discussed in various contexts by Tiao and Zellner (1964b), Olkin and Rubin (1964), Geisser (1965a), and Dickey (1967). For arbitrary centering of the distribution $\mathbf{T}$ could be replaced by $(\mathbf{T} - \boldsymbol{\Theta})$ in (6.2.6) and (6.2.7). It is easy to check that when $\mathbf{T}$ is replaced by $(\mathbf{T} - \boldsymbol{\Theta})$ in this distribution, $E(\mathbf{T}) = \boldsymbol{\Theta}$.

Theorem (6.2.2) (Dickey): Let $\mathbf{T}': q \times p$ be given by $\mathbf{T}' = \mathbf{X}'\mathbf{U}^{-1/2}$, where $\mathbf{U}: p \times p$, $\mathcal{L}(\mathbf{U}) = W(\mathbf{P}, p, m - q)$, $\mathbf{P} > 0$, $m > p + q - 1$, $\mathbf{U}$ is independent of $\mathbf{X}$, and the row vectors of $\mathbf{X}: p \times q$ are independently distributed as $N(\mathbf{0},\mathbf{Q})$, $\mathbf{Q} > 0$. Then, $\mathbf{T}$ has density $p(\mathbf{T})$ defined in (6.2.6) or (6.2.7).

Proof: First multiply the Wishart density of $\mathbf{U}$ by the joint density of the rows of $\mathbf{X}$ to get for the joint density of $\mathbf{U}$ and $\mathbf{X}$,

$$p(\mathbf{U},\mathbf{X}) \propto |\mathbf{U}|^{(m-q-p-1)/2} \exp\left[-\tfrac{1}{2} \operatorname{tr} (\mathbf{P}^{-1}\mathbf{U} + \mathbf{X}\mathbf{Q}^{-1}\mathbf{X}')\right].$$

Transform $\mathbf{X}$ to $\mathbf{T}$ letting $\mathbf{T} = (\mathbf{U}^{1/2'})^{-1}\mathbf{X}$. This gives

$$p(\mathbf{U},\mathbf{T}) \propto |\mathbf{U}|^{(m-p-1)/2} \exp\left[-\tfrac{1}{2} \operatorname{tr} (\mathbf{P}^{-1} + \mathbf{T}\mathbf{Q}^{-1}\mathbf{T}')\mathbf{U}\right].$$

The $\mathbf{U}$ matrix is easily integrated out using the properties of the Wishart distribution. The result is

$$p(\mathbf{T}) \propto |\mathbf{P}^{-1} + \mathbf{T}\mathbf{Q}^{-1}\mathbf{T}'|^{-m/2},$$

which is sufficient for the proof.

Theorem (6.2.3) Dickey: Suppose $\mathbf{T}: p \times q$ follows the distribution given in (6.2.6) and $\mathbf{T}' = (\mathbf{X}_1', \mathbf{X}_2')$. The *conditional* distribution of $\mathbf{X}_1$ given $\mathbf{X}_2$ is also a matrix T-distribution with m d.f., with density given by

$$p(\mathbf{x}_1|\mathbf{x}_2) \propto \frac{1}{|\mathbf{P}_{11}^{-1} + (\mathbf{x}_1 + \mathbf{P}_{11}^{-1}\mathbf{P}_{12}\mathbf{x}_2)(\mathbf{Q} + \mathbf{x}_2'\mathbf{P}_{22\cdot1}\mathbf{x}_2)^{-1}(\mathbf{x}_1 + \mathbf{P}_{11}^{-1}\mathbf{P}_{12}\mathbf{x}_2)'|^{m/2}},$$

where $\mathbf{P}_{22\cdot1} = \mathbf{P}_{22} - \mathbf{P}_{21}\mathbf{P}_{11}^{-1}\mathbf{P}_{12}$, $\mathbf{P}_{11} > 0$, and

$$\mathbf{P} = \begin{pmatrix} \mathbf{P}_{11} & \mathbf{P}_{12} \\ \mathbf{P}_{21} & \mathbf{P}_{22} \end{pmatrix},$$

where if $\mathbf{X}_1 : p_1 \times q$, then $\mathbf{P}_{11} : p_1 \times p_1$.

Proof: Write $p(\mathbf{T})$ in terms of the partitioned matrices which follow $\mathbf{T}' = (\mathbf{X}_1', \mathbf{X}_2')$. Then recognize the functional form shown in the theorem.

Theorem (6.2.4) Dickey: Suppose T follows the law given in (6.2.6) and $\mathbf{T}' = (\mathbf{X}_1', \mathbf{X}_2')$, $\mathbf{X}_i : p_i \times q$, $i = 1,2$. Then the *marginal* distribution of $\mathbf{X}_2$ is also matrix $\mathbf{T}$, but with density given by

$$p(\mathbf{x}_2) \propto \frac{1}{|\mathbf{P}_{22\cdot 1}^{-1} + \mathbf{x}_2 \mathbf{Q}^{-1} \mathbf{x}_2'|^{(m-p_1)/2}},$$

where $\mathbf{P}_{22\cdot 1}$ is defined in Theorem (6.2.3).

Proof: The proof follows immediately from Theorem (6.2.2).

Corollary (6.2.1) Dickey: If $\mathbf{T}$ follows a matrix $\mathbf{T}$-distribution, the marginal distribution of any row or any column of $\mathbf{T}$ is multivariate Student $\mathbf{t}$.

Proof: If P_{22} is a scalar, $p(\mathbf{x}_2)$ in Theorem (6.2.4) is clearly a multivariate Student $\mathbf{t}$-density. Such is the case when $p_1 = p - 1$, $p_2 = 1$. The rows may always be so arranged. When so arranged, $\mathbf{x}_2'$ is the last column of $\mathbf{T}'$.

Matrix $\mathbf{T}$-distributions have also been called generalized multivariate Student $\mathbf{t}$-distributions, and matrix variate $\mathbf{T}$-distributions. They arise, for example, as predictive distributions in problems involving prediction of variables in a multivariate regression (see, for instance, Geisser, 1965), and as the posterior distribution with respect to a diffuse prior of the coefficient matrix in a multivariate regression (Chapter 8).

Theorem (6.2.5): Suppose $\mathbf{T}$ (and $\mathbf{T}'$) follows the law given in (6.2.6), and let $\psi = \mathbf{C}_1 \mathbf{T}' \mathbf{C}_2$, where $\psi : r \times 1$, $\mathbf{C}_1 : r \times q$, $\mathbf{T}' : q \times p$, $\mathbf{C}_2 : p \times 1$. Then, the density of ψ is given by

$$p(\psi) \propto \{\mathbf{C}_2' \mathbf{P}^{-1} \mathbf{C}_2 + \psi' (\mathbf{C}_1 \mathbf{Q} \mathbf{C}_1')^{-1} \psi\}^{-m/2}.$$

Proof: Use the representation given in Theorem (6.2.2) that $\mathbf{T}' = \mathbf{X}' \mathbf{U}^{-1/2}$. Then, the rows of $\mathbf{X}$ are readily transformed by $\mathbf{C}_1$, so that if

$$\underset{(q \times p)}{\mathbf{X}'} \equiv (\mathbf{x}_1, \ldots, \mathbf{x}_p), \quad \mathcal{L}(\mathbf{C}_1 \mathbf{x}_j) = N(0, \mathbf{C}_1 \mathbf{Q} \mathbf{C}_1'), \qquad \text{for } j = 1, \ldots, p.$$

Moreover, if $v^2 = \mathbf{C}_2' \mathbf{U} \mathbf{C}_2$, where $\mathcal{L}(\mathbf{U}) = W(\mathbf{P}, q, m - q)$, $\mathcal{L}(v^2)$

$= W(\mathbf{C}_2'\mathbf{P}\mathbf{C}_2, 1, m - q)$. Since $\psi = \mathbf{C}_1\mathbf{X}'v^{-1/2}$, substitution gives the result.

Remark: This result is useful in Bayesian MANOVA (see Press, 1980a). Another result that is useful in Bayesian MANOVA involves setting posterior probability regions (Bayesian confidence sets). The result is given in the next theorem.

Theorem (6.2.6) (Geisser): Suppose $\mathbf{T}$ follows the law given in (6.2.6). Then, if

$$U \equiv \frac{|\mathbf{P}^{-1}|}{|\mathbf{P}^{-1} + \mathbf{T}\mathbf{Q}^{-1}\mathbf{T}'|}$$

where $\mathbf{P}: p \times p$, $\mathbf{P} > 0$, $\mathbf{Q}: q \times q$, $\mathbf{Q} > 0$, $\mathbf{T}: p \times q$,

$$\mathcal{L}(U) = \mathcal{L}(U_{p,q,m-q}),$$

where $U_{p,q,m-q}$ is defined in eqn. (8.4.15), and it is shown in Anderson, 1958, p. 194, that $U_{p,q,m-q}$ follows the distribution of a product of independent beta variates.

Proof: Note that $0 < U < 1$, and find its kth moment. It is found that its kth moment is easily evaluated as the normalizing constant of the matrix T-distribution and that it is the same as the kth moment of a product of independent beta variates. (See Geisser 1965a, p. 156).

6.2.4. The Inverted Matrix T-Distribution

The inverted matrix T-distribution is a generalization to matrix variates of the inverted multivariate Student t-distribution defined in (6.2.5). Its density is given by

$$p(\mathbf{T}) = \frac{|\mathbf{P}|^{(m-p-1)/2}}{k(m,q,p)|\mathbf{Q}|^{p/2}} \cdot |\mathbf{P}^{-1} - \mathbf{T}\mathbf{Q}^{-1}\mathbf{T}'|^{(m-p-q-1)/2}, \qquad (6.2.8)$$

for $\mathbf{P}^{-1} - \mathbf{T}\mathbf{Q}^{-1}\mathbf{T}' > 0$, $m > p + q - 1$, $\mathbf{P} > 0$, $\mathbf{Q} > 0$.

6.2.5 Multiple t-Distributions

Multiple t-distributions are those for which the density is proportional to a product of multivariate Student t-densities with different parameters. Thus, the random p-vector $\mathbf{t}: p \times 1$ is said to have a multiple t-distribution of order r if its density is expressible in the form (see Tiao and Zellner 1964a, p. 222)

$$f(\mathbf{t}) \propto \prod_{j=1}^{r} \frac{|\mathbf{\Sigma}_j|^{-1/2}}{[n_j + (\mathbf{t} - \boldsymbol{\theta}_j)'\mathbf{\Sigma}_j^{-1}(\mathbf{t} - \boldsymbol{\theta}_j)]^{(n_j+p)/2}}, \tag{6.2.9}$$

where $\mathbf{\Sigma}_j > 0$ for all j, and $\{n_j,\boldsymbol{\theta}_j,\mathbf{\Sigma}_j;j = 1,\ldots,r\}$ are fixed parameters of the distribution. Such a distribution represents, for example, the posterior distribution (under the assumption of diffuse priors) of the coefficient vector in a Normal multivariate regression model in which r independent samples of data are taken. In the case of $r = 2$, (6.2.9) is applicable to the problem of making inferences about a population mean when a sample is drawn from each of two multivariate Normal populations with common mean vectors and unequal but scalar covariance matrices (a positive constant times the identity matrix), see Sect. 2.2. In the form (6.2.9), the distribution does not readily yield to computation of fractiles.

A Normal approximation for (6.2.9) is easy to obtain, as follows, when all of the sample sizes are large. Rewrite (6.2.9) in the form

$$f(\mathbf{t}) \propto \exp\left(-\tfrac{1}{2}\right)\left\{\sum_{j=1}^{r} (n_j + p) \log\left[1 + (\mathbf{t} - \boldsymbol{\theta}_j)' \frac{\mathbf{\Sigma}_j^{-1}}{n_j} (\mathbf{t} - \boldsymbol{\theta}_j)\right]\right\}. \tag{6.2.10}$$

Since $\log(1 + x) \cong x$, approximately, when $-1 < x < 1$, for all n_j sufficiently large,

$$\log\left[1 + (\mathbf{t} - \boldsymbol{\theta}_j)' \frac{\mathbf{\Sigma}_j^{-1}}{n_j} (\mathbf{t} - \boldsymbol{\theta}_j)\right] \cong (\mathbf{t} - \boldsymbol{\theta}_j)' \frac{\mathbf{\Sigma}_j^{-1}}{n_j} (\mathbf{t} - \boldsymbol{\theta}_j).$$

Using this approximation in (6.2.10), completing the square in $\mathbf{t}$, and combining terms, gives approximately for large $n_1,\ldots,n_r$,

$$\mathcal{L}(\mathbf{t}) = N(\boldsymbol{\phi},\mathbf{\Omega}), \tag{6.2.11}$$

where

$$\boldsymbol{\phi} = \left[\sum_{j=1}^{r}\left(\frac{n_j + p}{n_j}\mathbf{\Sigma}_j^{-1}\right)\right]^{-1}\left[\sum_{j=1}^{r}\left(\frac{n_j + p}{n_j}\mathbf{\Sigma}_j^{-1}\boldsymbol{\theta}_j\right)\right],$$

and

$$\mathbf{\Omega} = \left[\sum_{j=1}^{r}\left(\frac{n_j + p}{n_j}\mathbf{\Sigma}_j^{-1}\right)\right]^{-1}.$$

6.3 BETA-TYPE DISTRIBUTIONS

6.3.1 The Univariate Beta Distribution

$X: 1 \times 1$ is said to follow a univariate beta distribution if its density is given by

$$f(x) = \frac{1}{B(a,b)} x^{a-1}(1-x)^{b-1}, \qquad a > 0, b > 0, 0 < x < 1, \quad (6.3.1)$$

and $f(x) = 0$, otherwise, where $B(a,b)$ denotes the beta function. It is assumed that this distribution is well known. It will be denoted by $\beta(a,b)$.

For convenience, we recall here that for the distribution in (6.3.1),

$$E(X) = \frac{a}{a+b}, \qquad \text{var}(X) = \frac{ab}{(a+b)^2(a+b+1)}.$$

The distribution of a product of independent beta variates arises in various contexts. Such a distribution is sometimes approximated by that of a single beta variate (see e.g. Press, May, 1980, pp. 37–38).

6.3.2 The Dirichlet Distribution

A multivariate generalization of the univariate beta distribution is the Dirichlet distribution. It is defined as follows.

Let $\mathbf{x}' \equiv (x_1, \ldots, x_p)$ be a p-variate vector whose components lie in the simplex (see for instance, Wilks, 1962, p. 177):

$$S_p: \{(x_1, \ldots, x_p): x_k > 0, k = 1, \ldots, p, \sum_1^p x_k < 1\}.$$

Then $\mathbf{x}$ follows the Dirichlet distribution if its density is given by

$$f(\mathbf{x}) = \frac{\Gamma(\nu_1 + \cdots + \nu_{p+1})}{\Gamma(\nu_1) \cdots \Gamma(\nu_{p+1})} x_1^{\nu_1 - 1} \cdots x_p^{\nu_p - 1}(1 - x_1 - \cdots - x_p)^{\nu_{p+1}-1}, \quad (6.3.2)$$

for $\mathbf{x}$ in S_p, and $f(\mathbf{x}) = 0$, otherwise; $\nu_k > 0$, $k = 1, \ldots, p + 1$. If $\mathbf{x}$ has density $f(\mathbf{x})$, then use the notation

$$\mathcal{L}(\mathbf{x}) = D(\nu_1, \ldots, \nu_p; \nu_{p+1}).$$

Note that (6.3.2) reduces to (6.3.1) if $p = 1$.

Properties of the Dirichlet Distribution

The following properties of the Dirichlet distribution are frequently useful.

1. Moments: If $\mathcal{L}(\mathbf{x}) = D(\nu_1,\ldots,\nu_p;\nu_{p+1})$, all of the mixed moments about the origin are given by

$$E(x_1^{r_1}\cdots x_p^{r_p}) = \frac{\Gamma(\nu_1+r_1)\cdots\Gamma(\nu_p+r_p)\Gamma(\nu_1+\cdots+\nu_{p+1})}{\Gamma(\nu_1)\cdots\Gamma(\nu_p)\Gamma(\nu_1+\cdots+\nu_{p+1}+r_1+\cdots+r_p)}.$$

Thus,

$$E(x_k) = \frac{\nu_k}{\nu_1+\cdots+\nu_{p+1}}, \qquad k = 1,\ldots,p,$$

$$\text{var}\,(x_k) = \frac{\nu_k(\nu_1+\cdots+\nu_{p+1}-\nu_k)}{(\nu_1+\cdots+\nu_{p+1})^2(\nu_1+\cdots+\nu_{p+1}+1)}, \qquad k = 1,\ldots,p,$$

and

$$\text{cov}\,(x_j,x_k) = -\frac{\nu_j\nu_k}{(\nu_1+\cdots+\nu_{p+1})^2(\nu_1+\cdots+\nu_{p+1}+1)},$$

for $j \neq k = 1,\ldots,p$.

This property follows directly by substituting in 6.3.2 and integrating Dirichlet densities with different degrees of freedom.

2. Relation to Gamma Distribution: A scalar X_j follows the one parameter univariate gamma distribution if its density is

$$g_j(x) = \frac{x^{j-1}e^{-x}}{\Gamma(j)}, \qquad x > 0,\, j > 0,$$

and $g_j(x) = 0$, otherwise. Suppose x_k has density $g_{\nu_k}(x)$ for

$$k = 1,\ldots,p+1,$$

and the x_k's are independent. Define

$$y_k = \frac{x_k}{x_1+\cdots+x_{p+1}}, \qquad k = 1,\ldots,p.$$

Then, if $\mathbf{y}' \equiv (y_1,\ldots,y_p)$,

$$\mathcal{L}(\mathbf{y}) = D(\nu_1,\ldots,\nu_p;\nu_{p+1}).$$

This property is easily established by using Theorem 3.1.1 on transformations.

An illustration of the use of this property is found in the representation of the von Neumann statistic for testing for serial correlation in time series and regression analysis. Define the (von Neumann) ratio of the two quadratic forms

$$P = \frac{\sum_1^p \lambda_i t_i^2}{\sum_1^{p+1} t_i^2}, \qquad \lambda_i > 0,$$

where $\mathbf{t}' \equiv (t_1,\ldots,t_{p+1})$, and $\mathcal{L}(\mathbf{t}) = N(\mathbf{0},\mathbf{I})$. Then, by setting

$$y_k = \frac{t_k^2}{\sum_1^{p+1} t_i^2}, \qquad k = 1,\ldots,p, \qquad \mathbf{y}' \equiv (y_1,\ldots,y_p),$$

and $\boldsymbol{\lambda}' \equiv (\lambda_1,\ldots,\lambda_p)$, it is not hard to see that

$$\mathcal{L}(P) = \mathcal{L}(\boldsymbol{\lambda}'\mathbf{y}),$$

where $\mathcal{L}(y_k) = \beta\left(\frac{1}{2}, \frac{p}{2}\right)$, for $k = 1,\ldots,p$. That is, $\mathcal{L}(\mathbf{y}) = D\left(\frac{1}{2},\ldots,\frac{1}{2}; \frac{1}{2}\right)$.

3. Marginals: Let $\mathbf{x}' \equiv (x_1,\ldots,x_p)$, with

$$\mathcal{L}(\mathbf{x}) = D(\nu_1,\ldots,\nu_p;\ \nu_{p+1});$$

then, for $\mathbf{y}' \equiv (x_1,\ldots,x_q)$, $q < p$,

$$\mathcal{L}(\mathbf{y}) = D(\nu_1,\ldots,\nu_q;\nu_{q+1} + \cdots + \nu_{p+1}).$$

The proof follows by direct integration of the undesired variables in (6.3.2).

4. Conditional Distributions: Let $\mathbf{x}' \equiv (x_1,\ldots,x_p)$ with

$$\mathcal{L}(\mathbf{x}) = D(\nu_1,\ldots,\nu_p;\nu_{p+1});$$

then

$$\mathcal{L}\left(\frac{x_p}{1 - x_1 - \cdots - x_{p-1}} \,\middle|\, x_1,\ldots,x_{p-1}\right) = \beta(\nu_p,\nu_{p+1}).$$

This property may be established by means of Theorem (3.1.1).

Other properties of the Dirichlet distribution may be found in Wilks (1962).

6.3.3 The Inverted Dirichlet Distribution

A distribution associated with the univariate beta $\beta(a,b)$, defined in (6.3.1), is the "inverted beta distribution" (see Raiffa and Schlaifer, 1961, p. 221), whose density is given by

$$f(x) = \frac{c^b}{B(a,b)} \cdot \frac{x^{a-1}}{(x + c)^{a+b}}, \qquad x \geq 0, \tag{6.3.3}$$

and $f(x) = 0$, otherwise, where a, b, $c > 0$.

Denote this distribution by

$$\mathcal{L}(x) = \beta^{-1}(a,b;c).$$

Thus, it is easy to check that if $\mathcal{L}(y) = \beta(a,b)$, and if $x = y(1-y)^{-1}$, $\mathcal{L}(x) = \beta^{-1}(a,b;1)$.

The β^{-1} distribution was well known in multivariate form but was used in a Bayesian context in 1965 by Tiao and Guttman (1965), who referred to it as the *inverted Dirichlet* distribution [see the remark following Property (2), below].

Let $\mathbf{x}' \equiv (x_1,\ldots,x_p)$. Then $\mathbf{x}$ is said to follow an inverted Dirichlet distribution, that is, $\mathcal{L}(\mathbf{x}) = D^{-1}(\nu_1,\ldots,\nu_p;\nu_{p+1})$, if its density is given by

$$f(\mathbf{x}) = \frac{\Gamma(\nu_1 + \cdots + \nu_{p+1})}{\Gamma(\nu_1)\cdots\Gamma(\nu_{p+1})} \cdot \frac{x_1^{\nu_1-1}\cdots x_p^{\nu_p-1}}{\left(1 + \sum_1^p x_k\right)^{\nu_1+\cdots+\nu_{p+1}}}, \tag{6.3.4}$$

for $0 < x_k < \infty$, $k = 1,\ldots,p$.

It may be checked that if $\mathcal{L}(\mathbf{y}) = D(\nu_1,\ldots,\nu_p;\nu_{p+1})$, and if

$$x_k = \frac{y_k}{1 - y_1 - \cdots - y_p}, \qquad \text{for } k = 1,\ldots,p,$$

and if $\mathbf{x}' \equiv (x_1,\ldots,x_p)$, then

$$\mathcal{L}(\mathbf{x}) = D^{-1}(\nu_1,\ldots,\nu_p;\nu_{p+1}).$$

Properties of the Inverted Dirichlet Distribution

1. **Moments:** If $\mathcal{L}(\mathbf{x}) = D^{-1}(\nu_1,\ldots,\nu_p;\nu_{p+1})$,

$$E(x_1^{r_1}\cdots x_p^{r_p}) = \frac{\Gamma\left(\nu_{p+1} - \sum_1^p r_j\right)}{\Gamma(\nu_{p+1})} \cdot \frac{\Gamma(\nu_1 + r_1)\cdots\Gamma(\nu_p + r_p)}{\Gamma(\nu_1)\cdots\Gamma(\nu_p)},$$

for $\nu_{p+1} > \Sigma_1^p r_j$. Thus,

$$Ex_k = \frac{\nu_k}{\nu_{p+1} - 1}, \qquad \text{var}(x_k) = \frac{\nu_k(\nu_k + \nu_{p+1} - 1)}{(\nu_{p+1} - 1)^2(\nu_{p+1} - 2)},$$

$$\text{cov}(x_j,x_k) = \frac{\nu_j\nu_k}{(\nu_{p+1} - 1)^2(\nu_{p+1} - 2)}.$$

2. **Marginals:** Let $\mathbf{x}' \equiv (x_1,\ldots,x_p)$, for $\mathcal{L}(\mathbf{x}) = D^{-1}(\nu_1,\ldots,\nu_p;\nu_{p+1})$. Then, if $\mathbf{y}' \equiv (x_1,\ldots,x_q)$, $q < p$, $\mathcal{L}(\mathbf{y}) = D^{-1}(\nu_1,\ldots,\nu_q;\nu_{p+1})$.

The preceding properties are easily checked by direct integration.

The D^{-1}-distribution was used by Tiao and Guttman (1965) to simplify the preposterior analysis of the error variance in a multivariate regression model using natural conjugate priors on the parameters.

A special case of the inverted Dirichlet distribution is the *multivariate* **F**-*distribution* (a generalization of the univariate F-distribution used in the analysis of variance).

Let $x_0, x_1, \ldots, x_p$ be independent, scalar random variables following the laws $\mathcal{L}(x_j) = \chi^2(a_j), j = 0, 1, \ldots, p$, where $\chi^2(n)$ denotes the chi-square distribution with n d.f. Define $\mathbf{y}' \equiv (y_1, \ldots, y_p)$, where $y_j = x_j/x_0$, $j = 1, \ldots, p$. The components of y are all correlated and follow the multivariate **F**-distribution with density

$$p(\mathbf{y}) = \frac{c \prod_1^p y_j^{(a_j/2)-1}}{\left[1 + \sum_1^p y_j\right]^{\Sigma_0^p a_j/2}}, \tag{6.3.5}$$

where

$$c = \frac{\Gamma(a_0 + a_1 + \cdots + a_p)}{\Gamma\left(\frac{a_0}{2}\right)\Gamma\left(\frac{a_1}{2}\right)\cdots\Gamma\left(\frac{a_p}{2}\right)}.$$

The multivariate **F**-distribution is one of the fundamental distributions required for testing hypotheses about means, variances, and covariances in Normal populations having covariance matrices with intraclass structure [see (2.2.4)]; also Wilks (1936), and Wilks (1946). Note that (6.3.5) is the same as (6.3.4) with $a_j = 2\nu_j$, $j = 1, \ldots, p$, and $a_0 = 2\nu_{p+1}$.

6.3.4 Multivariate Beta Distributions

Several multivariate analogues for the beta and inverted beta distributions have been developed for random matrices. The distributions are based upon functions of Wishart matrices. The next two theorems are generalizations of the Dirichlet and inverted Dirichlet distributions to matrix valued random variables. Proofs are given in Olkin and Rubin (1964).

Theorem (6.3.1): Let $\mathbf{S}_0, \mathbf{S}_1, \ldots, \mathbf{S}_r$ be independent with $\mathcal{L}(\mathbf{S}_j) = W(\mathbf{\Sigma}, p, n_j)$, $j = 0, 1, \ldots, r$. Then, if

$$\mathbf{W}_j = \left(\sum_0^r \mathbf{S}_i\right)^{-1/2} \mathbf{S}_j \left(\sum_0^r \mathbf{S}_i\right)^{-1/2}, \qquad j = 1, \ldots, r,$$

the joint distribution of the $\mathbf{W}_j$'s has the *matrix Dirichlet density*

$$p(\mathbf{W}_1, \ldots, \mathbf{W}_r) \propto \left[\prod_{j=1}^r |\mathbf{W}_j|^{(n_j-p-1)/2}\right] \left|\mathbf{I} - \sum_1^r \mathbf{W}_j\right|^{(n_0-p-1)/2}, \tag{6.3.6}$$

for $\mathbf{W}_j > 0$, $\mathbf{I} - \Sigma_1^r \mathbf{W}_j > 0$, $j = 1,\ldots,r$.

The distribution in (6.3.6) occurs as the joint natural conjugate prior for the sampling distribution of the elements of the sample transition probability matrix in a Markov Chain (see Martin, 1967, p. 141). The correspondence may be made by taking the $\mathbf{W}_j$'s in (6.3.6) as diagonal matrices and using as many of them as is required to fill out the dimensions of the Markov matrix.

Theorem (6.3.2): Let $\mathbf{S}_0, \mathbf{S}_1,\ldots,\mathbf{S}_r$ be independent with $\mathcal{L}(\mathbf{S}_j) = W(\mathbf{I},p,n_j)$, $j = 0,\ldots,r$, and define $\mathbf{S}_0^{1/2}$ by the equation $\mathbf{S}_0 = (\mathbf{S}_0^{1/2})^2$. If $\mathbf{V}_j \equiv \mathbf{S}_0^{-1/2}\mathbf{S}_j\mathbf{S}_0^{-1/2}$, $j = 1,\ldots,r$, the joint distribution of $\mathbf{V}_1,\ldots,\mathbf{V}_r$ is given by the *matrix inverted Dirichlet density*

$$p(\mathbf{V}_1,\ldots,\mathbf{V}_r) \propto \frac{\prod_{j=1}^{r} |\mathbf{V}_j|^{(n_j-p-1)/2}}{\left|\mathbf{I} + \sum_1^r \mathbf{V}_j\right|^{n/2}}, \tag{6.3.7}$$

for $\mathbf{V}_j > 0$, where $n = \Sigma_0^r n_j$.

Equations (6.3.6) and (6.3.7) are direct generalizations of (6.3.2) and (6.3.4), respectively. The normalizing constants are somewhat complicated and are given in Olkin and Rubin (1964), as are other matrix generalizations of beta type distributions. Note that (6.3.7) reduces to the density of the inverted Dirichlet distribution (6.3.4) when $p = 1$.

These distributions arise in applications in which it is desired to test for the equality of several Normal populations (see Anderson, 1958, p. 251), in the multivariate analysis of variance (see Roy and Gnanadesikan, 1959), and in multivariate slippage problems (see Karlin and Truax, 1960, p 321).

6.4 THE MULTIVARIATE LOG-NORMAL DISTRIBUTION

6.4.1 Density

For scalar random variables a variable is said to follow the log-Normal distribution if its logarithm follows the Normal distribution. This notion is of great interest in business and economic applications where many positive random variables appear to follow the log-Normal distribution quite closely.

For example, the Cobb-Douglas production function, frequently used in economics to relate an output variable Y to several input variables $X_1,\ldots,X_p$, has the form

$$Y = a_0 X_1^{a_1} X_2^{a_2} \cdots X_p^{a_p} U,$$

where $(a_0,a_1,\ldots,a_p)$ are unknown coefficients to be determined empirically [see, for instance, Nerlove (1965)], and U is a disturbance term whose logarithm is assumed to be Normally distributed. Taking logs on both sides of the equation yields a conventional univariate regression equation in the logarithms of the original variables. In many problems, however, an additional difficulty is created by demanding that the parameters be constrained in some way (such as requiring that $\Sigma_1^p a_j = 1$, $a_j \geq 0$). There are many questions that center around the implications of operations on the logged variables, on the original variables. The idea of the log-Normal distribution is easily generalized to vectors in the following way.

Let $\mathbf{y}: p \times 1$ follow the law $N(\boldsymbol{\theta},\boldsymbol{\Sigma})$. Define $\mathbf{y}' \equiv (y_1,\ldots,y_p)$, and let $y_j = \log x_j$. Define the "vector logarithm" (see Exercise 2.8).

$$\overline{\mathbf{log}}\ \mathbf{x} \equiv \begin{pmatrix} \log x_1 \\ \cdot \\ \cdot \\ \cdot \\ \log x_p \end{pmatrix}, \qquad \mathbf{x}' \equiv (x_1,\ldots,x_p).$$

Then $\mathbf{x}$ is said to follow the multivariate log-Normal distribution with density given by

$$p(\mathbf{x}) = \frac{1}{(2\pi)^{p/2}|\boldsymbol{\Sigma}|^{1/2}\Pi_1^p x_j} \exp\left[-\tfrac{1}{2}(\overline{\mathbf{log}}\ \mathbf{x} - \boldsymbol{\theta})'\boldsymbol{\Sigma}^{-1}(\overline{\mathbf{log}}\ \mathbf{x} - \boldsymbol{\theta})\right], \quad (6.4.1)$$

for $x_j > 0, j = 1,\ldots,p$, and $p(\mathbf{x}) = 0$, otherwise. Use the notation $\mathcal{L}(\mathbf{x}) = LN(\boldsymbol{\theta},\boldsymbol{\Sigma})$ to denote that $\mathbf{x}$ has density (6.4.1). The $LN(\boldsymbol{\theta},\boldsymbol{\Sigma})$ distribution is often applicable to the multivariate regression model.

Reference to Chapter 4 will readily show that under very general conditions (for example, when second moments of the logs of the original random variables exist), the joint distribution of products of N independent random variables tends to multivariate log-Normality as $N \to \infty$.

6.4.2 Moments

Let $\mathbf{z} \equiv (z_1,\ldots,z_p)'$, and let $\mathbf{z} \equiv \overline{\mathbf{log}}\ \mathbf{x}$. Some moments of the log-Normal distribution are given in the next theorem.

Theorem (6.4.1): If $\mathcal{L}(\mathbf{x}) = LN(\boldsymbol{\theta},\boldsymbol{\Sigma})$,

(a) $E(x_i)^k = \exp\,(k\theta_i + \tfrac{1}{2}k^2\sigma_i^2)$,

and

(b) $$\operatorname{cov}(x_i,x_j) = \exp\{\theta_i + \theta_j + \tfrac{1}{2}(\sigma_i^2 + \sigma_j^2) + \rho_{ij}\sigma_i\sigma_j\} - \exp\{\theta_i + \theta_j + \tfrac{1}{2}(\sigma_i^2 + \sigma_j^2)\},$$

for $k = 1, 2,\ldots,$ where $\boldsymbol{\theta} \equiv (\theta_1,\ldots,\theta_p)'$, $\boldsymbol{\Sigma} \equiv (\sigma_{ij})$, $\sigma_{ij} = \rho_{ij}\sigma_i\sigma_j$, $\sigma_{ii} = \sigma_i^2$, $\sigma_{jj} = \sigma_j^2$; $i, j = 1,\ldots,p$.

Proof: Suppose y is a scalar and $\mathcal{L}(y) = LN(\phi,\sigma^2)$. Then, if $t = \log y$, $\mathcal{L}(t) = N(\phi,\sigma^2)$, and

$$E(y^k) = E(\exp tk) = \int_{-\infty}^{\infty} \frac{e^{tk}}{\sigma\sqrt{2\pi}} \exp\left[-\frac{(t-\phi)^2}{2\sigma^2}\right] dt.$$

Then, if $v = (t - \phi)/\sigma$,

$$E(y^k) = e^{k\phi} \int_{-\infty}^{\infty} \frac{\exp\left[-\frac{1}{2}(v^2 - 2k\sigma v)\right]}{\sqrt{2\pi}} dv.$$

Thus,

$$E(y^k) = \exp\left(k\phi + \frac{k^2\sigma^2}{2}\right), \tag{6.4.2}$$

which establishes part (a) of the theorem.

To establish part (b), recall that

$$\text{cov}\,(x_i,x_j) = E(x_ix_j) - E(x_i)E(x_j). \tag{6.4.3}$$

But

$$E(x_ix_j) = E[\exp\,(z_i + z_j)],$$

and since $\mathcal{L}(z_i) = N(\theta_i,\sigma_i^2)$,

$$\mathcal{L}(z_i + z_j) = N(\theta_i + \theta_j,\ \sigma_i^2 + \sigma_j^2 + 2\rho_{ij}\sigma_i\sigma_j).$$

Hence, applying (6.4.2) to (x_ix_j) gives

$$E(x_ix_j) = \exp\,\{\theta_i + \theta_j + \tfrac{1}{2}(\sigma_i^2 + \sigma_j^2 + 2\rho_{ij}\sigma_i\sigma_j)\}.$$

Applying (6.4.2) to x_i and x_j separately and substituting into (6.4.3) establishes part (b).

Corollary (6.4.1): If $\mathcal{L}(\mathbf{x}) = LN(\boldsymbol{\theta},\boldsymbol{\Sigma})$,

$$\text{var}\,(x_i) = \exp\,(2\theta_i + 2\sigma_i^2) - \exp\,(2\theta_i + \sigma_i^2).$$

Proof: Let $k = 1$ in part (a) of the theorem, and let $\rho_{ij} = 1$ and $i = j$ in part (b).

6.5 STABLE DISTRIBUTIONS

6.5.1 The Univariate Stable Family

The stable family of distributions is of increasing interest in applications in business, economics, and the behavioral biological, physical, and social sciences in general. This is true for at least the following reasons.

(1) Previously, many processes were thought to be "approximately Normal" because several observations far out in the tails were discarded as outliers or as spurious. However, on retaining all obser-

vations, it was found that many processes have much more probability mass in their tails than the Normal distribution. Such "fat tail" behavior is typical of the stable non-Normal distributions, and of some other distributions as well (see Chapter 12). However, an advantage of the stable distribution explanation for the behavior of a process is that linear combinations of such variables are often easier to analyze under a stable distribution assumption (this will not always be true). When several dependent stable variables are involved, multivariate stable distributions are required. In Chapter 12, multivariate stable distributions will be used to structure a probabilistic theory of risk minimization for portfolio analysis (the optimum allocation of funds to various assets in an investment portfolio).

(2) Many relations among variables occurring in the behavioral and social sciences are of the structural type. For example, let

$$Y_1 = a_1Y^* + e_1, \qquad Y_2 = a_2Y^* + e_2, \qquad a_1 \neq 0, \qquad a_2 \neq 0,$$

where Y_1 and Y_2 are observable random variables, Y^* is an unobservable random variable, e_1 and e_2 are random disturbances or errors, a_1 and a_2 are unknown constants, and Y^*, e_1, and e_2 are mutually independent. It is well known [see Fix (1949)] that in order for the regression of Y_1 on Y_2 to be a linear function of Y_2 for any a_2 in some interval about the origin, it is necessary and sufficient that Y^* and e_2 should follow stable laws with finite means. This result is also true for the extension to many structural relations and many unobservable (latent) variables.

Linearity of regression and identifiability of the slope parameter in linear structural relations are notions that have been related to the underlying distributions.[1] Certain procedures involving linearity and identifiability which are used in factor analysis and latent structure analysis (see Chapter 10), and econometrics (errors-in-the-variables model), need to be examined in terms of whether or not the underlying distributions are stable.

Definition of the Univariate Stable Family

A distribution function $F(x)$ is said to be stable if to every $b_1 > 0$, $b_2 > 0$, and real c_1, c_2, there corresponds a positive number b and a real number c such that

$$F\left(\frac{x - c_1}{b_1}\right) * F\left(\frac{x - c_2}{b_2}\right) = F\left(\frac{x - c}{b}\right), \tag{6.5.1}$$

[1] See, for instance, Ferguson (1955), and also Lukacs and Laha (1964).

where the $*$ operation means "convolution." In general, if F_1 and F_2 are continuous cdf's with densities f_1 and f_2, respectively, their convolution is denoted as

$$F(x) = F_1(x) * F_2(x),$$

and is defined as

$$\begin{aligned} F(x) &= \int_{-\infty}^{\infty} F_1(x - t)f_2(t)dt, \\ &= \int_{-\infty}^{\infty} F_2(x - t)f_1(t)dt. \end{aligned} \tag{6.5.2}$$

A definition equivalent to that in (6.5.1) is the following: Let X and Y be independent with cdf. F. Then, F is stable if to every pair $b_1 > 0$, and $b_2 > 0$, there exists a $b_3 > 0$ and a c such that $Z \equiv (b_1X + b_2Y + c)/b_3$ also has cdf. F. Either of these definitions asserts that if sums of linear functions of independent and identically distributed random variables belong to the same family of distributions, the family is called stable. For example, such a property has already been verified for the Normal distribution so it must be a member of the stable family.

It is well known that stable distributions are continuous and have densities. But these densities are generally expressible in terms of infinite series only. However, a convenient way of studying the stable family is in terms of characteristic functions (see Section 3.1.8). Recall that the characteristic function of a continuous[2] random variable $X: 1 \times 1$ is defined as

$$\phi_X(t) = Ee^{itX} = \int_{-\infty}^{\infty} e^{itx}f(x)dx, \tag{6.5.3}$$

where $f(x)$ is the density function of X, and $i \equiv \sqrt{-1}$.

The univariate stable family was originally defined by Lévy (1924). In terms of characteristic functions, a family of distributions is said to be stable with characteristic function $\phi(t)$ if and only if $\phi(t)$ may be written in the form

$$\log \phi(t) = iat - \gamma|t|^\alpha \left[1 + i\beta \frac{t}{|t|} \omega(t,\alpha)\right], \tag{6.5.4}$$

where

$$\omega(t,\alpha) = \begin{cases} \tan \dfrac{\pi\alpha}{2}, & \text{if } \alpha \neq 1, \\ \dfrac{2}{\pi} \log |t|, & \text{if } \alpha = 1. \end{cases}$$

In this representation, a is a centering or location parameter, γ is a scale parameter, β is related to the skewness or symmetry of the distribution ($\beta = 0$ for symmetric distributions) and α is called the "characteristic

[2] If X is discrete with possible values $x_1, \ldots, x_N$ assumable with probabilities $p_1, \ldots, p_N$, $\phi_X(t) = \Sigma_1^N p_j \exp(itx_j)$; N might be infinite.

exponent" of the law ($\alpha = 2$ for Normal distributions). Thus, the univariate stable family is characterized by four parameters (a,γ,β,α), and they are confined to the domains

$$-\infty < a < \infty, \qquad \gamma \geq 0, \qquad -1 \leq \beta \leq +1, \qquad 0 < \alpha \leq 2.$$

If γ is limited to strictly positive values, the distributions will be nondegenerate. The quantity $(t/|t|)$, in (6.5.4), is defined to be zero at $t = 0$.

Elementary Results for Univariate Stable Distributions

(1) All nondegenerate stable distributions are continuous (and have continuously differentiable densities).
(2) For $\alpha = 1$, $\beta = 0$, $\phi(t)$ corresponds to Cauchy distributions.
(3) For $\alpha = 2$, $\beta = 0$, $\phi(t)$ corresponds to Normal distributions.
(4) For $0 < \alpha \leq 1$, the members of the stable family have no first- or higher-order integer moments; for $1 < \alpha < 2$, the stable distributions possess a first moment, and all moments of order δ, for all $\delta < \alpha$.
(5) The stable random variable is positive if and only if $\beta = -1$, $\alpha < 1$, and $a \geq 0$; it is negative if and only if $\beta = +1$, $\alpha < 1$, and $a \leq 0$.
(6) All stable distributions are unimodal.

For further discussion of univariate stable distributions see Lévy (1924), (1925), (1937), (1954), Gnedenko and Kolmogorov (1954), Ibragimov and Chernin (1959), Lukacs (1960), Loéve (1960), Ferguson (1962), Lukacs and Laha (1964), Feller II (1966), and Port (1966). Parameter estimation is discussed in Chapter 15.

Example (6.5.1): Suppose X is a scalar and $\mathfrak{L}(X) = N(\theta,\sigma^2)$. Find the characteristic function of X.

$$\phi_X(t) = \int_{-\infty}^{\infty} e^{itx} \cdot \frac{1}{\sigma\sqrt{2\pi}} \exp\left[-\frac{1}{2\sigma^2}(x-\theta)^2\right] dx.$$

Letting $y = \dfrac{x-\theta}{\sigma}$ gives

$$\phi_X(t) = e^{it\theta} \int_{-\infty}^{\infty} e^{it\sigma y} \cdot \frac{e^{-y^2/2}}{\sqrt{2\pi}}\, dy.$$

Expand the complex exponential in series and integrate termwise to get

$$\phi_X(t) = e^{it\theta} \sum_{k=0}^{\infty} \frac{(it\sigma)^k}{k!} \int_{-\infty}^{\infty} y^k \cdot \frac{e^{-y^2/2}}{\sqrt{2\pi}}\, dy.$$

Since the integral is just $E(Y^k)$ for $\mathcal{L}(Y) = N(0,1)$, and it is well known that $EY^k = 0$ for $k = 2l + 1$, and $EY^k = \frac{k!}{2^{k/2}(k/2)!}$ for $k = 2l$, where l is an integer, it follows that

$$\phi_X(t) = e^{it\theta} \sum_{k=0,2,\ldots}^{\infty} \frac{(it\sigma)^k}{k!} \cdot \frac{k!}{2^{k/2}\left(\frac{k}{2}\right)!} = e^{it\theta} \sum_{l=0}^{\infty} \left(\frac{it\sigma}{2^{1/2}}\right)^{2l} \cdot \frac{1}{l!}.$$

Summing the series gives

$$\phi_X(t) = \exp\left(it\sigma - \frac{\sigma^2 t^2}{2}\right).$$

That is, the log characteristic function of a univariate Normal distribution is given by

$$\log \phi_X(t) = it\theta - \frac{\sigma^2}{2} t^2. \tag{6.5.5}$$

Notice that $\phi_X(t)$ has the form given in (6.5.4) with $\theta = a$, $\gamma = \sigma^2/2$, and $\alpha = 2$.

Example (6.5.2): In (6.5.4), take $\beta = 0$ and $\alpha = 1$. Then

$$\log \phi(t) = iat - \gamma|t|.$$

This is the characteristic function of a univariate Cauchy distribution [the proof of this assertion is obtained by taking $p = 1$ in the proof of Theorem (6.5.6)]. The density of a Cauchy random variable X is given by

$$f(x) = \frac{\gamma}{\pi[\gamma^2 + (x - a)^2]},$$

for $-\infty < x < \infty$, $-\infty < a < \infty$, $\gamma > 0$. It is easy to check that $E(X)$ does not exist (and of course there are no higher-order moments). But this property was expected of a stable distribution with $\alpha = 1$. Note also that when $\beta = 0$, for any α, a is the median of the distribution; when $\beta = 0$ and $\alpha = 1$, the scale parameter γ is the semi-interquartile range. Moreover, whenever $\beta = 0$, the density is symmetric about the median.

6.5.2 Multivariate Stable Distributions

The univariate family of stable distributions was extended to the multivariate case by Lévy (1937). The family defined was one whose characteristic function could be represented as a certain integral whose form varies as the parameter α changes.

Recall from Section 3.1.8 that the characteristic function of a random vector $\mathbf{Y}: p \times 1$ is given by

$$\begin{aligned}\phi_{\mathbf{Y}}(\mathbf{t}) &\equiv \phi_{\mathbf{Y}}(t_1,\ldots,t_p) \equiv Ee^{i\mathbf{t}'\mathbf{Y}} \\ &= E\exp\left(i\sum_1^p t_jY_j\right) = \int e^{i\mathbf{t}'\mathbf{y}}f(\mathbf{y})d\mathbf{y},\end{aligned} \tag{6.5.6}$$

where $\mathbf{y}$ is $p \times 1$, $f(\mathbf{y})$ is the joint density function of $\mathbf{Y}$, and $\mathbf{t}' \equiv (t_1,\ldots,t_p)$; the p-fold integration is assumed to extend over all possible values of the components of $\mathbf{y}$.

Let $\phi_{\mathbf{Y}}(\mathbf{t})$ denote the characteristic function of a random p-vector following a multivariate stable law. Then Lévy's definition is that

$$\log \phi_{\mathbf{Y}}(\mathbf{t}) = iP_1(\mathbf{t}) - \tfrac{1}{2}P_2(\mathbf{t}) + \int\left[e^{i\mathbf{t}'\mathbf{u}} - 1 - \frac{i\mathbf{t}'\mathbf{u}}{1 + \mathbf{u}'\mathbf{u}}\right]\frac{d\rho}{\rho^{\alpha+1}}\,d\mathbf{\Phi}(\mathbf{u}),$$

where $P_1(\mathbf{t})$ and $P_2(\mathbf{t})$ are homogeneous polynomials of degree one and two, respectively; $\rho \equiv (\mathbf{u}'\mathbf{u})^{1/2}$ denotes the length of the vector $\mathbf{u}$; and $\mathbf{\Phi}(\mathbf{u})$ denotes a function proportional to a probability function (finite measure) and is defined on and integrable over the surface of the sphere $\rho = 1$. The integration (in polar coordinates) is assumed to be over the entire surface of the sphere ($d\mathbf{\Phi}$), and over all radii; otherwise $\mathbf{\Phi}(\mathbf{u})$ is arbitrary.

Lévy related the multivariate stable family defined above to the univariate stable family by showing (see Lévy, 1954, p. 221) that a necessary and sufficient condition for a random p-vector $\mathbf{Y}$ to follow a multivariate stable law is that every linear combination of the components of $\mathbf{Y}$ be stable. In particular, a symmetric multivariate stable law is one for which every linear combination of the variables is symmetric stable. It is well known that this result is true for the Normal distribution; that is, a necessary and sufficient condition for $\mathbf{Y}$ to follow a multivariate Normal law is that every linear combination of the components of $\mathbf{Y}$ be univariate Normal (this result may be proven by using characteristic functions). Using this definition of a multivariate stable law, the following theorem was established.

Theorem (6.5.1): The random vector $\mathbf{Y}: p \times 1$ follows a multivariate stable law if and only if its characteristic function, $\phi_{\mathbf{Y}}(\mathbf{t})$, is representable in the form

$$\begin{aligned}\log \phi_{\mathbf{Y}}(\mathbf{t}) &= \log \phi_{Y_1,\ldots,Y_p}(t_1,\ldots,t_p) \\ &= ia(t_1,\ldots,t_p) - \gamma(t_1,\ldots,t_p)[1 + i\beta(t_1,\ldots,t_p)\omega(1,\alpha)];\end{aligned}$$

or, more simply,

$$\log \phi_{\mathbf{Y}}(\mathbf{t}) = ia(\mathbf{t}) - \gamma(\mathbf{t})[1 + i\beta(\mathbf{t})\omega(1,\alpha)], \tag{6.5.7}$$

where $\omega(t,\alpha)$ is defined in (6.5.4), and where for every scalar u:

$$\text{(i)}\quad \gamma(t_1u,\ldots,t_pu) = |u|^\alpha\gamma(t_1,\ldots,t_p), \qquad \gamma > 0,$$

$$\text{(ii)}\quad \beta(t_1u,\ldots,t_pu) = \frac{u}{|u|}\beta(t_1,\ldots,t_p), \qquad -1 \le \beta \le +1,$$

$$\text{(iii)}\quad a(t_1u,\ldots,t_pu) = a(t_1,\ldots,t_p)u - \gamma(t_1,\ldots,t_p)\beta(t_1,\ldots,t_p)|u|^\alpha \frac{u}{|u|}[\omega(u,\alpha) - \omega(1,\alpha)];$$

that is,

$$a(\mathbf{t}u) = a(\mathbf{t})u - \gamma(\mathbf{t})\beta(\mathbf{t})|u|^\alpha \frac{u}{|u|}[\omega(u,\alpha) - \omega(1,\alpha)].$$

The distribution is multivariate symmetric stable if and only if $\beta(\mathbf{t}) \equiv \beta(t_1,\ldots,t_p) \equiv 0$.

Proof: See Ferguson (1955).

[*Remarks:* Functions that satisfy (i) of this theorem are called positive homogeneous of degree α. Note that a, γ, β are functions of $\mathbf{t}$ in this theorem, whereas in the univariate case, in (6.5.4), they were constants. Note also that not all functions satisfying (6.5.7) and (i)–(iii) will be logs of characteristic functions.]

Example (6.5.3): Let $\mathbf{X}: p \times 1$ follow the law $\mathcal{L}(\mathbf{X}) = N(\boldsymbol{\theta},\boldsymbol{\Sigma})$, for $\boldsymbol{\Sigma} > 0$. Find the characteristic function of $\mathbf{X}$.

For $\mathbf{t} \equiv (t_1,\ldots,t_p)'$,

$$\phi_\mathbf{X}(\mathbf{t}) = E(e^{i\mathbf{t}'\mathbf{x}}).$$

Let $\mathbf{Y} = \boldsymbol{\Sigma}^{-1/2}(\mathbf{X} - \boldsymbol{\theta})$. Then $\mathcal{L}(\mathbf{Y}) = N(\mathbf{0},\mathbf{I})$. So

$$\phi_\mathbf{X}(\mathbf{t}) = E\{\exp[i\mathbf{t}'(\boldsymbol{\theta} + \boldsymbol{\Sigma}^{1/2}\mathbf{Y})]\} = \exp(i\mathbf{t}'\boldsymbol{\theta}) \cdot E[\exp(i\mathbf{t}'\boldsymbol{\Sigma}^{1/2}\mathbf{Y})],$$

or

$$\phi_\mathbf{X}(\mathbf{t}) = \exp(i\mathbf{t}'\boldsymbol{\theta}) \cdot \phi_\mathbf{Y}(\mathbf{z}),$$

where $\mathbf{z}' = \mathbf{t}'\boldsymbol{\Sigma}^{1/2}$. Since by independence of the components of $\mathbf{Y} \equiv (Y_j)$, and by Example (6.5.1),

$$\phi_\mathbf{Y}(\mathbf{z}) = \prod_{j=1}^{p} E(e^{iz_jY_j}) = \prod_{j=1}^{p} e^{-z_j^2/2} = e^{-\mathbf{z}'\mathbf{z}/2},$$

$$\phi_\mathbf{X}(\mathbf{t}) = \exp(i\mathbf{t}'\boldsymbol{\theta} - \tfrac{1}{2}\mathbf{t}'\boldsymbol{\Sigma}\mathbf{t}).$$

That is, the log characteristic function for the multivariate Normal distribution is given by

$$\log \phi_\mathbf{X}(\mathbf{t}) = i\mathbf{t}'\boldsymbol{\theta} - \tfrac{1}{2}\mathbf{t}'\boldsymbol{\Sigma}\mathbf{t}. \tag{6.5.8}$$

Example (6.5.4): To appreciate the meaning of a function being positive homogeneous of degree α, compare (6.5.7) and (6.5.8). Clearly, $a(\mathbf{t}) = \mathbf{t}'\boldsymbol{\theta}$. That is,

$$a(t_1,t_2,\ldots,t_p) = t_1\theta_1 + t_2\theta_2 + \cdots + t_p\theta_p.$$

Then, for any u, $a(\mathbf{t}u) = u\mathbf{t}'\boldsymbol{\theta} = ua(\mathbf{t})$.

Also note that since the term $\frac{1}{2}\mathbf{t}'\boldsymbol{\Sigma}\mathbf{t}$ in (6.5.8) is real, it is necessary that $\beta(\mathbf{t}) \equiv 0$ in (6.5.7). Thus,

$$\gamma(\mathbf{t}) = \tfrac{1}{2}\mathbf{t}'\boldsymbol{\Sigma}\mathbf{t}.$$

Hence,

$$\gamma(\mathbf{t}u) = \tfrac{1}{2}u^2\mathbf{t}'\boldsymbol{\Sigma}\mathbf{t} = u^2\gamma(\mathbf{t}),$$

so that (i) of Theorem (6.5.1) is clearly satisfied. Condition (ii) is satisfied trivially, and condition (iii) was demonstrated above. Thus, $N(\boldsymbol{\theta},\boldsymbol{\Sigma})$ satisfies Theorem (6.5.1), and so is in the multivariate stable family (with characteristic exponent $\alpha = 2$). Of course this result also follows from the fact that all linear combinations of the components of a multivariate Normal vector follow univariate Normal distributions.

6.5.3 A Multivariate Symmetric Stable Family

This section introduces a family of multivariate stable distributions that is useful for the stable portifolio analysis model (see Chapter 12). This family and its properties, and the most general family, are discussed in Press, 1972; and Paulauskas, 1976. A family which generalizes the one presented below is discussed in Complement 6.3.

First restrict the family of distributions to those symmetric about a point in p-space. That is, in (6.5.7), take $\beta(\mathbf{t}) \equiv 0$, so that the log characteristic function of the vector $\mathbf{Y}: p \times 1$ is given by

$$\log \phi_{\mathbf{Y}}(\mathbf{t}) = ia(\mathbf{t}) - \gamma(\mathbf{t}), \qquad \gamma > 0,$$

where for every u, $\gamma(u\mathbf{t}) = |u|^\alpha\gamma(\mathbf{t})$, and $a(u\mathbf{t}) = ua(\mathbf{t})$. To satisfy this last condition on $a(\mathbf{t})$ it is necessary[3] to take $a(\mathbf{t}) = \mathbf{a}'\mathbf{t}$, where $\mathbf{a} \equiv (a_1,\ldots,a_p)'$ is a p-vector with $-\infty < a_j < \infty$. That is, take

$$\log \phi_{\mathbf{Y}}(\mathbf{t}) = i\mathbf{a}'\mathbf{t} - \gamma(\mathbf{t}), \qquad \gamma > 0,$$

with $\gamma(\mathbf{t})$ positive homogeneous of degree α. It still remains to choose $\gamma(\mathbf{t})$ subject to $\phi_{\mathbf{Y}}(\mathbf{t})$ remaining a characteristic function.

How to choose $\gamma(\mathbf{t})$ is, in part, the problem of how to generalize univariate distributions to multivariate distributions. There is no unique

[3] Suppose the distribution of $\mathbf{Y}$ is symmetric about $\mathbf{a}: p \times 1$. Then if $\mathbf{Z} = \mathbf{Y} - \mathbf{a}$, it is required that $\phi_{\mathbf{Z}}(\mathbf{t}) = \phi_{\mathbf{Z}}(-\mathbf{t})$. This condition, and (i) and (iii) of Theorem (6.5.1) require that $a(\mathbf{t}) = \mathbf{a}'\mathbf{t}$.

way to do it for dependent variables. (A well-defined multivariate distribution is the one obtained for the joint distribution of independent variables.) The best that can be done in the general case is to choose a family of multivariate distributions that produces the desired univariate distributions as marginals, and, in addition, is rich enough in interpretation to possess some properties of special interest.

Now consider the multivariate symmetric stable distributions in general (since the Normal is just a special case). A family which possesses the *ordinary* reproductive property required of multivariate stable distributions (that linear combinations of independent stable vectors belong to the same stable family), in addition to the *special* reproductive property that linear combinations of the *components* of independent stable vectors belong to the same stable family is proposed in Theorem (6.5.2), below.

First, for some integer $m \geq 1$, define the function

$$\log \phi_{\mathbf{Y}}(\mathbf{t}) = i\mathbf{a}'\mathbf{t} - \tfrac{1}{2}\sum_{j=1}^{m} (\mathbf{t}'\mathbf{\Omega}_j\mathbf{t})^{\alpha/2}, \tag{6.5.9}$$

where $\mathbf{\Omega}_j$ is a $p \times p$ positive semidefinite symmetric matrix of constants for every j, $j = 1,\ldots,m$, $0 < \alpha \leq 2$, and $\mathbf{a}$ is an arbitrary p-vector. Families defined by $\phi_{\mathbf{Y}}(\mathbf{t})$ will be said to be of order m. If $\Sigma_1^m\mathbf{\Omega}_j > 0$, it will be seen that $\phi_{\mathbf{Y}}(\mathbf{t})$ corresponds to a nonsingular distribution (that is, there is a density in p-space). Otherwise, the density is singular so that all the probability mass lies in a subspace of p-space. To avoid ambiguities, it will be assumed that no two of the Ω_j's are proportional. A review of stable distributions is given in Press (1975).

Theorem (6.5.2): $\phi_{\mathbf{Y}}(\mathbf{t})$ defined in (6.5.9) for $\Sigma_1^m\mathbf{\Omega}_j > 0$ is the characteristic function of a nonsingular multivariate symmetric stable distribution of order m, with characteristic exponent α.

Proof: See Complement 6.1.

Remark: While all members of the family given by eqn. (6.5.9) correspond to multivariate symmetric stable distributions, there are other such distributions not included in this family (see Complement 6.3).

Corollary (marginal characteristic functions): If $\mathbf{Y}: p \times 1$, $\mathbf{Y}_1: p_1 \times 1$, $\mathbf{Y}_2: (p - p_1) \times 1$, and $\mathbf{Y}' = (\mathbf{Y}_1', \mathbf{Y}_2')$, the log characteristic function of $\mathbf{Y}_1$ is given by

$$\log \phi_{\mathbf{Y}_1}(\mathbf{t}_1) = i\mathbf{a}_1'\mathbf{t}_1 - \tfrac{1}{2}\sum_{1}^{m} [\mathbf{t}_1'\mathbf{\Omega}_{11}(j)\mathbf{t}_1]^{\alpha/2},$$

where $\mathbf{t}_1 : p_1 \times 1$, $\mathbf{a}' = (\mathbf{a}_1', \mathbf{a}_2')$, $\mathbf{\Omega}_j = (\mathbf{\Omega}_{ik}(j)), i, k = 1, 2$.

Proof: Let $\mathbf{t}' = (\mathbf{t}_1', \mathbf{t}_2')$ in (6.5.9) and set $\mathbf{t}_2 = \mathbf{0}$.

If a random vector $\mathbf{Y}: p \times 1$ belongs to the family of multivariate symmetric stable distributions given in (6.5.9), the property will be denoted by

$$\mathcal{L}(\mathbf{Y}) = S_p(m, \mathbf{a}; \mathbf{\Omega}_1, \ldots, \mathbf{\Omega}_m; \alpha).$$

For many problems of interest, the number of parameters required to specify the distribution will be considerably fewer than that required for the general case. This fact will emerge more clearly in the examples to follow later in this section. Some properties of the family of laws described by (6.5.9) are given in the following theorems. Parameter estimation is discussed in Chapter 12.

Theorem (6.5.3): Suppose for $\mathbf{Y}: p \times 1$ and $\mathbf{Z}: p \times 1$,

$$\mathcal{L}(\mathbf{Y}) = S_p(m, \mathbf{a}_1; \mathbf{\Omega}_1, \ldots, \mathbf{\Omega}_m; \alpha),$$
$$\mathcal{L}(\mathbf{Z}) = S_p(n, \mathbf{a}_2; \boldsymbol{\psi}_1, \ldots, \boldsymbol{\psi}_n; \alpha),$$

and $\mathbf{Y}$ and $\mathbf{Z}$ are independent. Then if, for $\mathbf{A}_1 : q \times p$ and $\mathbf{A}_2 : q \times p$, $\mathbf{Y}^* = \mathbf{A}_1\mathbf{Y} + \mathbf{b}_1$ and $\mathbf{Z}^* = \mathbf{A}_2\mathbf{Z} + \mathbf{b}_2$, where $\mathbf{b}_1$ and $\mathbf{b}_2$ are $q \times 1$ vectors, and $q \leq p$,

$$\mathcal{L}(\mathbf{Y}^* + \mathbf{Z}^*) = S_q(\nu, \mathbf{a}; \mathbf{\Theta}_1, \ldots, \mathbf{\Theta}_\nu; \alpha),$$

where

$$\nu = m + n, \qquad \mathbf{a} = \mathbf{A}_1\mathbf{a}_1 + \mathbf{A}_2\mathbf{a}_2 + \mathbf{b}_1 + \mathbf{b}_2,$$

and

$$\mathbf{\Theta}_j = \begin{cases} \mathbf{A}_1\mathbf{\Omega}_j\mathbf{A}_1', & j = 1, \ldots, m \\ \mathbf{A}_2\boldsymbol{\psi}_{j-m}\mathbf{A}_2', & j = m+1, \ldots, m+n. \end{cases}$$

Proof: By independence,

$$\begin{aligned} \phi_{\mathbf{Y}^*+\mathbf{Z}^*}(\mathbf{t}) &= \phi_{\mathbf{Y}^*}(\mathbf{t})\phi_{\mathbf{Z}^*}(\mathbf{t}) \\ &= E \exp[i\mathbf{t}'(\mathbf{A}_1\mathbf{Y} + \mathbf{b}_1)] E \exp[i\mathbf{t}'(\mathbf{A}_2\mathbf{Z} + \mathbf{b}_2)] \\ &= \exp[i\mathbf{t}'(\mathbf{b}_1 + \mathbf{b}_2)]\phi_{\mathbf{Y}}(\mathbf{A}_1'\mathbf{t})\phi_{\mathbf{Z}}(\mathbf{A}_2'\mathbf{t}). \end{aligned}$$

Applying (6.5.9) gives the result.

[*Remark:* Theorem (6.5.3) contains the essence of the multivariate stable family defined in this chapter; that is, that the sum of independent linear combinations of the components of vectors in the same family gives rise to a new vector in the same family of distributions. Note that if all multivariate symmetric stable distributions were restricted to families with $m = 1$, Theorem (6.5.3) could not be satisfied (unless $\mathbf{A}_1$, $\mathbf{A}_2$, $\mathbf{\Omega}_j$, $\boldsymbol{\psi}_j$ were all scalars).]

Theorem (6.5.4): All multivariate stable laws are continuous and have continuous densities.

Proof: From (6.5.7), for any admissible choice of $a(\mathbf{t})$, $\beta(\mathbf{t})$, and $\gamma(\mathbf{t}) \geq 0$,

$$|\phi_{\mathbf{Y}}(\mathbf{t})| = e^{-\gamma(\mathbf{t})} \leq 1.$$

Since $\exp\{-\gamma(\mathbf{t})\}$ vanishes at infinity [this follows from (6.5.7), (i)] to an order greater than unity (the order is exponential), $\phi(\mathbf{t})$ is absolutely integrable. The result now follows from the fact that laws for which $|\phi(\mathbf{t})|$ is integrable must have continuous densities; see, for instance, Cramér (1946), p. 101.

Theorem (6.5.5): Let $\mathbf{Y}: p \times 1$ be a random vector following the law $S_p(m,\mathbf{a};\boldsymbol{\Omega}_1,\ldots,\boldsymbol{\Omega}_m;\alpha)$. Then $\mathbf{Y}$ is expressible in the form $\mathbf{Y} = \Sigma_1^m \mathbf{X}_j$, where the $\mathbf{X}_j$'s are mutually independent and $\mathcal{L}(\mathbf{X}_j) = S_p(1,\mathbf{a}/m;\boldsymbol{\Omega}_j;\alpha)$; and conversely.

[*Remark:* If a vector follows a multivariate symmetric stable law of the form given in (6.5.9), Theorem (6.5.5) asserts that the vector may be thought of as decomposable as a sum of factors (vectors), each of which follows a stable law.]

Proof: Note that (6.5.9) may be written in the form

$$\phi_{\mathbf{Y}}(\mathbf{t}) = \prod_{j=1}^{m} \exp\left[\frac{i\mathbf{a}'\mathbf{t}}{m} - \tfrac{1}{2}(\mathbf{t}'\boldsymbol{\Omega}_j\mathbf{t})^{\alpha/2}\right].$$

But if $\mathcal{L}(\mathbf{X}_j) = S_p(1,\mathbf{a}/m;\boldsymbol{\Omega}_j;\alpha)$, its log characteristic function is $[i\mathbf{a}'\mathbf{t}/m - \frac{1}{2}(\mathbf{t}'\boldsymbol{\Omega}_j\mathbf{t})^{\alpha/2}]$. Using independence gives the result. The converse follows directly from repeated applications of Theorem (6.5.3).

The following examples will illustrate special cases of the multivariate symmetric stable family.

Example (6.5.5): Let $\mathbf{Y}: p \times 1$ follow the law $\mathcal{L}(\mathbf{Y}) = S_p(m,\mathbf{a};\boldsymbol{\Omega}_1,\ldots,\boldsymbol{\Omega}_m;2)$. Then, from (6.5.9),

$$\begin{aligned}\log \phi_{\mathbf{Y}}(\mathbf{t}) &= i\mathbf{a}'\mathbf{t} - \tfrac{1}{2}\sum_{j=1}^{m}(\mathbf{t}'\boldsymbol{\Omega}_j\mathbf{t}) \\ &= i\mathbf{a}'\mathbf{t} - \tfrac{1}{2}\mathbf{t}'\left(\sum_1^m \boldsymbol{\Omega}_j\right)\mathbf{t}.\end{aligned}$$

Hence, if $\boldsymbol{\Sigma} \equiv \Sigma_1^m \boldsymbol{\Omega}_j$, $\mathcal{L}(\mathbf{Y}) = N(\mathbf{a},\boldsymbol{\Sigma})$. Thus,

$$S_p\left(1,\mathbf{a};\sum_1^m \boldsymbol{\Omega}_j;2\right) = S_p(m,\mathbf{a};\boldsymbol{\Omega}_1,\ldots,\boldsymbol{\Omega}_m;2) = N\left[\mathbf{a},\sum_1^m \boldsymbol{\Omega}_j\right], \quad (6.5.10)$$

the multivariate Normal distribution.

Example (6.5.6): For $\alpha = 1$, and $\mathbf{a} = \mathbf{0}$, (6.5.9) becomes

$$\log \phi_{\mathbf{Y}}(\mathbf{t}) = -\tfrac{1}{2} \sum_{j=1}^{m} (\mathbf{t}'\boldsymbol{\Omega}_j\mathbf{t})^{1/2}.$$

If $\mathbf{I}$ denotes the identity matrix of order p, and if $m = 1$, and $\boldsymbol{\Omega}_1 = 4\mathbf{I}$, the log characteristic function becomes

$$\log \phi_{\mathbf{Y}}(\mathbf{t}) = -(t_1^2 + \cdots + t_p^2)^{1/2}.$$

That is, $\mathbf{Y}$ follows a standard spherical Cauchy distribution.

Example (6.5.7): Let $\alpha = 1$ in (6.5.9), and take $\boldsymbol{\Omega}_j$ to be diagonal for all j. Moreover, if $\omega_i(j)$ denotes the ith diagonal element of $\boldsymbol{\Omega}_j$, and if $\gamma > 0$, take

$$\omega_i(j) = \begin{cases} \gamma^2, & \text{if } i = j,\ i = 1,\ldots,p, \qquad j = 1,\ldots,m, \\ 0, & \text{if } i \neq j. \end{cases}$$

Since in this case, $(\mathbf{t}'\boldsymbol{\Omega}_j\mathbf{t}) = t_j^2\gamma^2$, $j = 1,\ldots,m$, $m \leq p$, and $(\mathbf{t}'\boldsymbol{\Omega}_j\mathbf{t}) = 0$ for $p < m$ and $j > p$,

$$\log \phi_{\mathbf{Y}}(\mathbf{t}) = i\mathbf{a}'\mathbf{t} - \frac{\gamma}{2} \sum_{j=1}^{\nu} (t_j^2)^{1/2},$$

where $\nu \equiv \min(m,p)$. If $a_1 = a_2 = \cdots = a_p = a_0$, and if $m \equiv p$,

$$\log \phi_{\mathbf{Y}}(\mathbf{t}) = ia_0(t_1 + \cdots + t_p) - \frac{\gamma}{2}(|t_1| + \cdots + |t_p|).$$

That is, the components of $\mathbf{Y}$ are independent Cauchy variates each of which has median equal to a_0 and semi-interquartile range equal to $\gamma/2$. The location and scaling are standardized by choosing $a_0 = 0$ and $\gamma = 2$.

Example (6.5.8): Let $\alpha = 1$ in (6.5.9), and take each $\boldsymbol{\Omega}_j$ to be the special Green's matrix [see (2.2.7)]

$$\boldsymbol{\Omega}_j = \begin{bmatrix} \omega_1^2(j) & \omega_1(j)\omega_2(j) & \cdots & \omega_1(j)\omega_p(j) \\ \cdot & \cdot & & \cdot \\ \cdot & \cdot & \cdots & \cdot \\ \cdot & \cdot & & \cdot \\ \omega_p(j)\omega_1(j) & \omega_p(j)\omega_2(j) & \cdots & \omega_p^2(j) \end{bmatrix},$$

for $j = 1,\ldots,m$. Then, if $\boldsymbol{\omega}(j) \equiv [\omega_1(j),\ldots,\omega_p(j)]'$,

$$\mathbf{t}'\boldsymbol{\Omega}_j\mathbf{t} = \sum_{i=1}^{p}\sum_{k=1}^{p} t_i t_k \omega_i(j)\omega_k(j)$$
$$= \left[\sum_{i=1}^{p} t_i\omega_i(j)\right]^2 = [\mathbf{t}'\boldsymbol{\omega}(j)]^2.$$

Hence,

$$(\mathbf{t}'\boldsymbol{\Omega}_j\mathbf{t})^{1/2} = |\mathbf{t}'\boldsymbol{\omega}(j)|.$$

Therefore, if the components of $\mathbf{Y}$ all have median equal to zero,

$$\log \phi_{\mathbf{Y}}(\mathbf{t}) = -\tfrac{1}{2}\sum_{j=1}^{m} |\mathbf{t}'\boldsymbol{\omega}(j)|.$$

Now suppose that $X_1,\ldots,X_m$ are independent random variables following identical Cauchy distributions with median equal to zero, and semi-interquartile range equal to unity. Let $\mathbf{X}' \equiv (X_1,\ldots,X_m)$. The log characteristic function for X_j is

$$\log \phi_{X_j}(t_j) = -|t_j|.$$

Since the components of $\mathbf{X}$ are independent,

$$\phi_{\mathbf{X}}(\mathbf{t}) = E\exp(i\mathbf{t}'\mathbf{X}) = \prod_{j=1}^{m} \phi_{X_j}(t_j),$$

so that

$$\log \phi_{\mathbf{X}}(\mathbf{t}) = -\sum_{j=1}^{m} |t_j|.$$

Let $\mathbf{w}$ denote a $p \times 1$ vector each of whose components is a linear combination of the components of $\mathbf{X}$; specifically, let

$$\underset{(p\times 1)}{\mathbf{w}} = \underset{(p\times m)}{\boldsymbol{\psi}} \cdot \underset{(m\times 1)}{\mathbf{X}}, \qquad p \le m,$$

and $\boldsymbol{\psi} \equiv \left[\frac{\omega(1)}{2},\ldots,\frac{\omega(m)}{2}\right]$. Then, since

$$\phi_{\mathbf{w}}(\mathbf{t}) = E\exp(i\mathbf{t}'\mathbf{w}) = E\exp(i\mathbf{t}'\boldsymbol{\psi}\mathbf{X}) = \phi_{\mathbf{X}}(\boldsymbol{\psi}'\mathbf{t}),$$

and since the jth component of $\boldsymbol{\psi}'\mathbf{t}$ is $\frac{1}{2}\mathbf{t}'\boldsymbol{\omega}(j)$, $j = 1,\ldots,m$,

$$\log \phi_{\mathbf{w}}(\mathbf{t}) = -\tfrac{1}{2}\sum_{j=1}^{m} |\mathbf{t}'\boldsymbol{\omega}(j)|.$$

Thus, it is seen that a Green's matrix assumption on the $\boldsymbol{\Omega}_j$ parameters corresponds to the multivariate Cauchy distribution of sums of independent and identically distributed Cauchy variates.

Example (6.5.9): In (6.5.9) take $m = 1$, $\mathbf{\Omega}_1 = \mathbf{\Omega}$, and take $\mathbf{a} = \mathbf{0}$. Then

$$\log \phi_{\mathbf{Y}}(\mathbf{t}) = -\tfrac{1}{2}(\mathbf{t}'\mathbf{\Omega}\mathbf{t})^{\alpha/2}.$$

This characteristic function corresponds to the ellipsoidal stable distribution of characteristic exponent α, $0 < \alpha \leq 2$. Note that the standard spherical Cauchy distribution of Example (6.5.6) is a special case. The term ellipsoidal is used since the contours of $\phi_{\mathbf{Y}}(\mathbf{t})$ [the curves along which $\phi_{\mathbf{Y}}(\mathbf{t})$ is constant] are the convex curves (ellipsoids) belonging to the set $\{\mathbf{t}: (\mathbf{t}'\mathbf{\Omega}\mathbf{t}) \leq c\}$ for any constant $c > 0$.

Although most stable distributions, both univariate and multivariate, do not have densities expressible in simple closed form, the multivariate Normal and multivariate Cauchy distributions are exceptions. For the Cauchy distribution, a density is given in the following theorem.

Let $\phi_{\mathbf{Y}}(\mathbf{t})$ denote the characteristic function of a random p-vector $\mathbf{Y}$, $\mathbf{t}: p \times 1$, and suppose

$$\log \phi_{\mathbf{Y}}(\mathbf{t}) = i\mathbf{a}'\mathbf{t} - (\mathbf{t}'\mathbf{\Sigma}\mathbf{t})^{1/2}, \qquad \mathbf{\Sigma} \geq 0. \tag{6.5.11}$$

Then $\mathbf{Y}$ will be said to follow a *multivariate Cauchy distribution of order one*. If $\mathbf{\Sigma} > 0$, the distribution is nonsingular.

Theorem (6.5.6): The density of a nonsingular multivariate Cauchy vector of order one is given by

$$f(\mathbf{y}) = \frac{c|\mathbf{\Sigma}|^{-1/2}}{[1 + (\mathbf{y} - \mathbf{a})'\mathbf{\Sigma}^{-1}(\mathbf{y} - \mathbf{a})]^{(p+1)/2}}, \qquad \mathbf{\Sigma} > 0, \qquad -\infty < y_j < \infty,$$

where

$$c = \frac{\Gamma\left(\frac{p+1}{2}\right)}{\pi^{(p+1)/2}}, \qquad \mathbf{a} = (a_1, \ldots, a_p)', \qquad -\infty < a_j < \infty.$$

Proof: Complement 6.2.

[*Remarks:* If $\mathbf{Y}$ has the density given in Theorem (6.5.6),

$$\mathcal{L}(\mathbf{Y}) = S_p(1,\mathbf{a};4\mathbf{\Sigma};1).$$

Now let $\mathbf{Z} = \mathbf{Y}_1 + \cdots + \mathbf{Y}_m$, where the $\mathbf{Y}_j$'s are independent and

$$\mathcal{L}(\mathbf{Y}_j) = S_p\left(1, \frac{\mathbf{a}}{m}; \mathbf{\Omega}_j;1\right), \qquad j = 1, \ldots, m.$$

Then, $\mathbf{Z}$ follows a *multivariate Cauchy distribution of order m*, with log characteristic function

$$\log \phi_{\mathbf{Z}}(\mathbf{t}) = i\mathbf{a}'\mathbf{t} - \tfrac{1}{2}\sum_{j=1}^{m} (\mathbf{t}'\mathbf{\Omega}_j\mathbf{t})^{1/2},$$

corresponding to the law $S_p(m,\mathbf{a};\mathbf{\Omega}_1,\ldots,\mathbf{\Omega}_m;1)$, given in (6.5.9) for $\alpha = 1$; Examples (6.5.6), (6.5.7), and (6.5.8) are obtained as special cases.]

6.6 Generalized Multivariate Distributions–Zonal Polynomials

The structural form of the continuous multivariate distributions discussed so far is oftentimes too restrictive. For example, in the case of the normal distribution, specifying the first two moments fixes all higher order moments, so if we wish to specify a particular type of tail behavior, for instance, we cannot do it independently of the mean and covariance matrix and still remain within the normal family. We sometimes have need for such specifications, however (see, e.g., Kaufman and Press, 1973, 1976, and Press, 1981). There have been many efforts to generalize the known classes of multivariate distributions to make them richer in parameters (see e.g., Johnson and Kotz, 1972). We will discuss below what is probably the richest class proposed so far. This class that is rich in parameters is called the class of generalized multivariate distributions. These distributions will be introduced in terms of generalization of the hypergeometric function; it is called a hypergeometric function of matrix argument (or of several matrix arguments). These hypergeometric functions of matrix argument are in turn expressible in terms of zonal polynomials. These terms will be defined and discussed below.

6.6.1 Hypergeometric Functions of Matrix Argument

The hypergeometric function of a scalar variable z is defined as

$$F(a,b;c;z) = 1 + \sum_{n+1}^{\infty} \frac{(a)_n (b)_n}{(c)_n} \cdot \frac{z^n}{n!}, \tag{6.6.1}$$

where

$$(\alpha)_n \equiv \alpha(\alpha+1)(\alpha+2)\ldots(\alpha+n-1), \qquad n \geq 1,$$
$$(\alpha)_0 \equiv 1, \qquad \alpha \neq 0.$$

This series converges for $|z| < 1$ if c is neither zero nor a negative integer and if $c - a - b > 0$. For α a non-negative integer,

$$(\alpha)_n = \frac{\Gamma(\alpha+n)}{\Gamma(\alpha)}.$$

In integral form,

$$F(a,b;c;z) = \frac{\Gamma(c)}{\Gamma(b)\Gamma(c-b)} \int_0^1 t^{b-1}(1-t)^{c-b-1}(1-tz)^{-a}dt, \quad (6.6.2)$$

for $|z| < 1$, and if $c > b > 0$ (note that we are assuming all quantities are real). The case of $a = c$ and $b = 1$ yields the geometric series $\sum_0^\infty z^n$; hence the term hypergeometric function.

It is natural to extend the definition to any number of numerator and denominator parameters. We define the "generalized" hypergeometric function

$$_pF_q[a_1, \ldots, a_p; b_1, \ldots, b_q; z] = 1 + \sum_{n=1}^{\infty} \frac{\prod_{i=1}^{p} (a_i)_n}{\prod_{j=1}^{q} (b_j)_n} \frac{z^n}{n!}, \quad (6.6.3)$$

in which no denominator parameter b_j is allowed to be zero or a negative integer. If $p \leq q$, the series converges for all finite z; if $p = q + 1$ the series converges for $|z| < 1$ and diverges for $|z| > 1$; if $p > q + 1$ the series diverges for $z \neq 0$. A corresponding integral representation is

$$\begin{aligned} {}_pF_q&[a_1, a_2, \ldots, a_p; b_1, b_2, \ldots, b_q; z] \\ &= \frac{\Gamma(b_1)}{\Gamma(a_1)\Gamma(b_1 - a_1)} \int_0^1 t^{a_1-1}(1-t)^{b_1-a_1-1} \\ &\times {}_pF_{q-1}[a_2, \ldots, a_p; b_2, \ldots, b_q; zt]\, dt. \end{aligned} \quad (6.6.4)$$

For properties of these functions, see e.g., Rainville, 1960; Erdélyi, 1953, Chap. 2; Slater, 1960 (confluent hypergeometric functions, i.e., $_1F_1$).

The generalized hypergeometric function defined above is still a function of a scalar variable z. This definition has been extended to multiple arguments in various ways. A useful, up-to-date, summary is given in Exton, 1976, and a development of integral representations is given in Exton, 1978. The generalization to functions of matrix argument was made by Herz, 1955. Using Laplace transforms of matrix argument, he defines the hypergeometric functions of the $m \times m$ matrix $\mathbf{Z}$, iteratively, from:

$$
\begin{aligned}
{}_{p+1}F_q(a_1,\ldots,a_p,\gamma;b_1,\ldots,b_q;\mathbf{Z})
&= \frac{1}{\Gamma_m(\gamma)} \int_{\Lambda>0} \exp\{tr(-\Lambda)\} \\
&\quad \times {}_pF_q(a_1,\ldots,a_p;b_1,\ldots,b_q;\Lambda\mathbf{Z})\,|\Lambda|^{\gamma-p}\,d\Lambda, \qquad (6.6.5)
\end{aligned}
$$

for $\mathbf{Z}<0$, and

$$
\begin{aligned}
{}_pF_{q+1}(a_1,\ldots,a_p;b_1,\ldots,b_q,\gamma;\Lambda)
&= \frac{\Gamma_m(\gamma)}{(2\lambda i)^{m(m+1)/2}} \cdot \int_{\mathbf{Z}>0} \exp\{\mathrm{tr}(\mathbf{Z})\} \\
&\quad \times {}_pF_q(a_1,\ldots,a_p;b_1,\ldots,b_q;\Lambda\mathbf{Z}^{-1})\,|\mathbf{Z}|^{-\gamma}\,d\mathbf{Z}, \qquad (6.6.6)
\end{aligned}
$$

where the integral converges in some region of the complex plane, we assume $\mathbf{Z}$ is a positive definite symmetric matrix of order m, $\Gamma_m(\gamma)$ denotes the multivariate gamma function (see p. 101), and ${}_0F_0(\Lambda) \equiv \exp\{\mathrm{tr}(\Lambda)\}$.

For example, in the simple case of one numerator parameter and no denominator parameters, from (6.6.5) the function becomes

$$
\begin{aligned}
{}_1F_0(\gamma;\mathbf{Z}) &= \frac{1}{\Gamma_m(\gamma)} \int_{\Lambda>0} e^{-\mathrm{tr}\Lambda}\,{}_0F_0(\Lambda\mathbf{Z})\,|\Lambda|^{\gamma}\,d\Lambda \\
&= \frac{1}{\Gamma_m(\gamma)} \int_{\Lambda>0} |\Lambda|^{\gamma} e^{-\mathrm{tr}\Lambda(I-\mathbf{Z})}\,d\Lambda.
\end{aligned}
$$

It is straightforward to check by transforming the definition of the multivariate gamma function (p. 101), that

$$
{}_1F_0(\gamma;\mathbf{Z}) = |\mathbf{I}-\mathbf{Z}|^{-\gamma}.
$$

The hypergeometric function ${}_pF_q(\mathbf{Z})$ converges for $p > q+1$, only for $\mathbf{Z}=0$; if $p = q+1$, it converges if all the latent roots of $\mathbf{Z}$ are less than one; if $p \leq q$ it converges for all $\mathbf{Z}$.

In the next section we define zonal polynomials and show how to evaluate numerically hypergeometric functions of matrix argument, in terms of zonal polynomials.

6.6.2 Zonal Polynomials

Zonal polynomials derive from the theory of group representations but have been applied to statistical distribution theory and elsewhere (see Hua, 1959; and James, 1961, 1964). We will not derive them here, but will define them so that they may be used in applications.

Let $\mathbf{Z}$ denote an $m \times m$ matrix of real numbers. The zonal polynomial $C_{K(j)}(\mathbf{Z})$, corresponding to the partition $K(j)$ of the integer j, is a symmetric, homogeneous polynomial function of the latent roots of $\mathbf{Z}$, of degree j. A homogeneous polynomial of (x,y) is one such as

$$f(x,y) = x^n + a_1 x^{n-1} y + a_2 x^{n-2} y^2 + \ldots + y^n,$$

where each term has the same degree, and the a_j's are real constants; so $g(x,y) = (x^2 y + x)$ is not homogeneous. A symmetric polynomial in (x,y) is one whose value is unchanged when x and y are interchanged. Although no explicit general formula for the zonal polynomials has yet been given they are uniquely defined for the $m \times m$ matrix $\mathbf{Z}$ from

$$(\operatorname{tr} \mathbf{Z})^j = \sum_{K(j)} C_{K(j)}(\mathbf{Z}), \tag{6.6.7}$$

where the sum over $K(j)$ means sum over all partitions of the integer j, of weight r not exceeding m. A partition of the number j of weight r is a set of r positive integers $(j_1, j_2, \ldots, j_r)$ such that $\sum_{i=1}^{r} j_i = j$, $j_1 \geq j_2 \geq \ldots \geq j_r > 0$. For example, for the number 4, we can find partitions as shown below:

Partition	*Weight*
(1,1,1,1)	4
(2,1,1)	3
(3,1)	2
(2,2)	2
(4)	1

Let s_g denote the sum of the gth powers of the latent roots λ_j of $\mathbf{Z}$: i.e.,

$$s_g = \sum_{j=1}^{m} \lambda_j^g.$$

Below we give the first few zonal polynomials corresponding to $\mathbf{Z}$.

j	$K(j)$	$C_{K(j)}(\mathbf{Z})$
1	1	s_1
2	(1,1)	$2/3\,(s_1^2 - s_2)$
2	(2)	$1/3\,(s_1^2 + 2s_2)$
3	(1,1,1)	$1/3\,(s_1^3 - 3s_1 s_2 + 2s_3)$
3	(2,1)	$3/5\,(s_1^3 + s_1 s_2 - 2s_3)$
3	(3)	$1/15\,(s_1^3 + 6s_1 s_2 + 8s_3)$.

A listing of the zonal polynomials of orders one through twelve is given in Parkhurst and James (1974).

It follows from (6.6.7) that

$$e^{\text{tr}(\mathbf{Z})} = \sum_{j=0}^{\infty} \frac{(\text{tr}\,\mathbf{Z})^j}{j!} = \sum_{j=0}^{\infty} \sum_{K(j)} \frac{C_{K(j)}(\mathbf{Z})}{j!}$$

$$= \sum_{j=0}^{\infty} \sum_{K} \frac{C_K(\mathbf{Z})}{j!}, \tag{6.6.8}$$

since the j is usually supressed.

For example, if $\mathbf{x}:m \times 1$ and $\mathcal{L}(\mathbf{x}) = N(\mathbf{0},\Sigma)$, $\Sigma > 0$, its density is given by

$$f(\mathbf{x}) \propto \exp\{\text{tr}(\mathbf{Z})\},$$

where $\mathbf{Z} \equiv \mathbf{x}\mathbf{x}'\Sigma^{-1}$. So $f(\mathbf{x})$ may be expressed in terms of zonal polynomials by using eqn. (6.6.8).

Properties of Zonal Polynomials

(1) $C_K(\mathbf{Z})$ is invariant under orthogonal rotations, so that if $\Gamma\Gamma' = I$ where $\Gamma: m \times m$, $C_K(\Gamma\mathbf{Z}\Gamma') = C_K(\mathbf{Z})$;
(2) $C_K(\mathbf{Z}) \neq 0$;
(3) $C_K(\mathbf{Z}) > 0$ if $\mathbf{Z} > 0$;
(4) $C_{K(j)}(a\mathbf{Z}) = (a^j)C_{K(j)}(\mathbf{Z})$, for a positive constant a;
(5) For $m = 1$, $C_{K(j)}(Z) = Z^j$, where Z is now a scalar; thus, the zonal polynomials of the matrix $\mathbf{Z}$ are analogous to the simple powers of the variable Z in one dimension;
(6) For conformable matrices $\mathbf{A}$, $\mathbf{B}$, and $\mathbf{Z}$, $C_K(\mathbf{ABZ}) = C_K(\mathbf{BZA}) = C_K(\mathbf{ZAB})$;
(7) If $\mathbf{V}: p \times p$ and $\mathcal{L}(\mathbf{V}) = W(\Sigma, p, n)$, $E[C_{k(j)}(\mathbf{V})] = 2^j(n/2)_k C_{k(j)}(\Sigma)$; the result is due to Constantine, 1963.
(8) In general for $\mathbf{Z}:m \times m$, $C_K(\mathbf{Z})$ has the form: $C_{K(j)}(\mathbf{Z}) = d_K s_1^{j_1} s_2^{j_2} \ldots s_m^{j_m}$ + (lower order terms), where s_g = sum of gth powers of the latent roots of $\mathbf{Z}$, $(j_1, \ldots, j_m)$ denotes a partition of j, and d_K denotes a constant.
(9) The zonal polynomials are linearly independent and therefore serve as a basis for the symmetric functions of $\mathbf{Z}$.

Zonal Polynomials and Hypergeometric Functions

The hypergeometric function of matrix argument $\mathbf{Z}:m \times m$ is defined in terms of the zonal polynomials (see Constantine, 1963, and James, 1954) as

$$
{}_pF_q(a_1,\ldots,a_p;b_1,\ldots,b_q;\mathbf{Z}) = \sum_{k=0}^{\infty}\sum_{K(k)} \frac{(a_1)_K \ldots (a_p)_K}{(b_1)_K \ldots (b_q)_K};\frac{C_K(\mathbf{Z})}{k!}, \qquad (6.6.9)
$$

where $K(k)$ runs over all partitions of k of weight r not exceeding m. A hypergeometric function of two matrix arguments $\mathbf{Z}_1 : m \times m$, and $\mathbf{Z}_2 : m \times m$ is defined as

$$
{}_pF_q(a_1,\ldots,a_p;b_1,\ldots,b_q;\mathbf{Z}_1,\mathbf{Z}_2) = \sum_{k=0}^{\infty}\sum_{K(k)} \frac{(a_1)_K \ldots (a_p)_K}{(b_1)_K \ldots (b_q)_K} \cdot \frac{C_K(\mathbf{Z}_1)C_K(\mathbf{Z}_2)}{C_K(\mathbf{I})k!}. \qquad (6.6.10)
$$

Eqns. (6.6.9) and (6.6.10) can be used to approximate hypergeometric functions to any degree of accuracy required. Unfortunately, experience has shown, however, that the series do not converge rapidly.

Non-central Distributions

Many non-central and other distributions of Statistics are expressible in terms of hypergeometric functions. Some of them are given below for scalar arguments of the hypergeometric function.

(1) ${}_0F_0(z) = \exp(z)$;
(2) ${}_1F_1(a;z) = (1-z)^{-a}$;
(3) ${}_1F_1(a;b;z)$ = confluent hypergeometric function. This function is part of the density definition of the non-central F-distribution (see James, 1964; Fisher, 1928).
(4) ${}_0F_1(b;z)$ = Bessel function. This function is part of the density definition of the non-central chi-squared distribution (see James, 1964, p. 484; Fisher, 1928).
(5) ${}_2F_1(a_1,a_2;b;z)$ = Gaussian hypergeometric function. This function is part of the density of the sample multiple correlation coefficient (see James, 1964, p. 486; Fisher, 1928). It is also part of the density of the sample canonical correlation coefficients (see Constantine, 1963; and James, 1964).

Hypergeometric functions of matrix argument help define the density of the non-central Wishart distribution. It is defined in terms of ${}_0F_1$ functions of matrix argument, i.e., Bessel functions of matrix argument (see James, 1961). The non-central distribution of the sample latent roots of Wishart distributed matrices is given in terms of a ${}_1F_1$ function of matrix argument (see Constantine, 1963; and James, 1964). Population principle components from a Bayesian point of view also have a distribution whose density depends upon a ${}_1F_1$ function of matrix argument (see Geisser, 1965a; Constantine, 1963; and James, 1964). Hotelling's $T_0{}^2$-distribution was given in terms of zonal polynomials (see Sect. 6.1.3). So many of the important distributions of multivariate analysis are expressible in terms of hypergeometric functions of matrix argument.

6.6.3 Generalized Multivariate Distributions

Generalized multivariate distributions were developed by Roux, 1971. By tacking ${}_pF_q$ functions of matrix argument on to the usual multivariate densities and finding the appropriate normalizing constants, Roux was able to develop a generalized form for many multivariate distributions. Some of them are given below.

(1) *Wishart type:*

Let $\mathbf{Z}:m \times m$ denote a random matrix whose density is given by

$$f(\mathbf{Z}) = \frac{|B|^n {}_pF_q(a_1, \ldots, a_p; b_1, \ldots, b_q; \mathbf{BRBZ})}{\Gamma_m(n)\,{}_{(p+1)}F_q(a_1, \ldots, a_p, n; b_1, \ldots, b_q; \mathbf{BR})} \cdot |\mathbf{Z}|^{n-(m+1)/2} \exp\{\operatorname{tr}(-\mathbf{BZ})\}, \tag{6.6.11}$$

for $\mathbf{Z} > 0$, $\mathbf{R} > 0$, and $\mathbf{B} > 0$, and $f(\mathbf{Z}) = 0$, otherwise. The (a_j, b_j) are restricted to values such that $f(\mathbf{Z}) > 0$. If $\mathbf{R} = \mathbf{0}$, the density reduces to that of an ordinary Wishart distribution. If $p = 0$, $q = 1$, the density reduces to that of a non-central Wishart distribution. For an application of this distribution to a problem of Bayesian inference in group judgment formulation, see Press, 1981.

(2) *Beta Type:*

Let $\mathbf{Z}:m \times m$ denote a random matrix whose density is given by

$$f(\mathbf{Z}) = \frac{\Gamma_m(n_1 + n_2)}{\Gamma_m(n_1)\Gamma_m(n_2)} [{}_{p+1}F_q(a_1, \ldots, a_p, n_1; b_1, \ldots, b_q; n_1 + n_2; \mathbf{R}]^{-1} \cdot |\mathbf{Z}|^{n_1-(m+1)/2} |\mathbf{I} - \mathbf{Z}|^{n_2-(m+1)/2} \, {}_pF_q(a_1, \ldots, a_p; b_1, \ldots, b_q; \mathbf{RZ}), \tag{6.6.12}$$

for $\mathbf{Z} > 0$, and $f(\mathbf{Z}) = 0$, otherwise. If $\mathbf{R} = \mathbf{0}$ the central multivariate beta distribution results. There is of course a corresponding definition for a Type 2 beta distribution (see Roux, 1971).

(3) *Dirichlet Type:*

For $\mathbf{Z}_1, \ldots, \mathbf{Z}_k$ all $m \times m$, define the joint density

$$f(\mathbf{Z}_1, \ldots, \mathbf{Z}_k) = C \prod_{j=1}^{k} |\mathbf{X}_j|^{r_j - (m+1)/2} |\mathbf{I} - \sum_{j=1}^{k} \mathbf{Z}_k|^{n_2 - (m+1)/2}$$

$$\cdot\ {}_pF_q(a_1, \ldots, a_p; b_1, \ldots, b_q; \mathbf{R}\Sigma_1{}^k \mathbf{Z}_j), \quad (6.6.13)$$

for $0 < \mathbf{Z}_j < \mathbf{I}$, and $0 < \mathbf{I} - \sum_{j=1}^{k} \mathbf{Z}_j$. The normalizing constant is

$$C = \frac{\Gamma_m(n_1 + n_2)}{[\prod_{j=1}^{k} \Gamma_m(r_j)]\, \Gamma_m(n_2)\, {}_{(p+1)}F_{(q+1)}(a_1, \ldots, a_p, n_1; b_1, \ldots, b_q, n_1 + n_2; \mathbf{R})}$$

and $n_1 \equiv \sum_{j=1}^{k} r_j; f(\mathbf{Z}_1, \ldots, \mathbf{Z}_k) = 0$, otherwise. A Type 2 form of this distribution may be given as well (see Roux, 1971).

REFERENCES

Anderson (1958).
Ando and Kaufman (1963).
Bose and Roy (1938).
Bowker (1960).
Constantine (1963, 1966).
Cornish (1954).
Cramer (1946).
de Silva (1978).
Dickey (1967).
Dunnett and Sobel (1954).
Eisenhart, Hastay, and Wallis (1947).
Erdelyi (1953).
Exton (1976, 1978).
Feller II (1966).
Ferguson (1955, 1962).
Fisher (1928).
Fix (1949).
Geisser (1965a).
Gnedenko and Kolmogorov (1954).
Herz (1955).
Hotelling (1931, 1944, 1951).
Hsu (1938).
Hua (1959).
Ibragimov and Chernin (1959).
Ito (1962).
James (1954, 1961, 1964).
Johnson & Kotz (1972).
Karlin and Truax (1960).
Kaufman & Press (1973, 1976).
Kshirsagar (1960).
Lazarsfeld and Henry (1968).
Lévy (1924, 1925, 1937, 1954).
Loève (1960).
Lukacs (1960).
Lukacs and Laha (1964).
Mandelbrot (1964).
Nerlove (1965).
Olkin and Rubin (1964).
Parkhurst & James (1974).
Paulauskas (1976).
Pearson (1934).
Port (1966).

Press (1969a, 1969c, 1972, 1975, 1980d, 1981).
Raiffa and Schlaifer (1961).
Rainville (1960).
Roy and Gnanadesikan (1959).
Roux (1971).
Slater (1960).
Student (1908).
Tiao and Guttman (1965).
Tiao and Zellner (1964b, 1965).
Wilks (1932, 1936, 1946, 1962).

COMPLEMENT 6.1

Theorem (6.5.2): Log $\phi_{\mathbf{Y}}(\mathbf{t})$ defined in (6.5.9) is a multivariate stable characteristic function.

Proof: First note that since $\beta(\mathbf{t}) = 0$, (ii) of Theorem (6.5.1) is trivially satisfied. Also, since $\beta(\mathbf{t}) = 0$, from (6.5.9), for every u,

$$a(u\mathbf{t}) = (u\mathbf{t})'\mathbf{a} = u(\mathbf{t}'\mathbf{a}) = ua(\mathbf{t}),$$

so that (iii) of Theorem (6.5.1) is satisfied. Also, from (6.5.9), for every u,

$$\gamma(u\mathbf{t}) = \tfrac{1}{2}\sum_{j=1}^{m} (u^2\mathbf{t}'\mathbf{\Omega}_j\mathbf{t})^{\alpha/2}$$

$$= \frac{|u|^{\alpha}}{2}\sum_{j=1}^{m} (\mathbf{t}'\mathbf{\Omega}_j\mathbf{t})^{\alpha/2} = |u|^{\alpha}\gamma(\mathbf{t}).$$

Since $\Sigma_1^m\mathbf{\Omega}_j > 0$ by assumption, $\Sigma_1^m(\mathbf{t}'\mathbf{\Omega}_j\mathbf{t}) > 0$, and hence, $[\Sigma_1^m(\mathbf{t}'\mathbf{\Omega}_j\mathbf{t})]^{\alpha/2} > 0$, $0 < \alpha \leq 2$. Since $\gamma(\mathbf{t}) \equiv \Sigma_1^m(\mathbf{t}'\mathbf{\Omega}_j\mathbf{t})^{\alpha/2} > [\Sigma_1^m(\mathbf{t}'\mathbf{\Omega}_j\mathbf{t})]^{\alpha/2}$, $\gamma(\mathbf{t}) > 0$. Hence, (i) of Theorem (6.5.1) is satisfied. Thus, it only remains to show that $\phi_{\mathbf{Y}}(\mathbf{t})$ in (6.5.9) is in fact a characteristic function.

Let $\mathbf{X}: p \times 1$ denote a random vector with characteristic function

$$\phi_{\mathbf{X}}(\mathbf{t}) = \exp\,[-\tfrac{1}{2}(\mathbf{t}'\mathbf{t})^{\alpha/2}].$$

It was shown by Lévy (1954), p. 224, that $\phi_{\mathbf{X}}(\mathbf{t})$ is a multivariate stable characteristic function corresponding to a spherical probability law of characteristic exponent α. This fact also follows from Bochner's Theorem (see, for instance, Lukacs, 1960) and an application of the same inequality used in the last paragraph.

Let $\mathbf{Y}_j = \mathbf{A}_j\mathbf{X}$, where $\mathbf{A}_j: p \times p$. Then $\mathbf{Y}_j: p \times 1$ follows the law with characteristic function

$$\begin{aligned}\phi_{\mathbf{Y}_j}(\mathbf{t}) &= E \exp(i\mathbf{t}'\mathbf{Y}_j) = E \exp(i\mathbf{t}'\mathbf{A}_j\mathbf{X}) = \phi_{\mathbf{X}}(\mathbf{A}_j'\mathbf{t}) \\ &= \exp[-\tfrac{1}{2}(\mathbf{t}'\mathbf{A}_j\mathbf{A}_j'\mathbf{t})^{\alpha/2}].\end{aligned}$$

Note that $\mathbf{A}_j\mathbf{A}_j'$ must be a positive semidefinite matrix (see Chapter 2), and take $\mathbf{\Omega}_j \equiv \mathbf{A}_j\mathbf{A}_j'$. Then

$$\phi_{\mathbf{Y}_j}(\mathbf{t}) = \exp[-\tfrac{1}{2}(\mathbf{t}'\mathbf{\Omega}_j\mathbf{t})^{\alpha/2}].$$

Let $\mathbf{Y}^* \equiv \Sigma_1^m \mathbf{Y}_j$, where the $\mathbf{Y}_j$'s are mutually independent. Then the characteristic function of $\mathbf{Y}^*$ is

$$\phi_{\mathbf{Y}^*}(\mathbf{t}) = \exp\left[-\tfrac{1}{2}\sum_1^m (\mathbf{t}'\mathbf{\Omega}_j\mathbf{t})^{\alpha/2}\right].$$

If $\mathbf{Y} = \mathbf{Y}^* + \mathbf{a}$, the characteristic function of $\mathbf{Y}$ is given in (6.5.9).

COMPLEMENT 6.2

Theorem (6.5.6): The density of a nonsingular multivariate Cauchy vector, $\mathbf{Y}$, of order one, is given by

$$f(\mathbf{y}) = \frac{c|\mathbf{\Sigma}|^{-1/2}}{[1 + (\mathbf{y} - \mathbf{a})'\mathbf{\Sigma}^{-1}(\mathbf{y} - \mathbf{a})]^{(p+1)/2}}, \qquad \mathbf{\Sigma} > 0,\ -\infty < y_j < \infty,$$

where

$$c = \Gamma\left(\frac{p+1}{2}\right)\pi^{-(p+1)/2}, \qquad \mathbf{a} = (a_1, \ldots, a_p)', \qquad -\infty < a_j < \infty.$$

Proof:[5] It will be shown that the density in the theorem corresponds to the characteristic function given in (6.5.11).

Let $\mathbf{X}$ have density

$$g(\mathbf{x}) = \frac{c}{(1 + \mathbf{x}'\mathbf{x})^{(p+1)/2}},$$

for c defined in the theorem.

The characteristic function of $\mathbf{X}$ is given by

[5] I am grateful to M. L. Eaton for suggesting this proof. My original proof noted that $g(\mathbf{x})$ is a multivariate Student t-distribution and this results from a gamma mixture of multivariate Normal densities (Raiffa and Schlaifer, p. 256). Interchanging order of integration in $\phi(\mathbf{t})$ and evaluating the integrals gives the result. That proof relies upon a useful technique, but does not have the pedagogical virtue of reducing the multivariate problem to the analogous univariate problem.

$$\phi(\mathbf{t}) = \int \exp(i\mathbf{t}'\mathbf{x})g(\mathbf{x})d\mathbf{x} = c\int \frac{\exp(i\mathbf{t}'\mathbf{x})}{(1+\mathbf{x}'\mathbf{x})^{(p+1)/2}}\,d\mathbf{x}.$$

Let $\mathbf{\Gamma}: p \times p$ denote an orthogonal matrix whose first column is the vector $(\mathbf{t}'\mathbf{t})^{-1/2}\mathbf{t}$. Then, if $\mathbf{x} = \mathbf{\Gamma}\mathbf{z}$, since $\mathbf{\Gamma}'\mathbf{\Gamma} = \mathbf{I}$,

$$\phi(\mathbf{t}) = c\int \frac{\exp(i\mathbf{t}'\mathbf{\Gamma}\mathbf{z})}{(1+\mathbf{z}'\mathbf{z})^{(p+1)/2}}\,d\mathbf{z}.$$

Note that if $\boldsymbol{\gamma}_j$ denotes the jth column of $\mathbf{\Gamma}$, $j = 2,\ldots,p$,

$$\mathbf{t}'\mathbf{\Gamma} = \mathbf{t}'\left(\frac{\mathbf{t}}{(\mathbf{t}'\mathbf{t})^{1/2}}, \boldsymbol{\gamma}_2,\ldots,\boldsymbol{\gamma}_p\right) = \sqrt{\mathbf{t}'\mathbf{t}}\,(1,0,\ldots,0).$$

Hence, if $\mathbf{z} = (z_j)$,

$$\phi(\mathbf{t}) = c\int \frac{\exp[iz_1(\mathbf{t}'\mathbf{t})^{1/2}]}{\left(1+\sum_1^p z_j^2\right)^{(p+1)/2}}\,d\mathbf{z}.$$

Define

$$F_k \equiv \int \frac{dz_k}{(b_k + z_k^2)^{(k+1)/2}}, \qquad b_k \equiv 1 + \sum_1^{k-1} z_j^2,$$

for $k = 2,\ldots,p$. Now note that

$$\phi(\mathbf{t}) = c\int \exp[iz_1(\mathbf{t}'\mathbf{t})^{1/2}]\,dz_1 \int \cdots \int F_p\,dz_{p-1}.$$

It is easy to check (using the normalizing constant of a univariate Student t-density) that

$$F_k = \frac{\sqrt{\pi}\,\Gamma(k/2)}{\Gamma\left(\frac{k+1}{2}\right) b_k^{k/2}} \equiv \frac{g_k}{b_k^{k/2}},$$

where

$$g_k = \frac{\sqrt{\pi}\,\Gamma(k/2)}{\Gamma\left(\frac{k+1}{2}\right)}.$$

Since $b_k = b_{k-1} + z_{k-1}^2$, each integration results in a reduced power of b_k, with the result that

$$\phi(\mathbf{t}) = cg_pg_{p-1},\ldots,g_2\int \frac{\exp(iz_1\sqrt{\mathbf{t}'\mathbf{t}})}{1+z_1^2}\,dz_1.$$

The last integral is recognized as a form of the characteristic function of a univariate Cauchy distribution. It is evaluated by the theory of residues of complex variables (integrating around a closed contour consisting of a semicircle in the upper half plane, and the real axis, and letting the radius of the semicircle approach infinity). The result is

$$\int \frac{\exp\,(iz_1\sqrt{\mathbf{t}'\mathbf{t}})}{1+z_1^2}\,dz_1 = \pi \exp\,(-\sqrt{\mathbf{t}'\mathbf{t}}).$$

Combining terms gives

$$\phi(\mathbf{t}) = \exp\,(-\sqrt{\mathbf{t}'\mathbf{t}}).$$

Letting $\mathbf{X} = \mathbf{\Sigma}^{-1}(\mathbf{Y} - \mathbf{a})$ gives the desired result.

COMPLEMENT 6.3

A class of multivariate symmetric stable distributions more general than the one presented in eqn. (6.5.9) was given by de Silva (1978). This class is defined by a log-characteristic function for a random vector $\mathbf{Y}:p \times 1$, as

$$\log\phi(\mathbf{t}) = i\mathbf{a}'\mathbf{t} - \sum_{j=1}^{m}\Big(\sum_{k=1}^{r}|\mathbf{b}_{jk}'\mathbf{t}|^{\beta}\Big)^{\alpha/\beta},$$

for $\mathbf{t}:p \times 1$, $0 < \alpha \leq \beta \leq 2$, $p \leq r$, $\mathbf{b}_{jk}:p \times 1$, and no two of the sets $W_j = \{\mathbf{b}_{jk}:k = 1,\ldots,r\}$ are proportional. α is called the characteristic exponent, as defined previously, and (β,r,m) are other parameters of the distribution. It is straightforward to check the conditions in Theorem (6.5.1) to show that $\phi(\mathbf{t})$ is indeed a member of the general class (for $\beta(\mathbf{t}) \equiv 0$). Note, moreover, that for $\beta = 2$, and $\Omega_j \equiv \sum_{k=1}^{r} \mathbf{b}_{jk}\mathbf{b}_{jk}'$, $\phi(\mathbf{t})$ reduces to the characteristic function defined in eqn. (6.5.9).

EXERCISES

6.1 Suppose on the basis of interviews with 100 consumers and retailers, it is found that average price changes for two commodities are given by

$$\bar{\mathbf{x}}' = (4,-3),$$

and the sample covariance matrix for the 100 independent observations is given by

$$\mathbf{S} = \tfrac{1}{99}\sum_{1}^{100}(\mathbf{x}_k - \bar{\mathbf{x}})(\mathbf{x}_k - \bar{\mathbf{x}})',$$

$$\mathbf{S} = \begin{pmatrix} 2 & \frac{1}{2} \\ \frac{1}{2} & 1 \end{pmatrix}.$$

Supposing $\mathcal{L}(\mathbf{x}_k) = N(\boldsymbol{\theta},\mathbf{\Sigma})$ $\mathbf{\Sigma} > 0$, for independent observations

$\mathbf{x}_1, \ldots, \mathbf{x}_{100}$, test the hypothesis $H: \boldsymbol{\theta}' = \boldsymbol{\theta}_0' = (2, -2)$, versus the alternative hypothesis $A: \boldsymbol{\theta}' \neq \boldsymbol{\theta}_0'$, at the 1 percent level of significance.

6.2 Let $X_j(t)$ denote the change in the number of families considered to be in the poverty class (the "poverty" level changes annually) in State j, $j = 1, \ldots, p$, in the United States in two successive years, ending with year t. Let $\mathbf{X}(t) \equiv [X_1(t), \ldots, X_p(t)]$ and assume that $\mathbf{X}(t_1), \ldots, \mathbf{X}(t_N)$ are mutually independent and $\mathcal{L}[\mathbf{X}(t)] = N(\boldsymbol{\theta}, \boldsymbol{\Sigma})$, for $(\boldsymbol{\theta}, \boldsymbol{\Sigma})$ given. Adopt a diffuse prior for $(\boldsymbol{\theta}, \boldsymbol{\Sigma})$ and find the marginal posterior distribution for $\boldsymbol{\theta}$ given $\mathbf{X}(t_1), \ldots, \mathbf{X}(t_N)$.

6.3 In Exercise 6.2, find the predictive distribution for a new observation, $\mathbf{X}(t_{N+1})$.

6.4 Let $\mathbf{X}_j: p \times 1$ denote a vector representing the response of subject j to a question posed in a sample survey that has p possible response categories, $j = 1, \ldots, N$. The numbers of subjects in each of the response categories follow a multinomial distribution. Find the posterior distribution for the category probabilities given the observations, under the assumption of a natural conjugate prior distribution. [*Hint:* See Example (4.4.1).]

6.5 Let X and Y denote response times for the same individual in a pair of psychological experiments so that X and Y are correlated. Suppose it is found that log (X) and log (Y) follow the bivariate Normal distribution $N(\boldsymbol{\theta}, \boldsymbol{\Sigma})$, $\boldsymbol{\Sigma} > 0$, where

$$\boldsymbol{\theta} = \begin{pmatrix} -2 \\ 3 \end{pmatrix}, \qquad \boldsymbol{\Sigma} = \begin{pmatrix} 6 & 2 \\ 2 & 3 \end{pmatrix}.$$

Evaluate:
(a) $E(3X + 2Y)$,
(b) var $(3X + 2Y)$.

6.6 Let X denote the time between successive arrivals of Douglas DC-10 commercial airplanes at a busy airport, and let Y denote the time between successive arrivals of Boeing 747 commercial airplanes at the same airport. Suppose X and Y are independent and each follows the stable law with density given by

$$p(v) = \frac{(h/2\pi)^{1/2}}{v^{3/2}} e^{-h/2v}, \qquad v \geq 0,\ h > 0,$$

and $p(v) = 0$ for $v \leq 0$. Note that for this distribution, the characteristic exponent is $\frac{1}{2}$, $\beta = +1$, and the centering parameter $a = 0$.

Let $z \equiv c_1 X + c_2 Y$. Find
(a) the density of z,
(b) the characteristic function of z.

6.7 Let Y_j denote the return one period from now on the jth oil security in an investment portfolio, $j = 1,\ldots,4$ and $\mathbf{Y} \equiv (Y_1,\ldots,Y_4)'$; and let Z_j denote the one period return on the jth office machine security in the portfolio, $\mathbf{Z} \equiv (Z_1,\ldots,Z_4)'$. Suppose $\mathbf{Y}$ and $\mathbf{Z}$ are independent and (see Section 6.5.3)

$$\mathcal{L}(\mathbf{Y}) = S_4(1,\mathbf{a};\mathbf{\Omega};\tfrac{3}{2}),$$
$$\mathcal{L}(\mathbf{Z}) = S_4(1,\mathbf{b};\boldsymbol{\psi};\tfrac{3}{2}),$$

where

$$\mathbf{a}' \equiv (1,3,-2,2), \qquad \mathbf{b}' \equiv (1,-8,-3,4),$$

$$\underset{(4\times 4)}{\mathbf{\Omega}} = 3\begin{pmatrix} 1 & & & \\ & \cdot & \frac{1}{2} & \\ & & \cdot & \\ & \frac{1}{2} & \cdot & \\ & & & 1 \end{pmatrix}, \qquad \underset{(4\times 4)}{\boldsymbol{\psi}} = 9\begin{pmatrix} 1 & & & \\ & \cdot & \frac{3}{4} & \\ & & \cdot & \\ & \frac{3}{4} & \cdot & \\ & & & 1 \end{pmatrix}.$$

Find the distribution law of $(\mathbf{c}_1'\mathbf{Y} + \mathbf{c}_2'\mathbf{Z})$, where

$$\mathbf{c}_1 \equiv (\tfrac{2}{8},\tfrac{1}{8},0,0)', \qquad \mathbf{c}_2 \equiv (\tfrac{1}{8},\tfrac{1}{8},\tfrac{2}{8},\tfrac{1}{8})'.$$

6.8 Suppose that in an experiment on behavior, k mutually exclusive responses are possible on a single trial. Suppose that out of n independent multinomial type trials [see Example (4.4.1)], r_1 responses of the first type, r_2 responses of the second type, $\ldots$, and r_k responses of the kth type are obtained. Let p_j denote the probability of a response of type j on any trial, $j = 1,\ldots,k$.

Suppose the prior distribution for $(p_1,\ldots,p_k)$ is natural conjugate (the Dirichlet distribution of Section 6.3.2).

Find the predictive density for the responses of each type in a new sequence of N trials, given the results of the earlier experiment. This result will generalize the distribution obtained in Exercise 3.9, and is called the Dirichlet—multinomial distribution.

6.9 Suppose $\mathbf{Y}: p \times 1$, and define for $0 < \alpha \leq 2$, $\log \phi_{\mathbf{Y}}(\mathbf{t}) = ia(\mathbf{t}) - \gamma(\mathbf{t})[1 + i\beta(\mathbf{t})\omega(1,\alpha)]$, where $\omega(u,\alpha)$ is defined in (6.5.4); $\gamma(\mathbf{t}) \equiv \frac{1}{2}\Sigma_1^m(\mathbf{t}'\mathbf{\Omega}_j\mathbf{t})^{\alpha/2}$, $\mathbf{\Omega}_j \geq 0$ and $\mathbf{\Omega}_j: p \times p$, $m = 1, 2,\ldots$; $\beta(\mathbf{t}) \equiv \Sigma_1^p\beta_j t_j/|t_j|$, for $\Sigma_1^p|\beta_j| \leq 1$; for $\alpha \neq 1$, $a(\mathbf{t}) \equiv \mathbf{a}'\mathbf{t}$, and for $\alpha = 1$, $a(\mathbf{t}) = \mathbf{a}'\mathbf{t} - \gamma(\mathbf{t})\beta(\mathbf{t})(2/\pi)\log|\mathbf{g}'\mathbf{t}|$, where $\mathbf{a}$ and $\mathbf{g}$ are arbitrary p-vectors. Is $\phi_{\mathbf{Y}}(\mathbf{t})$ a multivariate stable characteristic function?

6.10 Let $\mathbf{Y}$ be a random p-vector, and let $\phi_{\mathbf{Y}}(\mathbf{t})$ denote its characteristic function. Suppose $\log \phi_{\mathbf{Y}}(\mathbf{t}) = -\frac{1}{2}\Sigma_1^m(\mathbf{t}'\mathbf{\Omega}_j\mathbf{t})^{\alpha/2}$, for some positive integer m, where $\mathbf{\Omega}_j: p \times p$, $\mathbf{\Omega}_j \geq 0$ for every j, and $0 < \alpha \leq 2$. Show that $\log \phi_{\mathbf{Y}}(\mathbf{t})$ is concave if and only if $1 \leq \alpha \leq 2$, and that for $0 < \alpha < 1$, $\log \phi_{\mathbf{Y}}(\mathbf{t})$ is neither concave nor convex. [*Hint:* Calculate the Hessian matrix and show that its latent roots are positive if and only if $1 < \alpha \leq 2$. Then apply Property (8) of

2.11.2, and the results on concavity in 2.14.3. This result will be useful in portfolio analysis (see Chapter 12).] Why must the case of $\alpha = 1$ be studied separately?

6.11 Let $\mathbf{Y}: 2 \times 1$ have log-characteristic function

$$\log \phi_{\mathbf{Y}}(t) = -|2t_1 + 3t_2| - |6t_1 - 5t_2|.$$

Show that $\phi_{\mathbf{Y}}(\mathbf{t})$ is the characteristic function of a multivariate symmetric stable distribution not included in the family defined by eqn. (6.5.9).

6.12 Let $\mathbf{T}$ denote a random $p \times q$ matrix following the law given by eqn. (6.2.7). Define

$$U = \frac{|\mathbf{Q}|}{|\mathbf{Q} + \mathbf{T}'\mathbf{X}'\mathbf{X}\mathbf{T}|}.$$

Find $E(U)$. [Hint: see Theorem (6.2.6).]

6.13 Let $\mathbf{Z}: m \times m$ denote a random matrix following the generalized Wishart distribution (see Sect. 6.6.3). Evaluate $E|\mathbf{Z}|^k$, $k \geq 1$.

7

BASIC MULTIVARIATE STATISTICS IN THE NORMAL DISTRIBUTION

Study of the foundations of multivariate analysis (Part I) culminates in this chapter with an introduction to estimation and hypothesis testing associated with the Normal distribution. The methods introduced may often be extended to problems of inference and decision making involving non-Normal distributions in an obvious way. The Normal distribution was selected both because of its central role and because of its tractability.

Section 7.1 is devoted to conventional aspects of estimation from both sampling theory and Bayesian points of view, while Section 7.2 briefly discusses the decision-theoretic difficulties associated with estimating means in multivariate populations. Section 7.3 is devoted to hypothesis testing involving means, variances, covariances, correlations, covariance matrices, and correlation matrices. Many problems of inference discussed in this chapter, and in the chapters of Part II can be handled by the general technique of analysis of covariance structures (see Jöreskog, 1970, 1973). In this approach, one may fix the structural form of the mean response vector and its covariance matrix, and may also constrain many of the parameters. Then, parameters are estimated and hypotheses are tested using a Fletcher-Powell optimization technique. Many multivariate models may be examined within this framework. Alternatively, we will study each model separately to bring out individual features more readily.

7.1 ESTIMATION

7.1.1 Sufficiency

Suppose $\mathbf{x}_1, \ldots, \mathbf{x}_N$ is a sample of $p \times 1$ mutually independent random vectors that are identically distributed as $N(\boldsymbol{\theta}, \boldsymbol{\Sigma})$. Their joint density (likelihood) is given by

$$p(\mathbf{x}_1, \ldots, \mathbf{x}_N | \boldsymbol{\theta}, \boldsymbol{\Sigma}) = \frac{1}{(2\pi)^{Np/2} |\boldsymbol{\Sigma}|^{N/2}} \exp\left[-\tfrac{1}{2} \sum_{j=1}^{N} (\mathbf{x}_j - \boldsymbol{\theta})' \boldsymbol{\Sigma}^{-1} (\mathbf{x}_j - \boldsymbol{\theta})\right]. \tag{7.1.1}$$

Recall that in the univariate case, sufficient statistics for $(\boldsymbol{\theta}, \boldsymbol{\Sigma})$ are found in practice by means of the Neyman-Fisher factorization criterion. That is, the likelihood function is factored into the product of a function of the observations, and a function of the parameters and summary statistics. Such a criterion is also applicable to multivariate distributions and is an immediate generalization of the univariate case.

Sufficiency Criterion for Multivariate Distributions

Let $p(\mathbf{x}_1, \ldots, \mathbf{x}_N | \boldsymbol{\phi})$ denote the joint density for N observations $\mathbf{x}_j : p \times 1$, given a parameter matrix $\boldsymbol{\phi} : q \times r$. Then $\mathbf{T}(\mathbf{x}_1, \ldots, \mathbf{x}_N) \equiv \mathbf{T}$ is sufficient for $\boldsymbol{\phi}$ if and only if there exist nonnegative functions f and g such that

$$p(\mathbf{x}_1, \ldots, \mathbf{x}_N | \boldsymbol{\phi}) = f(\mathbf{T}; \boldsymbol{\phi}) g(\mathbf{x}_1, \ldots, \mathbf{x}_N). \tag{7.1.2}$$

Now apply (7.1.2) to (7.1.1). Define $\mathbf{A} \equiv \Sigma_1^N (\mathbf{x}_j - \boldsymbol{\theta})(\mathbf{x}_j - \boldsymbol{\theta})'$. Then

$$p(\mathbf{x}_1, \ldots, \mathbf{x}_N | \boldsymbol{\theta}, \boldsymbol{\Sigma}) = \frac{1}{(2\pi)^{Np/2} |\boldsymbol{\Sigma}|^{N/2}} \exp\left(-\tfrac{1}{2} \operatorname{tr} \boldsymbol{\Sigma}^{-1} \mathbf{A}\right).$$

But if $\bar{\mathbf{x}}$ denotes the sample mean vector, $\bar{\mathbf{x}} \equiv N^{-1} \Sigma_1^N \mathbf{x}_\alpha$,

$$\mathbf{A} = \sum_1^N [(\mathbf{x}_j - \bar{\mathbf{x}}) + (\bar{\mathbf{x}} - \boldsymbol{\theta})][\mathbf{x}_j - \bar{\mathbf{x}}) + (\bar{\mathbf{x}} - \boldsymbol{\theta})]',$$

or, since the sum of the cross-product terms vanish,

$$\mathbf{A} = \sum_1^N (\mathbf{x}_j - \bar{\mathbf{x}})(\mathbf{x}_j - \bar{\mathbf{x}})' + N(\bar{\mathbf{x}} - \boldsymbol{\theta})(\bar{\mathbf{x}} - \boldsymbol{\theta})'.$$

Define

$$\mathbf{V} = \sum_1^N (\mathbf{x}_j - \bar{\mathbf{x}})(\mathbf{x}_j - \bar{\mathbf{x}})',$$

where $\mathbf{V}$ denotes the sample covariance matrix (up to a factor of proportionality). That is, the diagonal elements of $\mathbf{V}$ are proportional to sample variances and the off-diagonal elements are proportional to sample covariances. Thus,

$$\mathbf{A} = \mathbf{V} + N(\bar{\mathbf{x}} - \boldsymbol{\theta})(\bar{\mathbf{x}} - \boldsymbol{\theta})',$$

and

$$p(\mathbf{x}_1,\ldots,\mathbf{x}_N|\boldsymbol{\theta},\boldsymbol{\Sigma}) = \frac{1}{(2\pi)^{Np/2}|\boldsymbol{\Sigma}|^{N/2}} \exp\{-\tfrac{1}{2}\operatorname{tr}\boldsymbol{\Sigma}^{-1}[\mathbf{V} + N(\bar{\mathbf{x}} - \boldsymbol{\theta})(\bar{\mathbf{x}} - \boldsymbol{\theta})']\}. \quad (7.1.3)$$

Use of (7.1.2) with $g(\mathbf{x}_1,\ldots,\mathbf{x}_N) \equiv 1$, $\mathbf{T} \equiv (\bar{\mathbf{x}},\mathbf{V})$, $\boldsymbol{\phi} \equiv (\boldsymbol{\theta},\boldsymbol{\Sigma})$ gives the result summarized in the following theorem.

Theorem (7.1.1): If $\mathbf{x}_1,\ldots,\mathbf{x}_N$ are independent p-vector observations from $N(\boldsymbol{\theta},\boldsymbol{\Sigma})$, $(\bar{\mathbf{x}},\mathbf{V})$ is sufficient for $(\boldsymbol{\theta},\boldsymbol{\Sigma})$, where $\bar{\mathbf{x}}$ and $\mathbf{V}$ are the sample mean vector and sample covariance matrix, respectively.

Theorem (7.1.2): $\bar{\mathbf{x}}$ and $\mathbf{V}$ are independent.

Proof: In Complement 5.1 it is shown that if $\boldsymbol{\Gamma} = (\gamma_{ij})$, $i, j = 1,\ldots,N$, is any $N \times N$ orthogonal matrix with $\gamma_{Nj} = N^{-1/2}$, $j = 1,\ldots,N$, and if

$$\mathbf{Y}_i = \sum_{j=1}^{N} \gamma_{ij}\mathbf{x}_j, \qquad i = 1,\ldots,N,$$

the $\mathbf{Y}_i$'s are independent with mean zero for $i = 1,\ldots,N-1$, and $\mathbf{Y}_N = \sqrt{N}\,\bar{\mathbf{x}}$. However, it is now shown that the independence holds for $i = N$ as well. Accordingly, for $i \neq N$,

$$\begin{aligned}
\operatorname{cov}(\mathbf{Y}_N,\mathbf{Y}_i) &= E(\mathbf{Y}_N - \sqrt{N}\,\boldsymbol{\theta})\mathbf{Y}_i' = E(\mathbf{Y}_N\mathbf{Y}_i') \\
&= E\left(\sum_{j=1}^{N} \gamma_{Nj}\mathbf{x}_j\right)\left(\sum_{k=1}^{N} \gamma_{ik}\mathbf{x}_k'\right) \\
&= \sum_{j=1}^{N}\sum_{k=1}^{N} \gamma_{Nj}\gamma_{ik}E(\mathbf{x}_j\mathbf{x}_k') \\
&= (\boldsymbol{\Sigma} + \boldsymbol{\theta}\boldsymbol{\theta}')\left(\sum_{j=1}^{N} \gamma_{Nj}\gamma_{ij}\right) + \boldsymbol{\theta}\boldsymbol{\theta}'\left(\sum_{\substack{j=1 \\ j\neq k}}^{N}\sum_{k=1}^{N} \gamma_{Nj}\gamma_{ik}\right).
\end{aligned}$$

By orthogonality of $\boldsymbol{\Gamma}$ the first term vanishes. Moreover, since $\gamma_{Nj} = \gamma_{Nk} = N^{-1/2}$,

$$\operatorname{cov}(\mathbf{Y}_N,\mathbf{Y}_i) = \boldsymbol{\theta}\boldsymbol{\theta}'\left(\sum_{j=1}^{N}\sum_{k=1}^{N} \gamma_{Nk}\gamma_{ik}\right),$$

which must also vanish by orthogonality of $\boldsymbol{\Gamma}$. Hence,

$$\operatorname{cov}(\mathbf{Y}_N,\mathbf{Y}_i) = \mathbf{0}, \qquad i = 1,\ldots,N-1.$$

Since $\bar{\mathbf{x}}$ depends only on $\mathbf{Y}_N$, and, as is shown in Complement 5.1, $\mathbf{V}$ depends only on $\mathbf{Y}_1,\ldots,\mathbf{Y}_{N-1}$, $\bar{\mathbf{x}}$ and $\mathbf{V}$ must be independent.

7.1.2 Maximum Likelihood Estimation

Define, for $\boldsymbol{\Lambda} \equiv \boldsymbol{\Sigma}^{-1}$, $L(\boldsymbol{\theta},\boldsymbol{\Lambda}) \equiv \log p(\mathbf{x}_1,\ldots,\mathbf{x}_N|\boldsymbol{\theta},\boldsymbol{\Lambda})$. Then, from (7.1.3),

$$L(\boldsymbol{\theta},\boldsymbol{\Lambda}) = \frac{N}{2}\log|\boldsymbol{\Lambda}| - \frac{1}{2}\operatorname{tr}\boldsymbol{\Lambda}\mathbf{V} - \frac{N}{2}\operatorname{tr}\boldsymbol{\Lambda}(\bar{\mathbf{x}}-\boldsymbol{\theta})(\bar{\mathbf{x}}-\boldsymbol{\theta})' - \log(2\pi)^{Np/2}.$$

Differentiating with respect to $\boldsymbol{\theta}$ gives

$$\frac{\partial L(\boldsymbol{\theta},\boldsymbol{\Lambda})}{\partial\boldsymbol{\theta}} = -\frac{N}{2}\frac{\partial}{\partial\boldsymbol{\theta}}\operatorname{tr}\boldsymbol{\Lambda}(\bar{\mathbf{x}}-\boldsymbol{\theta})(\bar{\mathbf{x}}-\boldsymbol{\theta})'.$$

Equivalently,

$$\begin{aligned}\frac{\partial L(\boldsymbol{\theta},\boldsymbol{\Lambda})}{\partial\boldsymbol{\theta}} &= -\frac{N}{2}\frac{\partial}{\partial\boldsymbol{\theta}}(\bar{\mathbf{x}}-\boldsymbol{\theta})'\boldsymbol{\Lambda}(\bar{\mathbf{x}}-\boldsymbol{\theta}),\\ &= -\frac{N}{2}\cdot 2\boldsymbol{\Lambda}(\boldsymbol{\theta}-\bar{\mathbf{x}}).\end{aligned}$$

Setting the derivative equal to zero and solving gives $\hat{\boldsymbol{\theta}} = \bar{\mathbf{x}}$. Moreover, since

$$\frac{\partial^2 L(\boldsymbol{\theta},\boldsymbol{\Lambda})}{\partial\boldsymbol{\theta}\partial\boldsymbol{\theta}'} = -N\boldsymbol{\Lambda}$$

which is clearly negative definite (for nondegenerate distributions), $\bar{\mathbf{x}}$ corresponds to a maximum of $L(\boldsymbol{\theta},\boldsymbol{\Lambda})$.

Similarly [from (2.14.2)′ and (5) of Section 2.14.2],

$$\frac{\partial L(\boldsymbol{\theta},\boldsymbol{\Lambda})}{\partial\boldsymbol{\Lambda}} = \frac{N}{2}(2\boldsymbol{\Lambda}^{-1} - \operatorname{diag}\boldsymbol{\Lambda}^{-1}) - \frac{1}{2}(2\mathbf{A} - \operatorname{diag}\mathbf{A}),$$

where $\mathbf{A} \equiv \Sigma_1^N(\mathbf{x}_j-\boldsymbol{\theta})(\mathbf{x}_j-\boldsymbol{\theta})'$. [*Note:* This is the same $\mathbf{A}$ used in Section 7.1.1.] Setting the derivative equal to zero and solving gives the estimator

$$\hat{\boldsymbol{\Sigma}} = \frac{\hat{\mathbf{A}}}{N} = \frac{1}{N}\sum_1^N(\mathbf{x}_j-\hat{\boldsymbol{\theta}})(\mathbf{x}_j-\hat{\boldsymbol{\theta}})'.$$

Note from Bellman (1960, p. 125) that $\log|\Lambda|$ is concave, and since $\operatorname{tr}(\Lambda\mathbf{V})$ is linear in Λ, $L(\boldsymbol{\theta},\Lambda)$ is concave in Λ, so $\hat{\Sigma}$ must correspond to a maximum.

and this matrix is clearly negative definite. Summarizing:

Theorem (7.1.3): For $\mathbf{x}_1,\ldots,\mathbf{x}_N$ independent p-vectors from $N(\boldsymbol{\theta},\boldsymbol{\Sigma})$, the maximum likelihood estimators of $\boldsymbol{\theta}$, $\boldsymbol{\Sigma}$ are

$$\hat{\boldsymbol{\theta}} = \bar{\mathbf{x}},$$

and

$$\hat{\boldsymbol{\Sigma}} = \frac{1}{N}\sum_{1}^{N}(\mathbf{x}_j - \bar{\mathbf{x}})(\mathbf{x}_j - \bar{\mathbf{x}})'. \tag{7.1.4}$$

7.1.3 Distributions of Sufficient Statistics

To make interval estimates of the parameters of a multivariate Normal distribution or to test hypotheses, it is of interest to know the distributions of the sufficient statistics. The following two theorems give the basic results.

Theorem (7.1.4): If $\mathbf{x}_1,\ldots,\mathbf{x}_N$ are independent p-variate vector observations from $N(\boldsymbol{\theta},\boldsymbol{\Sigma})$ and $\bar{\mathbf{x}}$ is the sample mean vector,

$$\mathcal{L}(\bar{\mathbf{x}}) = N\left(\boldsymbol{\theta}, \frac{\boldsymbol{\Sigma}}{N}\right).$$

Proof: Since $\mathcal{L}(\mathbf{x}_j) = N(\boldsymbol{\theta},\boldsymbol{\Sigma})$, and the $\mathbf{x}_j$'s are independent, $\mathcal{L}(\Sigma_1^N \mathbf{x}_j) = N(N\boldsymbol{\theta},N\boldsymbol{\Sigma})$. Division of the sum by N gives the result.

Theorem (7.1.5): If $\mathbf{x}_1,\ldots,\mathbf{x}_N$ are independent p-variate vector observations from $N(\boldsymbol{\theta},\boldsymbol{\Sigma})$ and $\mathbf{V}$ is the sample covariance matrix,

$$\mathcal{L}(\mathbf{V}) = W(\boldsymbol{\Sigma}, p, N - 1 \equiv n).$$

Proof: See Complement 5.1.

Example (7.1.1): Suppose a holding company owns p firms. Let $\mathbf{x}_\alpha : p \times 1$ denote the vector of profits for each of its firms during week α, $\alpha = 1,\ldots,N$. Assume the $\mathbf{x}_\alpha$ vectors are mutually independent and Normally distributed according to $N(\boldsymbol{\theta},\boldsymbol{\Sigma}_0)$, where $\boldsymbol{\Sigma}_0$ is a known matrix (having been determined in the past). It is expected that $\boldsymbol{\theta}$ will be estimated each year, so what is really of interest is an estimator that will have merit when averaged over many samples. Find an interval estimate of θ_1, the first element of $\boldsymbol{\theta}$, which will provide 95 percent confidence.

The solution to this problem is obtained by immediately going to the sufficient statistic. Since $\boldsymbol{\Sigma}_0$ is known, $\bar{\mathbf{x}}$ is sufficient for $\boldsymbol{\theta}$, and $\mathcal{L}(\bar{\mathbf{x}}) =$

$N(\boldsymbol{\theta},N^{-1}\boldsymbol{\Sigma}_0)$. Moreover, if $\bar{x}_1$ denotes the first element of $\bar{\mathbf{x}}$, it is easy to check that $\bar{x}_1$ is sufficient for θ_1. Hence, if $\sigma_{11}^{(0)}$ denotes the variance of the first element of any $\mathbf{x}_\alpha$, $\alpha = 1,\ldots,N$, the distribution of the sufficient statistic for θ_1 is just the marginal distribution of $\bar{x}_1$, given by

$$\mathcal{L}(\bar{x}_1) = N(\theta_1,N^{-1}\sigma_{11}^{(0)}).$$

Thus, if $y \equiv N^{1/2}(\sigma_{11}^{(0)})^{-1/2}(\bar{x}_1 - \theta_1)$, $\mathcal{L}(y) = N(0,1)$, so that

$$.95 = P\{-1.96 \leq y \leq 1.96\}.$$

Therefore, a 95 percent confidence interval for θ_1 is given by

$$\left(\bar{x}_1 - 1.96\sqrt{\frac{\sigma_{11}^{(0)}}{N}},\ \bar{x}_1 + 1.96\sqrt{\frac{\sigma_{11}^{(0)}}{N}}\right).$$

Joint confidence statements on several parameters simultaneously may be made using the joint distribution.

7.1.4 Bayes Estimation Using Decision Theory

Suppose $\mathbf{X}: p \times 1$ has density $p(\mathbf{x}|\boldsymbol{\theta})$ and there is a loss function associated with estimating $\boldsymbol{\theta}$ by $\hat{\boldsymbol{\theta}}$, given by $L(\boldsymbol{\theta},\hat{\boldsymbol{\theta}})$. Then the risk, $\rho(\hat{\boldsymbol{\theta}}|\boldsymbol{\theta})$, of using $\hat{\boldsymbol{\theta}}$ is defined to be $E[L(\boldsymbol{\theta},\hat{\boldsymbol{\theta}})|\boldsymbol{\theta}]$, so that the risk depends upon which $\boldsymbol{\theta}$ is chosen. Thus,

$$\rho(\hat{\boldsymbol{\theta}}|\boldsymbol{\theta}) = \int L(\boldsymbol{\theta},\hat{\boldsymbol{\theta}})p(\mathbf{x}_1,\ldots,\mathbf{x}_N|\boldsymbol{\theta})\,d\mathbf{x}_1 \cdots d\mathbf{x}_N,$$

where $p(\mathbf{x}_1,\ldots,\mathbf{x}_N|\boldsymbol{\theta})$ denotes the likelihood function of a sample of N observations.

Let $p(\boldsymbol{\theta})$ denote a prior distribution for $\boldsymbol{\theta}$. Then, if $p(\boldsymbol{\theta}|\mathbf{x}_1,\ldots,\mathbf{x}_N)$ denotes the posterior distribution of $\boldsymbol{\theta}$ given the sample, the Bayes estimator of $\boldsymbol{\theta}$ is defined to be the estimator $\hat{\boldsymbol{\theta}}$ (rule for estimating) which minimizes the loss averaged over the posterior distribution; that is, the posterior risk is

$$\rho_p(\mathbf{x}_1,\ldots,\mathbf{x}_N) = \min_{\hat{\boldsymbol{\theta}}} \int L(\boldsymbol{\theta},\hat{\boldsymbol{\theta}})p(\boldsymbol{\theta}|\mathbf{x}_1,\ldots,\mathbf{x}_N)\,d\boldsymbol{\theta}.$$

The Bayes risk is

$$E[\rho_p(\mathbf{x}_1,\ldots,\mathbf{x}_N)] = \int \rho_p(\mathbf{x}_1,\ldots,\mathbf{x}_N)p(\mathbf{x}_1,\ldots,\mathbf{x}_N)d\mathbf{x}_1 \cdots d\mathbf{x}_N,$$

where $p(\mathbf{x}_1,\ldots,\mathbf{x}_N)$ is the unconditional density of the sample. Hence, the Bayes risk, for an arbitrary $\hat{\boldsymbol{\theta}}$, is

$$\int p(\mathbf{x}_1,\ldots,\mathbf{x}_N)\int \min_{\hat{\theta}} L(\boldsymbol{\theta},\hat{\boldsymbol{\theta}})p(\boldsymbol{\theta}\,|\,\mathbf{x}_1,\ldots,\mathbf{x}_N)d\boldsymbol{\theta}\,d\mathbf{x}_1 \ldots d\mathbf{x}_N$$

$$= \min_{\hat{\theta}} \int\{\int L(\boldsymbol{\theta},\hat{\boldsymbol{\theta}})p(\mathbf{x}_1,\ldots,\mathbf{x}_N|\boldsymbol{\theta})\,d\mathbf{x}_1,\ldots,d\mathbf{x}_N\}p(\boldsymbol{\theta})\,d\boldsymbol{\theta} = \int \rho(\hat{\boldsymbol{\theta}}|\boldsymbol{\theta})p(\boldsymbol{\theta})\,d\boldsymbol{\theta}.$$

Since the Bayes estimator also minimizes this integral, the Bayes estimator minimizes the Bayes risk, when it exists. That is, there are occasions when the Bayes risk does not exist, such as when improper priors are used for $\boldsymbol{\theta}$. Then, the multiple integral taken over both the sample and parameter spaces, defined above, diverges; but the Bayes estimator still may exist since the integral of the loss weighted by the posterior distribution may exist (the posterior distribution must be proper even with an improper prior).

Example (7.1.2): Let $\mathbf{X}: p \times 1$ have moments $E\mathbf{X} = \boldsymbol{\theta}$, var $\mathbf{X} = \boldsymbol{\Sigma}$, for $\boldsymbol{\Sigma} > 0$. Define the multivariate quadratic loss function

$$L(\boldsymbol{\theta},\hat{\boldsymbol{\theta}}) = (\boldsymbol{\theta} - \hat{\boldsymbol{\theta}})'(\boldsymbol{\theta} - \hat{\boldsymbol{\theta}}). \tag{7.1.5}$$

The loss averaged with respect to the posterior distribution is given by

$$E[L(\boldsymbol{\theta},\hat{\boldsymbol{\theta}})|\text{sample}] = E[(\boldsymbol{\theta} - \hat{\boldsymbol{\theta}})'(\boldsymbol{\theta} - \hat{\boldsymbol{\theta}})|\text{sample}],$$

or

$$E[L(\boldsymbol{\theta},\hat{\boldsymbol{\theta}})|\text{sample}] = E(\boldsymbol{\theta}'\boldsymbol{\theta}|\text{sample}) - 2\hat{\boldsymbol{\theta}}'E(\boldsymbol{\theta}|\text{sample}) + \hat{\boldsymbol{\theta}}'\hat{\boldsymbol{\theta}}.$$

To find the minimum risk conditional on the sample, differentiate with respect to $\hat{\boldsymbol{\theta}}$, set the result equal to zero and solve. This gives for the Bayes estimator

$$\hat{\boldsymbol{\theta}}_{\text{Bayes}} = E(\boldsymbol{\theta}|\text{sample}). \tag{7.1.6}$$

That is, the Bayes estimator is the mean of the posterior distribution.

Note that since

$$\frac{\partial^2 E[L(\boldsymbol{\theta},\hat{\boldsymbol{\theta}})|\text{sample}]}{\partial\hat{\boldsymbol{\theta}}\partial\hat{\boldsymbol{\theta}}'} = 2\mathbf{I},$$

a positive definite matrix, $\hat{\boldsymbol{\theta}}_{\text{Bayes}}$ actually corresponds to a minimum value.

7.1.5 Bayes Estimation without Decision Theory

When loss functions are unavailable or not deemed appropriate, the analyst is left with a choice of which measure of location of the posterior distribution he should use. Often, the mode is selected in such circumstances since it corresponds to maximum posterior probability (with minimum information available, a uniform prior density over the real line makes the mode of the posterior distribution correspond to the maximum likelihood estimator).

The importance of assessing parameters of prior distributions as precisely as possible has been pointed out by Dempster (1963) and (1966), in that in small samples, results widely different from sampling theory results can be obtained (for estimation of the parameters of a covariance matrix).

7.1.6 Bayes Estimates in the Multivariate Normal Distribution

Suppose $\mathbf{x}_1,\ldots,\mathbf{x}_N$ are independent and identically distributed as $N(\boldsymbol{\theta},\boldsymbol{\Sigma})$. Then if

$$\bar{\mathbf{x}} = \frac{1}{N}\sum_{j=1}^{N} \mathbf{x}_j, \text{ and } \mathbf{V} = \sum_{1}^{N} (\mathbf{x}_j - \bar{\mathbf{x}})(\mathbf{x}_j - \bar{\mathbf{x}})',$$

by Theorem (7.1.2), if $n \equiv N - 1$, the joint density of $\bar{\mathbf{x}}$ and $\mathbf{V}$ is given by

$$p(\bar{\mathbf{x}},\mathbf{V}|\boldsymbol{\theta},\boldsymbol{\Sigma}) \propto \frac{|\mathbf{V}|^{(n-p-1)/2}}{|\boldsymbol{\Sigma}|^{N/2}} \exp\left\{-\tfrac{1}{2} \operatorname{tr} \boldsymbol{\Sigma}^{-1}[\mathbf{V} + N(\bar{\mathbf{x}} - \boldsymbol{\theta})(\bar{\mathbf{x}} - \boldsymbol{\theta})']\right\}.$$

Case I. Diffuse Prior: Assume [see Equations (3.6.1) and (3.6.5)]

$$p(\boldsymbol{\theta},\boldsymbol{\Sigma}) \propto \frac{1}{|\boldsymbol{\Sigma}|^{(p+1)/2}}. \tag{7.1.7}$$

The posterior distribution becomes

$$p(\boldsymbol{\theta},\boldsymbol{\Sigma}|\bar{\mathbf{x}},\mathbf{V}) \propto \frac{1}{|\boldsymbol{\Sigma}|^{(N+p+1)/2}} \exp\left\{-\tfrac{1}{2} \operatorname{tr} \boldsymbol{\Sigma}^{-1}[\mathbf{V} + N(\bar{\mathbf{x}} - \boldsymbol{\theta})(\bar{\mathbf{x}} - \boldsymbol{\theta})']\right\}. \tag{7.1.8}$$

The marginal posterior distribution of $\boldsymbol{\theta}$ is found by using the normalizing constant of the inverted Wishart distribution [Equation (5.2.1)]. The result is

$$p(\boldsymbol{\theta}|\bar{\mathbf{x}},\mathbf{V}) \propto \frac{1}{|\mathbf{V} + N(\boldsymbol{\theta} - \bar{\mathbf{x}})(\boldsymbol{\theta} - \bar{\mathbf{x}})'|^{N/2}}.$$

Use (2.4.3) and simplify to get

$$p(\boldsymbol{\theta}|\bar{\mathbf{x}},\mathbf{V}) \propto \frac{1}{\{1 + N(\boldsymbol{\theta} - \bar{\mathbf{x}})'\mathbf{V}^{-1}(\boldsymbol{\theta} - \bar{\mathbf{x}})\}^{N/2}}. \tag{7.1.9}$$

Thus, $(\boldsymbol{\theta}|\bar{\mathbf{x}},\mathbf{V})$ follows a multivariate Student **t**-distribution. If there is a quadratic loss function as given in (7.1.5), the Bayes estimator of $\boldsymbol{\theta}$ is the mean of the multivariate **t**-distribution; that is,

$$\boldsymbol{\theta}_{\text{Bayes}} = \bar{\mathbf{x}}, \qquad N > 1.$$

This estimator coincides with both the mode and the maximum likelihood estimator.

Integrating (7.1.8) with respect to $\boldsymbol{\theta}$ ($\boldsymbol{\theta}|\boldsymbol{\Sigma}$ follows a multivariate Normal distribution) gives the marginal posterior density with respect to $\boldsymbol{\Sigma}$. The result is that

$$\mathcal{L}(\boldsymbol{\Sigma}|\bar{\mathbf{x}},\mathbf{V}) = W^{-1}(\mathbf{V}, p, N + p). \tag{7.1.10}$$

Hence, the mean of the posterior distribution is [see (5.2.4)]

$$E(\boldsymbol{\Sigma}|\bar{\mathbf{x}},\mathbf{V}) = \frac{1}{N - p - 2}\mathbf{V}.$$

It is not hard to check with the aid of (2.14.5) and (2.14.11) that this estimator is the Bayes estimator for a quadratic loss function of the form

$$L(\mathbf{\Sigma},\hat{\mathbf{\Sigma}}) = \text{tr } (\mathbf{\Sigma} - \hat{\mathbf{\Sigma}})^2. \tag{7.1.11}$$

[Merely differentiate $EL(\mathbf{\Sigma},\hat{\mathbf{\Sigma}})$ with respect to $\hat{\mathbf{\Sigma}}$, set the result equal to zero, and solve.] See Exercise 2.11.

Case II. Natural Conjugate Prior: Rewrite the likelihood function in the form

$$p(\bar{\mathbf{x}},\mathbf{V}|\boldsymbol{\theta},\mathbf{\Sigma}) \propto \frac{\exp(-\frac{1}{2}\text{tr}\,\mathbf{\Sigma}^{-1}\mathbf{V})}{|\mathbf{\Sigma}|^{(N-1)/2}} \cdot \frac{\exp\left[-\dfrac{N}{2}(\boldsymbol{\theta} - \bar{\mathbf{x}})'\mathbf{\Sigma}^{-1}(\boldsymbol{\theta} - \bar{\mathbf{x}})\right]}{|\mathbf{\Sigma}|^{1/2}}. \tag{7.1.12}$$

If in this equation, $\boldsymbol{\theta}$ and $\mathbf{\Sigma}$ are thought of as the random variables, with $\bar{\mathbf{x}}$ and $\mathbf{V}$ held fixed (see Section 3.6.4), $\boldsymbol{\theta}$ given $\mathbf{\Sigma}$ follows $N(\bar{\mathbf{x}},\mathbf{\Sigma}/N)$, and $\mathbf{\Sigma}$ follows an inverted Wishart distribution. Hence, by enriching the distributions and using $p(\boldsymbol{\theta},\mathbf{\Sigma}) = p(\boldsymbol{\theta}|\mathbf{\Sigma})p(\mathbf{\Sigma})$, the prior may be written as

$$p(\boldsymbol{\theta},\mathbf{\Sigma}) \propto \frac{1}{|\mathbf{\Sigma}|^{1/2}} \exp\left[-\tfrac{1}{2}(\boldsymbol{\theta} - \boldsymbol{\phi})'\mathbf{\Sigma}^{-1}(\boldsymbol{\theta} - \boldsymbol{\phi})\right] \cdot \frac{\exp(-\frac{1}{2}\text{ tr }\mathbf{\Sigma}^{-1}\mathbf{G})}{|\mathbf{\Sigma}|^{m/2}}, \tag{7.1.13}$$

where $\boldsymbol{\phi}$, $\mathbf{G}$, and m are parameters of the prior distribution, $m > 2p$ (see also (3.6.19)). The joint posterior distribution becomes

$$p(\boldsymbol{\theta},\mathbf{\Sigma}|\bar{\mathbf{x}},\mathbf{V}) \propto \frac{\exp\{-\frac{1}{2}\text{ tr }\mathbf{\Sigma}^{-1}[\mathbf{V} + \mathbf{G} + (\boldsymbol{\theta} - \boldsymbol{\phi})(\boldsymbol{\theta} - \boldsymbol{\phi})' + N(\boldsymbol{\theta} - \bar{\mathbf{x}})(\boldsymbol{\theta} - \bar{\mathbf{x}})']\}}{|\mathbf{\Sigma}|^{(N+m+1)/2}}. \tag{7.1.14}$$

The marginal posterior distribution of $\boldsymbol{\theta}$ is found in the same way as in the diffuse prior case (integrating an inverted Wishart distribution). The result is

$$p(\boldsymbol{\theta}|\bar{\mathbf{x}},\mathbf{V}) \propto \frac{1}{|\mathbf{V} + \mathbf{G} + (\boldsymbol{\theta} - \boldsymbol{\phi})(\boldsymbol{\theta} - \boldsymbol{\phi})' + N(\boldsymbol{\theta} - \bar{\mathbf{x}})(\boldsymbol{\theta} - \bar{\mathbf{x}})'|^{(N+m-p)/2}}.$$

Completing the square on $\boldsymbol{\theta}$ gives

$$p(\boldsymbol{\theta}|\bar{\mathbf{x}},\mathbf{V}) \propto \frac{1}{|\mathbf{C} + (\boldsymbol{\theta} - \mathbf{a})(\boldsymbol{\theta} - \mathbf{a})'|^{(N+m-p)/2}},$$

where

$$\mathbf{a} = \frac{\boldsymbol{\phi} + N\bar{\mathbf{x}}}{1 + N}, \qquad \mathbf{C} = \left[\frac{\mathbf{V} + \mathbf{G}}{1 + N} + \frac{N}{(1 + N)^2}(\boldsymbol{\phi} - \bar{\mathbf{x}})(\boldsymbol{\phi} - \bar{\mathbf{x}})'\right].$$

Equivalently, using (2.4.3),

$$p(\boldsymbol{\theta}|\bar{\mathbf{x}},\mathbf{V}) \propto \frac{1}{\{1 + (\boldsymbol{\theta} - \mathbf{a})'\mathbf{C}^{-1}(\boldsymbol{\theta} - \mathbf{a})\}^{(N+m-p)/2}},$$

a multivariate **t**-distribution. Hence, with respect to the quadratic loss function given in (7.1.5), the Bayes estimator is $E(\boldsymbol{\theta}|\bar{\mathbf{x}},\mathbf{V})$, or

$$\hat{\boldsymbol{\theta}}_{\text{Bayes}} = \frac{\boldsymbol{\phi} + N\bar{\mathbf{x}}}{1 + N}, \qquad N + m - p > 1.$$

Observe that $\hat{\boldsymbol{\theta}}_{\text{Bayes}}$ approaches the maximum likelihood estimator in large samples, as is predicted in (4.6.1). Next concentrate on $\boldsymbol{\Sigma}$.

The marginal posterior distribution of $\boldsymbol{\Sigma}$ is obtained by integrating (7.1.14) with respect to $\boldsymbol{\theta}$. This integration is carried out by first completing the square on $\boldsymbol{\theta}$ in the exponent of (7.1.14). This gives

$$p(\boldsymbol{\Sigma}|\bar{\mathbf{x}},\mathbf{V}) \propto \frac{\exp[-\frac{1}{2}\operatorname{tr}\boldsymbol{\Sigma}^{-1}(\mathbf{V} + \mathbf{G})]}{|\boldsymbol{\Sigma}|^{(N+m+1)/2}} \int \exp\left\{-\frac{(1+N)}{2}[(\boldsymbol{\theta} - \mathbf{a})'\boldsymbol{\Sigma}^{-1}(\boldsymbol{\theta} - \mathbf{a}) + B]\right\} d\boldsymbol{\theta},$$

where [compare with (3.6.20)]

$$B = \frac{\boldsymbol{\phi}'\boldsymbol{\Sigma}^{-1}\boldsymbol{\phi} + N\bar{\mathbf{x}}'\boldsymbol{\Sigma}^{-1}\bar{\mathbf{x}}}{1 + N} - \mathbf{a}'\boldsymbol{\Sigma}^{-1}\mathbf{a}.$$

The integral is now easily evaluated using the normalizing constant of a multivariate Normal distribution. The final result is that

$$\mathcal{L}(\boldsymbol{\Sigma}|\bar{\mathbf{x}},\mathbf{V}) = W^{-1}\left(\mathbf{V} + \mathbf{G} + \frac{N}{1+N}(\bar{\mathbf{x}} - \boldsymbol{\phi})(\bar{\mathbf{x}} - \boldsymbol{\phi})', p, N + m\right).$$

Consequently, for the loss function in (7.1.11), the Bayes estimator is $E(\boldsymbol{\Sigma}|\bar{\mathbf{x}},\mathbf{V})$, and is found from (5.2.4) to be

$$\hat{\boldsymbol{\Sigma}}_{\text{Bayes}} = \frac{\mathbf{V} + \mathbf{G} + \dfrac{N}{1+N}(\bar{\mathbf{x}} - \boldsymbol{\phi})(\bar{\mathbf{x}} - \boldsymbol{\phi})'}{N + m - 2p - 2}, \qquad N + m - 2p > 2.$$

7.2 ADMISSIBILITY PROBLEMS IN ESTIMATION

Let $\mathbf{x}_1,\ldots,\mathbf{x}_N$ be independent p-variate observations from $N(\boldsymbol{\theta},\mathbf{I}_p)$. If the population is univariate ($p = 1$), it is well known that for loss functions proportional to

$$L(\boldsymbol{\theta},\hat{\boldsymbol{\theta}}) = (\boldsymbol{\theta} - \hat{\boldsymbol{\theta}})'(\boldsymbol{\theta} - \hat{\boldsymbol{\theta}}), \tag{7.2.1}$$

$\hat{\boldsymbol{\theta}} = \bar{\mathbf{x}}$, the sample mean, is an admissible estimator of $\boldsymbol{\theta}$. That is, there is no estimator $\hat{\boldsymbol{\theta}}^*$ for which the risk

$$E[L(\hat{\boldsymbol{\theta}}^*,\boldsymbol{\theta})|\boldsymbol{\theta}] \leq E[L(\bar{\mathbf{x}},\boldsymbol{\theta})|\boldsymbol{\theta}]$$

for all $\boldsymbol{\theta}$, and

$$E[L(\hat{\boldsymbol{\theta}}^*,\boldsymbol{\theta}_0)|\boldsymbol{\theta}_0] < E[L(\bar{\mathbf{x}},\boldsymbol{\theta}_0)|\boldsymbol{\theta}_0]$$

for some $\boldsymbol{\theta} = \boldsymbol{\theta}_0$. It was generally believed that this optimality property of $\bar{\mathbf{x}}$ held for multivariate populations ($p > 1$) as well as for the univariate case. However, Stein (1956) showed that although $\bar{\mathbf{x}}$ does remain admissible for $p = 2$, it becomes inadmissible for $p \geq 3$. He demonstrated this by exhibiting a class of estimators that has the property attributed to $\hat{\boldsymbol{\theta}}^*$ above.

In 1961 James and Stein showed that for just one observation, $\mathbf{x}$, from $N(\boldsymbol{\theta},\mathbf{I}_p)$, for $p \geq 3$,

$$\hat{\boldsymbol{\theta}}^* = \left(1 - \frac{p-2}{\mathbf{x}'\mathbf{x}}\right)\mathbf{x}$$

has smaller expected loss (risk) than $\mathbf{x}$, for all $\boldsymbol{\theta}$; that is, $\hat{\boldsymbol{\theta}}^*$ "dominates" $\mathbf{x}$. More generally, if $\mathbf{x}_1,\ldots,\mathbf{x}_N$ are independent p-vectors following $N(\boldsymbol{\theta},\boldsymbol{\Sigma})$, $\boldsymbol{\Sigma} > 0$, and

$$L[\hat{\boldsymbol{\theta}};(\boldsymbol{\theta},\boldsymbol{\Sigma})] = (\boldsymbol{\theta} - \hat{\boldsymbol{\theta}})'\boldsymbol{\Sigma}^{-1}(\boldsymbol{\theta} - \hat{\boldsymbol{\theta}}), \tag{7.2.2}$$

if

$$\bar{\mathbf{x}} = \frac{1}{N}\sum_{1}^{N} \mathbf{x}_j, \qquad \mathbf{S} = \frac{1}{N}\sum_{1}^{N} (\mathbf{x}_j - \bar{\mathbf{x}})(\mathbf{x}_j - \bar{\mathbf{x}})',$$

and if

$$c = \frac{p-2}{N-(p-2)},$$

$$\hat{\boldsymbol{\theta}}^* = \left(1 - \frac{c}{\bar{\mathbf{x}}'\mathbf{S}^{-1}\bar{\mathbf{x}}}\right)\bar{\mathbf{x}} \tag{7.2.3}$$

will dominate $\bar{\mathbf{x}}$. Since it is possible for $\bar{\mathbf{x}}'\mathbf{S}^{-1}\bar{\mathbf{x}}$ to become smaller than c even if the components of $\boldsymbol{\theta}$ are positive, it is not unreasonable to try to improve upon $\hat{\boldsymbol{\theta}}^*$ by using (see Stein, 1962)

$$\hat{\boldsymbol{\theta}}^{**} = \left[\max\left(0,\ 1 - \frac{c}{\bar{\mathbf{x}}'\mathbf{S}^{-1}\bar{\mathbf{x}}}\right)\right]\bar{\mathbf{x}}. \tag{7.2.4}$$

Note that all measurements in the problem were taken with respect to an arbitrary origin at zero. If another origin (at $\mathbf{x}_0$ say) is more appropriate in a given problem, corrected observations $(\mathbf{x}_\alpha - \mathbf{x}_0)$, $\alpha = 1,\ldots,N$, should be used[1] in $\boldsymbol{\theta}^{**}$ instead of $\mathbf{x}_\alpha$.

[1] For example, if a Belgian and an American measure temperature, pressure, and humidity in the same place at the same time, the Belgian might find the component of $\bar{\mathbf{x}}$ for temperature is zero (centigrade) and so from (7.2.3) that component of $\hat{\boldsymbol{\theta}}^*$

A number of remarks relative to these results are in order.

(1) The general effect of the Stein estimator (7.2.3) or (7.2.4) is to draw $\bar{\mathbf{x}}$ closer in toward the origin.
(2) Non-Normal distributions and nonquadratic loss functions also give rise to estimators that should be modified when admissibility of estimators in high numbers of dimensions is of concern in a problem.
(3) For large sample sizes the inadmissibility problem disappears and is of no concern. Also, it will not be a troublesome problem if p is small. The major difficulties arise for large p and small N.
(4) A Bayesian with one observation, $\mathbf{x}$, on $N(\boldsymbol{\theta},\mathbf{I}_p)$, using a diffuse prior for a loss function given by (7.2.1), would be inclined to estimate $\boldsymbol{\theta}$ by $\hat{\boldsymbol{\theta}} = E(\boldsymbol{\theta}|\mathbf{x}) = \mathbf{x}$. Moreover, with a diffuse prior, $p(\boldsymbol{\theta})$ is constant over the entire p-dimensional space—an improper distribution. So if the risk is averaged over the parameter space, the result is not finite. In the case of a natural conjugate prior, or other proper prior, the Bayesian estimate is admissible. That is, the Bayesian procedure will not be in conflict with the decision-theoretic difficulties raised by the Stein results, as long as the prior distribution is proper. When the prior distribution is improper (diffuse), Bayesian estimates will still be useful so long as admissibility is not an overriding consideration. Moreover, by using appropriate priors, Bayesian estimators of the Stein type may be constructed (see, for example, Exercise 7.12).

Stein type estimators have been developed for the parameters of Poisson distributions, in Tsui and Press, 1977a, 1977b. For an account of more recent work relating to Stein-type estimation, see Judge and Bock, 1978. For an excellent descriptive discussion of the Stein shrinker phenomenon, see Efron and Morris, 1977.

7.3 HYPOTHESIS TESTING

7.3.1 Testing for a Given Mean with Known Covariances

Suppose $\mathbf{x}_1,\ldots,\mathbf{x}_N$ are independent p-vectors from $N(\boldsymbol{\theta},\boldsymbol{\Sigma}_0)$, where $\boldsymbol{\Sigma}_0$ is a known covariance matrix, $\boldsymbol{\Sigma}_0 > 0$ (perhaps many previous samples have established the errors (that is, the variances) and correlations very well, but the means are different from one situation to the next). It may

is not corrected. However, the American will measure in Fahrenheit and will find that the temperature component of $\bar{\mathbf{x}}$ is 32. Hence, for him the temperature component of $\hat{\boldsymbol{\theta}}^*$ will be corrected (reduced somewhat). Their final answers can never possibly agree, even after converting both to the same units, unless they used the same origin at the start.

be of interest to test the hypothesis $H: \boldsymbol{\theta} = \boldsymbol{\theta}_0$, a known vector, versus alternatives such as

$$A_1: \boldsymbol{\theta} \neq \boldsymbol{\theta}_0; \quad A_2: \theta_j > \theta_{j0}, \quad j = 1,\ldots,p; \quad A_3: \theta_j < \theta_{j0}, \quad j = 1,\ldots,p.$$

Since $\bar{\mathbf{x}}$ is sufficient it is only necessary to consider $\bar{\mathbf{x}}$, where $\mathcal{L}(\bar{\mathbf{x}}) = N[\boldsymbol{\theta},(1/N)\boldsymbol{\Sigma}_0]$. Since it is of interest to study whether $(\boldsymbol{\theta} - \boldsymbol{\theta}_0)$ deviates from zero, the test should be based on this quantity, and clearly, $\mathcal{L}(\bar{\mathbf{x}} - \boldsymbol{\theta}_0) = N[\boldsymbol{\theta} - \boldsymbol{\theta}_0, (1/N)\boldsymbol{\Sigma}_0]$. Moreover, since $(1/N)\boldsymbol{\Sigma}_0$ is a known quantity, the data can be easily transformed into a vector with independent components, and the problem can be reexamined in terms of the transformed sufficient statistic. Accordingly, if $\mathbf{y} = \sqrt{N}\,\boldsymbol{\Sigma}_0^{-1/2}(\bar{\mathbf{x}} - \boldsymbol{\theta}_0)$, and $\boldsymbol{\phi} = \sqrt{N}\,\boldsymbol{\Sigma}_0^{-1/2}(\boldsymbol{\theta} - \boldsymbol{\theta}_0)$, $\mathbf{y}$ is an observable quantity, $\mathcal{L}(\mathbf{y}) = N(\boldsymbol{\phi},\mathbf{I})$, and the transformed problem is $H: \boldsymbol{\phi} = \mathbf{0}$ versus, say, $A_1: \boldsymbol{\phi} \neq \mathbf{0}$. The components of $\mathbf{y}$ are all independent (since $\mathbf{y}$ has an identity covariance matrix) so that

$$\mathcal{L}(\mathbf{y}'\mathbf{y}|\boldsymbol{\phi}) = \chi_p'^2(\boldsymbol{\phi}'\boldsymbol{\phi}). \tag{7.3.1}$$

That is, conditional on $\boldsymbol{\phi}$, $\mathbf{y}'\mathbf{y}$ follows a noncentral chi-square distribution with p d.f. and noncentrality parameter $\boldsymbol{\phi}'\boldsymbol{\phi}$. Hence, under H, $\mathbf{y}'\mathbf{y}$ follows a central chi-square distribution.

From a Bayesian viewpoint with a diffuse prior on $\boldsymbol{\theta}$, it is immediate that $\mathcal{L}(\boldsymbol{\phi}|\mathbf{y}) = N(\mathbf{y},\mathbf{I})$, so that the Bayesian analogue of (7.3.1) is

$$\mathcal{L}(\boldsymbol{\phi}'\boldsymbol{\phi}|\mathbf{y}) = \chi_p'^2(\mathbf{y}'\mathbf{y}).$$

Using an informative prior distribution would generate a stronger result.

Example (7.3.1): Suppose $\mathbf{x} = (r,s)'$, where r and s are scalars denoting scores on a verbal examination, and on a quantitative examination for entrants to a graduate school, respectively. Observations $\mathbf{x}_1,\ldots,\mathbf{x}_{100}$ are taken and it is known that $\mathcal{L}(\mathbf{x}_\alpha) = N(\boldsymbol{\theta},\boldsymbol{\Sigma}_0)$, where for $\alpha = 1,\ldots,100$,

$$\boldsymbol{\Sigma}_0 = 4\begin{pmatrix} 1 & .5 \\ .5 & 1 \end{pmatrix}.$$

Test the hypothesis at the 5 percent level of significance that $\boldsymbol{\theta} = \boldsymbol{\theta}_0 \equiv (6,7)'$, versus the alternative that $\boldsymbol{\theta} \neq \boldsymbol{\theta}_0$, given that $\bar{\mathbf{x}} = (6.5,6.5)'$.

It is easy to check either from (2.8.3) or directly that the latent roots of $\boldsymbol{\Sigma}_0$ are 6 and 2. Hence, $\boldsymbol{\Sigma}_0$ may be written

$$\boldsymbol{\Sigma}_0 = \begin{pmatrix} \frac{1}{\sqrt{2}} & \frac{1}{\sqrt{2}} \\ \frac{1}{\sqrt{2}} & -\frac{1}{\sqrt{2}} \end{pmatrix} \begin{pmatrix} 6 & 0 \\ 0 & 2 \end{pmatrix} \begin{pmatrix} \frac{1}{\sqrt{2}} & \frac{1}{\sqrt{2}} \\ \frac{1}{\sqrt{2}} & -\frac{1}{\sqrt{2}} \end{pmatrix}$$

$$= \frac{1}{2}\begin{pmatrix}1 & 1\\ 1 & -1\end{pmatrix}\begin{pmatrix}6 & 0\\ 0 & 2\end{pmatrix}\begin{pmatrix}1 & 1\\ 1 & -1\end{pmatrix}.$$

Therefore, using (2.12.10),

$$\boldsymbol{\Sigma}_0^{-1/2} = \frac{1}{2}\begin{pmatrix}1 & 1\\ 1 & -1\end{pmatrix}\begin{pmatrix}\frac{1}{\sqrt{6}} & 0\\ 0 & \frac{1}{\sqrt{2}}\end{pmatrix}\begin{pmatrix}1 & 1\\ 1 & -1\end{pmatrix}$$

$$= \frac{1}{2}\begin{pmatrix}\frac{1}{\sqrt{6}} + \frac{1}{\sqrt{2}} & \frac{1}{\sqrt{6}} - \frac{1}{\sqrt{2}}\\ \frac{1}{\sqrt{6}} - \frac{1}{\sqrt{2}} & \frac{1}{\sqrt{6}} + \frac{1}{\sqrt{2}}\end{pmatrix}$$

$$\doteq \begin{pmatrix}.60 & -.15\\ -.15 & .60\end{pmatrix}.$$

So if $\mathbf{y} = \sqrt{N}\,\boldsymbol{\Sigma}_0^{-1/2}(\bar{\mathbf{x}} - \boldsymbol{\theta}_0)$, with $N = 100$,

$$\mathbf{y} = 3.75\begin{pmatrix}1\\ -1\end{pmatrix}, \qquad \text{and} \qquad \mathbf{y}'\mathbf{y} = 28.13.$$

But under H, $(\mathbf{y}'\mathbf{y})$ follows a central chi-square distribution with two degrees of freedom so that the 95 percent point is at 5.99. Hence, the hypothesis that $\boldsymbol{\theta} = \boldsymbol{\theta}_0$ is rejected. (See also Exercise 7.10.)

7.3.2 Testing for a Given Mean with Unknown Covariances

One-Sample Problem: In the typical case, if $\mathbf{x}_1, \ldots, \mathbf{x}_N$ are independent p-vectors from $N(\boldsymbol{\theta}, \boldsymbol{\Sigma})$, $\boldsymbol{\Sigma} > 0$, $\boldsymbol{\Sigma}$ is an unknown matrix. Then, to test $H: \boldsymbol{\theta} = \boldsymbol{\theta}_0$ versus say $A: \boldsymbol{\theta} \neq \boldsymbol{\theta}_0$, a procedure (likelihood ratio) with many optimal properties is based upon Hotelling's T^2-statistic (see Chapter 6) which, of course, is a function of the sufficient statistic

$$(\bar{\mathbf{x}}, \mathbf{V}) \equiv \left(\frac{1}{N}\sum_1^N \mathbf{x}_j, \sum_1^N (\mathbf{x}_j - \bar{\mathbf{x}})(\mathbf{x}_j - \bar{\mathbf{x}})'\right).$$

Define

$$\begin{aligned} T^2 &= N(N-1)(\bar{\mathbf{x}} - \boldsymbol{\theta}_0)'\mathbf{V}^{-1}(\bar{\mathbf{x}} - \boldsymbol{\theta}_0),\\ \lambda &= N(\boldsymbol{\theta} - \boldsymbol{\theta}_0)'\boldsymbol{\Sigma}^{-1}(\boldsymbol{\theta} - \boldsymbol{\theta}_0). \end{aligned} \tag{7.3.2}$$

Then

$$\mathcal{L}\left(\frac{T^2}{(N-1)}\,\frac{(N-p)}{p}\right) = F_{p,N-p}(\lambda), \tag{7.3.3}$$

where $F_{p,N-p}(\lambda)$ denotes an F-distribution with p and $(N-p)$ d.f., and noncentrality parameter λ. Note that a test of H versus A requires the use of the upper tail of $F_{p,N-p}(\lambda)$ since any deviation from H will cause T^2 to assume a larger value.

Example (7.3.2): Suppose that a bivariate sample of nine cities ($N = 9$) is taken from $N(\boldsymbol{\theta},\boldsymbol{\Sigma})$ to measure cigarette and liquor consumption. The data are summarized (in coded units) as

$$\bar{\mathbf{x}} = \begin{pmatrix} 1 \\ 4 \end{pmatrix}, \qquad \mathbf{V} = \begin{pmatrix} 9 & 0 \\ 0 & 4 \end{pmatrix}.$$

It is desired to test $H: \boldsymbol{\theta} = \boldsymbol{\theta}_0 = (2,3)'$, versus $A: \boldsymbol{\theta} \neq \boldsymbol{\theta}_0$. From (7.3.2), $T^2 = 26$. Hence, substitution in (7.3.3) gives an F-statistic of 11.375. But $F_{2,7}(0) = 4.47$ at the 5 percent level of significance. Thus, T^2 is significant at 5 percent, so H can be rejected.

It is being assumed, of course, that a 5 percent significance level is reasonable. If, in fact, 5 percent makes economic sense (in terms of risk preference and in terms of the power of the test) the problem is finished. If, however, the desired level of significance is not really well determined, or if it is really important to know whether $\boldsymbol{\theta}$ differs from $\boldsymbol{\theta}_0$ by a "substantial" amount, the posterior distribution of $\boldsymbol{\theta}$ will be most helpful (in fact, the posterior distribution of $\boldsymbol{\theta} - \boldsymbol{\theta}_0$ will be even more helpful).

Two-Sample Problem: Suppose $\mathbf{x}_1, \ldots, \mathbf{x}_{N_1}$ are independent p-vectors from $N(\boldsymbol{\theta}_1,\boldsymbol{\Sigma})$, $\mathbf{y}_1, \ldots, \mathbf{y}_{N_2}$ are independent p-vectors from $N(\boldsymbol{\theta}_2,\boldsymbol{\Sigma})$ and are independent of the $\mathbf{x}_j$'s, and it is of interest to test $H: \boldsymbol{\theta}_1 = \boldsymbol{\theta}_2$ versus $A: \boldsymbol{\theta}_1 \neq \boldsymbol{\theta}_2$. Let

$$\bar{\mathbf{x}} = \frac{1}{N_1}\sum_1^{N_1} \mathbf{x}_j, \qquad \bar{\mathbf{y}} = \frac{1}{N_2}\sum_1^{N_2} \mathbf{y}_j,$$

$$\mathbf{V}_1 = \sum_1^{N} (\mathbf{x}_j - \bar{\mathbf{x}})(\mathbf{x}_j - \bar{\mathbf{x}})'\Big). \qquad \mathbf{V}_2 = \sum_1^{N_2} (\mathbf{y}_j - \bar{\mathbf{y}})(\mathbf{y}_j - \bar{\mathbf{y}})',$$

$$\mathbf{S} = \frac{1}{N_1 + N_2 - 2}(\mathbf{V}_1 + \mathbf{V}_2).$$

Then

$$\mathcal{L}(\bar{\mathbf{x}}) = N\left(\boldsymbol{\theta}_1, \frac{\boldsymbol{\Sigma}}{N_1}\right), \qquad \mathcal{L}(\bar{\mathbf{y}}) = N\left(\boldsymbol{\theta}_2, \frac{\boldsymbol{\Sigma}}{N_2}\right),$$

$$\mathcal{L}(\mathbf{V}_j) = W(\boldsymbol{\Sigma}, p, N_j - 1), \qquad j = 1, 2.$$

Combining these results gives

$$\mathcal{L}(\bar{\mathbf{x}} - \bar{\mathbf{y}}) = N\left(\boldsymbol{\theta}_1 - \boldsymbol{\theta}_2, \left(\frac{1}{N_1} + \frac{1}{N_2}\right)\boldsymbol{\Sigma}\right),$$

$$\mathcal{L}(\mathbf{V}_1 + \mathbf{V}_2) = W(\boldsymbol{\Sigma}, p, N_1 + N_2 - 2).$$

Since $(\bar{\mathbf{x}} - \bar{\mathbf{y}})$ is independent of $(\mathbf{V}_1 + \mathbf{V}_2)$, use Hotelling's T^2-statistic.

Let

$$T^2 = \left(\frac{N_1N_2}{N_1 + N_2}\right)(\bar{\mathbf{x}} - \bar{\mathbf{y}})'\mathbf{S}^{-1}(\bar{\mathbf{x}} - \bar{\mathbf{y}}). \tag{7.3.4}$$

Then

$$\mathcal{L}\left(\frac{N_1 + N_2 - p - 1}{(N_1 + N_2 - 2)p}T^2\right) = F_{p,N_1+N_2-p-1}(\lambda), \tag{7.3.5}$$

where

$$\lambda = \frac{N_1N_2}{N_1 + N_2}(\boldsymbol{\theta}_1 - \boldsymbol{\theta}_2)'\boldsymbol{\Sigma}^{-1}(\boldsymbol{\theta}_1 - \boldsymbol{\theta}_2).$$

Example (7.3.3): Suppose $\mathbf{x}: p \times 1$ is a random vector whose ith component denotes the number of crimes of type i reported in a given police precinct in a large metropolitan area. Suppose $\mathbf{y}: p \times 1$ denotes the same crime statistic reported for a different police precinct in the same metropolitan area. Adopt the model $\mathcal{L}(\mathbf{x}_j) = N(\boldsymbol{\theta}_1,\boldsymbol{\Sigma})$, $\mathcal{L}(\mathbf{y}_k) = N(\boldsymbol{\theta}_2,\boldsymbol{\Sigma})$, where $\mathbf{x}_1,\ldots,\mathbf{x}_{N_1}$ and $\mathbf{y}_1,\ldots,\mathbf{y}_{N_2}$ denote independent observations of $\mathbf{x}$ and $\mathbf{y}$, respectively, and the $\mathbf{x}_j$'s and $\mathbf{y}_k$'s are independent. It is of interest to determine whether the p crime types have the same reported incidence in the two areas of the city represented by the two police precincts. That is, test $H: \boldsymbol{\theta}_1 = \boldsymbol{\theta}_2$ versus $A: \boldsymbol{\theta}_1 \neq \boldsymbol{\theta}_2$; and of course if $\boldsymbol{\theta}_1 \neq \boldsymbol{\theta}_2$, determine what are the true relationships.

The testing problem is solved by using the two-sample Hotelling T^2-statistic in (7.3.4) and (7.3.5). If H is rejected, further analysis is required.

[*Remark:* This problem would be approached in a Bayesian way by studying the posterior distribution of $(\boldsymbol{\theta}_1 - \boldsymbol{\theta}_2)$. In this problem, prior information is likely to be readily assessable.]

7.3.3 Testing for Equality of Covariance Matrices

Multivariate tests involving variances and covariances are generally natural extensions of univariate tests. It will be seen in multivariate regression and the multivariate analysis of variance and covariance (Chapter 8) that it is often important to determine whether the covariance matrices of several populations are equal. That is, the assumption of homoscedasticity (equality of variances) used in the univariate general linear model is extended to a multivariate homoscedasticity assumption by taking equal covariance matrices. To test whether this assumption is correct, even approximately, it is necessary to rely upon asymptotic results. It should be noted that since it is well known that tests of equality of variance are extremely sensitive to non-Normality in the univariate case, the same sensitivity should not be surprising in the multivariate case.

Let $\mathbf{x}_1(i),\ldots,\mathbf{x}_{N_i}(i)$ denote N_i independent p-variate observations from $N(\boldsymbol{\theta}_i,\boldsymbol{\Sigma}_i)$, $i = 1,\ldots,K$. It is desired to test the hypothesis

$$H: \boldsymbol{\Sigma}_1 = \boldsymbol{\Sigma}_2 = \cdots = \boldsymbol{\Sigma}_K.$$

Since the sample covariance matrix of each population is sufficient for each population covariance matrix, define

$$\mathbf{V}_i = \sum_{\alpha=1}^{N_i} [\mathbf{x}_\alpha(i) - \overline{\mathbf{x}(i)}][\mathbf{x}_\alpha(i) - \overline{\mathbf{x}(i)}]',$$

for $i = 1,\ldots,K$, where the bars denote sample means. Let $N \equiv \Sigma_1^K N_i$. It is easy to check that a test based upon the likelihood ratio statistic (replacing maximum likelihood estimates by unbiased estimates) can be based upon the statistic

$$G = c + \sum_{i=1}^{K} (N_i - 1) \log |\mathbf{V}_i| - (N - K) \log |\mathbf{V}_1 + \cdots + \mathbf{V}_K|, \tag{7.3.6}$$

where

$$c = p\,(N - K) \log (N - K) - \sum_{i=1}^{K} p\,(N_i - 1) \log (N_i - 1).$$

That is, e^G is the likelihood ratio statistic modified for unbiased estimates. This test statistic was suggested by Bartlett (1937); an alternative test statistic based upon union-intersection test procedures is given in Roy (1957), Chapter 11.

The test of H is to reject H if $G <$ constant, and the constant is determined at any preassigned significance level from the distribution of G under H. For $p = 1$, the test of H is based upon an F-distribution. For $p > 1$, the distribution is more complicated.

Define the constants

$$a = 1 - \left(\sum_{i=1}^{K} \frac{1}{N_i - 1} - \frac{1}{N - K}\right) \frac{2p^2 + 3p - 1}{6(p + 1)(K - 1)},$$

$$b = \frac{p(p + 1)}{48a^2} \left[(p - 1)(p + 2) \left(\sum_{i=1}^{K} \frac{1}{(N_i - 1)^2} - \frac{1}{(N - K)^2} \right) - 6(K - 1)(1 - a)^2 \right].$$

It was shown by Box (1949) that a close approximation to the distribution of G under H is given by

$$P\{-aG \le t\} = P\{\chi^2(f) \le t\} + b[P\{\chi^2(f + 4) \le t\} - P\{\chi^2(f) \le t\}] + O\left(\frac{1}{(N - K)^3}\right), \tag{7.3.7}$$

where $f = (p/2)(p + 1)(K - 1)$. That is, in large samples it is approximately true that under H,

$$\mathfrak{L}[-aG] = \mathfrak{L}\left\{\chi^2\left[\left(\frac{p}{2}\right)(p + 1)(K - 1)\right]\right\}. \tag{7.3.8}$$

This kind of approximation due to Box will be used again in Chapter 8 to test hypotheses about means.

Suppose it is found that the covariance matrices are not the same. Then many of the standard multivariate models will not apply. However, it is still possible to test equality of mean vectors even when the covariance matrices are unequal (this is called the Behrens-Fisher problem); see for example Bennett (1951), Anderson (1963), and Press (1967a). The methods used in the references cited are multivariate extensions of the univariate procedure of Scheffé (1943) for solving the Behrens-Fisher problem. Bennett directly extended the Scheffé procedure to two multivariate populations, Anderson extended Bennett's procedure to many multivariate populations, and Press extended the procedure to patterned covariance matrices such as the intraclass covariance matrix. In the univariate case, if the variances are unequal but not too different, the test procedures will still generally be applicable. An analogous robustness may also hold in the multivariate case (see Section 8.10).

7.3.4 Testing Correlation Coefficients and Matrices

Since correlation coefficients occur so frequently in applications, in this section methods of testing hypotheses concerning them are given. Also, to determine whether a correlation matrix computed from observed data actually reflects correlated components of a random vector, it is often desirable to examine whether the sample correlation matrix differs significantly from the identity matrix. These problems are discussed below.

Correlation Coefficients

Suppose that for each α, $\mathbf{z}_\alpha' = (x_\alpha, y_\alpha)$ is a two-dimensional random vector following the law $N(\boldsymbol{\theta}, \boldsymbol{\Sigma})$, $\boldsymbol{\Sigma} > 0$. Let $\mathbf{z}_1, \ldots, \mathbf{z}_N$ be independent observations from this distribution and form the sample correlation coefficient

$$r = \frac{\sum_1^N (x_i - \bar{x})(y_i - \bar{y})}{\left\{\left[\sum_1^N (x_i - \bar{x})^2\right]\left[\sum_1^N (y_i - \bar{y})^2\right]\right\}^{1/2}}.$$

Theorem (7.3.1): Let ρ denote the population correlation coefficient between the components of $\mathbf{z}$. Then:

(a) If $\rho = 0$,

$$\mathcal{L}\left(\frac{r\sqrt{N-2}}{\sqrt{1-r^2}}\right) = t_{N-2},$$

where t_k denotes the Student t-distribution with k d.f.

(b) If for all ρ, $-1 < \rho < 1$,

$$v \equiv \frac{1}{2}\log\left(\frac{1+r}{1-r}\right) \equiv \tanh^{-1} r,$$

and

$$v_0 = \frac{1}{2}\log\left(\frac{1+\rho}{1-\rho}\right) \equiv \tanh^{-1}\rho,$$
$$\lim_{N\to\infty} \mathcal{L}[(v - v_0)\sqrt{N-3}] = N(0,1).$$

Proof: See Fisher (1921).

Example (7.3.4): In a large sales organization it was believed that score on the examination in a company sponsored salesmanship course has about 50 percent correlation with sales of an individual. To test this hypothesis, observations were taken on many employees after the examination.

Let x_j denote test score for the jth employee, y_j denote sales for the jth employee, $j = 1,\ldots,103$. Let $\mathbf{z}_j' \equiv (x_j,y_j)$, and assume the model of this section is directly applicable (in an actual situation it might be necessary, for example, to first log the data to satisfy the model assumptions of Normality and constant variances of the underlying variables). It is desired to test $H: \rho = .5$ versus $A: \rho \neq .5$. Suppose that the data collected yield a sample correlation coefficient of $r = .40$ so that $v = .42$, where v is defined in Theorem (7.3.1). Since for $\rho = .5$, $v_0 = .55$, Theorem (7.3.1) asserts that, approximately,

$$\mathcal{L}(v) = N(.55,.01),$$

or

$$\mathcal{L}\left(\frac{v - .55}{.1}\right) = N(0,1).$$

Hence, a two tail test at the 5 percent level shows that

$$.95 = P\{.35 \leq v \leq .75\}.$$

Thus, $v = .42$ is not significant, so the hypothesis that $\rho = .5$ cannot be rejected at the 5 percent significance level. Note that if the hypothesis were that the correlation is at least .5, a one-tailed test would be appropriate.

Theorem (7.3.2): Suppose there are samples based upon $N_1, \ldots, N_q$ observations from q bivariate Normal populations with correlation coefficients $\rho_1, \ldots, \rho_q$. Let $Z_i \equiv \tanh^{-1} r_i$, where $r_1, \ldots, r_q$ are sample correlation coefficients from the q populations, respectively. Define

$$W = \sum_1^q (N_i - 3)(Z_i - \bar{Z})^2,$$

where

$$\bar{Z} = \frac{\sum_1^q (N_i - 3)Z_i}{\sum_1^q (N_i - 3)}.$$

Then it is approximately true (in large samples) that

$$\mathcal{L}(W) = \chi_{q-1}^2.$$

Proof: See, for instance, Graybill (1961).

[*Remark:* Theorem (7.3.2) is often used to test the hypothesis $H: \rho_1 = \rho_2 = \cdots = \rho_q$, against A: the ρ_j's are unequal. This problem arises, for example, in testing whether the teaching methods used in q schools are equivalent, where ρ_j denotes the correlation coefficient between, say, mathematics and sociology examination scores in School j.]

Correlation Matrices

Given the independent p-variate observations $\mathbf{x}_1, \ldots, \mathbf{x}_N$ from $N(\boldsymbol{\theta}, \boldsymbol{\Sigma})$, it is sometimes of interest to study whether the sample correlation matrix actually reflects correlation among the components of $\mathbf{x}$ or is merely the result of sampling variation. That is, it is desired to test the hypothesis $H: \mathbf{R} = \mathbf{I}$ versus $A: \mathbf{R} \neq \mathbf{I}$, where $\mathbf{R}$ denotes the population correlation matrix. Equivalently, it is desired to test that $\boldsymbol{\Sigma}$ is a diagonal covariance matrix, or that the components of $\mathbf{x}$ are independent.

Theorem (7.3.3): If $\hat{\mathbf{R}}$ denotes the sample correlation matrix based upon N observations from $N(\boldsymbol{\theta}, \boldsymbol{\Sigma})$, the likelihood ratio statistic for testing $H: \mathbf{R} = \mathbf{I}$ versus $A: \mathbf{R} \neq \mathbf{I}$ is $|\hat{\mathbf{R}}|^{N/2}$, so that the test is to reject H if $|\hat{\mathbf{R}}|^{N/2} < c$, where c is found for any given level of significance from

$$\lim_{N\to\infty} \mathcal{L}\left[-\left(N-1-\frac{2p+5}{6}\right)\log|\hat{\mathbf{R}}|\right] = \chi_{(p/2)(p-1)}{}^2.$$

Proof: The likelihood ratio statistic is found in a straightforward manner from first principles. The limiting distribution is found from Wilks' Theorem (see Chapter 8, and Wilks, 1938) and the fact that a correlation matrix has $p(p-1)/2$ unconstrained parameters. The asymptotic result of Wilks was modified somewhat by Bartlett (1954) to accelerate the convergence, and it is Bartlett's form that is given in the theorem.

REFERENCES

Anderson (1958).
Anderson (1963).
Bartlett (1954).
Bennett (1951).
Box (1949).
Dempster (1963).
Dempster (1966).
Efron and Morris (1977).
Fisher (1921).
James and Stein (1961).
Jöreskog (1970, 1973).
Judge and Bock (1978).
Press (1967a).
Roy (1957).
Scheffé (1943).
Stein (1956).
Stein (1962).
Tsui and Press (1977a, 1977b).
Wilks (1938).

EXERCISES

7.1 Let $\mathbf{x}_1,\ldots,\mathbf{x}_N$ denote N independent observations from the multivariate log-Normal distribution with density

$$p(\mathbf{y}) = \frac{1}{(2\pi)^{p/2}|\boldsymbol{\Sigma}|^{1/2}\prod_{j=1}^{p} y_j}\exp\{-\tfrac{1}{2}(\overline{\log}\,\mathbf{y}-\boldsymbol{\theta})'\boldsymbol{\Sigma}^{-1}(\overline{\log}\,\mathbf{y}-\boldsymbol{\theta})\}.$$

Find sufficient statistics for $\boldsymbol{\theta}$ and $\boldsymbol{\Sigma}$.

7.2 Let $\mathbf{x}_j: p \times 1$ denote the responses to p questions asked, of the jth person in a sample survey; $j = 1,\ldots,N$. Assume $\mathcal{L}(\mathbf{x}_j) = N(\boldsymbol{\theta},\boldsymbol{\Sigma})$ for all j. Suppose stratified random sampling has been used so that $\boldsymbol{\Sigma}$ follows the intraclass pattern of correlation,

$$\boldsymbol{\Sigma} = \sigma^2\begin{pmatrix} 1 & & & \\ & \cdot & \rho & \\ & & \cdot & \\ & \rho & \cdot & \\ & & & 1\end{pmatrix}, \qquad \boldsymbol{\Sigma} > 0.$$

Find maximum likelihood estimators of $\boldsymbol{\theta}$, σ^2, ρ.

7.3 Suppose $\mathbf{x}_\alpha$ denotes a vector of the three scores on a battery of three tests for the αth child in a class, and $\mathcal{L}(\mathbf{x}_\alpha) = N(\boldsymbol{\theta},\boldsymbol{\Sigma})$, $\boldsymbol{\Sigma} > 0$. On testing ten children ($\alpha = 1,\ldots,10$) it is found that

$$\bar{\mathbf{x}} = \tfrac{1}{10}\sum_{1}^{10} \mathbf{x}_\alpha = \begin{pmatrix}71\\84\\56\end{pmatrix},$$

and

$$\mathbf{S} = \tfrac{1}{9}\sum_{\alpha=1}^{10} (\mathbf{x}_\alpha - \bar{\mathbf{x}})(\mathbf{x}_\alpha - \bar{\mathbf{x}})' = \begin{pmatrix}4 & 1 & 1\\1 & 3 & 2\\1 & 2 & 2\end{pmatrix}.$$

Suppose it is believed a priori that $\mathcal{L}(\boldsymbol{\theta}|\boldsymbol{\Sigma}) = N(\boldsymbol{\phi},\boldsymbol{\Sigma})$, and $\mathcal{L}(\boldsymbol{\Sigma}) = W^{-1}(\mathbf{G},3,4)$, with [see (5.2.1)]

$$\boldsymbol{\phi} = \begin{pmatrix}70\\70\\70\end{pmatrix}, \qquad \mathbf{G} = \begin{pmatrix}4 & 2 & 2\\2 & 4 & 1\\2 & 1 & 3\end{pmatrix}.$$

(a) If $L(\boldsymbol{\theta},\hat{\boldsymbol{\theta}}) = (\boldsymbol{\theta} - \hat{\boldsymbol{\theta}})'(\boldsymbol{\theta} - \hat{\boldsymbol{\theta}})$ denotes a loss function for $\boldsymbol{\theta}$, find the Bayes estimate of $\boldsymbol{\theta}$.

(b) If $L(\boldsymbol{\Sigma},\hat{\boldsymbol{\Sigma}}) = \operatorname{tr}(\boldsymbol{\Sigma} - \hat{\boldsymbol{\Sigma}})^2$ denotes a loss function for $\boldsymbol{\Sigma}$, find the Bayes estimate for $\boldsymbol{\Sigma}$.

(c) Find the Bayes estimates of $\boldsymbol{\theta}$ and $\boldsymbol{\Sigma}$ assuming there were millions of children included in the sample estimates $\bar{\mathbf{x}}$ and $\mathbf{S}$ (asymptotic Bayes estimates).

7.4 For the observed data given in Exercise 7.3, compute the Stein estimator of $\boldsymbol{\theta}$ and compare it with the maximum likelihood estimator.

7.5 Suppose $\mathbf{x}_1,\ldots,\mathbf{x}_N$ are independent p-variate observations from $N(\boldsymbol{\theta},\boldsymbol{\Sigma}_0)$, where $\mathbf{x}_\alpha$ denotes a vector whose components are reported numbers of crimes of various types, and

$$\boldsymbol{\Sigma}_0 = \operatorname{diag}(4,2,7).$$

Suppose it is found that based upon ten observations ($N = 10$), $\bar{\mathbf{x}} = (6,3,2)'$. It is of interest to test $H: \boldsymbol{\theta} = \boldsymbol{\theta}_0 = (5,4,3)'$ versus $A: \boldsymbol{\theta} \neq \boldsymbol{\theta}_0$.

(a) Using a standard sampling theory approach, test H versus A at the 5 percent level of significance.

(b) Adopt a diffuse prior and compute the posterior distribution of $\boldsymbol{\phi}$, where

$$\boldsymbol{\phi} = \sqrt{N}\,\boldsymbol{\Sigma}_0^{-1/2}(\boldsymbol{\theta} - \boldsymbol{\theta}_0).$$

(c) Test, from the Bayesian viewpoint, whether the first component of $\boldsymbol{\theta}$ is 5, by accepting the hypothesis if a 95 percent credibility interval for ϕ_1, the first component of $\boldsymbol{\phi}$ in Part (b) includes 0, and rejecting if it does not.

7.6 Suppose $\mathbf{x}_1,\ldots,\mathbf{x}_N$ denote independent bivariate vectors of observations representing reported number of robberies and reported number of car thefts at N different times in a given police precinct. Let $\mathbf{y}_1,\ldots,\mathbf{y}_M$ denote independent observations of the same quantities after the number of patroling police in the precinct has doubled. Suppose the $\mathbf{x}_j$'s and $\mathbf{y}_j$'s are independent with $\mathcal{L}(\mathbf{x}_j) = N(\boldsymbol{\theta},\boldsymbol{\Sigma}_1)$, and $\mathcal{L}(\mathbf{y}_k) = N(\boldsymbol{\phi},\boldsymbol{\Sigma}_2)$, $j = 1,\ldots,N$, $k = 1,\ldots,M$. Assuming $\boldsymbol{\Sigma}_1 = \boldsymbol{\Sigma}_2$, $\bar{\mathbf{x}} = (4,7)'$, $\bar{\mathbf{y}} = (6,6)'$,

$$\mathbf{V}_1 = \sum_1^{10} (\mathbf{x}_j - \bar{\mathbf{x}})(\mathbf{x}_j - \bar{\mathbf{x}})' = \begin{pmatrix} 10 & 3 \\ 3 & 8 \end{pmatrix},$$

$$\mathbf{V}_2 = \sum_1^{20} (\mathbf{y}_k - \bar{\mathbf{y}})(\mathbf{y}_k - \bar{\mathbf{y}})' = \begin{pmatrix} 6 & 2 \\ 2 & 4 \end{pmatrix},$$

so that $N = 10$, $M = 20$, test (at the 5 percent level) the hypothesis $H: \boldsymbol{\theta} = \boldsymbol{\phi}$ versus $A: \boldsymbol{\theta} \neq \boldsymbol{\phi}$; that is, test the hypothesis that doubling police manpower has no effect on reported crime.

7.7 For the data given in Exercise 7.6, test the hypothesis (at the 5 percent level) that $\boldsymbol{\Sigma}_1 = \boldsymbol{\Sigma}_2$ assuming the samples are sufficiently large for Bayesian asymptotic theory to be a useful approximation. [*Hint:* See Exercise 7.5(c).]

7.8 Suppose $\mathbf{z}_\alpha \equiv (x_\alpha,y_\alpha)'$ denotes the scores of psychotics, and of normals, respectively, on a verbal achievement test, $\alpha = 1,\ldots,11$. Suppose also that $\mathcal{L}(\mathbf{z}_\alpha) = N(\boldsymbol{\theta},\boldsymbol{\Sigma})$. On the basis of the observed data it is found that

$$\mathbf{V} = \sum_1^{11} (\mathbf{z}_\alpha - \bar{\mathbf{z}})(\mathbf{z}_\alpha - \bar{\mathbf{z}})' = \begin{pmatrix} 40 & 30 \\ 30 & 60 \end{pmatrix}.$$

Test the hypothesis that scores of psychotics and those of normals are independent (at the 1 percent level).

7.9 For the sample covariance matrix given in Exercise 7.3, test the hypothesis at the 10 percent level that the components are independent (assuming Normality of the observations, and that 10 observations is a large sample).

7.10 How would a Bayesian handle the problem in Example (7.3.1) if he used a diffuse prior?

7.11 The following statement appeared in *Time* magazine: "As early as the 12th century, Hebrew scholars began to question whether the entire 'Book of Isaiah' was written by the same author. Liberal Scripture scholars have long agreed that there are at least two distinct collections in Isaiah, one comprising the first 39 chapters, the other the remaining 27. Now modern technology has ratified that thesis.

Using an Elliott 503 computer, Yehuda Radday, a lecturer in biblical studies and Hebrew in Haifa, produced a 175-page statistical linguistics analysis of Isaiah. He applied 18 standard tests to measure such features as word length and vocabulary eccentricity. An additional test devised by Radday measured Isaiah's war idioms and metaphors .... All 19 tests turned up significant differences between the two parts. Radday's conclusion: The probability that one prophet wrote the book is one in 100,000."[2]

Comment on this quotation from a statistical standpoint, as follows. What methodology would you have used, given the same data? What methodology is implied in the quotation? If the two statistical procedures are not the same, which is better and why? Assuming the quotation correctly described Radday's work, explain why you agree or disagree with his conclusions.

7.12 Suppose $\mathbf{x}: p \times 1$, and $\mathcal{L}(\mathbf{x}|\boldsymbol{\theta}) = N(\boldsymbol{\theta},\sigma_1^2\mathbf{I})$, for σ_1^2 known. Show that if $\mathcal{L}(\boldsymbol{\theta}|\phi) = N(\phi\mathbf{e},\sigma_2^2\mathbf{I})$, where ϕ is a scalar, $\mathbf{e}$ is a p-vector of ones, and σ_2^2 is known, and if the density of ϕ is proportional to a constant, the posterior density of $\boldsymbol{\theta}$ given $\mathbf{x}$ is Normal, $E(\boldsymbol{\theta}|\mathbf{x}) = \mathbf{Gx}$, where $\mathbf{G}$ denotes an intraclass covariance matrix, and in particular, for $\boldsymbol{\theta} = (\theta_i)$, $i = 1,\ldots, p$,

$$E(\theta_1|\mathbf{x}) = \frac{h_1x_1 + h_2\bar{x}}{h_1 + h_2},$$

where $h_i \equiv \sigma_i^{-2}$, $i = 1, 2$, and $\bar{x} \equiv \Sigma_1^p(x_i/p)$.

Note that the mean of θ_1 is a weighted average of the traditional maximum likelihood estimator, x_1, and the component mean, $\bar{x}$, and the weights are the precisions of the sampling and prior distributions. The effect of this averaging is to pull the estimator away from x_1 toward the component mean, $\bar{x}$. (See also the Lindley discussion of Stein, 1962.)

7.13 Score on a sales examination, X was claimed to be related to years of education, Y. The following data was collected.

X	80	60	50	75	85	77	75	82	85	45
Y	16	12	13	10	17	12	14	15	16	10

[2] *Time* magazine, "Isaiah and the Computer," April 13, 1970. Reprinted by permission from TIME, The Weekly Newsmagazine; Copyright Time, Inc., 1970.

Assume (X, Y) follows a bivariate normal distribution and test the hypothesis at the 5% level of significance that the population correlation coefficient is zero.

7.14 A pharmaceutical company claims to have developed a drug for simultaneously reducing blood vessel clogging cholesterol levels in the human body, and for inducing body weight reduction. To test this claim the Food and Drug Administration took a sample of 100 people and measured their cholesterol levels and body weights. A random half of the sample was given the new drug and the other half was given a placebo. After a suitable time cholesterol levels and body weights were recorded again. The data (in coded units) is as follows:

$$\bar{\mathbf{x}} = \begin{pmatrix} 4 \\ 10 \end{pmatrix}, \qquad \bar{\mathbf{y}} = \begin{pmatrix} 2 \\ 12 \end{pmatrix},$$

$$\mathbf{V}_1 = \begin{pmatrix} 16 & 6 \\ 6 & 9 \end{pmatrix}, \qquad \mathbf{V}_2 = \begin{pmatrix} 12 & 8 \\ 8 & 10 \end{pmatrix},$$

where $\bar{\mathbf{x}}$ denotes the average of the changes in the vector of cholesterol levels and body weights for the group that received the drug, and $\bar{\mathbf{y}}$ denotes the same quantity for the group that received the placebo; $\mathbf{V}_1$ and $\mathbf{V}_2$ denote the corresponding covariance matrices (unscaled).

Assume bivariate normality with equal covariance matrices for the underlying data. Test the hypothesis that the drug has no effect, at the 5% level of significance.

7.15 For the data given in Exercise 7.14, test the hypothesis that the covariance matrices are equal (at the 5% level).

part II

Models

8

REGRESSION AND THE ANALYSIS OF VARIANCE

8.1 INTRODUCTION

This chapter provides methods of analyzing the general linear model, including regression and the analysis of variance.

Five distinct regression models are defined in Section 8.2, and attention is focused on two of them. The univariate regression problem (there is only one dependent variable) is examined in Section 8.3 under the presence of heteroscedasticity (unequal variances of observations), and serial correlation (correlated observations), for a variety of special cases.

Multivariate regression (one dependent variable in each of p equations) is treated from the sampling theory point of view in Section 8.4, and from the Bayesian point of view in Section 8.6. In the standard multivariate regression model, the matrix of regressors is the same in each equation (one equation for each dependent variable). Multivariate regression in which the matrix of regressors is different for each dependent variable (generalized regression) is examined in Section 8.5.

The multivariate analysis of variance is treated for the one-and two-way layouts in Section 8.7, and the composite procedure for combining multivariate analysis of variance and multivariate regression (multivariate analysis of covariance) is described in Section 8.8. Techniques are provided in Section 8.9 for making multiple comparisons and simultaneous confidence intervals from both sampling theory and Bayesian points of view. Finally, in Section 8.10 there is some discussion of how to check for violations of the underlying assumptions of the general linear model, and also some discussion of how to correct for such violations, and the effects such violations have on statistical inferences.

8.2 UNIVARIATE REGRESSION AND MODEL CLASSIFICATION

There are many regression models which have been studied. They differ from one another according to how the basic random variables were generated, whether the basic random variables were observed or whether surrogates were used, whether the underlying relationships are among stochastic or deterministic variables, and according to the nature of the errors in the data. Violating the underlying assumptions of a model has the effect of vitiating the meaning of any regression relationships established.

In this section, various regression models and their underlying assumptions are delineated. To simplify the exposition, the models classified will all be univariate, although there are multivariate analogues for each case. It should be borne in mind in selecting a regression model that inferences drawn about the parameters may be different, depending upon the assumptions of the model.

8.2.1 Simple Univariate Multiple Regression

Suppose $(y_j, x_{j1}, \ldots, x_{j,k-1})$ are all observable scalar variables, $j = 1, \ldots, n$, and are related by the equation

$$y_j = a_0 + a_1 x_{j1} + a_2 x_{j2} + \cdots + a_{k-1} x_{j,k-1} + e_j, \tag{8.2.1}$$

where the e_j's are random errors and the a_j's are unknown coefficients. The model is more conveniently written in matrix form.

Let $\mathbf{y} \equiv (y_1, \ldots, y_n)'$, $\mathbf{e} \equiv (e_1, \ldots, e_n)'$, $\boldsymbol{\beta} \equiv (a_0, a_1, \ldots, a_{k-1})'$, and define the *regressor matrix*

$$\underset{(n \times k)}{\mathbf{X}} = \begin{pmatrix} 1 & x_{11} & x_{12} & \cdots & x_{1,k-1} \\ \cdot & \cdot & \cdot & & \cdot \\ \cdot & \cdot & \cdot & & \cdot \\ \cdot & \cdot & \cdot & & \cdot \\ 1 & x_{n1} & x_{n2} & \cdots & x_{n,k-1} \end{pmatrix}.$$

(The dimensions of a matrix will often be written under it for clarity.) Now, (8.2.1) may be written in the more compact form

$$\underset{(n \times 1)}{\mathbf{y}} = \underset{(n \times k)}{\mathbf{X}} \cdot \underset{(k \times 1)}{\boldsymbol{\beta}} + \underset{(n \times 1)}{\mathbf{e}}. \tag{8.2.2}$$

This equation represents the fundamental equation for many univariate models. These models differ according to the assumptions made about

($\mathbf{y}, \mathbf{X}, \boldsymbol{\beta}, \mathbf{e}$) and their interrelations. Some of the models are discussed below and are summarized for easy reference in Table 8.2.1. It should be noted that in practice, portions of several of the basic regression models may be present in a specific model.

8.2.2 Regression Model Classification

Functional Regression (Model I)

In this model $\mathbf{y}$ is assumed to be a vector of n independent observations on a random variable Y which is observed without any measurement error. $\mathbf{X}$ is a matrix of known constants which are values of deterministic mathematical variables such as calendar time, speed of a commercial airplane, or distance a rigid body will move in a frictionless vacuum under a preassigned initial speed. In most cases, variables are called deterministic or stochastic depending upon their degree of predictability. Thus, if a variable is highly predictable in repeated experiments, it is called deterministic; otherwise, stochastic. The error vector, $\mathbf{e}$, has n unobservable components which might include errors of measurement and other unexplainable variables in the relationship. It is always assumed that $E(\mathbf{e}) = \mathbf{0}$, and it is usually assumed for purposes of inference that $\mathbf{e}$ follows a multivariate Normal distribution. However, $\boldsymbol{\Sigma}$, the covariance matrix of $\mathbf{e}$, is sometimes assumed to be a scalar matrix (sphericity of the errors), in which case $\boldsymbol{\Sigma} = \sigma^2\mathbf{I}_n$ for σ^2 an unknown scalar, and sometimes $\boldsymbol{\Sigma}$ is assumed to be an $n \times n$ matrix with arbitrary components (this case is taken up in the next section). The assumption that $\boldsymbol{\Sigma} = \sigma^2\mathbf{I}_n$ implies that the errors are uncorrelated and that they all have the same variance (homoscedasticity). The coefficient vector $\boldsymbol{\beta}$ is assumed to be nonrandom (deterministic) so that $E(\mathbf{y}) = \mathbf{X}\boldsymbol{\beta}$, and var $(\mathbf{y})$ = var $(\mathbf{e}) = \boldsymbol{\Sigma}$.

Conditional Regression (Model II)

This model presupposes that there is a random variable Y and $(k - 1)$ random variables $(X_1, \ldots, X_{k-1})$ all of which are jointly distributed. The data of (8.2.1) are assumed to be generated by holding the X_α's fixed (or under control) at values $x_{j\alpha}$, $\alpha = 1, \ldots, k - 1$, $j = 1, \ldots, n$, and letting Y vary so that $(Y|X_1, \ldots, X_{k-1}) = y_j$ is observed. This is done n times. Thus, if an experiment could be carried out in which n families which may be considered homogeneous with regard to their spending habits are given certain preassigned annual incomes (fixed values), a postulated linear relationship between their total consumptions for the year and their annual incomes represents a case of this model.

Since Y is observed conditional on X, and since $\mathbf{y} = \mathbf{X}\boldsymbol{\beta} + \mathbf{e}$, $E(\mathbf{y}|\mathbf{X}) =$

Table 8.2.1 Regression Model Classification Summary

Model name	*Form of relationships*	*Assumptions about* $\mathbf{y}$	*Assumptions about* $\mathbf{X}$	*Assumptions about* $\boldsymbol{\beta}$	*Assumptions about errors*
Functional Regression (Model I)	$\mathbf{y} = \mathbf{X}\boldsymbol{\beta} + \mathbf{e}$	Random, uncontrolled but observable	Strictly deterministic mathematical variables, observable	Deterministic	$E(\mathbf{e}) = \mathbf{0}$; $\text{var}(\mathbf{e}) = \boldsymbol{\Sigma}$, or $\text{var}(\mathbf{e}) = \sigma^2\mathbf{I}$; $\mathfrak{L}(\mathbf{e})$ is Normal
Conditional Regression (Model II)	$\mathbf{y} \mid \mathbf{X} = \mathbf{X}\boldsymbol{\beta} + \mathbf{e}$	Random, conditioned on fixed $x_{\alpha j}$'s, and observable	$(Y, X_1, \ldots, X_{k-1})$ jointly distributed, $Y \mid X_1, \ldots, X_{k-1}$ is observed, and $X_j = x_j$ is observed	Deterministic	Same as Model I
Stochastic Regression (Model III)	$\mathbf{y} = \mathbf{X}\boldsymbol{\beta} + \mathbf{e}$	Random, uncontrolled, and observable	$x_{\alpha j}$ is stochastic and observed; $\mathbf{X}$ is independent of $\mathbf{e}$	Deterministic	Same as Model I; $\mathbf{e}$ is independent of $\mathbf{X}$
Errors in the Variables (Model IV)	$\mathbf{y} = \mathbf{X}\boldsymbol{\beta}$ $x_{\alpha j}^* = x_{\alpha j} + v_{\alpha j}$ $y_\alpha^* = y_\alpha + u_\alpha$	Unobservable and either random or not	$x_{\alpha j}$ is unobservable and either random or not	Deterministic	$\mathfrak{L}(\mathbf{u}) = N(0, \sigma^2\mathbf{I})$; $\mathfrak{L}(\mathbf{v}_j) = N(0, \tau_j^2\mathbf{I})$; $(\mathbf{u}, \mathbf{v}_1, \ldots, \mathbf{v}_{k-1})$ all independent
Random Parameters (Model V)	$\mathbf{y} = \mathbf{X}\boldsymbol{\beta} + \mathbf{e}$	Random, uncontrolled, and observable	Same as either Model I or Model II	Random and independent of $\mathbf{e}$	Same as Model I; $\mathbf{e}$ is independent of $\boldsymbol{\beta}$

$\mathbf{X}\boldsymbol{\beta}$, var $(\mathbf{y}|\mathbf{X})$ = var $(\mathbf{e}|\mathbf{X}) = \boldsymbol{\Sigma}$. The remarks about the assumptions associated with the error vector are identical with those made in the functional regression model. The coefficient vector $\boldsymbol{\beta}$ is again assumed to be nonrandom. Finally, $\mathbf{X}$ is assumed to be known, measured, or observed without error. Since in this model the values of the regressor variables may be designed or preselected, $\mathbf{X}$ is also called the *design matrix*. In other models $\mathbf{X}$ may be stochastic, and therefore not subject to design.

Stochastic Regression (Model III)

The term stochastic regression alludes to the fact that in this model, the independent variables, x_{jk}, in (8.2.1) are assumed to be stochastic; that is, $\mathbf{X}$ is a random matrix. The k jointly distributed random variables $(Y, X_1, \ldots, X_{k-1})$ are permitted to vary without control or conditioning, and are observed n times. Such simultaneous random behavior is typical of random variables that arise in the social sciences.

For example, suppose Y denotes household income, and X_j denotes proportion of income spent on the jth type of expenditure (rent, food, clothing, and so on). A random sample of n households is observed. Here, the variables are all stochastic and not under control so Model III is most appropriate.

Remarks made about the error vector are identical to those made in Model I. The $\boldsymbol{\beta}$ vector is again assumed to be nonrandom and unknown, and (Y, X) are assumed to be observed without error (although there is still an error $\mathbf{e}$ present to correct for omitted variables and imperfections in the assumed linear relationship). Finally, the stochastic matrix $\mathbf{X}$ is assumed to be independent of $\mathbf{e}$. It is an interesting circumstance that even if $\mathbf{X}$ is stochastic, inferences about the elements of $\boldsymbol{\beta}$ or $\boldsymbol{\Sigma}$ (point and interval estimates, for example) are the same as those obtained for the case of nonstochastic $\mathbf{X}$, so long as the distribution of $\mathbf{X}$ does not depend upon the regression parameters, $\boldsymbol{\beta}$ and $\boldsymbol{\Sigma}$. That is, since

$$p(\boldsymbol{\beta}, \boldsymbol{\Sigma} | \text{sample}) \propto p(\mathbf{y}, \mathbf{X} | \boldsymbol{\beta}, \boldsymbol{\Sigma}) p(\boldsymbol{\beta}, \boldsymbol{\Sigma}),$$

or

$$p(\boldsymbol{\beta}, \boldsymbol{\Sigma} | \text{sample}) \propto p(\mathbf{y} | \mathbf{X}, \boldsymbol{\beta}, \boldsymbol{\Sigma}) p(\mathbf{X} | \boldsymbol{\beta}, \boldsymbol{\Sigma}) p(\boldsymbol{\beta}, \boldsymbol{\Sigma}),$$

if $p(\mathbf{X} | \boldsymbol{\beta}, \boldsymbol{\Sigma})$ does not depend upon $(\boldsymbol{\beta}, \boldsymbol{\Sigma})$,

$$p(\boldsymbol{\beta}, \boldsymbol{\Sigma} | \text{sample}) \propto p(\mathbf{y} | \mathbf{X}, \boldsymbol{\beta}, \boldsymbol{\Sigma}) p(\boldsymbol{\beta}, \boldsymbol{\Sigma}).$$

In such a case, inferences about $\boldsymbol{\beta}$ or $\boldsymbol{\Sigma}$ may be made without regard to whether or not the regressor matrix is stochastic. (For this reason, in the sequel, the assumption of nonstochastic, or at least fixed, $\mathbf{X}$ will generally be made implicitly and interest will be focused on Models I and II.) However, the distribution of $\hat{\boldsymbol{\beta}}$, the maximum likelihood estimator of $\boldsymbol{\beta}$, is different from that in Models I and II when $\mathbf{X}$ is stochastic (it is no longer Normally distributed when the errors are Normally distributed). Therefore, predictions of $\hat{\boldsymbol{\beta}}$ for experimental design purposes must take the randomness of $\mathbf{X}$ into account. That is, a prediction interval for $\hat{\boldsymbol{\beta}}$ (or the Bayesian predictive distribution) must be based upon the non-Normal distribution of $\hat{\boldsymbol{\beta}}$. An analogous statement may be made for $\hat{\boldsymbol{\Sigma}}$.

Errors in the Variables (Model IV)

In this model assume $\mathbf{y} = \mathbf{X}\boldsymbol{\beta}$, where $\mathbf{X}$ may be either stochastic or deterministic. However, neither $\mathbf{y}$ nor $\mathbf{X}$ is directly observable without error. Specifically, it is only possible to observe $y_\alpha^* = y_\alpha + u_\alpha$, and $x_{\alpha j}^* = x_{\alpha j} + v_{\alpha j}$, where u_α and $v_{\alpha j}$ are errors of measurement or observation. It is often reasonable to assume that $\mathcal{L}(u_\alpha) = N(0,\sigma^2)$, $\mathcal{L}(v_{\alpha j}) = N(0,\tau_j^2)$, $\alpha = 1,\ldots,n$, $j = 1,\ldots,k-1$, $\mathbf{u} \equiv (u_1,\ldots,u_n)'$, $\mathbf{v}_j \equiv (v_{1j},v_{2j},\ldots,v_{nj})'$ are independent, and the components of $\mathbf{u}$ and $\mathbf{v}_j$ are all mutually independent. In fact, it is a rare instance in which there are no measurement errors, so that this model would seem appropriate in many situations. Because of theoretical difficulties, the model is not so easy to analyze without some additional simplifying assumptions. Fortunately, however, measurement errors are often small enough to be neglected relative to the linear relationship being studied so that one of the simpler models discussed above can be used instead. For additional discussion of this model, see, for instance, Kendall and Stuart (1961), Vol. II, and Zellner (1972), Chapter 5.

Random Parameters (Model V)

The basic relation (8.2.2) is again applicable, where $(\mathbf{y},\mathbf{X})$ can follow the assumptions of either functional or conditional regression (so that regardless, $\mathbf{X}$ is nonstochastic). However, the coefficient vector, $\boldsymbol{\beta}$, is assumed to be stochastic and independent of $\mathbf{e}$. The error vector, $\mathbf{e}$, complies with the assumptions laid down in Model I. Model V is discussed in some detail in Scheffé (1959) from the analysis of variance standpoint, and is considered from a regression standpoint by Swamy (1967; 1968).

8.2.3 Estimation in Simple Univariate Regression

In (8.2.2), solve for $\mathbf{e}$ and form the sum of squares of the errors

$$\mathcal{E} = \mathbf{e}'\mathbf{e} = (\mathbf{y} - \mathbf{X}\boldsymbol{\beta})'(\mathbf{y} - \mathbf{X}\boldsymbol{\beta}).$$

To estimate $\boldsymbol{\beta}$ by *ordinary least squares*, minimize $\mathcal{E}$ with respect to $\boldsymbol{\beta}$. Accordingly, differentiate $\mathcal{E}$ with respect to $\boldsymbol{\beta}$, set the result equal to zero, and solve for $\boldsymbol{\beta}$. Thus, $-2\mathbf{X}'(\mathbf{y} - \mathbf{X}\hat{\boldsymbol{\beta}}) = \mathbf{0}$, or

$$\mathbf{X}'\mathbf{X}\hat{\boldsymbol{\beta}} = \mathbf{X}'\mathbf{y}. \tag{8.2.3}$$

This vector equation summarizes k equations in the k components of $\hat{\boldsymbol{\beta}}$; the equations are called the *normal equations.* To solve the normal equations, $\mathbf{X}'\mathbf{X}$ must be eliminated from the left-hand side of (8.2.3). Since $\mathbf{X}$ is $n \times k$, $\mathbf{X}'\mathbf{X}$ is $k \times k$. If $k \leq n$ and rank $(\mathbf{X}'\mathbf{X}) = k$, $\mathbf{X}'\mathbf{X}$ is nonsingular. Then $\mathbf{X}'\mathbf{X}$ has an inverse, so that (8.2.3) is easily solved as

$$\hat{\boldsymbol{\beta}} = (\mathbf{X}'\mathbf{X})^{-1}\mathbf{X}'\mathbf{y}, \tag{8.2.4}$$

and $\hat{\mathbf{e}} \equiv (\mathbf{y} - \mathbf{X}\hat{\boldsymbol{\beta}})$ is the *vector of residuals.* However, if $k > n$ (more parameters than observations), or if $k \leq n$, but some of the underlying independent variables are perfectly correlated, there will be *multicollinearity* ($|\mathbf{X}'\mathbf{X}| = 0$) and an inverse will not exist. In either case, there are an infinite number of solutions to (8.2.3) and unless the extraneous variables are dropped by deleting the corresponding rows and columns in $\mathbf{X}'\mathbf{X}$ (and deleting the corresponding elements of $\boldsymbol{\beta}$), or prior information is brought to bear on the problem in some form to eliminate the ambiguity, a unique solution cannot be obtained. Prior information can be introduced from the sampling theory viewpoint by imposing side conditions on the parameters and using the formalism of a generalized inverse. Prior information enters the problem for a Bayesian when he assesses an informative prior distribution for the parameters. Use of a diffuse (noninformative) prior will not extricate the analyst from the multicollinearity problem, however, since such a prior does not add enough information.

Let $(\mathbf{X}'\mathbf{X})^*$ denote a generalized inverse of $(\mathbf{X}'\mathbf{X})$. Then

$$(\mathbf{X}'\mathbf{X})(\mathbf{X}'\mathbf{X})^*(\mathbf{X}'\mathbf{X}) = \mathbf{X}'\mathbf{X}$$

[see (2.5.11)]. The most general solution to the normal equations in (8.2.3) is

$$\hat{\boldsymbol{\beta}} = (\mathbf{X}'\mathbf{X})^*\mathbf{X}'\mathbf{y} + [(\mathbf{X}'\mathbf{X})^*(\mathbf{X}'\mathbf{X}) - \mathbf{I}]\mathbf{z}, \tag{8.2.5}$$

where $\mathbf{z}$ is an arbitrary $k \times 1$ vector that may be stochastic or deterministic. When $(\mathbf{X}'\mathbf{X})^{-1}$ exists, $(\mathbf{X}'\mathbf{X})^* = (\mathbf{X}'\mathbf{X})^{-1}$, and (8.2.5) reduces to (8.2.4). However, when $|\mathbf{X}'\mathbf{X}| = 0$, (8.2.5) still provides solutions to the normal equations.

For example, it is easy to check [see (2.12.5)] that when $(\mathbf{X}'\mathbf{X})^*$ is a pseudoinverse [so that $(\mathbf{X}'\mathbf{X})^*(\mathbf{X}'\mathbf{X})(\mathbf{X}'\mathbf{X})^* = (\mathbf{X}'\mathbf{X})^*$], by taking $\mathbf{z} = \mathbf{0}$, (8.2.5) yields the solution for which $\hat{\boldsymbol{\beta}}'\hat{\boldsymbol{\beta}}$ is minimum. However, taking $\mathbf{z} = \mathbf{0}$ rules out specific side conditions. This point is exemplified below.

Suppose $\boldsymbol{\beta}$ is 2×1, $\boldsymbol{\beta} = (\beta_1, \beta_2)'$,

$$[(\mathbf{X}'\mathbf{X})^*\mathbf{X}'\mathbf{X} - \mathbf{I}] \equiv \mathbf{A} = \begin{pmatrix} A_{11} & A_{12} \\ A_{21} & A_{22} \end{pmatrix};$$

$(\mathbf{X'X})^*\mathbf{X'y} \equiv (u_1,u_2)'$; and $\mathbf{z} = (z_1,z_2)'$. Then

$$\begin{pmatrix}\hat\beta_1\\ \hat\beta_2\end{pmatrix} = \begin{pmatrix}u_1\\ u_2\end{pmatrix} + \begin{pmatrix}A_{11} & A_{12}\\ A_{21} & A_{22}\end{pmatrix}\begin{pmatrix}z_1\\ z_2\end{pmatrix},$$

so that

$$\hat\beta_1 = u_1 + A_{11}z_1 + A_{12}z_2,$$
$$\hat\beta_2 = u_2 + A_{21}z_1 + A_{22}z_2.$$

Suppose it is known a priori that $\beta_1 + \beta_2 = 1$. Then z_1 and z_2 should be selected so that $\hat\beta_1 + \hat\beta_2 = 1$. Such a constraint rules out the possibility of taking $\mathbf{z} = \mathbf{0}$ since if $\mathbf{z} = \mathbf{0}$, $\hat\beta_1 = u_1$ and $\hat\beta_2 = u_2$. But u_1 and u_2 are functions of X and y, and there is no reason why it is necessary to have $u_1 + u_2 = 1$ (except in some special case). Thus, to satisfy the side condition that $\beta_1 + \beta_2 = 1$, it is necessary to impose a stochastic linear constraint upon z_1 and z_2; namely,

$$(A_{11} + A_{21})z_1 + (A_{12} + A_{22})z_2 = 1 - u_1 - u_2.$$

A specific generalized inverse often used (and used in the above illustration) is the pseudoinverse [see (2.12.5)] obtained by factoring $\mathbf{X'X}$ into its symmetric decomposition of latent vectors and latent roots and taking the reciprocals of the nonzero latent roots. In the sequel, solutions of normal equations will sometimes be written in terms of generalized inverses to include the case of multicollinearity.

8.2.4 Goodness of Fit in Univariate Regression

One measure of how well a regression model accounts for the observed variation in a set of univariate data is the *sample multiple correlation coefficient*, R; R^2 is sometimes referred to as the *sample coefficient of multiple determination*. By definition,

$$R^2 = \frac{\text{sum of squares due to regression}}{\text{sum of squares about the mean}}.$$

R^2 is defined algebraically as follows.

Let $\hat y_j$ denote the estimated value of y_j defined in (8.2.1), so that $\hat y_j = \hat a_0 + \hat a_1 x_{j,1} + \cdots + \hat a_{k-1}x_{j,k-1}$, where $\hat a_j$ is the least squares estimator of a_j. Define $\bar y$ by $n\bar y \equiv \Sigma_1^n y_j$. Now form the difference

$$y_j - \bar y = (y_j - \hat y_j) + (\hat y_j - \bar y).$$

Squaring both sides and summing from $j = 1$ to $j = n$ gives

$$\sum_1^n (y_j - \bar y)^2 = \sum_1^n (y_j - \hat y_j)^2 + \sum_1^n (\hat y_j - \bar y)^2 + 2\sum_1^n (y_j - \hat y_j)(\hat y_j - \bar y).$$

It is now shown that the cross product term must vanish. First represent the model of (8.2.1) in the form

$$y_j = a_0 + \mathbf{b}'\mathbf{x}_j + e_j,$$

where $\mathbf{b}$ and $\mathbf{x}_j$ denote $(k - 1) \times 1$ vectors. The normal equations for the parameters in this format are easily found to be (for $n\bar{\mathbf{x}} \equiv \Sigma_1^n \mathbf{x}_j$)

$$\hat{a}_0 = \bar{y} - \hat{\mathbf{b}}'\bar{\mathbf{x}}, \qquad \text{and} \qquad \sum_1^n (y_j - \hat{a}_0 - \hat{\mathbf{b}}'\mathbf{x}_j)\mathbf{x}_j = \mathbf{0}.$$

Since $\hat{y}_j = \hat{a}_0 + \hat{\mathbf{b}}'\mathbf{x}_j$, $\Sigma_1^n(y_j - \hat{y}_j) = 0$. Hence, the cross product term in the expansion of $\Sigma(y_j - \bar{y})^2$ is

$$\begin{aligned} 2\sum_1^n (y_j - \hat{y}_j)(\hat{y}_j - \bar{y}) = 2\sum_1^n (y_j - \hat{y}_j)\hat{y}_j &= 2\sum_1^n (y_j - \hat{y}_j)(\hat{a}_0 + \hat{\mathbf{b}}'\mathbf{x}_j) \\ &= 2\hat{\mathbf{b}}'\sum_1^n (y_j - \hat{y}_j)\mathbf{x}_j, \\ &= 2\hat{\mathbf{b}}'\sum_1^n (y_j - \hat{a}_0 - \hat{\mathbf{b}}'\mathbf{x}_j)\mathbf{x}_j = 0. \end{aligned}$$

Thus,

$$\sum_1^n (y_j - \bar{y})^2 = \sum_1^n (y_j - \hat{y}_j)^2 + \sum_1^n (\hat{y}_j - \bar{y})^2.$$

The left-hand side is the sum of squares about the mean, and since $(\hat{y}_j - \bar{y})$ is a deviation directly attributable to the regression, the ratio

$$R^2 = \frac{\sum_1^n (\hat{y}_j - \bar{y})^2}{\sum_1^n (y_j - \bar{y})^2} = 1 - \frac{\hat{\mathbf{e}}'\hat{\mathbf{e}}}{\sum_1^n (y_j - \bar{y})^2} \tag{8.2.6}$$

is close to unity when $\hat{y}_j$ is a good predictor of y_j, and is close to zero when $\hat{y}_j$ is a poor predictor.

In multivariate regression, the value of R^2 can be computed for each equation separately to study the effectiveness of each relationship in accounting for observed variation. A low value of R^2 often means some important variables have been omitted (see Exercise 8.16), or the assumed form of the regression relation is inadequate (perhaps quadratic or higher-order terms are required).

It is not hard to show that another interpretation of R which is sometimes useful is that R is the sample correlation coefficient between the scalars Y and $\hat{Y}$ (see Exercise 8.11).

The basis for the last assertion is found in the conditional regression model (see Section 8.2.2). There, the variables $(Y,X_1,\ldots,X_{k-1})$ are assumed to be joint Normally distributed, and if Y is regressed on $(X_1,\ldots,X_{k-1})' \equiv \mathbf{x}$,

$$\mathcal{L}(\mathbf{y}|\mathbf{X}) = N(\mathbf{X}\boldsymbol{\beta},\boldsymbol{\Sigma}),$$

where $\mathbf{y} \equiv (y_1,\ldots,y_n)'$ is the vector of observations of Y, and the design matrix, $\mathbf{X}$, is defined as in Section 8.2.1. Let $\boldsymbol{\alpha}$ denote any $(k-1) \times 1$ vector of parameters. Then the strength of the relationship between Y and $\mathbf{x}$ is measured by the *population multiple correlation coefficient*, P, defined as

$$P = \max_{\boldsymbol{\alpha}} \text{corr}\,(Y,\boldsymbol{\alpha}'\mathbf{x}). \tag{8.2.7}$$

That is, P is the largest possible correlation between Y and linear combinations of the components of $\mathbf{x}$. P^2 is the *population coefficient of multiple determination.* The sample version of (8.2.7) uses the fact that for maximum correlation, $\boldsymbol{\alpha}$ is estimated by the least squares estimator of $\boldsymbol{\beta}$. Thus, it is reasonable that R is the sample correlation between Y and $\hat{Y}$ (see Exercise 8.11).

R^2 in (8.2.6) may be intrinsically large because the assumed regression relationship is a close representation of the data, or it may be large because of sampling variation. The likelihood ratio test of the hypothesis $H: P = 0$ versus $A: P > 0$ is to reject H if

$$Q \equiv \frac{R^2}{1 - R^2} \cdot \frac{n-k}{k-1} > \text{constant},$$

where under H, $\mathcal{L}(Q) = F_{k-1,n-k}$. The upper tail of the F-distribution may thus be used to check whether there is a significant regression of Y on $\mathbf{x}$, or whether Y is really uncorrelated with the components of $\mathbf{x}$.

The distribution of R^2, for $P^2 > 0$, is not expressible in closed form in terms of elementary functions. In conditional regression models, for $P^2 > 0$, Q (above) follows a noncentral F-distribution. For stochastic regression models, the distribution of R^2 when $P^2 > 0$ and $(Y,X_1,\ldots,X_{k-1})$ follows a joint Normal distribution was derived by Fisher (1928), who gave the resulting density as an infinite series, and also in terms of a hypergeometric function. An approximate (large sample) relation suggested by Fisher for obtaining interval estimates of P^2 in stochastic regression, in large samples, for $P^2 > 0$, and $k > 2$, is given by

$$\mathcal{L}\{\sqrt{n}\,(\tanh^{-1} R - \tanh^{-1} P)\} \cong N(0,1).$$

From this relation, it follows that with confidence coefficient $(1 - \alpha)$,

$$\tanh^{-1} R - \frac{z_{\alpha/2}}{\sqrt{n}} \leq \tanh^{-1} P \leq \tanh^{-1} R + \frac{z_{\alpha/2}}{\sqrt{n}},$$

where $z_{\alpha/2}$ is the $\alpha/2$ fractile point of the $N(0,1)$ distribution. Since $\tanh(\cdot)$ is a monotone transformation, the last relation is equivalent to

$$\tanh\left[\tanh^{-1} R - \frac{z_{\alpha/2}}{\sqrt{n}}\right] \leq P \leq \tanh\left[\tanh^{-1} R + \frac{z_{\alpha/2}}{\sqrt{n}}\right].$$

For numerical convenience it is sometimes helpful to use the relation

$$\tanh^{-1} R \equiv \tfrac{1}{2} \log\left(\frac{1 + R}{1 - R}\right), \qquad \text{for } R < 1.$$

This interval estimator is useful for making quantitative comparisons of goodness of fit in regression. Often, R^2 will be used as a descriptive measure to assess the goodness of fit of an assumed regression model. Then a new model will be compared with the old model (perhaps another independent variable has been added) by comparing the associated values of R^2, which may have changed from $R^2 = .6$ to $R^2 = .7$. But what has happened to P^2? Interval estimates provide a much better insight into what has actually happened in the population by changing the model.

Bayesian approaches to comparison of regression models involve setting out several alternative models for the same data, assessing a prior distribution for the various alternative models (such as: a priori, all models being compared are equally likely), assessing a prior distribution for the regression parameters, evaluating the posterior distribution for model j and the regression parameters given the data, and then finding the marginal posterior density for model j by integrating out the regression parameters. The model with the largest marginal posterior probability is selected as the one that best fits the observed data. This Bayesian type of approach has the advantage of evaluating the goodness of fit of an entire model to a set of data, as opposed to the separate equation-by-equation evaluation described above (which does not really take correlations and interactions among equations into account). A Bayesian development of the distribution of the squared population multiple correlation coefficient is given in Press & Zellner, 1978.

The Use of R^2 as a Screening Device

Exploratory research often requires that a problem defined in many dimensions be reduced to one defined in many fewer dimensions. When some variables are believed to be considerably less important than others, the former variables are eliminated. The criterion for elimination is sometimes the ordinary correlation coefficient. That is, the matrix of all possible pairwise correlations is computed for all the variables and the smallest correlations are mistakenly used to dictate which variables to eliminate. The multiple correlation coefficient between one of the variables (the dependent variable) and all the rest is a function of all the pairwise correlations, and it is generally believed that if a pairwise correlation between a variable and the dependent variable is small, R^2 will not be changed appreciably if the associated variable is eliminated. However, such is not the case in general (see for instance Kendall and Stuart, Vol. II, p. 336), in that R^2 may be greatly affected even when the simple pairwise correlation is close to zero! Thus, it is not recommended that simple pairwise correlation be used as a screening device in this fashion. Instead, the effect of each variable on multiple R^2 can be examined. Those variables which contribute little to R^2 are eliminated.

8.3 WEIGHTED UNIVARIATE REGRESSION

In the previous section the underlying assumptions of various univariate regression models were discussed. In this and succeeding sections (which treat the multivariate regression problem) discussion will be confined to Models I and II assumptions, or a combination thereof, since the formalism is the same for both (in testing hypotheses in Models I and II, the powers of tests may differ—see Graybill, 1961). The simple, uncorrelated errors, univariate regression problem was discussed in Section 8.2. The correlated errors, univariate regression problem is taken up in this section because that problem is really one of applied multivariate analysis.

Consider the univariate model

$$\underset{(n\times 1)}{\mathbf{y}} = \underset{(n\times k)}{\mathbf{X}} \cdot \underset{(k\times 1)}{\boldsymbol{\beta}} + \underset{(n\times 1)}{\mathbf{u}}, \qquad \mathcal{L}(\mathbf{u}) = N(\mathbf{0},\sigma^2\mathbf{I}). \tag{8.3.1}$$

Here, the errors are independent and homoscedastic (all errors have the same variance). Minimizing $\mathbf{u}'\mathbf{u} = (\mathbf{y} - \mathbf{X}\boldsymbol{\beta})'(\mathbf{y} - \mathbf{X}\boldsymbol{\beta})$ with respect to $\boldsymbol{\beta}$ gives the least squares estimator [as in (8.2.4)]

$$\hat{\boldsymbol{\beta}} = (\mathbf{X}'\mathbf{X})^*\mathbf{X}'\mathbf{y} + [(\mathbf{X}'\mathbf{X})^*(\mathbf{X}'\mathbf{X}) - \mathbf{I}]\mathbf{z}, \tag{8.3.2}$$

and the corresponding unbiased estimator of σ^2,

$$\hat{\sigma}^2 = \frac{1}{n-k}(\mathbf{y} - \mathbf{X}\hat{\boldsymbol{\beta}})'(\mathbf{y} - \mathbf{X}\hat{\boldsymbol{\beta}}), \tag{8.3.3}$$

where $(\mathbf{X}'\mathbf{X})^*$ is any generalized inverse of $(\mathbf{X}'\mathbf{X})$, and $\mathbf{z}$ is an arbitrary $k \times 1$ vector. Assume that when necessary, there is sufficient prior information to estimate $\boldsymbol{\beta}$ uniquely. The well-known *Gauss-Markoff Theorem* (see, for instance, Scheffé, 1959, p. 14) and its extension for generalized inverses (see Rao, 1965, p. 182) assert that for the model in (8.3.1), $\hat{\boldsymbol{\beta}}$ is an unbiased estimator of $\boldsymbol{\beta}$ that is linear in $\mathbf{y}$, and among all linear unbiased estimators of $\boldsymbol{\beta}$, $\hat{\boldsymbol{\beta}}$ has minimum variance. Usually we can also assume that $\hat{\boldsymbol{\beta}}$ is consistent.[1]

Now suppose the errors are correlated, or heteroscedastic, or both, and adopt the model

$$\mathbf{y} = \mathbf{X}\boldsymbol{\beta} + \mathbf{u}, \qquad |\mathbf{X}'\boldsymbol{\Sigma}^{-1}\mathbf{X}| \neq 0, \qquad \boldsymbol{\Sigma} > 0, \qquad \mathcal{L}(\mathbf{u}) = N(\mathbf{0},\sigma^2\boldsymbol{\Sigma}). \quad (8.3.4)$$

Since the assumptions of (8.3.1) are no longer valid, the Gauss-Markoff Theorem need no longer apply. That is, if (8.3.2) and (8.3.3) are used under the assumptions of (8.3.4), inefficient estimators of $\boldsymbol{\beta}$ and σ^2 may result. Such is the problem of weighted univariate regression, and the assumptions of the model are defined by (8.3.4).

If (8.3.4) is multiplied throughout by $\boldsymbol{\Sigma}^{-1/2}$ in an effort to reduce this problem to the one of simple univariate regression, the model may be rewritten in the form

$$\mathbf{y}_0 = \mathbf{X}_0\boldsymbol{\beta} + \mathbf{u}_0, \qquad |\mathbf{X}_0'\mathbf{X}_0| \neq 0, \qquad \mathcal{L}(\mathbf{u}_0) = N(\mathbf{0},\sigma^2\mathbf{I}), \quad (8.3.5)$$

where

$$\mathbf{y}_0 = \boldsymbol{\Sigma}^{-1/2}\mathbf{y}, \qquad \mathbf{X}_0 = \boldsymbol{\Sigma}^{-1/2}\mathbf{X}, \qquad \mathbf{u}_0 = \boldsymbol{\Sigma}^{-1/2}\mathbf{u}.$$

But (8.3.5) is equivalent to (8.3.1). Therefore, the minimum variance linear unbiased estimator of $\boldsymbol{\beta}$ is given by

$$\tilde{\boldsymbol{\beta}} = (\mathbf{X}_0'\mathbf{X}_0)^{-1}\mathbf{X}_0'\mathbf{y}_0 = (\mathbf{X}'\boldsymbol{\Sigma}^{-1}\mathbf{X})^{-1}\mathbf{X}'\boldsymbol{\Sigma}^{-1}\mathbf{y}, \quad (8.3.6)$$

and the corresponding unbiased estimator of σ^2 is given by

$$\tilde{\sigma}^2 = \frac{1}{n-k}(\mathbf{y} - \mathbf{X}\tilde{\boldsymbol{\beta}})'\boldsymbol{\Sigma}^{-1}(\mathbf{y} - \mathbf{X}\tilde{\boldsymbol{\beta}}). \quad (8.3.7)$$

[1] Note that for $\mathbf{X}$ fixed, $\hat{\boldsymbol{\beta}}$ will be a consistent estimator of $\boldsymbol{\beta}$ if $\lim_{n\to\infty} (\mathbf{X}'\mathbf{X}/n) =$ constant; if $\mathbf{X}$ is stochastic, $\hat{\boldsymbol{\beta}}$ will be consistent if $\operatorname{plim}_{n\to\infty} (\mathbf{X}'\mathbf{X}/n) =$ constant. These assertions may be proved as follows:

Suppose $|\mathbf{X}'\mathbf{X}| > 0$. Then,

$$\hat{\boldsymbol{\beta}} = (\mathbf{X}'\mathbf{X})^{-1}\mathbf{X}'\mathbf{y} = (\mathbf{X}'\mathbf{X})^{-1}\mathbf{X}'(\mathbf{X}\boldsymbol{\beta} + \mathbf{u}) = \boldsymbol{\beta} + (\mathbf{X}'\mathbf{X})^{-1}\mathbf{X}'\mathbf{u},$$

or

$$\hat{\boldsymbol{\beta}} - \boldsymbol{\beta} = \left(\frac{\mathbf{X}'\mathbf{X}}{n}\right)^{-1}\left(\frac{\mathbf{X}'\mathbf{u}}{n}\right).$$

Now suppose $\mathbf{X} = (\mathbf{e},\mathbf{x})$, where $\mathbf{e}$ is an n-vector of ones and $\mathbf{x} = (\mathbf{x}_1,\ldots,\mathbf{x}_{k-1})$. Since $\operatorname{plim}_{n\to\infty}\left(\frac{\mathbf{X}'\mathbf{u}}{n}\right) = \left(\operatorname{plim}_{n\to\infty}\frac{\mathbf{X}'}{n}\right)\mathbf{u} = \mathbf{0}$, if $\operatorname{plim}_{n\to\infty}\left(\frac{\mathbf{X}'\mathbf{X}}{n}\right) =$ constant, an application of the result in Footnote 2, Section 4.5 will give the result.

Aitken (1935) obtained the result (8.3.6) by minimizing the weighted sum of squares $\mathbf{u}'\mathbf{\Sigma}^{-1}\mathbf{u}$ with respect to $\boldsymbol{\beta}$. This procedure is called the *method of generalized least squares.*

It is immediate from (8.3.6) that

$$\text{var }\tilde{\boldsymbol{\beta}} = \sigma^2(\mathbf{X}'\mathbf{\Sigma}^{-1}\mathbf{X})^{-1}. \tag{8.3.8}$$

Thus, if $\mathbf{\Sigma}$ is a known matrix, (8.3.6)–(8.3.8) may be used in the standard way to make inferences about $\boldsymbol{\beta}$ and σ^2 (Student t-tests, simultaneous confidence intervals, Bayesian estimation and hypothesis testing, and so on). In some situations, $\mathbf{\Sigma}$ is in fact known and $(\tilde{\boldsymbol{\beta}},\tilde{\sigma}^2)$ are computable [see Example (8.3.1)]. However, in many situations $\mathbf{\Sigma}$ is unknown and must be estimated from the data.

Suppose $\hat{\mathbf{\Sigma}}$ is a consistent[2] estimator of $\mathbf{\Sigma}$. Then define

$$\hat{\hat{\boldsymbol{\beta}}} = (\mathbf{X}'\hat{\mathbf{\Sigma}}^{-1}\mathbf{X})^{-1}\mathbf{X}'\hat{\mathbf{\Sigma}}^{-1}\mathbf{y} \tag{8.3.9}$$

and

$$\hat{\hat{\sigma}}^2 = \frac{1}{n-k}(\mathbf{y} - \mathbf{X}\hat{\hat{\boldsymbol{\beta}}})'\hat{\mathbf{\Sigma}}^{-1}(\mathbf{y} - \mathbf{X}\hat{\hat{\boldsymbol{\beta}}}). \tag{8.3.10}$$

In large samples, $\hat{\hat{\boldsymbol{\beta}}}$, $\hat{\hat{\sigma}}^2$ preserve the Gauss-Markoff optimality properties of $\tilde{\boldsymbol{\beta}}$, $\tilde{\sigma}^2$. However, in very small samples, there is no guarantee that (8.3.9) and (8.3.10) will yield more efficient estimates than $\hat{\boldsymbol{\beta}}$, and $\hat{\sigma}^2$, defined in (8.3.2) and (8.3.3).

The main problem remaining is how to find a consistent estimator of $\mathbf{\Sigma}$. A natural question at this point is why not use the residuals from a naive regression [using (8.3.1) and (8.3.2) when in fact (8.3.4) is true]? The proposed estimator would be

$$\hat{\mathbf{\Sigma}} = \frac{1}{n-k}(\mathbf{y} - \mathbf{X}\hat{\boldsymbol{\beta}})(\mathbf{y} - \mathbf{X}\hat{\boldsymbol{\beta}})'. \tag{8.3.11}$$

The reasons for not using this estimator are discussed below.

Note that $\hat{\mathbf{\Sigma}}$ in (8.3.11) is an $n \times n$ matrix so that as $n \to \infty$, $\hat{\mathbf{\Sigma}}$ becomes a matrix of infinite dimensions. Moreover, since $\mathbf{\Sigma}$ is an $n \times n$ symmetric matrix, it has $n(n + 1)/2$ distinct parameters. As $n \to \infty$, the number of distinct parameters becomes unbounded. Furthermore, for a

[2] Note that $\mathbf{\Sigma}$ is an $n \times n$ matrix, so that if $\mathbf{\Sigma}$ has a full complement of $n(n + 1)/2$ distinct parameters, the number of unknown parameters grows with sample size (which is the problem of *incidental parameters* in the errors-in-the-variables model; see Neyman and Scott, 1948). In this case, inconsistent estimators result. It is only when the number of parameters in $\mathbf{\Sigma}$ remains finite and sufficiently small while the sample size grows that consistent estimators of $\mathbf{\Sigma}$ are obtainable. This can occur only when additional information is introduced to reduce the number of parameters in $\mathbf{\Sigma}$ (such as by giving it a special structural form).

given n, there are a total of $[k + n(n + 1)/2]$ parameters to be estimated by a total of n observations, a clear impossibility. Finally, it is easy to check that for $\hat{\mathbf{\Sigma}}$ in (8.3.11), rank $(\hat{\mathbf{\Sigma}}) = 1$. Hence, $|\hat{\mathbf{\Sigma}}| = 0$, so that (8.3.9) cannot be computed.

Since the residuals cannot be used as in (8.3.11), it is necessary to learn more about the nature of the serial correlation or heteroscedasticity of the errors by studying the residuals from the standard regression (8.3.1), and then imposing enough additional assumptions on the problem to permit consistent estimation of $\mathbf{\Sigma}$.

Example (8.3.1)—Known Covariance Matrix: Demand for a certain product is assumed to be the sum of two random variables representing two independent processes:

(1) Y = demand induced by television advertising,
(2) Z = demand attributable to nontelevision advertising.

In City A, both forms of advertising are used and X_1 units of the product are demanded.

In City B, television is not used and X_2 units are demanded during the same time period.

In City C, only television is used and X_3 units are demanded in the same time period.

Note that X_1, X_2, X_3 are measurements on $Y + Z$, Z, and Y, respectively.

Suppose $EY = \lambda$, $EZ = \mu$, var $Y = \sigma^2$, var $Z = \sigma^2$, and X_1, X_2, X_3 are mutually independent. Estimate (μ,λ,σ^2).

Structure the problem by letting

$$\begin{aligned} X_1 &= \mu + \lambda + u_1, & E(u_1) &= 0, & \text{var}\,(u_1) &= 2\sigma^2, \\ X_2 &= \mu \qquad\; + u_2, & E(u_2) &= 0, & \text{var}\,(u_2) &= \sigma^2, \\ X_3 &= \qquad \lambda + u_3, & E(u_3) &= 0, & \text{var}\,(u_3) &= \sigma^2, \end{aligned} \tag{8.3.12}$$

where the u_α's are assumed to be uncorrelated errors. To write the model in compact form, define

$$\mathbf{y} = \begin{pmatrix} X_1 \\ X_2 \\ X_3 \end{pmatrix}, \quad \mathbf{u} = \begin{pmatrix} u_1 \\ u_2 \\ u_3 \end{pmatrix}, \quad \mathbf{X} = \begin{pmatrix} 1 & 1 \\ 1 & 0 \\ 0 & 1 \end{pmatrix}, \quad \boldsymbol{\beta} = \begin{pmatrix} \mu \\ \lambda \end{pmatrix}.$$

Then $\mathbf{y} = \mathbf{X}\boldsymbol{\beta} + \mathbf{u}$, $E(\mathbf{u}) = \mathbf{0}$, var $(\mathbf{u}) = \sigma^2\mathbf{\Sigma}$, where

$$\mathbf{\Sigma} = \begin{pmatrix} 2 & 0 & 0 \\ 0 & 1 & 0 \\ 0 & 0 & 1 \end{pmatrix}.$$

Thus, the errors are uncorrelated and heteroscedastic, but $\mathbf{\Sigma}$ is known; that is, the covariance matrix of $\mathbf{u}$ is known up to a scalar multiplier.[3] Therefore, $\boldsymbol{\beta}$ should not be estimated by ordinary least squares, but in fact should be estimated by weighted least squares, and the solution is immediately given by (8.3.6). In this case,

$$\mathbf{\Sigma}^{-1} = \begin{pmatrix} \frac{1}{2} & 0 & 0 \\ 0 & 1 & 0 \\ 0 & 0 & 1 \end{pmatrix},$$

$$\mathbf{X}'\mathbf{\Sigma}^{-1}\mathbf{X} = \frac{1}{2}\begin{pmatrix} 3 & 1 \\ 1 & 3 \end{pmatrix}, \qquad (\mathbf{X}'\mathbf{\Sigma}^{-1}\mathbf{X})^{-1} = \frac{1}{4}\begin{pmatrix} 3 & -1 \\ -1 & 3 \end{pmatrix},$$

so that

$$(\mathbf{X}'\mathbf{\Sigma}^{-1}\mathbf{X})^{-1}\mathbf{X}'\mathbf{\Sigma}^{-1} = \frac{1}{4}\begin{pmatrix} 1 & 3 & -1 \\ 1 & -1 & 3 \end{pmatrix}.$$

Hence,

$$\tilde{\mu} = \tfrac{1}{4}(X_1 + 3X_2 - X_3),$$
$$\tilde{\lambda} = \tfrac{1}{4}(X_1 - X_2 + 3X_3).$$

To obtain an unbiased estimate of σ^2, substitute in (8.3.7). Since $n = 3$ and $k = 2$,

$$\tilde{\sigma}^2 = \tfrac{1}{4}(X_2 + X_3 - X_1)^2.$$

For comparative purposes, note that if the heteroscedasticity had been ignored, an ordinary least squares estimator would have yielded

$$\hat{\mu} = \tfrac{1}{3}(X_1 + 2X_2 - X_3), \qquad \hat{\lambda} = \tfrac{1}{3}(X_1 - X_2 + 2X_3).$$

It is easy to check that $\operatorname{var}\hat{\mu} - \operatorname{var}\tilde{\mu} = \sigma^2/36$ and $\operatorname{var}\hat{\lambda} - \operatorname{var}\tilde{\lambda} = \frac{7}{8}\sigma^2$; that is, the generalized least squares estimators are superior, as expected.

Example (8.3.2)—Heteroscedasticity: Suppose it is determined that although there is no serial correlation, there is heteroscedasticity; that is,

$$\sigma^2\mathbf{\Sigma} = \begin{pmatrix} \sigma_1^2 & & & \\ & \cdot & 0 & \\ & & \cdot & \\ & 0 & \cdot & \\ & & & \sigma_n^2 \end{pmatrix}.$$

Suppose further that a plot of the observed dependent variables against the corresponding values of, say, X_2 indicates a relationship of the form (see Figure 8.3.1)

$$\sigma_\alpha^2 = cx_{\alpha 2}, \qquad \alpha = 1,\ldots,n, \tag{8.3.13}$$

[3] It should be noted that more generally, $\mathbf{\Sigma}$ will be known in any situation in which an observed process is the sum of several independent, observable processes with the same variance.

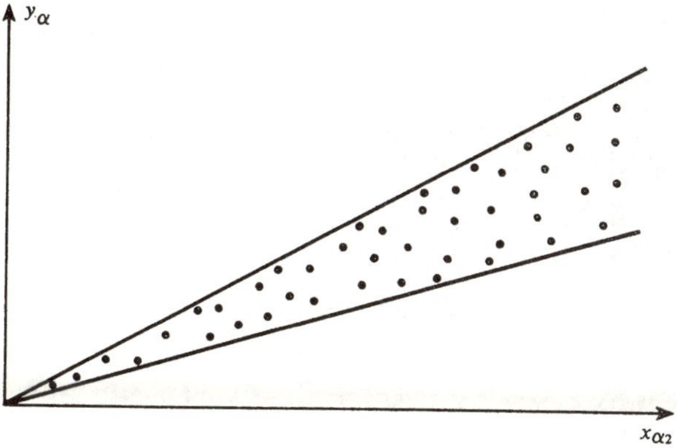

Figure 8.3.1. Heteroscedasticity

for some fixed c. That is, the variance of y_α is proportional to the corresponding value of X_2. Define $\mathbf{D}_2 \equiv \text{diag}\,(x_{1,2},x_{2,2},\ldots,x_{n,2})$. Then, from (8.3.6), the constant of proportionality cancels out, and

$$\tilde{\boldsymbol{\beta}} = (\mathbf{X}'\mathbf{D}_2^{-1}\mathbf{X})^{-1}\mathbf{X}'\mathbf{D}_2^{-1}\mathbf{y}. \qquad (8.3.14)$$

This case occurs frequently in practice and is immediately recognizable by the appearance of a funnel-shaped graph, as in Figure 8.3.1.

A very simple, but frequently occurring model is the one in which $y_i = \theta + u_i$, $i = 1,\ldots,n$, $E(u_i) = 0$, and the u_i's are uncorrelated with var $u_i = \sigma_i^2$, $i = 1,\ldots,n$. It is easy to check from (8.3.6) that the generalized least squares estimator of θ is

$$\hat{\theta} = \frac{\sum_1^n h_i y_i}{\sum_1^n h_i},$$

where $h_i \equiv \sigma_i^{-2}$ is the precision of u_i, $i = 1,\ldots,n$ (see Section 3.6.4). That is, if the variances of the disturbances are equal, $\hat{\theta} = \bar{y}$, whereas if the variances are unequal, the y_i's must be averaged with unequal weights, the weights being the corresponding precisions. Thus, this type of model is sometimes referred to as *weighted regression.*

Example (8.3.3)—First-Order Markoff (Autoregressive) Correlation: Consider the model

$$\underset{(n\times 1)}{\mathbf{y}} = \underset{(n\times k)}{\mathbf{X}} \cdot \underset{(k\times 1)}{\boldsymbol{\beta}} + \underset{(n\times 1)}{\mathbf{e}}, \quad \mathcal{L}(\mathbf{e}) = N(\mathbf{0},\boldsymbol{\Omega}), \qquad (8.3.15)$$

where, for any time t,

$$e_t = \rho e_{t-1} + u_t, \qquad |\rho| < 1, \qquad E(\mathbf{u}) = \mathbf{0}, \qquad \text{var}\ \mathbf{u} = \sigma_0^2\mathbf{I}, \qquad (8.3.16)$$

for some unknown constant σ_0^2. Assume the u_t process started at $t = -\infty$. The condition that $|\rho| < 1$ implies the corresponding system is stable in that e_t does not grow as $t \to \infty$ (var e_t is constant). The first-order difference equation is easily solved by iteration, so that for any k, and any integer α,

$$e_\alpha = \sum_{j=0}^{k} \rho^j u_{\alpha-j} + \rho^{k+1} e_{\alpha-k-1}.$$

Since $|\rho| < 1$, as $k \to \infty$, the second term approaches zero. Hence, for $\alpha = 1,\ldots,n$,

$$e_\alpha = \sum_{j=0}^{\infty} \rho^j u_{\alpha-j}.$$

Since var $(\mathbf{e}) = \mathbf{\Omega}$,

$$\text{cov}\,(e_\alpha, e_\beta) = Ee_\alpha e_\beta = E\left(\sum_{j=0}^{\infty} \rho^j u_{\alpha-j}\right)\left(\sum_{k=0}^{\infty} \rho^k u_{\beta-k}\right),$$

and var $\mathbf{u} = \sigma_0^2\mathbf{I}$, it is easily found that $\mathbf{\Omega} = \sigma^2\mathbf{\Sigma}$, where $\sigma^2 = \sigma_0^2/(1 - \rho^2)$, and

$$\mathbf{\Sigma} = \begin{pmatrix} 1 & \rho & \rho^2 & \cdots & \rho^{n-1} \\ \rho & 1 & \rho & \cdots & \rho^{n-2} \\ \cdot & \cdot & \cdot & & \cdot \\ \cdot & \cdot & \cdot & & \cdot \\ \cdot & \cdot & \cdot & & \cdot \\ \rho^{n-1} & \rho^{n-2} & \rho^{n-3} & \cdots & 1 \end{pmatrix}.$$

$\mathbf{\Sigma}$ is the same Toeplitz matrix given in Chapter 2, Equation (2.2.8); its inverse is given in Equation (2.5.10).

Thus, assuming that the regression disturbances are correlated according to a first-order Markoff scheme, $\mathbf{\Sigma}$ depends on only one parameter, ρ. The regression residuals $\hat{e}_\alpha$ can be used to estimate ρ. Using $\hat{e}_\alpha$ in place of e_α in (8.3.16) suggests the least squares estimator

$$\hat{\rho} = \frac{\sum_{1}^{n-1} \hat{e}_\alpha \hat{e}_{\alpha+1}}{\sum_{1}^{n-1} \hat{e}_\alpha^2}. \qquad (8.3.17)$$

The estimator given in (8.3.17) is shown to be a consistent estimator of ρ in Complement 8.1. $\hat{\hat{\boldsymbol{\beta}}}$ is easily evaluated from (8.3.9) by substituting $\hat{\rho}$ for ρ in $\mathbf{\Sigma}$.

Example (8.3.4)—Circular Correlation: Suppose that the nature of the problem, or of the residuals, indicates that the errors tend to follow a circular correlation model [see Equation (2.2.5)]. Then

$$\underset{(n\times 1)}{\mathbf{y}} = \underset{(n\times k)}{\mathbf{X}} \cdot \underset{(k\times 1)}{\boldsymbol{\beta}} + \underset{(n\times 1)}{\mathbf{u}}, \qquad \mathcal{L}(\mathbf{u}) = N(\mathbf{0},\sigma^2\boldsymbol{\Sigma}),$$

where

$$\left.\begin{aligned} &\boldsymbol{\Sigma} = (\sigma_{ij}),\ \sigma_{ij} = \rho_{j-1}, && \text{for } i \leq j,\\ &\text{and} \\ &\sigma_{ij} = \sigma_{ji}, && \text{for } i > j,\ i, j = 1,\ldots,n,\\ &\rho_0 \equiv 1, \text{ and } \rho_j = \rho_{n-j}, && j = 1,\ldots,n-1. \end{aligned}\right\} \qquad (8.3.18)$$

The form of $\sigma^2\boldsymbol{\Sigma}$ for the cases of $n = 4$ and $n = 5$ are shown below for illustrative purposes.

$$n = 4 \qquad\qquad\qquad\qquad n = 5$$

$$\sigma^2 \begin{pmatrix} 1 & \rho_1 & \rho_2 & \rho_1 \\ \rho_1 & 1 & \rho_1 & \rho_2 \\ \rho_2 & \rho_1 & 1 & \rho_1 \\ \rho_1 & \rho_2 & \rho_1 & 1 \end{pmatrix}, \qquad \sigma^2 \begin{pmatrix} 1 & \rho_1 & \rho_2 & \rho_2 & \rho_1 \\ \rho_1 & 1 & \rho_1 & \rho_2 & \rho_2 \\ \rho_2 & \rho_1 & 1 & \rho_1 & \rho_2 \\ \rho_2 & \rho_2 & \rho_1 & 1 & \rho_1 \\ \rho_1 & \rho_2 & \rho_2 & \rho_1 & 1 \end{pmatrix}.$$

Such covariance matrices are not only symmetric and positive semi-definite, but also imply that equally spaced temporal or spatial points are similarly correlated. These covariance matrices have been studied by Whittle (1951), Wise (1955), Press (1964), and Olkin and Press (1969). Note that $\boldsymbol{\Sigma}$ in (8.3.18) is easily inverted in analytical form using Property (10) of Section 2.12; as an example,

$$\boldsymbol{\Sigma}^{-1} = \boldsymbol{\Gamma}\mathbf{D}_\tau^{-1}\boldsymbol{\Gamma}', \qquad (8.3.19)$$

where $\mathbf{D}_\tau = \text{diag}(\tau_1,\ldots,\tau_n)$, and from (2.8.4), (2.8.8), and symmetry of $\boldsymbol{\Sigma}$,

$$\tau_j = \left[\sum_{k=1}^{n} \rho_{k-1} \cos\frac{2\pi}{n}(j-1)(k-1)\right], \qquad j = 1,\ldots,n, \qquad (8.3.20)$$

$\boldsymbol{\Gamma}$ is an orthogonal matrix defined by $\boldsymbol{\Gamma}\boldsymbol{\Gamma}' = \mathbf{I}_n$, and

$$\boldsymbol{\Gamma} = (\gamma_{jk}), \qquad \gamma_{jk} = \frac{1}{\sqrt{n}}\left[\cos\frac{2\pi}{n}(j-1)(k-1) + \sin\frac{2\pi}{n}(j-1)(k-1)\right]. \qquad (8.3.21)$$

Consistent estimation of $\boldsymbol{\Sigma}$ has been studied by Olkin and Press (1969). Substituting a consistent estimator of $\boldsymbol{\Sigma}$ into (8.3.9) gives the Aitken estimator of $\boldsymbol{\beta}$.

A frequently occurring circular correlation matrix which is a special case of (8.3.18), and which further restricts the parameters of the model is discussed below.

Suppose that on the basis of a residual analysis it is decided that a realistic model is

$$\boldsymbol{\Sigma} = \begin{pmatrix} 1 & & & \\ & \cdot & \rho & \\ & & \cdot & \\ & \rho & \cdot & \\ & & & 1 \end{pmatrix}, \qquad -\frac{1}{n-1} < \rho < 1, \tag{8.3.22}$$

the intraclass correlation matrix [see Equation (2.2.4)]. Clearly (8.3.18) reduces to (8.3.22) when all off-diagonal elements are equal. Then, no matter how large n is, the number of parameters to be estimated remains fixed at one; namely ρ.

It was shown by McElroy (1967) that if the regression contains a constant term, and if the correlation matrix of the disturbances has the intraclass form given in (8.3.22) (with equal variances), the ordinary least squares estimator coincides with the Aitken, best linear unbiased estimator. Hence, it is not necessary to estimate ρ to obtain an estimator of β that is best. (See also Section 8.4.2, Efficiency of Estimators.)

Example (8.3.5)—Distributed Lags: Distributed lag regression models appear frequently in econometric contexts. For example, Koyck (1954) discussed the use of distributed lag models in investment analysis, Nerlove (1958) discussed their use in agriculture, and Griliches (1967) provided an extensive review of the distributed lag literature. A very simple, but often occurring, distributed lag model is discussed below.

Assume the dependent variable observations y_j are related to observations on a single independent variable x_j, through the equation

$$y_j = \alpha_0 x_j + \alpha_0\alpha_1 x_{j-1} + \alpha_0\alpha_1^2 x_{j-2} + \cdots + u_j, \tag{8.3.23}$$

for $j = 1,\ldots,n$, $-\infty < \alpha_0 < \infty$, $0 \leq \alpha_1 < 1$. Assume there are observations x_0, y_0 available. Equation (8.3.23) is equivalent to

$$y_j = \alpha_0 \sum_{i=0}^{\infty} \alpha_1^i x_{j-i} + u_j.$$

In this model, the coefficients of lagged terms decay exponentially as the points become further removed in time. The above sum is sometimes referred to as an exponentially smoothed average of the x_j's. (A more general weighting scheme than this exponential one was described by Jorgenson, 1966.)

Suppose for $\mathbf{u} \equiv (u_1,\ldots,u_n)'$,

$$E(\mathbf{u}) = \mathbf{0}, \qquad \text{var}(\mathbf{u}) = \sigma_0^2\mathbf{I}_n. \tag{8.3.24}$$

By subtracting $\alpha_1 y_{j-1}$ from y_j, on both sides of (8.3.23), the equation may be rewritten in the form

$$y_j = \alpha_1 y_{j-1} + \alpha_0 x_j + u_j^*, \qquad j = 1, 2, \ldots, n, \tag{8.3.25}$$

where $u_j^* \equiv u_j - \alpha_1 u_{j-1}$. Note that $E(u_j^*) = 0$. Moreover,

$$\begin{aligned} E(u_j^* u_k^*) &= E(u_j - \alpha_1 u_{j-1})(u_k - \alpha_1 u_{k-1}) \\ &= E(u_j u_k - \alpha_1 u_{j-1} u_k - \alpha_1 u_j u_{k-1} + \alpha_1^2 u_{j-1} u_{k-1}) \\ &= \sigma_0^2[\delta_{j,k} - \alpha_1 \delta_{j-1,k} - \alpha_1 \delta_{j,k-1} + \alpha_1^2 \delta_{j-1,k-1}], \end{aligned}$$

where $\delta_{i,j}$ is zero for $i \neq j$ and one for $i = j$. Clearly, if $\mathbf{u}^* \equiv (u_1^*, \ldots, u_n^*)'$,

$$\text{var } \mathbf{u}^* = \sigma^2 \mathbf{\Sigma}, \tag{8.3.26}$$

where $\mathbf{\Sigma}$ (generally unknown) is given by

$$\mathbf{\Sigma} = \begin{pmatrix} 1 & \rho & & 0 \\ \rho & \ddots & \ddots & \\ & \ddots & \ddots & \rho \\ 0 & & \rho & 1 \end{pmatrix},$$

$$\sigma^2 \equiv \sigma_0^2(1 + \alpha_1^2), \qquad \rho \equiv -\frac{\alpha_1}{1 + \alpha_1^2}.$$

This is the tri-diagonal matrix defined in (2.2.9).

Define $\mathbf{y} \equiv (y_1, \ldots, y_n)'$, $\boldsymbol{\beta} = (\alpha_0, \alpha_1)'$, and

$$\mathbf{X} = \begin{pmatrix} x_1 & y_0 \\ \cdot & \cdot \\ \cdot & \cdot \\ \cdot & \cdot \\ x_n & y_{n-1} \end{pmatrix}.$$

The revised regression model becomes

$$\mathbf{y} = \mathbf{X}\boldsymbol{\beta} + \mathbf{u}^*, \qquad E(\mathbf{u}^*) = \mathbf{0}, \qquad \text{var }(\mathbf{u}^*) = \sigma^2 \mathbf{\Sigma}. \tag{8.3.27}$$

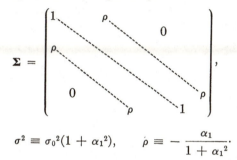

Superficial examination of (8.3.27) leads one to believe that the distributed lag model of (8.3.23) has been reduced to a simple univariate weighted regression model, so that if ρ in (8.3.26) is estimated consistently, the Aitken approach will provide optimal estimates. Closer scrutiny of (8.3.27), however, reveals that not only are the errors correlated, but also, because of the lagged structure in the dependent variables, the **X** matrix is stochastic, and not independent of $\mathbf{u}^*$. A general consequence of the dependence of **X** and $\mathbf{u}^*$ is that the least squares estimator of $\boldsymbol{\beta}$ need not be consistent, depending upon the nature of the dependence (see, for instance, Goldberger, 1964, p. 276). It is now shown how $\boldsymbol{\beta}$ can be estimated consistently and efficiently, at least asymptotically.

Koyck (1954) proposed a three-step procedure for consistent estimation of $\boldsymbol{\beta}$. The procedure subsumes the condition that for every element, *plim* $(\mathbf{X}'\mathbf{X}/n)$ remains bounded as n approaches infinity.

Koyck Procedure

Step (1): Compute the ordinary least squares estimator in (8.3.27), assuming $\boldsymbol{\Sigma} = \mathbf{I}$,

$$\hat{\boldsymbol{\beta}} \equiv \begin{pmatrix} \hat{\alpha}_0 \\ \hat{\alpha}_1 \end{pmatrix} = (\mathbf{X}'\mathbf{X})^{-1}\mathbf{X}'\mathbf{y}. \tag{8.3.28}$$

Step (2): Compute the residuals from this regression, $\hat{\mathbf{u}}^* \equiv \mathbf{y} - \mathbf{X}\hat{\boldsymbol{\beta}}$.

Step (3): Solve for $\bar{\alpha}_0$ and $\bar{\alpha}_1$ from

$$\bar{\boldsymbol{\beta}} \equiv \begin{pmatrix} \bar{\alpha}_0 \\ \bar{\alpha}_1 \end{pmatrix} = \hat{\boldsymbol{\beta}} + (\mathbf{X}'\mathbf{X})^{-1} \begin{pmatrix} 0 \\ \delta \end{pmatrix}, \tag{8.3.29}$$

where

$$\delta = \frac{\bar{\alpha}_1(\hat{\mathbf{u}}^*)'(\hat{\mathbf{u}}^*)}{1 + \bar{\alpha}_1\hat{\alpha}_1}.$$

This computation requires solving a quadratic equation in $\bar{\alpha}_1$.

The result of the above computations is $\bar{\boldsymbol{\beta}}$, a consistent estimator of $\boldsymbol{\beta}$, but not necessarily an efficient one. Now agument the Koyck procedure (using properties of tri-diagonal matrices) to obtain asymptotically efficient estimators (estimators whose variances are as small as possible in large samples).

Suppose that in (8.3.27), $\boldsymbol{\Sigma}$ were known. Then

$$\tilde{\boldsymbol{\beta}} \equiv \begin{pmatrix} \tilde{\alpha}_0 \\ \tilde{\alpha}_1 \end{pmatrix} = (\mathbf{X}'\boldsymbol{\Sigma}^{-1}\mathbf{X})^{-1}\mathbf{X}'\boldsymbol{\Sigma}^{-1}\mathbf{y} \tag{8.3.30}$$

would be an asymptotically efficient estimator. This is seen as follows. Multiply through by $\boldsymbol{\Sigma}^{-1/2}$ to get

$$\mathbf{y}_0 = \mathbf{X}_0\boldsymbol{\beta} + \mathbf{u}_0, \qquad \mathcal{L}(\mathbf{u}_0) = N(\mathbf{0},\sigma^2\mathbf{I}), \tag{8.3.31}$$

where $\mathbf{y}_0 = \boldsymbol{\Sigma}^{-1/2}\mathbf{y}$, $\mathbf{X}_0 = \boldsymbol{\Sigma}^{-1/2}\mathbf{X}$, $\mathbf{u}_0 = \boldsymbol{\Sigma}^{-1/2}\mathbf{u}^*$. Recall (and it is easy to check) that $\tilde{\boldsymbol{\beta}}$, given in (8.3.30), is a maximum likelihood estimator of $\boldsymbol{\beta}$. Hence, it is asymptotically efficient.

The problem remaining is to estimate $\boldsymbol{\Sigma}$. If there were available a consistent estimator of ρ, defined in (8.3.26), $\boldsymbol{\Sigma}$ could be estimated consistently (since $\boldsymbol{\Sigma}$ depends only upon the single parameter, ρ). Fortunately such is the case; to wit, $\bar{\alpha}_1$ defined in (8.3.29) is a consistent estimator of α_1, and therefore,

$$\bar{\rho} = -\frac{\bar{\alpha}_1}{1 + \bar{\alpha}_1^2} \tag{8.3.32}$$

is a consistent estimator of ρ. So the expanded Koyck procedure is to augment the above three steps with:

Step (4): Compute $\bar{\rho}$ from (8.3.32), and then by substituting $\bar{\rho}$ into (8.3.26) for ρ, compute

$$\bar{\mathbf{\Sigma}} = \begin{pmatrix} 1 & \bar{\rho} & & & & \\ & & \cdot & 0 & & \\ \bar{\rho} & 1 & \cdot & & & \\ & \cdot & \cdot & \cdot & & \\ & & \cdot & \cdot & \bar{\rho} & \\ & 0 & \cdot & \cdot & & \\ & & & \bar{\rho} & 1 & \end{pmatrix}. \tag{8.3.33}$$

Step (5): Compute $\bar{\bar{\boldsymbol{\beta}}}$ from

$$\bar{\bar{\boldsymbol{\beta}}} \equiv \begin{pmatrix} \bar{\bar{\alpha}}_0 \\ \bar{\bar{\alpha}}_1 \end{pmatrix} = (\mathbf{X}'\bar{\mathbf{\Sigma}}^{-1}\mathbf{X})^{-1}\mathbf{X}'\bar{\mathbf{\Sigma}}^{-1}\mathbf{y}, \tag{8.3.34}$$

where $\bar{\mathbf{\Sigma}}$ is defined by (8.3.32) and (8.3.33). The proposed estimator, $\bar{\bar{\boldsymbol{\beta}}}$, yields estimators of α_0 and α_1 which are not only consistent, but asymptotically have all the optimality and distributional properties of maximum likelihood estimators; specifically, they are asymptotically efficient. Note that $\bar{\mathbf{\Sigma}}^{-1}$ may be computed exactly by using (2.8.5) and (2.8.9) in (8.3.19).

8.4 MULTIVARIATE REGRESSION

Although there are multivariate analogues of all of the models discussed in Section 8.2, only the first two models are treated below (and they are studied together). However, in the next section, the univariate weighted regression model discussed in Section 8.3 is extended to the multivariate case.

The identifying notion of multivariate regression is that of the presence of several correlated dependent variables (p of them). In the standard multivariate regression model there is a separate univariate regression equation corresponding to each of the p-dependent variables, and N observations are taken jointly on the dependent and independent variables. In more general multivariate regression models (simultaneous equation systems), several dependent variables may appear in a single equation.

8.4.1 The Standard Multivariate Regression Model

Let $\mathbf{y}_1, \ldots, \mathbf{y}_p$ be $N \times 1$ vectors representing N independent observations on each of p correlated dependent random variables. Assume the linear model (linear in the parameters $\boldsymbol{\beta}_j$)

$$\underset{(N\times 1)}{\mathbf{y}_j} = \underset{(N\times q)(q\times 1)}{\mathbf{X} \cdot \boldsymbol{\beta}_j} + \underset{(N\times 1)}{\mathbf{u}_j}, \qquad j = 1, \ldots, p, \tag{8.4.1}$$

where $\mathbf{X}$ is known and may be thought of as arising either as a "functional" or a "conditional" *regressor matrix*[4] (see Section 8.2 for definition of these terms); $\boldsymbol{\beta}_j$ is an unknown parameter vector; and $\mathbf{u}_j$ is an error or disturbance vector. For a given j, (8.4.1) is a univariate regression.

The basic model equation may be written in a more compact form in the following way. Define

$$\mathbf{Y} \equiv (\mathbf{y}_1, \ldots, \mathbf{y}_p), \qquad \mathbf{U} \equiv (\mathbf{u}_1, \ldots, \mathbf{u}_p), \qquad \mathbf{B} \equiv (\boldsymbol{\beta}_1, \ldots, \boldsymbol{\beta}_p).$$

Then (8.4.1) becomes

$$\underset{(N\times p)}{\mathbf{Y}} = \underset{(N\times q)(q\times p)}{\mathbf{X} \cdot \mathbf{B}} + \underset{(N\times p)}{\mathbf{U}}. \tag{8.4.2}$$

Assumptions

To define the multivariate regression model completely, impose the following assumptions and constraints on the quantities in (8.4.2).

(1)
$$p + q \leq N. \tag{8.4.3}$$

This assumption is used so that there will be a sufficient number of observations available to estimate all of the parameters. Let N_0 denote the total number of observations, and let N_p denote the total number of parameters. Then, since there are N observations per equation, and p equations, $N_0 = Np$. Since a row of $\mathbf{U}$ has p correlated errors [and all rows will have the same covariance matrix, by assumption (3)], there will be a corresponding $p \times p$ covariance matrix with $p(p + 1)/2$ distinct elements in general, and fewer distinct elements if the covariance matrix is structured. Moreover, unless there are some a priori constraints on the coefficient matrix, $\mathbf{B}$ will have pq distinct parameters. Thus,

$$N_p \leq \frac{p(p+1)}{2} + pq.$$

[4] The term "regressor matrix" is used to include both the case in which $\mathbf{X}$ may be predetermined or designed; and also, the case in which $\mathbf{X}$ is stochastic and uncontrollable.

Hence, substitution gives

$$N_0 - N_p \geq Np - pq - \frac{p(p+1)}{2} = p\left[N - (p+q) + \frac{p-1}{2}\right].$$

To be certain that $(N_0 - N_p)$ will be nonnegative for all $p \geq 1$, it is sufficient if $N \geq p + q$.

(2) $$\operatorname{rank} \underset{(N\times q)}{(\mathbf{X})} = q. \tag{8.4.4}$$

This condition will be required in order to obtain a unique solution to the normal equations. If (8.4.4) is not satisfied, use a generalized inverse and introduce prior information for selecting a **z** vector in each equation, as in (8.2.5), to obtain a solution. An alternative is to use a completely Bayesian approach with a strictly informative prior distribution for the parameters.

(3) Each row of **U** has covariance matrix $\mathbf{\Sigma}$.

Define the rows of **U** by

$$\mathbf{U} \equiv (\mathbf{v}_1, \ldots, \mathbf{v}_N)',$$

where $\mathbf{v}_j$ is a $p \times 1$ vector, $j = 1, \ldots, N$. Assume for all j that

$$E(\mathbf{v}_j) = \mathbf{0}, \qquad \operatorname{var}(\mathbf{v}_j) = \mathbf{\Sigma} \equiv (\sigma_{ij}), \qquad \text{and} \qquad \mathbf{\Sigma} > 0. \tag{8.4.5}$$

[*Remark:* An alternative form of this assumption, which is often useful, is obtained by stringing out the columns of **U** into a long $Np \times 1$ vector **u**, where $\mathbf{u}' \equiv (\mathbf{u}_1', \mathbf{u}_2', \ldots, \mathbf{u}_p')$. Then (8.4.5) becomes

$$E(\mathbf{u}) = \mathbf{0}, \qquad \operatorname{var}(\mathbf{u}) = \mathbf{\Sigma} \otimes \mathbf{I}_N, \tag{8.4.6}$$

where the direct product $\otimes$ is defined in Equation (2.10.1).]

(4) $$\mathcal{L}(\mathbf{v}_j) = N(\mathbf{0}, \mathbf{\Sigma}), \qquad j = 1, \ldots, N. \tag{8.4.7}$$

This assumption is not needed for point estimation although it will be used for making inferences in interval estimation and testing.

The multivariate regression model is defined by (8.4.2)–(8.4.7).

8.4.2 Estimation

The unknown parameters which must be estimated are **B**, $\mathbf{\Sigma}$. They will be estimated by the methods of least squares and maximum likelihood.

Least Squares Estimation: Since each equation in (8.4.1) corresponds to a univariate regression, minimize $(\mathbf{u}_j'\mathbf{u}_j)$ with respect to $\boldsymbol{\beta}_j$, for each fixed j

and find, as in (8.3.2),

$$\underset{(q\times 1)}{\hat{\boldsymbol{\beta}}_j} = (\mathbf{X}'\mathbf{X})^{-1}\mathbf{X}'\mathbf{y}_j, \qquad j = 1,\ldots,p. \tag{8.4.8}$$

Note that $\hat{\mathbf{B}}$ does not depend upon Σ, which would not be true in the general case for $\operatorname{var}(\mathbf{u}) \neq \Sigma \otimes \mathbf{I}_N$.

Then if

$$\begin{aligned} \hat{\mathbf{B}} \equiv (\hat{\boldsymbol{\beta}}_1,\ldots,\hat{\boldsymbol{\beta}}_p) &= [(\mathbf{X}'\mathbf{X})^{-1}\mathbf{X}'\mathbf{y}_1,\ldots,(\mathbf{X}'\mathbf{X})^{-1}\mathbf{X}'\mathbf{y}_p], \\ \underset{(q\times p)}{\hat{\mathbf{B}}} &= (\mathbf{X}'\mathbf{X})^{-1}\mathbf{X}'\mathbf{Y}. \end{aligned} \tag{8.4.9}$$

It may be checked that this same result is obtained by minimizing $\operatorname{tr}(\mathbf{U}'\mathbf{U})$ or $|\mathbf{U}'\mathbf{U}|$ with respect to $\mathbf{B}$.

The residuals may be used to estimate $\boldsymbol{\Sigma}$. Thus, define the *matrix of residuals:*

$$\hat{\mathbf{U}} = \mathbf{Y} - \mathbf{X}\hat{\mathbf{B}}.$$

Then the residual sums of squares is $\hat{\mathbf{U}}'\hat{\mathbf{U}}$. Since each residual variance and covariance involves an estimate of a q-vector $\hat{\boldsymbol{\beta}}$, an unbiased estimator of $\boldsymbol{\Sigma}$ is given by

$$\hat{\boldsymbol{\Sigma}} = \frac{1}{N-q}\hat{\mathbf{U}}'\hat{\mathbf{U}} = \frac{1}{N-q}(\mathbf{Y} - \mathbf{X}\hat{\mathbf{B}})'(\mathbf{Y} - \mathbf{X}\hat{\mathbf{B}}). \tag{8.4.10}$$

That $\hat{\boldsymbol{\Sigma}}$ is unbiased may be seen as follows. Define the symmetric idempotent matrix.

$$\mathbf{A} = \mathbf{I} - \mathbf{X}(\mathbf{X}'\mathbf{X})^{-1}\mathbf{X}'.$$

Then $\hat{\mathbf{U}} = \mathbf{A}\mathbf{Y}$ and $\hat{\mathbf{u}}_j = \mathbf{A}\mathbf{y}_j$. Define $(N-q)\hat{\sigma}_{jk} \equiv \hat{\mathbf{u}}_j'\hat{\mathbf{u}}_k$. Taking expectations gives

$$\begin{aligned} E(\hat{\mathbf{u}}_j'\hat{\mathbf{u}}_k) = E\operatorname{tr}(\hat{\mathbf{u}}_k\hat{\mathbf{u}}_j') = E\operatorname{tr}[\mathbf{A}\mathbf{y}_k\mathbf{y}_j'\mathbf{A}'] &= E\operatorname{tr}[\mathbf{y}_k\mathbf{y}_j'\mathbf{A}'\mathbf{A}] \\ &= \operatorname{tr}[E(\mathbf{y}_k\mathbf{y}_j')\mathbf{A}'\mathbf{A}]. \end{aligned}$$

Since

$$\begin{aligned} E(\mathbf{y}_k\mathbf{y}_j') &= \sigma_{kj}\mathbf{I}_N + \mathbf{X}\boldsymbol{\beta}_k\boldsymbol{\beta}_j'\mathbf{X}', \\ E(\hat{\mathbf{u}}_j'\hat{\mathbf{u}}_k) &= \sigma_{jk}\operatorname{tr}\mathbf{A} + \operatorname{tr}(\mathbf{A}\mathbf{X}\boldsymbol{\beta}_k\boldsymbol{\beta}_j'\mathbf{X}'). \end{aligned}$$

But $\mathbf{AX} = \mathbf{0}$, and $\operatorname{tr}\mathbf{A} = N - q$. Hence $\hat{\sigma}_{jk}$ is unbiased for all j, k.

Maximum Likelihood Estimation: Use (8.4.7) to form the likelihood function for $\mathbf{U}$. Define $\boldsymbol{\Lambda} \equiv \boldsymbol{\Sigma}^{-1}$. Then

$$p(\mathbf{U}) = \prod_{j=1}^{N} p(\mathbf{v}_j) \propto |\boldsymbol{\Lambda}|^{N/2}\exp\left(-\tfrac{1}{2}\sum_{j=1}^{N}\mathbf{v}_j'\boldsymbol{\Lambda}\mathbf{v}_j\right),$$

or

$$p(\mathbf{U}) \propto |\boldsymbol{\Lambda}|^{N/2}\exp\left\{-\tfrac{1}{2}\operatorname{tr}\left[\boldsymbol{\Lambda}\left(\sum_{1}^{N}\mathbf{v}_j\mathbf{v}_j'\right)\right]\right\},$$

or

$$p(\mathbf{U}) \propto |\mathbf{\Lambda}|^{N/2} \exp\left(-\tfrac{1}{2} \operatorname{tr} \mathbf{\Lambda U'U}\right).$$

Differentiating $\log p(\mathbf{U})$ with respect to $\mathbf{\Lambda}$ and setting the result equal to zero gives[5]

$$\frac{N}{2}[2\mathbf{\Lambda}^{-1} - \operatorname{diag}\mathbf{\Lambda}^{-1}] - \frac{1}{2}[2\hat{\mathbf{U}}'\hat{\mathbf{U}} - \operatorname{diag}\hat{\mathbf{U}}'\hat{\mathbf{U}}]$$

or, if $\hat{\hat{\mathbf{\Sigma}}}$ denotes the maximum likelihood estimator of $\mathbf{\Sigma}$,

$$\hat{\hat{\mathbf{\Sigma}}} = \frac{1}{N}\hat{\mathbf{U}}'\hat{\mathbf{U}},$$

which is a biased version of (8.4.10).

Similarly, element-by-element differentiation of $\log p(\mathbf{U})$ with respect to B_{ij} [where $\mathbf{B} \equiv (B_{ij})$] yields (8.4.9). Thus, under the assumption of Normality, maximum likelihood estimation yields the same results as equation-by-equation least squares estimation.

Properties of Estimators

(1) Consistency. $\hat{\mathbf{B}}$ and $\hat{\hat{\mathbf{\Sigma}}}$ are consistent under the assumptions of the model defined above since they are maximum likelihood estimators and all required regularity conditions for maximum likelihood estimators are satisfied (for a statement of these conditions, see, for instance, Wald, 1949).

(2) Unbiasedness. That $\hat{\mathbf{\Sigma}}$ is unbiased was shown above. The proof for $\hat{\mathbf{B}}$ is immediate:

$$E\hat{\mathbf{B}} = (\mathbf{X}'\mathbf{X})^{-1}\mathbf{X}'E\mathbf{Y} = (\mathbf{X}'\mathbf{X})^{-1}\mathbf{X}'\mathbf{X}\mathbf{B} = \mathbf{B}.$$

So $\hat{\mathbf{B}}$ and $\hat{\mathbf{\Sigma}}$ are unbiased estimators.

(3) Efficiency. The least squares estimator, $\hat{\mathbf{B}}$, was derived from the equation-by-equation estimators which, under the above assumptions, are minimum variance linear unbiased estimators. Moreover, by permitting the observations in *all* equations to be used to estimate parameters in a given equation, and finding that only observations in the given equation are required, Rao (1965, p. 459) showed that the optimal multivariate least squares estimators are identical to the optimal equation-by-equation

[5] Use the same log-concavity argument as in Sect. 7.1.2 to show that a maximum value is obtained.

univariate least squares estimators; that is, $\hat{\mathbf{B}}$ has maximum efficiency, in that among all linear unbiased estimators, the covariance matrix of $\hat{\mathbf{B}}$ subtracted from that of any other estimator, results in a positive definite matrix.

It is also of interest to determine the extent to which the assumptions of the standard multivariate regression model made above can be relaxed and still have $\mathbf{B}$ retain the property of maximum efficiency. It was pointed out above [Example (8.3.4)] that if in a univariate regression with a constant term, the covariance matrix of the disturbances follows an intraclass pattern, the ordinary least squares estimators are also best linear unbiased. A more general result due to Rao (1967) and (1968) is the following: A necessary and sufficient condition in any univariate regression model that the ordinary least squares estimator of $\mathbf{B}$ be also a minimum variance linear unbiased estimator is that the covariance matrix of the disturbances be expressible in the form

$$\mathbf{\Sigma} = \mathbf{X\Theta X}' + \mathbf{Z\Gamma Z}' + \alpha\mathbf{I},$$

where $\mathbf{\Theta}$ and $\mathbf{\Gamma}$ are arbitrary matrices, α is an arbitrary nonnegative scalar, and $\mathbf{Z}$ is a matrix of maximum rank such that $\mathbf{Z}'\mathbf{X} = \mathbf{0}$. It may be checked that the intraclass covariance matrix is a member of this family by taking $\mathbf{\Theta} = \mathbf{0}$, $\alpha = \sigma^2(1 - \rho)$, $\rho < 1$, and $\mathbf{\Gamma} = \sigma^2\rho\mathbf{uu}'$, where $\mathbf{u}$ is chosen so that $\mathbf{u}'\mathbf{Z}' \equiv (1,\ldots,1)$, and $\mathbf{Z}'\mathbf{X} = \mathbf{0}$.

(4) **Orthogonality.** Note that the space of observations is orthogonal to the space of residuals; that is,

$$\mathbf{X}'\hat{\mathbf{U}} = \mathbf{0}, \qquad \hat{\mathbf{Y}}'\hat{\mathbf{U}} = (\mathbf{X}\hat{\mathbf{B}})'\hat{\mathbf{U}} = \mathbf{0}. \tag{8.4.11}$$

(5) **Covariance Matrix and Distribution of $\hat{\boldsymbol{\beta}}$.** From (8.4.8),

$$\operatorname{cov}(\hat{\boldsymbol{\beta}}_i,\hat{\boldsymbol{\beta}}_j) = \sigma_{ij}(\mathbf{X}'\mathbf{X})^{-1}.$$

Hence, for the strung out vectors

$$\underset{(1\times pq)}{\hat{\boldsymbol{\beta}}'} \equiv (\underset{(1\times q)}{\hat{\boldsymbol{\beta}}_1'},\ldots,\underset{(1\times q)}{\hat{\boldsymbol{\beta}}_p'}), \qquad \boldsymbol{\beta}' \equiv (\boldsymbol{\beta}_1',\ldots,\boldsymbol{\beta}_p'),$$

$$\operatorname{var}(\hat{\boldsymbol{\beta}}) = \mathbf{\Sigma} \otimes (\mathbf{X}'\mathbf{X})^{-1} \equiv \mathbf{\Phi},$$

which is a $pq \times pq$ matrix. Thus,

$$\mathcal{L}(\hat{\boldsymbol{\beta}}) = N(\boldsymbol{\beta},\mathbf{\Phi}). \tag{8.4.12}$$

Moreover, $\hat{\boldsymbol{\beta}}$ and $\hat{\mathbf{\Sigma}}$ are independent [see Theorem (7.1.2)].

(6) **Distribution of $\hat{\mathbf{\Sigma}}$.** It is easy to see from (8.4.10) that since $(N - q)\hat{\mathbf{\Sigma}}$ is a sum of terms of the form given in Theorem (5.1.1) (Chapter 5),

$$\mathcal{L}\{(N - q)\hat{\mathbf{\Sigma}}\} = W(\mathbf{\Sigma}, p, N - q). \tag{8.4.13}$$

Computation of Estimates

Appendix A contains a tabulation of some computer programs recommended for use in analyzing the various models discussed in Part II. The regression program of Stroud, Zellner, and Chau (1965) not only provides estimates of $\mathbf{B}$ and $\mathbf{\Sigma}$ for the multivariate regression model discussed in this section, but also provides estimates for more general models involving lagged values of the dependent and independent variables and possibly more than one dependent variable in each equation (simultaneous equation systems).

Example (8.4.1): A firm produces p distinct products. Define

$$
\begin{aligned}
y_{ij} &= \text{profit on product } j \text{ in year } i,\\
X_{i1} &= \text{warehousing costs in year } i,\\
X_{i2} &= \text{total sales in year } i,\\
X_{i3} &= \text{number of employees on payroll in year } i,\\
X_{i4} &= \text{total purchasing costs in year } i,\\
&\cdot\\
&\cdot\\
&\cdot\\
X_{iq} &= \text{number of college graduates in administrative positions in year } i,\\
\mathbf{X} &= (X_{ij}),\ \mathbf{X}: N \times q.
\end{aligned}
$$

Suppose, for illustration, that $p = 3$, $q = 2$, and $N = 5$; also, $\mathbf{y}_j = \mathbf{X}\boldsymbol{\beta}_j + \mathbf{u}_j$, $j = 1, 2, 3$. (Note that there is no constant term assumed since there can be no profits without sales.) Suppose further that the profit vectors for each product are given by

$$
\mathbf{y}_1 = \begin{pmatrix} 3\\ 2\\ 4\\ 4\\ 1 \end{pmatrix}, \qquad \mathbf{y}_2 = \begin{pmatrix} 4\\ 5\\ 4\\ 5\\ 2 \end{pmatrix}, \qquad \mathbf{y}_3 = \begin{pmatrix} 2\\ 3\\ 4\\ 3\\ 1 \end{pmatrix},
$$

so that with the regressor matrix, the data matrices are

$$
\underset{(N\times p)}{\mathbf{Y}} = \begin{pmatrix} 3 & 4 & 2\\ 2 & 5 & 3\\ 4 & 4 & 4\\ 4 & 5 & 3\\ 1 & 2 & 1 \end{pmatrix}, \qquad \underset{(N\times q)}{\mathbf{X}} = \begin{pmatrix} 12 & 10\\ 8 & 8\\ 8 & 10\\ 9 & 7\\ 12 & 12 \end{pmatrix}.
$$

To find least-squares estimates of $\mathbf{B} = (\boldsymbol{\beta}_1,\ldots,\boldsymbol{\beta}_p)$ and $\boldsymbol{\Sigma}$ (assuming a standard multivariate regression model is applicable), it is first necessary to evaluate $\mathbf{X}'\mathbf{X}$, and $(\mathbf{X}'\mathbf{X})^{-1}$. Thus,

$$\mathbf{X}'\mathbf{X} = \begin{pmatrix} 497 & 471 \\ 471 & 457 \end{pmatrix}, \qquad (\mathbf{X}'\mathbf{X})^{-1} = \begin{pmatrix} .0864 & -.089 \\ -.089 & .094 \end{pmatrix}.$$

Substituting into (8.4.9) and (8.4.10) gives

$$\hat{\mathbf{B}} = \begin{pmatrix} .185 & .390 & -.137 \\ .085 & -.011 & .397 \end{pmatrix}, \qquad \hat{\boldsymbol{\Sigma}} = \begin{pmatrix} 3.622 & 3.271 & 3.012 \\ 3.271 & 4.709 & 3.597 \\ 3.012 & 3.597 & 2.943 \end{pmatrix}.$$

8.4.3 Hypothesis Testing in Linear Models

Procedures for Testing Coefficients in a Single Equation

In the model $\mathbf{Y} = \mathbf{XB} + \mathbf{U}$, it may be of interest to test hypotheses about the coefficients in a single equation, or about the coefficients in several equations simultaneously. Suppose first that only the coefficients in a single equation (the αth) are of interest. Then, if $\hat{\boldsymbol{\beta}}_\alpha$ denotes the estimator of all coefficients in the αth equation,

$$\mathcal{L}(\hat{\boldsymbol{\beta}}_\alpha) = N(\boldsymbol{\beta}_\alpha, \sigma_{\alpha\alpha}(\mathbf{X}'\mathbf{X})^{-1}),$$

for $\alpha = 1,\ldots,p$. Thus, to test coefficients in the αth equation for significance (or to see if they are equal to preassigned values), Student t-tests should be used. For example, to test $H: \beta_{\alpha k} = \beta_{\alpha k}{}^*$, a known value, where $\boldsymbol{\beta}_\alpha = (\beta_{\alpha 1},\ldots,\beta_{\alpha q})'$, and $\hat{\boldsymbol{\beta}}_\alpha = (\hat{\beta}_{\alpha 1},\ldots,\hat{\beta}_{\alpha q})'$, test using the fact that under H,

$$\mathcal{L}[(\hat{\beta}_{\alpha k} - \beta_{\alpha k}{}^*)(w_{kk}\hat{\sigma}_{\alpha\alpha})^{-1/2}] = t_{N-q},$$

where $\hat{\sigma}_{\alpha\alpha} = (N-q)^{-1}(\mathbf{y}_\alpha - \mathbf{X}\hat{\boldsymbol{\beta}}_\alpha)'(\mathbf{y}_\alpha - \mathbf{X}\hat{\boldsymbol{\beta}}_\alpha)$, $\alpha = 1,\ldots,p$, $k = 1,\ldots,q$, w_{kk} is the kth diagonal element of $(\mathbf{X}'\mathbf{X})^{-1}$, and t_{N-q} denotes a Student t-distribution with $(N-q)$ d.f.

Now suppose it is of interest to study a given coefficient in each of the equations simultaneously. For example, suppose the system of equations involves present test score in mathematics as the dependent variable in the first equation, present test score in reading as the dependent variable in the second equation, and present test score in a different subject in each of the p equations. The independent variables might be intelligence quotient, high school grade point average, and so on. Suppose it is desired to test whether high school grade point average contributes to present examination scores. Then the hypothesis to test is that simultaneously, the coefficient in each equation corresponding to high school grade point average is zero. This involves a simultaneous test and the technique described below for carrying out the test is applicable to both regression and the analysis of variance (to be discussed in later sections of this chapter).

Procedure for Testing Coefficients in Several Equations

Step (1): Partition the **B** matrix into a part which focuses on the null hypothesis and another part which remains the same under the two hypotheses:

$$\underset{(q\times p)}{\mathbf{B}} = \begin{pmatrix}\mathbf{B}_1\\ \mathbf{B}_2\end{pmatrix}, \qquad \mathbf{B}_1: q_1 \times p, \qquad \mathbf{B}_2: q_2 \times p, \qquad q_1 + q_2 = q.$$

Step (2): Set up the linear model and the hypotheses so that under $H_0: \mathbf{B}_1 = \mathbf{B}_1^* =$ known matrix, and under $H_1: \mathbf{B}_1 \neq \mathbf{B}_1^*$.

Step (3): Compute the value of the likelihood ratio statistic λ, given by the determinantal ratio

$$\lambda^{2/N} = \frac{|\hat{\hat{\Sigma}}_{H_0+H_1}|}{|\hat{\hat{\Sigma}}_{H_0}|}, \tag{8.4.14}$$

where $\hat{\hat{\mathbf{\Sigma}}}_H$ denotes the value of $\hat{\hat{\mathbf{\Sigma}}}$ under hypothesis H, and $H_0 + H_1$ denotes the union of H_0 and H_1. Under $H_0 + H_1$, $\hat{\hat{\mathbf{\Sigma}}}$ is given by (8.4.10) with $(N - q)$ replaced by N. Under H_0, Σ is found easily by running the regression again assuming H_0 is true, and then computing Σ. The required Σ would be printed out in any computer regression routine.

Step (4): Carry out the standard likelihood ratio test; that is, reject H_0 if $\lambda < c$, where c is a constant determined by using the approximate procedure discussed below.

Box Approximation: An asymptotic expansion for the distribution of a monotone function of the likelihood ratio statistic was developed by Box (1949). The representation converges extremely rapidly and therefore significance levels derived from it will be quite accurate even for small values of N. For large N, the Box result is equivalent to the large sample result of Wilks (1938) that under H_0,

$$\lim_{N\to\infty} \mathcal{L}(-2\log\lambda) = \chi^2(r),$$

where $\chi^2(r)$ denotes a chi-square distribution with r degrees of freedom, and r is the difference between the number of d.f. of the likelihood under H_1 and the number of d.f. of the likelihood under H_0. Fixing a significance level will determine c. For the linear model defined above, $r = pq_1$. For the Box procedure, define the quantity

$$U_{p,q_1,n} \equiv \lambda^{2/N}, \tag{8.4.15}$$

where $n \equiv N - q$, and q_1 is the number of rows of $\mathbf{B}_1$. Assuming $p \leq q_1$, it is approximately true that under H_0,

$$P\{-m \log U_{p,q_1n,} \leq z\} = P\{\chi^2(pq_1) \leq z\} + \frac{\gamma}{m^2}[P\{\chi^2(pq_1 + 4) \leq z\} - P\{\chi^2(pq_1) \leq z\}], \tag{8.4.16}$$

where

$$m = N - q_2 - \tfrac{1}{2}(p + q_1 + 1),$$

$$\gamma = \frac{pq_1(p^2 + q_1^2 - 5)}{48}.$$

If just the first term of (8.4.16) is used (neglecting the term in γ) the total error in the approximation is $O(N^{-2})$; if both terms are used, the error is $O(N^{-4})$. If $q_1 < p$, use the result that under H_0, $\mathcal{L}(U_{p,q_1,n}) = \mathcal{L}(U_{q_1,p,n+q_1-p})$.

In order to obtain the power of the test the distribution of λ under H_1 is required. An approximate result may be obtained by using the theorems in Section 4.4.

[*Remark:* In the Mesa 97 computer program for testing in the general linear model, an approximate test based on an F-statistic is used as an alternative to the Box approximation approach (see Rao, 1952, p. 262).]

Example (8.4.2): To understand the use of the Box approximation, suppose $p = 1$, $q_1 = 2$, $N - q = 50$, $N - q_2 = N - q + q_1 = 52$. Note that $p \leq q_1$. It is easy to check that $m = 50$, $\gamma = 0$. Hence,

$$P\{-50 \log U_{1,2,50} \leq z\} \cong P\{\chi^2(2) \leq z\}.$$

At a 5 percent significance level (upper tail), $P\{\chi^2(2) \leq z\} = .95$ implies $z = 5.99$. Therefore,

$$P\{-50 \log U_{1,2,50} \leq 5.99\} \cong .95,$$

so that with 95 percent probability, $U_{1,2,50} \geq .887$. That is, reject H_0 with 95 percent confidence if $\lambda^{2/N} < .887$.

8.5 GENERALIZED MULTIVARIATE REGRESSION

This section presents a multivariate regression model that is more general than the one discussed in Section 8.4. The present model generalizes the earlier model in that the design matrices in the various regression equations for each dependent variable may be different from one another. The model is due to Zellner (1962), and he referred to the model as the *seemingly unrelated regressions* model.

8.5.1 Generalized Multivariate Regression Model

Suppose there are p univariate regression equations

$$\underset{(N\times 1)}{\mathbf{y}_j} = \underset{(N\times k_j)(k_j\times 1)}{\mathbf{X}_j \cdot \boldsymbol{\beta}_j} + \underset{(N\times 1)}{\mathbf{u}_j}, \qquad j = 1,\ldots,p, \tag{8.5.1}$$

in which $\mathbf{X}_j$ is the design matrix in the jth equation. Note that in the special case $\mathbf{X}_1 = \mathbf{X}_2 = \cdots = \mathbf{X}_p$, this model reduces to that of Section 8.4. Suppose the errors corresponding to the αth observation in each of the equations $(u_{1\alpha},\ldots,u_{p\alpha})$ are correlated, and have $p \times p$ covariance matrix $\boldsymbol{\Sigma} > 0$. The model may be written in compact (and seemingly univariate) form by defining

$$\mathbf{y} \equiv (\mathbf{y}_1',\ldots,\mathbf{y}_p')', \qquad \boldsymbol{\beta} \equiv (\boldsymbol{\beta}_1',\ldots,\boldsymbol{\beta}_p')',$$
$$\mathbf{u} \equiv (\mathbf{u}_1',\ldots,\mathbf{u}_p')', \qquad \mathbf{X} \equiv \text{diag}\,(\mathbf{X}_1,\ldots,\mathbf{X}_p).$$

Then,

$$\underset{(Np\times 1)}{\mathbf{y}} = \underset{(Np\times k)(k\times 1)}{\mathbf{X} \cdot \boldsymbol{\beta}} + \underset{(Np\times 1)}{\mathbf{u}}, \tag{8.5.2}$$

and

$$\underset{(Np\times 1)}{\mathcal{L}(\mathbf{u})} = N(\mathbf{0}, \boldsymbol{\Sigma} \otimes \mathbf{I}_N), \qquad \underset{(p\times p)}{\boldsymbol{\Sigma}} > 0, \tag{8.5.3}$$

where $k \equiv \Sigma_1^p k_j$, and $\mathbf{X}$ is a block diagonal matrix. Note that if $(\boldsymbol{\Sigma} \otimes \mathbf{I}_N)$ were a scalar matrix, (8.5.3) could be treated as a univariate regression and the Gauss-Markoff Theorem would apply. The case in which the error covariance matrix is not scalar was the problem in univariate weighted regression (8.3.4). The solution then and now is to use the Aitken estimator (generalized least squares) corresponding to (8.5.2) and (8.5.3); to wit

$$\begin{aligned}\tilde{\boldsymbol{\beta}} &= (\mathbf{X}'[\boldsymbol{\Sigma} \otimes \mathbf{I}_N]^{-1}\mathbf{X})^{-1}\mathbf{X}'[\boldsymbol{\Sigma} \otimes \mathbf{I}_N]^{-1}\mathbf{y} \\ &= (\mathbf{X}'[\boldsymbol{\Sigma}^{-1} \otimes \mathbf{I}_N]\mathbf{X})^{-1}\mathbf{X}'[\boldsymbol{\Sigma}^{-1} \otimes \mathbf{I}_N]\mathbf{y}.\end{aligned} \tag{8.5.4}$$

Thus, for known $\boldsymbol{\Sigma}$, $\tilde{\boldsymbol{\beta}}$ has minimum variance in the class of linear unbiased estimators. However, when $\boldsymbol{\Sigma}$ is unknown, a consistent estimator of $\boldsymbol{\beta}$ is obtained by replacing $\boldsymbol{\Sigma}$ with $\hat{\boldsymbol{\Sigma}}$, any consistent estimator of $\boldsymbol{\Sigma}$. As in univariate weighted regression, $\hat{\mathbf{u}}\hat{\mathbf{u}}'/(N - q)$ cannot be used as an estimator of $\boldsymbol{\Sigma}$ since that quantity has rank one, and therefore does not have an inverse (which is required for computation of $\tilde{\boldsymbol{\beta}}$).

8.5.2 Estimation of $\boldsymbol{\Sigma}$

In ordinary weighted regression, at this point it would be necessary to study the residuals to bring additional information to bear on the problem of estimating $\boldsymbol{\Sigma}$. However, in the case of generalized multivariate regression, a simpler formal procedure is available.

Define

$$\mathbf{Y} \equiv (\mathbf{y}_1, \ldots, \mathbf{y}_p), \qquad \mathbf{U} \equiv (\mathbf{u}_1, \ldots, \mathbf{u}_p),$$
$$\underset{(N\times k)}{\mathbf{Z}} \equiv (\underset{(N\times k_1)}{\mathbf{X}_1}, \ldots, \underset{(N\times k_p)}{\mathbf{X}_p}), \qquad \underset{(k\times p)}{\mathbf{B}} \equiv \text{diag}\,(\underset{(k_1\times 1)}{\boldsymbol{\beta}_1}, \ldots, \underset{(k_p\times 1)}{\boldsymbol{\beta}_p}).$$

Then (8.5.1) may be rewritten in the form

$$\underset{(N\times p)}{\mathbf{Y}} = \underset{(N\times k)(k\times p)}{\mathbf{Z} \cdot \mathbf{B}} + \underset{(N\times p)}{\mathbf{U}}. \tag{8.5.5}$$

Now note that symbolically, (8.5.5) looks just like (8.4.2), the more restrictive multivariate regression model. However, closer inspection reveals they differ in that in the generalized model, the off-block diagonal elements of the coefficient matrix, $\mathbf{B}$, are zero, a priori, and in the simpler model, $\mathbf{B}$ is a full matrix. Nevertheless, ignore the zero restrictions and estimate $\mathbf{B}$ (assuming $\mathbf{Z}'\mathbf{Z} > 0$) by

$$\hat{\mathbf{B}} = (\mathbf{Z}'\mathbf{Z})^{-1}\mathbf{Z}'\mathbf{Y}, \tag{8.5.6}$$

recognizing it will be an inefficient estimator since many coefficients are being estimated that are known to be zero. The virtue of doing this, however, is to obtain the $\boldsymbol{\Sigma}$ estimator,

$$\underset{(p\times p)}{\hat{\boldsymbol{\Sigma}}} = \frac{1}{N}\hat{\mathbf{U}}'\hat{\mathbf{U}} = \frac{1}{N}(\mathbf{Y} - \mathbf{Z}\hat{\mathbf{B}})'(\mathbf{Y} - \mathbf{Z}\hat{\mathbf{B}}), \tag{8.5.7}$$

where $\hat{\boldsymbol{\Sigma}}$ is a matrix of full rank, p, and is consistent, but biased.[6] Hence, $\hat{\boldsymbol{\Sigma}}^{-1}$ exists, and an estimator that asymptotically has the same properties as $\tilde{\boldsymbol{\beta}}$ is obtained by substituting $\hat{\boldsymbol{\Sigma}}$ given in (8.5.6) and (8.5.7) into

$$\hat{\tilde{\boldsymbol{\beta}}} = [\mathbf{X}'(\hat{\boldsymbol{\Sigma}}^{-1} \otimes \mathbf{I}_N)\mathbf{X}]^{-1}\mathbf{X}'(\hat{\boldsymbol{\Sigma}}^{-1} \otimes \mathbf{I}_N)\mathbf{y}. \tag{8.5.8}$$

The "updated estimator" of $\boldsymbol{\Sigma}$ [updated from that of (8.5.7)] is therefore given by

$$\hat{\hat{\boldsymbol{\Sigma}}} = \frac{1}{N}(\mathbf{Y} - \mathbf{Z}\hat{\hat{\mathbf{B}}})'(\mathbf{Y} - \mathbf{Z}\hat{\hat{\mathbf{B}}}), \tag{8.5.9}$$

where $\hat{\hat{\mathbf{B}}} \equiv \text{diag}\,(\hat{\tilde{\boldsymbol{\beta}}}_1, \ldots, \hat{\tilde{\boldsymbol{\beta}}}_p)$, and the $\hat{\tilde{\boldsymbol{\beta}}}_j$ are found from (8.5.8).

[6] There is no need to correct for bias since what is of greatest interest is rapid convergence of $\hat{\boldsymbol{\Sigma}}$ to $\boldsymbol{\Sigma}$.

8.5.3 Properties of Estimators

Some of the properties of $\hat{\tilde{\boldsymbol{\beta}}}$ were examined in Zellner (1962), Zellner (1963), and in Zellner and Huang (1962). For example, it has been shown for $\hat{\tilde{\boldsymbol{\beta}}}$ defined in (8.5.8), that

(1)
$$\hat{\tilde{\boldsymbol{\beta}}} = \tilde{\boldsymbol{\beta}} + \mathbf{O}\left(\frac{1}{N}\right), \text{ in probability,}^{7} \tag{8.5.10}$$

and

(2) *Bias:*
$$E\hat{\tilde{\boldsymbol{\beta}}} = \boldsymbol{\beta} + \mathbf{O}\left(\frac{1}{N}\right). \tag{8.5.11}$$

Moreover, it is easy to find from (8.5.4) that

(3)
$$\text{var } \tilde{\boldsymbol{\beta}} = [\mathbf{X}'(\boldsymbol{\Sigma}^{-1} \otimes \mathbf{I}_N)\mathbf{X}]^{-1}. \tag{8.5.12}$$

Then[7]

(4)
$$\text{var } \hat{\tilde{\boldsymbol{\beta}}} = \text{var } \tilde{\boldsymbol{\beta}} + \mathbf{o}\left(\frac{1}{N}\right). \tag{8.5.13}$$

Let $\hat{\mathbf{B}} \equiv \text{diag }(\hat{\boldsymbol{\beta}}_1,\ldots,\hat{\boldsymbol{\beta}}_p)$, and define $\underset{(k\times 1)}{\hat{\boldsymbol{\beta}}} \equiv (\hat{\boldsymbol{\beta}}_1',\ldots,\hat{\boldsymbol{\beta}}_p')'$, for $\hat{\mathbf{B}}$ given by (8.5.6). Then, for $\tilde{\boldsymbol{\beta}}$ defined by (8.5.4),

(5)
$$\text{var } \tilde{\beta}_{ij} \leq \text{var } \hat{\beta}_{ij}, \qquad i = 1,\ldots,k_j, \qquad j = 1,\ldots,p, \tag{8.5.14}$$

where $\tilde{\boldsymbol{\beta}}_j \equiv (\tilde{\beta}_{ij})$, $\hat{\boldsymbol{\beta}}_j = (\hat{\beta}_{ij})$.

(6) *Gain in Efficiency:* To examine the gain in efficiency in going to the generalized multivariate regression model from the standard multivariate regression model, it is illuminating to examine the two-equation system:

$$\begin{pmatrix} \mathbf{y}_1 \\ \mathbf{y}_2 \end{pmatrix} = \begin{pmatrix} \mathbf{X}_1 & \mathbf{0} \\ \mathbf{0} & \mathbf{X}_2 \end{pmatrix} \begin{pmatrix} \boldsymbol{\beta}_1 \\ \boldsymbol{\beta}_2 \end{pmatrix} + \begin{pmatrix} \mathbf{u}_1 \\ \mathbf{u}_2 \end{pmatrix}.$$

[7] Equation (8.5.10) implies that

$$\underset{N\to\infty}{\text{plim}} \{N(\hat{\tilde{\boldsymbol{\beta}}} - \tilde{\boldsymbol{\beta}})\} = \text{constant}.$$

Equation (8.5.13) implies that

$$\lim_{N\to\infty} \{N(\text{var } \hat{\tilde{\boldsymbol{\beta}}} - \text{var } \tilde{\boldsymbol{\beta}})\} = \mathbf{0}.$$

It was shown by Zellner and Huang (1962) that for the Aitken estimator of $\boldsymbol{\beta}_1$, given in (8.5.4), the generalized variance is

$$|\text{var}\,(\tilde{\boldsymbol{\beta}}_1)| = \frac{(1-\rho^2)^{k_1}}{\prod_{j=1}^{k_1}(1-\rho^2 r_j^2)}\,|\sigma^2(\mathbf{X}_1'\mathbf{X}_1)^{-1}|, \tag{8.5.15}$$

where ρ is the correlation coefficient between the αth disturbance in the first equation and the αth disturbance in the second equation; and r_j is the positive square root of the jth largest latent root of the equation.

$$|\mathbf{X}_1'\mathbf{X}_2(\mathbf{X}_2'\mathbf{X}_2)^{-1}\mathbf{X}_2'\mathbf{X}_1 - r^2(\mathbf{X}_1'\mathbf{X}_1)| = 0,$$

so that r_j^2 is the square of the jth canonical correlation [Chapter 11; compare with (11.2.7)]. As will be seen in Chapter 11, $0 \le r_j^2 \le 1$, so that $|\text{var}\,(\tilde{\boldsymbol{\beta}}_1)|$ is never larger than $|\sigma^2(\mathbf{X}_1'\mathbf{X}_1)^{-1}|$, the generalized variance of the single equation estimator of $\boldsymbol{\beta}_1$ (the estimator obtained by standard univariate regression). If $\mathbf{X}_1'\mathbf{X}_2 = \mathbf{0}$, $r_j = 0$ for all j. It is clear from (8.5.15) that this is the case of greatest gain in efficiency since then $|\text{var}\,(\tilde{\boldsymbol{\beta}}_1)|$ is as small as possible (for a given correlation between equations). Of course it is also true that the greater the correlation between the equations, the greater is the gain in efficiency (if $\rho = 0$, there is no gain). The results hold for the many-equation systems as well as for the two-equation systems. Moreover, in the case of a many-equation system, the generalized canonical correlations among many sets of variates should be applicable (Chapter 11). However, even in the two-equation system, it is not so clear as to whether the gain in efficiency is so large (or even whether there is always a gain) in the case when the Aitken estimator must be approximated because the $\boldsymbol{\Sigma}$ matrix is unknown (and must be estimated).

(7) *Hypothesis Testing:* For $\tilde{\boldsymbol{\beta}}$ defined in (8.5.4),

$$\mathcal{L}(\tilde{\boldsymbol{\beta}}) = N\{\boldsymbol{\beta},[\mathbf{X}'(\boldsymbol{\Sigma}^{-1}\otimes\mathbf{I}_N)\mathbf{X}]^{-1}\}.$$

Suppose it is of interest to test $H: \boldsymbol{\beta} = \boldsymbol{\beta}^*$, a known vector. Let $\boldsymbol{\Phi} = [\mathbf{X}'(\boldsymbol{\Sigma}^{-1}\otimes\mathbf{I}_N)\mathbf{X}]^{-1}$. Now use the results of Section 7.3. Let

$$\mathbf{y} \equiv \boldsymbol{\Phi}^{-1/2}(\tilde{\boldsymbol{\beta}} - \boldsymbol{\beta}^*).$$

Then

$$\mathcal{L}(\mathbf{y}) = N(\boldsymbol{\theta},\mathbf{I}),$$

where $\boldsymbol{\theta} \equiv \boldsymbol{\Phi}^{-1/2}(\boldsymbol{\beta} - \boldsymbol{\beta}^*)$. Hence,

$$\mathcal{L}(\mathbf{y}'\mathbf{y}|\boldsymbol{\theta}) = \chi_p'^2(\boldsymbol{\theta}'\boldsymbol{\theta}),$$

and under H, $\boldsymbol{\theta} = \mathbf{0}$ so that $\mathbf{y}'\mathbf{y}$ has a central chi-square distribution with p d.f. This result may be used directly when $\boldsymbol{\Sigma}$ is known.

When $\boldsymbol{\Sigma}$ is unknown but N is large, $\hat{\tilde{\boldsymbol{\beta}}}$ will be close to $\tilde{\boldsymbol{\beta}}$, and $\hat{\boldsymbol{\Sigma}}$ will be close to $\boldsymbol{\Sigma}$ in probability, so that the indicated chi-square test based upon $\hat{\mathbf{y}} = \hat{\boldsymbol{\Phi}}^{-1/2}(\hat{\tilde{\boldsymbol{\beta}}} - \boldsymbol{\beta}^*)$ is approximately valid ($\hat{\boldsymbol{\Phi}}$ is obtained from $\boldsymbol{\Phi}$ by replacing $\boldsymbol{\Sigma}$ by $\hat{\boldsymbol{\Sigma}}$).

Example (8.5.1): An example of the use of generalized multivariate regression which is cited in Zellner (1962) is based upon data appearing in Boot and deWitt (1960, p. 27). There are two firms, General Electric Company and Westinghouse Electric Company, and data are taken from 1935–1954 on the investment functions of the firms. Define

$$\begin{aligned} I(t) &= \text{current gross investment of the firm,} \\ C(t-1) &= \text{firm's beginning-of-the-year capital stock,} \\ F(t-1) &= \text{value of firm's outstanding shares at the beginning of the year.} \end{aligned}$$

Assume

$$I(t) = \alpha_1 C(t-1) + \alpha_2 F(t-1) + \alpha_3 + u(t),$$

for $t = 1,\ldots,N$, and for some unknown parameters α_1, α_2, and α_3. Renaming to conform with the earlier notation gives the result in Table 8.5.1. Let $\mathbf{y}_1$ and $\mathbf{y}_2$ be $N \times 1$ vectors and define

$$\underset{(2N\times 1)}{\mathbf{y}} \equiv \begin{pmatrix} \mathbf{y}_1 \\ \mathbf{y}_2 \end{pmatrix}, \qquad \mathbf{y}_j \equiv [y_j(1),\ldots,y_j(N)]', \qquad j = 1, 2,$$

$$\underset{(N\times 3)}{\mathbf{X}_j} \equiv \begin{bmatrix} x_{j1}(1) & x_{j2}(1) & x_{j3}(1) \\ \cdot & \cdot & \cdot \\ \cdot & \cdot & \cdot \\ \cdot & \cdot & \cdot \\ x_{j1}(N) & x_{j2}(N) & x_{j3}(N) \end{bmatrix}, \qquad j = 1, 2,$$

and

$$\underset{(3\times 1)}{\boldsymbol{\beta}_1} \equiv (\beta_{11},\beta_{12},\beta_{13})', \qquad \underset{(3\times 1)}{\boldsymbol{\beta}_2} \equiv (\beta_{21},\beta_{22},\beta_{23})',$$

Table 8.5.1 Regression Variables

	$I(t)$	$C(t-1)$	$F(t-1)$	*Constant term*
General Electric	$y_1(t)$	$x_{11}(t)$	$x_{12}(t)$	$x_{13}(t) \equiv 1$
Westinghouse	$y_2(t)$	$x_{21}(t)$	$x_{22}(t)$	$x_{23}(t) \equiv 1$

where the ordering corresponds to that used in Table 8.5.1. Now write the model as

$$\mathbf{y} = \begin{pmatrix} \mathbf{X}_1 & \mathbf{0} \\ \mathbf{0} & \mathbf{X}_2 \end{pmatrix} \begin{pmatrix} \boldsymbol{\beta}_1 \\ \boldsymbol{\beta}_2 \end{pmatrix} + \begin{pmatrix} \mathbf{u}_1 \\ \mathbf{u}_2 \end{pmatrix},$$

where $\boldsymbol{\beta}_1 = (\alpha_1,\alpha_2,\alpha_3)'$ for General Electric, and $\boldsymbol{\beta}_2 = (\alpha_1,\alpha_2,\alpha_3)'$ for Westinghouse. The raw data for $N = 20$ are given in Table 8.5.2.

Table 8.5.2 Investment Function Data on General Electric and Westinghouse

Year	*General Electric* $I(t)$	$C(t-1)$	$F(t-1)$	*Westinghouse* $I(t)$	$C(t-1)$	$F(t-1)$
1935	33.1	97.8	1170.6	12.93	1.8	191.5
36	45.0	104.4	2015.8	25.90	.8	516.0
37	77.2	118.0	2803.3	35.05	7.4	729.0
38	44.6	156.2	2039.7	22.89	18.1	560.4
39	48.1	172.6	2256.2	18.84	23.5	519.9
1940	74.4	186.6	2132.2	28.57	26.5	628.5
41	113.0	220.9	1834.1	48.51	36.2	537.1
42	91.9	287.8	1588.0	43.34	60.8	561.2
43	61.3	319.9	1749.4	37.02	84.4	617.2
44	56.8	321.3	1687.2	37.81	91.2	626.7
1945	93.6	319.6	2007.7	39.27	92.4	737.2
46	159.9	346.0	2208.3	53.46	86.0	760.5
47	147.2	456.4	1656.7	55.56	111.1	581.4
48	146.3	543.4	1604.4	49.56	130.6	662.3
49	98.3	618.3	1431.8	32.04	141.8	583.8
1950	93.5	647.4	1610.5	32.24	136.7	635.2
51	135.2	671.3	1819.4	54.38	129.7	723.8
52	157.3	726.1	2079.7	71.78	145.5	864.1
53	179.5	800.3	2371.6	90.08	174.8	1193.5
54	189.6	888.9	2759.9	68.60	213.5	1188.9

Use ordinary least squares estimation on these data, equation by equation (ignoring the correlations), to obtain the following estimates (note that the coefficient of the constant term is given last)

$$\hat{\boldsymbol{\beta}}_1 = \begin{pmatrix} .15 \\ .03 \\ -10.0 \end{pmatrix}, \qquad \hat{\boldsymbol{\beta}}_2 = \begin{pmatrix} .09 \\ .05 \\ -.51 \end{pmatrix},$$

and

$$\hat{\boldsymbol{\Sigma}} = \begin{pmatrix} 660.83 & 176.45 \\ 176.45 & 88.67 \end{pmatrix}.$$

The regression equations are

$$I(t) = .15C(t-1) + .03F(t-1) - 10.0\text{: General Electric,}$$
$$I(t) = .09C(t-1) + .05F(t-1) - .51\text{: Westinghouse Electric.}$$

Using the Aitken estimator (8.5.8) gives

$$I(t) = .14C(t-1) + .04F(t-1) - 27.7\text{: General Electric,}$$
$$I(t) = .06C(t-1) + .06F(t-1) - 1.3\text{: Westinghouse Electric.}$$

The estimated standard deviations of the regression coefficients obtained by the two methods are given in Table 8.5.3.

Table 8.5.3 Standard Deviations of Regression Coefficients

COEFFICIENT OF	*Ordinary least squares*		*Aitken estimator*	
	GENERAL ELECTRIC	WESTINGHOUSE ELECTRIC	GENERAL ELECTRIC	WESTINGHOUSE ELECTRIC
$C(t-1)$	.03	.06	.02	.05
$F(t-1)$	.02	.02	.01	.01
Constant term	31.3	8.0	27.0	7.1

It is seen from Table 8.5.3 that the Aitken estimator has the smaller standard deviation for all of the regression coefficients (and the percentage reductions are generally large). The correlation between the errors in the two equations is approximately .73. Note that since $N = 20$, the bias in $\hat{\boldsymbol{\beta}}_j$ is quite small so there is little error in comparing the Aitken estimator with the ordinary least squares estimator on a variance basis rather than a mean squared error basis.

8.5.4 Weighted Generalized Multivariate Regression

The generalized multivariate regression model was extended to the "weighted" case by Parks (1967). He considered multivariate regression in which not only can the design matrices be different from one equation to the next, but also the errors can be both serially (from one observation to the next) and contemporaneously (from one equation to the next) correlated with arbitrary and unequal variances serially and contemporaneously. The problem is described below.

Adopt the model of (8.5.1) and (8.5.2), $\mathbf{y} = \mathbf{X}\boldsymbol{\beta} + \mathbf{u}$, but instead of (8.5.3) use the less restrictive assumption

$$\underset{(Np \times 1)}{\mathcal{L}(\mathbf{u})} = N(\mathbf{0}, \sigma^2\boldsymbol{\Omega}), \qquad \underset{(Np \times Np)}{\boldsymbol{\Omega}} > 0. \tag{8.5.16}$$

For this case, the Aitken estimator is given by

$$\underset{(k \times 1)}{\tilde{\boldsymbol{\beta}}} = (\mathbf{X}'\boldsymbol{\Omega}^{-1}\mathbf{X})^{-1}\mathbf{X}'\boldsymbol{\Omega}^{-1}\mathbf{y}, \tag{8.5.17}$$

with covariance matrix

$$\operatorname{var} \tilde{\boldsymbol{\beta}} = \sigma^2(\mathbf{X}'\boldsymbol{\Omega}^{-1}\mathbf{X})^{-1}, \tag{8.5.18}$$

as in (8.3.6) and (8.3.8). Moreover,

$$\tilde{\sigma}^2 = \frac{1}{Np - k}(\mathbf{y} - \mathbf{X}\tilde{\boldsymbol{\beta}})'\boldsymbol{\Omega}^{-1}(\mathbf{y} - \mathbf{X}\tilde{\boldsymbol{\beta}}), \tag{8.5.19}$$

corresponding to (8.3.7). Since (8.5.17)–(8.5.19) are useful only if $\boldsymbol{\Omega}$ is known (or if $\boldsymbol{\Omega}$ can be consistently estimated), this is the same problem as in weighted univariate regression (Section 8.3). That is, by bringing sufficient additional information to bear on the form of $\boldsymbol{\Omega}$ so as to reduce the number of distinct parameters in $\boldsymbol{\Omega}$ to a small number (by studying the ordinary regression residuals) $\boldsymbol{\Omega}$ can be replaced by a consistent estimator to yield $\hat{\tilde{\boldsymbol{\beta}}}$, as in (8.3.9). In an effort to do this for one important class of problems, Parks (1967) examined the case of errors that are autocorrelated according to a first-order Markoff scheme. Errors correlated according to different structures can be similarly approached. The result is given below.

Define $\mathbf{v}_t \equiv (u_{1t},\ldots,u_{pt})'$, $\boldsymbol{\varepsilon}_t \equiv (\epsilon_{1t},\ldots,\epsilon_{pt})'$, $t = 1,\ldots,N$, and suppose the errors are correlated according to the first-order autoregressive scheme

$$u_{it} = \rho_i u_{i,t-1} + \epsilon_{it}, \qquad |\rho_i| < 1, \qquad i = 1,\ldots,p, \qquad t = 2,\ldots,N;$$

or in compact form,

$$\mathbf{v}_t = \operatorname{diag}(\rho_1,\ldots,\rho_p)\mathbf{v}_{t-1} + \boldsymbol{\varepsilon}_t, \qquad t = 2,\ldots,N,$$

where $|\rho_i| < 1$, $E(\boldsymbol{\varepsilon}_t) = \mathbf{0}$, $\operatorname{var}(\boldsymbol{\varepsilon}_t) = \boldsymbol{\Sigma}$, all $t = 1,\ldots,N$, and $\operatorname{cov}(\boldsymbol{\varepsilon}_t,\boldsymbol{\varepsilon}_\tau) = \mathbf{0}$, $t \neq \tau$. Then the estimator indicated by (8.5.8) [and analogous to (8.3.9)] is

$$\hat{\tilde{\boldsymbol{\beta}}} = (\mathbf{X}'\hat{\boldsymbol{\Omega}}^{-1}\mathbf{X})^{-1}\mathbf{X}'\hat{\boldsymbol{\Omega}}^{-1}\mathbf{y}, \tag{8.5.20}$$

where $\hat{\boldsymbol{\Omega}} = \hat{\mathbf{P}}(\hat{\boldsymbol{\Sigma}} \otimes \mathbf{I}_N)\hat{\mathbf{P}}'$, $\hat{\mathbf{P}} = \operatorname{diag}(\hat{\mathbf{P}}_1,\ldots,\hat{\mathbf{P}}_p)$,

$$\hat{\mathbf{P}}_i = \begin{pmatrix} (1-\hat{\rho}_i^2)^{-1/2} & 0 & 0 & \cdots & 0 \\ \hat{\rho}_i(1-\hat{\rho}_i^2)^{-1/2} & 1 & 0 & \cdots & 0 \\ \hat{\rho}_i^2(1-\hat{\rho}_i^2)^{-1/2} & \hat{\rho}_i & 1 & \cdots & 0 \\ \cdot & \cdot & \cdot & \cdot & \cdot \\ \cdot & \cdot & \cdot & \cdot & \cdot \\ \cdot & \cdot & \cdot & \cdot & \cdot \\ \hat{\rho}_i^{N-1}(1-\hat{\rho}_i^2)^{-1/2} & \hat{\rho}_i^{N-2} & \hat{\rho}_i^{N-3} & \cdots & 1 \end{pmatrix}, \qquad i = 1,\ldots,p,$$

$$\hat{\rho}_i = \frac{\sum_{t=2}^{N} \hat{u}_{it}\hat{u}_{i,t-1}}{\sum_{t=2}^{N} \hat{u}_{i,t-1}^2}, \qquad \hat{\mathbf{u}}_i \equiv (\hat{u}_{i1},\ldots,\hat{u}_{iN})' = \mathbf{y}_i - \mathbf{X}_i\hat{\boldsymbol{\beta}}_i,$$

$\hat{\boldsymbol{\beta}}_i$ is the ordinary least squares estimator of $\boldsymbol{\beta}_i$ (ignoring the serial correlation), $\hat{\boldsymbol{\Sigma}} = (s_{ij})$, and

$$s_{ij} = \frac{1}{N} (\mathbf{y}_i^* - \mathbf{X}_i^*\hat{\boldsymbol{\beta}}_i)'(\mathbf{y}_j^* - \mathbf{X}_j^*\hat{\boldsymbol{\beta}}_j),$$

where $\mathbf{y}_i^* = \hat{\mathbf{P}}_i^{-1}\mathbf{y}_i$, and $\mathbf{X}_i^* = \hat{\mathbf{P}}_i^{-1}\mathbf{X}_i$. That is, the $\hat{\mathbf{P}}$ matrix is the estimated transformation matrix between the u_{it}'s and the ϵ_{it}'s (using an estimated value of ρ_i). When (8.5.1) is multiplied through by $\mathbf{P}_j^{-1}$, the result is a regression equation whose disturbances have a scalar covariance matrix. The standard Aitken approach is then used to develop the solution given in (8.5.20).

8.6 BAYESIAN INFERENCE IN MULTIVARIATE REGRESSION

The multivariate regression model was analyzed from the sampling theory viewpoint in the last two sections. Point estimators were developed and methods were provided for testing simple hypotheses concerning the regression coefficients. In this section similar problems are approached from the Bayesian viewpoint, using both *noninformative* and *informative* prior densities on the parameters. Various characteristics of the posterior distributions to be developed can be used (with or without loss functions) to form Bayesian point estimators. Also, Bayesian interval estimators (credibility intervals) can be formulated, a priori information of various kinds can be incorporated into the analyses, and posterior odds can be computed. It will be seen that for regression coefficients, interval estimators based upon the data of a given sample are easily found using the Bayesian approach of this section. For experimental design purposes, confidence interval estimators obtainable from the sampling theory approach of the earlier sections are often used. For the Bayesian approach to experimental design problems (preposterior analysis), see Raiffa and Schlaifer (1961); and Ando and Kaufman (1965) for a multivariate Normal extension. The distribution theory for preposterior analysis is generally more complicated than that of confidence interval estimation. However, because an entire distribution for the posterior mean is provided (which does not depend upon the sample data), greater flexibility is available for design purposes.

Consider the standard multivariate regression model

$$\underset{(N\times p)}{\mathbf{Y}} = \underset{(N\times q)}{\mathbf{X}} \cdot \underset{(q\times p)}{\mathbf{B}} + \underset{(N\times p)}{\mathbf{U}}, \qquad N \geq p + q, \tag{8.4.2}$$

or, for $\mathbf{Y} \equiv (\mathbf{y}_1,\ldots,\mathbf{y}_p)$,

$$\underset{(N\times 1)}{\mathbf{y}_j} = \underset{(N\times q)}{\mathbf{X}} \cdot \underset{(q\times 1)}{\boldsymbol{\beta}_j} + \underset{(N\times 1)}{\mathbf{u}_j}, \qquad j = 1,\ldots,p. \tag{8.4.1}$$

Let $\underset{(p\times N)}{\mathbf{U}'} \equiv (\underset{(p\times 1)}{\mathbf{v}_1}, \ldots, \underset{(p\times 1)}{\mathbf{v}_N})$, so that $\mathbf{U}'\mathbf{U} = \Sigma_1^N \mathbf{v}_j\mathbf{v}_j'$. Assume

$$\mathcal{L}(\underset{(p\times 1)}{\mathbf{v}_j}) = N(\mathbf{0},\mathbf{\Sigma}), \quad j = 1,\ldots,p, \qquad \underset{(p\times p)}{\mathbf{\Sigma}} > 0. \tag{8.4.7}$$

The likelihood function for the error matrix $\mathbf{U}$ is given by

$$p(\mathbf{U}) = p(\mathbf{v}_1,\ldots,\mathbf{v}_N) = \prod_{j=1}^{N} p(\mathbf{v}_j),$$

or

$$p(\mathbf{U}) \propto \frac{1}{|\mathbf{\Sigma}|^{N/2}} \exp\left(-\frac{1}{2}\operatorname{tr}\mathbf{\Sigma}^{-1}\mathbf{U}'\mathbf{U}\right).$$

Equivalently,

$$p(\mathbf{Y}|\mathbf{X},\mathbf{B},\mathbf{\Sigma}) \propto \frac{\exp\{-\frac{1}{2}\operatorname{tr}(\mathbf{Y}-\mathbf{XB})'(\mathbf{Y}-\mathbf{XB})\mathbf{\Sigma}^{-1}\}}{|\mathbf{\Sigma}|^{N/2}}. \tag{8.6.1}$$

Using the orthogonality properties associated with least squares estimators, (8.6.1) may be simplified. Note that for $\hat{\mathbf{B}} = (\mathbf{X}'\mathbf{X})^{-1}\mathbf{X}'\mathbf{Y}$,

$$(\mathbf{Y}-\mathbf{XB})'(\mathbf{Y}-\mathbf{XB}) = (\mathbf{Y}-\mathbf{X}\hat{\mathbf{B}})'(\mathbf{Y}-\mathbf{X}\hat{\mathbf{B}}) + (\mathbf{B}-\hat{\mathbf{B}})'(\mathbf{X}'\mathbf{X})(\mathbf{B}-\hat{\mathbf{B}}), \tag{8.6.2}$$

since in expanding the square of $[\mathbf{Y}-\mathbf{XB}] = [(\mathbf{Y}-\mathbf{X}\hat{\mathbf{B}}) + \mathbf{X}(\hat{\mathbf{B}}-\mathbf{B})]$ the cross product terms vanish. Hence, if

$$\mathbf{V} \equiv (\mathbf{Y}-\mathbf{X}\hat{\mathbf{B}})'(\mathbf{Y}-\mathbf{X}\hat{\mathbf{B}}), \tag{8.6.3}$$

$\mathbf{V}$ is a sample statistic, and (8.6.1) becomes

$$p(\mathbf{Y}|\mathbf{X},\mathbf{B},\mathbf{\Sigma}) \propto \frac{1}{|\mathbf{\Sigma}|^{N/2}} \exp\left\{-\frac{1}{2}\operatorname{tr}\mathbf{\Sigma}^{-1}[\mathbf{V} + (\mathbf{B}-\hat{\mathbf{B}})'\mathbf{X}'\mathbf{X}(\mathbf{B}-\hat{\mathbf{B}})]\right\}. \tag{8.6.4}$$

8.6.1 Diffuse (Noninformative) Prior

Proceed with the analysis by putting prior densities on $\mathbf{B}$ and $\mathbf{\Sigma}$, and then applying Bayes' theorem. Take $\mathbf{B}$ independent of $\mathbf{\Sigma}$, and use Jeffreys' theory of invariant prior densities [see Equations (3.6.1) and (3.6.5)]. Accordingly, assume

$$\left.\begin{aligned} p(\mathbf{B},\mathbf{\Sigma}) &= p(\mathbf{B})p(\mathbf{\Sigma}), \\ p(\mathbf{B}) \propto \text{constant}, &\qquad p(\mathbf{\Sigma}) \propto \frac{1}{|\mathbf{\Sigma}|^{(p+1)/2}} \end{aligned}\right\}. \tag{8.6.5}$$

The joint posterior density of $\mathbf{B}$ and $\mathbf{\Sigma}$ is seen to be

$$p(\mathbf{B},\mathbf{\Sigma}|\mathbf{Y},\mathbf{X}) \propto \frac{\exp\{-\frac{1}{2}\operatorname{tr}\mathbf{\Sigma}^{-1}[\mathbf{V} + (\mathbf{B}-\hat{\mathbf{B}})'(\mathbf{X}'\mathbf{X})(\mathbf{B}-\hat{\mathbf{B}})]\}}{|\mathbf{\Sigma}|^{(N+p+1)/2}}. \tag{8.6.6}$$

Inferences Concerning B

To obtain the marginal posterior density of **B**, integrate (8.6.6) with respect to $\mathbf{\Sigma}$. For simplicity, let $\mathbf{G} \equiv \mathbf{V} + (\mathbf{B} - \hat{\mathbf{B}})'\mathbf{X}'\mathbf{X}(\mathbf{B} - \hat{\mathbf{B}})$. Then recall [see (5.2.1)] that if $\mathcal{L}(\mathbf{\Sigma}) = W^{-1}(\mathbf{G},p,m)$, $m > 2p$,

$$p(\mathbf{\Sigma}) \propto \frac{|\mathbf{G}|^{(m-p-1)/2}}{|\mathbf{\Sigma}|^{m/2}} \exp\left(-\frac{1}{2}\operatorname{tr} \mathbf{\Sigma}^{-1}\mathbf{G}\right).$$

Using this fact to integrate (8.6.6) with respect to $\mathbf{\Sigma}$ gives for the marginal posterior density of **B**,

$$p(\mathbf{B}|\mathbf{Y},\mathbf{X}) \propto \frac{1}{|\mathbf{V} + (\mathbf{B} - \hat{\mathbf{B}})'\mathbf{X}'\mathbf{X}(\mathbf{B} - \hat{\mathbf{B}})|^{N/2}}. \tag{8.6.7}$$

That is, $(\mathbf{B}|\mathbf{Y},\mathbf{X})$ follows a matrix **T**-distribution [see (6.2.6)]. Hence, the marginal vectors (columns of **B**) follow multivariate Student **t**-distributions. Thus, for $j = 1,\ldots,p$,

$$p(\boldsymbol{\beta}_j|\mathbf{Y},\mathbf{X}) \propto \frac{1}{\{v_{jj} + (\boldsymbol{\beta}_j - \hat{\boldsymbol{\beta}}_j)'\mathbf{X}'\mathbf{X}(\boldsymbol{\beta}_j - \hat{\boldsymbol{\beta}}_j)\}^{(N-p+1)/2}}, \tag{8.6.8}$$

where $\mathbf{V} \equiv (v_{ij})$. Moreover, it must follow that each scalar element of **B** follows a univariate Student t-distribution. To see this, let $\boldsymbol{\theta} \equiv \boldsymbol{\beta}_j - \hat{\boldsymbol{\beta}}_j$, $v_{jj} \equiv \nu s_j^2$, $\nu \equiv N - p - q + 1$, and $\mathbf{X}'\mathbf{X} \equiv \mathbf{H}s_j^2$. Then, from (8.6.8),

$$p(\boldsymbol{\theta}|\mathbf{Y},\mathbf{X}) \propto \frac{1}{\{\nu + \boldsymbol{\theta}'\mathbf{H}\boldsymbol{\theta}\}^{(\nu+q)/2}}.$$

If $\boldsymbol{\theta} \equiv (\theta_1,\boldsymbol{\theta}_2')'$, where θ_1 denotes the first element of $\boldsymbol{\theta}$, $\boldsymbol{\theta}_2: (q-1) \times 1$ denotes the remaining elements, and **H** is partitioned accordingly, as

$$\mathbf{H} = \begin{pmatrix} h_{11} & \mathbf{H}_{12} \\ \mathbf{H}_{21} & \mathbf{H}_{22} \end{pmatrix}, \qquad h_{11}: 1 \times 1,$$

the quadratic form in $\boldsymbol{\theta}$ may be written as

$$\boldsymbol{\theta}'\mathbf{H}\boldsymbol{\theta} = \theta_1^2(h_{11} - \mathbf{H}_{12}\mathbf{H}_{22}^{-1}\mathbf{H}_{21}) + (\boldsymbol{\theta}_2 + \mathbf{H}_{22}^{-1}\mathbf{H}_{21}\theta_1)'\mathbf{H}_{22}(\boldsymbol{\theta}_2 + \mathbf{H}_{22}^{-1}\mathbf{H}_{21}\theta_1).$$

Integrating with respect to $\boldsymbol{\theta}_2$ gives

$$p(\theta_1|\mathbf{Y},\mathbf{X}) = \int p(\theta_1,\boldsymbol{\theta}_2|\mathbf{Y},\mathbf{X})d\boldsymbol{\theta}_2 = \int p(\boldsymbol{\theta}|\mathbf{Y},\mathbf{X})d\boldsymbol{\theta}_2,$$

or

$$p(\theta_1|\mathbf{Y},\mathbf{X}) \propto \frac{1}{\{\nu + H_{11\cdot2}\theta_1^2\}^{(\nu+q)/2}} \int \frac{d\boldsymbol{\theta}_2}{\left\{1 + \dfrac{(\boldsymbol{\theta}_2 + \boldsymbol{\phi})'\mathbf{H}_{22}(\boldsymbol{\theta}_2 + \boldsymbol{\phi})}{\nu + H_{11\cdot2}\theta_1^2}\right\}^{(\nu+q)/2}},$$

where $H_{11\cdot2} \equiv h_{11} - \mathbf{H}_{12}\mathbf{H}_{22}^{-1}\mathbf{H}_{21}$, and $\boldsymbol{\phi} \equiv \mathbf{H}_{22}^{-1}\mathbf{H}_{21}\theta_1$.

Since the integrand in the last integral is proportional to the density of a multivariate Student **t**-variate, by (6.2.4),

$$\int \frac{d\boldsymbol{\theta}_2}{\left\{1 + \dfrac{(\boldsymbol{\theta}_2 + \boldsymbol{\phi})'\mathbf{H}_{22}(\boldsymbol{\theta}_2 + \boldsymbol{\phi})}{\nu + H_{11\cdot 2}\theta_1^2}\right\}^{(\nu+q)/2}} \propto \frac{1}{\{\nu + H_{11\cdot 2}\theta_1^2\}^{-(q-1)/2}}.$$

Hence,

$$p(\theta_1|\mathbf{Y},\mathbf{X}) \propto \frac{1}{\{\nu + \theta_1^2 H_{11\cdot 2}\}^{(\nu+1)/2}}.$$

That is, $\theta_1 \sqrt{H_{11\cdot 2}}$ follows a Student t-distribution with ν degrees of freedom. If $\mathbf{A} \equiv (\mathbf{X}'\mathbf{X})^{-1}$, use of (2.6.3) gives the alternative form,

$$\mathfrak{L}\left(\frac{\theta_1}{s_j\sqrt{a_{11}}}\right) = t_\nu.$$

But the same argument applies to any element of $\boldsymbol{\theta}$. This result is summarized in the next theorem.

Theorem (8.6.1): Under the assumptions of Section 8.6.1, if β_{ij} denotes the ith element of $\boldsymbol{\beta}_j$, β_{ij} follows the Student t-distribution given by

$$\mathfrak{L}\left[\frac{\beta_{ij} - \hat{\beta}_{ij}}{s_j\sqrt{a_{ii}}}\,\middle|\, \hat{\beta}_{ij}, s_j, \mathbf{X}\right] = t_\nu, \tag{8.6.9}$$

where $\nu s_j^2 \equiv v_{jj}$, $j = 1, \ldots, p$, $i = 1, \ldots, q$, and $\mathbf{A} \equiv (a_{ij}) = (\mathbf{X}'\mathbf{X})^{-1}$. That is, a_{ii} is the ith diagonal element of $(\mathbf{X}'\mathbf{X})^{-1}$.

Note that a simpler but less illuminating proof of Theorem (8.6.1) is obtained by adapting (2.6.3) to H and applying (6.2.4)′ for univariate marginals.

Bayesian confidence intervals may be computed from Theorem (8.6.1) by finding significance points of the t-distribution. Also, since the posterior densities are centered at the least squares estimators, Bayesian point estimators with respect to quadratic loss functions are given by the corresponding least squares estimators.

Joint inference statements about all of the components of a column of **B** may be made by analogy from the results stated in the following theorem. That is, all coefficients in a single equation of a multiequation system may be tested for significance simultaneously using the following theorem.

Theorem (8.6.2): If $\boldsymbol{\beta}_1$ is the first column of **B**, and if $\mathbf{X}'\mathbf{X} > 0$,

$$\mathfrak{L}\{(\boldsymbol{\beta}_1 - \hat{\boldsymbol{\beta}}_1)'\mathbf{G}(\boldsymbol{\beta}_1 - \hat{\boldsymbol{\beta}}_1)|\mathbf{Y},\mathbf{X}\} = F_{q+1,N-p-q}, \tag{8.6.10}$$

where "F" denotes the F-distribution, and

$$\mathbf{G} \equiv \frac{\mathbf{X}'\mathbf{X}(\nu - 1)}{\nu(q + 1)s_1^2}.$$

Proof: Let $\mathbf{W}'\mathbf{W} = \mathbf{X}'\mathbf{X}/s_1^2$, $\mathbf{W} > 0$, and let $\boldsymbol{\theta} \equiv \mathbf{W}(\boldsymbol{\beta}_1 - \hat{\boldsymbol{\beta}}_1)$. Then from (8.6.8) and (2.15.1),

$$p(\boldsymbol{\theta}|\mathbf{Y},\mathbf{X}) \propto \frac{1}{\{\nu + \boldsymbol{\theta}'\boldsymbol{\theta}\}^{(\nu+q)/2}}.$$

Make the polar transformation

$$\begin{aligned}
\theta_1 &= (\boldsymbol{\theta}'\boldsymbol{\theta})^{1/2} \sin \phi_1, \\
\theta_2 &= (\boldsymbol{\theta}'\boldsymbol{\theta})^{1/2} \cos \phi_1 \sin \phi_2, \\
&\ \ \vdots \\
\theta_q &= (\boldsymbol{\theta}'\boldsymbol{\theta})^{1/2} \cos \phi_1 \cos \phi_2 \cdots \cos \phi_{q-1}.
\end{aligned}$$

It is straightforward to check that the Jacobian of the transformation is

$$J[(\theta_1,\ldots,\theta_q) \rightarrow ((\boldsymbol{\theta}'\boldsymbol{\theta})^{1/2},\phi_1,\ldots,\phi_{q-1})] = (\boldsymbol{\theta}'\boldsymbol{\theta})^{(q-1)/2} \prod_{i=1}^{q-2} \cos^{q-i-1} \phi_i.$$

Then

$$p[(\boldsymbol{\theta}'\boldsymbol{\theta})^{1/2},\phi_1,\ldots,\phi_{q-1}|\mathbf{Y},\mathbf{X}] \propto \frac{(\boldsymbol{\theta}'\boldsymbol{\theta})^{(q-1)/2} \prod_{i=1}^{q-2} \cos^{q-i-1} \phi_i}{\{\nu + \boldsymbol{\theta}'\boldsymbol{\theta}\}^{(\nu+q)/2}}.$$

Define

$$\psi \equiv \boldsymbol{\theta}'\boldsymbol{\theta} = (\boldsymbol{\beta}_1 - \hat{\boldsymbol{\beta}}_1)' \frac{\mathbf{X}'\mathbf{X}}{s_1^2} (\boldsymbol{\beta}_1 - \hat{\boldsymbol{\beta}}_1).$$

Substituting ψ into the last density expression and using the Jacobian of this transformation gives

$$p(\psi,\phi_1,\ldots,\phi_{q-1}|\mathbf{Y},\mathbf{X}) \propto \frac{\psi^{(q-2)/2} \prod_{1}^{q-2} \cos^{q-i-1} \phi_i}{(\nu + \psi)^{(\nu+q)/2}}. \tag{8.6.11}$$

Integrating over all $\phi_1,\ldots,\phi_{q-1}$ gives the required F-density up to a scale constant.

Notice in (8.6.11) that $p(\psi|\mathbf{Y},\mathbf{X})$ is constant whenever ψ is constant (which is also true in any density) Since ψ is a positive semidefinite quadratic form in $\boldsymbol{\beta}_1$, a constant ψ corresponds to hyperellipsoids centered at $\hat{\boldsymbol{\beta}}_1$. Hence, the Bayesian confidence regions are hyperellipsoids.

Inferences Concerning $\boldsymbol{\Sigma}$

Inferences about $\boldsymbol{\Sigma}$ are made from the marginal posterior distribution of $\boldsymbol{\Sigma}|\mathbf{Y}, \mathbf{X}$, an inverted Wishart distribution. Specifically:

Theorem (8.6.3): If the joint posterior distribution of $\mathbf{B}$ and $\boldsymbol{\Sigma}$ is given by (8.6.6),

$$\mathcal{L}(\boldsymbol{\Sigma}|\mathbf{Y},\mathbf{X}) = W^{-1}(\mathbf{V}, p, N + p - q + 1). \tag{8.6.12}$$

Proof: For $\underset{(q\times p)}{\mathbf{B}} \equiv (\boldsymbol{\beta}_1,\ldots,\boldsymbol{\beta}_p)$, define $\boldsymbol{\beta}' \equiv (\boldsymbol{\beta}_1',\ldots,\boldsymbol{\beta}_p')$. Then, by expanding the elements it is easy to find that

$$\text{tr}\, \boldsymbol{\Sigma}^{-1}[(\mathbf{B} - \hat{\mathbf{B}})'\mathbf{X}'\mathbf{X}(\mathbf{B} - \hat{\mathbf{B}})] = (\boldsymbol{\beta} - \hat{\boldsymbol{\beta}})'[\boldsymbol{\Sigma}^{-1} \otimes \mathbf{X}'\mathbf{X}](\boldsymbol{\beta} - \hat{\boldsymbol{\beta}}). \tag{8.6.13}$$

Now write the posterior density (8.6.6) in the form $p(\mathbf{B},\boldsymbol{\Sigma}|\mathbf{Y},\mathbf{X}) = p(\boldsymbol{\Sigma}|\mathbf{Y},\mathbf{X})p(\mathbf{B}|\boldsymbol{\Sigma},\mathbf{Y},\mathbf{X})$. Then

$$p(\mathbf{B},\boldsymbol{\Sigma}|\mathbf{Y},\mathbf{X}) \propto \frac{\exp\left(-\frac{1}{2}\,\text{tr}\,\boldsymbol{\Sigma}^{-1}\mathbf{V}\right)}{|\boldsymbol{\Sigma}|^{(N+p+1)/2}} \cdot \exp\left\{-\frac{1}{2}(\boldsymbol{\beta} - \hat{\boldsymbol{\beta}})'[\boldsymbol{\Sigma} \otimes (\mathbf{X}'\mathbf{X})^{-1}]^{-1}(\boldsymbol{\beta} - \hat{\boldsymbol{\beta}})\right\}.$$

Identify the second term with $p(\mathbf{B}|\boldsymbol{\Sigma},\mathbf{Y},\mathbf{X})$ and normalize it by multiplying numerator and denominator by $|\boldsymbol{\Sigma} \otimes (\mathbf{X}'\mathbf{X})^{-1}|^{1/2} = |\boldsymbol{\Sigma}|^{q/2}|\mathbf{X}'\mathbf{X}|^{-p/2}$. Absorbing $|\boldsymbol{\Sigma}|^{q/2}$ into the first term and identifying the first term with $p(\boldsymbol{\Sigma}|\mathbf{Y},\mathbf{X})$ proves the theorem.

Since $(\boldsymbol{\Sigma}|\mathbf{Y},\mathbf{X})$ follows an inverted Wishart distribution, the diagonal elements of $(\boldsymbol{\Sigma}|\mathbf{Y},\mathbf{X})$, namely, $(\sigma_{ii}|\mathbf{Y},\mathbf{X})$ follow marginal inverted gamma distributions (see Section 5.2.2). Hence, inferences about the variances are based upon percentage points of the inverted gamma distribution, which are in turn related to the gamma function (see Raiffa and Schlaifer, 1961, p. 228).

8.6.2 Natural Conjugate Prior

Suppose there is prior information readily assessable about the parameters. It is now shown that a conventional natural conjugate approach does not yield a prior density that has sufficient flexibility to permit the specification of unconstrained prior parameters.

The likelihood function is expressible in the form [using (8.6.13)]

$$p(\mathbf{Y}|\mathbf{X},\mathbf{B},\boldsymbol{\Sigma}) \propto \frac{\exp\left(-\frac{1}{2}\,\text{tr}\,\boldsymbol{\Sigma}^{-1}\mathbf{V}\right)}{|\boldsymbol{\Sigma}|^{N/2}} \cdot \exp\left\{-\frac{1}{2}(\boldsymbol{\beta} - \hat{\boldsymbol{\beta}})'[\boldsymbol{\Sigma} \otimes (\mathbf{X}'\mathbf{X})^{-1}]^{-1}(\boldsymbol{\beta} - \hat{\boldsymbol{\beta}})\right\}. \tag{8.6.14}$$

Hence, by interchanging the roles of the observable and unobservable random variables, this density with the appropriate normalization would imply the natural conjugate prior

$$\mathcal{L}(\boldsymbol{\beta},\boldsymbol{\Sigma}) = N[\hat{\boldsymbol{\beta}}, \boldsymbol{\Sigma} \otimes (\mathbf{X}'\mathbf{X})^{-1}], \qquad \mathcal{L}(\boldsymbol{\Sigma}) = W^{-1}(\mathbf{V},p,N).$$

That is, if the densities are *enriched* by generalizing the prior parameters, the natural conjugate approach leads to the prior densities

$$\mathcal{L}(\boldsymbol{\beta}|\boldsymbol{\Sigma}) = N[\boldsymbol{\phi}, \boldsymbol{\Sigma} \otimes \mathbf{A}], \qquad \mathcal{L}(\boldsymbol{\Sigma}) = W^{-1}(\mathbf{G},p,m), \qquad m > 2p, \quad (8.6.15)$$

where $\boldsymbol{\phi}$, $\mathbf{A}$, $\mathbf{G}$, m are arbitrary parameters available to be predetermined by the analyst. The joint prior becomes

$$p(\boldsymbol{\beta},\boldsymbol{\Sigma}) = p(\boldsymbol{\beta}|\boldsymbol{\Sigma})p(\boldsymbol{\Sigma}).$$

Rothenberg (1963) pointed out that this prior distribution has the property that the ratios of the variances (and covariances) of certain pairs of elements of $(\boldsymbol{\beta}|\boldsymbol{\Sigma})$ are required to be identical. That is, the prior parameters are constrained. For example, if τ_{ij} denotes the prior variance (given $\boldsymbol{\Sigma}$) of the jth element of $\boldsymbol{\beta}_i$, as given in (8.6.15), it is clear that since var $(\boldsymbol{\beta}|\boldsymbol{\Sigma}) = \boldsymbol{\Sigma} \otimes \mathbf{A}$, this constraint implies that

$$\frac{\tau_{11}}{\tau_{21}} = \frac{\tau_{12}}{\tau_{22}}.$$

Similar relationships hold throughout the covariance matrix. Hence, the conventional natural conjugate prior density is not sufficiently rich to permit complete freedom of choice in assessing the prior parameters. Similar constraint problems sometimes occur in other problems involving distributions other than the Normal. For these reasons, the notion of a natural conjugate prior density is now generalized to a form that eliminates this problem in a wide variety of circumstances.

Generalized Natural Conjugate Priors

Suppose the likelihood function is indexed by several matrix (vector) parameters. Carrying out the following steps will yield a generalized natural conjugate prior.

Step (1): Assume all parameters in the likelihood function $\boldsymbol{\Omega}_1,\ldots,\boldsymbol{\Omega}_g$ are known except $\boldsymbol{\Omega}_1$. Then interchange the roles of the observable and unobservable random variables to determine the form of the prior density.

Step (2): Enrich the resulting distribution by permitting the prior parameters to be arbitrary.

Step (3): Normalize the resulting enriched density to make it proper.

Step (4): Assume a second parameter $\mathbf{\Omega}_2$ is unknown and all others are known (including $\mathbf{\Omega}_1$) and determine the resulting density. Repeat Steps (2) and (3) for $\mathbf{\Omega}_2$. Repeat the entire process for all $\mathbf{\Omega}_j$.

Step (5): Assume that the priors on each of the parameter matrices are independent so that their prior densities may be multiplied to find the joint prior density. If a problem warranted it, this step could be generalized to a case in which covariances between $\mathbf{\Omega}_i$ and $\mathbf{\Omega}_j$ were permitted. For the multivariate regression problem, the independence assumption is sufficient.

[*Note:* As was true for the ordinary natural conjugate prior, when the sampling distribution possesses sufficient statistics, the joint posterior distribution will be in the same class as the joint prior distribution.]

Now apply the above procedure to (8.6.14) to find a generalized natural conjugate prior for $\mathbf{B}$ and $\mathbf{\Sigma}$. Assuming $\mathbf{\Sigma}$ is known and interchanging the roles of $(\mathbf{Y},\mathbf{X})$ and $\underset{\sim}{\beta}$ gives, after enrichment and normalization,

$$\underset{(pq \times 1)}{\mathcal{L}(\underset{\sim}{\beta})} = N(\underset{\sim}{\phi},\mathbf{F}), \qquad \mathbf{F} > 0,$$

where $\underset{\sim}{\phi}$ and $\mathbf{F}$ are arbitrary. Similarly,

$$\mathcal{L}(\mathbf{\Sigma}) = W^{-1}(\mathbf{G},p,m), \qquad m > 2p \qquad \mathbf{G} > 0.$$

Combining these results gives for the joint generalized natural conjugate prior (a Normal-inverted Wishart prior),

$$p(\mathbf{B},\mathbf{\Sigma}) \propto \frac{\exp\{-\frac{1}{2}\operatorname{tr}[\mathbf{G}\mathbf{\Sigma}^{-1} + (\underset{\sim}{\beta} - \underset{\sim}{\phi})'\mathbf{F}^{-1}(\underset{\sim}{\beta} - \underset{\sim}{\phi})]\}}{|\mathbf{\Sigma}|^{m/2}}. \tag{8.6.16}$$

This prior density does not have the constraint problems on $\mathbf{B}$ associated with the ordinary natural conjugate prior for this problem. However, there are still constraints on $\mathbf{\Sigma}$ [see Remark following Theorem (5.2.2] and if all constraints are to be removed, the prior must be generalized still further.

The joint posterior density for $\mathbf{B}$, $\mathbf{\Sigma}$ is found by multiplying (8.6.14) and (8.6.16):

$$p(\mathbf{B},\mathbf{\Sigma}|\mathbf{Y},\mathbf{X}) \propto |\mathbf{\Sigma}|^{-(m+N)/2} \exp\{-\tfrac{1}{2}\operatorname{tr}\mathbf{\Sigma}^{-1}[\mathbf{V} + \mathbf{G} + (\mathbf{B} - \hat{\mathbf{B}})'\mathbf{X}'\mathbf{X}(\mathbf{B} - \hat{\mathbf{B}})] - \tfrac{1}{2}(\underset{\sim}{\beta} - \underset{\sim}{\phi})'\mathbf{F}^{-1}(\underset{\sim}{\beta} - \underset{\sim}{\phi})\}, \tag{8.6.17}$$

which is seen to be again a Normal-inverted Wishart density.

Inferences Concerning B

The marginal posterior density of $\mathbf{B}$ is found (as in the case of the diffuse prior) by integrating (8.6.17) with respect to $\mathbf{\Sigma}$, using the inverted Wishart normalizing constant, to get

$$p(\mathbf{B}|\mathbf{Y},\mathbf{X}) \propto \frac{\exp\left[-\frac{1}{2}(\boldsymbol{\beta}-\boldsymbol{\phi})'\mathbf{F}^{-1}(\boldsymbol{\beta}-\boldsymbol{\phi})\right]}{|\mathbf{V}+\mathbf{G}+(\mathbf{B}-\hat{\mathbf{B}})'\mathbf{X}'\mathbf{X}(\mathbf{B}-\hat{\mathbf{B}})|^{(N+m-p-1)/2}}. \tag{8.6.18}$$

Note from (8.6.18) that the posterior density of $\mathbf{B}$ is the product of a Normal density and a matrix $\mathbf{T}$-density. Therefore, the exact percentage points corresponding to (8.6.18), which are required for making inferences about the components of $\mathbf{B}$, are not easily obtained. However, if a "large" sample is available, there is a good Normal distribution approximation that can be used instead. The result is given in the next theorem.

Theorem (8.6.4): If N is large, it is approximately true that if $\boldsymbol{\beta}$ is the $pq \times 1$ vector whose elements (q at a time) are the columns of $\mathbf{B}$ in (8.6.18),

$$\mathcal{L}(\boldsymbol{\beta}|\mathbf{Y},\mathbf{X}) \cong N(\boldsymbol{\beta}_0,\mathbf{J}^{-1}), \tag{8.6.19}$$

where

$$\boldsymbol{\beta}_0 = [\mathbf{F}^{-1} + N(\mathbf{V}+\mathbf{G})^{-1} \otimes (\mathbf{X}'\mathbf{X})]^{-1}\{\mathbf{F}^{-1}\boldsymbol{\phi} + N[(\mathbf{V}+\mathbf{G})^{-1} \otimes (\mathbf{X}'\mathbf{X})]\hat{\boldsymbol{\beta}}\},$$

and

$$\mathbf{J} = \mathbf{F}^{-1} + N(\mathbf{V}+\mathbf{G})^{-1} \otimes (\mathbf{X}'\mathbf{X}).$$

Equivalently, if $\mathbf{K} = (\mathbf{F}\mathbf{J})^{-1}$,

$$\boldsymbol{\beta}_0 = \mathbf{K}\boldsymbol{\phi} + (\mathbf{I}-\mathbf{K})\hat{\boldsymbol{\beta}}.$$

That is, $\boldsymbol{\beta}_0$ is a weighted average of the prior mean and the MLE.

Proof: See Complement 8.2.

Assuming N is large, inferences pertaining to $\boldsymbol{\beta}$ may be made using the approximate Normal distribution in Theorem (8.6.4). For joint inferences on all components of $\mathbf{B}$ simultaneously, use the fact [which follows immediately from (8.6.19)] that

$$\mathcal{L}\{(\boldsymbol{\beta}-\boldsymbol{\beta}_0)'\mathbf{J}(\boldsymbol{\beta}-\boldsymbol{\beta}_0)\} = \chi^2(pq). \tag{8.6.20}$$

Suppose $\boldsymbol{\beta}$ is partitioned into two subvectors (with $\boldsymbol{\beta}_0$ and $\mathbf{J}^{-1}$ similarly partitioned):

$$\boldsymbol{\beta} = \begin{pmatrix}\dot{\boldsymbol{\beta}}\\ \ddot{\boldsymbol{\beta}}\end{pmatrix}, \qquad \boldsymbol{\beta}_0 = \begin{pmatrix}\dot{\boldsymbol{\beta}}_0\\ \ddot{\boldsymbol{\beta}}_0\end{pmatrix}, \qquad \mathbf{J}^{-1} \equiv \begin{pmatrix}\mathbf{K}_{11} & \mathbf{K}_{12}\\ \mathbf{K}_{21} & \mathbf{K}_{22}\end{pmatrix}.$$

Then, from (8.6.19), the marginal posterior distribution of $\dot{\boldsymbol{\beta}}$ is given by

$$(\dot{\boldsymbol{\beta}}|\mathbf{Y},\mathbf{X}) = \mathbf{N}(\dot{\boldsymbol{\beta}}_0,\mathbf{K}_{11}). \tag{8.6.21}$$

Hence,

$$\mathcal{L}\{(\dot{\beta} - \dot{\beta}_0)'\mathbf{K}_{11}^{-1}(\dot{\beta} - \dot{\beta}_0)\} = \chi^2(r), \tag{8.6.22}$$

where r is the number of components in $\dot{\beta}$. Thus, use of (8.6.22) permits joint inferences for any subvector of β.

Finally, since a special case of (8.6.21) is that in which $\dot{\beta}$ is a single (scalar) element of β, the percentage points of a standard univariate Normal distribution may be used to make inferences about any single regression coefficient in any of the multivariate regression equations. It is only necessary to assess $(\phi,\mathbf{F},\mathbf{G})$, and observe $(\mathbf{V},\mathbf{X},\hat{\beta})$.

Inferences Concerning Σ

With a generalized natural conjugate prior, the posterior density of $(\Sigma|\mathbf{Y},\mathbf{X})$ does not have a form from which exact percentage points may be easily obtained. This is seen as follows. Rewrite (8.6.17) using (8.6.13), as

$$p(\mathbf{B},\Sigma|\mathbf{Y},\mathbf{X}) \propto \frac{\exp[-\frac{1}{2}\operatorname{tr}\Sigma^{-1}(\mathbf{V}+\mathbf{G})]}{|\Sigma|^{(N+m)/2}}$$
$$\exp\left\{-\frac{1}{2}(\beta-\hat{\beta})'[\Sigma^{-1}\otimes\mathbf{X}'\mathbf{X}](\beta-\hat{\beta}) - \frac{1}{2}(\beta-\phi)'\mathbf{F}^{-1}(\beta-\phi)\right\}.$$

Completing the square in β gives

$$p(\beta,\Sigma|\mathbf{Y},\mathbf{X}) \propto \frac{\exp[-\frac{1}{2}\operatorname{tr}\Sigma^{-1}(\mathbf{V}+\mathbf{G})]}{|\Sigma|^{(N+m)/2}}\cdot\exp\left[-\frac{1}{2}(\beta-\bar{\beta})'\mathbf{M}(\beta-\bar{\beta})\right], \tag{8.6.23}$$

where

$$\bar{\beta} = [\mathbf{F}^{-1} + \Sigma^{-1}\otimes(\mathbf{X}'\mathbf{X})]^{-1}\{\mathbf{F}^{-1}\phi + [\Sigma^{-1}\otimes(\mathbf{X}'\mathbf{X})]\hat{\beta}\},$$

and

$$\mathbf{M} \equiv \mathbf{F}^{-1} + \Sigma^{-1}\otimes(\mathbf{X}'\mathbf{X}).$$

Since (8.6.23) involves a Normal density in β, it is easy to integrate with respect to β to obtain the marginal posterior density of Σ:

$$p(\Sigma|\mathbf{Y},\mathbf{X}) \propto \frac{\exp[-\frac{1}{2}\operatorname{tr}\Sigma^{-1}(\mathbf{V}+\mathbf{G})]}{|\Sigma|^{(N+m)/2}|\mathbf{F}^{-1}+\Sigma^{-1}\otimes(\mathbf{X}'\mathbf{X})|^{1/2}}. \tag{8.6.24}$$

The density in (8.6.24) does not readily lend itself to computation of fractile points. Fortunately, however, a simple approximation to (8.6.24) is available for large sample sizes. It is given in the next theorem.

Theorem (8.6.5): If N is large, it is approximately true for $\Sigma|(\mathbf{Y},\mathbf{X})$ distributed as in (8.6.24), that

$$\mathcal{L}(\Sigma|\mathbf{Y},\mathbf{X}) \cong W^{-1}(\mathbf{V}+\mathbf{G}, p, N+m-q). \tag{8.6.25}$$

Proof: See Complement 8.3.

Note that since $\boldsymbol{\Sigma}$ is distributed approximately in the inverted Wishart distribution form, the diagonal elements (variances) are marginally distributed in the inverted gamma distribution form. Therefore, inferences concerning the individual variances are easily made by applying the remark made following Theorem (8.6.3) concerning percentage points of the inverted gamma distribution.

8.6.3 Forecasting and Prediction

In many situations, after a sample has been taken and the regression function has been estimated, it is desired to predict the value of the dependent variable for preassigned values of the independent variables. In the context of (8.4.1), suppose $\mathbf{y}_j = \mathbf{X}\boldsymbol{\beta}_j + \mathbf{u}_j$, and $\hat{\boldsymbol{\beta}}_j$ denotes the maximum likelihood estimator of $\boldsymbol{\beta}_j$, assuming $\mathcal{L}(\mathbf{u}_j) = N(\mathbf{0},\sigma^2\mathbf{I}_N)$. Let y^* denote a new scalar observation of the jth dependent variable and suppose (for fixed j),

$$y^* = z_1\beta_{1j} + \cdots + z_q\beta_{qj} + u^*, \tag{8.6.26}$$

where u^* is the associated disturbance term, the z_i's are assigned values of the independent variables, and $\boldsymbol{\beta}_j \equiv (\beta_{1j},\ldots,\beta_{qj})'$. Thus, if $\mathbf{z} \equiv (z_1,\ldots,z_q)'$, $\mathcal{L}(y^*|\beta,\sigma^2) = N(\mathbf{z}'\boldsymbol{\beta}_j,\sigma^2)$. If not much is known about the parameters, a priori, the following result is sometimes useful.

Theorem (8.6.6): Under the assumptions of the simpler linear regression model given in (8.4.1), if a new scalar observation, y^*, is given by (8.6.26), and if a diffuse prior is adopted for the parameters so that if $h \equiv 1/\sigma^2$,

$$p(\beta_j,h) \propto \frac{1}{h},$$

the predictive distribution of y^* given the observed data is

$$\mathcal{L}\left[\frac{y^* - \mathbf{z}'\hat{\boldsymbol{\beta}}_j}{\hat{\sigma}\sqrt{1 + \mathbf{z}'(\mathbf{X}'\mathbf{X})^{-1}\mathbf{z}}}\,\middle|\,\text{(sample)}\right] = t_{N-q}. \tag{8.6.27}$$

[*Remark:* A related result for the multivariate case is developed in Section 14.2.2.]

Example (8.6.1): Adopt the simple model

$$y_i|x_i = a + bx_i + u_i, \qquad i = 1,\ldots,N, \qquad \mathcal{L}(u_i) = N(0,\sigma^2),$$

and the u_i's are mutually uncorrelated. Then with

$$\boldsymbol{\beta} = \begin{pmatrix} a \\ b \end{pmatrix}, \qquad \mathbf{X}' = \begin{pmatrix} 1 & \cdots & 1 \\ x_1 & \cdots & x_N \end{pmatrix}, \qquad \mathbf{z} = \begin{pmatrix} 1 \\ x^* \end{pmatrix},$$
$$y^* = a + bx^* + u^*, \qquad \mathbf{y} = \mathbf{X}\boldsymbol{\beta} + \mathbf{u},$$

and substitution in (8.6.27) gives

$$\mathcal{L}\left[\frac{y^* - \hat{a} - \hat{b}x^*}{\hat{\sigma}\sqrt{1 + \frac{1}{N} + \frac{(x^* - \bar{x})^2}{\Sigma_1^N (x_i - \bar{x})^2}}} \middle| \text{(sample)}\right] = t_{N-2}.$$

Proof of Theorem (8.6.6):

$$p(y^*|\text{sample}) = \int p(y^*|\boldsymbol{\beta}_j,h)p(\boldsymbol{\beta}_j,h|\text{sample})d\boldsymbol{\beta}_j dh.$$

Since $\mathcal{L}(y^*|\boldsymbol{\beta}_j,h) = N(\mathbf{z}'\boldsymbol{\beta}_j,h^{-1})$,

$$p(\boldsymbol{\beta}_j,h|\text{sample}) \propto p(\boldsymbol{\beta}_j,h)p(\hat{\boldsymbol{\beta}}_j|\boldsymbol{\beta}_j,h)p(\hat{\sigma}^2|\sigma^2),$$
$$\mathcal{L}(\hat{\boldsymbol{\beta}}_j|\boldsymbol{\beta}_j,\sigma^2) = N(\boldsymbol{\beta}_j,\sigma^2(\mathbf{X}'\mathbf{X})^{-1}),$$

and

$$\mathcal{L}\left[\frac{(N-q)\hat{\sigma}^2}{\sigma^2}\middle|\sigma^2\right] = \chi^2_{N-q}$$

$$p(\boldsymbol{\beta}_j,h|\text{sample}) \propto h^{N/2-1}\exp\left\{-\frac{h}{2}[(N-q)\hat{\sigma}^2 + (\boldsymbol{\beta}_j - \hat{\boldsymbol{\beta}}_j)'(\mathbf{X}'\mathbf{X})(\boldsymbol{\beta}_j - \hat{\boldsymbol{\beta}}_j)]\right\}.$$

Thus,

$$p(y^*|\text{sample}) \propto \int h^{(N+1)/2-1}\exp\left\{-\frac{h}{2}[(\boldsymbol{\beta}_j - \hat{\boldsymbol{\beta}}_j)'(\mathbf{X}'\mathbf{X})(\boldsymbol{\beta}_j - \hat{\boldsymbol{\beta}}_j) + (N-q)\hat{\sigma}^2 + (y^* - \mathbf{z}'\boldsymbol{\beta}_j)^2]\right\} d\boldsymbol{\beta}_j dh.$$

Integrating with respect to h gives

$$p(y^*|\text{sample}) \propto \int \frac{d\boldsymbol{\beta}_j}{[(N-q)\hat{\sigma}^2 + (y^* - \mathbf{z}'\boldsymbol{\beta}_j)^2 + (\boldsymbol{\beta}_j - \hat{\boldsymbol{\beta}}_j)'\mathbf{X}'\mathbf{X}(\boldsymbol{\beta}_j - \hat{\boldsymbol{\beta}}_j)]^{(N+1)/2}}.$$

Completing the square on $\boldsymbol{\beta}_j$ gives

$$p(y^*|\text{sample}) \propto \int \frac{d\boldsymbol{\beta}_j}{[w + (\boldsymbol{\beta}_j - \mathbf{Q})'(\mathbf{X}'\mathbf{X} + \mathbf{z}\mathbf{z}')(\boldsymbol{\beta}_j - \mathbf{Q})]^{(N+1)/2}},$$

where

$$\mathbf{Q} = (\mathbf{X}'\mathbf{X} + \mathbf{z}\mathbf{z}')^{-1}(\hat{\boldsymbol{\beta}}'\mathbf{X}'\mathbf{X} + y^*\mathbf{z}')',$$

and

$$w = y^{*2} + \hat{\boldsymbol{\beta}}'\mathbf{X}'\mathbf{X}\hat{\boldsymbol{\beta}} + (N-q)\hat{\sigma}^2 - (\hat{\boldsymbol{\beta}}'\mathbf{X}'\mathbf{X} + y^*\mathbf{z}')(\mathbf{X}'\mathbf{X} + \mathbf{z}\mathbf{z}')^{-1}(\hat{\boldsymbol{\beta}}'\mathbf{X}'\mathbf{X} + y^*\mathbf{z}')'.$$

Integrating the multivariate Student t-density gives

$$p(y^*|\text{sample}) \propto \frac{1}{w^{(N-q+1)/2}}.$$

Applying (2.5.6) to $(\mathbf{X}'\mathbf{X} + \mathbf{zz}')^{-1}$ gives the result.

8.7 MULTIVARIATE ANALYSIS OF VARIANCE (MANOVA)

The analysis of variance (ANOVA) is a technique that was designed to ferret out differences in means of various populations. It may be treated as a special case of the regression model although it is usually approached by resolving the sum of squares of the observable variables about their means into several component parts. Moreover, the technique is specifically aimed at coping with variables that are qualitative rather than quantitative (for example, one might be analyzing the differential effects of using Feeds A, B, and C on the growth of hogs; Feed Type is a qualitative variable). For each model, the univariate case is first reviewed. Then the multivariate case is described. The discussion will be confined to the one- and two-way layouts (Sections 8.7.1 and 8.7.2) although results for higher-way layouts are immediately obtainable by analogy. Tests of hypotheses are generally discussed in terms of likelihood ratio criteria (assuming Normality). Alternative criteria are indicated in Section 8.7.3. Analysis of covariance and comparisons of means are discussed in Sections 8.8 and 8.9. All analysis of variance and covariance models to be discussed assume fixed effects; that is, rather than assuming the effects are representative of some larger population, interest will always center about the particular effects selected.

8.7.1 One-Way Layout

Univariate ANOVA—One-Way Layout

Suppose a company producing structural materials for construction purposes decides to see if it can produce a stronger product. Several different techniques are compared experimentally:

Technique No. 1—use an alloy of 50 percent of metal A and 50 percent of metal B.
Technique No. 2—apply a surface coating of metal C to metal B.
Technique No. 3—use an alloy of 25 percent of metal A and 75 percent of metal B.
Technique No. 4—use an alloy of three metals.

In ANOVA jargon, these techniques are called *treatments*.

Define:

y_{ij} = strength of the jth metal rod using Technique i, as measured by a strain gauge,

β_i = increased strength of an experimental rod under Technique i (the β_i are called the *effects*),

μ = expected strength of a rod without any new treatment,

e_{ij} = errors due to measurement problems, variations in the ambient environment, and so on,

for $j = 1, 2, \ldots, J_i$, and $i = 1, 2, 3, 4$. Let y_{ij} differ from its mean by an error term; that is, adopt the linear model

$$y_{ij} = \mu + \beta_i + e_{ij}, \tag{8.7.1}$$

or

$$y_{ij} = \theta_i + e_{ij}, \tag{8.7.2}$$

where θ_i is the expected strength of a rod using Technique i; $\theta_i \equiv \mu + \beta_i$. Assume for inferential purposes that the e_{ij}'s are all independent with $\mathcal{L}(e_{ij}) = N(0,\sigma^2)$. The problem is to test the hypothesis of "no effects," $H_0\colon \beta_1 = \beta_2 = \beta_3 = \beta_4 = 0$ (or equivalently, that the θ_i's are equal), against the alternative hypothesis H_1 that the effects are unequal; and if the effects are different, how do they compare? That is, it is desired to test whether the means of four normal populations are identical, given that they have the same variances (the fact that the y_{ij}'s have the same variance is called the assumption of homoscedasticity).

The parameters θ_i, $i = 1, \ldots, 4$ may be estimated by minimizing $\Sigma_1^4 \Sigma_1^{J_i} e_{ij}^2$ with respect to θ_i (least squares), or equivalently by maximum likelihood. The results are that under the alternative hypothesis,

$$\hat{\theta}_i = y_{i\cdot} \equiv \frac{1}{J_i} \sum_{j=1}^{J_i} y_{ij}, \tag{8.7.3}$$

and under the null hypothesis H_0, $\theta_i = \theta$, for all i,

$$\hat{\theta}_i = \hat{\theta} = \bar{y} \equiv \frac{1}{n} \sum_{i=1}^{4} \sum_{j=1}^{J_i} y_{ij}, \tag{8.7.4}$$

where $n = \Sigma_1^4 J_i$ is the total number of observations. The dot notation refers to the fact that an averaging has taken place over that subscript. The total sample variance between populations is $\frac{1}{3}\Sigma_1^4 J_i(y_{i\cdot} - \bar{y})^2$ and the sample variance within populations is

$$\frac{1}{n-4}\sum_{i=1}^{4}\sum_{j=1}^{J_i}(y_{ij}-y_{i\cdot})^2.$$

It is easily found that a likelihood ratio test of H_0 versus H_1 is carried out by forming the ratio of the two total sample variances and comparing the ratio with the percentage points of the $F_{3,n-4}$ distribution. The generalization to more than four populations is direct.

Multivariate ANOVA—One-Way Layout

The generalization of the ANOVA to the multivariate case is straightforward in terms of the form of the model, but special care is required for the testing of hypotheses and the application of related methods of simultaneous inference. The basis for this special care is discussed in Section 8.7.3.

Define the p-variate vector observations of population α,

$$\mathbf{z}_\alpha(1), \qquad \mathbf{z}_\alpha(2),\ldots,\mathbf{z}_\alpha(T_\alpha); \qquad \alpha = 1, 2,\ldots,q.$$

That is, there are q populations, each of which is p-variate, and there are T_α observations of the αth population. Suppose

$$\mathcal{L}[\mathbf{z}_\alpha(t)] = N(\boldsymbol{\theta}_\alpha,\boldsymbol{\Sigma}), \qquad \boldsymbol{\Sigma} > 0, \qquad t = 1,\ldots,T_\alpha,$$

and that all $\mathbf{z}_\alpha(t)$ are independent; thus, the assumption of homoscedasticity in the multivariate case is that each population has the same covariance matrix, $\boldsymbol{\Sigma}$. By analogy with (8.7.2), write

$$\underset{(p\times 1)}{\mathbf{z}_\alpha(t)} = \boldsymbol{\theta}_\alpha + \mathbf{v}_\alpha(t), \qquad \alpha = 1,\ldots,q. \tag{8.7.5}$$

The problem is to test

$$H_0\colon \boldsymbol{\theta}_1 = \boldsymbol{\theta}_2 = \cdots = \boldsymbol{\theta}_q$$

versus

$$H_1\colon \boldsymbol{\theta}_\alpha\text{'s are not all equal.}$$

An observation vector $\mathbf{z}_\alpha(t)$ might correspond to a $p \times 1$ vector of sales for a multiproduct firm; the jth component of $\mathbf{z}_\alpha(t)$ would correspond to sales of product number j, $j = 1, 2,\ldots,p$. Observations could be taken on a random day, one day each week, for many weeks.

Suppose there are q different techniques or treatments for improving sales and it is desired to compare their effects. The errors for a given α and t are all correlated since the same firm and sales conditions are involved.

To solve this problem, the observations will be rearranged so as to conform to the pattern of the standard multivariate regression model. Then the hypothesis testing techniques developed for that model will be applied. Accordingly, define for $N \equiv \Sigma_1^q T_\alpha$,

$$\underset{(N\times p)}{\mathbf{Y}} = [\mathbf{z}_1(1),\ldots,\mathbf{z}_1(T_1);\mathbf{z}_2(1),\ldots,\mathbf{z}_2(T_2);\ldots;\mathbf{z}_q(1),\ldots,\mathbf{z}_q(T_q)]',$$

$$\underset{(N\times p)}{\mathbf{U}} = [\mathbf{v}_1(1),\ldots,\mathbf{v}_1(T_1);\ldots;\mathbf{v}_q(1),\ldots,\mathbf{v}_q(T_q)]',$$

$$\underset{(N\times q)}{\mathbf{X}} = \left[\begin{array}{ccccc}
1 & 0 & \cdots & 0 & 1 \\
\vdots & \vdots & & \vdots & \vdots \\
1 & 0 & \cdots & 0 & 1 \\
\hline
0 & 1 & 0\ \cdots & 0 & 1 \\
\vdots & \vdots & \vdots & \vdots & \vdots \\
0 & 1 & 0\ \cdots & 0 & 1 \\
\hline
 & & \vdots & & \\
\hline
0 & & \cdots & 1 & 1 \\
\vdots & & & \vdots & \vdots \\
0 & & \cdots & 1 & 1 \\
\hline
0 & & \cdots & 0 & 1 \\
\vdots & & & \vdots & \vdots \\
0 & & \cdots & 0 & 1
\end{array}\right]
\begin{array}{l}
T_1 \times q \\ \\ T_2 \times q \\ \\ \vdots \\ \\ T_{q-1} \times q \\ \\ T_q \times q
\end{array},$$

$$\underset{(q\times p)}{\mathbf{B}} \equiv \begin{pmatrix} \mathbf{B}_1 \\ \mathbf{B}_2 \end{pmatrix}, \qquad \underset{(q_1\times p)}{\mathbf{B}_1} = \begin{pmatrix} \boldsymbol{\theta}_1' - \boldsymbol{\theta}_q' \\ \vdots \\ \boldsymbol{\theta}_{q-1}' - \boldsymbol{\theta}_q' \end{pmatrix}, \qquad \underset{(q_2\times p)}{\mathbf{B}_2} = \boldsymbol{\theta}_q',$$

where q_j is the number of rows in $\mathbf{B}_j$, $j = 1, 2$, $q_1 + q_2 = q$, and $q_2 \equiv 1$. The model (8.7.5) may now be rewritten in the form

$$\mathbf{Y} = \mathbf{XB} + \mathbf{U}, \tag{8.7.6}$$

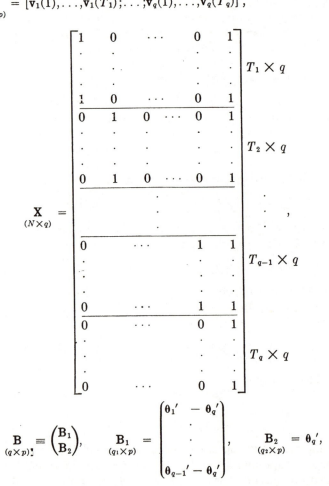

and it is desired to test $H_0\colon \mathbf{B}_1 = \mathbf{0}$ (which is equivalent to testing for equality of all q means). The test is again (as in the univariate case) carried out by comparing the appropriate variances. However, in the multivariate case this means comparing the generalized variances (determinants of the appropriate covariance matrices), as in Equation (8.4.14).

Denote the maximum likelihood estimated error covariance matrix under $H_0 + H_1$ by $\hat{\mathbf{\Sigma}}_1$ (the plus denotes union). Then

$$N\hat{\mathbf{\Sigma}}_1 = \hat{\mathbf{U}}'\hat{\mathbf{U}} = (\mathbf{Y} - \mathbf{X}\hat{\mathbf{B}})'(\mathbf{Y} - \mathbf{X}\hat{\mathbf{B}}), \tag{8.7.7}$$

where, by definition, $|\mathbf{X}'\mathbf{X}| \neq 0$, and we may write

$$\mathbf{Y} - \mathbf{X}\hat{\mathbf{B}} = \mathbf{AY}, \qquad \mathbf{A} = \mathbf{I} - \mathbf{X}(\mathbf{X}'\mathbf{X})^{-1}\mathbf{X}'.$$

It is easy to check that $\mathbf{A}$ is an idempotent matrix (that is, $\mathbf{A}^2 = \mathbf{A}$). Hence,

$$N\hat{\mathbf{\Sigma}}_1 = \mathbf{Y}'\mathbf{AY} = \sum_{\alpha=1}^{q}\sum_{t=1}^{T_\alpha} \mathbf{z}_\alpha(t)\mathbf{z}_\alpha'(t) - \sum_{\alpha=1}^{q} T_\alpha\bar{\mathbf{z}}_\alpha\bar{\mathbf{z}}_\alpha', \tag{8.7.8}$$

where

$$\bar{\mathbf{z}}_\alpha \equiv \frac{1}{T_\alpha}\sum_{t=1}^{T_\alpha} \mathbf{z}_\alpha(t).$$

Under H_0, $\mathbf{B}_1 = \mathbf{0}$, so that $\mathbf{B}' = (\mathbf{0},\mathbf{B}_2')$. Hence, if $\mathbf{X} \equiv (\mathbf{X}_1,\mathbf{X}_2)$ is partitioned according to the partitioning of $\mathbf{B}$,

$$\mathbf{Y} = (\mathbf{X}_1,\mathbf{X}_2)\begin{pmatrix}\mathbf{0}\\ \mathbf{B}_2\end{pmatrix} + \mathbf{U} = \mathbf{X}_2\mathbf{B}_2 + \mathbf{U},$$

and therefore (since $\mathbf{X}_2$ is a column of ones),

$$\hat{\mathbf{B}}_2 = (\mathbf{X}_2'\mathbf{X}_2)^{-1}\mathbf{X}_2'\mathbf{Y} = \frac{1}{N}\mathbf{X}_2'\mathbf{Y}. \tag{8.7.9}$$

Hence, the maximum likelihood estimated error covariance matrix under H_0 is

$$N\hat{\mathbf{\Sigma}}_0 = \hat{\mathbf{U}}_0'\hat{\mathbf{U}}_0 = (\mathbf{Y} - \mathbf{X}_2\hat{\mathbf{B}}_2)'(\mathbf{Y} - \mathbf{X}_2\hat{\mathbf{B}}_2) = \mathbf{Y}'\mathbf{CY}, \tag{8.7.10}$$

where

$$\mathbf{C} = \mathbf{I} - \mathbf{X}_2(\mathbf{X}_2'\mathbf{X}_2)^{-1}\mathbf{X}_2'$$

is idempotent, and zero subscripts refer to values of the matrices under H_0. It is not hard to check that (8.7.10) simplifies to

$$N\hat{\mathbf{\Sigma}}_0 = \sum_{\alpha=1}^{q}\sum_{t=1}^{T_\alpha} \mathbf{z}_\alpha(t)\mathbf{z}_\alpha'(t) - N\bar{\mathbf{z}}\bar{\mathbf{z}}', \tag{8.7.11}$$

where

$$\bar{\mathbf{z}} = \frac{1}{N}\sum_{\alpha=1}^{q}\sum_{t=1}^{T_\alpha} \mathbf{z}_\alpha(t).$$

The likelihood ratio test of H_0 versus H_1 is to reject H_0 if

$$\frac{|N\hat{\mathbf{\Sigma}}_1|}{|N\hat{\mathbf{\Sigma}}_0|} \equiv U_{p,q-1,N-q} < \text{constant}, \tag{8.7.12}$$

where the fractile points of $U_{p,q-1,N-q}$ are evaluated for large N, as in Section 8.4, by means of the Box approximation [see Equation (8.4.16)].

Example (8.7.1): Consider a two-dimensional one-way layout with equal numbers of observations on each of three populations; that is, suppose $p = 2$, $q_1 = 2$, $q = 3$, $T_1 = T_2 = T_3 = 10$, $q_2 = 1$, and suppose the observation vectors at times $t_1, \ldots, t_{10}$ are

$$\begin{array}{llllllllllll} & t_1 & t_2 & t_3 & t_4 & t_5 & t_6 & t_7 & t_8 & t_9 & t_{10} \\ \mathbf{z}_1(t) = & \begin{pmatrix}1\\-1\end{pmatrix}, & \begin{pmatrix}2\\1\end{pmatrix}, & \begin{pmatrix}0\\0\end{pmatrix}, & \begin{pmatrix}1\\1\end{pmatrix}, & \begin{pmatrix}2\\2\end{pmatrix}, & \begin{pmatrix}0\\1\end{pmatrix}, & \begin{pmatrix}-1\\-1\end{pmatrix}, & \begin{pmatrix}2\\0\end{pmatrix}, & \begin{pmatrix}3\\1\end{pmatrix}, & \begin{pmatrix}1\\0\end{pmatrix}; \\ \mathbf{z}_2(t) = & \begin{pmatrix}3\\1\end{pmatrix}, & \begin{pmatrix}3\\0\end{pmatrix}, & \begin{pmatrix}2\\1\end{pmatrix}, & \begin{pmatrix}2\\2\end{pmatrix}, & \begin{pmatrix}2\\3\end{pmatrix}, & \begin{pmatrix}2\\-1\end{pmatrix}, & \begin{pmatrix}0\\0\end{pmatrix}, & \begin{pmatrix}3\\2\end{pmatrix}, & \begin{pmatrix}4\\1\end{pmatrix}, & \begin{pmatrix}2\\2\end{pmatrix}; \\ \mathbf{z}_3(t) = & \begin{pmatrix}4\\2\end{pmatrix}, & \begin{pmatrix}1\\2\end{pmatrix}, & \begin{pmatrix}3\\1\end{pmatrix}, & \begin{pmatrix}3\\0\end{pmatrix}, & \begin{pmatrix}2\\3\end{pmatrix}, & \begin{pmatrix}1\\2\end{pmatrix}, & \begin{pmatrix}3\\2\end{pmatrix}, & \begin{pmatrix}4\\3\end{pmatrix}, & \begin{pmatrix}3\\3\end{pmatrix}, & \begin{pmatrix}2\\1\end{pmatrix}. \end{array}$$

It is easily found that

$$N = \sum_1^3 T_\alpha = 30, \qquad \bar{\mathbf{z}} = \begin{pmatrix}2\\1.13\end{pmatrix},$$

$$\bar{\mathbf{z}}_1 = \begin{pmatrix}1.1\\.4\end{pmatrix}, \qquad \bar{\mathbf{z}}_2 = \begin{pmatrix}2.3\\1.1\end{pmatrix}, \qquad \bar{\mathbf{z}}_3 = \begin{pmatrix}2.6\\1.9\end{pmatrix},$$

$$\mathbf{Y}'\mathbf{Y} = \begin{pmatrix}166 & 87\\87 & 80\end{pmatrix},$$

$$N\hat{\mathbf{\Sigma}}_1 = \begin{pmatrix}33.4 & 7.9\\7.9 & 30.2\end{pmatrix}, \qquad N\hat{\mathbf{\Sigma}}_0 = \begin{pmatrix}46 & 19.2\\19.2 & 41.6\end{pmatrix}.$$

Hence,

$$\tilde{U} \equiv \frac{|N\hat{\mathbf{\Sigma}}_1|}{|N\hat{\mathbf{\Sigma}}_0|} = .614.$$

It is now necessary to compare $\tilde{U}$ with an appropriate fractile of $U_{2,2,27}$.

Taking just the first term of the Box approximation, for illustration, gives (see Section 8.4)

$$P\{-m \log U_{p,q_1,N-q} \leq v\} \cong P\{\chi^2(pq_1) \leq v\},$$

where

$$m = N - q_2 - \tfrac{1}{2}(p + q_1 + 1) = 26.5.$$

Substituting the parameter values into the approximation gives

$$P\{-26.5 \log U_{2,2,27} \leq v\} \cong P\{\chi^2(4) \leq v\}.$$

For a 5 percent significance level (just to be specific), $v = 9.49$ (see the Chi-Square table in Appendix B). Therefore, with 95 percent confidence,

$$-26.5 \log U_{2,2,27} \leq 9.49,$$

or equivalently, $U_{2,2,27} \geq .7$. Since $\tilde{U} = .614$, reject H_0 at the 5 percent level of significance and conclude that the means of the three populations are significantly different. A discussion of how to make inferences concerning the extent to which the means actually differ is deferred until Section 8.9, where the subject of multiple comparisons is considered.

See Appendix A for listings of computer programs which will carry out the computations for the one-way MANOVA.

8.7.2 Two-Way Layout

The study will be restricted to the simplest case of equal numbers of observations per cell. However, interaction between the row and column effects is permitted, although interpretation of results in the presence of interaction effects is generally more difficult. As with the one-way layout, the univariate case is reviewed first.

Univariate ANOVA—Two-Way Layout—*K* Observations per Cell

Suppose a company is concerned with the problem of how to increase sales of a nationally sold and advertised product. They design and carry out an experiment involving two factors: advertising medium and packaging technique. Several types of each are considered. Suppose K sample observations on each possible combination are taken ($K > 1$). Let

y_{ijk} = sales of the product during the period when advertising medium i is used, $i = 1,\ldots,I$ ($i = 1$ corresponds to radio advertising, $i = 2$ corresponds to Saturday night television advertising, $i = 3$ corresponds to *New York Times* advertising, and so on); and when packaging technique j is used, $j = 1,\ldots,J$ ($j = 1$ corresponds to a red box, $j = 2$ corresponds to a red jar, $j = 3$ corresponds to a blue box, and so on); and for the kth observation, $k = 1,\ldots,K$.

Consider the model

$$y_{ijk} = \mu + \alpha_i + \beta_j + \gamma_{ij} + e_{ijk}, \tag{8.7.13}$$

where $i = 1,\ldots,I$, $j = 1,\ldots,J$, $k = 1,\ldots,K$, μ is the overall mean, α_i is the effect on sales attributable to the ith advertising medium, β_j is the effect on sales attributable to the jth packaging technique, γ_{ij} is the interaction effect between the ith advertising medium and the jth packaging technique, and e_{ijk} is the error term. Note that since (8.7.13) may be rewritten

$$y_{ijk} = (\mu + \alpha_. + \beta_. + \gamma_{i.} + \gamma_{.j}) + (\alpha_i - \alpha_.) + (\beta_j - \beta_.) + (\gamma_{ij} - \gamma_{i.} - \gamma_{.j}) + e_{ijk},$$

or

$$y_{ijk} = \mu^* + \alpha_i^* + \beta_j^* + \gamma_{ij}^* + e_{ijk},$$

where

$$\mu^* = \mu + \alpha_. + \beta_. + \gamma_{i.} + \gamma_{.j}, \qquad \alpha_i^* = \alpha_i - \alpha_., \qquad \beta_j^* = \beta_j - \beta_.,$$

$$\gamma_{ij}^* = \gamma_{ij} - \gamma_{i.} - \gamma_{.j}, \qquad \alpha_. = \frac{1}{I}\sum_1^I \alpha_i, \qquad \beta_. = \frac{1}{J}\sum_1^J \beta_j,$$

$$\gamma_{i.} = \frac{1}{J}\sum_{j=1}^J \gamma_{ij}, \qquad \gamma_{.j} = \frac{1}{I}\sum_{i=1}^I \gamma_{ij},$$

there is some ambiguity in the model parameters. However, if the model is augmented with the constraints

$$\alpha_. = 0, \qquad \beta_. = 0, \qquad \gamma_{i.} = 0, \qquad \gamma_{.j} = 0, \tag{8.7.14}$$

the solutions to the normal equations for the estimators of α_i, β_j, and γ_{ij}, will be unique. It is of interest to test the hypotheses

$$H_0\text{: all } \alpha_i = 0, \text{ versus } H_1\text{: not all } \alpha_i = 0,$$
$$H_0^*\text{: all } \beta_j = 0, \text{ versus } H_1^*\text{: not all } \beta_j = 0,$$

and

H_0^{**}: row and column effects are additive,[8] versus H_1^{**}: effects interact; that is, it is desired to know if the advertising medium has any effect on sales, if the packaging technique has any effect on sales, and whether or not the effects are additive. These hypotheses are tested by comparing the appropriate variance ratios with fractile points of the F-distribution, assuming that $\mathcal{L}(e_{ijk}) = N(0,\sigma^2)$, for all i, j, k, and the errors are mutually independent.

[8] The hypothesis of additivity is equivalent to the hypothesis of no interaction between effects.

Multivariate ANOVA—Two-Way Layout—*K* Observations per Cell

The MANOVA model for a two-way layout with K observations per cell ($K > 1$) is a simple generalization of (8.7.13) and (8.7.14) to the case of p dimensions. That is, let all variables denote p-vectors. Then the model becomes

$$\underset{(p\times 1)}{\mathbf{y}_{ijk}} = \underset{(p\times 1)}{\boldsymbol{\mu}} + \underset{(p\times 1)}{\boldsymbol{\alpha}_i} + \underset{(p\times 1)}{\boldsymbol{\beta}_j} + \underset{(p\times 1)}{\boldsymbol{\gamma}_{ij}} + \underset{(p\times 1)}{\mathbf{e}_{ijk}}, \tag{8.7.15}$$

where $i = 1,\ldots,I$, $j = 1,\ldots,J$, $k = 1,\ldots,K$, $K > 1$, and $\boldsymbol{\alpha}_{\cdot} = \boldsymbol{\beta}_{\cdot} = \boldsymbol{\gamma}_{i\cdot} = \boldsymbol{\gamma}_{\cdot j} = \mathbf{0}$. Also, assume

$$\mathcal{L}(\mathbf{e}_{ijk}) = N(\mathbf{0},\boldsymbol{\Sigma}), \qquad \boldsymbol{\Sigma} > 0, \tag{8.7.16}$$

and all $\mathbf{e}_{ijk}$ are mutually independent.

Define the covariance matrices

$$\mathbf{G} = \sum_{i=1}^{I}\sum_{j=1}^{J}\sum_{k=1}^{K} [\mathbf{y}_{ijk} - \mathbf{y}_{ij\cdot}][\mathbf{y}_{ijk} - \mathbf{y}_{ij\cdot}]', \tag{8.7.17}$$

$$\mathbf{H}_J = IK \sum_{j=1}^{J} [\mathbf{y}_{\cdot j\cdot} - \mathbf{y}_{\cdots}][\mathbf{y}_{\cdot j\cdot} - \mathbf{y}_{\cdots}]',$$

$$\mathbf{H}_I = JK \sum_{i=1}^{I} [\mathbf{y}_{i\cdot\cdot} - \mathbf{y}_{\cdots}][\mathbf{y}_{i\cdot\cdot} - \mathbf{y}_{\cdots}]', \tag{8.7.18}$$

$$\mathbf{H}_{IJ} = K \sum_{i=1}^{I}\sum_{j=1}^{J} [\mathbf{y}_{ij\cdot} - \mathbf{y}_{i\cdot\cdot} - \mathbf{y}_{\cdot j\cdot} + \mathbf{y}_{\cdots}][\mathbf{y}_{ij\cdot} - \mathbf{y}_{i\cdot\cdot} - \mathbf{y}_{\cdot j\cdot} + \mathbf{y}_{\cdots}]', \tag{8.7.19}$$

where $\mathbf{G}$ corresponds to total sample variance within populations (error sum of squares). $\mathbf{H}_J$ and $\mathbf{H}_I$ correspond to total sample variances between columns or between rows of the data tableau, and $\mathbf{H}_{IJ}$ is the interaction sample variance.

Likelihood ratio test procedures require that the condition

$$p \le IJ(K-1)$$

be satisfied, so that $\mathbf{G} > 0$. This may be seen as follows. It is shown in Complement 5.1 that $\Sigma_1^K(\mathbf{y}_{ijk} - \mathbf{y}_{ij\cdot})(\mathbf{y}_{ijk} - \mathbf{y}_{ij\cdot})'$ may always be transformed to $\Sigma_1^{K-1}\mathbf{Y}_{ijk}\mathbf{Y}_{ijk}'$. Then $\mathbf{G}$ may be rewritten as $\mathbf{G} = \mathbf{W}\mathbf{W}'$, where $\mathbf{W}: p \times IJ(K-1)$. But rank $(\mathbf{W}) = \min\,[p,\ IJ(K-1)] =$ rank $(\mathbf{G})$. By analogy, it is easy to see that rank $(\mathbf{H}_I) = \min\,[p,\ I-1]$, rank $(\mathbf{H}_J) = \min\,[p,\ J-1]$, rank $(\mathbf{H}_{IJ}) = \min\,[p,\ (I-1)(J-1)]$.

The likelihood ratio test of H_0: all $\boldsymbol{\alpha}_i = \mathbf{0}$, versus H_1: not all $\boldsymbol{\alpha}_i = \mathbf{0}$, is to reject H_0 if

$$\frac{|\mathbf{G}|}{|\mathbf{G} + \mathbf{H}_I|} < \text{constant},$$

and the constant is determined from the Box approximation [Equation (8.4.16)] using the fact that under H_0,

$$\mathcal{L}\left\{\frac{|\mathbf{G}|}{|\mathbf{G} + \mathbf{H}_I|}\right\} = \mathcal{L}\{U_{p,q_1,n}\}, \tag{8.7.20}$$

where the number of degrees of freedom for the row effects vector is $q_1 \equiv I - 1$, and $n \equiv IJ(K - 1)$ denotes the number of error degrees of freedom. The distribution theory follows from (8.7.16).

Similarly, to test H_0^*: all $\boldsymbol{\beta}_j = \mathbf{0}$, versus H_1^*: not all $\boldsymbol{\beta}_j = \mathbf{0}$, reject H_0^* if

$$\frac{|\mathbf{G}|}{|\mathbf{G} + \mathbf{H}_J|} < \text{constant},$$

where under H_0^*,

$$\mathcal{L}\left\{\frac{|\mathbf{G}|}{|\mathbf{G} + \mathbf{H}_J|}\right\} = \mathcal{L}\{U_{p,q_1^*,n}\}, \tag{8.7.21}$$

and the degrees of freedom for column effects is $q_1^* = J - 1$, with $n = IJ(K - 1)$.

If H_0 or H_0^* is rejected, it is concluded there are main effects, so long as there are no interaction effects (H_0^{**} is true). However, if H_0^{**} is rejected, the interpretation is more complicated (see for instance, Scheffé, 1959).

The test for interaction effects (H_0^{**} versus H_1^{**}) is to reject H_0^{**} if

$$\frac{|\mathbf{G}|}{|\mathbf{G} + \mathbf{H}_{IJ}|} < \text{constant},$$

where under H_0^{**},

$$\mathcal{L}\left\{\frac{|\mathbf{G}|}{|\mathbf{G} + \mathbf{H}_{IJ}|}\right\} = \mathcal{L}\{U_{p,q_1^{**},n}\}, \tag{8.7.22}$$

and the interaction degrees of freedom is $q_1^{**} = (I - 1)(J - 1)$, with $n = IJ(K - 1)$.

For purposes of inference on the variances, note that

$$\mathcal{L}(\mathbf{G}) = W[\mathbf{\Sigma}, p, IJ(K - 1)], \qquad \text{if } p \leq IJ(K - 1), \tag{8.7.23}$$

under H_0,

$$\mathcal{L}(\mathbf{H}_I) = W(\mathbf{\Sigma}, p, I - 1), \qquad \text{if } p \leq I - 1, \tag{8.7.24}$$

under H_0^*,

$$\mathcal{L}(\mathbf{H}_J) = W(\mathbf{\Sigma}, p, J-1), \qquad \text{if } p \leq J-1. \tag{8.7.25}$$

If the inequalities are not satisfied, singular distributions will result.

Multivariate ANOVA—Two-Way Layout—One Observation per Cell

If there is only one observation per cell available, the two-way layout MANOVA model is given by

$$\underset{(p\times 1)}{\mathbf{y}_{ij}} = \underset{(p\times 1)}{\boldsymbol{\mu}} + \underset{(p\times 1)}{\boldsymbol{\alpha}_i} + \underset{(p\times 1)}{\boldsymbol{\beta}_j} + \underset{(p\times 1)}{\mathbf{e}_{ij}}, \tag{8.7.26}$$

where $i = 1,\ldots,I$, $j = 1,\ldots,J$, $\boldsymbol{\alpha}_\cdot = \boldsymbol{\beta}_\cdot = \mathbf{0}$, and the $\mathbf{e}_{ij}$'s are mutually independent with $\mathcal{L}(\mathbf{e}_{ij}) = N(\mathbf{0},\mathbf{\Sigma})$, $\mathbf{\Sigma} > 0$.

Define the covariance matrices

$$\mathbf{G} = \sum_{i=1}^{I}\sum_{j=1}^{J} [\mathbf{y}_{ij} - \mathbf{y}_{i\cdot} - \mathbf{y}_{\cdot j} + \mathbf{y}_{\cdot\cdot}][\mathbf{y}_{ij} - \mathbf{y}_{i\cdot} - \mathbf{y}_{\cdot j} + \mathbf{y}_{\cdot\cdot}]', \tag{8.7.27}$$

$$\mathbf{H}_J = I\sum_{j=1}^{J} [\mathbf{y}_{\cdot j} - \mathbf{y}_{\cdot\cdot}][\mathbf{y}_{\cdot j} - \mathbf{y}_{\cdot\cdot}]', \tag{8.7.28}$$

$$\mathbf{H}_I = J\sum_{i=1}^{I} [\mathbf{y}_{i\cdot} - \mathbf{y}_{\cdot\cdot}][\mathbf{y}_{i\cdot} - \mathbf{y}_{\cdot\cdot}]', \tag{8.7.29}$$

where $\mathbf{G}$ is the multivariate error sum of squares, and $\mathbf{H}_J$ and $\mathbf{H}_I$ are the total sample variances between columns and between rows, respectively. It is assumed that $p \leq (I-1)(J-1)$, so that $\mathbf{G}$ will be positive definite. The proof that this condition is necessary is analogous to that for the case of K observations per cell. Note that there is no interaction term since with $K = 1$ it is assumed that effects are additive (since there are not enough degrees of freedom available to provide a good justification for assuming interaction).

The likelihood ratio test of H_0: all $\boldsymbol{\alpha}_i = \mathbf{0}$, versus H_1: not all $\boldsymbol{\alpha}_i = \mathbf{0}$ is to reject H_0 if

$$\frac{|\mathbf{G}|}{|\mathbf{G} + \mathbf{H}_I|} < \text{constant},$$

and the constant is determined from the Box approximation [Equation (8.4.16)] using the fact that under H_0,

$$\mathcal{L}\left\{\frac{|\mathbf{G}|}{|\mathbf{G} + \mathbf{H}_I|}\right\} = \mathcal{L}\{U_{p,q_1,n}\}, \tag{8.7.30}$$

where $q_1 = I - 1$, and $n = (I-1)(J-1)$.

Similarly, the test of H_0^*: all $\boldsymbol{\beta}_j = \mathbf{0}$, versus H_1^*: not all $\boldsymbol{\beta}_j = \mathbf{0}$, is to reject H_0^* if

$$\frac{|\mathbf{G}|}{|\mathbf{G} + \mathbf{H}_J|} < \text{constant},$$

where, under H_0^*,

$$\mathcal{L}\left\{\frac{|\mathbf{G}|}{|\mathbf{G} + \mathbf{H}_J|}\right\} = \mathcal{L}\{U_{p,q_1^*,n}\} \tag{8.7.31}$$

where $q_1^* = J - 1$, and $n = (I - 1)(J - 1)$.

Inferences concerning variances may be made with nonsingular distributions by noting that for $p \leq (I - 1)(J - 1)$,

$$\mathcal{L}(\mathbf{G}) = W[\mathbf{\Sigma},p,(I - 1)(J - 1)]. \tag{8.7.32}$$

Under H_0, if $p \leq I - 1$,

$$\mathcal{L}(\mathbf{H}_I) = W(\mathbf{\Sigma}, p, I - 1); \tag{8.7.33}$$

and under H_0^*, if $p \leq J - 1$,

$$\mathcal{L}(\mathbf{H}_J) = W(\mathbf{\Sigma}, p, J - 1). \tag{8.7.34}$$

Computer program listings for the two-way MANOVA are given in Appendix A.

Unequal Numbers of Observations per Cell

If it is possible to control which observations are to be collected, an effort should be made to secure equal numbers of observations per cell. In such a case the analyses given above in this section are directly applicable. If unequal numbers of observations per cell are presented, the analysis is more complicated involving the general solution of a system of m linear algebraic equations where $m \equiv \min(I - 1, J - 1)$. Nevertheless, exact procedures used in the univariate case may be extended to the multivariate case in a straightforward manner (least squares estimators are obtained directly). Note that a simplifying assumption often made in MANOVA with unequal numbers of observations per cell (and all cells have "approximately" the same number) is to assign each cell a frequency equal to the "mean" number of observations; denote this number by K. Then the ordinary least squares estimation analysis is carried out using a K-observations-per-cell model. This approach, though simpler than using the exact estimation method, is less efficient, so that tests based upon this approximation cannot be expected to be as powerful as those based upon the exact estimators. Note that since the covariance matrix of the mean vector of the (i,j)th cell is proportional to $1/n_{ij}$, where n_{ij} denotes the number of observations in cell (i,j), the mean of the recipricals of the n_{ij}'s is required rather than the mean of the n_{ij}'s. Hence, it is the *harmonic mean* of the n_{ij}'s rather than the arithmetic mean that should be used as the approximate mean cell frequency. Accordingly, take

$$K = \frac{IJ}{\sum_{i=1}^{I}\sum_{j=1}^{J}\frac{1}{n_{ij}}}.$$

If the "mean" cell frequency is not used, the estimated cell means have different covariance matrices, so that corresponding components of the vectors of cell means have different standard errors. If an analysis is large, it is recommended that the exact solution be used since the computer is carrying out all of the tedious calculations. However, the approximate solution is often useful in small analyses.

Missing Observations

It frequently happens in a MANOVA that one or more observations are missing, possibly because subjects with the required characteristics could not be obtained, or because acquired data was lost, because experimental animals accidentally died, or whatever. (Note, it is assumed that the treatments did not cause the data to be missing. The case in which observations are missing in a systematic way is excluded.) If such missing observations are really intrinsic to the study, it is desirable to estimate the missing values and use the estimates in the analysis. Accordingly, suppose the data follow a two-way layout multivariate ANOVA with one observation per cell. Assume the model is given by

$$\underset{(p\times 1)}{\mathbf{y}_{ij}} = \underset{(p\times 1)}{\boldsymbol{\mu}} + \underset{(p\times 1)}{\boldsymbol{\alpha}_i} + \underset{(p\times 1)}{\boldsymbol{\beta}_j} + \underset{(p\times 1)}{\mathbf{e}_{ij}}, \tag{8.7.26}$$

where $i = 1,\ldots,I$, $j = 1,\ldots,J$, $\boldsymbol{\alpha}. = \boldsymbol{\beta}. = \mathbf{0}$, and the $\mathbf{e}_{ij}$'s are mutually independent with $\mathcal{L}(\mathbf{e}_{ij}) = N(\mathbf{0},\boldsymbol{\Sigma})$, $\boldsymbol{\Sigma} > 0$ (note that there is no interaction term). Let $\mathbf{y}_{00} \equiv \mathbf{z}$ denote a particular $\mathbf{y}_{ij}$, and suppose $\mathbf{z}$ is a missing observation. The estimator of $\mathbf{y}_{00}$ based upon least squares is clearly

$$\hat{\mathbf{z}} = \hat{\boldsymbol{\mu}} + \hat{\boldsymbol{\alpha}}_0 + \hat{\boldsymbol{\beta}}_0, \tag{8.7.35}$$

where $\hat{\boldsymbol{\mu}}$, $\hat{\boldsymbol{\alpha}}_0$, and $\hat{\boldsymbol{\beta}}_0$ are the least squares estimators of $\boldsymbol{\mu}$, $\boldsymbol{\alpha}_0$, and $\boldsymbol{\beta}_0$ (they are also maximum likelihood estimators under the Normality assumption). Since

$$\hat{\boldsymbol{\mu}} = \mathbf{y}_{\cdot\cdot}, \qquad \hat{\boldsymbol{\alpha}}_0 = \mathbf{y}_{0\cdot} - \mathbf{y}_{\cdot\cdot}, \qquad \hat{\boldsymbol{\beta}}_0 = \mathbf{y}_{\cdot 0} - \mathbf{y}_{\cdot\cdot},$$

where the averages are all taken assuming no missing observations, if a prime denotes a summation taken over the actual observations only,

$$\hat{\mu} = \frac{1}{IJ}\left\{\left(\sum_i \sum_j\right)' \mathbf{y}_{ij} + \hat{\mathbf{z}}\right\},$$

$$\hat{\alpha}_0 = \frac{1}{J}\left\{\sum_j{}' \mathbf{y}_{0j} + \hat{\mathbf{z}}\right\} - \frac{1}{IJ}\left\{\left(\sum_i \sum_j\right)' \mathbf{y}_{ij} + \hat{\mathbf{z}}\right\},$$

$$\hat{\beta}_0 = \frac{1}{I}\left\{\sum_i{}' \mathbf{y}_{i0} + \hat{\mathbf{z}}\right\} - \frac{1}{IJ}\left\{\left(\sum_i \sum_j\right)' \mathbf{y}_{ij} + \hat{\mathbf{z}}\right\}.$$

Substituting these estimators into (8.7.35) gives

$$\begin{aligned}\hat{\mathbf{z}} = & \frac{1}{IJ}\left\{\left(\sum_i \sum_j\right)' \mathbf{y}_{ij} + \hat{\mathbf{z}}\right\} + \frac{1}{J}\left\{\sum_j{}' \mathbf{y}_{0j} + \hat{\mathbf{z}}\right\} \\ & - \frac{1}{IJ}\left\{\left(\sum_i \sum_j\right)' \mathbf{y}_{ij} + \hat{\mathbf{z}}\right\} + \frac{1}{I}\left\{\sum_i{}' \mathbf{y}_{i0} + \hat{\mathbf{z}}\right\} \\ & - \frac{1}{IJ}\left\{\left(\sum_i \sum_j\right)' \mathbf{y}_{ij} + \hat{\mathbf{z}}\right\}. \end{aligned} \tag{8.7.36}$$

Solving (8.7.36) for $\hat{\mathbf{z}}$ gives the missing observation estimator

$$\hat{\mathbf{z}} = \frac{I \sum_j{}' \mathbf{y}_{0j} + J \sum_i{}' \mathbf{y}_{i0} - \left(\sum_i \sum_j\right)' \mathbf{y}_{ij}}{(I-1)(J-1)}. \tag{8.7.37}$$

The result in (8.7.37) may now be substituted into the data array and used as if it were an additional observation. Least squares estimators may be obtained using $\hat{\mathbf{z}}$ as just another data point although the total number of degrees of freedom is reduced by one, to $IJ - 1$.

If there is more than one missing observation, the same analysis may be applied to each missing data point. Moreover, the same technique is applicable to layouts of order greater than two, and to layouts with more than one observation per cell. If some elements of some data vectors are present and others are missing there is considerable question as to the most appropriate method of analysis to use. There is a large literature on this subject which has been summarized and extended (for both regression and ANOVA) in a series of four papers (see Afifi and Elashoff, 1966; 1967; 1969a; 1969b). For a Bayesian approach to this problem, see Press and Scott (1975, 1976).

8.7.3 Alternative Hypothesis Testing Criteria

In univariate ANOVA, the optimal testing criterion is generally agreed to be an F test of the sample variance ratio. However, in MANOVA, alternative criteria have been proposed for testing the same hypotheses and these criteria do not always yield the same results in p dimensions. For $p = 1$, however, all the criteria become identical.

One criterion is, of course, that of likelihood ratio, which is the one recommended above. Other criteria are compared in some detail (on a power basis) by Smith, Gnanadesikan, and Hughes (1962), and by Gabriel (1968), and the implications of these criteria for programming are discussed in Bock (1963).

Some of the various testing criteria are compared below using $\mathbf{G}$ and $\mathbf{H}_I$ defined in (8.7.17) and (8.7.18) for the two-way layout. However, for the following discussion $\mathbf{H}_I$ should be thought of as either $\mathbf{H}_I$, $\mathbf{H}_J$, or $\mathbf{H}_{IJ}$. The case of K observations per cell will be understood.

(1) **Likelihood Ratio Criterion (Product of Roots):** Examine

$$\frac{|\mathbf{G}|}{|\mathbf{G} + \mathbf{H}_I|} = \frac{1}{|\mathbf{H}_I\mathbf{G}^{-1} + \mathbf{I}|} = \frac{1}{\prod_{j=1}^{p} (1 + \lambda_j)},$$

where λ_j is the jth latent root[9] of $\mathbf{H}_I\mathbf{G}^{-1}$.

(2) **Trace Criterion (Sum of Roots):** Examine

$$T_0^2 \equiv \text{tr}\,(\mathbf{H}_I\mathbf{G}^{-1}) = \sum_{j=1}^{p} \lambda_j,$$

where λ_j is the jth latent root of $\mathbf{H}_I\mathbf{G}^{-1}$.

(3) **Largest Characteristic Root Criterion:** Examine

$$\frac{\lambda}{1 + \lambda},$$

where λ is the largest latent root of $\mathbf{H}_I\mathbf{G}^{-1}$.

[9] Note from (2.4.2) that λ_j is also the jth latent root of the symmetric matrix $\mathbf{G}^{-1/2}\mathbf{H}_I\mathbf{G}^{-1/2}$. Thus, if $p \leq IJ(K - 1)$, $\mathbf{G}^{-1/2}$ is nonsingular (as was shown in Section 8.7.2), and by (2.7.4), rank $(\mathbf{H}_I\mathbf{G}^{-1})$ = rank $(\mathbf{G}^{-1/2}\mathbf{H}_I\mathbf{G}^{-1/2})$ = rank $(\mathbf{H}_I)$ = number of λ_j's different from zero. Recall that rank $(\mathbf{H}_I) = \min\,(p,q)$, where $q = q_1$, q_1^*, or q_1^{**} depending on whether $\mathbf{H}_I$ corresponds to rows, columns, or interactions. Thus, if $q = 1$, the three criteria are equivalent.

For the likelihood ratio criterion, exact tests are available for all p, although tables are available only for special cases (such as for testing for independence of normal variates), and for small p. Moreover, the Box approximation is available for any p (for very small sample sizes, additional terms of the Box expansion should be used), which is why this approach was recommended. The test is unbiased.

The trace criterion involves the use of Hotelling's T_0^2 statistic (see Section 6.1.3); however, it cannot be used for small sample sizes and appreciable p since only a result asymptotic in sample size is available. For additional information, see Anderson (1958), p. 224, and Pillai (1954).

The largest characteristic root criterion (Roy, 1957) can be applied in general although percentage point tables are available only for restricted values of the parameters. Appropriate tables were prepared by Foster and Rees (1957), Foster (1957; 1958), Heck (1960), and Pillai (1960); the Heck tables are reproduced in Appendix B. The test is unbiased.

Various possible hypothesis testing techniques (including nonparametric techniques) were compared according to their power functions by Gabriel (1968). He showed that from the sampling theory point of view, tests based upon the largest characteristic root criterion are best if tests involving all linear combinations of parameters are of interest. However, if interest is focused on tests involving all pairs of parameters, the use of many Student t-tests is a better procedure than any other that has been proposed. Thus, the optimal answer to MANOVA hypothesis testing (and confidence interval estimation) is that the best procedure to use varies depending upon the objective of the analyst. The maximum characteristic root criterion will be used for simultaneous confidence interval estimation (see Section 8.8.2) for cases in which an experiment can be designed for optimality over many samples, since simple results and tables are available for this case (and not for the likelihood ratio criterion). For inferences based upon a given sample, the Bayesian approach outlined in Section 8.9 should prove more useful than any of the confidence interval approaches discussed above. It is often useful to test hypotheses using all three criteria discussed above. If contradictory results are obtained, extreme caution should be exercised in interpreting the results.

Example (8.7.2): The multivariate analysis of variance was used to explore possible differences in grammatical usage among groups of adult English language speakers who differed in sex or in educational level (see Jones, 1966). Data were collected by tape recording responses from each of 54 adult speakers to the Thematic Apperception Test. Recordings were transcribed and each word spoken was classified into one of a set of 19 mutually exclusive grammatical categories. The relative frequency of words produced in each category was determined for each speaker. These proportions were determined for each of the following seven categories (which served as the dependent variables) after transforming the data (see Section 8.10) to make them behave more Normally distributed with constant variances.

$$\underset{(7\times 1)}{\mathbf{y}} = \begin{pmatrix} y_1 \\ \cdot \\ \cdot \\ \cdot \\ y_7 \end{pmatrix} = \begin{bmatrix} \text{personal subject pronouns} \\ \text{personal possessive pronouns} \\ \text{nouns (excluding high frequency nouns)} \\ \text{prepositions} \\ \text{indefinites (something, everywhere, anyplace, and so on)} \\ \text{quantifying modifiers} \\ \text{conjunctions} \end{bmatrix}.$$

The seven-dimensional problem may be analyzed as a two-way layout MANOVA with unequal numbers of observations per cell. The two factors are education and sex, with education analyzed with three levels, and sex with two levels. The numbers of observations in each cell are shown below (and of course the total number of observations is 54).

Cell Frequencies

Education	*Sex*	
	MALE	FEMALE
Failed to complete high school	8	10
Graduated from high school	11	8
Two or more years of college	9	8

The problem is to examine whether there are differences in grammatical usage attributable to differing levels of education (row effects); also, are there differences attributable to differing sex levels (column effects); and are there interaction effects? The sample correlation matrix of **y** is

$$\operatorname{corr}(\mathbf{y}) = \begin{bmatrix} 1.000 & & & & & & \\ -.042 & 1.000 & & & & & \\ -.619 & .257 & 1.000 & & & & \\ -.475 & .324 & .165 & 1.000 & & & \\ .152 & -.299 & -.541 & .112 & 1.000 & & \\ -.263 & -.071 & .134 & -.050 & -.025 & 1.000 & \\ .064 & -.024 & -.039 & .035 & .036 & -.157 & 1.000 \end{bmatrix},$$

and of course it is symmetrical. The sample standard deviations of the components of **y** are given by

$$(.068,.038,.063,.054,.045,.038,.046).$$

The summary statistics required for hypothesis testing are given in Table 8.7.1. The analysis is based upon a model using an average of 9 observations per cell (the harmonic mean is easily computed to be 8.8).

Table 8.7.1 Manova Summary ($p = 7$)

	Effect		
	EDUCATION	SEX	INTERACTION
Degrees of freedom under hypothesis	2	1	2
Degrees of freedom for error	48	48	48
Nonzero latent roots of ($\mathbf{HG}^{-1}$) λ_1	23.828	45.283	6.086
Nonzero latent roots of ($\mathbf{HG}^{-1}$) λ_2	3.826		1.846
Significance point of likelihood ratio	.433	.485	.741
Degrees of freedom	14, 84	7, 42	14, 84
Significance level	<.01	<.01	>.50
Significance point of trace ($\mathbf{HG}^{-1}$) $= T_0^2$	1.152	.943	.331
Significance level	<.01		>.05
Significance point of largest root criterion	.498	.515	.202
Significance level	<.01		>.05

The first result noted is that the hypothesis of additivity cannot be rejected; that is, there does not appear to be any interaction between effects.

Since there is only one degree of freedom for the sex effect, there is only one nonzero latent root of ($\mathbf{HG}^{-1}$), and in this case all three hypothesis testing criteria are equivalent. The result is a significant sex effect.

For the education effect there are two latent roots of ($\mathbf{HG}^{-1}$) so that the three criteria are different. However, it is seen that all three criteria yield the same conclusion; namely, that there is a significant education effect.

The data must now be studied further using multiple comparisons (Section 8.9) to determine the sources and magnitudes of the sex and education effects.

8.8 MULTIVARIATE ANALYSIS OF COVARIANCE (MANOCOVA)

The analysis of covariance is a technique designed to accommodate both the generally quantitative variables of regression and the generally qualitative variables of the analysis of variance. That is, it is sometimes the case that while there may be significant main effects in the analysis of variance of a set of data, when one or more continuous variables (called *concomitant variables*) is added to the model to help account for the variation in the dependent variables, the significance of the main effects disappears. Thus, it may be of interest to study whether there are significant main effects, over and above the variation in the data attributable to concomitant variables. There is a MANOCOVA model for every type of MANOVA model. For example, for a two-way layout with one observation per cell, the model may be written

$$\underset{(p\times 1)}{\mathbf{y}_{ij}} = \underset{(p\times 1)}{\boldsymbol{\mu}} + \underset{(p\times 1)}{\boldsymbol{\alpha}_i} + \underset{(p\times 1)}{\boldsymbol{\beta}_j} + \underset{(p\times k)(k\times 1)}{\boldsymbol{\Lambda}\cdot\mathbf{x}_{ij}} + \underset{(p\times 1)}{\mathbf{e}_{ij}}, \tag{8.8.1}$$

where $\boldsymbol{\mu}$, $\boldsymbol{\alpha}_i$, $\boldsymbol{\beta}_j$, and $\mathbf{e}_{ij}$ have the same interpretations as in Section 8.7; however, $\mathbf{x}_{ij}$ is a vector of k concomitant variables (different for each i and j), and $\boldsymbol{\Lambda}$ is a matrix of unknown coefficients that are assumed to be the same for every i and j. Note that the interaction term is assumed to be zero.

Equation (8.8.1) may easily be put into the form of the general linear model as in Section 8.7. Then the coefficients are estimable by least squares or maximum likelihood, as before. Moreover, the following hypotheses may now be of interest.

H_1: $\boldsymbol{\Lambda} = \mathbf{0}$; H_2: $\lambda_{mn} = 0$, for some given m and n, where $\boldsymbol{\Lambda} = (\lambda_{mn})$; H_3: $\lambda_{m_1n_1} = \lambda_{m_2n_2} = \cdots = \lambda_{m_rn_r}$ for some set of r elements of $\boldsymbol{\Lambda}$, $r = 1,\ldots,(pk)$; H_4: $\boldsymbol{\alpha}_i = \mathbf{0}$ for all $i = 1,\ldots,I$; H_5: $\boldsymbol{\beta}_j = \mathbf{0}$ for all $j = 1,\ldots,J$. Hypotheses may be tested as in Section 8.4.3.

8.9 MULTIPLE COMPARISONS

In most MANOVA problems, the rejection of the null hypothesis is only the first step in the problem (if the null hypothesis of "equal means" or of "no effects" is accepted, that will often be the last step). After rejection, it is usually of interest to know the reason the null hypothesis was rejected and to establish quantitative relationships among the unequal means (in most cases, the relationships are suggested by the data themselves). For example, if, in a one-way layout MANOVA, the population means of one population seemed to differ by 3 from the population means of another population, it would be of interest to test that hypothesis.

What is usually of interest is to make probability statements about the population means, given the observed data for the situation at hand. That is, Bayesian credibility intervals are what is required in most situations. For this reason a Bayesian approach to multiple comparisons is discussed in Section 8.9.1.

There are some situations, such as in experimental design, in which interest centers on the comparison of population means for the many different sets of data that might be obtained in sampling under a variety of circumstances. For this kind of a situation, the confidence interval approach of sampling theory is appropriate. A sampling theory approach to multivariate simultaneous confidence interval estimation is described in Section 8.9.2. (An alternative Bayesian approach of preposterior analysis is discussed in Raiffa and Schlaifer, 1961, for the univariate case.)

8.9.1 Bayesian Credibility Intervals in MANOVA

Adopt the one-way layout MANOVA model of (8.7.5), accompanied by

$$\mathcal{L}[\mathbf{z}_\alpha(t)] = N(\boldsymbol{\theta}_\alpha,\boldsymbol{\Sigma}), \tag{8.9.1}$$

$t = 1,\ldots,T_\alpha$, $\alpha = 1,\ldots,q$. Suppose it is of interest to examine simple comparisons between say, the first mean $\boldsymbol{\theta}_1$, and any other mean, say $\boldsymbol{\theta}_i$, $i \neq 1$. That is, interest is focused on $(\boldsymbol{\theta}_1 - \boldsymbol{\theta}_i)$ and on $\mathbf{a}'(\boldsymbol{\theta}_1 - c\boldsymbol{\theta}_i)$, for some prespecified $p \times 1$ constant vector $\mathbf{a}$ (which might be a unit vector as a special case) and some prespecified scalar c. The required distribution is found by reformulating the general linear model

$$\underset{(N\times p)}{\mathbf{Y}} = \underset{(N\times q)}{\mathbf{X}} \cdot \underset{(q\times p)}{\mathbf{B}} + \underset{(N\times p)}{\mathbf{U}}, \qquad N = \Sigma_\alpha^q T_\alpha \tag{8.9.2}$$

so that it fits the above requirements, and then using results from the Bayesian analysis of the multivariate regression model in Section 8.7.

Define

$$\underset{(p\times N)}{\mathbf{Y}'} = [\mathbf{z}_1(1),\ldots,\mathbf{z}_1(T_1);\ldots;\mathbf{z}_q(1),\ldots,\mathbf{z}_q(T_q)], \tag{8.9.3}$$

$$\underset{(p\times N)}{\mathbf{U}'} = [\mathbf{v}_1(1),\ldots,\mathbf{v}_1(T_1);\ldots;\mathbf{v}_q(1),\ldots,\mathbf{v}_q(T_q)], \tag{8.9.4}$$

$$\underset{(p\times q)}{\mathbf{B}'} = [\boldsymbol{\theta}_1 - c\boldsymbol{\theta}_i,\ \boldsymbol{\theta}_2 - c\boldsymbol{\theta}_i,\ldots,\boldsymbol{\theta}_{i-1} - c\boldsymbol{\theta}_i,\ c\boldsymbol{\theta}_i,\ \boldsymbol{\theta}_{i+1} - c\boldsymbol{\theta}_i,\ldots,\boldsymbol{\theta}_q - c\boldsymbol{\theta}_i]. \tag{8.9.5}$$

The associated design matrix is defined as follows. Let $\mathbf{e}_{T_\alpha}$ denote a vector of ones of dimension T_α. Then define

$$\underset{(N\times q)}{\mathbf{X}} = \begin{pmatrix} \mathbf{e}_{T_1} & & & & \mathbf{e}_{T_1} & & & \\ & \mathbf{e}_{T_2} & & & \mathbf{e}_{T_2} & \mathbf{0} & & \\ & & \cdot & & \cdot & & & \\ & & & \cdot & \cdot & & & \\ & & & & \mathbf{e}_{T_i} & & & \\ & & & & \cdot & \cdot & & \\ & & & \mathbf{0} & \cdot & & \cdot & \\ & & & & \mathbf{e}_{T_q} & & & \mathbf{e}_{T_q} \end{pmatrix}, \tag{8.9.6}$$

so that $\mathbf{X}$ is block diagonal except for ones running down the ith column; that is, except for ones in the ith column, there are vectors of ones centered along the principal diagonal, and zeros elsewhere.

Now recall from (8.6.7) that if a diffuse prior on $\mathbf{B}$ and $\mathbf{\Sigma}$ is assumed, the posterior distribution of $(\mathbf{B}|\mathbf{Y},\mathbf{X})$ is matrix $\mathbf{T}$ [see (6.2.6)]; that is,

$$p(\mathbf{B}|\mathbf{Y},\mathbf{X}) \propto \frac{1}{|\mathbf{V} + (\mathbf{B} - \hat{\mathbf{B}})'\mathbf{X}'\mathbf{X}(\mathbf{B} - \hat{\mathbf{B}})|^{N/2}}, \tag{8.9.7}$$

where $\mathbf{V} \equiv (\mathbf{Y} - \mathbf{X}\hat{\mathbf{B}})'(\mathbf{Y} - \mathbf{X}\hat{\mathbf{B}})$, and $\hat{\mathbf{B}} = (\mathbf{X}'\mathbf{X})^{-1}\mathbf{X}'\mathbf{Y}$. But (8.6.7) implies the rows of $\mathbf{B}$ follow a multivariate t-distribution. In particular, the first row of $\mathbf{B}$ follows the density

$$p(\boldsymbol{\theta}_1 - c\boldsymbol{\theta}_i)|\mathbf{Y},\mathbf{X}) \tag{8.9.8}$$

$$\propto \frac{1}{\{g_{11} + c^2 g_{ii} + [(\boldsymbol{\theta}_1 - c\boldsymbol{\theta}_i) - (\hat{\boldsymbol{\theta}}_1 - c\hat{\boldsymbol{\theta}}_i)]'\mathbf{V}^{-1}[(\boldsymbol{\theta}_1 - c\boldsymbol{\theta}_i) - (\hat{\boldsymbol{\theta}}_1 - c\hat{\boldsymbol{\theta}}_i)]\}^{(N-q+1)/2}},$$

where $(\boldsymbol{\theta}_1 - c\boldsymbol{\theta}_i)'$ is the first row of $\mathbf{B}$, $G \equiv (g_{ij}) \equiv (\mathbf{X}'\mathbf{X})^{-1}$, and $\mathbf{V} \equiv (v_{ij})$, $i,j = 1, \ldots, p$. Thus, (8.9.8) may be used to draw inferences concerning the difference between the first and ith mean vectors (by taking $c \equiv 1$).

To find the posterior distribution of $\mathbf{a}'(\boldsymbol{\theta}_1 - c\boldsymbol{\theta}_i)$ for known $\mathbf{a}$ and c, the distribution of a linear combination[10] of the elements of a vector that follows a multivariate Student t-distribution is required. The required result is given in Theorem (6.2.1). Accordingly, let $w \equiv \mathbf{a}'(\boldsymbol{\theta}_1 - c\boldsymbol{\theta}_i)$, and

[10] Such a linear combination will make sense only when it involves components measured in the same units.

$\hat{w} \equiv \mathbf{a}'(\hat{\boldsymbol{\theta}}_1 - c\hat{\boldsymbol{\theta}}_i)$. Then the scalar w follows the univariate Student t-density

$$p(w|\mathbf{Y},\mathbf{X}) \propto \frac{1}{\{\nu + \gamma(w - \hat{w})^2\}^{(\nu+1)/2}}, \tag{8.9.9}$$

where

$$\nu = N - p - q + 1, \qquad \gamma = \frac{\nu}{(g_{11} + c^2 g_{ii})(\mathbf{a}'\mathbf{V}\mathbf{a})},$$

that is,

$$\mathcal{L}\{\gamma^{1/2}(w - \hat{w})\,|\,\mathbf{Y},\mathbf{X}\} = t_\nu$$

For example, if it is desired to compare the first components of $\boldsymbol{\theta}_1$ and $\boldsymbol{\theta}_2$ ($i = 2$), take $\mathbf{a} = (1,0,\ldots,0)'$. Then $w = \theta_{11} - c\theta_{12}$, where θ_{11} and θ_{12} are the first elements of $\boldsymbol{\theta}_1$ and $\boldsymbol{\theta}_2$, respectively. Suppose the sample data suggest that $\theta_{11} = 2\theta_{12}$. Such a relation can be examined by letting $c = 2$.

Results analogous to (8.9.9) may be derived in an obvious way for a comparison of the jth mean $\boldsymbol{\theta}_j$ with the ith mean $\boldsymbol{\theta}_i$ for any $i, j, i \neq j$. Credibility intervals are established using significance points of the distribution in (8.9.9).

A more extensive discussion of this topic is given in Press (1980a).

8.9.2 Simultaneous Confidence Intervals in MANOVA

Confidence intervals associated with the maximum characteristic root criterion have been established for the means in MANOVA by Roy and Bose (1953), and by Roy (1957). Another approach which has been developed is the multivariate extension of the univariate Scheffé procedure for multiple comparisons. However, the former approach is the one discussed below because of its relation to the maximum root method of hypothesis testing (see Section 8.7.3) and the attendant power considerations.

Simultaneous Confidence Intervals in One-Way Layout MANOVA

Suppose $\mathbf{a}$ is a given $p \times 1$ vector of weights, ϵ is a desired level of significance, and the model given by (8.7.5) applies. Then a confidence interval for simple comparisons between the αth and $(\alpha + 1)$th means, at confidence level $(1 - \epsilon)$, is given by

$$1 - \epsilon = P\{\mathbf{a}'(\bar{\mathbf{z}}_\alpha - \bar{\mathbf{z}}_{\alpha+1}) - h \le \mathbf{a}'(\boldsymbol{\theta}_\alpha - \boldsymbol{\theta}_{\alpha+1}) \le \mathbf{a}'(\bar{\mathbf{z}}_\alpha - \bar{\mathbf{z}}_{\alpha+1}) + h\}, \tag{8.9.10}$$

where

$$h = \left\{\frac{\phi_\epsilon}{1 - \phi_\epsilon}\,\mathbf{a}'(N\hat{\boldsymbol{\Sigma}}_1)\mathbf{a}\left(\frac{1}{T_\alpha} + \frac{1}{T_{\alpha+1}}\right)\right\}^{1/2},$$

$N\hat{\boldsymbol{\Sigma}}_1$ is defined in (8.7.8), and ϕ_ϵ is the 100ϵ percent point on the charts established by Heck (1960) (see Appendix B). Note that the Heck charts are indexed by three parameters, designated as (s,m,n). For the one-way layout MANOVA, the parameters are given by

$$s \equiv \min(q-1, p), \qquad 2m \equiv |q-p-1| - 1,$$

and $2n = N - q - p - 1$. Also note that (8.9.10) may be applied to all possible linear comparisons among the means simultaneously, at confidence level $1 - \epsilon$.

Example (8.9.1): Suppose a 95 percent confidence interval is desired ($\epsilon = .05$), and $q = 3$, $p = 2$, $T_\alpha = T_{\alpha+1} = 20$, $\bar{z}_{1,\alpha} - \bar{z}_{1,\alpha+1} = 1$, for $\alpha = 1$; that is, the difference between the first components of $\boldsymbol{\theta}_1$ and $\boldsymbol{\theta}_2$ is of interest. The total number of observations is $N = \Sigma_1^q T_\alpha = 60$. The Heck chart parameter is

$$\phi_\epsilon(s,m,n) \equiv \phi_{.05}(2, -\tfrac{1}{2}, 27).$$

Choose $\mathbf{a} = (1,0)'$. Suppose the 1–1 element of $N\hat{\boldsymbol{\Sigma}}_1$ is unity. Then, since reference to the Heck chart shows that

$$\phi_{.05}(2, -\tfrac{1}{2}, 27) = .143,$$

$h = .129$. Hence, from (8.9.10),

$$.95 = P\{1 - .129 \le \theta_{1,1} - \theta_{1,2} \le 1 + .129\},$$

or

$$.95 = P\{.871 \le \theta_{1,1} - \theta_{1,2} \le 1.129\}.$$

Simultaneous Confidence Intervals in Two-Way Layout MANOVA

Suppose there are K observations per cell, but there is no interaction term associated with (8.7.15). Then

$$\underset{(p\times 1)}{\mathbf{y}_{ijk}} = \boldsymbol{\mu} + \boldsymbol{\alpha}_i + \boldsymbol{\beta}_j + \underset{(p\times 1)}{\mathbf{e}_{ijk}},$$

for $i = 1,\ldots,I$; $j = 1,\ldots,J$; $k = 1,\ldots,K$. Then, simultaneously, for all possible comparisons between row effects $\boldsymbol{\alpha}_\gamma$ and $\boldsymbol{\alpha}_\delta$, at confidence level $1 - \epsilon$,

$$\mathbf{a}'(\mathbf{y}_{\gamma\cdot\cdot} - \mathbf{y}_{\delta\cdot\cdot}) - r_I \le \mathbf{a}'(\boldsymbol{\alpha}_\gamma - \boldsymbol{\alpha}_\delta) \le \mathbf{a}'(\mathbf{y}_{\gamma\cdot\cdot} - \mathbf{y}_{\delta\cdot\cdot}) + r_I, \quad (8.9.11)$$

where

$$\mathbf{y}_{\gamma\cdot\cdot} = \frac{1}{K}\sum_{k=1}^{K} \mathbf{y}_{\gamma\cdot k} = \frac{1}{JK}\sum_{k=1}^{K}\sum_{j=1}^{J} \mathbf{y}_{\gamma jk},$$

a is an arbitrary fixed $p \times 1$ vector,

$$r_I = \left\{ \frac{2\phi_\epsilon \mathbf{a}'\mathbf{Q}\mathbf{a}}{(1 - \phi_\epsilon)JK} \right\}^{1/2},$$

$$IJ(K-1)\mathbf{Q} = \sum_{k=1}^{K} \sum_{j=1}^{J} \sum_{i=1}^{I} \mathbf{y}_{ijk}\mathbf{y}_{ijk}' - K \sum_{j=1}^{J} \sum_{i=1}^{I} \mathbf{y}_{ij\cdot}\mathbf{y}_{ij\cdot}',$$

$$\mathbf{y}_{ij\cdot} = \frac{1}{K} \sum_{k=1}^{K} \mathbf{y}_{ijk},$$

and ϕ_ϵ is the 100ϵ percent point on the Heck charts for $\phi_\epsilon(s,m,n)$, where $s = \min(I - 1, p)$, $2m = |I - 1 - p| - 1$, and

$$2n = IJ(K - 1) - p - 1.$$

Analogous to (8.9.11) for all possible comparisons between column effects, it can be asserted that

$$\mathbf{a}'(\mathbf{y}_{\cdot\gamma\cdot} - \mathbf{y}_{\cdot\delta\cdot}) - r_J \leq \mathbf{a}'(\boldsymbol{\beta}_\gamma - \boldsymbol{\beta}_\delta) \leq \mathbf{a}'(\mathbf{y}_{\cdot\gamma\cdot} - \mathbf{y}_{\cdot\delta\cdot}) + r_J, \quad (8.9.12)$$

where

$$r_J = \left\{ \frac{2\phi_\epsilon \mathbf{a}'\mathbf{Q}\mathbf{a}}{(1 - \phi_\epsilon)IK} \right\}^{1/2},$$

and $\phi_\epsilon(s,m,n)$ has for its parameters $s \equiv \min(J - 1, p)$, $2m = |J - 1 - p| - 1$, and $2n = IJ(K - 1) - p - 1$.

8.10 CHECKING AND VIOLATING THE ASSUMPTIONS IN THE GENERAL LINEAR MODEL (Including Discrete Dependent Variable Regression and Log-linear Models)

It is of paramount importance to the user of the techniques of multivariate analysis to know whether the models used are sensitive to violations of the assumptions of the model; that is, are the associated tests of hypotheses and interval estimates (including credibility intervals) *robust?* Unfortunately, there has not yet been very much study of this question for the case of the multivariate general linear model although some preliminary results have been obtained.

For example, Ito (1968) has shown that when sample sizes are very large, the effect of violation of the multivariate Normality assumption is slight on testing hypotheses about the mean vectors but can be serious for testing covariance matrices [this is the analogue of the result obtained by Scheffé (1959), Chapter 10, in the univariate case].

When it is determined that one or more of the assumptions of the model is not valid, methods of correcting for these violations must be established. Tests and corrections for such violations are discussed below.

8.10.1 Checking for Normality

A quick and frequently adequate[11] way of checking whether or not the components of the disturbance vectors follow univariate Normal distributions is to plot the residuals of the regression (or ANOVA) on Normal probability paper. Let $x_1, \ldots, x_n$ denote a general sample of independent and identically distributed univariate observations, and order the observations so that $x_{(1)} \leq x_{(2)} \leq \cdots \leq x_{(n)}$; that is, $x_{(j)}$ denotes the value of the jth smallest observation. Define $y_j = j/(n+1)$; y_j is an unbiased estimator of the cdf corresponding to $x_{(j)}$. The points $(x_{(j)}, y_j)$, $j = 1 \ldots n$ are now plotted on Normal probability paper. In the case of regression, these x_j's correspond to the regression residuals (in large samples). If the residuals fall along a straight line (approximately), the assumption of Normality is roughly acceptable. If the plot tends to "tail off" toward the horizontal (see Figure 8.10.1) at one or both extremities (even though there might be a pretty good straight line fit in the central portion of the plot) the underlying distributions are "fat-tailed" (relative to the Normal) and might be stable non-Normal (see Section 6.5). If the plot tends to tail off toward the vertical (Figure 8.10.2), the underlying distributions are "thin-tailed." In either case, Normality is not justified and some type of "correction" must be imposed.

A test for multivariate normality was proposed by Andrews et al. (1973). This test generalized the Box-Cox transformation (see Box, Cox, 1964) to the multivariate case.

8.10.2 Univariate Heteroscedasticity

In other situations the disturbance vectors are thought to have components with unequal variances (heteroscedasticity), another violation of a basic assumption generally made in the general linear model. Again some type of "correction" is necessary.

In the case of non-Normality, if the actual underlying distribution can be established, it may be possible to make inferences based upon the

[11] In the standard linear regression model (univariate) with $\mathbf{y} = \mathbf{X}\boldsymbol{\beta} + \mathbf{e}$, and $\hat{\mathbf{e}}$ denoting the residual vector, it is easy to check that $\hat{\mathbf{e}} = \mathbf{A}\mathbf{e}$, where $\mathbf{A}$ denotes the idempotent matrix $\mathbf{I}_n - \mathbf{X}(\mathbf{X}'\mathbf{X})^{-1}\mathbf{X}'$. As a result, with $\mathbf{X}$ fixed, if $\mathcal{L}(\mathbf{e}) = N(\mathbf{0}, \sigma^2\mathbf{I}_n)$, $\mathcal{L}(\hat{\mathbf{e}}) = N(\mathbf{0}, \sigma^2\mathbf{A})$, and so the variances of the $\hat{e}_\alpha$'s are different in general. Because their distributions are different it is not meaningful to plot the $\hat{e}_\alpha$'s on Normal paper when n is small. Nevertheless, because $\hat{e}_\alpha$ converges to e_α in probability [see remark following (C8.1.2)], its distribution also converges to that of e_α, which is the same for all α. Moreover, although the $\hat{e}_\alpha$'s are correlated, when the e_α's are uncorrelated, the $\hat{e}_\alpha$'s become less correlated as the sample size increases. Hence, in large samples, a plot of the residuals on Normal probability paper should provide a good check on the Normality of the disturbances.

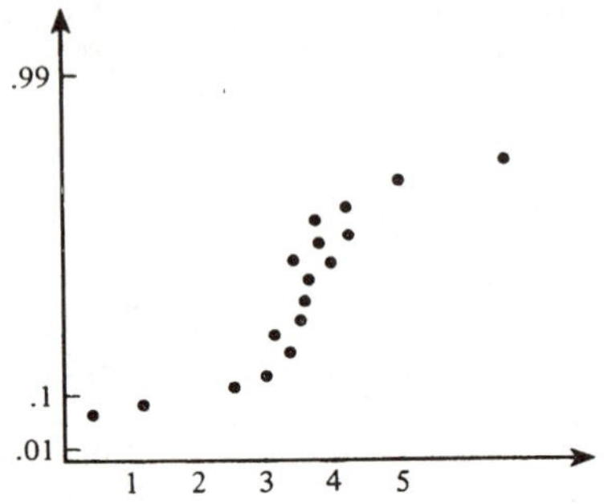

Figure 8.10.1. Normal Probability Paper Plot of Fat Tail Distribution

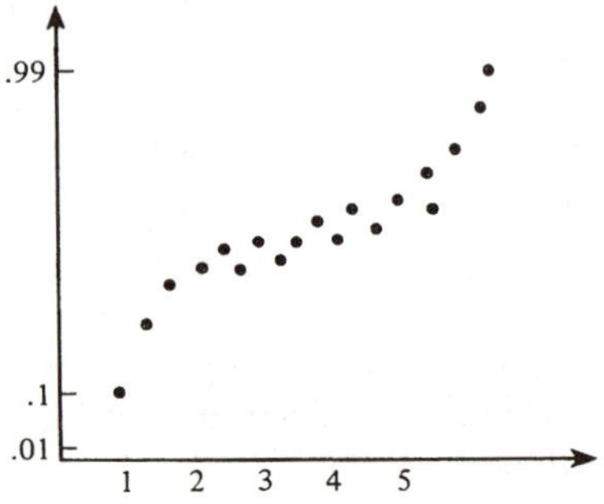

Figure 8.10.2. Normal Probability Paper Plot of Thin Tail Distribution

actual distribution. In the case of heteroscedasticity, if a simple relationship can be established among the variances of the components (such as var $(e_\alpha) \equiv \sigma_\alpha^2 = kx_\alpha$, where k is a constant and x_α is the αth observation of the independent variable in a simple linear regression), generalized least squares can be used and the problem can be obviated [see Example (8.3.2)]. Often, however, non-Normality, or heteroscedasticity, or both, are not resolvable as above. In such cases, *transformations of the data* might be carried out. In practice it often happens that both of the above problems will thereby be resolved simultaneously.

8.10.3 Transformations

For example, suppose that in a one-way layout univariate analysis of variance with q groups, the dependent variables, y_{ij}, $i = 1,\ldots,q$, $j = 1,\ldots,n$, are all proportions (numbers between zero and unity) following binomial distributions. If the means of the groups, θ_i, are different, the variances of the observations for the groups, $\theta_i(1 - \theta_i)$, are also different; that is, since the group variances are functions of the group means, unequal means implies heteroscedasticity. This dependence of the variance on the mean can be eliminated (this is called variance stabilization) by means of a transformation.

The appropriate variance stabilizing transformation is a function, f, such that, in large samples, if $z_{ij} = f(y_{ij})$, where y_{ij} is the original dependent variable, $z_{ij} = f(\theta_i) + e_{ij}$, where e_{ij} is an error term whose variance is a numerical constant for all i and j. Since in practice, the standardized asymptotic distribution is usually Normal, by working with large samples and by adjusting the asymptotic variance appropriately, non-Normality and heteroscedasticity can be resolved simultaneously (see Section 4.5 for the asymptotic distribution of a function).

In the general case, it is desired to find a function $f(\theta)$ which will make the asymptotic variance in (4.5.3) equal to a constant. That is, reference to (4.5.3) with $p = 1$ shows that if $\Sigma \equiv \sigma^2(\theta)$,

$$\mathbf{\Phi}'\mathbf{\Sigma}\mathbf{\Phi} \equiv \left(\frac{df(\theta)}{d\theta}\right)^2 \sigma^2(\theta),$$

and if this expression is to be constant, it is necessary that for some constant c (it may be taken to be any convenient value),

$$f(\theta) = \int \frac{c\,d\theta}{\sigma(\theta)}.$$

To illustrate, consider the case of the frequently occurring binomial distribution, cited above. There, $\sigma(\theta) = [\theta(1 - \theta)]^{1/2}$, and therefore, for c chosen appropriately, $f(\theta) = \arcsin \sqrt{\theta}$; also, $z_{ij} = \arcsin \sqrt{y_{ij}}$.

In the multivariate case, if each observation has p components but the components are uncorrelated, at least asymptotically, the $f(\theta)$ transformation may be applied to each component separately. If, however, the components are correlated, the optimal transformation procedure for stabilizing the asymptotic covariance matrix is not known. In this case, if there is reason to suspect that inferences will not be robust with respect to departures from multivariate Normality, large samples must be used with data that is aggregated and standardized so that Normality will be assured (Central Limit Theorem). An added complexity is that if sample sizes are small or even moderately large, it is often not known whether inference procedures are robust.

8.10.4 Discrete Dependent Variable Regression

If large samples are available and the dependent variable is discrete, a transformation of the dependent variable followed by conventional least squares estimation, as described in the last section, is a reasonable solution to the regression problem. In this section it is shown that when samples are small or moderately large and when the dependent variable can assume one of only several possible values, conventional estimation techniques are inappropriate, but alternatives are available.

For example, suppose n people are asked whether or not they favor integrated housing. Their responses are classifiable as binary, or dichotomous (excluding such categories as "no response," or "no opinion"). If it is desired to relate people's responses to their socioeconomic and demographic characteristics, a discrete dependent variable regression technique, as described below, may be used.

Attention will first be directed to the case in which the dependent variable is dichotomous. The analysis will then be extended to the case of dependent variables which can assume one of several (more than two) possible values. The multivariate case of several polytomous dependent variables is discussed in Sect. 8.10.5.

The theory of discrete dependent variable regression has its genesis in bioassay (see, for example, Finney, 1952). In that context, let y_i denote the ith dependent variable (a response for the ith animal injected with a drug); $i = 1,\ldots,n$. Let $y_i = a_2$ only if the dosage of the drug administered to the ith animal exceeds a certain (random) threshold level for that animal. Otherwise, take $y_i = a_1$, $a_1 < a_2$ (a_1 and a_2 are often taken equal to zero and one, respectively). Thus, if z_i denotes the threshold response level for animal i (the z_i's are assumed independently distributed), and if $\mathbf{x}_i'\boldsymbol{\beta}$ denotes the dosage level for that animal, where $\mathbf{x}_i$ is a vector of say q fixed regressor variables, and $\boldsymbol{\beta}$ is a vector of unknown coefficients, $y_i = a_2$ if $\mathbf{x}_i'\boldsymbol{\beta} \geq z_i$, and $y_i = a_1$ if $\mathbf{x}_i'\boldsymbol{\beta} < z_i$. It will also be necessary to relate y_i to z_i probabilistically.

Suppose first that y_i were regressed on the components of $\mathbf{x}_i$ in the conventional way. Then $y_i = \mathbf{x}_i'\boldsymbol{\beta} + u_i$; $i = 1,\ldots, n$, where u_i denotes the ith uncorrelated disturbance term and $Eu_i = 0$. But

$$\text{var}\,(y_i) = \text{var}\,(u_i) = \mathbf{x}_i'\boldsymbol{\beta}(1 - \mathbf{x}_i'\boldsymbol{\beta}),$$

since y_i is a Bernoulli random variable. Since var (u_i) depends on i, the resulting heteroscedasticity will generate inefficient estimators of $\boldsymbol{\beta}$. Another problem inherent in this model (ordinary least squares estimation) is that estimators of $\mathbf{x}_i'\boldsymbol{\beta}$ can have *any* numerical value, in spite of the fact that since $Ey_i = \mathbf{x}_i'\boldsymbol{\beta}$ and $a_1 \leq y_i \leq a_2$, $a_1 \leq \mathbf{x}_i'\boldsymbol{\beta} \leq a_2$. That is, ordinary least squares estimators of $\mathbf{x}_i'\boldsymbol{\beta}$ are often unreasonable. A possible alternative is the use of restricted least squares estimation in which $\boldsymbol{\beta}$ is estimated by ordinary least squares, subject to the constraint that $a_1 \leq \mathbf{x}_i'\boldsymbol{\beta} \leq a_2$. However, when the heteroscedasticity difficulty is accounted for (by generalized least squares) the resulting programming problem becomes especially complex.

In the bioassay type formulation of the problem the inequality constraint encountered above is obviated, but at the cost of assuming a probabilistic functional form between y_i and z_i. There, let $F(t)$ denote the cdf of z_i, given $\mathbf{x}_i$. Then

$$P\{y_i = a_2\} = P\{z_i \leq \mathbf{x}_i'\boldsymbol{\beta}|\mathbf{x}_i\} = F(\mathbf{x}_i'\boldsymbol{\beta}).$$

Also,

$$P\{y_i = a_1\} = 1 - F(\mathbf{x}_i'\boldsymbol{\beta}).$$

To estimate $\boldsymbol{\beta}$ it will be necessary to specify F. Note that if $a_1 = 0$ and $a_2 = 1$, the model may be written as

$$y_i = F(\mathbf{x}_i'\boldsymbol{\beta}) + u_i, \qquad i = 1,\ldots,n,$$

where the u_i's are uncorrelated and $E(u_i) = 0$. Since F is a cdf, F is monotone nondecreasing, $0 \leq F \leq 1$, and the inequality constraint is satisfied automatically.

The parameter vector $\boldsymbol{\beta}$ will be estimated by maximum likelihood. The likelihood function is given by

$$L = \prod_{i=1}^{n} [F(\mathbf{x}_i'\boldsymbol{\beta})]^{y_i}[1 - F(\mathbf{x}_i'\boldsymbol{\beta})]^{1-y_i}.$$

When $F(t)$ is taken to be the standard Normal cdf, the method is called *probit analysis*. The estimation by maximum likelihood is straightforward but computationally difficult since $F(t)$ is then an integral and the simultaneous equations which result from differentiation of $\log L$ are then extremely difficult to solve. A more prudent alternative is to select a cdf which has a simple algebraic form. For this purpose the logistic distribution suggested by Berkson (1944), with cdf $F(t) = (1 + e^{-t})^{-1}$, $-\infty < t < \infty$, serves admirably well. Moreover, it is easily checked that the Normal and logistic cdf's are numerically quite close to one another, except in their extreme tails (see Cox, 1970, p. 28 for a tabular comparison). Thus, unless interest must focus on the extreme tails, there is no reason to prefer the Normal over the logistic functional form.

An added advantage of the logistic distribution is that sufficient statistics are available (it is easy to check that if F denoted the Normal cdf in the above, there would be no summarization of the data in terms of sufficient statistics). Thus, when $F(t)$ denotes the logistic cdf, the likelihood function becomes

$$L = \prod_{i=1}^{n} \left[\frac{1}{1 + \exp(-\mathbf{x}_i'\boldsymbol{\beta})}\right]^{y_i} \left[1 - \frac{1}{1 + \exp(-\mathbf{x}_i'\boldsymbol{\beta})}\right]^{1-y_i}$$

$$= \frac{\left\{\exp \boldsymbol{\beta}' \sum_{i=1}^{n} \mathbf{x}_i y_i\right\}}{\prod_{i=1}^{n} \{1 + \exp(\mathbf{x}_i'\boldsymbol{\beta})\}}.$$

Hence, $\Sigma_1^n \mathbf{x}_i y_i \equiv \mathbf{t}^*$ is sufficient for $\boldsymbol{\beta}$, for given $\mathbf{x}_1,\ldots,\mathbf{x}_n$. Note that for $a_2 = 1$, $\mathbf{t}^*$ is merely the sum of those independent variable vectors for which a response was obtained ($y_i = a_2$).

A maximum likelihood estimator of $\boldsymbol{\beta}$, $\hat{\boldsymbol{\beta}}$, is obtained by differentiating $\log L$, setting the result equal to zero, and solving for $\boldsymbol{\beta}$. Since

$$L^* \equiv \log L = \boldsymbol{\beta}'\mathbf{t}^* - \sum_{i=1}^{n} \log\,[1 + \exp\,(\mathbf{x}_i'\boldsymbol{\beta})],$$

$\boldsymbol{\beta}$ must satisfy the equation

$$\sum_{i=1}^{n} \frac{1}{1 + \exp\,(-\mathbf{x}_i'\boldsymbol{\beta})}\,\mathbf{x}_i = \mathbf{t}^*.$$

This vector equation must be solved for $\boldsymbol{\beta}$ numerically, but the problem is simpler than the one of probit analysis (which involves Normal integrals). Solving regression problems in this manner using the logistic cdf is sometimes referred to as *logit analysis*. (It is easy to check that there is a unique solution since $\boldsymbol{\beta}$ is a q-vector and there are q implied equations to solve.)

It is sometimes of interest in the logit model to predict the value of the dependent variable corresponding to a new vector of independent variables. To see how this prediction is effected, suppose that on the basis of observing $(\mathbf{x}_1,y_1),\ldots,(\mathbf{x}_n,y_n)$, the logit model is used to determine $\hat{\boldsymbol{\beta}}$, and let $\mathbf{x}^*$ denote a new vector of independent variables. If y^* corresponds to $\mathbf{x}^*$, its value is predicted as

$$\hat{y}^* = a_1[1 - F(\hat{\boldsymbol{\beta}}'\mathbf{x}^*)] + a_2[F(\hat{\boldsymbol{\beta}}'\mathbf{x}^*)],$$

or if $a_1 = 0$ and $a_2 = 1$,

$$\hat{y}^* = F(\hat{\boldsymbol{\beta}}'\mathbf{x}^*).$$

The logit regression method is often used to classify observations in the classification and discrimination problem (see Chap. 13). For a discussion of this use of the model, see Press and Wilson (Dec. 1978).

In the foregoing, the only regressions considered were those for which the dependent variable is discrete with only two possible values. More generally, suppose the dependent variable for the ith subject, y_i, may have p possible values, $a_1,\ldots,a_p$. The logit model described above may be extended to cover this polychotomous case by building on some work of Mantel (1966).

Define $p_{ij} \equiv P\{y_i = a_j\}$, $i = 1,\ldots, n$, $j = 1,\ldots, p$, and take

$$p_{ij} = \frac{\exp\,(z_{ij})}{\sum_{k=1}^{p} \exp\,(z_{ik})},$$

where $\Sigma_1^p p_{ij} = 1$, $-\infty < z_{ij} < \infty$. Take

$$z_{ij} = \mathbf{x}_i'\boldsymbol{\gamma}_j \equiv \sum_{k=1}^{q} x_{ik}\gamma_{kj},$$

so that if $\mathbf{Z} = (z_{ij})$ is an $n \times p$ matrix, $\mathbf{X} = (x_{ik})$ is an $n \times q$ matrix, and $\mathbf{\Gamma} = (\gamma_{kj})$ is a $q \times p$ matrix, $\mathbf{Z} = \mathbf{X\Gamma}$. The $\mathbf{X}$ matrix is assumed to consist of elements which are observed values of independent regressor variables (and possibly, a row of ones). The $\mathbf{\Gamma}$ matrix contains the unknown coefficients to be estimated. An equivalent representation is

$$p_{ij} = \frac{1}{1 + \sum_{k=1}^{p}{}^{*} \exp\left[\mathbf{x}_i'(\boldsymbol{\gamma}_k - \boldsymbol{\gamma}_j)\right]}$$

or

$$p_{ij} = F\left(-\log\left[\sum_{k=1}^{p}{}^{*} \exp\,\{\mathbf{x}_i'(\boldsymbol{\gamma}_k - \boldsymbol{\gamma}_j)\}\right]\right),$$

where $F(t) = [1 + \exp(-t)]^{-1}$ denotes the logistic cdf, and the asterisk implies that $k \neq j$ in the summation.

In the special case of $p = 2$,

$$p_{i1} = \frac{\exp(z_{i1})}{\exp(z_{i1}) + \exp(z_{i2})}, \qquad p_{i2} = \frac{\exp(z_{i2})}{\exp(z_{i1}) + \exp(z_{i2})},$$

so that

$$p_{i1} = \frac{\exp(\mathbf{x}_i'\boldsymbol{\gamma}_1)}{\exp(\mathbf{x}_i'\boldsymbol{\gamma}_1) + \exp(\mathbf{x}_i'\boldsymbol{\gamma}_2)}, \qquad p_{i2} = \frac{\exp(\mathbf{x}_i'\boldsymbol{\gamma}_2)}{\exp(\mathbf{x}_i'\boldsymbol{\gamma}_1) + \exp(\mathbf{x}_i'\boldsymbol{\gamma}_2)}.$$

Now note that by using the parameterization $\boldsymbol{\beta} = 2\boldsymbol{\gamma}_1 = -2\boldsymbol{\gamma}_2$, the problem reduces to the case of regression with dichotomous dependent variables, discussed earlier.

The polytomous case of $y_i = a_1, \ldots, a_p$, with the logit model assumption, can be characterized in terms of sufficient statistics, just as in the dichotomous logit model case. The parameter matrix $\mathbf{\Gamma}$ is estimated below using the method of constrained maximum likelihood.

Let $v_{ij} = 1$ if $y_i = a_j$, and let $v_{ij} = 0$ otherwise. Then the likelihood function is given by

$$L = \prod_{i=1}^{n} p_{i1}^{v_{i1}} p_{i2}^{v_{i2}} \ldots p_{ip}^{v_{ip}},$$

where $\Sigma_1^p p_{ij} = 1$ and $\Sigma_1^p v_{ij} = 1$. Substituting the values of p_{ij} from above, and taking logs, gives

$$L^* \equiv \log L = \sum_{i=1}^{n}\sum_{j=1}^{p} v_{ij} \log \left\{ \frac{\exp(\mathbf{x}_i'\boldsymbol{\gamma}_j)}{\sum_{k=1}^{p} \exp(\mathbf{x}_i'\boldsymbol{\gamma}_k)} \right\}$$

$$= \sum_{i=1}^{n}\sum_{j=1}^{p} v_{ij}\mathbf{x}_i'\boldsymbol{\gamma}_j - \sum_{i=1}^{n}\sum_{j=1}^{p} v_{ij} \log \left[\sum_{k=1}^{p} \exp(\mathbf{x}_i'\boldsymbol{\gamma}_k) \right]$$

$$= \sum_{j=1}^{p} \mathbf{t}_j'\boldsymbol{\gamma}_j - \sum_{i=1}^{n} \log \left[\sum_{k=1}^{p} \exp(\mathbf{x}_i'\boldsymbol{\gamma}_k) \right],$$

where $\mathbf{t}_j \equiv \Sigma_1^n \mathbf{x}_i v_{ij}$ and $(\mathbf{t}_1,\ldots,\mathbf{t}_p)$ is sufficient for $\boldsymbol{\Gamma}$ (for given $\mathbf{X}$). There is a logit regression relationship to be estimated for each response category, but since the categories are exhaustive, the regressions must be related. Thus, the regressor variables are to be weighted by $\boldsymbol{\gamma}_j$ when estimating the likelihood of a response in category j, $j = 1,\ldots,p$. However, because a response not found in any of $p - 1$ categories must lie in the remaining category, the coefficient vectors must be constrained (with the result that their maximum likelihood estimators will be unique). It is simple to make this formulation of the model represent an extension of the dichotomous case by imposing the condition that $\Sigma_1^p \boldsymbol{\gamma}_j = \mathbf{0}$ (for $p = 2$, the model required that $\boldsymbol{\gamma}_1 = -\boldsymbol{\gamma}_2$). Thus, L^* is to be maximized with respect to $\boldsymbol{\Gamma} = (\boldsymbol{\gamma}_1,\ldots,\boldsymbol{\gamma}_p)$, subject to the constraint that $\Sigma_1^p \boldsymbol{\gamma}_j = \mathbf{0}$. Denote the Lagrangian by

$$\mathcal{L} = \sum_{j=1}^{p} \mathbf{t}_j'\boldsymbol{\gamma}_j - \sum_{i=1}^{n} \log \left[\sum_{k=1}^{p} \exp(\mathbf{x}_i'\boldsymbol{\gamma}_k) \right] - \boldsymbol{\lambda}' \sum_{j=1}^{p} \boldsymbol{\gamma}_j,$$

where $\boldsymbol{\lambda}$ denotes a q-vector of Lagrangian multipliers. Differentiating with respect to $\boldsymbol{\gamma}_l$ and setting the result equal to zero gives

$$\boldsymbol{\lambda} = \mathbf{t}_l - \sum_{i=1}^{n} \left[\frac{\exp(\mathbf{x}_i'\hat{\boldsymbol{\gamma}}_l)}{\sum_{k=1}^{n} \exp(\mathbf{x}_i'\hat{\boldsymbol{\gamma}}_k)} \right] \mathbf{x}_i,$$

or

$$\boldsymbol{\lambda} = \mathbf{t}_l - \sum_{i=1}^{n} \hat{p}_{il}\mathbf{x}_i, \qquad l = 1, 2,\ldots,p,$$

where the carets denote maximum likelihood estimators. Summing this equation over l gives

$$\boldsymbol{\lambda} p = \sum_{l=1}^{p} \mathbf{t}_l - \sum_{l=1}^{p}\sum_{i=1}^{n} \hat{p}_{il}\mathbf{x}_i.$$

Since $\Sigma_1^p \hat{p}_{il} = 1$, $\boldsymbol{\lambda} p = \Sigma_1^p \mathbf{t}_l - \Sigma_1^n \mathbf{x}_i$. But by definition,

$$\sum_{l=1}^{p} \mathbf{t}_l = \sum_{l=1}^{p} \sum_{i=1}^{n} \mathbf{x}_i v_{il} = \sum_{i=1}^{n} \mathbf{x}_i.$$

Hence, $\boldsymbol{\lambda} = \mathbf{0}$. Thus, the maximum likelihood estimators of the columns of $\mathbf{\Gamma}$ must satisfy

$$\sum_{i=1}^{n} \left[\frac{\exp(\mathbf{x}_i' \hat{\boldsymbol{\gamma}}_l)}{\sum_{k=1}^{p} \exp(\mathbf{x}_i' \hat{\boldsymbol{\gamma}}_k)} \right] \mathbf{x}_i = \sum_{i=1}^{n} \mathbf{x}_i v_{il}, \tag{1}$$

for $l = 1, 2, \ldots, p$, and

$$\sum_{j=1}^{p} \hat{\boldsymbol{\gamma}}_j = \mathbf{0}. \tag{2}$$

(It is straightforward to check that if any constraint of the form $\Sigma_1^p \delta_j \boldsymbol{\gamma}_j = \boldsymbol{\theta}$ is used, with $\Sigma_1^p \delta_j \neq 0$, $\delta_1, \ldots, \delta_p$ fixed scalars, and $\boldsymbol{\theta}$ a fixed vector, it would turn out that $\boldsymbol{\lambda} = \mathbf{0}$, and the same final result would be obtained.) These equations yield a maximum value, a fact which may be checked by evaluating $\partial^2 L^* / \partial \boldsymbol{\gamma}_l \, \partial \boldsymbol{\gamma}_l'$ and noting that the resulting Hessian matrix is negative definite. Equations (1) and (2) above may be solved by standard Newton-Raphson methods of successive iteration.

Note that if it is desired to predict the value of a new dependent variable, y^*, corresponding to a new $\mathbf{x}^*$, it is reasonable to use the predictor (of Ey^*)

$$\hat{y}^* = \sum_{j=1}^{p} \frac{a_j \exp(\hat{\boldsymbol{\gamma}}_j' \mathbf{x}^*)}{\sum_{k=1}^{p} \exp(\hat{\boldsymbol{\gamma}}_k' \mathbf{x}^*)}.$$

As an illustration of the case of univariate regression with $y_i = 0$ or 1, consider the following problem. A college entrance administrator wants to determine the relationship between the amount of money offered a potential student in scholarship and fellowship stipends, and the likelihood of the student's accepting the money and coming to the college. The administrator plans to examine n cases in which money was offered and to determine whether or not the person accepted. Let $y_i = 1$ if the ith awardee accepted the offer and let $y_i = 0$ if he declined. Let s_i denote the dollar amount of his offer; then y_i may now be related to s_i by means of the logit analysis model above, with $q = 2$ (to permit a constant term). That is, logit analysis is being used to relate y_i to $\beta_1 + \beta_2 s_i$. If $\boldsymbol{\beta}' = (\beta_1, \beta_2)$, and $\mathbf{x}_i' = (1, s_i)$, $\beta_1 + \beta_2 s_i = \mathbf{x}_i' \boldsymbol{\beta}$, $i = 1, \ldots, n$.

In another instance, participants in a government sponsored job training program are evaluated to determine the merits and degree of success of the program. It is of interest, for example, to determine whether participants have experienced a "positive" change in attitude while going through the program, and it is desired to predict which participants will experience such a change. Let n represent the number of graduates of the program who are studied. Let $y_i = 1$ if the ith participant experienced a positive attitude change (as indicated by counsellor's ratings), and $y_i = 0$ if he did not. Let $\mathbf{x}_i$ denote a vector of independent variables for the ith graduate in the group. These variables might include age, race or ethnic background, family income, or other socioeconomic or demographic characteristics of the individual. Logit analysis may again be used to regress y_i on $\mathbf{x}_i'\boldsymbol{\beta}$. Then, for a new individual entering the program who has a given vector of personal characteristics, $\mathbf{x}^*$, the estimated value of y^*, $\hat{y}^*$, for this person is $F(\hat{\boldsymbol{\beta}}'\mathbf{x}^*)$, where F denotes the cdf of the logistic distribution. If $\hat{y}^*$ is "close" to unity, the individual is likely to experience a positive attitude change in the program; otherwise, he is not.

For some purposes, it may be of interest to divide the program participants of the last example into those who experienced a "negative" change in attitude ($y_i = -1$), those who experienced no change ($y_i = 0$), and those who experienced a positive change ($y_i = +1$). The relationship between y and the regressor variables for this case may be estimated by the polytomous logit regression model set forth above. Note that the categories have been assumed to be unordered. If in fact they are ordered but the ordering is ignored, the same procedure may be used but inefficient estimators may result. The ordering may be accounted for by using the same model form and imposing constraints on the parameters to conform with the ordering. Such an ordered model is included within the unordered models described above.

8.10.5 Log-Linear Models and Analysis of Discrete Data

The polytomous response models described in Sect. 8.10.4 have been extended to the multivariate case of several jointly varying discrete dependent variables (see Nerlove and Press, 1973, 1976, 1978, 1980a, 1980b). There and elsewhere, there has been a steady development of methods of analyzing multivariate discrete data using log-linear models. This approach was introduced by Birch, 1963. In this type of model, the logarithm of the probability of falling into a given cell in a contingency table is expanded as an ANOVA model using as many cross classifications as there are jointly dependent discrete variables. In the Nerlove/Press approach, each of the "effects" is expressed as a function (a linear function is used in an associated computer program called Log-Lin, described in Appendix A) of exogenously determined variables. Parameters are estimated by constrained maximum likelihood, and hypotheses are tested using the likelihood ratio test procedure (and the Wilk's approximation to the asympotic distribution of the likelihood ratio statistic). These models are also discussed extensively in Bishop, Fienberg, and Holland, 1975; Fienberg, 1977; Goodman, 1978; Haberman, 1974, 1978, 1979; and Plackett, 1974.

The model is given as follows. Let $p_{i_1,\ldots,i_q}$ denote the probability of falling into cell $(i_1,\ldots,i_q)$ of a q-dimensional contingency table where $i_1 = 1,2,\ldots,I_1$, denote the I_1 categories of factor $1,\ldots$, $i_q = 1,\ldots,I_q$, denote the I_q categories of factor q. Now write

$$\begin{aligned}\log p_{i_1,\ldots,i_q} &= \mu + \alpha_1(i_1) + \cdots + \alpha_q(i_q)\\ &\quad + \beta_{12}(i_1,i_2) + \cdots + \beta_{q-1,q}(i_{q-1},i_q)\\ &\quad + \cdots\\ &\quad + \omega_{1,\ldots,q}(i_1,\ldots,i_q).\end{aligned}$$

Here μ denotes an overall (grand mean) effect; $\alpha_1(i_1)$ denotes an effect due to factor 1 at level $i_1,\ldots,\alpha_q(i_q)$ denotes an effect due to factor q at level i_q; $\beta_{12}(i_1,i_2)$ denotes a second order interaction effect between factors 1 and 2, at levels i_1 and i_2, respectively ; . . . ; $\omega_{1,\ldots,q}(i_1,\ldots,i_q)$ denotes a qth order interactions effect among all the q variables at levels $i_1,\ldots,i_q$. The condition that the cell probabilities must sum to unity requires that

$$\mu \equiv -\log \Sigma \ldots \Sigma \exp\{\theta_{i_1,\ldots,i_q}\},$$

where

$$\theta_{i_1,\ldots,i_q} \equiv \alpha_1(i_1) + \cdots + \omega_{1,\ldots,q}(i_1,\ldots,i_q).$$

Since all parameters in the log-linear model above are not estimable, some constraints must be imposed. We adopt the same ones used in the ANOVA model, viz.,

$$\begin{aligned}&\alpha_1(\cdot) = \alpha_2(\cdot) = \cdots = \alpha_q(\cdot) = 0;\\ &\beta_{12}(i_1,\cdot) = 0, \beta_{12}(\cdot,i_2) = 0, \ldots, \beta_{q-1,q}(\cdot,i_q) = 0;\\ &\vdots\\ &\omega_{1,\ldots,q}(i_1,\ldots,i_{q-1},\cdot) = 0, \ldots, \omega_{1,\ldots,q}(\cdot,i_2,\ldots,i_q) = 0\end{aligned}$$

The dot used in place of an index denotes summation over that index.

In many situations, we can assume that all main effects and all interaction effects depend upon a vector of exogenous variables $\mathbf{x}: r \times 1$ in a linear way, so that

$$\begin{aligned}\alpha_1(i_1) &= \mathbf{x}'\alpha_1^*(i_1); \ldots;\\ \omega_{1,\ldots,q}(i_1,\ldots,i_q) &= \mathbf{x}'\omega_1^*,\ \ldots,_q(i_1,\ldots,i_q).\end{aligned}$$

Here; all starred quantities are r-vectors; $\mathbf{x}$ is a vector that can be measured for each subject in the analysis. All starred vectors are estimable by maximum likelihood (and by other methods as well). A test of independence of all factors is a test that all interaction terms vanish. A likelihood ratio test is readily used. When all terms in the log-linear model are present a priori, the model is called *saturated*; if some terms are set equal to zero a priori, the model is called *unsaturated*. In any case the Log-Lin programs in Appendix A can be used to estimate all parameters by MLE in the presence of exogenous variables. When no exogenous variables are present other computer software such as ECTA may be used for estimation and hypothesis testing.

8.10.6 Serial Correlation

If univariate results can be used as a guide, it is to be expected that a serious violation of assumptions in the general linear model is violation of the independence-of-the-errors assumption. That is, if the errors are serially correlated (as determined from plots of the residuals for each reduced form equation separately) estimators of the mean vector elements may be extremely inefficient (large variances). This problem can be minimized by determining the nature of the serial correlation by residual analysis, and correcting for it by estimating parameters using generalized least squares.

8.10.7 Multivariate Heteroscedasticity

Heteroscedasticity in the multivariate case implies unequal covariance matrices, which give rise to the multivariate Behrens-Fisher problem. Methods of testing for and treating heteroscedasticity were discussed in Section 7.3.3 and in Example (8.3.2). Mild heteroscedasticity is probably not serious for comparisons of mean vectors in MANOVA or for estimating coefficients in regression. However, most assertions about robustness in the multivariate case are hardly more than speculation and need to be examined further.

8.10.8 Multicollinearity

In the regression model it is assumed that $|\mathbf{X}'\mathbf{X}| \neq 0$, where $\mathbf{X}$ is the regressor matrix. If there is *near multicollinearity* so that $|\mathbf{X}'\mathbf{X}|$ is close to zero, the variances of the coefficient estimators will be extremely large. Every effort should be made to avoid this situation, such as omitting appropriate variables (those which are most closely correlated) when the situation is suspected. If steps are not taken to minimize the effect of near multicollinearity, standard errors of coefficient estimators will be so large that Student t-ratios will not be large enough to be significant.

REFERENCES

Aitken (1935).
Anderson (1958).
Ando and Kaufman (1965).
Andrews, Gnanadesikan, and Warner (1973).
Berkson (1944).
Birch (1963).
Bishop, Fienberg, and Holland (1975).
Bock (1963).
Boot and de Witt (1960).
Box (1949).
Box, and Cox (1964).
Cox (1970).
Fienberg (1977).
Finney (1952).
Fisher (1928).
Foster (1957, 1958).
Foster and Rees (1957).
Gabriel (1968).
Goldberger (1964).
Goodman (1978).
Graybill (1961).
Griliches (1967).
Haberman (1974, 1978, 1979).
Heck (1960).
Ito (1968).
Johnston (1963).
Jones (1966).
Jorgenson (1966).
Kendall and Stuart (1961).
Koyck (1954).
Lindley (1965).
Mantel (1966).
McElroy (1967).
Morrison (1967).
Nerlove (1958).
Nerlove and Press (1973, 1976, 1978, 1980a, 1980b)
Neyman and Scott (1948).
Olkin and Press (1969).
Parks (1967).
Pillai (1954, 1960).
Plackett (1974).
Prais and Houthakker (1955).
Press (1964, 1980a).
Press and Scott (1975, 1976).
Press and Wilson (1978).
Press and Zellner (1978).
Raiffa and Schlaifer (1961).
Rao (1952, 1965, 1967, 1968).
Roberts (1969).
Rothenberg (1963).
Roy (1957).
Roy and Bose (1953).
Scheffé (1959).
Smith, Gnanadesikan, and Hughes (1962).
Stroud, Zellner, and Chau (1965).
Swamy (1967, 1968).
Tobin (1955).
Wald (1949).
Whittle (1951).
Wilks (1938).
Wise (1955).
Zellner (1962, 1963, 1967, 1972).
Zellner and Huang (1962).
Zellner and Lee (1965).

COMPLEMENT 8.1

Proof of the Consistency of $\hat{\varrho}$ in the First-Order Markov Regression Model [Example (8.3.3)]: Let

$$\underset{(n\times 1)}{\mathbf{y}} = \underset{(n\times k)}{\mathbf{X}} \cdot \underset{(k\times 1)}{\boldsymbol{\beta}} + \underset{(n\times 1)}{\mathbf{e}}, \qquad \mathcal{L}(\mathbf{e}) = N(\mathbf{0},\boldsymbol{\Omega}), \tag{8.3.15}$$

where for any t

$$e_t = \rho e_{t-1} + u_t, \qquad |\rho| < 1, \qquad E(\mathbf{u}) = \mathbf{0}, \qquad \text{var}(\mathbf{u}) = \sigma_0{}^2\mathbf{I}, \tag{8.3.16}$$

and the process is assumed to have started at $t = -\infty$. ρ is estimated by

$$\hat{\rho} = \frac{\sum_2^n \hat{e}_\alpha \hat{e}_{\alpha-1}}{\sum_2^n \hat{e}_{\alpha-1}{}^2}. \tag{8.3.17}$$

Define the residual vector

$$\hat{\mathbf{e}} = \mathbf{y} - \hat{\mathbf{y}} = \mathbf{y} - \mathbf{X}\hat{\boldsymbol{\beta}} = \mathbf{X}(\boldsymbol{\beta} - \hat{\boldsymbol{\beta}}) + \mathbf{e}, \tag{C8.1.1}$$

where $\hat{\boldsymbol{\beta}}$ denotes the least squares estimator of $\boldsymbol{\beta}$. Then, if $\mathbf{X} \equiv (\mathbf{x}_1,\ldots,\mathbf{x}_n)'$, where $\mathbf{x}_\alpha$ is a $k \times 1$ vector, (C8.1.1) becomes

$$\hat{e}_\alpha = \mathbf{x}_\alpha'(\boldsymbol{\beta} - \hat{\boldsymbol{\beta}}) + e_\alpha, \qquad \alpha = 1,\ldots,n. \tag{C8.1.2}$$

Since $\hat{\boldsymbol{\beta}}$ is assumed to be a consistent estimator of $\boldsymbol{\beta}$, that is, it is assumed

$$\underset{n\to\infty}{\text{plim}} \left(\frac{1}{n}\mathbf{X}'\mathbf{X}\right) = \text{constant},$$

it follows from (C8.1.2) that

$$\underset{n\to\infty}{\text{plim}}\ \hat{e}_\alpha = e_\alpha, \qquad \text{for } \alpha = 1,\ldots,n.$$

Hence, from (8.3.17),

$$\underset{n\to\infty}{\text{plim}}\ \hat{\rho} = \underset{n\to\infty}{\text{plim}} \frac{\sum_2^n e_\alpha e_{\alpha-1}}{\sum_2^n e_{\alpha-1}{}^2}. \tag{C8.1.3}$$

Substituting from (8.3.16) gives

$$\underset{n\to\infty}{\text{plim}}\ \hat{\rho} = \underset{n\to\infty}{\text{plim}} \frac{\sum_2^n (\rho e_{\alpha-1} + u_\alpha)e_{\alpha-1}}{\sum_2^n e_{\alpha-1}{}^2},$$

or

$$\operatorname*{plim}_{n\to\infty} \hat{\rho} = \rho + \operatorname*{plim}_{n\to\infty} \frac{\sum_{2}^{n} u_\alpha e_{\alpha-1}}{\sum_{2}^{n} e_{\alpha-1}^2}. \tag{C8.1.4}$$

It is now shown that the last term in (C8.1.4) is zero. Note that

$$\operatorname*{plim}_{n\to\infty} \frac{\sum_{2}^{n} u_\alpha e_{\alpha-1}}{\sum_{2}^{n} e_{\alpha-1}^2} = \frac{\operatorname*{plim}_{n\to\infty} \dfrac{1}{n-1} \sum_{2}^{n} u_\alpha e_{\alpha-1}}{\operatorname*{plim}_{n\to\infty} \dfrac{1}{n-1} \sum_{2}^{n} e_{\alpha-1}^2}.$$

Moreover, since [see Example (8.3.3)]

$$e_\alpha = \sum_{j=0}^{\infty} \rho^j u_{\alpha-j},$$

$$\operatorname*{plim}_{n\to\infty} \frac{1}{n-1} \sum_{\alpha=2}^{n} e_{\alpha-1}^2 = \operatorname*{plim}_{n\to\infty} \frac{1}{n-1} \sum_{\alpha=2}^{n} \sum_{j=0}^{\infty} \sum_{k=0}^{\infty} \rho^{j+k} u_{\alpha-1-j} u_{\alpha-1-k}$$

$$= \sum_{j=0}^{\infty} \sum_{k=0}^{\infty} \rho^{j+k} \left\{ \operatorname*{plim}_{n\to\infty} \frac{1}{n-1} \sum_{\alpha=2}^{n} u_{\alpha-1-j} u_{\alpha-1-k} \right\}.$$

Since the u_α's are uncorrelated, when $j \neq k$,

$$\operatorname*{plim}_{n\to\infty} \frac{1}{n-1} \sum_{\alpha=2}^{n} u_{\alpha-1-j} u_{\alpha-1-k} = 0.$$

Also, when $j = k$,

$$\operatorname*{plim}_{n\to\infty} \frac{1}{n-1} \sum_{\alpha=2}^{n} u_{\alpha-1-j}^2 = \sigma_0^2.$$

Hence,

$$\operatorname*{plim}_{n\to\infty} \frac{1}{n-1} \sum_{\alpha=2}^{n} e_{\alpha-1}^2 = \sigma_0^2 \sum_{j=0}^{\infty} \rho^{2j} = \frac{\sigma_0^2}{1-\rho^2} \equiv \sigma^2.$$

Thus,

$$\operatorname*{plim}_{n\to\infty} \hat{\rho} = \rho + \frac{1}{\sigma^2} \operatorname*{plim}_{n\to\infty} \frac{1}{n-1} \sum_2^n u_\alpha e_{\alpha-1}.$$

Now define for $\alpha = 1,\ldots,n-1$,

$$v_\alpha = u_{\alpha+1} e_\alpha.$$

Then, since e_α depends only upon $u_\alpha, u_{\alpha-1},\ldots$, $Ev_\alpha = E(u_{\alpha+1})E(e_\alpha) = 0$, for all $\alpha = 1,\ldots, n-1$. Also, for integer $k \geq 1$ and $2 \leq \alpha + k \leq n$, $E(v_\alpha v_{\alpha+k}) = E(u_{\alpha+1}e_\alpha u_{\alpha+k+1}e_{\alpha+k})$. Since e_α depends only on $u_\alpha, u_{\alpha-1},\ldots$, it follows that $u_{\alpha+k+1}$ is uncorrelated with $(u_{\alpha+1}e_\alpha e_{\alpha+k})$. Hence, $E(v_\alpha v_{\alpha+k}) = 0$; that is, the v_α's are mutually uncorrelated.

Now consider the variances of the v_α's.

$$\begin{aligned}\operatorname{Var}(v_\alpha) = E(v_\alpha^2) = E(u_{\alpha+1}^2 e_\alpha^2) &= E(u_{\alpha+1}^2)E(e_\alpha^2)\\ &= \operatorname{var}(u_{\alpha+1}) \operatorname{var}(e_\alpha) = \sigma_0^2\sigma^2 = \frac{\sigma_0^4}{1-\rho^2}.\end{aligned}$$

It is well known from the Weak Law of Large Numbers (see for instance, Rao, 1965, p. 92) that since the v_α's are uncorrelated, if

$$\lim_{n\to\infty} \frac{1}{n^2} \sum_1^n \operatorname{var}(v_\alpha) = 0,$$

$$\operatorname*{plim}_{n\to\infty} \frac{1}{n-1} \sum_1^{n-1} v_\alpha = E(v_\alpha) = 0.$$

But $\operatorname{var}(v_\alpha) = \text{constant}$. Hence, $\operatorname*{plim}_{n\to\infty} \hat{\rho} = \rho$.

COMPLEMENT 8.2

Proof of Theorem (8.6.4): This theorem is established by approximating the determinant in (8.6.18) for large N.

Suppose that for $\mathbf{X}'\mathbf{X}$ nonstochastic,

$$\lim_{N\to\infty} \frac{\mathbf{X}'\mathbf{X}}{N} = \text{constant},$$

and recall that $\mathbf{V}/N$ and $\hat{\mathbf{B}}$ are consistent estimators of $\mathbf{\Sigma}$ and $\mathbf{B}$, respectively.

Define

$$\mathbf{Q} = (\mathbf{B} - \hat{\mathbf{B}})'\left(\frac{\mathbf{X}'\mathbf{X}}{N}\right)(\mathbf{B} - \hat{\mathbf{B}}), \qquad \mathbf{H} = \frac{\mathbf{V} + \mathbf{G}}{N}, \qquad \mathbf{W} = \mathbf{H}^{-1/2}\mathbf{Q}\mathbf{H}^{-1/2}.$$

Then

$$\left|\frac{\mathbf{V}+\mathbf{G}}{N}+(\mathbf{B}-\hat{\mathbf{B}})\left(\frac{\mathbf{X}'\mathbf{X}}{N}\right)(\mathbf{B}-\hat{\mathbf{B}})\right|^{(N+m-p-1)/2}$$
$$\begin{aligned}&\equiv|\mathbf{H}+\mathbf{Q}|^{(N+m-p-1)/2}\\ &=|\mathbf{H}|^{(N+m-p-1)/2}|\mathbf{I}+\mathbf{W}|^{(N+m-p-1)/2}\\ &=|\mathbf{H}|^{(N+m-p-1)/2}\cdot|\mathbf{I}+\mathbf{W}|^{N/2}\cdot|\mathbf{I}+\mathbf{W}|^{(m-p-1)/2}.\end{aligned}\qquad\text{(C8.2.1)}$$

Note that since $\mathbf{H}$ does not depend upon $\boldsymbol{\beta}$ it will be treated as a constant of proportionality. Also note that $\underset{N\to\infty}{plim}\ \hat{\mathbf{B}}=\mathbf{B}$, and $\mathbf{Q}=\mathbf{O}_p(N^{-1})$; that is, $\underset{N\to\infty}{plim}\ (NQ)=$ a constant matrix. Hence, as $N\to\infty$ the latent roots of $\mathbf{W}$ approach zero, in probability. So.

$$\underset{N\to\infty}{plim}\ |\mathbf{I}+\mathbf{W}|^{(m-p-1)/2}=1.\qquad\text{(C8.2.2)}$$

Now consider $|\mathbf{I}+\mathbf{W}|^{N/2}$. Let $\lambda_1,\ldots,\lambda_p$ denote the latent roots of $\mathbf{W}$, and let $\mathbf{D}_\lambda\equiv\operatorname{diag}(\lambda_1,\ldots,\lambda_p)$. Then, if $\mathbf{W}=\boldsymbol{\Gamma}\mathbf{D}_\lambda\boldsymbol{\Gamma}'$, $\boldsymbol{\Gamma}\boldsymbol{\Gamma}'=\mathbf{I}$,

$$\begin{aligned}|\mathbf{I}+\mathbf{W}|^{N/2}&=\exp\left(\frac{N}{2}\log|\mathbf{I}+\mathbf{W}|\right)=\exp\left(\frac{N}{2}\log|\mathbf{I}+\mathbf{D}_\lambda|\right),\\ &=\exp\left[\frac{N}{2}\log\prod_{i=1}^{p}(1+\lambda_i)\right]=\exp\left[\frac{N}{2}\sum_{i=1}^{p}\log(1+\lambda_i)\right],\\ &=\exp\left\{\frac{N}{2}\sum_{i=1}^{p}\left[\lambda_i-\frac{\lambda_i^2}{2}+\frac{\lambda_i^3}{3}-\cdots\right]\right\},\\ &=\exp\left\{\frac{N}{2}\left[\operatorname{tr}\mathbf{W}-\frac{1}{2}\operatorname{tr}\mathbf{W}^2+\frac{1}{3}\operatorname{tr}\mathbf{W}^3-\cdots\right]\right\},\\ &=\exp\left(\frac{N}{2}\operatorname{tr}\mathbf{W}\right)\cdot\exp\left(-\frac{N}{4}\operatorname{tr}\mathbf{W}^2+\frac{N}{6}\operatorname{tr}\mathbf{W}^3-\cdots\right).\end{aligned}$$

It is easy to see that since $\mathbf{Q}=\mathbf{O}_p(N^{-1})$, $\operatorname{tr}\mathbf{W}^k=\mathbf{O}_p(N^{-k})$, $k=1,2,\ldots.$ Therefore,

$$|\mathbf{I}+\mathbf{W}|^{N/2}=\exp\left(\frac{N}{2}\operatorname{tr}\mathbf{W}\right)\cdot\left[1+\mathbf{O}_p\left(\frac{1}{N}\right)\right].\qquad\text{(C8.2.3)}$$

Substituting (C8.2.1) into (8.6.18), using (C8.2.2) and (C8.2.3) assuming N is large, and neglecting terms of order N^{-1} in probability, gives approximately,

$$p(\boldsymbol{\beta}|\mathbf{Y},\mathbf{X})\propto\exp\left\{-\frac{1}{2}(\boldsymbol{\beta}-\boldsymbol{\phi})'\mathbf{F}^{-1}(\boldsymbol{\beta}-\boldsymbol{\phi})\right\}\exp\left\{-\frac{N}{2}\operatorname{tr}\mathbf{W}\right\},$$

or

$$p(\boldsymbol{\beta}|\mathbf{Y},\mathbf{X}) \propto \exp\left\{-\frac{1}{2}(\boldsymbol{\beta}-\boldsymbol{\phi})'\mathbf{F}^{-1}(\boldsymbol{\beta}-\boldsymbol{\phi}) - \frac{N}{2}\operatorname{tr}\left(\frac{\mathbf{V}+\mathbf{G}}{N}\right)^{-1}(\mathbf{B}-\hat{\mathbf{B}})'\left(\frac{\mathbf{X}'\mathbf{X}}{N}\right)(\mathbf{B}-\hat{\mathbf{B}})\right\}.$$

Using (8.6.13) gives

$$p(\boldsymbol{\beta}|\mathbf{Y},\mathbf{X}) \propto \exp\left\{-\frac{1}{2}(\boldsymbol{\beta}-\boldsymbol{\phi})'\mathbf{F}^{-1}(\boldsymbol{\beta}-\boldsymbol{\phi}) - \frac{N}{2}(\boldsymbol{\beta}-\hat{\boldsymbol{\beta}})'\left[\left(\frac{\mathbf{V}+\mathbf{G}}{N}\right)^{-1}\otimes\left(\frac{\mathbf{X}'\mathbf{X}}{N}\right)\right](\boldsymbol{\beta}-\hat{\boldsymbol{\beta}})\right\}.$$

Completing the square on $\boldsymbol{\beta}$ proves the theorem.

COMPLEMENT 8.3

Proof of Theorem (8.6.5): This theorem may be proven by approximating the larger order determinant in (8.6.24).

Define

$$\mathbf{A} \equiv \boldsymbol{\Sigma}^{-1}\otimes(\mathbf{X}'\mathbf{X}) = N\boldsymbol{\Sigma}^{-1}\otimes\left(\frac{\mathbf{X}'\mathbf{X}}{N}\right), \tag{C8.3.1}$$

and assume $\lim_{N\to\infty}\left(\frac{X'X}{N}\right)$ = constant. For large N, $(A+F^{-1}) \cong A$. Then, if $Z \equiv A^{-1/2}F^{-1}A^{-1/2}$, $|\mathbf{F}^{-1}+\boldsymbol{\Sigma}^{-1}\otimes(\mathbf{X}'\mathbf{X})| = |\mathbf{F}^{-1}+\mathbf{A}| = |\mathbf{A}|\cdot|\mathbf{I}+\mathbf{Z}|$.

$$|\mathbf{F}^{-1}+\boldsymbol{\Sigma}^{-1}\otimes(\mathbf{X}'\mathbf{X})| = |\mathbf{F}^{-1}+\mathbf{A}| = |\mathbf{A}|\cdot|\mathbf{I}+\mathbf{Z}|. \tag{C8.3.2}$$

Note from (C8.3.1) that the latent roots of $\mathbf{Z}$ approach zero in probability, as N approaches infinity. Therefore,

$$\operatorname*{plim}_{N\to\infty}|\mathbf{I}+\mathbf{Z}| = 1. \tag{C8.3.3}$$

Now recall that [see (2.10.5)] if $\mathbf{S}_1: p\times p$, $|\mathbf{S}_1|\neq 0$, and $\mathbf{S}_2: q\times q$, $|\mathbf{S}_2|\neq 0$, $|\mathbf{S}_1\otimes\mathbf{S}_2| = |\mathbf{S}_1|^q\cdot|\mathbf{S}_2|^p$. Using this result on $\mathbf{A}$ gives

$$|\mathbf{A}| = |\boldsymbol{\Sigma}|^{-q}|\mathbf{X}'\mathbf{X}|^p. \tag{C8.3.4}$$

Combining (C8.3.2)–(C8.3.4) gives for large N (neglecting terms of order N^{-1} in probability),

$$|\mathbf{F}^{-1}+\boldsymbol{\Sigma}^{-1}\otimes(\mathbf{X}'\mathbf{X})|^{1/2} \cong |\boldsymbol{\Sigma}|^{-q/2}|\mathbf{X}'\mathbf{X}|^{p/2}. \tag{C8.3.5}$$

Substituting (C8.3.5) into (8.6.24) proves the theorem.

EXERCISES

8.1 Consider a Model III multivariate regression model (stochastic regression).

(a) Formulate the model giving appropriate assumptions.

(b) Are least squares estimators unbiased? Why? [*Hint:* See Table 8.2.1.]

8.2 Let y_j denote the change in the number of people residing in Chicago from year j to year $(j+1)$ who would prefer to live in integrated neighborhoods; $j = 1, 2, 3$. Suppose $\mathbf{y} = (y_1, y_2, y_3)'$, with

$$\underset{(3\times 1)}{\mathbf{y}} = \underset{(3\times 2)(2\times 1)}{\mathbf{X} \cdot \boldsymbol{\beta}} + \underset{(3\times 1)}{\mathbf{e}}, \qquad \text{and} \qquad \mathbf{X} = \begin{pmatrix} 1 & 0 \\ 0 & 1 \\ 1 & 1 \end{pmatrix},$$

where $\boldsymbol{\beta} = (\mu, \lambda)'$, and $\mathbf{e} = (e_1, e_2, e_3)'$. If $E(\mathbf{e}) = \mathbf{0}$, and var $(\mathbf{e}) =$

$$\sigma^2\boldsymbol{\Sigma}, \text{ where } \boldsymbol{\Sigma} = \begin{pmatrix} 2 & 1 & 2 \\ 1 & 3 & 1 \\ 2 & 1 & 4 \end{pmatrix}, \text{ and } \mathbf{y} = (-2,3,1)',$$

(a) estimate $\boldsymbol{\beta}$,

(b) estimate σ^2.

8.3 In Exercise 8.2 suppose instead of the $\boldsymbol{\Sigma}$ shown, $\boldsymbol{\Sigma}$ were unknown but with the structure

$$\boldsymbol{\Sigma} = \begin{pmatrix} 1 & \rho & \rho \\ \rho & 1 & \rho \\ \rho & \rho & 1 \end{pmatrix}, \qquad -\frac{1}{2} < \rho < 1.$$

Estimate $\boldsymbol{\beta}$.

8.4 For the data given in Example (8.4.1), test the hypothesis that $\beta_{11} = \beta_{12} = \beta_{13} = 0$, where $\mathbf{B} = (\beta_{ij})$, $i = 1, 2$, $j = 1, 2, 3$. (Use the likelihood ratio criterion and the 5 percent level of significance.)

8.5 Explain the conditions under which generalized multivariate regression methods yield more efficient slope coefficient estimators than those obtainable by standard multivariate regression techniques. (Consider a p equation system.)

8.6 For the data given in Example (8.4.1), assume a diffuse prior on $(\mathbf{B}, \boldsymbol{\Sigma})$ in the Normal standard multivariate regression model and test the hypothesis that $H: \beta_{11} = 0$ (use a 1 percent level of significance and accept H if a 99 percent credibility interval includes the origin).

8.7 Let $\pi_j \equiv N(\boldsymbol{\theta}_j,\boldsymbol{\Sigma})$, $j = 1, 2, 3$ denote three p-variate Normal populations and let $\{\mathbf{x}_1(j),\ldots,\mathbf{x}_{N_j}(j)\}$ denote samples of p-variate observations of size N_j for each of the populations. Explain any differences in the methods you would use to test $H_1: \boldsymbol{\theta}_1 = \boldsymbol{\theta}_2$, versus $A_1: \boldsymbol{\theta}_1 \neq \boldsymbol{\theta}_2$, as opposed to testing $H_2: \boldsymbol{\theta}_1 = \boldsymbol{\theta}_2 = \boldsymbol{\theta}_3$, versus $A_2: \boldsymbol{\theta}_1, \boldsymbol{\theta}_2, \boldsymbol{\theta}_3$ not all equal.

8.8 The country of Moralia has decided that to defend itself from its enemies it will need to purchase some fighter airplanes. Three different aircraft types are in contention for possible purchase (Type A, Type B, and Type C). Suppose data are available giving the results of test dive bombing for each aircraft type under a wide variety of different conditions of altitude, speed, and dive angle. Many different pilots have flown the airplanes (selected randomly) so that results do not reflect a "pilot experience bias." Suppose that for each test bomb drop run for a given airplane and given bomb release conditions of altitude, speed, and dive angle, there is available the target miss distance in feet in some two-dimensional coordinate system on the ground (and these miss distances are correlated).

Select a model (and explain the basis for your selection) for analyzing the above data to determine whether there are significant differences among the three aircraft types; also explain how you would compare the magnitude of the differences among aircraft, explaining why you have chosen this procedure. Assume that altitude, speed, and dive angle are all continuous variables that may assume any value within wide ranges.

8.9 Explain why the likelihood ratio, trace, and maximum root criteria for hypothesis testing in MANOVA are all equivalent when rank $(\mathbf{HG}^{-1}) = 1$, where $\mathbf{G}$ and $\mathbf{H}$ are defined in Section 8.7.3, and $\mathbf{H}$ denotes any main effects or interaction effects.

8.10 Parameterize the Normal multivariate regression model in (8.6.14) in terms $\boldsymbol{\Lambda} \equiv \boldsymbol{\Sigma}^{-1}$. Then, using the generalized natural conjugate prior approach of Section 8.6.2, find the joint generalized natural conjugate prior density for $(\mathbf{B},\boldsymbol{\Lambda})$.

8.11 Show that if $\mathbf{y} \equiv (y_j)$ denotes a vector of n observations on a dependent variable Y, and if for $k = \text{rank}\ (\mathbf{X})$, $k < n$,

$$\underset{(n\times 1)}{\mathbf{y}} = \underset{(n\times k)}{\mathbf{X}} \cdot \underset{(k\times 1)}{\boldsymbol{\beta}} + \underset{(n\times 1)}{\mathbf{u}}, \qquad E(\mathbf{u}) = \mathbf{0}, \qquad \text{var}\ (\mathbf{u}) = \sigma^2\mathbf{I}_n,$$

denotes a standard univariate regression model, the sample multiple correlation coefficient is given by

$$R = \widehat{\text{corr}}\,(Y,\hat{Y}) = \frac{\sum_1^n (y_j - \bar{y})(\hat{y}_j - \bar{\hat{y}})}{\left\{\sum_1^n (y_j - \bar{y})^2 \sum_1^n (\hat{y}_j - \bar{\hat{y}})^2\right\}^{1/2}},$$

where $\hat{\mathbf{y}} \equiv (\hat{y}_j) = \mathbf{X}\hat{\boldsymbol{\beta}}$, and $\bar{\hat{y}} \equiv \frac{1}{n}\sum_1^n \hat{y}_j$. [*Note:* When there is a constant term in the regression model, $\bar{\hat{y}} = \bar{y}$.]

8.12 Let $y_i = \alpha + \beta x_i + e_{1i}$, $z_i = a + bx_i + e_{2i}$, $i = 1,\ldots,n$, denote standard regressions of y on x and z on x, respectively; that is, $E(e_{1i}) = E(e_{2i}) = 0$, all errors are independent and homoscedastic. Define the new variables "corrected for x" as

$$y_i^* = y_i - \hat{\alpha} - \hat{\beta}x_i, \qquad z_i^* = z_i - \hat{a} - \hat{b}x_i,$$

where carets denote least squares estimators. The (sample) *partial correlation coefficient* of y and z "holding x fixed" is defined as the correlation between y^* and z^*; that is, since $Ey_i^* = Ez_i^* = 0$,

$$\rho_{yz|x} = \frac{\sum_1^n y_i^* z_i^*}{\left[\sum_1^n y_i^{*2} \sum_1^n z_i^{*2}\right]^{1/2}}.$$

Now suppose (X,Y,Z) follow a trivariate Normal distribution and x_i, y_i, z_i are their observed values. Find the conditional distribution of Y and Z given X and relate the correlation coefficient in this distribution to that found above.

8.13 When the cells have unequal numbers of observations in a MANOVA, the number of replications for each cell might be taken to be the harmonic average of the numbers of observations for all the cells. When might this be a poor practice? When would this procedure be reasonable?

8.14 Suppose $\mathbf{x}_1,\ldots,\mathbf{x}_N$ are independent bivariate vectors denoting changes in verbal and quantitative scores on achievement tests for people in a job training center who take the pair of tests before a training program, and then again after the training program. Suppose for $j = 1,\ldots,N$, $\mathcal{L}(\mathbf{x}_j|\boldsymbol{\theta},\boldsymbol{\Sigma}) = N(\boldsymbol{\theta},\boldsymbol{\Sigma})$.

In another such training program involving 300 trainees it was estimated that $\boldsymbol{\theta} = \boldsymbol{\theta}_0 \equiv (25,30)'$, $\boldsymbol{\Sigma} = \boldsymbol{\Sigma}_0 \equiv \begin{pmatrix} 100 & 50 \\ 50 & 100 \end{pmatrix}$. Moreover, verbal score changes were found to vary from 0 to 50 and quantitative score changes varied from 20 to 40. Verbal score changes were predictable about half the time from quantitative score changes.

Use this information on the other training program to assess a generalized natural conjugate prior for $\boldsymbol{\theta}$ and $\boldsymbol{\Sigma}$ (make some "reasonable" assumptions). Then show that the posterior distribution of $\boldsymbol{\theta}$ given the sample data is proportional to the product of a Normal density and a multivariate Student t-density centered at the sample mean.

[Hint (1): Ranges of variables might represent, say, plus or minus three standard deviations, so, for example,

$$3\{\mathrm{var}(\theta_1)\}^{1/2} = 25;$$

predictability might be interpreted as a correlation, so that

$$\mathrm{Cor}(\theta_1, \theta_2) = 0.5;$$

you might also take $E(\Sigma) = \Sigma_0$. Other interpretations might include order statistics, or other descriptive statistics.]

[*Hint (2):* Use the likelihood function of Section 7.1.6 and the prior density in (8.6.16), and eliminate $\boldsymbol{\Sigma}$.]

8.15 Suppose the production of a certain model automobile is summarized by the data below.

Year	*Cars* (10,000)
1971	2
1972	4
1973	8
1974	9

(a) Fit a regression line to the data by ordinary least squares, adopting the model $y_t = a + bt + e_t$, $t = 1, \ldots, 4$, where y_t = production at time t, and t = (year − 1970). This should be done without mechanical aids.

(b) Assume $e_t = \rho e_{t-1} + u_t$, $|\rho| < 1$, where the u_t's are mutually uncorrelated with mean zero and constant variance. Give an estimate of ρ.

(c) Use a computer program such as BMDO1R (UCLA Biomed) to carry out Part (a) and compare results. Plot the residuals. Would you expect $\hat{\rho}$ to have the sign found in Part (b) based upon your plot of the residuals? Why?

8.16 Show that the addition of a variable to a linear regression cannot decrease the value of the squared multiple correlation coefficient.

8.17 Show that at least one of the latent roots of a correlation matrix, $\mathbf{R}: p \times p$, must exceed one for $\mathbf{R} \neq \mathbf{I}$.

8.18 Let a system of linear simultaneous stochastic equations be denoted by the model (see Sect. 5.3.2)

$$\mathbf{Y\Gamma} = \mathbf{XB} + \mathbf{U},$$

where $\mathbf{Y}: N \times p$, $\mathbf{\Gamma}: p \times p$, $\mathbf{X}: N \times q$, $\mathbf{B}: q \times p$, $\mathbf{U}: N \times p$, and $|\mathbf{\Gamma}| \neq 0$. Assume $\mathbf{U}$ satisfies the usual assumptions of the multivariate regression model. Consider the problem of identifying and estimating $(\mathbf{\Gamma}, \mathbf{B})$. Is there an identification problem? Explain the nature of the identification problem. How would you estimate $(\mathbf{\Gamma}, \mathbf{B})$? [Hint: see, e.g., Goldberger, 1964.]

8.19 Theorem (8.6.6) gives the predictive distribution for a new observation y^*. Find the predictive distribution for $E(y^*)$ under the same assumptions.

8.20 Generalize the two-way layout MANOVA model of Sect. 8.7.2 to the multivariate three-way layout with K-observations per cell. Give the model, the restrictions, the sum-of-squares matrices, and the likelihood ratio tests for the usual hypotheses.

8.21 For the polytomous response model in Sect. (8.10.4), adopt a Dirichlet prior distribution on the category probabilities and develop the posterior distribution for the category probabilities. Find the posterior mean of p_1.

9

PRINCIPAL COMPONENTS

9.1 INTRODUCTION

The principal components model of multivariate analysis accounts for the variance within a set of data by providing those linear combinations of correlated variables that maximize the variance of the weighted sum. Data vectors taken from multivariate populations involve repeated observations on p possibly correlated random variables. It is sometimes the case that $p = 50$, 100, or even 1000. For this reason, it is natural to seek ways of rearranging or summarizing the data so that with as little loss of information as possible, the dimension of the problem is reduced. One approach toward a more parsimonious data representation is suggested by the notion that in the early stages of research, interest usually focuses on those variables that tend to exhibit the greatest variation from observation to observation. That is, since a variable which does not change much (relative to other variables) in repeated experiments may be treated approximately as a constant, by discarding low variance variables or groups of variables, and centering attention on high variance variables, the phenomenon of interest may be more conveniently studied in a subspace of dimension less than p; that is, less than that defined by the complete set of correlated variables. Although some information about the relationship among the variables is clearly lost by such a procedure, nevertheless, in many problems there is much more to gain than to lose by such an approach. More general procedures for reducing the dimension of a covariance matrix by a suitable approximation are given in Rao (1980).

The principal components approach to parsimony was introduced by Karl Pearson (1901) who studied the problem for the case of nonstochastic variables, and in a different context. The technique was generalized by Hotelling (1933) to the case of stochastic variables. Hotelling considered weighted sums of all the distinct random variables available and attempted to find a set of weights that would maximize the variance of the sum. The new variable (the weighted sum) is called the first principal component and oftentimes, its variance is such a large proportion of the total variance that only the first principal component is considered for further analyses which hinge on total variance. In other problems, other weighted sums that are orthogonal to the first sum are also considered.

The weights in the principal components associated with a vector of correlated attributes (or variables) are exactly the normalized latent vectors of the covariance matrix of the vector of attributes, and the latent roots of the covariance matrix are the variances of the principal components (the largest root is the variance of the first component).

It will be seen that principal components are sensitive to the units used in the analysis so that different sets of weights are obtained for different sets of units. Sometimes, sample correlation matrices are used instead of sample covariance matrices (by standardizing the variables relative to the sample); the problem of units is thereby avoided, since the "principal components" are then invariant with respect to changes in units. However, since the new variables are not really standardized relative to the population (they do not have unit variance except asymptotically) there is then introduced a problem of interpreting what has actually been computed. That is, in small or even moderate samples, the objective of finding linear functions of the original variables or even their standardized versions (relative to the population) with maximum variance is not achieved. Hence, the practice of computing principal components from the sample correlation matrix is not recommended unless samples are very large.

Large sets of data should not be subjected to a principal components analysis merely to obtain fewer variables to work with, without regard to an overall objective. Rather, an objective should be established first, and then the principal components model should be used only if it complements the objective. For example, in some problems where correlation rather than variance is of central interest or where there are likely to be important nonlinear functions of the observations of interest, much information about relationships may be lost if all but the first few principal components are dropped. However, for those problems in which principal components are appropriate, tremendous simplifications of the data can often be effected.

Bayesian results in principal components are quite sparse, and those that exist are complicated. For example, the posterior distribution of the variances of the principal components, using a diffuse prior density, was given by Geisser (1965a). The distribution is expressible in terms of zonal polynomials (see, for instance, James, 1961).

Section 9.2 contains the results of principal components analysis applied to random variables with known parameters, whereas in Section 9.3, the parameters are estimated, and asymptotic sampling theory results are given so that inferences can be made for the case of unknown parameters. Section 9.4 discusses the use of principal components analysis in conjunction with other models such as regression, and Section 9.5 presents some applications of principal components analysis in quality control, in the analysis of the prices of securities, and in the study of crime.

9.2 POPULATION PRINCIPAL COMPONENTS

Suppose in a problem, $\mathbf{X}: p \times 1$ denotes a vector of attributes, and $\boldsymbol{\phi} \equiv E(\mathbf{X})$, $\boldsymbol{\Sigma} \equiv \text{var}(\mathbf{X})$. It is assumed that $\boldsymbol{\Sigma}$ is a known quantity.

Let $\boldsymbol{\alpha} \equiv (\alpha_i)$ denote a p-variate vector of unknown weights for each of the components of $\mathbf{X}$, and let z_1 denote the scalar

$$z_1 = \boldsymbol{\alpha}'\mathbf{X} = \sum_{i=1}^{p} \alpha_i X_i.$$

Each weight α_i is a measure of the importance to be placed on that component. If it is required that $\boldsymbol{\alpha}'\boldsymbol{\alpha} = 1$, unique solutions for the principal components will result.

Assume that the elements of $\mathbf{X}$ are measured in the same units; otherwise, $\boldsymbol{\alpha}'\boldsymbol{\alpha} = 1$ is not a sensible requirement. If it is not reasonable in a particular problem for all of the components of $\mathbf{X}$ to be measured in the same units, and if the sample size is large so that the final results of a principal components analysis will be meaningful, new standardized variables should be formed from the X_i's by subtracting $\bar{X}_i$ and dividing them by their standard deviations. The corresponding covariance matrix will then be a sample correlation matrix, and all components will be measured in the same units.

Now maximize var (z_1). Since $z_1 = \boldsymbol{\alpha}'\mathbf{X}$, var $(z_1) = \boldsymbol{\alpha}'\boldsymbol{\Sigma}\boldsymbol{\alpha}$. Thus, the problem of finding those weights that will maximize var (z_1) is the problem of finding

$$\max_{\boldsymbol{\alpha}} (\boldsymbol{\alpha}'\boldsymbol{\Sigma}\boldsymbol{\alpha})$$

subject to the condition that $\boldsymbol{\alpha}'\boldsymbol{\alpha} = \Sigma_1^p \alpha_i^2 = 1$. The algebraic solution found from (2.11.9) shows that the largest possible value of var (z_1) is the largest latent root of $\boldsymbol{\Sigma}$. An alternative approach is to form the Lagrangian

$$t = \boldsymbol{\alpha}'\boldsymbol{\Sigma}\boldsymbol{\alpha} - \theta(\boldsymbol{\alpha}'\boldsymbol{\alpha} - 1),$$

where θ is a Lagrangian multiplier, and maximize t using the constraint. Differentiating t with respect to $\boldsymbol{\alpha}$, setting the result equal to zero, and solving, gives

$$\frac{\partial t}{\partial \boldsymbol{\alpha}} = 2\boldsymbol{\Sigma}\boldsymbol{\alpha} - 2\theta\boldsymbol{\alpha} = \mathbf{0},$$

or

$$(\boldsymbol{\Sigma} - \theta\mathbf{I})\boldsymbol{\alpha} = \mathbf{0}. \tag{9.2.1}$$

Since $\boldsymbol{\alpha} \not\equiv \mathbf{0}$ (since $\boldsymbol{\alpha}'\boldsymbol{\alpha} = 1$), there can be a solution only if

$$|\boldsymbol{\Sigma} - \theta\mathbf{I}| = 0. \tag{9.2.2}$$

That is, θ is a latent root of $\boldsymbol{\Sigma}$, and $\boldsymbol{\alpha}$ is a normalized latent vector of $\boldsymbol{\Sigma}$. Since $\boldsymbol{\Sigma}$ is $p \times p$, there are p values of θ that will satisfy the determinantal equation. However, the largest latent root is selected. The reasoning is as follows.

From (9.2.1), $\boldsymbol{\Sigma}\boldsymbol{\alpha} = \theta\boldsymbol{\alpha}$. Premultiplication by $\boldsymbol{\alpha}'$ gives

$$\boldsymbol{\alpha}'\boldsymbol{\Sigma}\boldsymbol{\alpha} = \theta\boldsymbol{\alpha}'\boldsymbol{\alpha} = \theta = \text{var}\,(z_1).$$

To maximize var (z_1) take θ as large as possible; that is, choose the largest latent root. Denote the ordered latent roots of $\boldsymbol{\Sigma}$ by $\theta_1 \geq \theta_2 \geq \cdots \geq \theta_p \geq 0$, and denote the corresponding normalized latent vectors by $(\boldsymbol{\alpha}_1,\ldots,\boldsymbol{\alpha}_p)$. The solutions to (9.2.1) and (9.2.2) are $(\boldsymbol{\alpha}_1,\theta_1)$; $z_1 = \boldsymbol{\alpha}_1'\mathbf{X}$ is called the *first principal component*. The result is summarized in the following theorem.

Theorem (9.2.1): Suppose for $\mathbf{X}: p \times 1$, $E\mathbf{X} = \boldsymbol{\phi}$, and var $\mathbf{X} = \boldsymbol{\Sigma}$; let $\theta_1 \geq \theta_2 \geq \cdots \geq \theta_p \geq 0$ denote the latent roots of $\boldsymbol{\Sigma}$, and let $\boldsymbol{\alpha}_1, \boldsymbol{\alpha}_2, \ldots, \boldsymbol{\alpha}_p$ denote the corresponding normalized latent vectors of $\boldsymbol{\Sigma}$; let $z_1 = \boldsymbol{\alpha}'\mathbf{X}$, and constrain $\boldsymbol{\alpha}$ so that $\boldsymbol{\alpha}'\boldsymbol{\alpha} = 1$. Then, if $\boldsymbol{\alpha} = \boldsymbol{\alpha}_1$, z_1 is called the first principal component of $\mathbf{X}$, and var $(z_1) = \theta_1$.

[*Remarks:* It should be noted that no distributional form has been assumed for $\mathbf{X}$ in Theorem (9.2.1), although there are interesting interpretations possible for these results when $\mathbf{X}$ is Normally distributed. Thus, if $\mathcal{L}(\mathbf{X}) = N(\boldsymbol{\phi},\boldsymbol{\Sigma})$, $\boldsymbol{\Sigma} > 0$, the contours of the distribution (surfaces of constant density) are ellipsoids, and z_1 represents the principal axis of the ellipsoid.

Note also that it is not required that $\boldsymbol{\Sigma}$ be nonsingular. Thus, some of the latent roots of $\boldsymbol{\Sigma}$ may be zeros; in addition, some of the roots may have multiplicities greater than one.]

Define $z_2 = \boldsymbol{\alpha}'\mathbf{X}$, where now not only is it required that $\boldsymbol{\alpha}'\boldsymbol{\alpha} = 1$, but in addition, it is required that z_2 be uncorrelated with z_1. Thus, the set of weights corresponding to a maximum variance for z_2 will be different from those found earlier for z_1. In fact, it is straightforward to show by using two Lagrangian multipliers for the two constraints, that var (z_2) is maximized for $\boldsymbol{\alpha} = \boldsymbol{\alpha}_2$, and the maximum variance is the second largest latent root, θ_2.

Analogously, define $z_3 = \boldsymbol{\alpha}'\mathbf{X}$, and require that for $\boldsymbol{\alpha}'\boldsymbol{\alpha} = 1$, z_3 be uncorrelated with both z_1 and z_2; moreover, sequentially defining $z_1, \ldots, z_p$, generate a set of p orthogonal axes that are obtained by using the normalized latent vectors of $\boldsymbol{\Sigma}$, and the variances of the z_j's will be the latent roots of $\boldsymbol{\Sigma}$. Summarizing:

Theorem (9.2.2): For the same definitions as in Theorem (9.2.1), the jth principal component of $\mathbf{X}$ is given by $z_j \equiv \boldsymbol{\alpha}_j'\mathbf{X}$, and var $(z_j) = \theta_j$, $j = 1, \ldots, p$. Moreover, corr$(z_1, z_2) = 0$, for all $i \neq j$.

Note that under the assumption of Normally distributed $\mathbf{X}$, the principal components represent a rotation of the coordinate axes of the X_j's to the principal axes of the ellipsoid defining the contours of the distribution. If there are multiple roots, the axes are not uniquely defined, however.

9.3 SAMPLE PRINCIPAL COMPONENTS

The results developed in the last section are directly applicable only if $\boldsymbol{\Sigma}$ is known. Unfortunately, in applied research this parameter is usually unknown and must be estimated. Typically, therefore, the decisions to be made about which principal components have sufficiently small variance to be ignored must be made on the basis of sample data. To answer such questions of inference, distributional assumptions will be made about the observed data (Normality), and about the structure of the covariance matrix, and results useful in large samples will be presented.

9.3.1 Point Estimation

Let $\mathbf{X}$ be a $p \times 1$ vector of attributes with $E(\mathbf{X}) = \boldsymbol{\phi}$, var $(\mathbf{X}) = \boldsymbol{\Sigma}$, both unknown. Suppose $\mathbf{x}_1, \ldots, \mathbf{x}_N$, $N > p$, are independent $p \times 1$ vector observations of $\mathbf{X}$. The sample mean and sample covariance matrix are given by

$$\hat{\boldsymbol{\phi}} = \bar{\mathbf{x}} = \frac{1}{N}\sum_{1}^{N} \mathbf{x}_j, \qquad \hat{\boldsymbol{\Sigma}} = \frac{1}{N}\sum_{1}^{N} (\mathbf{x}_j - \bar{\mathbf{x}})(\mathbf{x}_j - \bar{\mathbf{x}})'.$$

Recall from Theorem (7.1.3) that if $\mathcal{L}(\mathbf{X}) = N(\boldsymbol{\phi}, \boldsymbol{\Sigma})$, $\hat{\boldsymbol{\phi}}$, and $\hat{\boldsymbol{\Sigma}}$ are maximum likelihood estimators of $\boldsymbol{\phi}$ and $\boldsymbol{\Sigma}$.

Theorem (9.3.1): If $\hat{\boldsymbol{\phi}}$, $\hat{\boldsymbol{\Sigma}}$ denote moment estimators of $\boldsymbol{\phi}$ and $\boldsymbol{\Sigma}$ [maximum likelihood estimators of $\boldsymbol{\phi}$ and $\boldsymbol{\Sigma}$, for $\mathcal{L}(\mathbf{X}) = N(\boldsymbol{\phi},\boldsymbol{\Sigma})$], $\boldsymbol{\Sigma} > 0$, then moment estimators (maximum likelihood estimators) of the weights of the principal components of $\mathbf{X}$, and of the variances of the principal components of $\mathbf{X}$ are obtained by substituting $\hat{\boldsymbol{\Sigma}}$ for $\boldsymbol{\Sigma}$ in Theorem (9.2.2).

The theorem was originally proved by Girshick (1936); the proof was modernized by Anderson (1958).

An implication of Theorem (9.3.1) is that the sample latent roots and vectors are moment estimators (maximum likelihood estimators) as long as the population latent roots are distinct. Since there are many computer programs available for extracting latent roots and vectors of a matrix (see for instance, Appendix A), the sample principal components are easily computed.

Example (9.3.1): Consider a marketing problem associated with a certain cigarette brand, and let $\mathbf{X}: 4 \times 1$ denote a vector of four attributes of cigarette smokers. Suppose N smokers are observed and their attribute vectors are combined to yield the sample covariance matrix

$$\hat{\boldsymbol{\Sigma}} = \begin{pmatrix} 2.53 & 3.50 & 2.06 & 1.45 \\ 3.50 & 5.05 & 2.86 & 2.02 \\ 2.06 & 2.86 & 1.68 & 1.19 \\ 1.45 & 2.02 & 1.19 & .86 \end{pmatrix}.$$

Direct computation shows that $\hat{\boldsymbol{\Sigma}}$ is diagonalized (approximately) by the orthogonal matrix.

$$\hat{\boldsymbol{\Gamma}} = \begin{pmatrix} \frac{1}{2} & \frac{1}{2} & \frac{1}{2} & \frac{1}{2} \\ \frac{1}{\sqrt{2}} & -\frac{1}{\sqrt{2}} & 0 & 0 \\ \frac{1}{\sqrt{6}} & \frac{1}{\sqrt{6}} & 0 & \frac{2}{\sqrt{6}} \\ \frac{1}{2\sqrt{3}} & \frac{1}{2\sqrt{3}} & -\frac{3}{2\sqrt{3}} & \frac{1}{2\sqrt{3}} \end{pmatrix},$$

and that the latent roots of $\hat{\boldsymbol{\Sigma}}$ are 10, .1, .01, .02, respectively. Let $\hat{\mathbf{D}} \equiv \text{diag}\,(\hat{\lambda}_1,\hat{\lambda}_2,\hat{\lambda}_3,\hat{\lambda}_4) \equiv (10,.1,.02,.01)$; then, $\hat{\boldsymbol{\Sigma}} = \hat{\boldsymbol{\Gamma}}\hat{\mathbf{D}}\hat{\boldsymbol{\Gamma}}'$. Denote the columns of $\hat{\boldsymbol{\Gamma}}$ by $(\hat{\boldsymbol{\alpha}}_1,\hat{\boldsymbol{\alpha}}_2,\hat{\boldsymbol{\alpha}}_3,\hat{\boldsymbol{\alpha}}_4)$. The latent roots of $\hat{\boldsymbol{\Sigma}}$, $\hat{\lambda}_j$, are moment estimates of λ_j, and the columns of $\hat{\boldsymbol{\Gamma}}$, $\hat{\boldsymbol{\alpha}}_j$, are moment estimates of $\boldsymbol{\alpha}_j$, $j = 1,\ldots,4$. Note that the columns of $\hat{\boldsymbol{\Gamma}}$ may always be rearranged so as to conform with the ordering of the latent roots of $\hat{\boldsymbol{\Sigma}}$.

Suppose it is of interest to study those linear combinations of the components of $\mathbf{X}$ that have the greatest variance. The sample principal components are given by

$$\begin{aligned}
\hat{z}_1 &= \hat{\alpha}_1'\mathbf{x} = \frac{1}{2}x_1 + \frac{1}{\sqrt{2}}x_2 + \frac{1}{\sqrt{6}}x_3 + \frac{1}{2\sqrt{3}}x_4, \\
\hat{z}_2 &= \hat{\alpha}_2'\mathbf{x} = \frac{1}{2}x_1 - \frac{1}{\sqrt{2}}x_2 + \frac{1}{\sqrt{6}}x_3 + \frac{1}{2\sqrt{3}}x_4, \\
\hat{z}_3 &= \hat{\alpha}_3'\mathbf{x} = \frac{1}{2}x_1 \qquad\qquad\qquad\qquad\quad - \frac{3}{2\sqrt{3}}x_4. \\
\hat{z}_4 &= \hat{\alpha}_4'\mathbf{x} = \frac{1}{2}x_1 \qquad\qquad - \frac{2}{\sqrt{6}}x_3 + \frac{1}{2\sqrt{3}}x_4,
\end{aligned}$$

The next question of interest is: Is it really necessary to study all four principal components? If so, not much was gained by the transformation (there may be some advantages of interpretation in the new 4-dimensional coordinate system). Recall that the estimates of the variances of the principal components are the diagonal elements of $\hat{\mathbf{D}}$; moreover, the total variance of the components of $\mathbf{X}$ is the sum of the diagonal elements of $\mathbf{\Sigma}$; that is, if T denotes the total variance of the components of $\mathbf{X}$, $T = \text{tr}\,\mathbf{\Sigma}$, and $\hat{T} = \text{tr}\,\hat{\mathbf{\Sigma}}$. Therefore,

$$\hat{T} = \text{tr}\,\hat{\mathbf{D}} = \sum_{j=1}^{p} \hat{\lambda}_j.$$

In Example (9.3.1), $p = 4$, and $\hat{T} = 10.13$. Hence, $\hat{\lambda}_1$ accounts for 98.7 percent of the estimated total variance. Thus, it would appear that all principal components except the first may be dropped. However, in some problems, $(\hat{\lambda}_1/\text{tr}\,\hat{\mathbf{\Sigma}})$ is large mostly because of sampling variation. That is, interval estimates may show that the true proportion of variance removed by λ_1 is much smaller. Such problems of inference are discussed in the next section.

9.3.2 Inference in Principal Components

If the elements of $\mathbf{X}$ are standardized relative to the sample before the principal components are extracted, the distribution theory associated with the sample principal components and their variances is mostly intractable. In this case a simple recourse open to the researcher is to drop principal components only when they add but a few percent to the total estimated variance. Such a procedure is certainly valid in very large samples, and for many problems, it may not be too unreasonable even in moderate sized samples, assuming that variance maximization is a satisfactory criterion. Asymptotic distributional results are available for problems in which the variates have not been standardized (but hopefully, the variates have all been measured in the same units). Thus, if large samples are available it is usually possible to find interval estimates for the variances of the population principal components, and for the coefficients of the population principal components. Results are summarized in the following theorems.

Theorem (9.3.2): Let $\hat{\boldsymbol{\Sigma}}$ denote the maximum likelihood estimator for $\boldsymbol{\Sigma}$ based on a sample of size N from $N(\boldsymbol{\phi},\boldsymbol{\Sigma})$, and define $\hat{\boldsymbol{\lambda}} \equiv (\hat{\lambda}_1,\ldots,\hat{\lambda}_p)'$, $\boldsymbol{\lambda} \equiv (\lambda_1,\ldots,\lambda_p)'$, and $\hat{\lambda}_j$ and λ_j are the latent roots of $\hat{\boldsymbol{\Sigma}}$ and $\boldsymbol{\Sigma}$, respectively. Let $\mathbf{D} \equiv \text{diag}\,(\lambda_1,\ldots,\lambda_p)$. Then, if $\boldsymbol{\Sigma} > 0$, and all roots are distinct, so that $\lambda_1 > \lambda_2 > \cdots > \lambda_p > 0$,

$$\lim_{N\to\infty} \mathcal{L}\{\sqrt{N-1}\,(\hat{\boldsymbol{\lambda}} - \boldsymbol{\lambda})\} = N(\mathbf{0},2\mathbf{D}^2); \tag{9.3.1}$$

that is, $\hat{\lambda}_1,\ldots,\hat{\lambda}_p$ are asymptotically independent, unbiased, and Normally distributed with the variance of $\hat{\lambda}_j = 2\lambda_j^2/(N-1)$.

Theorem (9.3.3): For the same definitions and conditions as in Theorem (9.3.2), let $\hat{\boldsymbol{\alpha}}_j$ and $\boldsymbol{\alpha}_j$ denote the normalized latent vectors of $\hat{\boldsymbol{\Sigma}}$ and $\boldsymbol{\Sigma}$, respectively. Then, for $j = 1,\ldots,p$,

$$\lim_{N\to\infty} \mathcal{L}\{\sqrt{N-1}\,(\hat{\boldsymbol{\alpha}}_j - \boldsymbol{\alpha}_j)\} = N(\mathbf{0},\boldsymbol{\Lambda}_j), \tag{9.3.2}$$

where

$$\boldsymbol{\Lambda}_j = \lambda_j \sum_{\substack{i=1\\ j\neq i}}^{p} \frac{\lambda_i}{(\lambda_i - \lambda_j)^2}\, \boldsymbol{\alpha}_i\boldsymbol{\alpha}_i'.$$

Proof: Equations (9.3.1) and (9.3.2) are both proved in Girshick (1939).

[*Remarks:* It is emphasized that although the results given in Theorems (9.3.2) and (9.3.3) are approximately valid for large samples, they are valid only if the latent roots of $\boldsymbol{\Sigma}$ are distinct and if $\boldsymbol{\Sigma}$ is positive definite. The latter requirement is not restrictive, however, since singular covariance matrices do not arise too often in applications. Asymptotic distribution theory associated with singular $\boldsymbol{\Sigma}$'s is discussed in Anderson (1963) and the asymptotic moments are given in Girshick (1939). An important case in which the roots are not distinct is treated in Theorem (9.3.4).]

It is seen from (9.3.1) that for distinct roots, the sample latent roots of $\hat{\boldsymbol{\Sigma}}$ are distributed independently in large samples. Moreover, because of the simple distributional form, it is easy to obtain interval estimates of the variances of the principal components.

Note that because the sample mean and covariance matrix in Theorem (9.3.3) are not independent, Hotelling's T^2-distribution cannot be used to test hypotheses about $\boldsymbol{\alpha}_j$.

Example (9.3.2): For the same sample data as in Example (9.3.1), it is easy to compute an approximate interval estimate for the variance of the first principal component λ_1, for a sample of size $N = 101$. Recall that a point estimate is $\hat{\lambda}_1 = 10$. From (9.3.1), it is approximately true that

$$\mathcal{L}\left\{\frac{10(\hat{\lambda}_1 - \lambda_1)}{\lambda_1 \sqrt{2}}\right\} = N(0,1).$$

Hence, an equal tail 95 percent confidence interval for λ_1 is given approximately by

$$\left(\frac{\hat{\lambda}_1}{1 + \frac{2\sqrt{2}}{10}}, \frac{\hat{\lambda}_1}{1 - \frac{2\sqrt{2}}{10}}\right) = (7.7,13.9).$$

Note the large length of confidence interval possible even for N as large as 101. Since this type of situation in which confidence intervals are wide is not uncommon in applications, it is clear that large sources of variance corresponding to important relationships may be ignored if principal components are dropped merely on the basis of a superficial inspection of the sample variances (rather than upon an examination of interval estimates). Thus, if from Theorem (9.3.2) it is found that $1 - \alpha = P\{a_j < \lambda_j < b_j\}$, in large samples it is approximately true that

$$\frac{a_j}{\sum_1^p \hat{\lambda}_j} \leq \frac{\lambda_j}{\sum_1^p \lambda_j} \leq \frac{b_j}{\sum_1^p \hat{\lambda}_j}.$$

In Example (9.3.2) these bounds would imply $.76 \leq \dfrac{\lambda_j}{\Sigma\lambda_j} \leq 1$.

Theorem (9.3.4): Let $\hat{\mathbf{\Sigma}}: p \times p$ denote the sample covariance matrix for $\mathbf{\Sigma}$ based upon a sample of size N from $N(\boldsymbol{\phi},\mathbf{\Sigma})$, and define $\hat{\boldsymbol{\lambda}} \equiv (\hat{\lambda}_1,\ldots,\hat{\lambda}_p)'$, $\boldsymbol{\lambda} \equiv (\lambda_1,\ldots,\lambda_p)'$, where $\hat{\lambda}_j$ and λ_j are the latent roots of $\hat{\mathbf{\Sigma}}$ and $\mathbf{\Sigma}$, respectively. A likelihood ratio test of $H: \lambda_{j+1} = \cdots = \lambda_{j+r}$ versus $A: \lambda_{j+1},\ldots,\lambda_{j+r}$ are not all equal, is to reject H if $Q > c$, where for

$1 \leq j \leq p - r,\ 2 \leq r \leq p - 1,$

$$Q \equiv r(N-1)\log\bar{\lambda}_j - (N-1)\sum_{k=1}^{r}\log\hat{\lambda}_{j+k}, \tag{9.3.3}$$

where

$$\bar{\lambda}_j \equiv r^{-1}\sum_{k=1}^{r}\hat{\lambda}_{j+k},$$

and c is determined for any preassigned level of significance from the Wilks result that (see Section 8.4.3)

$$\lim_{N\to\infty} \mathcal{L}(Q) = \chi^2\left[\frac{r(r+1)}{2} - 1\right].$$

Proof: See Anderson (1963), p. 132.

This theorem provides a large sample procedure for testing whether the variances of any set of r successive principal components are equal. A useful consequence of this theorem is the following.

Corollary (9.3.1): For the conditions and definitions given in Theorem (9.3.4), the likelihood ratio procedure for testing

$$H: \mathbf{\Sigma} = \sigma^2 \begin{pmatrix} 1 & & & \\ & \cdot & \rho & \\ & & \cdot & \\ & \rho & \cdot & \\ & & & 1 \end{pmatrix}, \qquad -\frac{1}{p-1} < \rho < 1,$$

versus $A: \mathbf{\Sigma}$ is general, is to reject H if $Q > c$, where

$$Q = (p-1)(N-1)\log\bar{\lambda}_1 - (N-1)\sum_{k=1}^{p-1}\log\hat{\lambda}_{k+1}, \tag{9.3.4}$$

$$\bar{\lambda}_1 = (p-1)^{-1}\sum_{k=1}^{p-1}\hat{\lambda}_{k+1},$$

and c is determined from

$$\lim_{N\to\infty} \mathcal{L}(Q) = \chi^2\left[\frac{p(p-1)}{2} - 1\right].$$

Proof: Under H, $\mathbf{\Sigma}$ has an intraclass covariance pattern. Since under H, $\mathbf{\Sigma}$ may be written as $\mathbf{\Sigma} = \mathbf{\Gamma}\mathbf{D}_\lambda\mathbf{\Gamma}'$, where $\mathbf{D}_\lambda = \text{diag}(\lambda_1, \lambda_2, \ldots, \lambda_2)$, and where $\mathbf{\Gamma}$ does not depend upon the elements of $\mathbf{\Sigma}$ (see Section 2.8), to test H versus A is equivalent to testing $H^*: \mathbf{\Sigma}^* \equiv \mathbf{\Gamma}'\mathbf{\Sigma}\mathbf{\Gamma} = \mathbf{D}_\lambda$ versus $A^*: \mathbf{\Sigma}^*$ is general. Since the latent roots of $\mathbf{\Sigma}^*$ are the same as those of $\mathbf{\Sigma}$, the test is given by Theorem (9.3.4) with $j = 1$ and $r = p - 1$.

[*Remark:* If under H, $\mathbf{\Sigma}$ has intraclass covariance structure, a maximum likelihood estimator of λ_2 is easily found to be

$$\lambda_2^* = \frac{1}{p-1}\sum_{k=1}^{p-1}\hat{\lambda}_{1+k},$$

where $\hat{\lambda}_j$ denotes the jth latent root of $\hat{\mathbf{\Sigma}}$, the sample covariance matrix. Moreover, the asymptotic distribution of λ_2^* is, of course, Normal; specifically,

$$\lim_{N\to\infty} \mathcal{L}\{\sqrt{N-1}\,(\lambda_2^* - \lambda_2)\} = N\left(0, \frac{2\lambda_2^2}{p-1}\right). \tag{9.3.5}$$

Hence, in large samples it is straightforward to find confidence intervals for λ_2 from (9.3.5).]

9.4 PRINCIPAL COMPONENTS AND OTHER MODELS

Principal components analysis is often used in applied research as merely a first step in studying the reasonableness of various hypotheses about the phenomenon under investigation. Typically, the data will first be subjected to a principal components analysis to expose those linear combinations of variables with the greatest variance. Those with relatively small variance are dropped and the remaining subset of transformed variables is then used as the set of regressors in a regression, in conjunction with some dependent variables; or they may be further rotated according to some criterion in a factor analysis model (see Chapter 10); or they may be used in a clustering analysis (see Rao, 1964); or in a multivariate analysis of variance. For simplicity, in the remainder of this section, attention is confined to the use of principal components in regression. Its use in other models is completely analogous.

9.4.1 Principal Components in Regression

For regression problems in which principal component variates are useful, the following procedure is indicated (see, for instance, Massy, 1965). Let $\mathbf{X}$ denote a vector of independent variables that are candidates for the regressors in a regression. Let $\mathbf{\Gamma}$ denote the matrix of normalized latent vectors of $\mathbf{\Sigma}$, the covariance matrix of $\mathbf{X}$, and let $\lambda_1 \geq \lambda_2 \geq \cdots \geq \lambda_p$ denote the latent roots of $\mathbf{\Sigma}$.

Step (1): Subject the observation vectors to a principal components analysis to find the estimated matrix of weights $\hat{\mathbf{\Gamma}}$, and the estimated variances of the principal components, $\hat{\lambda}_j$, $j = 1,\ldots,p$.

Step (2): Rotate $\mathbf{X}$ into $\mathbf{X}^*$ by the transformation $\mathbf{X}^* = \hat{\mathbf{\Gamma}}\mathbf{X}$. The elements of $\mathbf{X}^*$ are the estimated principal components.

Step (3): Drop the elements of $\mathbf{X}^*$ with the smallest estimated variances, $\hat{\lambda}_j$. That is, use Theorem (9.3.2) to isolate the most important principal components. Denote the remaining principal components by $X_1^*, \ldots, X_r^*$, $r \leq p$.

Step (4): Let y denote a dependent scalar variable that is to be regressed on the independent variables. Regress y on $(X_1^*, \ldots, X_r^*)$.

[*Remark:* There has been implicit in the thinking of this chapter a basic notion of linearity. That is, the correlations discussed have simple interpretations only when the variables are linearly related. If the underlying relationships are in fact nonlinear, mathematical constructs such as principal components and correlation are difficult to interpret.]

9.5 APPLICATIONS OF PRINCIPAL COMPONENTS ANALYSIS

9.5.1 An Example of Principal Components in Quality Control

J. E. Jackson and R. H. Morris, working at the Eastman Kodak Company, were given the problem of determining whether or not the Company's development processes were under "control" (see Jackson and Morris, 1957). There are many variables involved and typically each of them is checked periodically (simultaneously) to ascertain whether or not that variable is still under control. The basic variables include color, tint, grain density, and so on, and they are correlated.

To measure picture quality they used optical filters and took sample data on 9 variables, using test strips of film. The 9 variables included

Covariance Matrix

High density			*Medium density*			*Low density*		
RED	GREEN	BLUE	RED	GREEN	BLUE	RED	GREEN	BLUE
177	179	95	96	53	32	−7	−4	−3
	419	245	131	181	127	−2	1	4
		302	60	109	142	4	4	11
			158	102	42	4	3	2
				137	96	4	5	6
					128	2	2	8
						34	31	33
							39	39
								48

Correlation Matrix

High density			Medium density			Low density		
RED	GREEN	BLUE	RED	GREEN	BLUE	RED	GREEN	BLUE
1	.66	.41	.57	.34	.21	−.09	−.05	−.03
	1	.69	.51	.75	.55	−.02	.01	.03
		1	.27	.54	.72	.04	.04	.09
			1	.69	.30	.06	.04	.02
				1	.73	.06	.06	.07
					1	.02	.02	.10
						1	.85	.81
							1	.89
								1

3 in the high optical density part of the film corresponding to red, green, and blue; 3 in the medium optical density part of the film (red, green, and blue); and 3 in the low optical density part of the film. They computed the covariance matrix and the associated correlation matrix, shown below, and tried to determine how to reduce the dimension of the problem so that only a few variables need to be examined for control. Note that in these matrices, the omitted elements below the diagonal are the symmetric images of those above the diagonal.

Using the sample covariance matrix, the sample principal components were computed. It was found that for their sample variances, $\hat{\lambda}_i$, $i = 1,\ldots,9$,

$$\left(\frac{100\hat{\lambda}_1}{\sum_1^9 \hat{\lambda}_i},\ldots,\frac{100\hat{\lambda}_5}{\sum_1^9 \hat{\lambda}_i}\right) = (61,14,9,7,6);$$

that is, the first sample principal component accounts for 61 percent of the total sample variance, and so on. Since the first five principal components accounted for 97 percent of the total sample variance, the other four components were dropped. The remaining five principal components were interpreted physically, and each of the five orthogonal variables was used as a control variable. In this case, not only was a more parsimonious representation achieved, but also, it was found that the new variables were easier to interpret than the original variables in terms of the chemical processes that are disarrayed when a picture is out of control. Unfortunately, although it is satisfying to have the interpretation simplified, it is not usually the case that the principal components have simpler interpretations than the original variables, as in this illustration.

9.5.2 An Application of Principal Components Analysis to Stock Market Indices

Feeney and Hester (1967) questioned the use of the equal weighting scheme used in computing the Dow Jones Index of 30 industrial stocks. They considered the alternative of a principal components weighting scheme, since such a price change index will be most sensitive if the weights are assigned in a way that captures the maximum variance of the set of reference stock prices over time. The principal components analysis was carried out for 3 different functions of the same set of variables.

First, they considered Index (a) which was based upon raw stock prices (adjusted only for splits and stock dividends).

Second, they considered Index (b) in which each stock price was standardized (the sample mean was subtracted and the result was divided by the sample standard deviation) before carrying out a principal components analysis. Thus, the latent roots and vectors are extracted from the correlation matrix instead of the covariance matrix.

The third set of variables studied were obtained by first standardizing, and then subtracting off a linear trend (just a term proportional to time). This was called Index (c).

The period covered includes the 50 consecutive quarters subsequent to December 31, 1950 (until June 30, 1963). All data were adjusted for stock dividends and splits.

The first two principal components for each of the three indices are shown in Table 9.5.1.

Comparison of this table with raw data shows that stocks which were high-priced in 1963 tend to be weighted heavily in Index (a). See for instance, duPont, General Foods, AT&T, Eastman Kodak, Owens-Illinois, and Union Carbide. Moreover, the behavior of the Dow Jones Index is highly correlated with that of the behavior of the first principal component for Index (a) showing that the Dow Jones Index is very informative if investors focus their attention on raw stock prices.

The weights for Index (b) seem to be dominated by time trends so that stocks that had a low rate of price appreciation during the 12-year period such as Chrysler, Swift, United Aircraft, American Can, Anaconda, and Westinghouse, have low weights. The other weights are approximately uniformly distributed over the remaining stocks.

It is seen that one-third of the weights of Index (c) are negative and the first principal component accounts for only 40 percent of the total variance, whereas those for Index (b) account for about 66 percent of the total variance. Hence, the time trends appear to account for the difference.

In general, it is seen that the initial form of the data must be carefully selected with the end objective in mind, since results of the attendant principal components analyses can differ substantially.

Table 9.5.1 Weights for First Two Principal Components of Price Indices *a*, *b*, and *c*

	a		*b*		*c*	
	(1)	(2)	(1)	(2)	(1)	(2)
Allied Chemical	.137	.039	.820	−.251	.837	−.049
Alcoa	.405	.308	.757	−.596	.424	−.851
American Can	.067	.035	.647	−.399	.895	.340
AT & T	.361	−.272	.800	.559	−.773	.013
American Tobacco	.123	−.081	.792	.498	.485	.487
Anaconda	.101	.168	.423	−.734	.911	−.164
Bethlehem	.217	.104	.849	−.416	.761	−.566
Chrysler	−.042	.025	−.305	−.199	.765	.493
duPont	.951	.217	.931	−.235	−.255	−.865
Eastman Kodak	.587	−.231	.900	.334	−.779	−.505
General Electric	.376	.047	.932	−.141	−.015	−.896
General Foods	.391	−.254	.841	.511	−.841	−.126
General Motors	.236	.009	.922	−.080	.510	−.405
Goodyear	.222	−.012	.958	.044	.562	−.453
International Harvester	.123	−.051	.836	.326	.636	.390
International Nickel	.297	−.061	.927	.159	−.123	−.465
International Paper	.156	.054	.917	−.335	.943	−.156
Johns-Manville	.150	−.004	.872	−.001	.732	.008
Owens-Illinois	.374	−.012	.943	.025	−.104	−.811
Procter and Gamble	.342	−.238	.823	.532	−.733	−.030
Sears	.319	−.185	.845	.455	−.694	−.030
Standard Oil (Cal.)	.195	.007	.863	−.089	.646	−.075
Esso	.198	.068	.793	−.334	.680	−.270
Swift	.037	.019	.344	−.194	.813	.360
Texaco	.251	−.090	.889	.271	−.371	−.024
Union Carbide	.422	.131	.903	−.286	.193	−.910
United Aircraft	.186	.155	.636	−.592	.721	−.321
U. S. Steel	.407	.136	.866	−.283	.207	−.886
Westinghouse	.145	.004	.670	−.006	.585	−.016
Woolworth	.175	−.124	.786	.537	.117	.348
Percent Variance	75.76	13.93	65.67	13.70	39.90	23.20

9.5.3 A Principal Components Analysis of Crime

Ahamad (1967) analyzed annual crime statistics for England and Wales by means of a principal components model. His avowed purpose was "to determine to what extent the variation in the numbers of crimes from year to year may be explained by a small number of unrelated factors." Since the method used for classifying crimes changed in 1950, the study period was restricted to the years 1950–1963. Moreover, since the true number of crimes committed is never known, the reported number of crimes was used as a surrogate. Eighteen crime types were studied and the 18×18 sample correlation matrix was used as the basis for extracting principal components (based upon 14 observations). The raw data (Y_j) is shown in Table 9.5.2, and the sample correlation matrix is shown in Table 9.5.3. The first three principal components that resulted from the standardized variables, Z_j, $j = 1, \ldots, 18$, are shown in Table 9.5.4.

It was found that 92 percent of the total variance was "explained" by the first three principal components. By correlating changes in population over the same time period with principal components, the author concluded that increases in crime are largely traceable to population growth. However, Walker (1967) in a critique of Ahamad's work pointed out that "time is the most important factor, and that if time had been eliminated from all the correlations a very different picture would have emerged, which might have revealed more interesting relationships between the numbers of crimes." Thus, it is clear again, as in the previous illustration, that the results and interpretation of a principal components analysis depend strongly upon the form of the initial data. In the stock market illustration, Feeney and Hester did just what Walker suggests when they considered Index (c), the standardized data corrected for a linear time trend; and the results were strikingly different.

[*Remarks:* It should be noted that the form of the initial data and how to interpret them are not the only problems of concern in studying the principal components associated with a set of multivariate data. For example, the data vectors in the above illustrations have been assumed to be independent over time. This is probably a poor assumption and it is more likely that corrections need to be made for serial correlation (in each component of the data vectors).

There has been no discussion of inference in the above illustrations. In the crime illustration there were only 14 observations. Many of the coefficients in the principal components may not be significantly different from zero. Is the sample size large enough for asymptotic theory to be valid?

The limiting distributions may not always be Normal. For example, in the stock market illustration, the distributions may be multivariate stable (see Chapter 6) but non-Normal.]

Table 9.5.2 Recorded Number of Offences for Groups of Offences Used in the Study, 1950–1963

Variate	1950	1951	1952	1953	1954	1955	1956	1957	1958	1959	1960	1961	1962	1963
Y_1	529	455	555	456	487	448	477	491	453	434	492	459	504	510
Y_2	5,258	5,619	5,980	6,187	6,586	7,076	8,433	9,774	10,945	12,707	14,391	16,197	16,430	18,655
Y_3	4,416	4,876	5,443	5,680	6,357	6,644	6,196	6,327	5,471	5,732	5,240	5,605	4,866	5,435
Y_4	8,178	9,223	9,026	10,107	9,279	9,953	10,505	11,900	11,823	13,864	14,304	14,376	14,788	14,722
Y_5	92,839	95,946	97,941	88,607	75,888	74,907	85,768	105,042	131,132	133,962	151,378	164,806	192,302	219,138
Y_6	1,021	800	1,002	980	812	823	965	1,194	1,692	1,900	2,014	2,349	2,517	2,483
Y_7	301,078	355,407	341,512	308,578	285,199	295,035	323,561	360,985	409,388	445,888	489,258	531,430	588,566	635,627
Y_8	25,333	27,216	27,051	27,763	26,267	22,966	23,029	26,235	29,415	34,061	36,049	39,651	44,138	45,923
Y_9	7,586	9,716	9,188	7,786	6,468	7,016	7,215	8,619	10,002	10,254	11,696	13,777	15,783	17,777
Y_{10}	4,518	4,993	5,003	5,309	5,251	2,184	2,559	2,965	3,607	4,083	4,802	5,606	6,256	6,935
Y_{11}	3,790	3,378	4,173	4,649	4,903	4,086	4,040	4,689	5,376	5,598	6,590	6,924	7,816	8,634
Y_{12}	118	74	120	108	104	92	119	121	164	160	241	205	250	257
Y_{13}	20,844	19,963	19,056	17,772	17,379	17,329	16,677	17,539	17,344	18,047	18,801	18,525	16,449	15,918
Y_{14}	9,477	10,359	9,108	9,278	9,176	9,460	10,997	12,817	14,289	14,118	15,866	16,399	16,852	17,003
Y_{15}	24,616	21,122	23,339	19,919	20,585	19,197	19,064	19,432	24,543	26,853	31,266	29,922	34,915	40,434
Y_{16}	49,007	55,229	55,635	55,688	57,011	57,118	63,289	71,014	69,864	69,751	74,336	81,753	89,709	89,149
Y_{17}	2,786	2,739	2,598	2,639	2,587	2,607	2,311	2,310	2,371	2,544	2,719	2,820	2,614	2,777
Y_{18}	3,126	4,595	4,145	4,551	4,343	4,836	5,932	7,148	9,772	11,211	12,519	13,050	14,141	22,896

Table 9.5.3 Correlation Matrix

Variate	1	2	3	4	5	6	7	8	9	10	11	12	13	14	15	16	17
1																	
2	−.041																
3	−.371	−.133															
4	−.174	.969	−.070														
5	.128	.947	−.379	.881													
6	.013	.970	−.315	.940	.963												
7	.074	.961	−.339	.909	.992	.966											
8	.099	.923	−.377	.870	.974	.950	.975										
9	.157	.900	−.406	.822	.981	.915	.982	.967									
10	.347	.469	−.522	.371	.648	.550	.623	.751	.701								
11	.089	.954	−.173	.896	.948	.942	.942	.957	.909	.645							
12	.183	.941	−.322	.895	.943	.956	.936	.922	.887	.565	.950						
13	.175	−.502	−.534	−.488	−.385	−.367	−.389	−.354	−.346	−.027	−.535	−.389					
14	−.082	.972	−.234	.962	.930	.964	.944	.885	.875	.412	.893	.922	−.397				
15	.264	.876	−.471	.782	.961	.910	.943	.954	.945	.723	.920	.932	−.272	.834			
16	−.022	.976	−.075	.949	.926	.932	.942	.892	.891	.439	.931	.900	−.590	.958	.816		
17	.173	.201	−.545	.109	.339	.282	.335	.448	.414	.677	.304	.286	.428	.144	.498	.074	
18	.015	.957	−.167	.893	.959	.916	.954	.919	.926	.542	.944	.905	−.523	.909	.915	.927	.256

Table 9.5.4 Coefficients of Principal Components

Variate		*First component*	*Second component*	*Third component*
Homicide	Z_1	.085	.540	.797
Woundings	Z_2	.971	−.199	−.054
Homosexual offences	Z_3	−.311	−.809	.118
Heterosexual offences	Z_4	.917	−.294	−.162
Breaking and entering	Z_5	.992	.051	.021
Robbery	Z_6	.976	−.041	−.091
Larceny	Z_7	.992	.013	−.030
Frauds and false pretenses	Z_8	.982	.115	−.030
Receiving	Z_9	.966	.126	.024
Malicious injuries to property	Z_{10}	.642	.566	.112
Forgery	Z_{11}	.974	−.078	.085
Blackmail	Z_{12}	.961	.007	.083
Assault	Z_{13}	−.422	.734	−.350
Malicious damage	Z_{14}	.943	−.176	−.139
Revenue laws	Z_{15}	.953	.233	.075
Intoxication laws	Z_{16}	.945	−.280	.040
Indecent exposure	Z_{17}	.337	.751	−.302
Taking motor vehicle without consent	Z_{18}	.962	−.129	.016

The above illustrations are intended to emphasize the care that should be exercised in applying the principal components model to a matrix of multivariate data.

REFERENCES

Ahamad (1967).
Anderson (1958, 1963).
Feeney and Hester (1967).
Geisser (1965a).
Girshick (1936, 1939).
Hotelling (1933).
Jackson and Morris (1957).
James (1961).
Massy (1965).
Okamoto (1968).
Pearson (1901).
Rao (1964, 1980).
Walker (1967).

EXERCISES

In the exercises below, assume the sample size is large enough for asymptotic theory to be applicable, and where inferences are required, assume the observations are Normally distributed.

9.1 Suppose that for a sample of size 101 the sample covariance matrix for a set of three variables which are purported to measure administrative ability (in the same units) is given by

$$\hat{\boldsymbol{\Sigma}} = 4.1\begin{pmatrix} 1 & .50 & .49 \\ .50 & 1.10 & .48 \\ .49 & .48 & .90 \end{pmatrix}.$$

(a) Find the sample principal components and their sample variances.

(b) Test the hypothesis that the variance of the first principal component is 8 (use a 5 percent level of significance).

9.2 In Exercise 9.1 compute the sample correlation matrix and determine the answer to Part (a) for the sample correlation matrix. Why is there a difficulty in interpreting this result unless the sample size is large?

9.3 In Exercise 9.1, assume the population covariance matrix has intraclass covariance structure.

(a) Find the maximum likelihood estimates of the distinct latent roots.

(b) Find a 95 percent confidence interval for the multiple latent root.

9.4 Find 95 percent confidence intervals for the variances of the principal components in Exercise 9.1.

9.5 For the covariance matrix given in Example (8.4.1), find the principal components and their variances.

9.6 Suppose $\mathbf{X}:p \times 1$ follows a general multivariate **t**-distribution. Show that the principal components have the same geometric interpretation as they would if $\mathbf{X}$ were normally distributed.

9.7 Let $\mathbf{X}_1, \ldots, \mathbf{X}_N$ denote bivariate observations from a covariance stationary time series, so of course, the observations are not independent (in general). If we wish to find the principal components, how might a correction be made for the dependence of the observations? [Hint: see eqn. (8.3.16), and assume observations follow a first order Markov process.]

9.8 Let $\boldsymbol{\Sigma}:p \times p$ have latent roots $\lambda_1 \geq \lambda_2 \geq \ldots \geq \lambda_p > 0$. Explain why

$$g \equiv \frac{1}{p}\sum_{1}^{p} (\lambda_i - \bar{\lambda})^2$$

where $\bar{\lambda} = \frac{1}{p}\sum_1^p \lambda_i$ is a good measure of whether or not a principal components analysis would be useful in a given situation; the higher the value of g, the better.

10

FACTOR ANALYSIS AND LATENT STRUCTURE ANALYSIS

10.1 INTRODUCTION

Factor analysis is a technique of multivariate analysis that attempts to account for the correlation pattern in a set of observable random variables in terms of a minimal number of unobservable or latent random variables called factors. It is an approach that might be used as the first step in a sequence of investigations which are aimed at developing insight into the relationships among variables. A second step might involve more direct input-response analyses of the data (such as the regression model analyses discussed in Chapter 8).

In the factor analytic approach, a matrix of N observations of p correlated random variables is examined to see if the data could have been generated by a linear model involving a minimum number of unobservable random variables that are in a certain sense fundamental to the data generating process. (Problems of interpretation are compounded when nonlinear models[1] are used.) These fundamental variables (factors) and linear combinations of them are used to "explain" the observed data. The principal output of a factor analysis is usually the choice of weights that should be used in the linear combinations of factors. The overall effect of the analysis is a regrouping of the data into patterns which can sometimes be interpreted meaningfully.

The formulation and early development of factor analysis models had its genesis in the field of psychology, and is generally attributed to Spearman (1904). His efforts were extended and generalized by many

[1] McDonald (1962) has considered a nonlinear model of factor analysis.

contributors to the field. Applications of the various models used in factor analysis are now found throughout the behavioral and social sciences. For example, the models have proven useful in growth economics (see Megee, 1965); in marketing (see Harper and Baron, 1951; Harper, 1956; Pilgrim, 1959; Stoetzel, 1960; Vincent, 1962; Massy, 1963; Mukherjee, 1965); in consumer expenditure models (see Millican, 1959); in finance (see King, 1966); in personnel management (see Ewart, Seashore, and Tiffin, 1941; Lawshe and Satter, 1944; Lawshe and Alessi, 1946; Lawshe and Maleski, 1946; Lawshe, Dudek, and Wilson, 1948; Howard and Schutz, 1952; Schreiber, Smith, and Harrell, 1952; Glickman, 1958); in advertising (see Twedt, 1952; Twery, 1958; Kirsch and Banks, 1962; Ekeblad and Stasch, 1967); in education (see Carrol and Schweiker, 1951; Walsh, 1964); in geography (see Berry, 1961; Ahmad, 1965); in psychology (see Bartlett, 1953; Harman 1976). There are numerous other examples in political science, sociology, anthropology, and law. Some recent references summarizing the latest developments in factor analysis are: Cattell, 1978; Harman, 1976; Jöreskog and Sörbom, 1979; and Mulaik, 1972.

If insufficient attention is paid to the underlying assumptions of a factor analysis model and to the multiplicities of solutions possible, not surprisingly, interpretation of the resulting factor patterns may be equivocal. Armstrong (1967) wrote a useful expository article which should be viewed as required reading by anyone planning to use factor analysis on real data. He showed by example that by ignoring some of the underlying assumptions and theory, absurd conclusions can be reached. The assumptions that underlie factor analysis models were detailed in Anderson and Rubin (1956). When the underlying model assumptions are viewed critically, factor analysis can often provide great insight into the associative patterns underlying a set of multivariate data.

Section 10.2 presents a discussion of the basic factor analysis models used in this chapter. The models yield unique solutions up to a rotation of the coordinate system. Their parameters are estimated in Section 10.3. In Section 10.4, it is shown how to carry out statistical tests for the appropriateness of the models. Section 10.5 discusses an example involving the interpretation of factor analysis results. Section 10.6 discusses the indeterminacies of a factor model that are associated with arbitrary rotations. Section 10.7 relates the factor analysis model to the principal components model of Chapter 9.

Section 10.8 presents the latent structure analysis model and shows how it is related to the factor analysis model. The latent structure analysis model was developed in sociology (and, to some extent, psychology as well), as a technique for making inferences from discrete, cross classified data, which might be collections of responses of individuals to dichotomous questions in a questionnaire or sample survey. It is often of interest to try to explain the variations and associations in the response pattern by showing that there are really some underlying fundamental subpopulations into which the individuals can be classified that accounts for the observed pattern in a simple way. That is, the original population may be subdivided by the analysis into homogeneous groups (latent classes) that are interpretable as different levels of intelligence, or perhaps different levels of cultural sophistication, and so on. Although the probabilities of belonging to each of the groups, and of responding affirmatively (rather than negatively) to a specific question in a sample survey given that an individual belongs to a specific group (latent class), are initially not known (or observable), the latent class model of latent structure analysis provides a means of inferring these probabilities from the observed data (assuming the unknown parameters are identified). It will be seen that the model bears a strong relationship to the factor analysis model.

10.2 FACTOR ANALYSIS MODELS

10.2.1 The Orthogonal Factor Model

Let $\mathbf{X}: p \times 1$ be an observable random vector, with $E(\mathbf{X}) \equiv \boldsymbol{\theta}$, var $(\mathbf{X}) \equiv \boldsymbol{\Sigma}$. Suppose that each element of $\mathbf{X}$ may be thought of as having been generated by a linear combination of orthogonal, unobservable factors upon which some error has been superimposed. In matrix form, the basic model is given by

$$\underset{(p\times 1)}{\mathbf{X}} = \underset{(p\times m)}{\boldsymbol{\Lambda}} \cdot \underset{(m\times 1)}{\mathbf{f}} + \underset{(p\times 1)}{\boldsymbol{\theta}} + \underset{(p\times 1)}{\mathbf{e}}, \tag{10.2.1}$$

where $\mathbf{f}$ denotes the vector of factors, $\boldsymbol{\Lambda}$ is a matrix of coefficient parameters and is called the factor loading matrix, and $\mathbf{e}$ denotes the error or disturbance term.

The assumptions associated with (10.2.1) are enumerated and explicated below.

Assumptions of the Orthogonal Factor Model

(1) $\mathcal{L}(\mathbf{f}) = N(\mathbf{0},\mathbf{I}_m), \qquad m \leq p;$ (10.2.2)

(2) $\mathcal{L}(\mathbf{e}) = N(\mathbf{0},\mathbf{D}_\psi), \qquad$ where $\mathbf{D}_\psi \equiv \text{diag}\,(\psi_1,\ldots,\psi_p);$ (10.2.3)

(3) $\mathbf{e}$ and $\mathbf{f}$ are independent; (10.2.4)

(4) $\boldsymbol{\Lambda}'\mathbf{D}_\psi^{-1}\boldsymbol{\Lambda} \equiv \mathbf{D}_J: m \times m = \text{diag}\,(J_1,\ldots,J_m),$
where $J_1 > J_2 > \cdots > J_m.$ (10.2.5)

From (10.2.1)–(10.2.4) it is seen that $\boldsymbol{\Sigma} = \boldsymbol{\Lambda}\boldsymbol{\Lambda}' + \mathbf{D}_\psi$, so that $\boldsymbol{\Sigma}$ will be positive definite except for degenerate solutions.

The requirement of (10.2.2) that $m \leq p$ implies that the "fundamental structure" of the data is no more complicated than that of the observed data. If this were not true, there would be little to gain from the factor analysis. Moreover, the usual goal is to find a suitable m which is strictly less than p, and as small as possible, consistent with the data.

Assumption (1) provides the distribution required for maximum likelihood estimation of the parameters. Assumptions (1)–(3) imply that the observed data must be Normally distributed for this method of estimation to be applicable. If, as is often the case, the data are not Normal, the data could first be subjected to a Normality inducing transformation (see for instance, Section 8.10.3, or Rao, 1965, p. 357) before carrying out a maximum likelihood estimation factor analysis. The factors are taken to have zero mean since any nonzero mean could always be subtracted off and absorbed into $\boldsymbol{\theta}$. The factors are taken to have unit variance since, as will be seen in Section 10.3, the estimation equations are invariant with respect to scale changes of the factors. Finally, the factors are assumed to be orthogonal, or mutually uncorrelated. This idea is consistent with the notion of an elemental or simplest latent structure. In the next subsection, this assumption is relaxed to one which permits $\mathbf{f}$ to have an arbitrary correlation matrix, implying oblique rather than orthogonal axes.

It is seen in Assumption (2) that the errors are mutually independent, but are permitted to be heteroscedastic (differing variances); moreover, Assumption (3) requires that they be independent of the factors. Assumption (3) is imposed to ensure consistency of the estimators.

Assumption (4) is imposed in order to obtain a unique solution to the parameter estimation equations. That is, without Assumption (4), an infinity of solutions would be possible, and they would all be related through orthogonal transformations (see Section 10.6). By using Assumption (4), a particular solution is always assured. In some situations, it may seem inappropriate to use the solution implied by Assumption (4). In other cases the solution in a rotated coordinate system may be preferable. It will be seen that such solutions may always be obtained by multiplying the matrix of parameter estimates of the factor loadings by an appropriate orthogonal matrix. An interpretation of (10.2.5) is that the columns of $\mathbf{\Lambda}$ are orthogonal with respect to a weighting function. The weights, ψ_j, $j = 1,\ldots,p$, are called the *specific variances* (or specificities), and are defined in (10.2.3). The unknown parameters $(\mathbf{\Lambda},\mathbf{D}_\psi)$ are estimated in Section 10.3.

10.2.2 The Oblique Factor Model

In some situations there is no reason to believe that the factors are orthogonal; in fact, it is often believed a priori that they are nonorthogonal, or oblique.

For example, suppose the observable vectors represent socioeconomic characteristics of buyers of a certain type of automobile. The latent factors, though different from one another, probably all depend in some complicated way upon the utility function of the buyer. Therefore, it is quite likely that the factor structure is composed of mutually correlated factors.

It will be seen in Section 10.3 that the estimation equations are invariant with respect to changes in scale of the factors. Therefore, to generalize (10.2.2), it is sufficient to permit the factors to have an arbitrary correlation matrix; that is, it will be assumed that

$$\mathcal{L}(\mathbf{f}) = N(\mathbf{0},\mathbf{R}), \qquad \mathbf{R} > 0, \tag{10.2.6}$$

where $\mathbf{R}: m \times m$ denotes the correlation matrix of $\mathbf{f}$. Thus, all factors are assumed to be standardized (zero mean and unit variance), but correlated. The oblique factor model is given by (10.2.1)–(10.2.5), with (10.2.2) replaced by (10.2.6).

10.3 ESTIMATION (Including Factor Scores)

The procedure customarily used in factor analysis is to fix m, the number of factors, beforehand, and then estimate $\mathbf{\Lambda}$ and $\mathbf{D}_\psi$. Estimation is followed by hypothesis testing which, in turn, is followed by interpretation of the factor loadings. The parameters are estimated below by the method of maximum likelihood developed for the factor analysis model by Lawley (1940).

10.3.1 Estimation in the Orthogonal Factor Model

First note from (10.2.1)–(10.2.4) that

$$\text{var}(\mathbf{X}) \equiv \mathbf{\Sigma} = \mathbf{\Lambda\Lambda}' + \mathbf{D}_\psi. \tag{10.3.1}$$

Suppose $\mathbf{x}_1,\dots,\mathbf{x}_N$ are independent, p-variate observations of $\mathbf{X}$. The maximum likelihood estimators of the mean and covariance matrix are

$$\hat{\boldsymbol{\theta}} = \bar{\mathbf{x}} = \frac{1}{N}\sum_{1}^{N} \mathbf{x}_j,$$

$$\hat{\mathbf{\Sigma}} = \mathbf{S} \equiv \frac{\mathbf{A}}{N} \equiv \frac{1}{N}\sum_{1}^{N} (\mathbf{x}_j - \bar{\mathbf{x}})(\mathbf{x}_j - \bar{\mathbf{x}})'. \tag{10.3.2}$$

Since $\mathbf{X}$ is Normally distributed, it follows from Theorem (5.1.2) that $\mathcal{L}(\mathbf{A}) = W(\mathbf{\Sigma},p,n)$, $n \equiv N - 1$, with density for $p \leq n$ given by

$$p(\mathbf{A}) \propto \frac{|\mathbf{A}|^{(n-p-1)/2}}{|\mathbf{\Sigma}|^{n/2}} \exp\left(-\frac{1}{2} \operatorname{tr} \mathbf{\Sigma}^{-1}\mathbf{A}\right). \qquad (10.3.3)$$

Recall that $\mathcal{L}(\bar{\mathbf{x}}) = N(\boldsymbol{\theta},N^{-1}\mathbf{\Sigma})$, and $\bar{\mathbf{x}}$ is independent of $\mathbf{A}$. Hence, multiplying the densities of $\bar{\mathbf{x}}$ and $\mathbf{A}$, taking logs, and making use of (10.3.1) gives the log-likelihood function

$$L = c + \left(\frac{n-p-1}{2}\right) \log |\mathbf{A}| - \frac{N}{2} \log |\mathbf{\Lambda\Lambda}' + \mathbf{D}_\psi| - \frac{1}{2} \operatorname{tr} (\mathbf{\Lambda\Lambda}' + \mathbf{D}_\psi)^{-1}\mathbf{A}$$
$$- \frac{N}{2} (\bar{\mathbf{x}} - \boldsymbol{\theta})'(\mathbf{\Lambda\Lambda}' + \mathbf{D}_\psi)^{-1}(\bar{\mathbf{x}} - \boldsymbol{\theta}), \qquad (10.3.4)$$

where c is a numerical constant. The last term in (10.3.4) may be set equal to zero by noting that it vanishes if $\hat{\boldsymbol{\theta}} = \bar{\mathbf{x}}$. It is now of interest to estimate $\mathbf{\Lambda}$ and $\mathbf{D}_\psi$. First note that the estimators will not depend on the units of the factors. This may be seen as follows.

Suppose $\mathcal{L}(\mathbf{f}) = N(\mathbf{0},\mathbf{D}_\sigma^2)$, where $\mathbf{D}_\sigma = \operatorname{diag}(\sigma_1,\ldots,\sigma_m)$, $\sigma_j^2 = \operatorname{var}(f_j)$, $j = 1,\ldots,m$. Thus, from (10.2.1),

$$\mathbf{\Sigma} = \operatorname{var}(\mathbf{X}) = (\mathbf{\Lambda D}_\sigma)(\mathbf{\Lambda D}_\sigma)' + \mathbf{D}_\psi,$$

and if $\mathbf{\Lambda}_0 \equiv \mathbf{\Lambda D}_\sigma$,

$$\mathbf{\Sigma} = \mathbf{\Lambda}_0\mathbf{\Lambda}_0' + \mathbf{D}_\psi.$$

That is, no generality is lost by taking the factors to have unit variances and then estimating $\mathbf{\Lambda}$. Henceforth, it will be assumed that $\mathcal{L}(\mathbf{f}) = N(\mathbf{0},\mathbf{R})$, where $\mathbf{R}$ denotes the correlation matrix of $\mathbf{f}$ and in the orthogonal factor model, $\mathbf{R} = \mathbf{I}$. Estimation results for the orthogonal factor model are given in the following lemma. A discussion of the number of estimable parameters in this model is given in Section 10.4.1. See also the discussion of Assumption (4) at the end of Section 10.2.1.

Lemma (10.3.1): The maximum likelihood estimation equations for $\mathbf{\Lambda}$ and $\psi_1,\ldots,\psi_p$ in the orthogonal factor model are given by

$$\operatorname{diag}(\mathbf{D}_{\hat{\psi}} + \hat{\mathbf{\Lambda}}\hat{\mathbf{\Lambda}}') = \operatorname{diag}(\mathbf{S}), \qquad (10.3.5a)$$
$$\mathbf{S}\mathbf{D}_{\hat{\psi}}^{-1}\hat{\mathbf{\Lambda}} = \hat{\mathbf{\Lambda}}(\mathbf{I} + \hat{\mathbf{\Lambda}}'\mathbf{D}_{\hat{\psi}}^{-1}\hat{\mathbf{\Lambda}}), \qquad (10.3.5b)$$

where $\hat{\mathbf{\Lambda}}$ and $\mathbf{D}_{\hat{\psi}}$ are the maximum likelihood estimators of $\mathbf{\Lambda}$ and $\mathbf{D}_\psi$, respectively.

Proof: Setting $\hat{\boldsymbol{\theta}} = \bar{\mathbf{x}}$, differentiate (10.3.4) with respect to $\mathbf{\Lambda}$ and $\psi_1,\ldots,\psi_p$, and set the results equal to zero. The differentiation is somewhat complicated algebraically, but details may be found in Lawley and Maxwell

(1963), and in Anderson and Rubin (1956). It may be verified that t estimators yield a maximum value.

Equations (10.3.5a) and (10.3.5b) do not have a unique solution. H ever, if the constraints of (10.2.5) are imposed, a unique set of solutioı guaranteed. These equations were programmed using a rapidly conv ing algorithm devised by Jöreskog (1967). Moreover, a computer progr is available for solutions of these equations when the parameters subject to constraints, as is sometimes the case in certain experimen design problems; see Appendix A, routines UMLFA and RMLFA.

[*Remark:* It has been shown by Anderson and Rubin (1956) th $\sqrt{N}(\hat{\mathbf{\Lambda}} - \mathbf{\Lambda})$ has a limiting Normal distribution with zero mean. Ho ever, the asymptotic covariance matrix is extremely complicated.]

10.3.2 Estimation in the Oblique Factor Model

In the oblique factor model, it is assumed that

$$\text{corr}\,(\mathbf{f}) = \mathbf{R}.$$

Otherwise the model is identical to that of the orthogonal factor mode The maximum likelihood estimation equations are somewhat more com plicated than those in the orthogonal factor model as a consequence The results are given in the following lemma.

Lemma (10.3.2): The maximum likelihood estimation equations for $\mathbf{\Lambda}$, $\mathbf{R}$, and $\psi_1, \ldots, \psi_p$ in the oblique factor model are given by

$$\begin{aligned} \hat{\mathbf{R}}\hat{\mathbf{\Lambda}}'(\hat{\mathbf{\Lambda}}\hat{\mathbf{\Lambda}}' + \mathbf{D}_{\hat{\psi}})^{-1}[\mathbf{I}_p - \mathbf{S}(\hat{\mathbf{\Lambda}}\hat{\mathbf{\Lambda}}' + \mathbf{D}_{\hat{\psi}})^{-1}]& \\ = \hat{\mathbf{R}}\hat{\mathbf{\Lambda}}'[\mathbf{I}_p - (\hat{\mathbf{\Lambda}}\hat{\mathbf{\Lambda}}' + \mathbf{D}_{\hat{\psi}})^{-1}\mathbf{S}]\mathbf{D}_{\hat{\psi}}^{-1};& \quad (10.3.6a) \\ \mathbf{D}_{\hat{\psi}} = \text{diag}\,(\mathbf{S} - \hat{\mathbf{\Lambda}}\hat{\mathbf{R}}\hat{\mathbf{\Lambda}}');& \quad (10.3.6b) \\ \hat{\mathbf{R}}\hat{\mathbf{\Lambda}}'\mathbf{D}_{\hat{\psi}}^{-1}\hat{\mathbf{\Lambda}} + \mathbf{I}_m = (\hat{\mathbf{\Lambda}}'\mathbf{D}_{\hat{\psi}}^{-1}\hat{\mathbf{\Lambda}})^{-1}(\hat{\mathbf{\Lambda}}'\mathbf{D}_{\hat{\psi}}^{-1}\mathbf{S}\mathbf{D}_{\hat{\psi}}^{-1}\hat{\mathbf{\Lambda}}),& \quad (10.3.6c) \end{aligned}$$

where $\hat{\mathbf{\Lambda}}$, $\hat{\mathbf{R}}$, and $\mathbf{D}_{\hat{\psi}}$ are the maximum likelihood estimators of $\mathbf{\Lambda}$, $\mathbf{R}$, and $\mathbf{D}_{\psi}$, respectively.

Proof: Differentiate (10.3.4) (after setting $\hat{\boldsymbol{\theta}} = \bar{\mathbf{x}}$) with respect to $\mathbf{\Lambda}$, $\mathbf{R}$, and $\psi_1, \ldots, \psi_p$, and set the results equal to zero. Algebraic simplification gives the results in (10.3.6). It is straightforward to show that a maximum is actually obtained.

10.3.3 Estimation of Factor Scores

Suppose each of N people is given a battery of p tests, and $\mathbf{x}_1, \ldots, \mathbf{x}_N$ are independent p-vectors denoting the observed scores. The orthogonal factor model then becomes

$$\mathbf{x}_j = \Lambda \mathbf{f}_j + \theta + \mathbf{e}_j, \qquad j = 1, \ldots, N, \tag{10.3.7}$$

where

$$\mathcal{L}(\mathbf{f}_j) = N(\mathbf{0},\mathbf{I}), \qquad \mathcal{L}(\mathbf{e}_j) = N(\mathbf{0},\mathbf{D}_\psi),$$

$\mathbf{e}_j$ and $\mathbf{f}_j$ are independent and $\mathbf{D}_\psi \equiv \text{diag}(\psi_1,\ldots,\psi_p)$, $\Lambda: p \times m$, and $m < p$. Although the $\mathbf{f}_j$'s are unobserved factors, it is often of interest to estimate them.

For example, suppose the tests were achievement tests given to high school students, $m = 2$, $p = 100$, and the two latent factors correspond to verbal and quantitative abilities. A quantitatively oriented college (an engineering school, say) looking at the "raw" scores, $\mathbf{x}_j$, might have a difficult time determining which students were the most quantitatively oriented, and therefore, most likely to be academically successful were they to attend that engineering college. The "factor scores," $\mathbf{f}_j$, however, immediately yield the desired information.

There is, of course, a basic identifiability problem; the factor scores are not identified by the data, given this model. By making some additional structural assumptions about the model, however, it becomes possible to estimate the factor scores.

First note that, as before,

$$\text{var}\,\mathbf{x}_j = \Sigma = \Lambda\Lambda' + \mathbf{D}_\psi. \tag{10.3.8}$$

Now assume that $E(\mathbf{f}_j)$ is a linear function of the raw scores, $\mathbf{x}_j$; i.e., assume

$$(\underset{(m\times 1)}{\mathbf{f}_j}|\mathbf{x}_j) = \underset{(m\times p)}{\mathbf{A}}\underset{(p\times 1)}{\mathbf{x}_j} + \underset{(m\times 1)}{\mathbf{u}_j}, \qquad j = 1,\ldots,N, \tag{10.3.9}$$

where

$$E(\mathbf{u}_j) = 0, \qquad \text{var}(\mathbf{u}_j) = \sigma^2\mathbf{I}_m.$$

Thus, if $\mathbf{F} \equiv (f_1,\ldots,f_N)$, $\mathbf{U} \equiv (\mathbf{u}_1,\ldots,\mathbf{u}_N)$, and $\mathbf{X} \equiv (\mathbf{x}_1,\ldots,\mathbf{x}_N)$, the matrix form of the model becomes

$$(\mathbf{F}|\mathbf{X}) = \mathbf{AX} + \mathbf{U}.$$

If $\mathbf{F}$ were observable, and therefore known, we could estimate $\mathbf{A}$ by ordinary least squares, and find

$$\hat{\mathbf{A}} = (\mathbf{FX}')(\mathbf{XX}')^{-1}. \tag{10.3.10}$$

It is straightforward to show by putting (10.3.7) into matrix form involving all the observations that

$$E(\mathbf{FX}') = N\Lambda'.$$

Since by the law of large numbers $(\mathbf{FX}')$ will behave like $E(\mathbf{FX}')$ in large samples, it will be asymptotically true (from 10.3.10) that

$$\hat{\mathbf{A}} \approx N\hat{\Lambda}'(\mathbf{XX}')^{-1} = \hat{\Lambda}'\left(\frac{\mathbf{XX}'}{N}\right)^{-1},$$

where $\hat{\Lambda}$ denotes the estimated factor loading matrix. But denoting the sample covariance matrix by

$$\mathbf{S} = \frac{1}{N}\sum_{1}^{N} (\mathbf{x}_j - \mathbf{x})(\mathbf{x}_j - \mathbf{x})' = \frac{1}{N}\mathbf{XX}' - \hat{\theta}\hat{\theta}'$$

gives

$$\hat{\mathbf{A}} \approx \Lambda'(\mathbf{S} + \hat{\theta}\hat{\theta}')^{-1}.$$

Since from (10.3.9), $\hat{\mathbf{F}} = \hat{\mathbf{A}}\mathbf{X}$, or $\hat{\mathbf{f}}_j = \hat{\mathbf{A}}\mathbf{x}_j$,

$$\hat{\mathbf{f}}_j \approx \hat{\Lambda}'(\mathbf{S} + \hat{\theta}\hat{\theta}')^{-1}\mathbf{x}_j, \qquad j = 1, \ldots, N. \tag{10.3.11}$$

Factor scores can be estimated from eqn. (10.3.11).

10.4 TESTING FOR THE APPROPRIATE NUMBER OF FACTORS

This section provides a procedure (involving a sequence of likelihood ratio tests) for testing the appropriateness of the preselected number of factors in a factor analysis model. Such testing is of interest in Confirmatory Factor Analysis in which we begin the analysis with an a priori belief in a particular model that has a given number of factors. We then test the model to see if the data actually confirm the hypothesis. In Exploratory Factor Analysis, by contrast, we don't have a model in mind a priori, and we let the data determine the model.

10.4.1 Likelihood Ratio Testing

The likelihood ratio method may be used to develop a test of whether the orthogonal factor model hypothesis H_1: $\mathbf{\Sigma} = \mathbf{\Lambda\Lambda}' + \mathbf{D}_\psi$ is more consistent with the data than a general covariance matrix hypothesis H': $\mathbf{\Sigma} > 0$. Also, a procedure may be developed to test the oblique factor model hypothesis H_2: $\mathbf{\Sigma} = \mathbf{\Lambda R \Lambda}' + \mathbf{D}_\psi$ versus H': $\mathbf{\Sigma} \neq \mathbf{\Lambda R \Lambda}' + \mathbf{D}_\psi$.

In both models, under hypothesis H', $\mathbf{\Sigma}$ is estimated by $\mathbf{S} = \mathbf{A}/N$. Under H_1 or H_2, $\mathbf{\Sigma}$ is estimated by its maximum likelihood estimator $\hat{\mathbf{\Sigma}}$, which is obtained by solving either (10.3.5) or (10.3.6). The notation $\hat{\mathbf{\Sigma}}$ will be used without regard to which model has been selected.

It is straightforward to find that if λ denotes the likelihood ratio statistic for H_1 versus H', or H_2 versus H',

$$-2\log\lambda = n\log\frac{|\hat{\mathbf{\Sigma}}|}{|\mathbf{S}|} + N(\mathrm{tr}\mathbf{S}\hat{\mathbf{\Sigma}}^{-1} - p), \tag{10.4.1}$$

where $n \equiv N - 1$. The likelihood ratio test procedure dictates that H_1 or H_2 be rejected if $-2 \log \lambda > k$, where k is some fixed constant. However, since the distribution of $-2 \log \lambda$ is too complicated in small samples to permit the determination of k, the test cannot be carried out. In large samples, however, Wilks' Theorem [see Section 8.4.3, Step (4), page 217] asserts that under H_1 or H_2,

$$\lim_{N \to \infty} \mathfrak{L}(-2 \log \lambda) = \chi^2(r), \tag{10.4.2}$$

where $\chi^2(r)$ denotes a chi-square distribution with r degrees of freedom, and r must now be evaluated. Note first, however, that if $\hat{\mathbf{\Sigma}}$ is evaluated exactly, from (10.3.5a) and (10.3.6b) it will be true in either model that $\operatorname{tr}(\mathbf{S}\hat{\mathbf{\Sigma}}^{-1}) = \operatorname{tr}(\mathbf{I}_p) = p$. Hence, when $\hat{\mathbf{\Sigma}}$ is found very precisely, (10.4.1) reduces to

$$-2 \log \lambda = n \log \frac{|\hat{\mathbf{\Sigma}}|}{|\mathbf{S}|}. \tag{10.4.3}$$

However, if the computing algorithm has not converged rapidly to a precise solution for the factor loadings, (10.4.1) should be used instead [if the convergence has been very slow, it is even possible that use of (10.4.3) could yield a negative value for chi-square].

The degrees of freedom parameter, r, is defined to be the difference between the number of unconstrained parameters under H' and the number under H_1 or H_2. Consider first H_1 versus H'. Here, under H', there are $p(p+1)/2$ distinct unconstrained parameters in $\mathbf{\Sigma}$. Under H_1, there are mp parameters in $\mathbf{\Lambda}$ to be estimated, and p parameters in $\mathbf{D}_\psi$ to be estimated, for a total of $(mp + p)$. However, $\mathbf{\Lambda}'\mathbf{D}_\psi^{-1}\mathbf{\Lambda}$ which has $m(m+1)/2$ distinct parameters is constrained to be diagonal by (10.2.7). Hence, this condition imposes $(m(m+1)/2) - m$ more constraints on the problem. Effectively, under H_1, $\mathbf{\Sigma}$ has $[(mp + p) - (m(m+1)/2) + m]$ degrees of freedom. Thus,

$$r = \frac{p(p+1)}{2} - \left[(mp + p) - \frac{m(m+1)}{2} + m\right],$$

or, if $r = r_1$ for the orthogonal factor model, simplification gives

$$r_1 = \tfrac{1}{2}\{(p - m)^2 - (p + m)\}. \tag{10.4.4}$$

In the oblique factor model, under H', the result is the same. However, under H_2, the number of parameters to be estimated increases by the number of distinct elements in R, namely $m(m-1)/2$. That is, there are $m(m-1)/2$ fewer degrees of freedom available in this model. Subtracting $m(m-1)/2$ from r_1 gives r_2, the number of degrees of freedom in the oblique factor model; that is,

$$r_2 = \frac{p}{2}(p - 2m - 1). \tag{10.4.5}$$

Bartlett Correction

It was pointed out by Bartlett (1954) that if in (10.4.1), n is replaced by

$$n' = n - \frac{2p + 5}{6} - \frac{2}{3}m, \tag{10.4.6}$$

the convergence of $(-2 \log \lambda)$ to its limiting chi-square distribution will be more rapid.

Factor Selection by MAICE Criterion

We note in passing that determining the appropriate number of factors by carrying out a sequence of likelihood ratio tests (which are mutually dependent), leaves much to be desired. An alternative criterion developed by Akaike, 1973 and 1977, has an information theoretic basis. Akaike suggests that we compute the value of the Akaike Information Criterion (AIC) for models with different numbers of factors, and then select that model which yields the minimum AIC estimate (MAICE). The definition is:

$$\begin{aligned} \text{AIC} = &-2 \log (\text{maximized likelihood}) \\ &+2\,(\text{number of independent parameters}). \end{aligned}$$

In a numerical example, Akaike presents the data taken from a maximum likelihood factor analysis:

Number of Factors	AIC
1	123.46
2	−3.65
3	−17.85*
4	−14.68

and points out that the AIC attains its minimum value (*) for 3 factors; so we should conclude that a 3 factor model is the best fit to the data.

The MAICE criterion is also useful in other contexts, such as selecting among regression models, and model fitting in time series analysis.

Example (10.4.1): In studying the Los Angeles Standard Metropolitan Statistical Area, Harman (1967) collected data on five socioeconomic variables. The correlation matrix for the five variables was computed on the basis of 12 census tracts ($N = 12$); it is shown in Table 10.4.1.

Table 10.4.1 Correlation Matrix for Census Tract Data

Variable	(1)	(2)	(3)	(4)	(5)
(1) Total Population	1.00	.01	.97	.44	.02
(2) Median School Years	.01	1.00	.15	.69	.86
(3) Total Employment	.97	.15	1.00	.51	.12
(4) Miscellaneous Professional Services	.44	.69	.51	1.00	.78
(5) Median House Value	.02	.86	.12	.78	1.00

Assume the data are Normally distributed. We now carry out a maximum likelihood factor analysis on the correlation matrix. Note that since the sample correlation matrix *does not follow* a Wishart distribution (under H_1, H_2, or H'), it is not immediately clear that the estimation procedures of Sect. 10.3 are applicable. [The exact distribution of the sample correlation matrix is extremely complicated and is expressible in terms of zonal polynomials (defined in Sect. 6.6). Since correlations are simple functions of moments, methods of Section 4.5 may be used to calculate the asymptotic distribution of the sample correlation matrix, and functions of it (see Olkin and Siotani, 1964)]. The sample correlation matrix may still be used for maximum likelihood factor analysis, however, because of the scale invariance property given below.

It may be checked that by setting $\hat{\mathbf{\Lambda}} = \mathbf{D}_s\hat{\mathbf{\Lambda}}_0$, and $\mathbf{D}_{\hat{\psi}} = \mathbf{D}_s\mathbf{D}_{\hat{\psi}_0}\mathbf{D}_s$, with $\mathbf{S} = \mathbf{D}_s\hat{\mathbf{R}}\mathbf{D}_s$, where $\hat{\mathbf{R}}$ is the estimated data correlation matrix and $\mathbf{D}_s \equiv \operatorname{diag}(s_1, \ldots, s_p)$, with s_j the estimated standard deviation of X_j, the same equations as in (10.3.5a) and (10.3.5b) are obtained. For this reason, the sample correlation matrix (instead of the sample covariance matrix) may be used in the analysis. MLE's of the factor loadings and error variances based upon the sample covariance matrix may then be obtained simply, by means of the above transformations. For example, if $(\hat{\lambda}_{ij}^{(0)}, \hat{\psi}_i^{(0)})$ denote the MLE's based upon $\hat{\mathbf{R}}$, $(\hat{\lambda}_{ij}, \hat{\psi}_i) \equiv (s_i\hat{\lambda}_{ij}^{(0)}, s_i^2\hat{\psi}_i^{(0)})$ are the MLE's corresponding to $\hat{\Sigma}$, $i = 1, \ldots, p$, $j = 1, \ldots, m$.

Example (10.4.2): Suppose in an oblique factor analysis involving ten attributes of 200 known drug abusers (obtained from police records), an effort is made to "fit the data" with two factors. It is desired to test whether the two-factor model provides an acceptable fit at the 5 percent level of significance.

To focus attention on the testing problem, assume the likelihood ratio statistic (using the Bartlett Correction) is given by $\lambda = .36$. Note that in the notation of this chapter, $p = 10$, $N = 200$, $m = 2$. Moreover, since

N is large, the distribution of $(-2 \log \lambda)$ is approximately chi-square with r_2 df, where r_2 is defined in (10.4.5). That is,

$$\mathcal{L}(-2 \log \lambda) \cong \chi^2(25).$$

Moreover, when $\lambda = .36$, $-2 \log \lambda \cong 2$. Reference to Appendix B shows that the 95 percent fractile point of $\chi^2(25)$ is 37.7. Hence, H_2: $\mathbf{\Sigma} = \mathbf{\Lambda R \Lambda}' + \mathbf{D}_\psi$ cannot be rejected. That is, the data do not seem incompatible with a two-factor oblique model. Note, however, that it is still possible that a single factor model would be sufficient.

10.5 INTERPRETATION OF A FACTOR ANALYSIS

An extremely important, but often troublesome, aspect of factor analysis is the interpretaion of results. In a typical analysis, $\mathbf{\Lambda}$ and $\mathbf{D}_\psi$ (and possibly $\mathbf{R}$) are estimated, and then significance tests are made on the appropriateness of the model. Next, an effort is made to find significance in patterns detected in the factor loading matrix.

Since correlations (between observable random variables and factors) are often of greater interest in a factor analysis than variances, it is frequently reasoned that the underlying patterns should not be too sensitive, in such analyses, to location and scale changes (changes of units) and such changes might even bring the patterns into closer focus. For this reason, the sample covariance matrix is sometimes converted into the sample correlation matrix prior to subjecting the data to a factor analysis. Thus, if $\mathbf{Y} \equiv (Y_j)$, and

$$Y_j = \frac{X_j - \theta_j}{(\text{var } X_j)^{1/2}}, \qquad j = 1, \ldots, p,$$

where $\mathbf{X}$ is defined in (10.2.1), and if var $(\mathbf{Y}) = \mathbf{Q}$, the correlation matrix, the standardized (orthogonal) model becomes

$$\mathbf{Y} = \mathbf{\Lambda f} + \mathbf{e}, \qquad \mathbf{Q} = \mathbf{\Lambda\Lambda}' + \mathbf{D}_\psi. \tag{10.5.1}$$

Then, if $\mathbf{\Lambda} \equiv (\lambda_{ij})$,

$$\text{cov }(\mathbf{Y},\mathbf{f}) = E\mathbf{Y}\mathbf{f}' = E(\mathbf{\Lambda f} + \mathbf{e})\mathbf{f}' = \mathbf{\Lambda} E\mathbf{f}\mathbf{f}' = \mathbf{\Lambda},$$

so that λ_{ij} may be interpreted as the covariance of Y_i with f_j. Since Y_i and f_j are both standardized, λ_{ij} is the correlation between Y_i and f_j. Recall that, as in Example (10.4.1), factor analysis of the model in (10.5.1) cannot be carried out by maximum likelihood. However, the method of principal components (Chapter 9), or some other method, might be used, assuming the sample size is large enough for the standardization to be interpretable relative to the population (see Section 9.1).

Example (10.5.1): Table 10.5.1 shows the estimated factor loading matrix for a seven-factor analysis carried out by King (1966) on the prices of 63 securities. Seven factors were selected to test the hypothesis that seven collections of stocks move in groups. Observations taken from June 1927 to December 1960 were standardized so that the analysis was carried out using the estimated correlation matrix. The securities were rearranged in the table so as to focus attention on particular groupings.

Reference to the table shows that the first column of the $\mathbf{\Lambda}$ matrix contains relatively large entries only; that is, regardless of which security is studied, the loading is fairly large. The interpretation is that there is a basic market factor operating which has a major effect on all securities.

Scrutiny of the remaining columns of $\mathbf{\Lambda}$ shows that large entries tend to group the stocks by industry. Thus, the largest entries in the second column tend to correspond to tobacco stocks, those in the third column correspond to oil and oil related stocks, and so forth. Thus, of the seven factors in the model, the first is a market factor and the remaining six are industry factors.

Of course the fact that seven factors were selected for this model and seven easily interpretable factors resulted is not coincidence; rather, it is an example of how important it is to carry out a study by first formulating a hypothesis, and then using the data to test the hypothesis. Sometimes, the data will suggest new hypotheses. In such cases, the hypotheses suggested should then be tested with new sets of data.

[*Remark:* When large samples are available, some of the arbitrariness in isolating factors may be eliminated by splitting the sample into several parts and carrying out the factor analysis on each subsample. If a particular factor pattern persists through each subsample, greater confidence may then be placed on the results.]

10.6 THE ROTATION PROBLEM OF FACTOR ANALYSIS

A fundamental problem of factor analysis which leads to multiplicities of solutions is the one of rotation. That is, since

$$\mathbf{\Sigma} = \mathbf{\Lambda}\mathbf{\Lambda}' + \mathbf{D}_\psi,$$

if $\mathbf{\Delta}$ is any orthogonal matrix (conformable with $\mathbf{\Lambda}$),

$$\mathbf{\Sigma} = (\mathbf{\Lambda}\mathbf{\Delta})(\mathbf{\Lambda}\mathbf{\Delta})' + \mathbf{D}_\psi = \mathbf{\Lambda}^*\mathbf{\Lambda}^{*\prime} + \mathbf{D}_\psi,$$

where $\mathbf{\Lambda}^* \equiv \mathbf{\Lambda}\mathbf{\Delta}$. Thus, regardless of which factor loading estimate is used, it is always possible to rotate $\mathbf{\Lambda}$ by an orthogonal matrix to yield a new estimate, $\mathbf{\Lambda}^*$, that will have the same associated $\mathbf{\Sigma}$. Note also that while $\mathbf{\Sigma}$ provides $p(p + 1)/2$ distinct numerical values, there are a generally

Table 10.5.1 Factor Loading Matrix for Securities

Security	*Factor* 1	*Factor* 2	*Factor* 3	*Factor* 4	*Factor* 5	*Factor* 6	*Factor* 7
American Snuff	.550	.360	−.155	−.120	−.040	.072	.031
American Tobacco	.666	.468	−.168	−.150	−.039	.009	.067
Bayuk Cigars	.567	.168	−.038	−.102	−.041	−.039	.133
Consolidated Cigar	.699	.099	−.097	−.043	−.010	.037	.079
General Cigar	.624	.200	−.042	−.084	−.024	−.000	.022
G. W. Helme	.392	.344	−.122	−.209	.039	.103	−.037
Liggett & Myers	.657	.450	−.111	−.092	−.062	−.026	−.001
P. Lorrilard	.620	.405	−.054	−.135	−.158	−.000	.100
Philip Morris	.514	.268	−.014	−.062	−.126	−.081	.125
Reynolds Tobacco	.625	.416	−.140	−.133	−.094	.033	.093
U. S. Tobacco	.376	.346	−.077	−.126	−.139	.034	.109
Continental Oil	.787	−.079	.443	−.044	−.156	−.200	−.093
Standard Oil (N. J.)	.753	−.115	.389	−.054	−.186	−.079	−.075
Texaco Inc.	.756	−.138	.466	−.060	−.176	−.132	−.114
Atlantic Refining	.709	−.092	.390	−.071	−.186	−.113	−.036
Pure Oil	.738	−.146	.464	−.020	−.104	−.198	−.166
Shell Oil	.736	−.165	.366	−.015	−.065	−.145	−.123
Skelly Oil	.700	−.173	.493	−.052	−.181	−.139	−.118
Socony Mobil Oil	.745	−.083	.306	−.084	−.066	−.088	−.090
Sun Oil	.545	−.101	.286	−.061	−.118	−.037	−.067
Tidewater Oil Co.	.722	−.096	.437	−.101	−.114	−.152	−.114
Union Oil of Calif.	.730	−.073	.351	−.100	−.114	−.085	−.086
Republic Steel	.871	−.117	−.008	.264	−.036	−.098	−.038
Amer. Smelting & Ref.	.832	−.007	−.018	.205	−.031	−.115	−.035
Amer. Steel Foundries	.856	−.049	−.009	.183	.025	−.130	−.041
Bethlehem Steel	.828	−.151	−.051	.315	−.013	−.102	−.033
Calumet and Hecla	.762	−.077	−.037	.239	.019	−.120	−.046
Inland Steel	.772	−.090	−.042	.239	−.053	−.103	.045
Inspirat. Cons. Copper	.771	−.028	−.068	.254	.053	−.146	−.075
Interlake Iron Corp.	.736	−.192	−.108	.261	.134	−.109	−.033
Magma Copper	.716	−.023	.024	.241	−.085	−.149	−.018
U. S. Steel	.814	−.177	−.045	.297	.031	−.095	−.065
Vanadium Corp. Amer.	.821	−.068	−.016	.190	−.039	−.024	−.065
Chesapeake & Ohio	.794	.004	−.125	−.014	.222	.046	−.144
Southern Pacific	.857	−.077	−.151	.016	.348	−.063	−.107
Atch. Topeka & S. Fe	.832	−.088	−.093	.010	.320	−.059	−.147
Louisville & Nashville	.771	−.045	−.116	.032	.295	−.085	−.112
Kansas City Southern	.788	−.065	−.068	−.051	.296	−.007	−.160
Missouri Kansas Tex.	.680	−.027	−.117	−.045	.331	.033	−.219
Northern Pacific	.833	−.058	−.077	−.024	.325	−.066	−.153
Union Pacific	.811	−.073	−.077	.003	.252	−.041	−.106
New York Central	.828	−.081	−.126	.072	.310	−.077	−.142
Reading Co.	.771	−.149	−.036	.085	.287	−.130	−.139
Allegheny Power Syst.	.785	.000	−.107	−.142	−.022	.357	−.068
Amer. & Foreign Power	.708	−.036	−.102	−.017	−.051	.225	.000
Brooklyn Union Gas	.653	.065	−.118	−.224	−.065	.376	.023
Columbia Gas System	.756	−.016	−.107	−.083	−.062	.346	−.057
Consolidated Edison	.757	.048	−.146	−.207	−.043	.400	.005
Laclede Gas Co.	.519	−.034	−.067	−.195	.100	.195	−.001
Peoples Gas	.711	−.025	−.126	−.148	−.030	.305	.062
Southern Calif. Ed.	.689	.077	−.091	−.242	−.139	.445	.011
Detroit Edison	.657	.034	−.080	−.164	−.090	.306	.037
Pacific Gas & Elect.	.698	.092	−.091	−.249	−.138	.454	−.006
Montgomery Ward	.796	.090	−.089	−.005	−.154	.010	.246
City Stores	.513	.032	−.088	−.077	−.043	−.028	.282
Arnold Constable	.709	.023	−.042	−.008	−.167	−.013	.283
Associated Dry Goods	.846	.080	−.087	.004	−.105	−.040	.237
Gimbel Bros.	.790	−.001	−.002	−.009	−.120	−.052	.229
S. S. Kresge	.705	.084	−.109	−.063	−.169	.107	.255
S. H. Kress	581	.156	−.073	−.112	−.146	.085	.196
May Dept. Stores	.802	.112	−.070	.014	−.175	−.052	.278
Outlet Co.	.423	.118	−.028	−.205	−.010	.009	.186
Sears Roebuck	.783	.053	−.050	−.038	−.189	.035	.276

greater number of $(mp + p)$ parameters to be estimated, so that a unique solution is generally impossible. This is an identification problem.

10.6.1 Sampling Approaches for Obtaining Unique Solutions

Various procedures have been proposed for eliminating the ambiguity in the factor analysis solution due to rotation. The assumption in (10.2.5) was one approach. Another involves the five rules for *simple structure* suggested by Thurstone (1947). In this procedure, it is specified dogmatically that zeros should appear in specific frequencies, in each of the rows and columns of the factor loading matrix, $\mathbf{\Lambda}$. The implication of this approach is that by so doing, a simple interpretation for the data will result. The fact that there is considerable room for subjectivity in assigning the zero elements of the $\mathbf{\Lambda}$ matrix has led to some criticism of the method. However, in this respect, the technique is somewhat similar to the Bayesian approach to introducing subjectivity into an analysis (see Section 10.6.2). Suppose that by some assessment scheme, enough of the elements of $\mathbf{\Lambda}$ can be preassigned so that the identification problem is eliminated (most of the preassigned elements will usually be taken as zero). The method of maximum likelihood can now be applied to the remaining elements of $\mathbf{\Lambda}$ (and those of $\mathbf{D}_\psi$) by differentiating the log-likelihood function with respect to unpreassigned elements only. The solution may be obtained numerically using the same computer routine developed by Jöreskog (1967) for the case in which all elements of $\mathbf{\Lambda}$ are unknown (the routine has been modified to permit some elements to be preassigned).

Other sampling theory suggestions for eliminating the rotational ambiguity in a factor analysis solution have involved various criteria which depend in some way on the variation or dispersion among the λ_{ij}^2's that results after an orthogonal transformation of $\mathbf{\Lambda}$. These approaches are typified by that of the *varimax rotation* procedure of Kaiser (1958).

Kaiser defined the *simplicity of a factor* as the variance of its squared factor loadings. That is, the simplicity corresponding to the kth element of the factor vector is the variance of the squared loadings in the kth column of $\mathbf{\Lambda}$. Accordingly, define the simplicity of the kth factor as

$$s_k = \frac{1}{p}\sum_{j=1}^{p}(\lambda_{jk}^2)^2 - \left(\frac{1}{p}\sum_{j=1}^{p}\lambda_{jk}^2\right)^2, \qquad (10.6.1)$$

for $k = 1,\ldots,m$. The *total simplicity* is $s \equiv \Sigma_1^m s_k$, and Kaiser originally suggested in his doctoral dissertation that the elements of the orthogonal rotation matrix be selected so that s is maximized. This is called the *raw varimax criterion.* In 1958, however, he suggested that the rotation result would be more satisfying if the squared loadings were normalized, or weighted, prior to computing the column variances. The weight suggested for any coefficient in the jth row of $\mathbf{\Lambda}$ is that part of the total variance (associated with the jth element of the observation vector) which is attributable to the common factors. In factor analysis jargon, this is called the jth *communality*.[2] Thus, if c_j^2 denotes the jth communality, and if $\mathbf{\Sigma} \equiv (\sigma_{ij})$,

$$c_j^2 \equiv \sigma_{jj} - \psi_j. \tag{10.6.2}$$

Moreover, in the orthogonal factor model, since $\mathbf{\Sigma} = \mathbf{\Lambda\Lambda}' + \mathbf{D}_\psi$,

$$c_j^2 = \sum_{\alpha=1}^{m} \lambda_{j\alpha}^2, \qquad j = 1,\ldots,p. \tag{10.6.3}$$

From (10.6.2) it is clear that c_j^2 is invariant under orthogonal rotations of $\mathbf{\Lambda}$. Then the *normalized total simplicity* becomes

$$s^* = \sum_{k=1}^{m} s_k^*,$$

where

$$s_k^* = \frac{1}{p}\sum_{j=1}^{p}\left(\frac{\lambda_{jk}^2}{c_j^2}\right)^2 - \left(\frac{1}{p}\sum_{j=1}^{p}\frac{\lambda_{jk}^2}{c_j^2}\right)^2. \tag{10.6.4}$$

If the elements of an orthogonal rotation matrix are selected so as to maximize s^*, the rotation procedure is called the *normal varimax rotation* criterion. Factor analysis computer programs that incorporate the normal varimax rotation criterion as a final rotation after the maximum likelihood solution is obtained are listed in Appendix A.

10.6.2 A Bayesian Approach to Unique Solutions

Since the maximum likelihood factor analysis solution is unique only up to a final rotation of the factor axes, it is tempting to try to eliminate the ambiguity in the solution by using subjective information in Bayesian fashion. Conceptually, this approach is sound, although there are some technical difficulties in obtaining numerical solutions.

The likelihood function for the observations given the parameters is given in (10.3.3) as $p(\mathbf{A}|\mathbf{\Lambda},\psi_1,\ldots,\psi_p)$ for the orthogonal factor model, and $p(\mathbf{A}|\mathbf{\Lambda},\mathbf{R},\psi_1,\ldots,\psi_p)$ for the oblique factor model, where in (10.3.3),

[2] It is straightforward to check, using (2.5.1), that if R_j^2 denotes the squared multiple correlation coefficient between the jth attribute and all the others,

$$R_j^2 = 1 - (r^{jj})^{-1}$$

where r^{jj} denotes the jj element of $\mathbf{R}^{-1}$, and where $\mathbf{R}$ is the attribute correlation matrix. It may be shown that R_j^2 is a lower bound for the jth communality.

$\boldsymbol{\Sigma} = \boldsymbol{\Lambda\Lambda}' + \mathbf{D}_\psi$ or $\boldsymbol{\Sigma} = \boldsymbol{\Lambda R \Lambda}' + \mathbf{D}_\psi$, respectively. For simplicity, adopt the orthogonal factor model. Then

$$p(\mathbf{A}|\boldsymbol{\Lambda},\psi_1,\ldots,\psi_p) \propto \frac{|\mathbf{A}|^{(n-p-1)/2}}{|\boldsymbol{\Lambda\Lambda}' + \mathbf{D}_\psi|^{n/2}} \exp\left[-\tfrac{1}{2}\operatorname{tr}\mathbf{A}(\boldsymbol{\Lambda\Lambda}' + \mathbf{D}_\psi)^{-1}\right]. \quad (10.6.5)$$

Now assume there are q elements of $\boldsymbol{\Lambda}$ about which there is informative prior information. Let $\boldsymbol{\omega}_0$ denote the $q \times 1$ vector of these elements of $\boldsymbol{\Lambda}$, and suppose the prior distribution of $\boldsymbol{\omega}_0$ is given[3] by $\mathcal{L}(\boldsymbol{\omega}_0) = N(\boldsymbol{\phi},\mathbf{G})$, $\mathbf{G} > 0$, where $\boldsymbol{\phi}$ and $\mathbf{G}$ are preassigned. It is often reasonable to take $\boldsymbol{\phi}$ as the zero vector, and $\mathbf{G}$ to be diagonal with small diagonal elements. Now string out the remaining elements of $\boldsymbol{\Lambda}$ into a $(pm - q) \times 1$ vector, $\boldsymbol{\omega}_1$. A noninformative, or diffuse prior on $\boldsymbol{\omega}_1$ implies $p(\boldsymbol{\omega}_1) \propto$ constant. Adopt an improper prior on $(\psi_1,\ldots,\psi_p)$ of the form $p(\psi_1,\ldots,\psi_p) \propto |\mathbf{D}_\psi|^{-\nu}$, $\nu < 1$. Hence, if all parameters are taken to be independent, a priori, the joint prior density is given by

$$\begin{aligned} p(\boldsymbol{\Lambda},\mathbf{D}_\psi) &= p(\boldsymbol{\omega}_0,\boldsymbol{\omega}_1,\psi_1,\ldots,\psi_p) \\ &= p(\boldsymbol{\omega}_0)p(\boldsymbol{\omega}_1)p(\psi_1,\ldots,\psi_p) \\ &\propto \frac{\exp\left\{-\tfrac{1}{2}(\boldsymbol{\omega}_0 - \boldsymbol{\phi})'\mathbf{G}^{-1}(\boldsymbol{\omega}_0 - \boldsymbol{\phi})\right\}}{|\mathbf{D}_\psi|^{\nu}}. \end{aligned} \quad (10.6.6)$$

Multiplying (10.6.5) by (10.6.6) and absorbing some unnecessary constants into the proportionality constant gives the posterior density

$$p(\boldsymbol{\Lambda},\mathbf{D}_\psi|\mathbf{A}) \propto \frac{p(\boldsymbol{\omega}_0)}{|\mathbf{D}_\psi|^{\nu} \cdot |\boldsymbol{\Lambda\Lambda}' + \mathbf{D}_\psi|^{n/2}} \exp\left[-\tfrac{1}{2}\operatorname{tr}\mathbf{A}(\boldsymbol{\Lambda\Lambda}' + \mathbf{D}_\psi)^{-1}\right]. \quad (10.6.7)$$

To find the marginal posterior density for $\boldsymbol{\Lambda}$, it is necessary to integrate this result with respect to $\psi_1,\ldots,\psi_p$. For $p = 1$, the integral is transformable to an integral in the hypergeometric function family. The problem is more complicated for general p. Of course an approximation which is asymptotic in sample size (n) may prove useful in some situations. For further developments of the Bayesian Factor Analysis model, see Kaufman and Press, 1973, 1976.

10.7 FACTOR ANALYSIS AND PRINCIPAL COMPONENTS

In this section the factor analysis model is related to the principal components model of Chapter 9. (See also, Kruskal, 1978.)

[3] An alternative assumption when the λ_{ij}'s are correlations is that $\boldsymbol{\omega}_0$ follows a transformed Dirichlet distribution whose elements are transformed beta variates defined on the range $(-1,1)$. Note that if X follows a beta distribution (so that $0 < X < 1$), the appropriate transformation so that $-1 < Y < 1$ is $Y = 2X - 1$.

In the orthogonal factor analysis model of (10.2.1) with $\mathbf{Y} \equiv \mathbf{X} - \boldsymbol{\theta}$, $\mathbf{Y} = \boldsymbol{\Lambda}\mathbf{f} + \mathbf{e}$ and

$$\boldsymbol{\Sigma} = \boldsymbol{\Lambda}\boldsymbol{\Lambda}' + \mathbf{D}_\psi.$$

Now suppose the errors are small enough to be ignored (so that $\mathbf{D}_\psi \cong \mathbf{0}$). A factorization of $\boldsymbol{\Sigma}$ is sought[4] in which $\boldsymbol{\Sigma} = \boldsymbol{\Lambda}\boldsymbol{\Lambda}'$.

One such factorization (see Section 10.6 for a discussion of the infinity of possible solutions) is the one provided by the principal components solution of Chapter 9. That is, if $\boldsymbol{\Sigma}$ were known, and if a principal components analysis were carried out, $\boldsymbol{\Sigma}$ would be factored as

$$\underset{(p\times p)}{\boldsymbol{\Sigma}} = \underset{(p\times p)}{\boldsymbol{\Gamma}}\,\underset{(p\times p)}{\mathbf{D}_\lambda}\,\underset{(p\times p)}{\boldsymbol{\Gamma}'} = (\boldsymbol{\Gamma}\mathbf{D}_\lambda^{1/2})(\boldsymbol{\Gamma}\mathbf{D}_\lambda^{1/2})',$$

where $\boldsymbol{\Gamma}$ is an orthogonal matrix whose columns are the normalized latent vectors of Σ, and $\mathbf{D}_\lambda \equiv \text{diag}(\lambda_1, \ldots, \lambda_p)$, where the λ_j's are the latent roots of Σ. So that if by definition,

$$\underset{(p\times p)}{\boldsymbol{\Lambda}} \equiv \underset{(p\times p)}{\boldsymbol{\Gamma}}\,\underset{(p\times p)}{\mathbf{D}_\lambda^{1/2}}, \tag{10.7.1}$$

(10.7.1) provides one solution to the factor analysis problem of factoring $\boldsymbol{\Sigma}$ in the form $\boldsymbol{\Lambda}\boldsymbol{\Lambda}'$. Moreover, if $\lambda_1 > \lambda_2 > \cdots > \lambda_p > 0$, this solution is unique. However, note that in (10.7.1), $\boldsymbol{\Lambda}$ is a $p \times p$ matrix, as indeed it must be since nothing more than a rotation of axes to principal axes has taken place. Next, examine the latent roots of $\boldsymbol{\Sigma}$ to determine which ones account for the largest proportion of tr $\boldsymbol{\Sigma}$ [using Theorem (9.3.2)]. Retain only the first m roots $\lambda_1 > \lambda_2 > \cdots > \lambda_m$; and delete the latent vectors corresponding to the deleted latent roots. Let $\mathbf{D}_\lambda^* \equiv \text{diag}\,(\lambda_1, \ldots, \lambda_m)$, and let $\boldsymbol{\Gamma}^*$ denote the corresponding matrix of latent vectors ($\boldsymbol{\Gamma}^*$ is obtained from $\boldsymbol{\Gamma}$ by deleting the $p - m$ columns of $\boldsymbol{\Gamma}$ corresponding to the $p - m$ smallest latent roots). Define

$$\underset{(p\times m)}{\boldsymbol{\Lambda}^*} \equiv \underset{(p\times m)}{\boldsymbol{\Gamma}^*}\,\underset{(m\times m)}{(\mathbf{D}_\lambda^*)^{1/2}}. \tag{10.7.2}$$

If $m < p$, some degree of parsimony has been achieved in the explanation of the data since now only m factors are required to explain data vectors expressed in p dimensions. The solution given in (10.7.2) is called the *principal factor solution.* Note that although no distributional assump-

[4] In some applications, instead of ignoring the errors, the communalities, $c_j^2 = \sigma_{jj} - \psi_j$, are estimated and $\hat{\boldsymbol{\Sigma}} - \hat{\mathbf{D}}_\psi$ is factored. However, there is no general agreement on the optimal way of estimating c_j^2. One frequently adopted approach is to standardize the data first, and then do a principal components type factor analysis on $(\hat{\mathbf{P}} - \hat{\mathbf{D}}_\psi^*)$, where $\hat{\mathbf{P}}$ denotes the sample correlation matrix, and $\hat{\mathbf{D}}^*$, the corresponding diagonal matrix of estimated specific variances. To obtain $\hat{\mathbf{D}}_\psi^*$, $c_j^{*2} = 1 - \psi_j^*$ is estimated by the sample squared multiple correlation coefficient between variable j and all the others (see Section 8.2.4).

tion was required to obtain this solution, it was necessary to estimate the communalities in an arbitrary way.

If $\mathbf{\Sigma}$ is estimated by its sample value $\hat{\mathbf{\Sigma}}$ (under Normality, $\hat{\mathbf{\Sigma}}$ is also its maximum likelihood estimator), the estimated principal factor solution is given by

$$\underset{(p\times m)}{\hat{\mathbf{\Lambda}}^*} = \underset{(p\times m)(m\times m)}{\hat{\mathbf{\Gamma}}^* (\hat{\mathbf{D}}_\lambda{}^*)^{1/2}}, \tag{10.7.3}$$

where $\hat{\mathbf{\Gamma}}^*$ and $\hat{\mathbf{D}}_\lambda{}^*$ are the estimated values of $\hat{\mathbf{\Gamma}}^*$ and $\hat{\mathbf{D}}_\lambda{}^*$, obtainable from the factorization $\hat{\mathbf{\Sigma}} = \hat{\mathbf{\Gamma}}\hat{\mathbf{D}}_\lambda\hat{\mathbf{\Gamma}}'$ by appropriate deletions. Of course, the diagonal elements of $\hat{\mathbf{D}}_\lambda{}^*$ are the estimators of the variances of the principal components, and $\hat{\mathbf{\Gamma}}^*$ is the matrix of estimators of the weights in the first m principal components of $\hat{\mathbf{\Sigma}}$. The principal factor and maximum likelihood solutions are but two solutions to the factor analysis problem.

10.8 LATENT STRUCTURE ANALYSIS

10.8.1 Introduction

Suppose a sample of people is taken and each individual is asked to respond yes or no as to whether or not he has read each of K magazines. Let π_j denote the probability of a positive response for magazine j. For each person there are 2^K categories of responses and the π_j's define a multinomial distribution for the categories [see Example (4.4.1)].

In the latent class model of latent structure analysis it is assumed that inherently, the population is stratified into m mutually exclusive and exhaustive groups ($m > 1$). In the example with magazines, the stratification might be by intelligence level of the individual. There might be three groups ($m = 3$) with group one including individuals with "low intelligence," group two including those with "normal intelligence," and group three those with "high intelligence."

Let ν_α denote the probability that an individual comes from the αth group, $\alpha = 1,\ldots,m$, and let $\lambda_{\alpha i}$ denote the probability that an individual from group α responds "yes" for magazine i. Let π_i denote the probability an individual responds "yes" for magazine i; π_{ij} denote the probability an individual responds "yes" for both magazines i and j; π_{ijk} for a "yes" for magazines i, j, k; and so on. Assume independence within each group so that given an individual belongs to group α, the joint probability he responds "yes" for magazines i, j, $k,\ldots$, is equal to the product of the marginal probabilities for each magazine separately. The model may now be written (using the rules of conditional probability)

$$\pi_i = \sum_{\alpha=1}^{m} \nu_\alpha \lambda_{\alpha i},$$

$$\pi_{ij} = \sum_{\alpha=1}^{m} \nu_\alpha \lambda_{\alpha i} \lambda_{\alpha j}, \qquad i \neq j,$$

$$\pi_{ijk} = \sum_{\alpha=1}^{m} \nu_\alpha \lambda_{\alpha i} \lambda_{\alpha j} \lambda_{\alpha k}, \qquad i \neq j,\ j \neq k,\ k \neq i.$$

$$\vdots \qquad \vdots$$

Similar equations may be written for the negative response categories.

Suppose that the group to which an individual belongs is not observed. The above equations can often be used to infer the group to which an individual belongs from the individual's pattern of responses. Thus, in the latent class model of latent structure analysis, the π_i's, π_{ij}'s, π_{ijk}'s,. . ., are assumed known (or estimable from samples), and ν_α, and the $\lambda_{\alpha i}$'s, are unknown parameters to be determined. In the jargon of latent structure analysis, the π variables are called the *manifest probabilities*, and are directly estimable using sample proportions. The π_i's are called the first-order manifest probabilities, the π_{ij}'s are called the second-order manifest probabilities, the π_{ijk}'s are the third-order manifest probabilities, and so on. The $\lambda_{\alpha i}$'s are called the *latent probabilities*, and the ν_α's are called the *recruitment probabilities* (of a respondent falling into, or being recruited from, the αth latent group). The totality of equations that relate the π's to the λ's and ν's are called the *accounting equations*, and the K contexts of the questions (magazine j, $j = 1,\ldots,K$ in the above illustration) are called *items*.

The latent structure model with the assumption of independence of response within each latent class was developed by Lazarsfeld (1950) and is described in detail in Lazarsfeld and Henry (1968). The formulation used above was given by Anderson (1954) who provided a formal matrix solution to the problem (in terms of the roots of a determinantal equation) using only the first-, second-, and third-order manifest probabilities. This solution to the latent class model of latent structure analysis is given below.

10.8.2 Solution to the Accounting Equations

Let $\mathbf{\Lambda}$ denote the matrix of unknown latent parameters (latent class probabilities)

$$\underset{m \times (K+1)}{\mathbf{\Lambda}} = \begin{pmatrix} 1 & \lambda_{11} & \lambda_{12} & \cdots & \lambda_{1K} \\ 1 & \lambda_{21} & \lambda_{22} & \cdots & \lambda_{2K} \\ \cdot & \cdot & \cdot & & \cdot \\ \cdot & \cdot & \cdot & & \cdot \\ \cdot & \cdot & \cdot & & \cdot \\ 1 & \lambda_{m1} & \lambda_{m2} & \cdots & \lambda_{mK} \end{pmatrix}. \tag{10.8.1}$$

Define

$$\underset{(m \times m)}{\mathbf{N}} = \text{diag}\,(\nu_1,\ldots,\nu_m), \tag{10.8.2}$$

and

$$\underset{(m \times m)}{\mathbf{\Delta}} = \text{diag}\,(\lambda_{1k},\ldots,\lambda_{mk}), \tag{10.8.3}$$

where $k > 2m - 2$. That is, the items have an arbitrary ordering, and to obtain a solution by this technique there must be at least $(2m - 1)$ items[5] of which k are selected to define $\mathbf{\Delta}$. Thus, if there are three latent subgroups (say low, medium, and high intelligence levels), and six items (say six different magazines), so that $m = 3$, and $K = 6$, $\mathbf{\Delta}$ may be chosen as diag $(\lambda_{15},\lambda_{25},\lambda_{35})$, or diag $(\lambda_{16},\lambda_{26},\lambda_{36})$ for any ordering of the items. It is not known how to choose $\mathbf{\Delta}$ so that parameter estimators will be most efficient.

Now define $\mathbf{\Lambda}_1$: $m \times m$ as the matrix consisting of the first m columns of $\mathbf{\Lambda}$, and $\mathbf{\Lambda}_2$ as the matrix consisting of the first column of $\mathbf{\Lambda}$, and the next $(m - 1)$ columns of $\mathbf{\Lambda}$ not contained in $\mathbf{\Lambda}_1$, so that

$$\underset{(m \times m)}{\mathbf{\Lambda}_2} = \begin{pmatrix} 1 & \lambda_{1m} & \cdots & \lambda_{1,2m-2} \\ \cdot & \cdot & & \cdot \\ \cdot & \cdot & & \cdot \\ \cdot & \cdot & & \cdot \\ 1 & \lambda_{mm} & \cdots & \lambda_{m,2m-2} \end{pmatrix}. \tag{10.8.4}$$

So far, only unobservable (latent) parameters have been defined. Now define the matrix of third-order manifest probabilities (observable)

$$\underset{(m \times m)}{\mathbf{\Pi}} = \begin{bmatrix} \pi_{00k} & \pi_{0mk} & \pi_{0,m+1,k} & \cdots & \pi_{0,2m-2,k} \\ \pi_{10k} & \pi_{1mk} & \pi_{1,m+1,k} & \cdots & \pi_{1,2m-2,k} \\ \cdot & \cdot & \cdot & & \cdot \\ \cdot & \cdot & \cdot & & \cdot \\ \cdot & \cdot & \cdot & & \cdot \\ \pi_{m-1,0,k} & \pi_{m-1,m,k} & \pi_{m-1,m+1,k} & \cdots & \pi_{m-1,2m-2,k} \end{bmatrix}, \tag{10.8.5}$$

where any π with a zero subscript is defined to be the value of π with the zero suppressed. Thus, $\pi_{00k} \equiv \pi_k$, and $\pi_{0jk} \equiv \pi_{jk}$. Take $\pi_{000} = 1$. Let $\mathbf{\Pi}^*$

[5] For any preassigned number of latent classes m, there must be at least $(2m - 1)$ items for the parameters λ_{ij}, and ν_j to be estimable by this technique.

be defined the same as $\mathbf{\Pi}$, except that each element π_{ijk} is replaced by π_{ij} (so that its entries correspond to joint probabilities for two items instead of three). Thus, $\mathbf{\Pi}^*$ is the matrix of second-order manifest probabilities.

The basic accounting equations relating the observable variables to the latent variables may now be written in the compact form

$$\mathbf{\Pi} = \mathbf{\Lambda}_1'\mathbf{N}\mathbf{\Delta}\mathbf{\Lambda}_2, \qquad \mathbf{\Pi}^* = \mathbf{\Lambda}_1'\mathbf{N}\mathbf{\Lambda}_2. \tag{10.8.6}$$

Suppose the manifest probabilities are all known, and that $\mathbf{\Lambda}_1$, $\mathbf{N}$, $\mathbf{\Lambda}_2$ are all nonsingular (so that, for example, all ν_α must be positive). It is easy to find $\mathbf{\Delta}$ when $\mathbf{\Pi}$ and $\mathbf{\Pi}^*$ are known. Since $\mathbf{\Delta}$ is diagonal, its nonzero elements are the roots of the determinantal equation

$$|\mathbf{\Delta} - \theta\mathbf{I}_m| = 0. \tag{10.8.7}$$

Multiplying (10.8.7) on the left by $|\mathbf{\Lambda}_1'\mathbf{N}|$ and on the right by $|\mathbf{\Lambda}_2|$ shows that the θ's satisfy the equation

$$|\mathbf{\Pi} - \theta\mathbf{\Pi}^*| = 0. \tag{10.8.8}$$

Moreover, since $\mathbf{\Pi}$ and $\mathbf{\Pi}^*$ are known, $\mathbf{\Delta}$ is determined. It only remains to show how to find $\mathbf{\Lambda}_1$, $\mathbf{\Lambda}_2$, and $\mathbf{N}$ in terms of $\mathbf{\Pi}$, $\mathbf{\Pi}^*$, and the θ's.

From (10.8.8) it follows that there is a latent vector $\mathbf{x}_\alpha: m \times 1$, corresponding to $\theta = \theta_\alpha$ such that

$$\mathbf{\Pi}\mathbf{x}_\alpha = \theta_\alpha\mathbf{\Pi}^*\mathbf{x}_\alpha, \qquad \alpha = 1,\ldots,m. \tag{10.8.9}$$

If the θ_α's are distinct, the $\mathbf{x}_\alpha$'s are unique up to a multiplicative constant (see Chapter 2). Let the θ_α's be ordered so that $\theta_1 > \theta_2 > \cdots > \theta_m$. Define $\underset{(m\times m)}{\mathbf{X}} = (\mathbf{x}_1,\ldots,\mathbf{x}_m)$, and $\mathbf{D}_\theta \equiv \text{diag}\,(\theta_1,\ldots,\theta_m)$. Now (10.8.9) may be written in the compact form

$$\mathbf{\Pi}\mathbf{X} = \mathbf{\Pi}^*\mathbf{X}\mathbf{D}_\theta. \tag{10.8.10}$$

Note that $\mathbf{D}_\theta = \mathbf{\Delta}$. It is easy to check by substituting (10.8.6) into (10.8.10) that one solution for $\mathbf{X}$ is $\mathbf{X} = \mathbf{\Lambda}_2^{-1}$. Now $\mathbf{X}$ is uniquely determined up to a multiplication on the right by an arbitrary nonsingular diagonal matrix; that is, each column of $\mathbf{X}$ can still be multiplied by an arbitrary constant. Thus, the general solution is

$$\mathbf{X} = \mathbf{\Lambda}_2^{-1}\mathbf{D}_x,$$

where $\mathbf{D}_x$ is diagonal. The $\mathbf{D}_x$ matrix is easily determined by recalling that since the first column of $\mathbf{\Lambda}_2$ must be unity, and since

$$\mathbf{\Lambda}_2 = \mathbf{D}_x\mathbf{X}^{-1}, \tag{10.8.11}$$

each diagonal element of $\mathbf{D}_x$ must be the reciprocal of the first element of the corresponding row of $\mathbf{X}^{-1}$. Thus, $\mathbf{\Lambda}_2$ is uniquely determined.

Now return to (10.8.8) and consider the corresponding $1 \times m$ latent row vectors $\mathbf{y}_\alpha'$ given by

$$\mathbf{\Pi}'\mathbf{y}_\alpha = \theta_\alpha(\mathbf{\Pi}^*)'\mathbf{y}_\alpha, \qquad \alpha = 1,\ldots,m. \tag{10.8.12}$$

Apply the same argument as used for $\mathbf{x}_\alpha$ in (10.8.9), since everything is analogous with $\mathbf{y}_\alpha$ replacing $\mathbf{x}_\alpha$, and $\mathbf{\Lambda}_1$ replacing $\mathbf{\Lambda}_2$. Accordingly, let $\underset{(m\times m)}{\mathbf{Y}} \equiv (\mathbf{y}_1,\ldots,\mathbf{y}_m)$. Then, by analogy with (10.8.11),

$$\mathbf{\Lambda}_1 = \mathbf{D}_y\mathbf{Y}^{-1}, \tag{10.8.13}$$

where $\mathbf{D}_y$ is the diagonal matrix whose diagonal elements are the reciprocals of the first elements in the corresponding rows of $\mathbf{Y}^{-1}$.

The solution for $\mathbf{N}$ is obtainable directly from the accounting equations. Thus, solving the second equation in (10.8.6) gives

$$\mathbf{N} = (\mathbf{\Lambda}_1')^{-1}\mathbf{\Pi}^*\mathbf{\Lambda}_2^{-1}. \tag{10.8.14}$$

A computer routine (LASY) for solving the equations of the latent class model by the method of this section is described in Appendix A. It will accommodate up to ten latent classes. There is another program available (called BAN) that generates maximum likelihood estimators of the parameters (see Lazarsfeld and Henry, 1968).

10.8.3 Inference in the Latent Class Model

In most real situations, the manifest probabilities (the π's) will not be known and will have to be estimated. Let a sample of M respondents be taken, and let n_i denote the number of people in the sample who respond "yes" for "item i"; n_{ij} the number who respond "yes" for items i and j; and n_{ijk} the number who respond "yes" for items i, j, and k. Then the sample estimators of the manifest probabilities π_i (they are also maximum likelihood estimators in the associated multinomial distribution) are n_i/M, those of π_{ij} are n_{ij}/M, and those of π_{ijk} are n_{ijk}/M. The unknown parameters in the latent class model may now be consistently estimated (assuming they are identified) using the solution developed in Section 10.8.2, with the first-, second-, and third-order manifest probabilities replaced by their sample values. Just as in the estimation problem of factor analysis (Section 10.3), the number of latent classes (or factors). m, must be preassigned. Then, a fit is found to the m-class model.

Let $\hat{\mathbf{\Lambda}}_1$, $\hat{\mathbf{\Lambda}}_2$, $\hat{\mathbf{N}}$, $\hat{\mathbf{\Delta}}$, $\hat{\mathbf{\Pi}}$, $\hat{\mathbf{\Pi}}^*$ denote the estimated values of $\mathbf{\Lambda}_1$, $\mathbf{\Lambda}_2$, $\mathbf{N}$, $\mathbf{\Delta}$, $\mathbf{\Pi}$, $\mathbf{\Pi}^*$, respectively. Since the entries of $\hat{\mathbf{\Pi}}$ and $\hat{\mathbf{\Pi}}^*$ are the sample frequencies corresponding to the manifest probabilities, $\hat{\mathbf{\Pi}}$ and $\hat{\mathbf{\Pi}}^*$ are consistent estimators of $\mathbf{\Pi}$ and $\mathbf{\Pi}^*$. Since $\hat{\mathbf{\Lambda}}_1$, $\hat{\mathbf{\Lambda}}_2$, $\hat{\mathbf{N}}$, and $\hat{\mathbf{\Delta}}$ are the same continuous functions of $\hat{\mathbf{\Pi}}$ and $\hat{\mathbf{\Pi}}^*$ that $\mathbf{\Lambda}_1$, $\mathbf{\Lambda}_2$, $\mathbf{N}$, and $\mathbf{\Delta}$ are of $\mathbf{\Pi}$ and $\mathbf{\Pi}^*$, $\hat{\mathbf{\Lambda}}_1$, $\hat{\mathbf{\Lambda}}_2$, $\hat{\mathbf{N}}$, and $\hat{\mathbf{\Delta}}$ are consistent estimators of $\mathbf{\Lambda}_1$, $\mathbf{\Lambda}_2$, $\mathbf{N}$, $\mathbf{\Delta}$, respectively.[6]

In an Appendix to Lazarsfeld and Henry (1968), Anderson showed that in large samples, $\hat{\mathbf{\Lambda}}_1$, $\hat{\mathbf{\Lambda}}_2$, $\hat{\mathbf{N}}$, and $\hat{\mathbf{\Delta}}$ are all unbiased and follow multivariate Normal distributions.

Just as in the factor analysis model where it is of interest to test whether the number of factors selected is appropriate or consistent with the data (Section 10.4), in the latent class model of latent structure analysis it is of interest to test whether the number of latent classes selected is consistent with the data. When large samples are available, a chi-square, goodness-of-fit test may be used.

Suppose the manifest probabilities have been estimated based upon a sample of size M as $\hat{\mathbf{\Pi}}$ and $\hat{\mathbf{\Pi}}^*$, and a solution $\hat{\mathbf{\Lambda}}_1$, $\hat{\mathbf{\Lambda}}_2$, $\hat{\mathbf{N}}$, and $\hat{\mathbf{\Delta}}$ has been found. These estimates of the latent parameters may now be used to generate estimates for the Kth-order manifest probabilities. That is, from basic conditional probability theory, since

$$\pi_{i_1,\ldots,i_K} = \sum_{\alpha=1}^{m} \nu_\alpha \lambda_{\alpha i_1} \lambda_{\alpha i_2} \cdots \lambda_{\alpha i_K}, \tag{10.8.15}$$

for $i_j \neq i_k$ and for all j and k, the implied Kth-order manifest probabilities (called *pseudo-manifest probabilities*) are

$$\tilde{\pi}_{i_1,\ldots,i_K} = \sum_{\alpha=1}^{m} \hat{\nu}_\alpha \hat{\lambda}_{\alpha i_1} \cdots \hat{\lambda}_{\alpha i_K}. \tag{10.8.16}$$

Since there are only two possible responses for each item, there are 2^K possible pseudo-manifest probabilities. The $\tilde{\pi}$'s may now be compared with the observed Kth-order manifest probabilities, $\hat{\pi}_{i_1,\ldots,i_K} \equiv n_{i_1,\ldots,i_K}/M$, to see whether they are in close agreement. To do so, compute the statistic

$$C_M = \sum \left[\frac{(n_{i_1,\ldots,i_K} - M\tilde{\pi}_{i_1,\ldots,i_K})^2}{M\tilde{\pi}_{i_1,\ldots,i_K}} \right], \tag{10.8.17}$$

where the summation is to be taken over all 2^K possible combinations of subscripts. For example, if there are three items ($K = 3$), there will be eight possible triples of responses. That is, if a "one" denotes a positive response and a "zero" a negative response, the possible responses are (0,0,0), (0,0,1), (0,1,0), (0,1,1), (1,0,0), (1,0,1), (1,1,0), (1,1,1).

Now test the hypothesis H: the data are consistent with m latent classes, versus the alternative hypothesis A: there are not m latent

[6] The property of consistency implies the existence of reasonable estimators in very large samples. However, in small samples the estimators are not even guaranteed to be real, and if they are real, they might not lie in the unit interval.

classes. Since in large samples the estimators are Normally distributed, it follows that

$$\lim_{M \to \infty} \mathfrak{L}\{C_M\} = \chi_p^2, \tag{10.8.18}$$

where $p = 2^K - m(K + 1) - 1$. The degrees of freedom, p, is obtained from the fact that p = number of categories minus one $(2^K - 1)$, minus the number of estimated parameters (mK latent probabilities, and m recruitment probabilities).

Example (10.8.1): Lazarsfeld and Henry (1968) analyzed some data from a study by Kadushin (1966) of people who have decided to undertake psychotherapy. Nine items were used as indicators of sophistication. An index of sophistication was sought for the purposes of comparing reasons for seeking psychiatric help, and comparing the people who applied for help at the various clinics studied. The items and their first-order frequency response estimates are given in Table 10.8.1 for a sample of $M = 1089$.

Table 10.8.1

Item	n_i/M
1. Go to plays several times a year	.569
2. Go to concerts several times a year	.411
3. Go to cocktail parties several times a year	.299
4. Go to museums several times a year	.597
5. Read psychology books and articles	.280
6. Know others with similar problems	.438
7. Know close friends or relatives who went to a psychiatrist	.679
8. Told a few people about coming for therapy	.553
9. Asked friends for referral	.421

Using sample frequency estimates for the first-, second-, and third-order manifest probabilities, and applying the solution of Section 10.8.2 to the case of a simple latent class dichotomy ($m = 2$) of "sophisticated" and "unsophisticated" gives the estimated latent probabilities shown in Table 10.8.2.

The estimated recruitment probabilities (ν_1 and ν_2) of belonging to each of the latent classes are

$$\hat{\nu}_1 = .54, \qquad \hat{\nu}_2 = .46.$$

The chi-square statistic of (10.8.17) was computed as $C_M = 772$, and from (10.8.18), $p = 491$ (since $K = 9$ and $m = 2$). It is easily checked

Table 10.8.2 Latent Probabilities for a Two-Class Model

Latent class	*Item*								
	1	2	3	4	5	6	7	8	9
Sophisticated	.83	.56	.43	.81	.35	.61	.92	.75	.68
Unsophisticated	.27	.23	.14	.35	.21	.23	.39	.53	.11

[from the fact that for large p, $\mathcal{L}\{\sqrt{2C} - \sqrt{2p-1}\} \cong N(0,1)$] that C is significant at all standard levels of significance. Hence, H is rejected (that is, the data are not consistent with two latent classes).

A reexamination of the problem led to the notion that there are really two types of sophistication present, a "cultural sophistication" and a "psychiatric sophistication." Moreover, it was decided to discard item 5. An alternative model was tried in which there were two classes of each type of sophistication for a total of four latent classes. The latent probability estimates are given in Table 10.8.3. Class (1) corresponds to high

Table 10.8.3 Latent Probabilities for a Four-Class Model

Latent class	*Item*							
	1	2	3	4	6	7	8	9
Class (1)	.91	.73	.45	.92	.63	.92	.79	.72
Class (2)	.82	.55	.30	.67	.25	.45	.28	.14
Class (3)	.34	.06	.30	.28	.59	.90	.72	.59
Class (4)	.03	.07	.07	.25	.25	.35	.28	.09

cultural and high psychiatric sophistication, Class (2) to high cultural and low psychiatric, Class (3) to low cultural and high psychiatric, and Class (4) to low cultural and low psychiatric.

The recruitment probabilities were estimated[7] as

$$\hat{\nu}_1 = .38, \qquad \hat{\nu}_2 = .19, \qquad \hat{\nu}_3 = .16, \qquad \hat{\nu}_4 = .27.$$

[7] The actual estimation procedure used here was maximum likelihood although the estimation procedure described in this section could have been used equally well.

For this model the C_M-statistic was found to be $C_M = 232$ and $p = 219$ (since now $K = 8$, and $m = 4$). Hence, now a test of H versus A would not be able to reject H at any reasonable significance level.

10.8.4 Relationship between the Latent Structure and Factor Analysis Models

The subject of latent structure analysis actually includes many different models with differing underlying assumptions. When the manifest probabilities are discrete, and there are a finite number of latent classes that are supposed to account for the relationships among the observable variables, the latent class model of latent structure analysis is the one under consideration (and this is the only model discussed above in this chapter). However, the latent structure in a problem may permit the manifest probabilities to be representable in terms of polynomials in several continuous variables (instead of in terms of latent classes). As a special case, the polynomials may be taken as linear in each latent variable. Now the latent structure is identical with the latent structure of factor analysis. Other latent structure analysis models include a generalization to the case of continuous manifest variables. If a latent structure analysis model has continuous manifest variables, and if the latent structure is linear in several continuous latent variables, the model is identical with the factor analysis model. While the factor analysis model has continuous manifest (response) variables and continuous latent factors, and the latent structure analysis model has discrete manifest variables and discrete latent factors or classes, models have recently been developed that fall in between the two, in that they have discrete manifest variables, but continuous latent factors. These models are in the spirit of Probit Analysis (see Sect. 8.10.4), and may be thought of as a discrete factor analysis (see Muthén, 1978 and 1979).

A major difference between the latent class model of latent structure analysis and the factor analysis model is that the former has no associated rotation problem leading to a multiplicity of solutions. The factor analysis model (with maximum likelihood estimation) leans heavily on multivariate Normality of the underlying variables, so that only second-order moments (variances, covariances, and correlations) are used. However, the latent class model uses not only first- and second-order moments, but third- and higher-order moments as well. In the MLE approach the entire cross classified structure is used. This procedure eliminates ambiguity in the solution. A more detailed account of the relationship between factor analysis and latent structure analysis is given in Green (1952).

10.8.5 Numerical Considerations

Those interested in actually carrying out a numerical latent structure analysis of a set of real, cross classified, categorical data should note that for problems involving four or more items, solutions based upon the first-, second-, and third-order manifest probabilities estimated by sample moments, are, in general, different from one another according to which three items are selected as the basis for computation. Results could be computed for all possible combinations, and then averaged. Alternatively, the maximum likelihood estimation approach yields unique solutions (see the computer software package developed by Clifford Clogg), without such a choice having to be made.

Users of the MLE approach should be aware of the severe hardware limitation on the size of a problem that can be analyzed. For example, suppose there are just thirteen items, and each one has five possible response categories. Then the number of cells in the corresponding contingency table of cross classifications is 5^{13}, or 1,220,703,125; that is, over one billion cells! Such a table would exceed the space limitations of known computers. But the problem seems small superficially with only thirteen items. In such cases, it is necessary to reduce the number of items, and if possible, to collapse the number of response categories, to yield a problem of more manageable size.

REFERENCES

Ahmad (1965).
Akaike (1973, 1977).
Anderson (1954).
Anderson and Rubin (1956).
Armstrong (1967).
Bartlett (1953, 1954).
Berry (1961).
Carrol and Schweiker (1951).
Cattell (1978).
Ekeblad and Stasch (1967).
Ewart, Seashore, and Tiffin (1951).
Galton (1888).
Green (1952).
Harman (1976).
Harper (1956).
Harper and Baron (1948, 1951).
Harrison (1960).
Howard and Schutz (1952).
Joreskog (1967).
Joreskog and Sorbom (1979).
Kadushin (1966).
Kaiser (1958).
Kaufman and Press (1973, 1976).
King (1966).
Kirsch and Banks (1962).
Kruskal (1978).
Lawley (1940).
Lawley and Maxwell (1963).
Lawshe and Alessi (1946).
Lawshe, Dudek, and Wilson (1948).
Lawshe and Maleski (1946).
Lawshe and Satter (1944).

Lawshe and Wilson (1946).
Lazarsfeld (1950).
Lazarsfeld and Henry (1968).
Massy (1963).
McDonald (1962).
Megee (1965).
Miller (1966).
Millican (1959).
Mukherjee (1965).
Mulaik (1972).
Olkin and Siotani (1964).
Pilgrim (1959).
Rao (1955, 1964, 1965).
Schreiber, Smith, and Harrell (1952).
Spearman (1904).
Stoetzel (1960).
Thurstone (1947).
Twedt (1952).
Twery (1958).
Vincent (1962).
Walsh (1964).
Wilks (1938).

EXERCISES

10.1 Suppose the fractions of "yes" votes and "no" votes on a congressional bill are recorded for every bill voted upon in a particular session of the United States Congress. It is of interest to see whether there are underlying constructs such as an agricultural bloc, or a civil rights bloc, or a liberal way of voting, and so on, which can account for the observed variation in the data in a simple way. Select a multivariate model you feel would be most appropriate for studying this problem, and formulate the model in the context of this problem stating the underlying assumptions of the model selected.

10.2 In a factor analysis problem, compare the advantages and disadvantages in terms of required assumptions, interpretation, and so on, of carrying out the analysis on the data correlation matrix instead of the data covariance matrix.

10.3 Compare the reasons for using a maximum likelihood method of solution of a factor analysis problem with those for using a principal factor solution. Are there conditions under which one might be used and not the other?

10.4 Suppose it is desired to carry out a Bayesian analysis of the latent class model of latent structure analysis. Find the joint natural conjugate prior density for the latent parameters.

10.5 Explain the circumstances under which the orthogonal factor model of factor analysis should be used, as opposed to the circumstances under which the oblique factor model should be used.

10.6 By consulting some of the references, suggest two distinct methods for estimating communalities in a factor analysis model.

10.7 Suppose in the latent class model of latent structure analysis it is assumed there is lack of correlation among the response variables for each item instead of independence. How would the model change?

10.8 Devise a group of five questions (each with two possible answers) one might ask (items) of each of N subjects in order to classify them into their degree of discrimination against people of other races. Assuming there are three classes of discrimination (low, medium, and high), how would you analyze the results using the latent class model of latent structure analysis?

10.9 Suppose on the basis of a sample of data from M subjects the latent probabilities and recruitment probabilities of a latent class model of latent structure analysis are estimated. Assume there are m latent classes and that, a priori, each subject has equal likelihood of belonging to each of the classes. Devise a Bayesian procedure for classifying each subject into one of the latent classes, a posteriori.

10.10 It is hypothesized that there are two distinct schools of thought about strategic policy of superpowers in the nuclear age: those who feel a superpower should hold a "flexible response" posture, so that whatever weapons the enemy develops, the superpower should develop a capability to respond to these weapons; and those who feel the superpower should maintain an "assured destruction" posture, so that the enemy will be assured that they will also be destroyed if they choose to attack.

To test this hypothesis 100 experts in military strategy were given a questionnaire with four items scored dichotomously as zero or one. Responses to the survey instrument are shown below.

	Response Pattern	Frequency		Response Pattern	Frequency
1.	(0,0,0,0)	0	9.	(1,0,0,0)	0
2.	(0,0,0,1)	0	10.	(1,0,0,1)	0
3.	(0,0,1,0)	4	11.	(1,0,1,0)	2
4.	(0,0,1,1)	38	12.	(1,0,1,1)	2
5.	(0,1,0,0)	6	13.	(1,1,0,0)	35
6.	(0,1,0,1)	2	14.	(1,1,0,1)	5
7.	(0,1,1,0)	2	15.	(1,1,1,0)	2
8.	(0,1,1,1)	1	16.	(1,1,1,1)	1

Carry out a latent structure analysis of this data, and test the hypothesis that the data are consistent with the postulated two latent class model (use a 5% level of significance, and the MLE method). Estimate the recruitment probabilities and the latent probabilities. Assume the sample size is large enough for large sample results to be applicable.

10.11 Explain the difference between the Factor Analysis and Latent Structure Analysis models.

10.12 Using the data contained in the covariance matrix in Example (9.3.1):

(a) Carry out a maximum likelihood factor analysis separately for a one-factor model, two-factor model, three-factor model and four-factor model.

(b) Adopt the MAICE approach to determine how many factors is most appropriate (see Section 10.4.1).

10.13 Principal components analysis is often used by researchers instead of factor analysis. What would be the advantages?

10.14 It is often useful to carry out a factor analysis with no distributional assumptions on the data. MINRES is such a procedure. Explain the MINRES procedure. [Hint: see Harman, 1976.]

10.15 Suppose you carry out a factor analysis and find that the factors have no natural interpretation. Would you still defend the results? Why?

11

CANONICAL CORRELATIONS

11.1 INTRODUCTION

The canonical correlations model was proposed by Hotelling (1935; 1936) for studying linear relationships between two sets of variates. To simplify the relationships he suggested that a few uncorrelated linear combinations of the variables in each set be studied instead of the two complete sets. Such a parsimonious representation of the original data is often useful in applied research.

In education, it may be desirable to consider a certain linear combination of scores obtained by a student in high school in order to predict a linear combination of his scores in college (or scores on measures of success in the working career of the individual). By studying many sets of such scores, simple relationships may be established. For specific examples of the use of canonical correlations in educational research see Cooley and Lohnes (1962), Chapter 3.

In finance it may prove useful to use linear combinations of easily measurable economic quantities to study the behavior of the prices of a group of securities (for example, a certain linear combination of the prices of 5 judiciously selected grain and beef futures might be a useful surrogate for the future security prices of a group of retail supermarket chains).

The method of canonical correlations has been applied in economics (see Waugh, 1942; Hooper, 1959; and Stone, 1947); in marketing (see Green, Halbert, and Robinson, 1966) and in most of the behavioral and social sciences. Canonical correlations have already appeared in Chapter 8 as a measure of efficiency in generalized multivariate regression.

The canonical correlations model selects weighted sums of variables from each of two sets to form new variables in each of the sets, so that the correlation between the new variables in different sets is maximized while the new variables within each set are constrained to be uncorrelated with mean zero and variance one. The ordinary (product moment) correlation between two random variables is by now very familiar. The generalization of this simple idea to a measure of association between one random variable and a vector of others was discussed in Chapter 8 in terms of the multiple correlation coefficient. The canonical correlation coefficient generalizes the notion even further to correlation between two random vectors.

The sampling theory approach is used throughout this chapter. The Bayesian solution proposed by Geisser (1965a) employed a diffuse prior density on the parameters and the assumption of multivariate Normality for the observed data. In even this simple case, the results he obtained are expressed in terms of zonal polynomials (see Sect. 6.6).

The canonical correlations model for populations with known parameters is discussed in Section 11.2. The case of unknown parameters and the accompanying problems of inference associated with observed data are discussed in Section 11.3. In Section 11.4 there is a summary of the extension of the canonical correlations model to more than two sets of random variables.

11.2 KNOWN PARAMETERS

Let $\mathbf{X}: p \times 1$ be a random vector with $E(\mathbf{X}) = \boldsymbol{\theta}$, var $(\mathbf{X}) = \boldsymbol{\Sigma}$, assumed to be positive definite. Assume $\boldsymbol{\Sigma}$ is known. Partition $\mathbf{X}$ into two sets of variables (and partition $\boldsymbol{\Sigma}$ accordingly) so that

$$\mathbf{X} = \begin{pmatrix} \mathbf{Y} \\ \mathbf{Z} \end{pmatrix}, \qquad \boldsymbol{\Sigma} = \begin{pmatrix} \boldsymbol{\Sigma}_{11} & \boldsymbol{\Sigma}_{12} \\ \boldsymbol{\Sigma}_{21} & \boldsymbol{\Sigma}_{22} \end{pmatrix},$$

where $\mathbf{Y}: p_1 \times 1$, $\mathbf{Z}: p_2 \times 1$, $p_1 + p_2 = p$; without loss of generality, assume $p_1 \leq p_2$. Let $\boldsymbol{\alpha}: p_1 \times 1$ and $\boldsymbol{\gamma}: p_2 \times 1$ be unknown vectors to be determined so that the correlation between $\boldsymbol{\alpha}'\mathbf{Y}$ and $\boldsymbol{\gamma}'\mathbf{Z}$ is as large as possible.

Define $U_1 = \boldsymbol{\alpha}'\mathbf{Y}$, and $V_1 = \boldsymbol{\gamma}'\mathbf{Z}$, and let $\rho(U_1, V_1)$ denote the correlation between U_1 and V_1. Next recall that correlations are invariant under linear transformations.

Lemma (11.2.1): Let $\rho(s,t)$ denote the correlation between the random variables s and t. If $s^* = as + b$, and $t^* = ct + d$, where a, b, c, d are fixed constants ($a \neq 0$, $c \neq 0$), and a and c are of the same sign,

$$\rho(s^*,t^*) = \rho(s,t).$$

Proof: Substitute in the definition of correlation.

A certain arbitrariness arises because of the invariance of $\rho(U_1, V_1)$ under appropriate linear transformations of the variables. Uniqueness of solutions can be achieved, however, by requiring that $\text{var}(U_1) = \text{var}(V_1) = 1$; that is,

$$\text{var}\,(U_1) = \boldsymbol{\alpha}'[\text{var}\,(\mathbf{Y})]\boldsymbol{\alpha} = \boldsymbol{\alpha}'\boldsymbol{\Sigma}_{11}\boldsymbol{\alpha} = 1, \tag{11.2.1}$$

$$\text{var}\,(V_1) = \boldsymbol{\gamma}'[\text{var}\,(\mathbf{Z})]\boldsymbol{\gamma} = \boldsymbol{\gamma}'\boldsymbol{\Sigma}_{22}\boldsymbol{\gamma} = 1. \tag{11.2.2}$$

Thus, the problem is to maximize (we can take $EU_1 = EV_1 = 0$ without loss of generality),

$$\rho(U_1,V_1) = \frac{EU_1V_1}{\sqrt{(\text{var}\,U_1)(\text{var}\,V_1)}} = \frac{\boldsymbol{\alpha}'\boldsymbol{\Sigma}_{12}\boldsymbol{\gamma}}{[(\boldsymbol{\alpha}'\boldsymbol{\Sigma}_{11}\boldsymbol{\alpha})(\boldsymbol{\gamma}'\boldsymbol{\Sigma}_{22}\boldsymbol{\gamma})]^{1/2}}, \tag{11.2.3}$$

subject to (11.2.1) and (11.2.2).

Let $\boldsymbol{\alpha}_0 \equiv \boldsymbol{\alpha}(\boldsymbol{\alpha}'\boldsymbol{\Sigma}_{11}\boldsymbol{\alpha})^{-1/2}$, $\boldsymbol{\gamma}_0 \equiv \boldsymbol{\gamma}(\boldsymbol{\gamma}'\boldsymbol{\Sigma}_{22}\boldsymbol{\gamma})^{-1/2}$. It is easy to see the problem is equivalent to that of maximizing $\boldsymbol{\alpha}_0'\boldsymbol{\Sigma}_{12}\boldsymbol{\gamma}_0$, subject to $\boldsymbol{\alpha}_0'\boldsymbol{\Sigma}_{11}\boldsymbol{\alpha}_0 = 1$, $\boldsymbol{\gamma}_0'\boldsymbol{\Sigma}_{22}\boldsymbol{\gamma}_0 = 1$. Thus, it is not necessary to normalize EU_1V_1, and the problem may be solved in terms of covariances.

Define the Lagrangian

$$\mathcal{L} = \boldsymbol{\alpha}'\boldsymbol{\Sigma}_{12}\boldsymbol{\gamma} - \phi_1(\boldsymbol{\alpha}'\boldsymbol{\Sigma}_{11}\boldsymbol{\alpha} - 1) - \phi_2(\boldsymbol{\gamma}'\boldsymbol{\Sigma}_{22}\boldsymbol{\gamma} - 1),$$

where ϕ_1 and ϕ_2 are Lagrangian multipliers. Differentiating $\mathcal{L}$ with respect to $\boldsymbol{\alpha}$ and $\boldsymbol{\gamma}$ separately, and setting the results equal to zero gives

$$\lambda \equiv 2\phi_1 = 2\phi_2 = \rho(U_1,V_1), \tag{11.2.4}$$

and

$$\begin{pmatrix} -\lambda\boldsymbol{\Sigma}_{11} & \boldsymbol{\Sigma}_{12} \\ \boldsymbol{\Sigma}_{21} & -\lambda\boldsymbol{\Sigma}_{22} \end{pmatrix}\begin{pmatrix} \boldsymbol{\alpha} \\ \boldsymbol{\gamma} \end{pmatrix} = 0. \tag{11.2.5}$$

In order that (11.2.5) possess a nontrivial solution (that is, $\boldsymbol{\alpha}$ and $\boldsymbol{\gamma}$ not both identically zero) it is necessary that

$$\begin{vmatrix} -\lambda\boldsymbol{\Sigma}_{11} & \boldsymbol{\Sigma}_{12} \\ \boldsymbol{\Sigma}_{21} & -\lambda\boldsymbol{\Sigma}_{22} \end{vmatrix} = 0. \tag{11.2.6}$$

It is not hard to find that (11.2.6) has p roots of which $(p_2 - p_1)$ are zero; the remaining $(2p_1)$ nonzero roots are of the form $\lambda = \pm\theta_i$, $i = 1,\ldots,p_1$; choose the largest positive root, $\lambda \equiv \lambda_1$, to correspond to maximum $\rho(U_1,V_1)$, and find the corresponding $\boldsymbol{\alpha} = \boldsymbol{\alpha}_1$, and $\boldsymbol{\gamma} = \boldsymbol{\gamma}_1$. Use of (2.6.7) and (11.2.6) gives the equivalent formulation

$$|\boldsymbol{\Sigma}_{12}\boldsymbol{\Sigma}_{22}^{-1}\boldsymbol{\Sigma}_{21} - \lambda^2\boldsymbol{\Sigma}_{11}| = 0, \quad \text{and} \quad (\boldsymbol{\Sigma}_{12}\boldsymbol{\Sigma}_{22}^{-1}\boldsymbol{\Sigma}_{21} - \lambda^2\boldsymbol{\Sigma}_{11})\boldsymbol{\alpha} = \mathbf{0}, \tag{11.2.7}$$

which has p_1 solutions for λ^2, and p_1 solutions for $\boldsymbol{\alpha}$; also from (2.6.8),

$$|\boldsymbol{\Sigma}_{21}\boldsymbol{\Sigma}_{11}^{-1}\boldsymbol{\Sigma}_{12} - \lambda^2\boldsymbol{\Sigma}_{22}| = 0, \quad \text{and} \quad (\boldsymbol{\Sigma}_{21}\boldsymbol{\Sigma}_{11}^{-1}\boldsymbol{\Sigma}_{12} - \lambda^2\boldsymbol{\Sigma}_{22})\boldsymbol{\gamma} = \mathbf{0}, \tag{11.2.8}$$

which has p_2 solutions for λ^2, and p_2 solutions for $\boldsymbol{\gamma}$. It may be found from (2.4.2) that the nonzero solutions of (11.2.7) for λ^2 are identical to those of (11.2.8). $U_1 \equiv \boldsymbol{\alpha}_1'\mathbf{Y}$ and $V_1 \equiv \boldsymbol{\gamma}_1'\mathbf{Z}$ are called the *first canonical variates*, and the correlation of U_1 and V_1, $\lambda_1 \equiv \rho(U_1, V_1)$, is called the *first canonical correlation*. Applying (2.14.16) to $\mathcal{L}$ shows that a maximum has actually been obtained.

Next define $U_2 = \boldsymbol{\alpha}_2'\mathbf{Y}$ and $V_2 = \boldsymbol{\gamma}_2'\mathbf{Z}$ so that var U_2 = var $V_2 = 1$, U_2 and V_2 are uncorrelated with U_1 and V_1 respectively, and $\rho(U_2, V_2)$ is as large as possible. The result is that $\rho(U_2, V_2) = \lambda_2$, the second largest root of (11.2.6). If $\boldsymbol{\alpha}_2$ and $\boldsymbol{\gamma}_2$ are solutions of (11.2.5) corresponding to λ_2, U_2 and V_2 are called the second canonical variates and λ_2 is called the second canonical correlation.

Continuing as above, define successive canonical variates as variates which are uncorrelated with all previous variates and are standardized. However, the number of pairs of positively correlated canonical variates is limited to p_1, as seen in (11.2.7). The results may be conveniently summarized in matrix form. Let $\mathbf{U} \equiv (U_1, \ldots, U_{p_1})'$, $\mathbf{V} \equiv (V_1, \ldots, V_{p_1})'$, $\mathbf{W} = (W_k)$, $\mathbf{W} \equiv (V_{p_1+1}, \ldots, V_{p_2})'$, where $U_j = \boldsymbol{\alpha}_j'\mathbf{Y}$, $V_j = \boldsymbol{\gamma}_j'\mathbf{Z}$, $j = 1, \ldots, p_1$, and $W_k = \boldsymbol{\gamma}_k'\mathbf{Z}$, $k = p_1 + 1, \ldots, p_2, p_1 \leq p_2$. Let $\mathbf{D}_\lambda = \text{diag}(\lambda_1, \ldots, \lambda_{p_1})$, $\lambda_1 \geq \lambda_2 \geq \cdots \geq \lambda_{p_1}$. That is, λ_1 is the largest positive square root of the latent roots of (11.2.6).

Theorem (11.2.1): For $\mathbf{U}$, $\mathbf{V}$, and $\mathbf{W}$ defined as above, if $\mathbf{W}$ is also normalized so that var $(\mathbf{W}) = \mathbf{I}$,

$$\text{var}\begin{pmatrix}\mathbf{U}\\ \mathbf{V}\\ \mathbf{W}\end{pmatrix} = \begin{pmatrix}\mathbf{I} & \mathbf{D}_\lambda & \mathbf{0}\\ \mathbf{D}_\lambda & \mathbf{I} & \mathbf{0}\\ \mathbf{0} & \mathbf{0} & \mathbf{I}\end{pmatrix}, \tag{11.2.9}$$

where $\lambda_1 \geq \lambda_2 \geq \cdots \geq \lambda_{p_1} \geq 0$.

Proof: See, for instance, Hotelling (1936) or Anderson (1958), Chapter 12.

The implications of this theorem are that $\mathbf{U}$ and $\mathbf{W}$ are uncorrelated, $\mathbf{V}$ and $\mathbf{W}$ are uncorrelated, $\mathbf{U}$, $\mathbf{V}$, $\mathbf{W}$ are all standardized, and that the jth pair of elements of $\mathbf{U}$ and $\mathbf{V}$, (U_j, V_j) have nonnegative correlation λ_j. The elements of $\mathbf{U}$, $\mathbf{V}$, and $\mathbf{W}$ are the canonical variates and the λ_j's are the corresponding canonical correlations.

[*Remark:* It is easy to check [in light of Lemma (11.2.1)] that the canonical correlations and the canonical variates are invariant under changes in location and scale so that in this model the units of the variables are not of any concern.]

In applications it is often the case that the first canonical correlation is large relative to the remaining canonical correlations, which are all near zero. In such studies often only the first canonical variates are retained for further analysis. In other applications the first few pairs of canonical variates have appreciably large correlation, so that a large reduction in dimension of the problem may still be achieved by retaining only those first few pairs.

11.3 UNKNOWN PARAMETERS AND INFERENCE IN SAMPLING

In this section it is assumed that all parameters are unknown. Point estimates are given for the canonical variates and correlations, and procedures are provided for testing for correlation between the components of a pair of estimated canonical variates, and for correlation between entire sets of canonical variates.

11.3.1 Point Estimation

Suppose, as in Section 11.2, that $\mathbf{T}: p \times 1$ is a random vector with $E(\mathbf{T}) = \boldsymbol{\theta}$, and var $(\mathbf{T}) = \boldsymbol{\Sigma} > 0$. However, now assume $\boldsymbol{\theta}$ and $\boldsymbol{\Sigma}$ are unknown. Let $\mathbf{t}_1, \mathbf{t}_2, \ldots, \mathbf{t}_N$ be independent, $p \times 1$ vector observations of $\mathbf{T}$. Partition $\mathbf{t}_j$ and $\boldsymbol{\Sigma}$ as

$$\mathbf{t}_j = \begin{pmatrix} \mathbf{y}_j \\ \mathbf{z}_j \end{pmatrix}, \qquad \boldsymbol{\Sigma} = \begin{pmatrix} \boldsymbol{\Sigma}_{11} & \boldsymbol{\Sigma}_{12} \\ \boldsymbol{\Sigma}_{21} & \boldsymbol{\Sigma}_{22} \end{pmatrix}, \qquad j = 1, \ldots, N,$$

where $\mathbf{y}_j: p_1 \times 1$ and $\mathbf{z}_j: p_2 \times 1$, $p_1 + p_2 = p$, $p_1 \leq p_2$. Define the maximum likelihood estimators of $\boldsymbol{\Sigma}$ and $\boldsymbol{\theta}$, under normality,

$$\hat{\boldsymbol{\Sigma}} = \frac{1}{N} \sum_1^N (\mathbf{t}_j - \bar{\mathbf{t}})(\mathbf{t}_j - \bar{\mathbf{t}})', \qquad \hat{\boldsymbol{\theta}} \equiv \bar{\mathbf{t}} \equiv \frac{1}{N} \sum_1^N \mathbf{t}_j.$$

Let $\hat{\boldsymbol{\Sigma}}$ be partitioned in the same way as $\boldsymbol{\Sigma}$, so that

$$\hat{\boldsymbol{\Sigma}} = \begin{pmatrix} \hat{\boldsymbol{\Sigma}}_{11} & \hat{\boldsymbol{\Sigma}}_{12} \\ \hat{\boldsymbol{\Sigma}}_{21} & \hat{\boldsymbol{\Sigma}}_{22} \end{pmatrix}$$

$$\equiv \begin{bmatrix} \frac{1}{N} \sum_1^N (\mathbf{y}_j - \bar{\mathbf{y}})(\mathbf{y}_j - \bar{\mathbf{y}})' & \frac{1}{N} \sum_1^N (\mathbf{y}_j - \bar{\mathbf{y}})(\mathbf{z}_j - \bar{\mathbf{z}})' \\ \frac{1}{N} \sum_1^N (\mathbf{z}_j - \bar{\mathbf{z}})(\mathbf{y}_j - \bar{\mathbf{y}})' & \frac{1}{N} \sum_1^N (\mathbf{z}_j - \bar{\mathbf{z}})(\mathbf{z}_j - \bar{\mathbf{z}})' \end{bmatrix}. \tag{11.3.1}$$

It is easy to check that using $\hat{\boldsymbol{\Sigma}}$ in place of $\boldsymbol{\Sigma}$ to estimate canonical variates and correlations gives maximum likelihood results. Specifically:

Lemma (11.3.1): Replacing $\boldsymbol{\Sigma}_{ij}$ by $\hat{\boldsymbol{\Sigma}}_{ij}$ in (11.2.5) and (11.2.6), $i, j = 1, 2$, yields maximum likelihood estimators of $\boldsymbol{\alpha}$, $\boldsymbol{\gamma}$, and λ.

Thus, all the canonical variates and correlations are obtainable by substituting sample estimates for population values in the defining equations. Computer programs for extracting the canonical variates and correlations are referenced in Appendix A.

11.3.2 Testing Canonical Correlations for Significance

Let $\mathbf{T}' \equiv (\mathbf{Y}',\mathbf{Z}')$, $\mathbf{t}_j' \equiv (\mathbf{y}_j',\mathbf{z}_j')$, and let $\mathbf{t}_1,\ldots,\mathbf{t}_N$ be independent p-variate observations of $\mathbf{T}$ with var $(\mathbf{t}_j) = \boldsymbol{\Sigma}$. Suppose it is desired to determine whether $\mathbf{Y}$ and $\mathbf{Z}$ are correlated, and whether any of the canonical correlations are significantly different from zero. To carry out such tests first make the assumption, not yet required, that $\mathcal{L}(\mathbf{t}_j) = N(\boldsymbol{\theta},\boldsymbol{\Sigma})$, $\boldsymbol{\Sigma} > 0$, $j = 1,\ldots,N$. If $\mathbf{t}_j$ does not seem to be Normally distributed, the data can sometimes be transformed to approximate Normality before proceeding (see, for instance, Rao, 1965; also, Section 8.10.3).

Consider testing the hypotheses

$$H_1\colon \boldsymbol{\Sigma}_{12} = \mathbf{0} \text{ versus } A_1\colon \boldsymbol{\Sigma}_{12} \neq \mathbf{0}. \tag{11.3.2}$$

Acceptance of H_1 will also imply that $\mathbf{U} \equiv (U_1,\ldots,U_{p_1})'$ and $\mathbf{V} \equiv (V_1,\ldots,V_{p_1})'$ are uncorrelated (independent) since $\lambda_j \equiv \rho(U_j,V_j) = \boldsymbol{\alpha}_j'\boldsymbol{\Sigma}_{12}\boldsymbol{\gamma}_j$, $j = 1,\ldots,p_1$. That is, a test of H_1 is a test for the joint significance of the first p_1 canonical correlations. Direct computation shows that the likelihood ratio statistic, l, for testing H_1 versus A_1 is expressible as

$$l^{2/N} \equiv \Lambda_1 = \frac{|\hat{\boldsymbol{\Sigma}}|}{|\hat{\boldsymbol{\Sigma}}_{11}|\,|\hat{\boldsymbol{\Sigma}}_{22}|}, \tag{11.3.3}$$

where $\hat{\boldsymbol{\Sigma}}$, and $\hat{\boldsymbol{\Sigma}}_{ij}$, $i, j = 1, 2$ are defined in (11.3.1). The exact distribution of Λ_1 was studied by Hotelling (1936), by Girshick (1939), and by Anderson (1958), p. 237. Moreover, it was shown by Narain (1950) that a test based upon (11.3.3) has the desirable property of being unbiased. Because the exact distribution of Λ_1 is complicated, an approximate (large sample) distribution was given by Bartlett (1938; 1939; 1941). In this connection it is preferable to express Λ_1 given in (11.3.3) in an alternative form.

Accordingly, since

$$\hat{\boldsymbol{\Sigma}} = \begin{vmatrix} \hat{\boldsymbol{\Sigma}}_{11} & \hat{\boldsymbol{\Sigma}}_{12} \\ \hat{\boldsymbol{\Sigma}}_{21} & \hat{\boldsymbol{\Sigma}}_{22} \end{vmatrix},$$

use of (2.6.7) gives for $\hat{\boldsymbol{\Sigma}}_{11} > 0$, and $\hat{\boldsymbol{\Sigma}}_{22} > 0$,

$$\begin{aligned} |\hat{\boldsymbol{\Sigma}}| &= |\hat{\boldsymbol{\Sigma}}_{22}|\,|\hat{\boldsymbol{\Sigma}}_{11} - \hat{\boldsymbol{\Sigma}}_{12}\hat{\boldsymbol{\Sigma}}_{22}^{-1}\hat{\boldsymbol{\Sigma}}_{21}| \\ &= |\hat{\boldsymbol{\Sigma}}_{22}|\,|\hat{\boldsymbol{\Sigma}}_{11}|\,|\mathbf{I} - \hat{\boldsymbol{\Sigma}}_{11}^{-1}\hat{\boldsymbol{\Sigma}}_{12}\hat{\boldsymbol{\Sigma}}_{22}^{-1}\hat{\boldsymbol{\Sigma}}_{21}|. \end{aligned}$$

Substitution in (11.3.3) shows that

$$\Lambda_1 = |\mathbf{I} - \hat{\mathbf{\Sigma}}_{11}^{-1}\hat{\mathbf{\Sigma}}_{12}\hat{\mathbf{\Sigma}}_{22}^{-1}\hat{\mathbf{\Sigma}}_{21}|.$$

Since it is clear from (11.2.7) that $\hat{\lambda}^2$ is a latent root of $\hat{\mathbf{\Sigma}}_{11}^{-1}\hat{\mathbf{\Sigma}}_{12}\hat{\mathbf{\Sigma}}_{22}^{-1}\hat{\mathbf{\Sigma}}_{21}$, it follows that

$$\Lambda_1 = \prod_{j=1}^{p_1} (1 - \hat{\lambda}_j^2). \tag{11.3.4}$$

Using the large sample result of Wilks [see Section 8.4.3, Step (4), page 217] for the distribution of a likelihood ratio statistic under H_1, and modifying it a bit to improve the rate of convergence, Bartlett gives the result

$$\lim_{N\to\infty} \mathcal{L}\{-\nu \log \Lambda_1\} = \chi^2(p_1p_2), \tag{11.3.5}$$

where $\nu = N - \frac{1}{2}(p_1 + p_2 + 1)$. That is, for N large, $-\nu \log \Lambda_1$ is distributed approximately as a chi-square variate.

Suppose H_1 is rejected in favor of A_1; that is, it is concluded that $\mathbf{\Sigma}_{12} \neq 0$. Bartlett suggests the use of

$$\Lambda_2 \equiv \frac{\Lambda_1}{1 - \hat{\lambda}_1^2} = \prod_{j=2}^{p_1} (1 - \hat{\lambda}_j^2), \tag{11.3.6}$$

with the associated distributional result that under H_2,

$$\lim_{N\to\infty} \mathcal{L}\{-\nu \log \Lambda_2\} = \chi^2[(p_1 - 1)(p_2 - 1)]. \tag{11.3.7}$$

That is, for large N, it is now possible to test for the joint significance of the remaining $(p_1 - 1)$ canonical correlations; that is, test H_2: $(\lambda_2,\ldots,\lambda_{p_1})$ are jointly not significant, versus A_2: $(\lambda_2,\ldots,\lambda_{p_1})$ are jointly significant. If H_1 is rejected but H_2 is accepted, λ_1 is the only significant canonical correlation.

If H_2 is rejected the test procedure can be continued with a third test, and then if necessary, with a fourth, and so on. At the rth stage, the test involves H_r: $(\lambda_r,\lambda_{r+1},\ldots,\lambda_{p_1})$ are jointly insignificant, versus A_r: $(\lambda_r,\lambda_{r+1},\ldots,\lambda_{p_1})$ are jointly significant; the test statistic is

$$\Lambda_r = \frac{\Lambda_{r-1}}{1 - \hat{\lambda}_{r-1}^2} = \prod_{j=r}^{p_1} (1 - \hat{\lambda}_j^2), \tag{11.3.8}$$

and under H_r,

$$\lim_{N\to\infty} \mathcal{L}\{-\nu \log \Lambda_r\} = \chi^2[(p_1 - r + 1)(p_2 - r + 1)]. \tag{11.3.9}$$

Note that after a sequence of tests has been carried out, the overall confidence level of the joint probability statement about the tests may be considerably lower than that for any single test (since the tests are interdependent).

Example (11.3.1): Suppose $\mathbf{Y}: 10 \times 1$ is a vector of socioeconomic indicators characterizing a neighborhood in a large metropolitan area, and $\mathbf{Z}: 12 \times 1$ is a vector reflecting the mean number of crimes reported in that neighborhood during a one month period for each of 12 types of crime. A sample of observations covering a 10 year period (120 months) is obtained and the data is preadjusted for seasonal effects (there is more crime in the warmer weather), long-term trends in crime, gross changes of character of the neighborhood, and so on. It may often be assumed that such effects are additive. Suppose these effects are removed by subtraction. A canonical correlations model is then applied to the resulting data to determine whether some simple socioeconomic indices can be used as predictors of crime indices. In this problem, $p_1 = 10$, $p_2 = 12$, $N = 120$. Suppose the squares of the sample canonical correlations that result from a canonical correlations analysis are given by $\hat{\lambda}_j^2 = (.810, .630, .440, .180, .100, .080, .060, .010, .002, .001)$, $j = 1, \ldots, 10$, respectively.

It is tempting to use just the first pair of canonical variates since $\hat{\lambda}_1^2$ is so large. However, $\hat{\lambda}_2^2$ and $\hat{\lambda}_3^2$ are also fairly large so that it would not be too surprising if all three population canonical correlations were significantly different from zero.

Since N is fairly large, Bartlett's test is applicable for testing for significance of the canonical correlations. The test data are summarized in Table 11.3.1. Since $\nu \doteq 108.5$, it is easy to check from (11.3.9) that H_1

Table 11.3.1 Summary of Crime Test Data

	Λ_1	Λ_2	Λ_3
Statistic	.025	.129	.349
Degrees of freedom	120	99	80

and H_2 are unquestionably rejected at the 1 percent level of significance, whereas H_3 is just barely rejected at that level. Moreover, simple computation shows that all tests of H_r for $r > 3$ will accept the hypothesis of insignificance of the canonical correlations. That is, $\lambda_4, \ldots, \lambda_{10}$ are not significantly different from zero.

Example (11.3.2): Green, Halbert, and Robinson (1966) were interested in relating various personality characteristics of individuals to their behavior regarding the extent to which they would be willing to let observational data modify their subjective (prior) feelings about how to make a decision.

An experimental game was constructed in which 36 subjects (students in a graduate school of business) were given prior information about the likelihood that a set of ten card decks belonged to one of two possible classes. The subjects had the option, on the basis of prior information alone, to guess which class the deck they selected for betting purposes belonged to, or, for a fixed cost per card, they could see some or all of the cards before placing their bet. Thus, the risk taking tendencies of the subjects were being measured. Two measures of this risk taking tendency were defined and scored for the subjects. The first measure was called "sensitivity," and measured the degree to which a subject varied the amount of information purchased in the experiment, as a function of his prior uncertainty. The scoring for sensitivity was a zero or a one. The second measure was called "bias," and it measured the amount of information a subject purchased during the whole game. The sample correlation coefficient between sensitivity and bias for the subjects was .09.

Personality of the subjects was measured by a battery of 20 tests that took two hours for the subjects to complete (such as the Allport-Vernon-Lindzey Study-of-Values Test which is supposed to classify subjects according to "theoretical," "economic," "aesthetic," "social," "political," and "religious" categories). The correlation matrix for the twenty personality scores and the two risk measures were computed and those personality variables which had zero correlation (at the 15 percent significance level) with both risk taking measures were dropped from further analysis. This resulted in five personality variables and the two risking taking variables. The 7 × 7 correlation matrix for these variables is given in Table 11.3.2 (note that the covariance matrix is not required since the results of a canonical correlations analysis will not depend on the scale used).

Table 11.3.2 Correlation Matrix for Personality and Risk

Risk taking variables	Sensitivity	1.00	.09	.27	−.36	.19	.44	−.23
	Bias		1.00	−.29	−.23	−.09	.05	−.29
Personality variables	(1) Intelligence			1.00	.22	−.23	−.23	.07
	(2) Responsibility				1.00	.18	−.17	.07
	(3) Open mindedness					1.00	.24	.01
	(4) Control over one's environment						1.00	−.25
	(5) Sociability							1.00

To relate risk taking to personality, a canonical correlations analysis was carried out on the correlation matrix of Table 11.3.2. The weights for the two canonical variates are given in Table 11.3.3.

Table 11.3.3

	Risk taking variables		*Personality variables*				
Weights for first canonical variate	.84	.47	−.34	−.52	.12	.36	−.38
Weights for second canonical variate	−.55	.89	−.58	.12	−.50	−.67	−.50

The sample canonical correlations of the two canonical variates were .61 and .35, respectively. The large sample test for significance of the canonical correlations was carried out [Equation (11.3.5)] and both canonical correlations were found to be significant at the 5 percent level of significance. However, since the sample size was not really large ($N = 36$), and since one of the variables (sensitivity to risk taking) was a Bernoulli (zero-one) random variable, the data vectors were not really drawn from a multivariate Normal population. Since the assumptions of the significance test were not satisfied, the resulting inferences must remain speculative. However, if results can be interpreted at least descriptively from the correction matrix, it is clear that Sensitivity and Unbiasedness tend to increase with higher IQ responsibility, and sociability, and to decrease with higher scores on attitude rigidity and degree of fatalism with respect to uncontrollable events.

11.4 CANONICAL CORRELATIONS WITH SEVERAL SETS OF VARIABLES

There have been several extensions of Hotelling's original canonical correlations model to more than two sets of variates. Anderson (1958), p. 306 suggested that the canonical variates be found by minimizing the generalized variance of the vector of transformed variates.

Horst (1961) proposed maximizing the sum of the correlations of pairs of linear combinations of variables (one linear combination from each of two sets form a pair).

These and three other procedures were studied by Kettenring(1971). He concluded that although all the methods reduce to Hotelling's model when there are only two sets of variates, with more than two sets, they are each designed to detect a different form of linear relationship among the sets of variables. Therefore, in exploratory studies it is often prudent to use several of the generalizations of the model in order not to exclude some underlying relationships which might prove important. For a recent inferential, non-iterative, generalization to several sets of variaxes, see Sen Gupta (1981).

REFERENCES

Anderson (1958).
Bartlett (1938, 1939, 1941).
Cooley and Lohnes (1962).
Geisser (1965a).
Girshick (1939).
Green, Halbert, and Robinson (1966).
Hooper (1959).
Horst (1962).
Hotelling (1935, 1936).
Kettenring (1971).
Narain (1950).
Rao (1965).
Sen Gupta (1981).
Stone (1947).
Waugh (1942).

EXERCISES

11.1 The population covariance matrix for changes in the numbers of users of four different illicit drugs was given in Exercise 3.1. Suppose it is desired to try to predict changes in marijuana and opium use from changes in the use of LSD and heroin. Determine the two canonical variates and their correlations, and interpret the results.

11.2 Consider carrying out the analysis of Exercise 11.1 on the implied correlation matrix instead of on the covariance matrix. To what extent is the difference in results of the two analyses predictable?

11.3 Suppose the covariance matrix used in Exercise 11.1 were a *sample* covariance matrix based upon a sample of size 100 (instead of a population covariance matrix).
   (a) Test the first canonical correlation for significance (at the 5 percent level);
   (b) test the second canonical correlation for significance.

11.4 How would you apply the ideas of canonical correlations analysis to the problem of testing hypotheses in the multivariate analysis of variance model?

11.5 How would you carry out a canonical correlations analysis on a set of data vectors collected weekly for N weeks, and which exhibits serial correlation?

11.6 Relate the notion of a canonical correlation coefficient to that of a multiple correlation coefficient.

11.7 How could inferential methods of canonical correlations analysis be used to study the meaningfulness of a small change in an index number such as the Department of Labor's "Cost of Living" index?

11.8 Compare the distribution of the sample canonical correlations with that of the population canonical correlations given the sample, with respect to a diffuse (vague) prior distribution. [Hint: You will need to examine the relevant journal articles.]

11.9 Discuss the meaning of canonical correlations applied to jointly dependent discrete random variables.

11.10 Explain the use of canonical correlations in studying goodness-of-fit in generalized regression.

12

STABLE PORTFOLIO ANALYSIS

12.1 INTRODUCTION

Portfolio analysis refers to the study of a portfolio of investments with the objective of maximizing some measure of portfolio performance, such as the expected return, while minimizing the risk. Stable portfolio analysis assumes that price changes of the assets in the portfolio follow symmetric stable distributions. If the prices are associated or related to one another, they are assumed to follow a multivariate symmetric stable distribution (see Chapter 6). It should be clear that the portfolio might contain human resources and social structures as well as cash and securities such as stocks and bonds. To avoid confusion the model will be discussed in a financial setting, although a potentially broader context should be understood.

The subject of portfolio analysis involves study of the utility function of the investor. The problem is to allocate the investor's resources to his assets in proportions that will maximize the expected utility of his portfolio. There is also some discussion of the simpler problem of how to minimize portfolio "risk" through diversification, by acquiring assets in optimal proportions (at some sacrifice to expected return). The measure of risk will be the scale parameter of an appropriate multivariate symmetric stable distribution. Thus, the underlying multivariate mathematical model is applicable to any situation requiring the minimization of the scale parameter of the distribution of a linear function of the elements of a random vector following a multivariate symmetric stable probability distribution, subject to a linear equality constraint and a non-negativity constraint on the coefficients of the elements. For an extensive discussion of the *Finance* theory behind the models discussed in this chapter, see Fama and Miller, 1972.

Section 12.2 contains some background of the problem, definition of the terms, and a basis for the model in its more general setting. Section 12.3 defines the *general* (unrestricted) stable portfolio analysis problem and presents an implicit solution which must be evaluated numerically. Section 12.4 proposes a policy for allocating scarce resources in the *restricted* stable portfolio analysis problem for the case in which prices of all assets in the portfolio are jointly distributed and follow a multivariate symmetric stable distribution (the marginal distributions all have the same characteristic exponent). Optimization under this assumption will be referred to as a Model I stable portfolio analysis. The restricted problem will be seen to be the one in which risk is minimized at some sacrifice to return.

Section 12.5 proposes a method of allocating resources in the more complicated *restricted* stable portfolio analysis problem (Model II) in which groups of assets are permitted to follow multivariate symmetric stable distributions with a different characteristic exponent for each group.

Section 12.6 is directed to the problem of parameter estimation in stable distributions. Several methods are proposed for estimating the parameters of univariate stable distributions, and multivariate symmetric stable distributions. The methods proposed are all based upon the sample characteristic function and they all provide estimators which are consistent, but not necessarily efficient. Hence, these methods should have greatest utility when large samples are available. However, at this time it is not known which of the methods is to be preferred.

12.2 BACKGROUND AND DEVELOPMENT OF THE MODEL

Let $p_j(\tau)$ denote the price per share of asset j at time τ, one period from now in the future, $p_j(\tau - 1)$ is the current price per share of asset j, and the current price is assumed known. The "return" on one share of this asset is given by

$$= \frac{p_j(\tau) - p_j(\tau - 1)}{p_j(\tau - 1)} \tag{12.2.1}$$

Thus, Y_j is a linear function of the random variable $p_j(\tau)$ for each j. Define the vector $\mathbf{Y} \equiv (Y_1, \ldots, Y_p)'$ representing the per share returns on all assets in the investment portfolio.

It will be assumed that $\mathbf{Y}: p \times 1$ follows a multivariate stable law. Empirical justification for this assumption has been offered by Mandelbrot (1964), Fama (1965b, p. 68), and Roll (1968). That is, evidence is not inconsistent with the notion of the Y_j's following distributions that are sometimes Normal, but that often have more probability mass in their tails than the Normal distribution. The compound Poisson distribution with Normal compounding would explain such behavior (see Press, 1968, and 1970), and so would a variety of other "fat tail" distributions. However, the property that sums of linear functions of stable variables with the same characteristic exponent are also stable makes this explanation for the variation of prices of assets particularly attractive for portfolio analysis.

To permit positive and negative price changes to be weighted in the same way, only the subclass of multivariate symmetric stable distributions is considered. The symmetry need not be around the origin, however, and only needs to be around any point in p-space. As a result, the family of distributions under consideration can be described by the log-characteristic function given in (6.5.9), Chapter 6; namely, for some m,

$$\log \phi_{\mathbf{Y}}(\mathbf{t}) = i\mathbf{a}'\mathbf{t} - \frac{1}{2}\sum_{j=1}^{m} (\mathbf{t}'\mathbf{\Omega}_j\mathbf{t})^{\alpha/2}, \tag{12.2.2}$$

where the $\mathbf{\Omega}_j$'s are scale matrices (assume no two are proportional) with the property that for every j, $\mathbf{\Omega}_j \geq 0$; $\mathbf{a} \equiv (a_1,\ldots,a_p)'$ is the location vector of the distribution (median) and is the mean vector if $\alpha > 1$, $-\infty < a_k < \infty$; more generally, α is restricted to the range $0 < \alpha \leq 2$. To avoid degenerate distributions, it is assumed that $\Sigma_1^m \mathbf{\Omega}_j > 0$. The law is denoted by

$$\mathcal{L}(\mathbf{Y}) = S_p(m,\mathbf{a};\mathbf{\Omega}_1,\ldots,\mathbf{\Omega}_m;\alpha). \tag{12.2.3}$$

It will be assumed generally that $1 < \alpha \leq 2$, unless indicated otherwise. There are several reasons for this assumption. First, for an investment setting, it is convenient to be able to speak of "expected returns" (assuming a finite first moment). Moreover, under a stable distribution assumption, the assumption that $\alpha > 1$ is in general agreement with empirical evidence. Finally, the assumption implies that although the distributions sometimes depart from Normality, they do not deviate "too much."

Let c_j denote the fraction of available funds to be invested in asset j, and let $\mathbf{c}' \equiv (c_1,\ldots,c_p)$. Assume the portfolio is to consist of p assets whose prices follow a multivariate symmetric stable law.

Let Q denote the return on the investment portfolio; that is,

$$Q = \mathbf{c}'\mathbf{Y}.$$

The dollar return on the portfolio is QW, where W denotes the total number of dollars invested in the portfolio. We will assume $W = 1$ in the sequel. The expected return on the portfolio is $E(Q) = \mathbf{c}'E(\mathbf{Y})$, and if $\mathbf{Y}$ follows the law (12.2.2) with $\alpha > 1$, $E(Q) = \mathbf{c}'\mathbf{a}$. One basic problem of a portfolio analysis is achieving the largest possible value of $E(Q)$ while keeping the "risk" of the portfolio as small as possible. That is, among all possible collections of assets that can be acquired, there are various combinations of collections and sets of proportions of each asset in the collection (portfolio) for which $E(Q)$ is as large as possible, while the risk associated with Q is as small as possible, subject to the constraints that $c_j \geq 0$ for all $j = 1, \ldots, p$, and $\Sigma_1^p c_j = 1$. These various combinations will be examined in the next section. In many situations there are additional physical and budgetary equality and inequality constraints that must be superimposed

Other attempts to analyze risk in portfolios constituted of assets whose prices follow stable distributions were made by Fama (1965a), Fama (1968), and Jensen (1969). These studies involve the use of univariate stable distributions only and attempt to account for the association of prices of different assets by postulating a model (called the Diagonal Model) in which return on an asset is a linear function of an overall market factor, and the parameters of the linear function differ for each asset. Thus, prices of different assets are related through the same market factor. By contrast, in the models described below, prices are assumed to be jointly distributed following a multivariate symmetric stable law. Moreover, in Model II, the price laws of different assets will no longer be required to have the same characteristic exponent.

There are many possible measures of portfolio risk. If α were equal to 2, the variance of Q could be used (or some function of variance). However, since for $\alpha < 2$, the variance does not exist, the scale parameter of the distribution will be used instead. Adopt the definition that the *risk of a portfolio* is equal to the scale parameter of the distribution of the return on the portfolio, Q. The scale parameter will be defined in terms of the parameters of the characteristic function of Q, which is developed below.

The characteristic function of the scalar Q is given by

$$\phi_Q(v) = Ee^{ivQ},$$

for a scalar variable v. Since

$$\phi_{\mathbf{Y}}(\mathbf{t}) = Ee^{i\mathbf{t}'\mathbf{Y}},$$

if $\mathbf{t} \equiv v\mathbf{c}$,

$$\phi_{\mathbf{Y}}(v\mathbf{c}) = E \exp (iv\mathbf{c}'\mathbf{Y}) = \phi_Q(v). \tag{12.2.4}$$

Substitution in (12.2.2) gives for the log characteristic function of the return on the portfolio,

$$\log \phi_Q(v) = iv(\mathbf{c}'\mathbf{a}) - \tfrac{1}{2}|v|^\alpha \sum_{j=1}^{m} (\mathbf{c}'\mathbf{\Omega}_j\mathbf{c})^{\alpha/2}. \tag{12.2.5}$$

The scale parameter or *risk* of this distribution is defined as

$$r(\mathbf{c}) \equiv \frac{1}{2} \sum_{j=1}^{m} (\mathbf{c}'\boldsymbol{\Omega}_j\mathbf{c})^{\alpha/2}.$$

That is, the scale parameter of a stable law is taken to be the coefficient of $-|v|^\alpha$ in (12.2.5).

12.3 THE GENERAL STABLE PORTFOLIO ANALYSIS PROBLEM

An *efficient portfolio* was defined by Markowitz (1959) and Sharpe (1963) as one which cannot achieve greater expected return without increasing risk. Thus, the portfolio analysis problem can be thought of as consisting of two parts: (a) the finding of all possible efficient portfolios, and (b) the selection of one portfolio from among the set of efficient ones. The second problem depends for its solution upon an individual's attitude toward risk taking. Attention is now focused upon problem (a). The efficient portfolios will be selected from the set of feasible portfolios, that is, from the set of combinations of expected return and risk which are possible. The feasible portfolios are shown below to lie in a closed region of the form shown in Figure 12.3.1. That is, for a portfolio composed of p assets, the boundary of the region of feasible solutions is defined by p nodal points which are arranged so that adjacent nodes are connected by convex curves of generally different shapes. Moreover, the boundary is continuous at the nodes so that the resulting simplex is closed. These assertions regarding the feasible portfolios are demonstrated below.

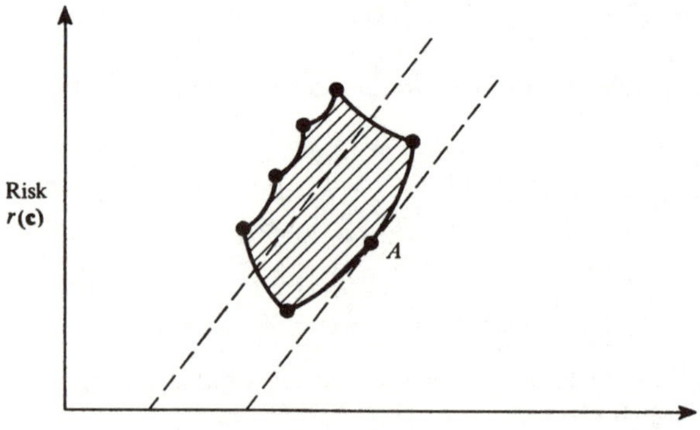

Figure 12.3.1. Feasible Portfolio Analysis Solutions

For notational simplicity, denote the expected return on the portfolio by $E \equiv EQ = \mathbf{c}'\mathbf{a}$, and denote the risk associated with the portfolio by $r \equiv r(\mathbf{c})$. Now compute $\partial^2 r/\partial E^2$ for α in the range $1 < \alpha \leq 2$, for all points where the second derivative may be evaluated (between two adjacent nodes). First note that by the chain rule,

$$\frac{\partial r}{\partial E} = \left(\frac{\partial r}{\partial \mathbf{c}'}\right)\left(\frac{\partial \mathbf{c}}{\partial E}\right).$$

Hence, by the product rule,

$$\frac{\partial^2 r}{\partial E^2} = \left(\frac{\partial^2 r}{\partial E \partial \mathbf{c}'}\right)\left(\frac{\partial \mathbf{c}}{\partial E}\right) + \left(\frac{\partial r}{\partial \mathbf{c}'}\right)\left(\frac{\partial^2 \mathbf{c}}{\partial E^2}\right).$$

Note that since $E = \mathbf{c}'\mathbf{a}$, $(\partial^2 \mathbf{c}/\partial E^2) = \mathbf{0}$. Applying the chain rule to $(\partial^2 r/\partial E \partial \mathbf{c}')$ gives

$$\frac{\partial^2 r}{\partial E^2} = \left(\frac{\partial \mathbf{c}'}{\partial E}\right)\mathbf{H}\left(\frac{\partial \mathbf{c}}{\partial E}\right),$$

where $\mathbf{H} = \partial^2 r/\partial \mathbf{c} \partial \mathbf{c}'$ denotes the Hessian matrix. But it was seen in Exercise 6.10 that $\mathbf{H} > 0$. Hence, the quadratic form $\partial^2 r/\partial E^2$ is positive definite; that is, r is convex in E. Note that if one of the assets is excluded from the portfolio (say the pth, by taking $c_p = 0$), the argument remains the same, except the resulting second derivative is different. Thus, the curvature of the simplex changes only at points at which assets are added or deleted. Continuity now requires that the simplex be closed.

From the set of feasible portfolios shown in Figure 12.3.1, the subset of efficient ones must now be selected. Following Sharpe (1963), consider the objective function $[\lambda E - r]$, where λ is assumed fixed. The dashed line passing through the interior of Figure 12.3.1 is a plot of the line $\lambda E - r =$ constant. The slope of the line is λ. For this value of λ it is clearly most desirable to have $[\lambda E - r]$ as large as possible. Since E increases to the right, it follows that for this value of λ, the best of the feasible portfolios is the one corresponding to point A in the figure; that is, the one corresponding to the point of tangency of the line $[\lambda E - r]$ with the set of feasible portfolios. The various lines $[\lambda E - r]$ are obtained by varying the vector $\mathbf{c}$, subject to the constraints that $\mathbf{c}'\mathbf{e} = 1$, and $c_j \geq 0$, $j = 1, \ldots, p$, where $\mathbf{e}$ denotes a p-vector of ones. It is clear that varying λ from $\lambda = +\infty$ to $\lambda = 0$ will yield all feasible portfolios. Hence, the subset of efficient portfolios may be obtained by fixing λ and then choosing c to satisfy

$$\max_{\mathbf{c}} \left[\lambda \mathbf{c}'\mathbf{a} - \frac{1}{2} \sum_{j=1}^{m} (\mathbf{c}'\boldsymbol{\Omega}_j \mathbf{c})^{\alpha/2}\right],$$

subject to $\mathbf{c}'\mathbf{e} = 1$, $c_j \geq 0$, $j = 1,\ldots,p$; and then changing λ and repeating the maximization. One point in the efficient set is obtained for each λ. If $\alpha = 2$, the maximization is a quadratic programming problem. For $1 < \alpha < 2$ it is a problem in concave programming since then the objective function is a concave function of $\mathbf{c}$ (and the return has an expectation).[1] The concavity may be demonstrated by computing the Hessian of (2.14.16) and showing its latent roots are nonnegative (see Exercise 6.10).

As an example, suppose the nonnegativity constraints, $c_j \geq 0$, to be inactive. Then the problem may be solved in principle by Lagrange multipliers. Accordingly, define the Lagrangian

$$\psi = \lambda\mathbf{c}'\mathbf{a} - \frac{1}{2}\sum_{j=1}^{m} (\mathbf{c}'\mathbf{\Omega}_j\mathbf{c})^{\alpha/2} - \phi(\mathbf{c}'\mathbf{e} - 1),$$

where $\mathbf{e}$ denotes a p-vector of ones and ϕ is a Lagrangian multiplier. Differentiating ψ with respect to $\mathbf{c}$ and setting the result equal to zero gives

$$\lambda\mathbf{a} - \frac{\alpha}{2}\sum_{j=1}^{m} (\mathbf{c}'\mathbf{\Omega}_j\mathbf{c})^{(\alpha/2)-1}\mathbf{\Omega}_j\mathbf{c} = \phi\mathbf{e}. \tag{12.3.1}$$

To find ϕ, multiply through by $\mathbf{c}'$ and use the condition that $\mathbf{c}'\mathbf{e} = 1$. Then

$$\phi = \lambda\mathbf{c}'\mathbf{a} - \frac{\alpha}{2}\sum_{j=1}^{m} (\mathbf{c}'\mathbf{\Omega}_j\mathbf{c})^{\alpha/2}.$$

Substituting this result into (12.3.1) and combining terms gives the vector equation

$$\lambda(\mathbf{a} - \mathbf{c}'\mathbf{a}\mathbf{e}) = \frac{\alpha}{2}\sum_{j=1}^{m} (\mathbf{c}'\mathbf{\Omega}_j\mathbf{c})^{\alpha/2}\left(\frac{\mathbf{\Omega}_j\mathbf{c}}{\mathbf{c}'\mathbf{\Omega}_j\mathbf{c}} - \mathbf{e}\right). \tag{12.3.2}$$

The value of $\mathbf{c}$ satisfying (12.3.2) yields the efficient portfolio for that λ when the nonnegativity constraints are inactive, when all securities in the portfolio jointly follow a multivariate symmetric stable distribution with characteristic exponent α, and when all distributional parameters are known. When the nonnegativity constraints are active, or when parameters must be estimated, the computer must be depended upon for the final solution. Of course, when parameters are unknown and estimated as in Section 12.6, estimates should be substituted for parameters in

[1] For $\alpha = 1$, maximizing *median* return could be used as a criterion instead of the one of maximizing *expected* return.

(12.3.2). Results obtained from this procedure will be appropriate in large samples.

It may be noted that when $\alpha = 2$, efficient portfolio combinations depend upon expected return and variance combinations. But a given efficient portfolio can equally well be described by expected return and the second moment of return (rather than variance of return). In this case $[\lambda EQ - EQ^2]$ is the function being maximized (subject to constraints). This function is the expectation of a quadratic utility function, and it is different for each investor, depending upon his value of λ.

An interesting alternative approach is suggested by the work of Pratt (1964). There, instead of representing an investor's utility function by the unbounded (in EQ) function used in the last paragraph, he suggests that one simple and more realistic functional form is

$$U(Q) = 1 - \omega e^{-g_1 Q} - (1 - \omega)e^{-g_2 Q},$$

where $0 \leq \omega \leq 1$, $g_1 > 0$, and $g_2 > 0$. This function is bounded, and has the property of decreasing "risk aversion," that is,

$$\frac{d}{dQ}\left[-\frac{U''(Q)}{U'(Q)}\right] < 0,$$

where the primes denote derivatives with respect to Q. For many purposes it may be desirable to choose $\mathbf{c}$ to maximize $E[U(Q)]$, subject to $\mathbf{c}'\mathbf{e} = 1$, and $c_j \geq 0$, $j = 1, \ldots, p$. Note that for this representation we must have $Q \geq 0$.

In the next two sections, a more narrow definition of portfolio analysis will be used. The advantages of working with this more narrowly defined sense of portfolio analysis are (1) it is feasible under certain conditions to obtain explicit solutions for individuals who prefer certain assets in their portfolios, and wish to incur a minimum risk penalty; and (2) insight is provided under certain conditions into how various patterns of association among prices of assets affect risk. The concern will not be with how to select assets so as to maximize $E(Q)$. Rather, it will be assumed that the assets have been selected (a special case would include all available securities on an exchange), and it is now a matter of determining how to allocate funds to each asset in such a way that the risk associated with the portfolio is minimized subject to the constraint that the allocation fractions total unity. All other constraints are ignored (so that by minimizing risk, expected return may be simultaneously reduced).

12.4 MODEL I—STABLE PORTFOLIO ANALYSIS (RESTRICTED PROBLEM)

12.4.1 Optimization for Efficient Portfolios

The scale parameter of the distribution represented in (12.2.5) was defined as $\frac{1}{2}\Sigma_1^m(\mathbf{c}'\mathbf{\Omega}_j\mathbf{c})^{\alpha/2}$. In the restricted problem considered in this section, the scale parameter must be minimized subject to the restriction that the fractions of funds allocated to each asset must total unity. All parameters m, $\mathbf{a}$, $\mathbf{\Omega}_1,\ldots,\mathbf{\Omega}_m$, and α are assumed known (estimation is discussed in Section 6). The problem can be formalized as follows.

Find $c: p \times 1$ with nonnegative components that minimizes the risk

$$r(\mathbf{c}) \equiv \frac{1}{2}\sum_{j=1}^{m}(\mathbf{c}'\mathbf{\Omega}_j\mathbf{c})^{\alpha/2},$$

subject to the constraint that $\mathbf{c}'\mathbf{e} = 1$, where $\mathbf{e}$ denotes a p-vector of ones. Since for $1 < \alpha \leq 2$, $r(\mathbf{c})$ is a convex function of $\mathbf{c}$ (see Exercise 6.10), the solution is obtained by convex programming. However, if the nonnegativity constraints are inactive, an implicit solution may be obtained by the method of Lagrange multipliers as the solution of a system of algebraic equations. The objective function is

$$\psi = \frac{1}{2}\sum_{j=1}^{m}(\mathbf{c}'\mathbf{\Omega}_j\mathbf{c})^{\alpha/2} - \phi(\mathbf{c}'\mathbf{e} - 1),$$

where ϕ is a Lagrange multiplier. Differentiating ψ with respect to $\mathbf{c}$, setting the result equal to zero, and solving gives

$$\phi\mathbf{e} = \frac{\alpha}{2}\sum_{j=1}^{m}(\mathbf{c}'\mathbf{\Omega}_j\mathbf{c})^{(\alpha/2)-1}\mathbf{\Omega}_j\mathbf{c}.$$

Imposing the constraint that $\mathbf{c}'\mathbf{e} = 1$ (by multiplying through by $\mathbf{c}'$) yields ϕ. The final result is the system of p equations

$$\sum_{j=1}^{m}(\mathbf{c}'\mathbf{\Omega}_j\mathbf{c})^{(\alpha/2)-1}\mathbf{\Omega}_j\mathbf{c} = \sum_{j=1}^{m}(\mathbf{c}'\mathbf{\Omega}_j\mathbf{c})^{\alpha/2}\mathbf{e}. \tag{12.4.1}$$

These equations cannot be solved explicitly for $\mathbf{c}$, for arbitrary $\mathbf{\Omega}_j$'s (although they can be solved in principle, by numerical methods). However, under restrictive assumptions about the patterns among the $\mathbf{\Omega}_j$'s and about their internal structures, explicit solutions can be obtained, as will be seen in the examples below. In this section, an efficient portfolio will be one satisfying the above constraints. If the solution found from (12.4.1) yields negative values of the c_i's, this approach is not applicable, and the problem must be solved by convex programming.

Example (12.4.1): A very important and realistic special case corresponding to a wide variety of multivariate symmetric stable distributions is the one in which $m = 1$ and $\mathbf{\Omega}_1 \equiv \mathbf{\Sigma}$ where $\mathbf{\Sigma} > 0$. The efficient portfolio may be found exactly and explicitly for this case by substituting in (12.4.1) and solving the system of equations. An alternative approach is to start from the beginning and minimize the risk for this case. This alternative route will be taken because the setting will prove basic to the sequel.

When $\mathbf{\Omega}_1 = \mathbf{\Sigma}$, the risk for the return on the portfolio is $(\frac{1}{2})(\mathbf{c}'\mathbf{\Sigma}\mathbf{c})^{\alpha/2}$. Therefore, the $\mathbf{c}$ that minimizes the risk is the same $\mathbf{c}$ that minimizes $[\frac{1}{2}\mathbf{c}'\mathbf{\Sigma}\mathbf{c}]$. The objective function is

$$\psi = \tfrac{1}{2}\mathbf{c}'\mathbf{\Sigma}\mathbf{c} - \phi(\mathbf{c}'\mathbf{e} - 1),$$

where ϕ is a Lagrange multiplier. Differentiating ψ with respect to $\mathbf{c}$, setting the result equal to zero, and solving for $\mathbf{c}$ gives $\mathbf{c} = \phi\mathbf{\Sigma}^{-1}\mathbf{e}$. Since $\mathbf{c}'\mathbf{e} = 1$, $\phi = (\mathbf{e}'\mathbf{\Sigma}^{-1}\mathbf{e})^{-1}$. Hence, the minimum is attained at

$$\mathbf{c} = \frac{\mathbf{\Sigma}^{-1}\mathbf{e}}{\mathbf{e}'\mathbf{\Sigma}^{-1}\mathbf{e}}. \tag{12.4.2}$$

To see that the result in (12.4.2) corresponds to a minimum, examine the second derivative of ψ with respect to $\mathbf{c}'$. The result is the positive definite matrix $\mathbf{\Sigma}$; that is,

$$\frac{\partial^2\psi}{\partial\mathbf{c}\partial\mathbf{c}'} = \mathbf{\Sigma} > 0.$$

This interpretation of the second derivative was discussed in Section 2.14.3. The result is particularly important in light of the lack of dependence of the solution on the characteristic exponent α. That is, in a restricted Model I portfolio with asset prices following any multivariate symmetric stable law of order one, to achieve an efficient portfolio, allocate funds according to (12.4.2) without regard to the value of α.

Example (12.4.2): Suppose the p assets in a (restricted) Model I portfolio are securities whose multivariate symmetric stable price law is of order one, as in Example (12.4.1), and its scale matrix follows the intraclass pattern of association. Then

$$\mathbf{\Sigma} = \sigma^2 \begin{pmatrix} 1 & & & \\ & \cdot & \rho & \\ & & \cdot & \\ & \rho & \cdot & \\ & & & 1 \end{pmatrix}, \qquad -\frac{1}{p-1} < \rho < 1,$$

where $\mathbf{\Sigma} : p \times p$ is the scale matrix. Then, since [see (2.5.8)]

$$\mathbf{\Sigma} = \sigma^2[(1-\rho)\mathbf{I}_p + \rho\mathbf{e}\mathbf{e}'],$$
$$\mathbf{\Sigma}^{-1} = \frac{\mathbf{I}_p}{\sigma^2(1-\rho)} - \frac{\rho\mathbf{e}\mathbf{e}'}{\sigma^2(1-\rho)[1+(p-1)\rho]},$$

as was shown in (2.5.9). Therefore,

$$\mathbf{\Sigma}^{-1}\mathbf{e} = \frac{\mathbf{e}}{\sigma^2[1+(p-1)\rho]},$$

and

$$\mathbf{e}'\mathbf{\Sigma}^{-1}\mathbf{e} = \frac{p}{\sigma^2[1+(p-1)\rho]}.$$

Substituting into (12.4.2) gives

$$\mathbf{c} = \frac{1}{p}\mathbf{e} = \begin{pmatrix} \frac{1}{p} \\ \cdot \\ \cdot \\ \cdot \\ \frac{1}{p} \end{pmatrix}.$$

That is, choose a portfolio with equal proportions of shares (fractions of the investment budget) in each asset. Note that the solution does not depend upon the value of ρ. Attention is focused upon this fact (by taking $\rho = 0$) in the next example.

Example (12.4.3): Suppose $\mathbf{\Sigma} = \sigma^2\mathbf{I}_p$, where $\mathbf{\Sigma}$ denotes the scale matrix in a Model I portfolio of the type described in Example (12.4.1). Substitution in (12.4.2) gives the same result as in Example (12.4.2); that is,

$$\mathbf{c} = \frac{1}{p}\mathbf{e},$$

the equal proportions solution. The fact that the solutions in the two examples are the same can be understood from the heuristic argument that in both cases, the scaling of all assets in the portfolio is the same, and the measures of association among all assets is the same. So it is not surprising that no asset should be held in any greater proportion than any other asset.

Example (12.4.4): Suppose the scale matrix in a Model I portfolio of the type described in Example (12.4.1) is given by $\mathbf{\Sigma} = \text{diag}\,(\sigma_1^2,\ldots,\sigma_p^2)$. Since $\mathbf{\Sigma}^{-1} = \text{diag}\,(1/\sigma_1^2,\ldots,1/\sigma_p^2)$, and

$$\mathbf{e}'\mathbf{\Sigma}^{-1}\mathbf{e} = \sum_{j=1}^{p} \frac{1}{\sigma_j^2},$$

from (12.4.2), the minimum risk allocation is

$$c_j = \frac{1/\sigma_j^2}{\sum_{j=1}^{p} \frac{1}{\sigma_i^2}}, \qquad j = 1,\ldots,p.$$

That is, if $h_j = 1/\sigma_j^2$ denotes the generalized precision parameter (the term "generalized" is used because variances do not exist unless $\alpha = 2$) and $h \equiv \Sigma_1^p h_j$ denotes the aggregate generalized precision (precision parameters are discussed in Chapter 3),

$$c_j = \frac{h_j}{h}, \qquad j = 1,\ldots,p.$$

12.5 MODEL II—STABLE PORTFOLIO ANALYSIS (RESTRICTED PROBLEM)

In Model II portfolio analysis, the restriction that all assets in the portfolio should have the same characteristic exponent α is removed. Now suppose that in the portfolio there are to be p_j assets whose prices follow a multivariate symmetric stable law with characteristic exponent α_j, $j = 1,\ldots,K$, and $\Sigma_1^K p_j = p$. Let $\mathbf{C}_j$ denote the $p_j \times 1$ vector of proportions of shares of assets of Type α_j to be acquired, $j = 1,\ldots,K$. Define

$$\mathbf{Y}' = (\mathbf{Z}_1',\ldots,\mathbf{Z}_K'),$$

where $\mathbf{Y}: p \times 1$ is the vector of returns per share for each asset, $\mathbf{Z}_j$ is the $p_j \times 1$ subvector of $\mathbf{Y}$ corresponding to assets whose price changes follow a stable law with characteristic exponent α_j, $j = 1,\ldots,K$, and suppose that all the $\mathbf{Z}_j$'s are independent. Define

$$Q = \sum_{j=1}^{K} \mathbf{C}_j'\mathbf{Z}_j$$

as the one period return on the portfolio, as in Model I. The characteristic function of Q is given by

$$\phi_Q(v) = E \exp(ivQ) = \phi_{\mathbf{Y}}(v\mathbf{C}), \tag{12.5.1}$$

where $\mathbf{C}' = (\mathbf{C}_1', \ldots, \mathbf{C}_K')$. This results from reasoning as in (12.2.4), and noting that $Q = \mathbf{C}'\mathbf{Y}$. Thus, if $\mathbf{t}' \equiv (\mathbf{t}_1', \ldots, \mathbf{t}_K')$, where the $\mathbf{t}_k'$'s are appropriate subvectors, and if

$$\log \phi_{\mathbf{Z}_k}(\mathbf{t}_k) = i\mathbf{t}_k'\mathbf{\Theta}_k - \frac{1}{2}\sum_{j=1}^{m} (\mathbf{t}_k'\mathbf{\Omega}_{jk}\mathbf{t}_k)^{\alpha_k/2},$$

where $\boldsymbol{\theta}' \equiv (\mathbf{\Theta}_1', \ldots, \mathbf{\Theta}_K')$, $\mathbf{\Theta}_k : p_k \times 1$, and $\mathbf{\Omega}_{jk} : p_k \times p_k$ are the location vector and the scale matrices, respectively, for the law of $\mathbf{Z}_k$, the law for Q becomes

$$\log \phi_Q(v) = iv(\mathbf{C}'\boldsymbol{\theta}) - \frac{1}{2}\sum_{k=1}^{K}\sum_{j=1}^{m} |v|^{\alpha_k/2}(\mathbf{C}_k'\mathbf{\Omega}_{jk}\mathbf{C}_k)^{\alpha_k/2}. \tag{12.5.2}$$

Thus, the resulting variable Q has a law that does not belong to any stable family. That is, its law is a mixture of stable laws, and the mixture is itself not a stable law (other than for the degenerate case in which all α_k's are equal).

Behavior of the prices of such a combination of assets can be explained, for the case of holdings of securities, in the following way. Prices of securities representing firms in the same industry or type of business follow a multivariate symmetric stable law with the same characteristic exponent, while prices for securities in different industries follow *independent* stable laws with different characteristic exponents. This is one kind of a model that gives rise to the mixed law in (12.5.2). Note, however, that although price changes may be pairwise independent across industries, they still can be affected by an overall market (or economy) factor that produces changes in most assets in related ways.

Now consider the risk for a Model II portfolio. The risk for the portfolio is defined to be the sum of the scale parameters corresponding to the independent subvectors $\mathbf{Z}_k$; that is,

$$r(\mathbf{C}) = \frac{1}{2}\sum_{k=1}^{K}\sum_{j=1}^{m} (\mathbf{C}_k'\mathbf{\Omega}_{jk}\mathbf{C}_k)^{\alpha_k/2}.$$

To find the efficient portfolio assuming the nonnegativity constraints are inactive, form the Lagrangian

$$\psi = \frac{1}{2}\sum_{k=1}^{K}\sum_{j=1}^{m} (\mathbf{C}_k'\mathbf{\Omega}_{jk}\mathbf{C}_k)^{\alpha_k/2} - \phi\left(\sum_{k=1}^{K} \mathbf{C}_k'\mathbf{e}_k - 1\right),$$

differentiate with respect to $\mathbf{C}_\delta$, $\delta = 1,\ldots,K$, and set the result equal to zero. Note that $\mathbf{e}_k$ is a p_k-vector of ones. Then,

$$\frac{\partial\psi}{\partial\mathbf{C}_\delta} = \frac{\alpha_\delta}{2}\sum_{j=1}^{m}(\mathbf{C}_\delta'\mathbf{\Omega}_{j\delta}\mathbf{C}_\delta)^{(\alpha_\delta/2)-1}\mathbf{\Omega}_{j\delta}\mathbf{C}_\delta - \phi\mathbf{e}_\delta = \mathbf{0}.$$

Multiplying through by $\mathbf{C}_\delta'$ and summing over δ gives (using $\Sigma_1^K\mathbf{C}_k'\mathbf{e}_k = 1$)

$$\phi = \sum_{\delta=1}^{K}\frac{\alpha_\delta}{2}\sum_{j=1}^{m}(\mathbf{C}_\delta'\mathbf{\Omega}_{j\delta}\mathbf{C}_\delta)^{\alpha_\delta/2}.$$

Substituting ϕ back into $\partial\psi/\partial\mathbf{C}_\delta = \mathbf{0}$ gives the system of equations

$$\sum_{k=1}^{K}\sum_{j=1}^{m}\alpha_k(\mathbf{C}_k'\mathbf{\Omega}_{jk}\mathbf{C}_k)^{\alpha_k/2}\mathbf{e}_\delta = \sum_{j=1}^{m}\alpha_\delta(\mathbf{C}_\delta'\mathbf{\Omega}_{j\delta}\mathbf{C}_\delta)^{\alpha_\delta/2}\left(\frac{\mathbf{\Omega}_{j\delta}\mathbf{C}_\delta}{\mathbf{C}_\delta'\mathbf{\Omega}_{j\delta}\mathbf{C}_\delta}\right). \tag{12.5.3}$$

Suppose, as in Example (12.4.1), that $m = 1$ and $\mathbf{\Omega}_{1k} \equiv \mathbf{\Sigma}_k$. Then

$$r(\mathbf{C}) = \frac{1}{2}\sum_{k=1}^{K}(\mathbf{C}_k'\mathbf{\Sigma}_k\mathbf{C}_k)^{\alpha_k/2}.$$

Therefore, the $\mathbf{C}_k$'s that minimize $r(\mathbf{C})$ are the same as the minimizers of $(\mathbf{C}_k'\mathbf{\Sigma}_k\mathbf{C}_k)$, separately, for each k.

Define the objective function

$$\psi = \frac{1}{2}\sum_{j=1}^{K}\mathbf{C}_j'\mathbf{\Sigma}_j\mathbf{C}_j - \phi(\mathbf{C}'\mathbf{e} - 1),$$

where ϕ is a Lagrangian multiplier, and $\mathbf{C}'\mathbf{e} = 1$ is a subsidiary condition that must be satisfied. Define the block diagonal matrix

$$\mathbf{\Sigma} = \begin{pmatrix} \mathbf{\Sigma}_1 & & & 0 \\ & \cdot & & \\ & & \cdot & \\ 0 & & & \mathbf{\Sigma}_K \end{pmatrix}. \tag{12.5.4}$$

Then

$$\psi = \tfrac{1}{2}\mathbf{C}'\mathbf{\Sigma}\mathbf{C} - \phi(\mathbf{C}'\mathbf{e} - 1).$$

But this is now the same problem solved in Model I, Example (12.4.1). Hence, a minimum exists at

$$\mathbf{C} = \frac{\mathbf{\Sigma}^{-1}\mathbf{e}}{\mathbf{e}'\mathbf{\Sigma}^{-1}\mathbf{e}}. \tag{12.5.5}$$

Substituting from (12.5.4) gives [since $\mathbf{\Sigma}^{-1} = \operatorname{diag}(\mathbf{\Sigma}_1^{-1},\ldots,\mathbf{\Sigma}_K^{-1})$],

$$\mathbf{C}_j = \frac{\mathbf{\Sigma}_j^{-1}\mathbf{e}_j}{\mathbf{e}'\mathbf{\Sigma}^{-1}\mathbf{e}}, \qquad j = 1,\ldots,K, \tag{12.5.6}$$

where $\mathbf{e}_j$ is a $p_j \times 1$ vector of ones (recall that $\mathbf{e}$ is a $p \times 1$ vector of ones, where $p = \Sigma_1^K p_j$).

Example (12.5.1): Suppose for $-1/(p_1 - 1) < \rho_1 < 1$, $-1/(p_2 - 1) < \rho_2 < 1$, $\mathbf{\Sigma}_1$ and $\mathbf{\Sigma}_2$ are the (order one) scale matrices in a Model II portfolio (restricted), where $\mathbf{\Sigma}_1$ and $\mathbf{\Sigma}_2$ each follow the intraclass patterns of association,

$$\underset{(p_1 \times p_1)}{\mathbf{\Sigma}_1} = \sigma_1^2 \begin{pmatrix} 1 & & & \\ & \cdot & \rho_1 & \\ & & \cdot & \\ \rho_1 & & \cdot & \\ & & & 1 \end{pmatrix}, \qquad \underset{(p_2 \times p_2)}{\mathbf{\Sigma}_2} = \sigma_2^2 \begin{pmatrix} 1 & & & \\ & \cdot & \rho_2 & \\ & & \cdot & \\ \rho_2 & & \cdot & \\ & & & 1 \end{pmatrix},$$

and $p_1 + p_2 = p$. Thus, $K = 2$. As in Example (12.4.1),

$$\mathbf{\Sigma}_j^{-1} = \frac{\mathbf{I}_{p_j}}{\sigma_j^2(1 - \rho_j)} - \frac{\rho_j \mathbf{e}_j \mathbf{e}_j'}{\sigma_j^2(1 - \rho_j)[1 + (p_j - 1)\rho_j]}, \qquad j = 1, 2.$$

Since

$$\mathbf{\Sigma}^{-1} = \begin{pmatrix} \mathbf{\Sigma}_1^{-1} & \mathbf{0} \\ \mathbf{0} & \mathbf{\Sigma}_2^{-1} \end{pmatrix},$$

$$\begin{aligned} \mathbf{e}'\mathbf{\Sigma}^{-1}\mathbf{e} &= \mathbf{e}_1'\mathbf{\Sigma}_1^{-1}\mathbf{e}_1 + \mathbf{e}_2'\mathbf{\Sigma}_2^{-1}\mathbf{e}_2 \\ &= \frac{p_1}{\sigma_1^2[1 + (p_1 - 1)\rho_1]} + \frac{p_2}{\sigma_2^2[1 + (p_2 - 1)\rho_2]}. \end{aligned}$$

Hence, for the $p_j \times 1$ vector of ones, $\mathbf{e}_j$,

$$\underset{(p_j \times 1)}{\mathbf{C}_j} = \frac{\left\{\dfrac{\mathbf{e}_j}{\sigma_j^2[1 + (p_j - 1)\rho_j]}\right\}}{\left\{\dfrac{p_1}{\sigma_1^2[1 + (p_1 - 1)\rho_1]} + \dfrac{p_2}{\sigma_2^2[1 + (p_2 - 1)\rho_2]}\right\}}, \qquad j = 1, 2.$$

That is, for all p_1 assets whose price law has characteristic exponent α_1 (and scale matrix $\mathbf{\Sigma}_1$), choose the same fraction g_1, where

$$g_1 \equiv \frac{\left\{\dfrac{1}{\sigma_1^2[1 + (p_1 - 1)\rho_1]}\right\}}{\left\{\dfrac{p_1}{\sigma_1^2[1 + (p_1 - 1)\rho_1]} + \dfrac{p_2}{\sigma_2^2[1 + (p_2 - 1)\rho_2]}\right\}},$$

and for all p_2 assets whose price law has characteristic exponent α_2 (and scale matrix $\mathbf{\Sigma}_2$), choose the same fraction g_2 where

$$p_1g_1 + p_2g_2 = 1.$$

Example (12.5.2): Suppose the future prices of two oil companies are represented by $\mathbf{Z}_1' = (Z_{11}, Z_{12})$, and the future prices of two mining companies are represented by $\mathbf{Z}_2' = (Z_{21}, Z_{22})$, with associated log-characteristic functions given by

$$\log \phi_{\mathbf{Z}_1}(\mathbf{t}_1) = i\mathbf{t}_1'\boldsymbol{\theta}_1 - \tfrac{1}{2}(\mathbf{t}_1'\boldsymbol{\Sigma}_1\mathbf{t}_1)^{3/4},$$
$$\log \phi_{\mathbf{Z}_2}(\mathbf{t}_2) = i\mathbf{t}_2'\boldsymbol{\theta}_2 - \tfrac{1}{2}(\mathbf{t}_2'\boldsymbol{\Sigma}_2\mathbf{t}_2)^{5/6},$$

respectively, where $\mathbf{t}_1' \equiv (t_{11}, t_{12})$, $\mathbf{t}_2' \equiv (t_{21}, t_{22})$, $\boldsymbol{\theta}_1' \equiv (\theta_{11}, \theta_{12})$, $\boldsymbol{\theta}_2' \equiv (\theta_{21}, \theta_{22})$, and

$$\boldsymbol{\Sigma}_1 = \begin{pmatrix} 3 & 1 \\ 1 & 2 \end{pmatrix}, \qquad \boldsymbol{\Sigma}_2 = \begin{pmatrix} 4 & 2 \\ 2 & 3 \end{pmatrix}.$$

Let $\mathbf{Y}' = (\mathbf{Z}_1', \mathbf{Z}_2')$. Find the best portfolio of these four securities.

From (12.5.6), it is necessary first to evaluate $\mathbf{e}'\boldsymbol{\Sigma}^{-1}\mathbf{e}$, where

$$\boldsymbol{\Sigma} = \begin{pmatrix} \boldsymbol{\Sigma}_1 & \mathbf{0} \\ \mathbf{0} & \boldsymbol{\Sigma}_2 \end{pmatrix}.$$

Thus,

$$\boldsymbol{\Sigma}^{-1} = \begin{pmatrix} \boldsymbol{\Sigma}_1^{-1} & \mathbf{0} \\ \mathbf{0} & \boldsymbol{\Sigma}_2^{-1} \end{pmatrix},$$

and since

$$\boldsymbol{\Sigma}_1^{-1} = \frac{1}{5}\begin{pmatrix} 2 & -1 \\ -1 & 3 \end{pmatrix}, \qquad \boldsymbol{\Sigma}_2^{-1} = \frac{1}{8}\begin{pmatrix} 3 & -2 \\ -2 & 4 \end{pmatrix},$$

$$\begin{aligned} \mathbf{e}'\boldsymbol{\Sigma}^{-1}\mathbf{e} &= \frac{1}{5}(2 + 3 - 1 - 1) + \frac{1}{8}(3 + 4 - 2 - 2) \\ &= \frac{39}{40}. \end{aligned}$$

Since

$$\boldsymbol{\Sigma}_1^{-1}\mathbf{e}_1 = \begin{pmatrix} \frac{1}{5} \\ \frac{2}{5} \end{pmatrix}, \qquad \boldsymbol{\Sigma}_2^{-1}\mathbf{e}_2 = \begin{pmatrix} \frac{1}{8} \\ \frac{2}{8} \end{pmatrix},$$

substituting in (12.5.6) gives

$$\mathbf{C}_1 = \frac{8}{39}\begin{pmatrix} 1 \\ 2 \end{pmatrix}, \qquad \mathbf{C}_2 = \frac{5}{39}\begin{pmatrix} 1 \\ 2 \end{pmatrix}.$$

Thus, the four securities should be purchased in the fractions $\left(\frac{8}{39}, \frac{16}{39}, \frac{5}{39}, \frac{10}{39}\right)$.

12.6 ESTIMATION IN STABLE DISTRIBUTIONS

The problem of parameter estimation in stable distributions is fundamental to portfolio analysis. The models discussed in this chapter have required the assumption that all parameters of the underlying stable distributions are known, at least approximately. Therefore, it is important to develop methods for estimating these parameters. Unfortunately, the effort that has been devoted to parameter estimation in stable distributions has been small. Some of this work will be reviewed now. Then, some additional analytical procedures for estimating the parameters of stable distributions are proposed. While the procedures should be useful in large samples, no claims of optimality are intended for any of the methods.

The case of $\alpha = 2$ for both univariate and multivariate distributions does not need to be discussed (this is the Normal distribution and is discussed in Chapters 3 and 7). Estimation in the Cauchy distribution ($\alpha = 1$) has received a modicum of attention in the univariate case (see, for instance, Rothenberg, Fisher, and Tilanus, 1964; Bloch, 1966; and Barnett, 1966). More generally, parameters of the univariate symmetric stable distributions were studied[2] by means of simulation for the case of $1 \leq \alpha \leq 2$. It was shown there that if $\delta \equiv \gamma^{1/\alpha}$, where γ is the "scale parameter," while for $\alpha = 1$, δ is exactly the semi-interquartile range of the distribution, δ remains approximately the semi-interquartile range even when $1 < \alpha \leq 2$. This approximate relation was used as a basis for estimating α when $1 < \alpha \leq 2$. First, let $y_1, \ldots, y_n$ denote a sample from the distribution defined in (12.6.1) with $\beta = 0$. Suppose $1 < \alpha \leq 2$, so that a may be estimated by $\bar{y}$, the sample mean. Then define the new observations $x_j \equiv y_j - \bar{y}, j = 1, \ldots, n$. Define the sample cdf $F_n(x)$, where $nF_n(x)$ denotes the number of x_j's $\leq x$. Fama and Roll suggest in this case the estimator of δ,

$$\hat{\delta} = \frac{1}{2(.827)} [\hat{x}_{72} - \hat{x}_{28}],$$

where $\hat{x}_h$ is defined by $F_n(\hat{x}_h) = h/100$. Next, they note that since the x_j's follow a stable distribution, for any p, $\sum_1^p x_j$ also follows a stable distribution, but with parameter $\delta^* = p^{1/\alpha}\delta$. Solving for α and replacing the parameters by their estimators gives

$$\hat{\alpha} = \frac{\log p}{\log \hat{\delta}^* - \log \hat{\delta}}.$$

[2] See Fama and Roll (1968; 1969).

The parameter δ is estimable directly, as indicated above, while δ^* may be estimated from the same relation by using aggregated subsamples (for example, suppose $n = 10$; then if $z_1 = x_1 + x_2$, $z_2 = x_3 + x_4, \ldots, z_5 = x_9 + x_{10}$, the z_j's form a sample of size 5 with parameter δ^*, which is estimable using the z_j's replaced by the x_j's; α is then estimable using $p = 2$).

Another approach for estimating α was suggested by Mandelbrot (1964). His work suggested that because a stable variable X satisfies the relation [see, for example, Feller (1966) p. 547]

$$\lim_{x \to \infty} x^\alpha P\{X > x\} = \text{constant},$$

it is approximately true that when x is large

$$P\{X > x\} \cong \frac{\text{constant}}{x^\alpha},$$

so that a plot of log $P\{X > x\}$ against log x should yield a straight line with slope α. Hence, if the sample cdf is used to estimate $P\{X > x\}$, in large samples, the slope will be an estimator of α. This approach is useful only for large x and only in large samples.

Several more methods of estimating parameters of stable distributions are suggested in the following sections. Moreover, all parameters will be estimable simultaneously, and the estimation techniques will be extended to the multivariate case. One method that generalizes some work of Kleinman (1965) to multiple parameter estimation in both univariate and multivariate distributions yields explicit estimators. Other methods do not have this virtue. Consider first the univariate case.

12.6.1 Estimation in Univariate Stable Distributions

Let the log-characteristic function for a scalar random variable Y following a stable law be given by

$$\log \phi(t) = iat - \gamma|t|^\alpha \left[1 + i\beta \frac{t}{|t|}\, \omega(t,\alpha)\right], \tag{12.6.1}$$

where $\omega(t,\alpha) = \tan(\pi\alpha/2)$, for $\alpha \neq 1$, and $\omega(t,\alpha) = (2/\pi) \log |t|$, for $\alpha = 1$. Let N independent observations $y_1, \ldots, y_N$ be taken. The problem is to estimate (a,γ,α,β). Consistent estimators are developed using sample characteristic functions.

Denote the sample characteristic function by

$$\hat{\phi}(t) = \frac{1}{N} \sum_{j=1}^{N} \exp(ity_j). \tag{12.6.2}$$

Thus, $\hat{\phi}(t)$ is computable for all values of t. Note that $\{\hat{\phi}(t), -\infty < t < \infty\}$

is a stochastic process, and for each t, $|\hat{\phi}(t)|$ is bounded above by unity. Hence, all moments of $\hat{\phi}(t)$ are finite, and $\hat{\phi}(t)$ for any fixed t, is the sample average of independent and identically distributed random variables. Hence, by the law of large numbers, $\hat{\phi}(t)$ is a consistent estimator of $\phi(t)$.

Estimation Method I (Minimum Distance)

Define[3]

$$g(a,\gamma,\alpha,\beta) \equiv \sup_t |\phi(t) - \hat{\phi}(t)|.$$

Then the minimum distance estimators of (a,γ,α,β) are the values of these parameters which minimize $g(a,\gamma,\alpha,\beta)$.

Estimation Method II (Minimum *r*th Mean Distance)

Define

$$h(a,\gamma,\alpha,\beta) \equiv \int_{-\infty}^{\infty} |\phi(t) - \hat{\phi}(t)|^r \, W(t)dt,$$

where $W(t)$ denotes a suitable convergence factor, such as $W(t) = (2\pi)^{-1/2} \exp(-t^2/2)$. Then the minimum rth mean estimators are those values of (a,γ,α,β) which minimize $h(a,\gamma,\alpha,\beta)$ for a fixed r. It is not clear which r should be selected. Both of the above methods yield consistent estimators, although no claim of efficiency is made. Moreover, numerous other norms (distances) involving $[\phi(t) - \hat{\phi}(t)]$ may be defined and used as estimation procedures, and it is not clear how to choose the best one. Paulson, Holcomb, and Leitch, 1975, adopted a numerical search procedure which minimizes the mean squared distance with respect to the four parameters of a univariate stable distribution. They used Hermitian quadrature to effect the integration, taking the weighting function, $W(t) = \exp(-t^2)$. The approach was applied to the problem of estimating the parameters for 20 randomly selected securities on the N.Y. Stock Exchange, in Leitch and Paulson, 1975.

An analytical estimation procedure that yields explicit estimators and involves minimal computation is a version of the method of moments.

Estimation Method III (Method of Moments)

From (12.6.1)

$$|\phi(t)|^2 = \exp(-2\gamma|t|^\alpha). \tag{12.6.3}$$

Hence, $\gamma|t|^\alpha = -\log|\phi(t)|$. Now choose two nonzero values of t; say t_1 and t_2, $t_1 \neq t_2$. Assume $\alpha \neq 1$. Then

$$\gamma|t_1|^\alpha = -\log|\phi(t_1)|, \qquad \gamma|t_2|^\alpha = -\log|\phi(t_2)|.$$

Solving these two equations simultaneously for α and γ gives [for $\phi(t)$ estimated by $\hat{\phi}(t)$] estimators for α and γ; namely,

[3] The term "sup" denotes supremum, or least upper bound.

$$\hat{\alpha} = \frac{\log \left[\dfrac{\log |\hat{\phi}(t_1)|}{\log |\hat{\phi}(t_2)|}\right]}{\log \left|\dfrac{t_1}{t_2}\right|}, \tag{12.6.4}$$

and

$$\log \hat{\gamma} = \frac{\log |t_1| \log [-\log |\hat{\phi}(t_2)|] - \log |t_2| \log [-\log |\hat{\phi}(t_1)|]}{\log \left|\dfrac{t_1}{t_2}\right|}. \tag{12.6.5}$$

To estimate β and a, define $u(t) \equiv \text{Im}\,[\log \phi(t)]$, where $\text{Im}\,[\psi(t)]$ denotes the imaginary part of any complex-valued function $\psi(t)$. Then, from (12.6.1),

$$u(t) = at - \gamma |t|^{\alpha-1} \beta t \omega(t,\alpha).$$

Choose two (possibly new[4]) nonzero values of t; say t_3 and t_4, $t_3 \neq t_4$. Then, for $\alpha \neq 1$,

$$a - \left[\gamma |t_3|^{\alpha-1} \tan \frac{\pi\alpha}{2}\right] \beta = \frac{u(t_3)}{t_3}, \tag{12.6.6}$$

and

$$a - \left[\gamma |t_4|^{\alpha-1} \tan \frac{\pi\alpha}{2}\right] \beta = \frac{u(t_4)}{t_4}. \tag{12.6.7}$$

Since

$$\hat{\phi}(t) = \left(\frac{1}{N} \sum_{j=1}^{N} \cos t y_j\right) + i \left(\frac{1}{N} \sum_{j=1}^{N} \sin t y_j\right),$$

in polar coordinates, $\hat{\phi}(t)$ is expressible as

$$\hat{\phi}(t) = \rho(t) e^{i\theta(t)},$$

where

$$\rho^2(t) = \left(\frac{1}{N} \sum_{j=1}^{N} \cos t y_j\right)^2 + \left(\frac{1}{N} \sum_{j=1}^{N} \sin t y_j\right)^2,$$

and

$$\tan \theta(t) = \left(\frac{\sum_{j=1}^{N} \sin t y_j}{\sum_{j=1}^{N} \cos t y_j}\right).$$

[4] It may be that the same pair of values, t_1 and t_2, which is used to estimate α and γ will also serve well to estimate a and β. However, this question requires further study.

Hence,

$$\log \hat{\phi}(t) = \rho(t) + i\theta(t),$$

and

$$\hat{u}(t) = \text{Im}\,[\log \hat{\phi}(t)] = \theta(t).$$

That is,

$$\hat{u}(t) = \tan^{-1}\left[\frac{\sum_{j=1}^{N} \sin ty_j}{\sum_{j=1}^{N} \cos ty_j}\right]. \tag{12.6.8}$$

Replacing $u(t)$ in (12.6.6) and (12.6.7) by the value of $\hat{u}(t)$ given in (12.6.8), and replacing the parameters by their estimators, and then solving the two linear equations simultaneously for β and a gives the estimators

$$\hat{\beta} = \frac{\left[\dfrac{\hat{u}(t_3)}{t_3} - \dfrac{\hat{u}(t_4)}{t_4}\right]}{[|t_4|^{\hat{\alpha}-1} - |t_3|^{\hat{\alpha}-1}]\hat{\gamma} \tan \dfrac{\pi\hat{\alpha}}{2}}, \tag{12.6.9}$$

and

$$\hat{a} = \frac{|t_4|^{\hat{\alpha}-1}\dfrac{\hat{u}(t_3)}{t_3} - |t_3|^{\hat{\alpha}-1}\dfrac{\hat{u}(t_4)}{t_4}}{|t_4|^{\hat{\alpha}-1} - |t_3|^{\hat{\alpha}-1}}. \tag{12.6.10}$$

Case of $\alpha = 1$: For $\alpha = 1$, (12.6.1) gives

$$u(t) = at - \frac{2\gamma\beta t}{\pi} \log |t|.$$

That is, for two nonzero values of t, say t_3 and t_4, $t_3 \neq t_4$,

$$a - \left[2\gamma \frac{\log |t_3|}{\pi}\right]\beta = \frac{u(t_3)}{t_3}, \tag{12.6.11}$$

and

$$a - \left[2\gamma \frac{\log |t_4|}{\pi}\right]\beta = \frac{u(t_4)}{t_4}. \tag{12.6.12}$$

Solving (12.6.11) and (12.6.12) simultaneously for β and a, with $u(t)$ replaced by $\hat{u}(t)$ and the parameters replaced by their estimators, gives the moment estimators for $\alpha = 1$,

$$\hat{\beta} = \frac{\left[\dfrac{\hat{u}(t_3)}{t_3} - \dfrac{\hat{u}(t_4)}{t_4}\right]}{\dfrac{2\hat{\gamma}}{\pi} \log \left|\dfrac{t_4}{t_3}\right|}, \tag{12.6.13}$$

and

$$\hat{a} = \frac{\dfrac{\log |t_4|}{t_3}\hat{u}(t_3) - \dfrac{\log |t_3|}{t_4}\hat{u}(t_4)}{\log |t_4| - \log |t_3|}. \tag{12.6.14}$$

The equations (12.6.4), (12.6.5), (12.6.9), and (12.6.10) yield moment estimators for (α,γ,β,a) for the case of $\alpha \neq 1$; for $\alpha = 1$, β and a are estimated from (12.6.13) and (12.6.14) with

$$\hat{\gamma} = -\frac{\log |\hat{\phi}(t_1)|}{|t_1|}.$$

The estimators given above are consistent since they are based upon the sample moments of cos tY and sin tY [see (12.6.2)], which are known to be consistent. However, convergence to the population values may not be too rapid, depending upon the choices of $t_1,\ldots,t_4$. Optimal selection of the t_j's requires further study. Fortunately, with security prices, observations are usually available inexpensively, so that the possibly large samples required for small estimation error are realistic. Interval estimation may be based upon the asymptotic Normal distributions for the above moment estimators. Thus, at least in large samples, the univariate estimation problem may be solved for the case of independently observed stable variables.

12.6.2 Estimation in Multivariate Symmetric Stable Distributions

For dependent variables following multivariate symmetric stable laws, the situation is more complicated, but clearly solvable, at least in large samples. For example, suppose $\mathbf{Y}: p \times 1$ follows a multivariate symmetric stable law

$$\mathcal{L}(\mathbf{Y}) = S_p(m,\mathbf{a};\mathbf{\Omega}_1,\ldots,\mathbf{\Omega}_m;\alpha),$$

so that its log characteristic function has the form given in (12.2.2). Consider the case in which $m = 1$ and $\mathbf{\Omega}_1 \equiv \mathbf{\Omega}$, so that

$$\log \phi_{\mathbf{Y}}(\mathbf{t}) = i\mathbf{a}'\mathbf{t} - \tfrac{1}{2}(\mathbf{t}'\mathbf{\Omega}\mathbf{t})^{\alpha/2}.$$

The problem is to estimate α, $\mathbf{a}$, and $\mathbf{\Omega}$, given a sample of p-variate observations, $\mathbf{y}_1,\ldots,\mathbf{y}_N$, where all $\mathbf{y}_j$ are $p \times 1$ vectors. By analogy with the univariate case, compute the sample characteristic function

$$\hat{\phi}_{\mathbf{Y}}(\mathbf{t}) = \frac{1}{N}\sum_{j=1}^{N} \exp\,(i\mathbf{t}'\mathbf{y}_j), \tag{12.6.15}$$

and use it as an estimator of $\phi_{\mathbf{Y}}(\mathbf{t})$. Now the minimum distance or the minimum rth mean distance estimators can be computed just as in the univariate case. Alternatively, the moment estimators will be found.

Since

$$\log |\phi_{\mathbf{Y}}(\mathbf{t})| = -\tfrac{1}{2}(\mathbf{t}'\mathbf{\Omega}\mathbf{t})^{\alpha/2}, \tag{12.6.16}$$

for $\mathbf{t} = (s_1,0,\ldots,0)' \equiv \mathbf{t}_1$, and $\mathbf{\Omega} \equiv (\omega_{ij})$, $i, j = 1,\ldots,p$,

$$\log |\phi_{\mathbf{Y}}(\mathbf{t}_1)| = -\tfrac{1}{2}(\omega_{11})^{\alpha/2}|s_1|^{\alpha},$$

or

$$|s_1|^{\alpha}(\omega_{11})^{\alpha/2} = -2 \log |\phi_{\mathbf{Y}}(\mathbf{t}_1)|.$$

Now take $\mathbf{t} = (s_2,0,\ldots,0)' \equiv \mathbf{t}_2$. Substituting in (12.6.16) gives

$$|s_2|^{\alpha}(\omega_{11})^{\alpha/2} = -2 \log |\phi_{\mathbf{Y}}(\mathbf{t}_2)|.$$

Taking the ratio of the results for $\mathbf{t}_1$ and $\mathbf{t}_2$ gives

$$\left|\frac{s_2}{s_1}\right|^{\alpha} = \frac{\log |\phi_{\mathbf{Y}}(\mathbf{t}_2)|}{\log |\phi_{\mathbf{Y}}(\mathbf{t}_1)|}.$$

Taking logs and replacing $\phi_{\mathbf{Y}}(\mathbf{t}_j)$ by $\hat{\phi}_{\mathbf{Y}}(\mathbf{t}_j)$, $j = 1, 2$, gives the moment estimator for α,

$$\hat{\alpha} = \frac{\log \left[\dfrac{\log |\hat{\phi}_{\mathbf{Y}}(\mathbf{t}_2)|}{\log |\hat{\phi}_{\mathbf{Y}}(\mathbf{t}_1)|}\right]}{\log \left|\dfrac{s_2}{s_1}\right|}. \tag{12.6.17}$$

$\hat{\alpha}$ is consistent since $\hat{\phi}_{\mathbf{Y}}(\mathbf{t})$ is consistent.

Now estimate $\mathbf{\Omega} \equiv (\omega_{ij})$ by selecting values of $\mathbf{t}$. First rewrite (12.6.16) as

$$\mathbf{t}'\mathbf{\Omega}\mathbf{t} = [-2 \log |\phi_{\mathbf{Y}}(\mathbf{t})|]^{2/\alpha}.$$

Since $\mathbf{t}'\mathbf{\Omega}\mathbf{t}$ is a quadratic form, if $\mathbf{t} \equiv (\tilde{t}_i)$,

$$\mathbf{t}'\mathbf{\Omega}\mathbf{t} = \sum_{i=1}^{p} \sum_{j=1}^{p} \tilde{t}_i\tilde{t}_j\omega_{ij} = [-2 \log |\phi_{\mathbf{Y}}(\mathbf{t})|]^{2/\alpha}.$$

Since there are $p(p + 1)/2$ distinct ω_{ij}'s, $p(p + 1)/2$ distinct nonzero choices[5] of the $\mathbf{t}$-vector will be required. Denote them by $\boldsymbol{\tau}_1, \boldsymbol{\tau}_2,\ldots,\boldsymbol{\tau}_M$, where $M \equiv p(p + 1)/2$. Then

$$\boldsymbol{\tau}_j'\hat{\mathbf{\Omega}}\boldsymbol{\tau}_j = [-2 \log |\hat{\phi}_{\mathbf{Y}}(\boldsymbol{\tau}_j)|]^{2/\alpha}, \tag{12.6.18}$$

[5] One possible choice that gives a simple solution is to take $\boldsymbol{\tau}_j$ to be a vector of zeros, except for a one in the jth place. Then (12.6.18) immediately yields $\hat{\omega}_{jj}$. Then, choosing $\boldsymbol{\tau}_j$ to be a vector of zeros, except for ones in the ith and jth places, gives $(\hat{\omega}_{ii} + \hat{\omega}_{jj} + 2\hat{\omega}_{ij})$ for this left-hand side of (12.6.18). Since $\hat{\omega}_{ii}$ and $\hat{\omega}_{jj}$ are already known, this procedure yields $\hat{\omega}_{ij}$, $i \neq j$.

for $j = 1,\ldots,M$ provides a system of M linear equations in M unknowns which is solvable by determinants [$\hat{\alpha}$ is assumed to be known from (12.6.17) above]; thus, the solutions of (12.6.18) are uniquely determined, and provide the moment estimator of $\mathbf{\Omega}$.

The location parameter of the distribution is easily estimated consistently by noting that if

$$u(\mathbf{t}) \equiv \text{Im}\,[\log \phi_{\mathbf{Y}}(\mathbf{t})],$$
$$u(\mathbf{t}) = \mathbf{a}'\mathbf{t} = \sum_{j=1}^{p} a_j \tilde{t}_j.$$

Then, since the multivariate estimation procedure is analogous to the univariate procedure, it is clear from (12.6.8) that if

$$\hat{u}(\mathbf{t}) = \text{Im}\,[\log \hat{\phi}_{\mathbf{Y}}(\mathbf{t})],$$
$$\hat{u}(\mathbf{t}) = \tan^{-1}\left[\frac{\sum_{j=1}^{N} \sin \mathbf{t}'\mathbf{y}_j}{\sum_{j=1}^{N} \cos \mathbf{t}'\mathbf{y}_j}\right].$$

Then, if $\boldsymbol{\tau}_1^*,\ldots,\boldsymbol{\tau}_p^*$ are distinct, nonzero values of the $\mathbf{t}$ vector, the system of equations

$$\hat{\mathbf{a}}'\boldsymbol{\tau}_j^* = \hat{u}(\boldsymbol{\tau}_j^*), \qquad j = 1,\ldots,p,$$

provides a basis for estimating $\mathbf{a}$. Thus, if $\mathbf{T}' \equiv (\boldsymbol{\tau}_1^*,\ldots,\boldsymbol{\tau}_p^*)$ is the $p \times p$ matrix of assigned $\mathbf{t}$ vectors, and $\hat{u}(\boldsymbol{\tau}_j^*)$ denotes the jth element of the $p \times 1$ vector $\mathbf{u}$, the moment estimator of $\mathbf{a}$ is given explicitly by

$$\hat{\mathbf{a}} = \mathbf{T}^{-1}\mathbf{u}, \qquad |\mathbf{T}| \neq 0. \tag{12.6.19}$$

The remarks made above about consistency, asymptotic Normality, and efficiency of the estimators of univariate stable distributions apply to the estimators of multivariate stable distributions as well. Hence, the estimators are consistent and sometimes, possibly not very efficient. However, the large samples required for a small estimation error are often available cheaply for portfolio analysis applications.

More general classes of symmetric stable distributions such as those involving multiple scale matrices $\mathbf{\Omega}_1,\ldots,\mathbf{\Omega}_m$ can be treated in much the same way; however, the estimation problem is clearly considerably larger in dimension, requiring, in this case, m systems of M simultaneous equations to be solved simultaneously. The properties of these estimators (both large and small sample) have not yet been established. Moreover, other estimation methods may be found which are superior, so that much research remains to be done in this area. At least it may be said at this time that if large samples are available, the methods outlined in this chapter should provide reasonable estimates of the parameters required for stable portfolio analysis.

REFERENCES

Barnett (1966).
Bloch (1966).
Fama (1965a, 1965b, 1968).
Fama and Miller (1972).
Fama and Roll (1968, 1969).
Jensen (1969).
Kleinman (1965).
Leitch and Paulson (1975).
Paulson, Holcomb, and Leitch (1975).
Mandelbrot (1965)
Markowitz (1959).
Press (1968, 1970).
Roll (1968).
Rothenberg, Fisher, and Tilanus (1964).
Sharpe (1963).

EXERCISES

12.1 A simple investment portfolio is to consist of shares of American Snuff, General Cigar, and P. Lorrilard Company. Let $\mathbf{Y} \equiv (Y_1, Y_2, Y_3)'$, where Y_j denotes the return on one share of Company j, and $j = 1, 2, 3$ corresponds to the three companies named. Suppose (hypothetically) it is found that $\mathbf{Y}$ follows the stable law [see (12.2.3)],

$$\mathcal{L}(\mathbf{Y}) = S_3(1,\mathbf{a};\mathbf{\Sigma};1.5),$$

where

$$\mathbf{\Sigma} = \begin{pmatrix} 4 & 2 & 3 \\ 2 & 7 & 1 \\ 3 & 1 & 9 \end{pmatrix}.$$

Find the allocation of funds among the three assets which makes the portfolio efficient in a restricted Model I sense.

12.2 Suppose the vector of single share returns on the portfolio in Exercise 12.1 followed the law

$$\mathcal{L}(\mathbf{Y}) = S_3(1,\mathbf{a};\mathbf{\Sigma};1.5),$$

where

$$\mathbf{\Sigma} = 4\begin{pmatrix} 1 & .5 & .5 \\ .5 & 1 & .5 \\ .5 & .5 & 1 \end{pmatrix}.$$

Find the efficient portfolio in a restricted Model I sense.

12.3 The single share returns vector (hypothetical) for Associated Dry Goods, Gimbel Brothers, and Sears, Roebuck and Company is found to follow a multivariate symmetric stable law of order 1, with scale matrix (see Section 12.4.2)

$$\mathbf{\Sigma} = \begin{pmatrix} 3 & 0 & 0 \\ 0 & 2 & 0 \\ 0 & 0 & 1 \end{pmatrix}.$$

Find the efficient portfolio in a restricted Model I sense.

12.4 Let $\mathbf{X} \equiv (X_1,X_2,X_3)'$ denote the single share vector of returns on American Tobacco Company, Consolidated Cigar, and Liggett and Myers; let $\mathbf{Y} \equiv (Y_1,Y_2)'$ denote the single share returns vector for Standard Oil of New Jersey, and Socony Mobil Oil; and let $\mathbf{Z} \equiv (Z_1,Z_2)'$ denote the single share returns vector for Consolidated Edison Company and Laclede Gas Company. Assume (hypothetically) that **X**, **Y**, and **Z** are mutually independent, and

$$\begin{aligned} \mathcal{L}(\mathbf{X}) &= S_3(1,\mathbf{\Theta}_\mathbf{X};\mathbf{\Omega}_\mathbf{X};1.5), \\ \mathcal{L}(\mathbf{Y}) &= S_2(1,\mathbf{\Theta}_\mathbf{Y};\mathbf{\Omega}_\mathbf{Y};1.8), \\ \mathcal{L}(\mathbf{Z}) &= S_2(1,\mathbf{\Theta}_\mathbf{Z};\mathbf{\Omega}_\mathbf{Z};2), \end{aligned}$$

where

$$\mathbf{\Omega}_\mathbf{X} = \begin{pmatrix} 4 & 1 & 0 \\ 1 & 3 & 2 \\ 0 & 2 & 2 \end{pmatrix},$$

$$\mathbf{\Omega}_\mathbf{Y} = \begin{pmatrix} 4 & 3 \\ 3 & 3 \end{pmatrix},$$

and

$$\mathbf{\Omega}_\mathbf{Z} = \begin{pmatrix} 2 & 1 \\ 1 & 3 \end{pmatrix},$$

Assuming the seven securities listed above constitute a prospective portfolio, find the allocation of funds that will make the portfolio efficient in a restricted Model II sense.

12.5 Explain the difference between an efficient portfolio in a restricted sense and an unrestricted efficient portfolio.

12.6 Suppose it is desired to minimize risk in a portfolio while maximizing median return, and the returns for prospective assets follow a multivariate symmetric stable distribution with characteristic exponent α in the range $0 < \alpha < 1$. Explain the numerical (and conceptual) difficulties associated with programming a solution to this problem. [*Hint:* See Exercise 6.10.]

12.7 Explain the difference between Models I and II portfolio analysis.

12.8 What is meant by the *restricted problem* of stable portfolio analysis?

12.9 Suppose you had a large sample of stable distribution data available. How would you estimate the characteristic exponent graphically?

13

CLASSIFICATION AND DISCRIMINATION MODELS

13.1 INTRODUCTION

Classification models are used to categorize an object on the basis of a profile of its characteristics. For example, suppose it is desired to select a market area for expansion of sales of a consumer product. Clearly, before expanding into a new area (such as a particular metropolitan area), it would be prudent to classify the area into either an area of likely buyers of the product, or an area not likely to contain buyers of the product. Such a classification might be based upon a profile of characteristics of the inhabitants of the area. Each of the metropolitan areas being considered for expansion could be classified on this basis.

More generally, let $\mathbf{z} : p \times 1$ denote a vector of observed attributes of an object or individual, and suppose it is known that $\mathbf{z}$ is an observation from exactly one of the mutually exclusive populations $\pi_1, \pi_2, \ldots, \pi_K$. The problem is to discriminate among the populations; that is, a decision must be made as to which population $\mathbf{z}$ belongs.

There is a vast literature on classification and discrimination. In order to classify $\mathbf{z}$ into one of the populations, Fisher (1936) suggested as a basis for classification decisions the use of a discriminant function linear in the components of $\mathbf{z}$. Other bases for classification have included likelihood ratio tests (see for instance, Anderson, 1958, Chapter 6), information theory (see Kullback, 1959) and Bayesian techniques (see, for instance, Geisser, 1964), to mention but a few. Only two general techniques are discussed in this chapter. However, it is believed that these techniques will prove useful for most situations of interest.

Section 13.2 sets forth a decision-theoretic model of the classification problem, assuming there are losses, a priori probabilities for each possible population, and known probability distributions for the random variables in each of the populations. A formal classification procedure is developed for this case. Unfortunately, the distributions can hardly ever be specified exactly because the parameters are usually unknown. In Section 13.3 the probability distributions for the random variables in each of the populations are assumed to have unknown parameters. A full Bayesian approach (with prior distributions on the unknown parameters) is suggested for this case which occurs so frequently. The final results are expressed in terms of the predictive density that $\mathbf{z}$ belongs to any particular one of the π_j's. Section 13.4 discusses a chi-square test for checking the ability of the classification procedure to correctly classify observations at a rate better than chance. In the sequel we will generally assume that our populations are continuous. In fact, in many situations of interest, the underlying populations are discrete, and the observed data may be taken to be multinomial; for example, suppose the data are multivariate scores on ten items in a questionnaire, and each item is polytomously scored. Classification with discrete populations has been nicely summarized in Goldstein and Dillon, 1978. In a logit model type approach, Press and Wilson, 1978 show that logit models are often better than classical approaches for discrimination even when the populations are continuous, but non-normal.

13.2 KNOWN PARAMETERS

It is sometimes the case that the populations of interest are well established; for example, perhaps so many observations have been taken in the past that for all intents and purposes the parameters may be thought of as known; alternatively, the population parameters might be known on theoretical grounds. In this section, for this case of known parameters, rules are developed for classifying observations on the basis that the rules should have minimum risk.

Suppose there are K continuous multivariate populations π_i, $i = 1,\ldots,K$, with corresponding probability densities $f_i(\mathbf{z}|\boldsymbol{\Theta}_i)$, where $\mathbf{z}$ is an observable p-vector and $\boldsymbol{\Theta}_i$ is a matrix of known parameters, and further that there are known a priori probabilities, p_i, that $\mathbf{z}\varepsilon\pi_i$, and that $\mathbf{z}$ is an observation from a particular population, π_i. Let $\{a_i,\ i = 1,\ldots,K\}$ denote the set of possible actions, where a_i is the action of classifying $\mathbf{z}$ into π_i. Finally, let c_{ij} denote the loss (sometimes called the *regret* or *opportunity loss;* $c_{ij} \geq 0$) associated with classifying $\mathbf{z}$ into π_i when in fact the correct decision should be to classify $\mathbf{z}$ into π_j; that is, c_{ij} is the loss associated with taking action a_i when π_j is the true state of nature. The loss matrix is given in Table 13.2.1. Note that all diagonal entries are zero since a correct classification implies no regret.

Table 13.2.1 Loss Matrix for Classification

Actions	*States of nature*			
	$\mathbf{z}\epsilon\pi_1$	$\mathbf{z}\epsilon\pi_2$	$\cdots$	$\mathbf{z}\epsilon\pi_K$
a_1	0	c_{12}	$\cdots$	c_{1K}
a_2	c_{21}	0	$\cdots$	c_{2K}
.	.	.		.
.	.	.		.
.	.	.		.
a_K	c_{K1}	c_{K2}	$\cdots$	0

13.2.1 Classification into Two Populations

Suppose $K = 2$ and it is desired to establish a decision rule for classifying $\mathbf{z}$ into π_1 or π_2. The decision-theoretic approach to making decisions is often based upon the axioms of utility theory (see, for instance, Savage, 1954) which dictate that that action should be taken which minimizes the average loss (that is, the risk). Let $L(a_i,\pi_j)$ denote the loss associated with taking action a_i when state of nature π_j prevails. Then, if ρ denotes the risk,

$$\begin{aligned}\rho = E[L(a,\pi)] &= \sum_{i=1}^{2}\sum_{j=1}^{2} L(a_i,\pi_j)P\{a = a_i,\ \mathbf{z}\epsilon\pi_j\}\\ &= c_{12}P\{a = a_1,\ \mathbf{z}\epsilon\pi_2\} + c_{21}P\{a = a_2,\ \mathbf{z}\epsilon\pi_1\}\\ &= c_{12}P\{a = a_1|\mathbf{z}\epsilon\pi_2\}p_2 + c_{21}P\{a = a_2|\mathbf{z}\epsilon\pi_1\}p_1.\end{aligned}$$

Define classification regions R_1 and R_2 so that if $\mathbf{z}$ falls in R_j, classify $\mathbf{z}$ into π_j (that is, take $a = a_j$), $j = 1, 2$. Then

$$\begin{aligned}\rho &= c_{12}p_2P\{\mathbf{z}\epsilon R_1|\mathbf{z}\epsilon\pi_2\} + c_{21}p_1P\{\mathbf{z}\epsilon R_2|\mathbf{z}\epsilon\pi_1\}\\ &= c_{12}p_2\int_{R_1} f_2(\mathbf{z}|\boldsymbol{\Theta}_2)d\mathbf{z} + c_{21}p_1\int_{R_2} f_1(\mathbf{z}|\boldsymbol{\Theta}_1)d\mathbf{z}.\end{aligned}$$

The problem then is to determine the regions R_1 and R_2 that minimize ρ. Since

$$\int_{R_1} f_j(\mathbf{z})d\mathbf{z} + \int_{R_2} f_j(\mathbf{z})d\mathbf{z} = 1,$$

$$\rho = c_{12}p_2\int_{R_1} f_2(\mathbf{z})d\mathbf{z} + c_{21}p_1\left[1 - \int_{R_1} f_1(\mathbf{z})d\mathbf{z}\right],$$

or

$$\rho = \int_{R_1}[c_{12}p_2f_2(\mathbf{z}|\boldsymbol{\Theta}_2) - c_{21}p_1f_1(\mathbf{z}|\boldsymbol{\Theta}_1)]d\mathbf{z} + c_{21}p_1.$$

It may now be concluded that ρ will be minimized if R_1 is selected to include all those $\mathbf{z}$'s for which

$$c_{12}p_2f_2(\mathbf{z}|\mathbf{\Theta}_2) - c_{21}p_1f_1(\mathbf{z}|\mathbf{\Theta}_1) \leq 0,$$

and R_1 excludes those $\mathbf{z}$'s for which the reverse inequality holds.[1] The result is summarized below.

Theorem (13.2.1): Suppose $\mathbf{z}: p \times 1$ is an observation either from a population π_1 with density $f_1(\mathbf{z}|\mathbf{\Theta}_1)$, or from a population π_2 with density $f_2(\mathbf{z}|\mathbf{\Theta}_2)$, with prior probabilities for π_j of p_j, and costs of misclassification c_{ij}. If $\mathbf{\Theta}_1$ and $\mathbf{\Theta}_2$ are known, the minimum risk rule is to classify $\mathbf{z}$ into π_1 if

$$\frac{f_1(\mathbf{z}|\mathbf{\Theta}_1)}{f_2(\mathbf{z}|\mathbf{\Theta}_2)} \geq \left(\frac{c_{12}}{c_{21}}\right)\left(\frac{p_2}{p_1}\right) \equiv \text{constant}; \tag{13.2.1}$$

otherwise classify $\mathbf{z}$ into π_2.

[*Remark:* Since the populations are assumed to be continuous, equality in (13.2.1) occurs with probability zero.]

13.2.2 Classification into Many Populations

If $K > 2$, the minimum risk decision rule for known parameter matrices $\mathbf{\Theta}_1,\ldots,\mathbf{\Theta}_K$ is to classify $\mathbf{z}$ into π_k if

$$\sum_{\substack{i=1 \\ (i\neq k)}}^{K} p_i c_{ki} f_i(\mathbf{z}|\mathbf{\Theta}_i) < \sum_{\substack{i=1 \\ (i\neq j)}}^{K} p_i c_{ji} f_i(\mathbf{z}|\mathbf{\Theta}_i), \tag{13.2.2}$$

for all $j = 1, 2,\ldots,K$; $j \neq k$. The extension of (13.2.1) to (13.2.2) is direct and straightforward to prove.

Example (13.2.1)—Two Normal Populations: Suppose $\pi_j = N(\boldsymbol{\theta}_j,\mathbf{\Sigma}_j)$, $\mathbf{\Sigma}_j > 0$, where $(\boldsymbol{\theta}_j,\mathbf{\Sigma}_j) \equiv \mathbf{\Theta}_j$ is known, $j = 1, 2$. Then, from (13.2.1), classify $\mathbf{z}$ into $N(\boldsymbol{\theta}_1,\mathbf{\Sigma}_1)$ if

$$\frac{|\mathbf{\Sigma}_1|^{-1/2}\exp\{-\frac{1}{2}(\mathbf{z}-\boldsymbol{\theta}_1)'\mathbf{\Sigma}_1^{-1}(\mathbf{z}-\boldsymbol{\theta}_1)\}}{|\mathbf{\Sigma}_2|^{-1/2}\exp\{-\frac{1}{2}(\mathbf{z}-\boldsymbol{\theta}_2)'\mathbf{\Sigma}_2^{-1}(\mathbf{z}-\boldsymbol{\theta}_2)\}} \geq \left(\frac{c_{12}}{c_{21}}\right)\left(\frac{p_2}{p_1}\right).$$

Equivalently, classify $\mathbf{z}$ into π_1 if

$$[(\mathbf{z}-\boldsymbol{\theta}_2)'\mathbf{\Sigma}_2^{-1}(\mathbf{z}-\boldsymbol{\theta}_2) - (\mathbf{z}-\boldsymbol{\theta}_1)'\mathbf{\Sigma}_1^{-1}(\mathbf{z}-\boldsymbol{\theta}_1)] \geq 2\log\left[\frac{|\mathbf{\Sigma}_1|^{1/2}}{|\mathbf{\Sigma}_2|^{1/2}}\frac{p_2}{p_1}\frac{c_{12}}{c_{21}}\right]. \tag{13.2.3}$$

[1] This result follows from the Neyman Pearson lemma (see, for instance, Kendall and Stuart, 1966).

For illustration, suppose the misclassification costs are equal, the prior probabilities are equal, and the covariance matrices are equal ($\mathbf{\Sigma}_1 = \mathbf{\Sigma}_2 = \mathbf{\Sigma}$) so that the populations differ only in location. The right-hand side of the last inequality becomes zero and the left-hand side simplifies, to give the rule: Classify $\mathbf{z}$ into $N(\boldsymbol{\theta}_1,\mathbf{\Sigma})$ if

$$[(\boldsymbol{\theta}_1 - \boldsymbol{\theta}_2)'\mathbf{\Sigma}^{-1}]\mathbf{z} \geq [\tfrac{1}{2}(\boldsymbol{\theta}_1'\mathbf{\Sigma}^{-1}\boldsymbol{\theta}_1 - \boldsymbol{\theta}_2'\mathbf{\Sigma}^{-1}\boldsymbol{\theta}_2)]. \tag{13.2.4}$$

Since the parameters are all known, both pairs of brackets in (13.2.4) can be computed and the inequality tested. Note that the discriminant function in (13.2.4) is linear in $\mathbf{z}$ in that it is of the form $\mathbf{a}'\mathbf{z} \geq b$, where $\mathbf{a}$ is a known vector and b is a known scalar. A related quantity is $D^2 \equiv (\boldsymbol{\theta}_1 - \boldsymbol{\theta}_2)'\mathbf{\Sigma}^{-1}(\boldsymbol{\theta}_1 - \boldsymbol{\theta}_2)$; D is known as the Mahalanobis distance between the two populations.

Example (13.2.2)—Three (or More) Normal Populations: Suppose $K = 3$, and all parameters are known, $j = 1, 2, 3$. The rule in (13.2.2) becomes: Classify $\mathbf{z}$ into π_3 if

$$p_1c_{31}f_1(\mathbf{z}|\boldsymbol{\Theta}_1) + p_2c_{32}f_2(\mathbf{z}|\boldsymbol{\Theta}_2) < p_2c_{12}f_2(\mathbf{z}|\boldsymbol{\Theta}_2) + p_3c_{13}f_3(\mathbf{z}|\boldsymbol{\Theta}_3),$$

and

$$p_1c_{31}f_1(\mathbf{z}|\boldsymbol{\Theta}_1) + p_2c_{32}f_2(\mathbf{z}|\boldsymbol{\Theta}_2) < p_1c_{21}f_1(\mathbf{z}|\boldsymbol{\Theta}_1) + p_3c_{23}f_3(\mathbf{z}|\boldsymbol{\Theta}_3).$$

A similar result is found for classifying $\mathbf{z}$ into π_1 and π_2 by permuting the subscripts.

Now consider the special case in which all misclassification costs c_{ij} are equal ($i \neq j$). Then the last two equations reduce to: Classify $\mathbf{z}$ into π_3 if

$$p_1f_1(\mathbf{z}|\boldsymbol{\Theta}_1) < p_3f_3(\mathbf{z}|\boldsymbol{\Theta}_3),$$

and (13.2.5)

$$p_2f_2(\mathbf{z}|\boldsymbol{\Theta}_2) < p_3f_3(\mathbf{z}|\boldsymbol{\Theta}_3).$$

Since the posterior probability density that $\mathbf{z}\epsilon\pi_j$, for given $\mathbf{z}$, is proportional to $p_jf_j(\mathbf{z}|\boldsymbol{\Theta}_j)$, this result shows that when the misclassification costs are equal and the population parameters are known, the classification rule becomes: Choose that π_j which maximizes the posterior probability density associated with π_j.

The Bayesian approach toward classification when all parameters are known and misclassification costs are equal, would begin with an evaluation of the posterior probability that $\mathbf{z}\epsilon\pi_j$ given $\mathbf{z}$, for each $j = 1,\ldots,K$. Then posterior odds might be computed for each pair of populations; alternatively, with $K > 2$, the population with the greatest posterior probability density can be selected.

When the costs of misclassification are unequal, the Bayesian would select the population that produced a minimum cost when averaged with respect to the posterior distribution. But this is equivalent to the sampling theory result obtained above.

Thus, the Bayesian and sampling theory approaches lead to the same classification rule when the parameters are known. Moreover, this result is clearly valid for all $K \geq 2$.

Now suppose the three populations are multivariate Normal; that is, $\pi_j = N(\boldsymbol{\theta}_j, \boldsymbol{\Sigma}_j)$, $\boldsymbol{\Sigma}_j > 0$, $j = 1, 2, 3$, and all parameters are known. Moreover, suppose all prior probabilities are equal, and all misclassification costs are equal. Then the rule becomes: Classify $\mathbf{z}$ into π_3 if

$$f_3(\mathbf{z}|\boldsymbol{\theta}_3,\boldsymbol{\Sigma}_3) > f_1(\mathbf{z}|\boldsymbol{\theta}_1,\boldsymbol{\Sigma}_1),$$

and

$$f_3(\mathbf{z}|\boldsymbol{\theta}_3,\boldsymbol{\Sigma}_3) > f_2(\mathbf{z}|\boldsymbol{\theta}_2,\boldsymbol{\Sigma}_2);$$

that is, classify $\mathbf{z}$ into π_3 if

$$(\mathbf{z} - \boldsymbol{\theta}_j)'\boldsymbol{\Sigma}_j^{-1}(\mathbf{z} - \boldsymbol{\theta}_j) - (\mathbf{z} - \boldsymbol{\theta}_3)'\boldsymbol{\Sigma}_3^{-1}(\mathbf{z} - \boldsymbol{\theta}_3) > \log \frac{|\boldsymbol{\Sigma}_3|}{|\boldsymbol{\Sigma}_j|}, \quad (13.2.6)$$

for $j = 1$ and $j = 2$. The subscripts would merely be permuted for classification into π_1 and π_2.

For classification of $\mathbf{z}$ into one of K known multivariate Normal populations, the rule is merely to classify $\mathbf{z}$ into the population with the largest density (for the case of equal p_j's and equal c_{ij}'s). Thus, classify $\mathbf{z}$ into π_K if

$$(\mathbf{z} - \boldsymbol{\theta}_j)'\boldsymbol{\Sigma}_j^{-1}(\mathbf{z} - \boldsymbol{\theta}_j) - (\mathbf{z} - \boldsymbol{\theta}_K)'\boldsymbol{\Sigma}_K^{-1}(\mathbf{z} - \boldsymbol{\theta}_K) > \log \frac{|\boldsymbol{\Sigma}_K|}{|\boldsymbol{\Sigma}_j|}, \quad (13.2.7)$$

for every $j = 1, 2, \ldots, K - 1$.

[*Remark:* Note that the discriminant functions given in (13.2.6) and (13.2.7) are quadratic in $\mathbf{z}$. However, if it may be assumed that the covariance matrices are equal, the discriminant functions become linear.]

13.3 UNKNOWN PARAMETERS

In this section the population parameters are assumed to be unknown, as is usually the case. Classification procedures are developed first for observations obtained from arbitrary, continuous, multivariate distributions and then for observations assumed to follow multivariate Normal distributions.

13.3.1 Arbitrary Distributions

Suppose π_j has an associated density $f_j(\mathbf{z}|\boldsymbol{\Theta}_j)$, $j = 1, 2, \ldots, K$, where $\boldsymbol{\Theta}_j$ is unknown. Suppose further that independent p-variate observations $\{\mathbf{x}_1(j), \ldots, \mathbf{x}_{N_j}(j)\}$ are available for each population j, $j = 1, \ldots, K$. Then, if these samples are used to form maximum likelihood estimates of the population parameters, and if the estimates are substituted for the parameter values in (13.2.2), large sample classification rules will be obtained. If sample sizes are sufficiently large, results obtained by this technique should be quite good. However, with moderate or small samples, the results could be quite poor.

13.3.2 Normally Distributed Observations

Sampling Theory Background

Now suppose that $\pi_j = N(\boldsymbol{\theta}_j, \boldsymbol{\Sigma}_j)$, $j = 1, \ldots, K$, and $(\boldsymbol{\theta}_j, \boldsymbol{\Sigma}_j)$ are unknown. Also suppose that the independent p-variate observations from each population $\{\mathbf{x}_1(j), \ldots, \mathbf{x}_{N_j}(j)\}$ are available, $j = 1, \ldots, K$. If $\boldsymbol{\Sigma}_1 = \boldsymbol{\Sigma}_2 = \cdots = \boldsymbol{\Sigma}_K$, likelihood ratio and similar procedures may be found easily although the distributions required to use these procedures in small samples are quite complicated (see, for instance, Wald, 1944; Anderson, 1951; and Sitgreaves, 1952). Asymptotic results were given for the general case of unequal means and unequal covariance matrices by Press (1964). A large variety of other techniques have been suggested from the sampling theory viewpoint; none of them are very simple. However, the Bayesian approach provides a useful and simple alternative in this case.

Bayesian Approach

Bayesian approaches to the classification problem in the case of Normally distributed observations with unknown parameters were discussed by Geisser (1964; 1966; and 1967) and Dunsmore (1966). The results are extremely simple to apply and there is no complicated distribution theory. The results are summarized below.

Define the sample mean and covariance matrix (unbiased estimator) for the jth population as

$$\bar{\mathbf{x}}(j) = \frac{1}{N_j} \sum_{i=1}^{N_j} \mathbf{x}_i(j), \qquad \mathbf{S}_j = \frac{1}{N_j - 1} \sum_{i=1}^{N_j} [\mathbf{x}_i(j) - \bar{\mathbf{x}}(j)][\mathbf{x}_i(j) - \bar{\mathbf{x}}(j)]',$$

and recall that p_j is the prior probability of classifying $\mathbf{z}$ into π_j, $j = 1, \ldots, K$.

Theorem (13.3.1): Let $\mathbf{z}: p \times 1$ be an observation from one of the populations $\pi_j = N(\boldsymbol{\theta}_j, \boldsymbol{\Sigma}_j)$, $j = 1, \ldots, K$ where the parameters $(\boldsymbol{\theta}_j, \boldsymbol{\Sigma}_j)$ are unknown. If the prior distribution of the parameters is diffuse, the predictive probability density (see Section 3.7) for classifying $\mathbf{z}$ into π_j is given by the multivariate Student t-density

$$p(\mathbf{z} \mid \text{data}, j) = \frac{k_j}{\left[1 + \dfrac{N_j}{N_j^2 - 1} (\mathbf{z} - \bar{\mathbf{x}}(j))' \mathbf{S}_j^{-1} (\mathbf{z} - \bar{\mathbf{x}}(j))\right]^{N_j/2}}, \tag{13.3.1}$$

where k_j is a constant not depending upon $\mathbf{z}$, and given by

$$k_j = \left[\frac{N_j}{(N_j+1)\pi}\right]^{p/2} \frac{\Gamma\left(\frac{N_j}{2}\right) p_j}{\Gamma\left(\frac{N_j - p}{2}\right) |(N_j - 1)\mathbf{S}_j|^{1/2}}.$$

Proof: The proof of this theorem is given in Complement 13.1.

[*Remark (1):* From (13.3.1) it follows that the *predictive odds ratio* for classifying $\mathbf{z}$ into π_i, as compared with π_j, becomes the ratio of the associated multivariate Student **t**-densities

$$\frac{p(\mathbf{z}|\text{data},i)}{p(\mathbf{z}|\text{data},j)} = L_{ij} \frac{\left\{1 + \frac{N_j}{N_j^2 - 1}(\mathbf{z} - \bar{\mathbf{x}}(j))'\mathbf{S}_j^{-1}(\mathbf{z} - \bar{\mathbf{x}}(j))\right\}^{N_j/2}}{\left\{1 + \frac{N_i}{N_i^2 - 1}(\mathbf{z} - \bar{\mathbf{x}}(i))'\mathbf{S}_i^{-1}(\mathbf{z} - \bar{\mathbf{x}}(i))\right\}^{N_i/2}}, \tag{13.3.2}$$

where L_{ij} is a constant given by

$$L_{ij} = \left(\frac{p_i}{p_j}\right)\left(\frac{|(N_j - 1)\mathbf{S}_j|}{|(N_i - 1)\mathbf{S}_i|}\right)^{1/2} \left(\frac{\Gamma\left(\frac{N_i}{2}\right)\Gamma\left(\frac{N_j - p}{2}\right)}{\Gamma\left(\frac{N_j}{2}\right)\Gamma\left(\frac{N_i - p}{2}\right)}\right) \left[\frac{N_i(N_j+1)}{N_j(N_i+1)}\right]^{p/2},$$

$i, j = 1, 2, \ldots, K$.]

[*Remark (2):* It is seen that the sample sizes need not be large for (13.3.1) to be applicable, as was the case in the sampling theory approach of Section (13.3.1). Therefore, the result is more generally applicable. Moreover, the covariance matrices need not be equal, which is usually required in sampling approaches.]

[*Remark (3):* If a natural conjugate rather than a vague prior is used, it is straightforward to check that the result of (13.3.2) is again obtained, except that in this case, the location, scale, and degrees of freedom parameters are different.]

[*Remark (4):* For an application of this technique to the problem of classifying undecided respondents in a sample survey, see Press and Yang, 1974.]

Example (13.3.1): The question of how similar are the audiences of two or more advertising vehicles may be answered, in part, by an appeal to discrimination analysis. Massy (1965) used this approach to evaluate the similarities among the audiences of 5 FM radio stations in the Boston metropolitan area. The data were collected from a sample of families who owned at least one FM radio receiver, and a mail questionnaire was used to obtain information on current station selections and some 47 socioeconomic and consumption variables. Respondents were given a series of scales simulating the markings on a typical FM dial and asked to note the position of the dial on each FM receiver in the home, as of the time the questionnaire was filled out. The result of this approach was a sample of 239 families for whom the station tuned to at response time could be unambiguously determined. This sample was used to estimate the parameters of the distributions (as in Section 13.3.1) for each population (all populations were assumed to be multivariate Normal).

Since the dimension of this problem was very large ($p = 47$), the data were first subjected to a factor analysis (see Chapter 10). This resulted in a set of 12 new variates which were used as summary or index variables for the original set. (Clearly some information was thereby lost.) The 12 remaining variables were then used in a 5-population discrimination analysis to establish a basis for classifying survey respondents into listeners of one of the 5 radio stations on the basis of their socioeconomic characteristics. The analysis resulted in a profile of socioeconomic characteristics of the listeners to each of the FM radio stations.

To carry out the analysis, Equation (13.2.7) was used assuming the a priori probabilities for each of the five populations were equal, and assuming the costs of misclassification were equal. Moreover, the densities were all taken to be Normal densities with equal covariance matrices and with parameters equal to the estimated sample values (as explained in Section 13.3.1). The 12 classification variables and their multiplying coefficients used to form the linear discriminant functions for the 5 FM stations are given in Table 13.3.1. This table was interpreted by Massy to provide the following audience profiles for each of the stations.

Station A: Ownership of a bigger or newer car, or more than one car, contributes most strongly to classification in A's audience. Families that seldom "go out" to movies, sports, or cultural events also are disproportionately likely to be A's. The younger the family, the higher its probability of being in the A audience.

Station B: The probability of classification in B increases as the family rises in occupational status. It is highest if the family did not send in for A's program guide. Younger families, and families that indicate a preference for opera over jazz, are more likely to be assigned to B.

Station C: Respondents assigned to C tend to be much older than average, and own fewer and/or older and smaller automobiles, and prefer jazz and popular music to opera. "Going out" contributes more to C's classification probability than to any other station. The same is true for sending in for A's program guide.

Table 13.3.1 Discriminant Function Coefficients

Variables	*Stations*				
	A	*B*	*C*	*D*	*E*
1. Durables ownership (high scorers more likely to own dishwashers, freezers, washers, dryers, second cars)	−.18	−.53	+ .27	− .74	−1.01
2. Age—older (+)	−.89	−.79	+1.18	+ .38	−1.72
3. Social class I—higher occupational status	+.41	+.90	+ .21	+1.22	+1.11
4. Musical preference I—classical and opera (+) versus popular (−)	+.03	+.06	− .26	+ .11	− .01
5. Social class II—"lower middle class" (high scorers use credit, have low income and assets, tend to have older cars)	+.20	+.29	− .34	+ .57	+ .94
6. Automobile ownership (high scorers own newer cars, tend toward foreign, lower priced, and larger models)	+.96	−.01	−1.27	−1.04	+1.10
7. Music preference II—folk (+) versus popular (−)	−.04	+.19	+ .06	+ .18	− .27
8. Source of entertainment (high scorers seldom "go out")	+.58	+.21	− .48	+ .24	+1.04
9. Wife's status—working wife (+)	+.36	−.09	− .06	− .69	− .49
10. Music preference III—opera (+) versus jazz (−)	+.28	+.35	− .55	− .13	+ .77
11. "Individualism" (high scorers tend to like folk music, dislike trading stamps, and not own TV set or shop in discount houses)	−.27	+.20	+ .71	+1.98	+ .20
12. Program guide—sent in for Station *A*'s guide (−)	+.31	+.43	+ .18	+ .25	+ .22
Constant	−.26	−.19	− .38	− .38	− .33

Station D: Individualism contributes most strongly to the probability of classification in this audience. Next in importance is occupational status. Affluence in automobile ownership strongly inhibits the chances of being so classified.

Station E: High classification probabilities for *E* are strongly related to occupational status and automobile affluence, and inversely related to "going out" and durables ownership. Younger people are much more likely to be classified in this audience. The group is most likely to exhibit "lower middle class" values (Social Class II). The extreme positive coefficient for opera versus jazz might be regarded as a dislike for jazz.

On the basis of the above audience profiles determined from the classification analysis, a potential advertiser would find it easy to select a particular FM station for advertising his product if he could establish the "type" of individual most likely to buy his product. Thus, if buyers of his product were given questionnaires to determine their socioeconomic backgrounds, all buyers could be classified by the above linear discriminant functions to see if most buyers would be likely to be drawn toward a particular radio station, and therefore, most likely to hear the commercial message.

Note that there has been no discussion of the validity of the assumptions of multivariate Normality, equality of covariance matrices, use of large sample results of parameter estimation, and so on. Violation of any of these assumptions would vitiate the results described above.

Example (13.3.2): Suppose $\mathbf{z}$ has two components ($p = 2$), and $\mathbf{z}$ is to be classified into one of two Normal populations ($K = 2$). Assume there is equal likelihood, a priori, of classifying $\mathbf{z}$ into the 2 populations so that $p_1 = p_2 = \frac{1}{2}$. A full Bayesian approach will be used for the classification.

Suppose that on the basis of 10 bivariate observations from each population ($N_1 = N_2 = N = 10$), the sufficient statistics are

$$\bar{\mathbf{x}}(1) = \begin{pmatrix} 1 \\ 1 \end{pmatrix}, \qquad \bar{\mathbf{x}}(2) = \begin{pmatrix} 0 \\ 0 \end{pmatrix},$$

$$\mathbf{S}_1 = \begin{pmatrix} 1 & \frac{1}{2} \\ \frac{1}{2} & 1 \end{pmatrix}, \qquad \mathbf{S}_2 = \begin{pmatrix} 1 & \frac{1}{4} \\ \frac{1}{4} & \frac{1}{2} \end{pmatrix}.$$

Since the sample sizes are equal and the prior probabilities are equal, from (13.3.2)

$$L_{ij} = \frac{|\mathbf{S}_j|^{1/2}}{|\mathbf{S}_i|^{1/2}}.$$

Since $|\mathbf{S}_1| = \frac{3}{4}$, and $|\mathbf{S}_2| = \frac{7}{16}$, $L_{12} = .76$. From (13.3.2),

$$\frac{p(\mathbf{z}|\text{data}, i = 1)}{p(\mathbf{z}|\text{data}, i = 2)} = \frac{.76\{1 + \frac{10}{99}(\mathbf{z} - \bar{\mathbf{x}}(2))'\mathbf{S}_2^{-1}(\mathbf{z} - \bar{\mathbf{x}}(2))\}^5}{\{1 + \frac{10}{99}(\mathbf{z} - \bar{\mathbf{x}}(1))'\mathbf{S}_1^{-1}(\mathbf{z} - \bar{\mathbf{x}}(1))\}^5}.$$

Suppose the observed vector to be classified is given by $\mathbf{z} = (\frac{1}{2}, \frac{1}{2})'$. Then

$$\mathbf{z} - \bar{\mathbf{x}}(1) = \begin{pmatrix} -\frac{1}{2} \\ -\frac{1}{2} \end{pmatrix}, \qquad \mathbf{z} - \bar{\mathbf{x}}(2) = \begin{pmatrix} \frac{1}{2} \\ \frac{1}{2} \end{pmatrix},$$

and

$$\mathbf{S}_1^{-1} = \begin{pmatrix} \frac{4}{3} & -\frac{2}{3} \\ -\frac{2}{3} & \frac{4}{3} \end{pmatrix}, \qquad \mathbf{S}_2^{-1} = \begin{pmatrix} \frac{8}{7} & -\frac{4}{7} \\ -\frac{4}{7} & \frac{16}{7} \end{pmatrix}.$$

Hence,

$$[\mathbf{z} - \bar{\mathbf{x}}(1)]'\mathbf{S}_1^{-1}[\mathbf{z} - \bar{\mathbf{x}}(1)] = \tfrac{1}{3},$$
$$[\mathbf{z} - \bar{\mathbf{x}}(2)]'\mathbf{S}_2^{-1}[\mathbf{z} - \bar{\mathbf{x}}(2)] = \tfrac{4}{7}.$$

Substitution of these results into the ratio of densities gives for the predictive odds ratio,

$$\frac{p(\mathbf{z}|\text{data}, i = 1)}{p(\mathbf{z}|\text{data}, i = 2)} = .92.$$

That is, the predictive odds are slightly in favor of π_2, but not much more than the *prior odds ratio* of 1:1. This result was to be expected in light of the observed data. That is, since the sample variances are so large relative to the distances between the sample means, and since the prior information sheds little additional light, the "boundary" between the two populations remains quite blurred.

It might be of interest to see what would have been the result of applying the sampling theory procedure used in Example (13.3.1) to this example. Since there are two populations, (13.2.3) would be applied if the parameters were all known, or if the sample sizes were very large. In this case, the sample sizes are each 10, so that asymptotic theory should not be expected to be applicable. However, what happens if the approach is used regardless?

Letting $p_1 = p_2$, and $c_{12} = c_{21}$, and substituting into (13.2.3) gives the rule: Classify $\mathbf{z}$ into π_1 if

$$(\mathbf{z} - \boldsymbol{\theta}_2)'\boldsymbol{\Sigma}_2^{-1}(\mathbf{z} - \boldsymbol{\theta}_2) - (\mathbf{z} - \boldsymbol{\theta}_1)'\boldsymbol{\Sigma}_1^{-1}(\mathbf{z} - \boldsymbol{\theta}_1) \geq \log \frac{|\boldsymbol{\Sigma}_1|}{|\boldsymbol{\Sigma}_2|}.$$

Now replace $[\boldsymbol{\theta}_1,\boldsymbol{\theta}_2,\boldsymbol{\Sigma}_1,\boldsymbol{\Sigma}_2]$ by their sample estimates $[\bar{\mathbf{x}}(1),\bar{\mathbf{x}}(2),\mathbf{S}_1,\mathbf{S}_2]$, given above. Then classify $\mathbf{z}$ into π_1 if

$$[\mathbf{z} - \bar{\mathbf{x}}(2)]'\mathbf{S}_2^{-1}[\mathbf{z} - \bar{\mathbf{x}}(2)] - [\mathbf{z} - \bar{\mathbf{x}}(1)]'\mathbf{S}_1^{-1}[\mathbf{z} - \bar{\mathbf{x}}(1)] \geq \log \frac{|\mathbf{S}_1|}{|\mathbf{S}_2|}.$$

But these quantities were evaluated numerically above. The left-hand side is $\frac{4}{7} - \frac{1}{3}$. Hence the rule is to classify $\mathbf{z}$ into π_1 if

$$\log \frac{|\mathbf{S}_1|}{|\mathbf{S}_2|} \leq \frac{4}{7} - \frac{1}{3} = \frac{5}{21} = .238.$$

But it was also found above that $|\mathbf{S}_1| = \frac{3}{4}$, and $|\mathbf{S}_2| = \frac{7}{16}$. Hence,

$$\log \frac{|\mathbf{S}_1|}{|\mathbf{S}_2|} = \log \frac{12}{7} = \log 1.71 = .536.$$

Therefore, $\mathbf{z}$ should be classified into π_2, the same conclusion reached by the Bayesian approach. However, in this case there is great uncertainty about the decision since large sample theory was used (and there was no reason to suppose it was valid to do so). Moreover, while the Bayesian approach provided a complete predictive distribution (or a continuum of "risks") for placing $\mathbf{z}$ in π_2 (in addition to the predictive odds), the sampling theory result provided only a decision. Sampling theory procedures that attempt to cope with the problem of risk associated with the classification decision in the absence of well-defined loss functions still fall short in that it is then required that sample sizes be large, and that the covariance matrices be equal, assumptions that are not always justified.

13.4 TEST FOR DISCRIMINATORY POWER

After a discrimination procedure has been established, it is of considerable interest to determine whether the discriminator is really useful. A method for studying the discriminatory power of a procedure involves the use of *Confusion matrices*, which were defined by Massy (1965) for comparing the similarities among populations.

A Confusion matrix provides a convenient method of summarizing the number of correct and incorrect classifications made by the discrimination procedure. Suppose there are K populations and N_j observations have been taken from π_j to estimate its parameters, $j = 1,\ldots,K$. Since the origins of all these observations are known, by applying the discrimination procedure to these observations, it is possible to score the fraction of successful classifications, and to test whether the procedure is significantly better than a purely random partitioning of the decision space.

Let n_{ij} denote the number of observations known to belong to population π_i, but which were classified into π_j. Then the Confusion matrix for the classification problem is defined to be the $K \times K$ matrix $\mathbf{C} \equiv (n_{ij})$ depicted below.

$$
\underset{(K\times K)}{\mathbf{C}} = \begin{array}{c} \\ \pi_1 \\ \pi_2 \\ \cdot \\ \cdot \\ \cdot \\ \pi_K \end{array} \begin{array}{c} \begin{array}{cccc} \pi_1 & \pi_2 & \cdots & \pi_K \end{array} \\ \left[\begin{array}{cccc} n_{11} & n_{12} & \cdots & n_{1K} \\ n_{21} & n_{22} & \cdots & n_{2K} \\ \cdot & \cdot & & \cdot \\ \cdot & \cdot & & \cdot \\ \cdot & \cdot & & \cdot \\ n_{K1} & n_{K2} & \cdots & n_{KK} \end{array}\right] \end{array} \quad \text{True } \pi_j\text{'s}
$$

Confusion matrix

Predicted π_j's

Diagonal elements of **C** denote the numbers of correct classifications (hits), and the off-diagonal elements denote the numbers of incorrect classifications (misses). The *normalized Confusion matrix*, $\mathbf{C}_0$, is easier to interpret than **C**. By definition,

$$\mathbf{C}_0 = (c_{ij}), \qquad c_{ij} = \frac{n_{ij}}{\sum_{j=1}^{K} n_{ij}}.$$

That is, the elements of the normalized Confusion matrix are fractions of correct and incorrect classifications.

To test the discriminatory power of the procedure use a chi-square test. Accordingly, define

$$Q = \frac{(n - e)^2}{e} + \frac{(\bar{n} - \bar{e})^2}{\bar{e}}, \tag{13.4.1}$$

where n and $\bar{n}$ denote the number of correct and incorrect classifications made by the discrimination procedure, respectively, and e and $\bar{e}$ denote the expected numbers of correct and incorrect classifications that would be made if the classifications were made at random. Then, if $N \equiv \Sigma_1^K N_j$ denotes the total number of observations classified, and the probability of a successful random classification is $1/K$,

$$n = \sum_{j=1}^{K} n_{jj}, \qquad \bar{n} = N - n, \qquad e = \frac{N}{K}, \qquad \bar{e} = N - \frac{N}{K}. \tag{13.4.2}$$

It is easy to check by substituting the relations in (13.4.2) into (13.4.1) that the test statistic is expressible as

$$Q = \frac{(N - nK)^2}{N(K - 1)}, \tag{13.4.3}$$

a form more convenient for numerical evaluation. Thus, to test H: hits took place at random versus A: the discrimination procedure did better than just chance, use the fact that under H,

$$\mathcal{L}(Q) = \chi_1^2. \tag{13.4.4}$$

It should be noted that since the same data is being used to rate the procedure as to define the procedure, the test for discriminatory power is not strictly appropriate. A correct test would be obtained by splitting the sample into one part which is used to establish the discrimination procedure and another which is used to test the procedure. However, if the sample is small, this approach is not recommended since then, inefficient estimators of the parameters would result at the expense of obtaining a good power testing procedure.

Example (13.4.1): Consider Example (13.3.1) in which audiences of 5 FM radio stations in Boston were classified according to 12 socioeconomic characteristics. The Confusion matrix and the normalized Confusion matrix for this problem are given in Tables 13.4.1 and 13.4.2. Results are based upon 239 observations with known classifications. Adding the diagonal elements of Table 13.4.1 shows that the total number of hits was 88, or 36.8 percent. Evaluating Q from (13.4.3) for $N = 239$, $n = 88$, and $K = 5$ gives $Q = 42.2$. Since at the 1 percent level, $\chi_1^2 = 6.63$, Q is certainly significant, and H must be rejected. Thus, the classification procedure does better than chance.

Table 13.4.1 Confusion Matrix for Radio Audiences

Actual audience	*Predicted audience*					*Totals*
	A	*B*	*C*	*D*	*E*	
A	43	13	8	21	14	99
B	16	15	15	13	13	72
C	3	5	14	5	4	31
D	2	3	5	9	4	23
E	2	1	0	4	7	14

Table 13.4.2 Normalized Confusion Matrix for Radio Audiences

Actual audience	*Predicted audience*				
	A	*B*	*C*	*D*	*E*
A	.43	.13	.08	.21	.14
B	.22	.21	.21	.18	.18
C	.10	.16	.45	.16	.13
D	.08	.13	.22	.39	.17
E	.14	.07	.00	.29	.50

Other examples of the use of discrimination analysis in marketing and business were given by Banks (1958), Evans (1959), and Frank and Massy (1963).

REFERENCES

Anderson (1951, 1958).
Banks (1958).
Dunsmore (1966).
Evans (1959).
Frank and Massy (1963).
Fisher (1936).
Geisser (1964, 1966, 1967).
Goldstein and Dillon (1978).
Kendall and Stuart (1966).
Kullback (1959).
Massy (1965).
Press (1964).
Press and Wilson (1978).
Press and Yang (1974).
Savage (1954).
Sitgreaves (1952).
Wald (1944).

COMPLEMENT 13.1

Proof of Theorem (13.3.1): Define $\mathbf{V}_j \equiv (N_j - 1)\mathbf{S}_j$, $j = 1,\ldots,K$. Then the posterior density of the parameters in the jth population, given the observed data, is

$$p(\boldsymbol{\theta}_j,\boldsymbol{\Lambda}_j|\bar{\mathbf{x}}(j),\mathbf{V}_j,\pi_j) \propto p(\bar{\mathbf{x}}(j),\mathbf{V}_j|\boldsymbol{\theta}_j,\boldsymbol{\Lambda}_j,\pi_j)p(\boldsymbol{\theta}_j,\boldsymbol{\Lambda}_j),$$

where $\boldsymbol{\Lambda}_j \equiv \boldsymbol{\Sigma}_j^{-1}$, $j = 1,\ldots,K$. That is, the posterior density is proportional to the product of the likelihood and the prior densities.

Since $\bar{\mathbf{x}}(j)$ and $\mathbf{V}_j$ are independent, and since

$$\mathcal{L}[\bar{\mathbf{x}}(j)|\boldsymbol{\theta}_j,\boldsymbol{\Lambda}_j] = N(\boldsymbol{\theta}_j,\boldsymbol{\Lambda}_j^{-1}N_j^{-1})$$

and

$$\mathcal{L}(\mathbf{V}_j|\boldsymbol{\Lambda}_j) = W(\boldsymbol{\Lambda}_j^{-1}, p, N_j - 1),$$

the likelihood is given by

$$p[\bar{\mathbf{x}}(j),\mathbf{V}_j|\boldsymbol{\theta}_j,\boldsymbol{\Lambda}_j,\pi_j] \propto |\mathbf{V}_j|^{(N_j-p-2)/2}|\boldsymbol{\Lambda}_j|^{N_j/2} \exp\{-\tfrac{1}{2}\operatorname{tr}\boldsymbol{\Lambda}_j[\mathbf{V}_j + N_j(\bar{\mathbf{x}}(j) - \boldsymbol{\theta}_j)(\bar{\mathbf{x}}(j) - \boldsymbol{\theta}_j)']\}.$$

If informative prior information is available, it may be brought in at this point either in the form of a natural conjugate prior density, or in the form of a specific preconceived prior density. In the absence of such prior information, adopt the diffuse prior density [see (3.6.5)]

$$p(\boldsymbol{\theta}_j,\boldsymbol{\Lambda}_j) \propto |\boldsymbol{\Lambda}_j|^{-(p+1)/2}.$$

Multiplying the likelihood and prior densities gives the posterior density

$$p(\boldsymbol{\theta}_j,\boldsymbol{\Lambda}_j|\bar{\mathbf{x}}(j),\mathbf{V}_j,\pi_j) \propto |\boldsymbol{\Lambda}_j|^{(N_j-p-1)/2} \exp\{-\tfrac{1}{2}\operatorname{tr}\boldsymbol{\Lambda}_j[\mathbf{V}_j + N_j(\bar{\mathbf{x}}(j) - \boldsymbol{\theta}_j)(\bar{\mathbf{x}}(j) - \boldsymbol{\theta}_j)']\}. \quad \text{(C13.1)}$$

The next step is to use this posterior density to obtain the predictive distribution for the observable vector $\mathbf{z}$. The result is given below.

Lemma (C13.1): If the prior density on the parameters is diffuse so that the posterior density is given by (C13.1), the predictive distribution of $\mathbf{z}$, given the observed data, is multivariate Student $\mathbf{t}$, and is centered at $\bar{\mathbf{x}}(j)$.

Proof: By definition, the predictive density of $\mathbf{z}$ for a classification into the jth population, is given by

$$p(\mathbf{z}|\bar{\mathbf{x}}(j),\mathbf{V}_j,\pi_j) = \int\int p(\mathbf{z}|\boldsymbol{\theta}_j,\boldsymbol{\Lambda}_j,\pi_j)p(\boldsymbol{\theta}_j,\boldsymbol{\Lambda}_j|\bar{\mathbf{x}}(j),\mathbf{V}_j,\pi_j)d\boldsymbol{\theta}_j d\boldsymbol{\Lambda}_j. \quad \text{(C13.2)}$$

That is, it is the sampling density of $\mathbf{z}$ averaged over all parameters with respect to the posterior density. Since

$$\mathfrak{L}(\mathbf{z}|\boldsymbol{\theta}_j,\boldsymbol{\Lambda}_j,\pi_j) = N(\boldsymbol{\theta}_j,\boldsymbol{\Lambda}_j^{-1}), \quad \text{(C13.3)}$$

substitution of (C13.1) and (C13.3) into (C13.2) gives

$$p(\mathbf{z}|\bar{\mathbf{x}}(j),\mathbf{V}_j,\pi_j) \propto \int\int |\boldsymbol{\Lambda}_j|^{(N_j-p)/2} \exp\left(-\tfrac{1}{2} \operatorname{tr} \boldsymbol{\Lambda}_j\mathbf{A}\right)d\boldsymbol{\theta}_j d\boldsymbol{\Lambda}_j,$$

where $\mathbf{A} \equiv \mathbf{V}_j + N_j(\bar{\mathbf{x}}(j) - \boldsymbol{\theta}_j)(\bar{\mathbf{x}}(j) - \boldsymbol{\theta}_j)' + (\mathbf{z} - \boldsymbol{\theta}_j)(\mathbf{z} - \boldsymbol{\theta}_j)'$.

Noting that if the integrand is viewed as a function of $\boldsymbol{\Lambda}_j$ it is proportional to a Wishart density [see (5.1.1)], the integration over $\boldsymbol{\Lambda}_j$ is easily found to result in

$$p(\mathbf{z}|\bar{\mathbf{x}}(j),\mathbf{V}_j,\pi_j) \propto \int \frac{d\boldsymbol{\theta}_j}{|\mathbf{A}|^{(N_j+1)/2}}.$$

Completing the square in $\boldsymbol{\theta}_j$ and simplifying gives

$$p(\mathbf{z}|\bar{\mathbf{x}}(j),\mathbf{V}_j,\pi_j) \propto \int \frac{d\boldsymbol{\theta}_j}{|\mathbf{B}_j + (\boldsymbol{\theta}_j - \bar{\boldsymbol{\theta}}_j)(\boldsymbol{\theta}_j - \bar{\boldsymbol{\theta}}_j)'|^{(N_j+1)/2}}, \quad \text{(C13.4)}$$

where

$$\bar{\boldsymbol{\theta}}_j = \frac{N_j\bar{\mathbf{x}}(j) + \mathbf{z}}{N_j + 1},$$

and

$$(N_j + 1)\mathbf{B}_j = \mathbf{V}_j + N_j\bar{\mathbf{x}}(j)\bar{\mathbf{x}}(j)' + \mathbf{z}\mathbf{z}' - \frac{(N_j\bar{\mathbf{x}}(j) + \mathbf{z})(N_j\bar{\mathbf{x}}(j) + \mathbf{z})'}{N_j + 1}.$$

Since (C13.4) is equivalent to

$$p(\mathbf{z}|\bar{\mathbf{x}}(j),\mathbf{V}_j,\pi_j) \propto \int \frac{d\boldsymbol{\theta}_j}{|\mathbf{B}_j|^{(N_j+1)/2}|\mathbf{I} + \mathbf{B}_j^{-1}(\boldsymbol{\theta}_j - \bar{\boldsymbol{\theta}}_j)(\boldsymbol{\theta}_j - \bar{\boldsymbol{\theta}}_j)'|^{(N_j+1)/2}},$$

use of the determinantal relation (2.4.3) gives

$$p(\mathbf{z}|\bar{\mathbf{x}}(j),\mathbf{V}_j,\pi_j) \propto \frac{1}{|\mathbf{B}_j|^{(N_j+1)/2}} \int \frac{d\boldsymbol{\theta}_j}{[1 + (\boldsymbol{\theta}_j - \bar{\boldsymbol{\theta}}_j)'\mathbf{B}_j^{-1}(\boldsymbol{\theta}_j - \bar{\boldsymbol{\theta}}_j)]^{(N_j+1)/2}}.$$

Note that the integrand is proportional to a multivariate Student **t**-density. Integrating gives

$$p(\mathbf{z}|\bar{\mathbf{x}}(j),\mathbf{V}_j,\pi_j) \propto \frac{|\mathbf{B}_j|^{1/2}}{|\mathbf{B}_j|^{(N_j+1)/2}} = |\mathbf{B}_j|^{-N_j/2}.$$

Using the definition of $\mathbf{B}_j$ and simplifying gives

$$p(\mathbf{z}|\bar{\mathbf{x}}(j),\mathbf{V}_j,\pi_j) \propto \frac{1}{\left|\mathbf{V}_j + \dfrac{N_j}{N_j+1}(\mathbf{z}-\bar{\mathbf{x}}(j))(\mathbf{z}-\bar{\mathbf{x}}(j))'\right|^{N_j/2}}.$$

Using (2.4.3) again, and substituting the relation $\mathbf{V}_j = (N_j - 1)\mathbf{S}_j$, gives the result asserted in the lemma; namely,

$$p(\mathbf{z}|\bar{\mathbf{x}}(j),\mathbf{V}_j,\pi_j) \propto \frac{1}{\left[1 + \dfrac{N_j}{N_j^2 - 1}(\mathbf{z}-\bar{\mathbf{x}}(j))'\mathbf{S}_j^{-1}(\mathbf{z}-\bar{\mathbf{x}}(j))\right]^{N_j/2}}. \qquad \text{(C13.5)}$$

Now apply Bayes' theorem to the predictive probability density for classifying a new observation $\mathbf{z}$ into π_j. This gives

$$p(\mathbf{z}|\text{data}, j) \propto p(\mathbf{z}|\bar{\mathbf{x}}(j),\mathbf{V}_j,\pi_j)p_j,$$

which is equivalent to (13.3.1), and therefore proves the theorem.

EXERCISES

13.1 Use the data of Exercise 7.6 to establish a sampling theory classification procedure for classifying new observations into $\pi_1 = N(\boldsymbol{\theta},\boldsymbol{\Sigma}_1)$, or $\pi_2 = N(\boldsymbol{\phi},\boldsymbol{\Sigma}_2)$. [*Hint:* Use the approach of Section 13.3.1 and assume sample sizes are sufficiently large.]

13.2 Using the classification procedure established in Exercise 13.1, classify the vectors

$$\mathbf{z}_1 = \begin{pmatrix}3\\4\end{pmatrix}, \quad \mathbf{z}_2 = \begin{pmatrix}3\\8\end{pmatrix}, \quad \mathbf{z}_3 = \begin{pmatrix}2\\5\end{pmatrix}, \quad \mathbf{z}_4 = \begin{pmatrix}7\\5\end{pmatrix}.$$

(Assume equal misclassification costs, and equal prior probabilities.)

13.3 Use the Bayesian approach of Section 13.3.2 with equal misclassification costs and equal prior probabilities and apply it to the data of Exercise 7.6. Develop the predictive probability density of classifying a new observation into one of the two populations.

13.4 Classify the vectors in Exercise 13.2 by means of the procedure developed in Exercise 13.3.

13.5 Compare the results in Exercises 13.2 and 13.4. Would you prefer one procedure over the other? Why?

13.6 Consider the problem about radio station audiences in Example (13.4.1). Suppose instead of using the equal likelihood prior probabilities, the sample frequencies (obtained from specific populations) were used as prior probabilities. What would you predict would happen to the discriminatory power of the procedure? How would the Confusion matrix be affected?

13.7 Suppose it were known that one of two authors wrote a particular article, but a decision as to which was the correct author was difficult. Suppose further that N_1 previous articles of author number (1) were available, and N_2 previous articles of author number (2) were available. Assume that all authors can be distinguished from one another on the basis of the frequencies with which they use certain words in their writings. Describe the procedure you would use to determine which of the two authors in question wrote the article. (See also Mosteller and Wallace, 1964.)

13.8 Use the logistic discrimination approach to develop a classification procedure for the data in Exercise 7.6 and the populations in Exercise 13.1. Classify the vectors in Exercise 13.2 by this procedure. Compare the results with those in Exercises 13.2 and 13.4. [Hint: see Press & Wilson, 1978.]

13.9 Suppose multivariate data were arriving sequentially in time (say, velocity information about an arriving airplane, and we wish to classify the airplane as *friend* or *foe*). How should a classification procedure be modified to account for this steadily increasing amount of data to be classified?

13.10 Explain the advantages of using a Bayesian classification procedure over those of a frequentist procedure.

13.11 Suppose you were attempting to diagnose a patient's disease on the basis of a vector of symptoms that included some variables that were continuous, and some that were discrete. Explain why the logistic regression approach to classification might be the preferred method of analysis.

13.12 Explain how you might use the "faces" approach to classification to classify the patient's disease in Exercise 13.11. [Hint: See Chernoff (1973).]

14

CONTROL IN THE MULTIVARIATE LINEAR MODEL

14.1 INTRODUCTION

In many applications it is often desirable not only to establish relationships among correlated dependent and independent variables (multivariate regression or analysis of variance), but also to attempt to modify or control these relationships so that some criterion function is optimized. This is the multivariate control problem.

Suppose some "system" can be characterized by a set of correlated equations each of which relates an output of the system (dependent variable) to an estimable function of the inputs (independent variables) to the system. Assume the relationship between inputs and outputs is described by a multivariate regression (see Chapter 8). More general relationships (nonlinear models) may be approached by approximating them by linear models, or by using the framework of dynamic programming. Unknown parameters complicate the problem, however. The "system" might be a business, a sector of an economy, a social system such as an educational or health program, a physical or biological unit, or any one of many possible constructs. It is desired to exercise some control over the output variables by appropriate adjustment of some of the input variables. It is often the case that there is some desired or target level of system output (goal or achievement level), and any deviation of the actual system output from the target output may be thought of as an error to be made as small as possible.

There is usually also an element of prediction involved. That is, in many applications, such as in planning of social systems, the problem of interest is how the planner should set variables under his control at the present time, so that the outputs of the system at some future time will be as close as possible to target values.

The independent variables of the system are typically of two types: controllable and uncontrollable. For example, a variable that can be controlled in inventory problems is the amount of a product that a firm should stock. In portfolio analysis, the number of shares of a given company which should be added to the portfolio is a controllable variable. However, the prices of the securities to be added to the portfolio cannot be controlled and are therefore uncontrollable variables. In the *uncontrollable* independent variable category, variables are also of two types: deterministic and stochastic. Thus, time might be an independent uncontrollable variable (deterministic) whose value is just accepted (such as in a system whose input is time in years and whose output is annual value of a budget), while the prices of securities to be added to a portfolio are uncontrollable, but stochastic.

It is often necessary to consider the economics of controlling a system. In particular, there may be an economic penalty associated with changing the controllable variables from their present values. For example, if the system is an educational construct and an important criterion variable is number of students per year educated up to a certain level, there is clearly a cost of increasing the number of existing schools from current levels to higher levels.

In this chapter it is assumed that the system is representable by the multivariate general linear model and that the parameters of the model have already been estimated according to the principles discussed in Chapter 8. The only problem remaining then is that of control. A Bayesian approach is used to develop the predictive distribution for the system output at some time in the future. Optimal control is established by minimizing a loss function with respect to the predictive distribution. The loss function assumed is comprised of two parts: one part that is quadratic in the difference between actual and target system output, and one part that is quadratic in the difference between present and future levels of the controllable variables. The latter part of the loss function is called the cost of control.

The general subject of controlling stochastically described systems is discussed, for example, in Aoki (1967), Bellman (1961), Box and Jenkins (1962), Sworder (1966), and Tou (1965). Work on specific classes of systems and on particular models was carried out, for example, by Bather and Chernoff (1967), Fisher (1962), Florentin (1962), Freund (1956), Kihlstrom (1967), Simon (1956), and Theil (1957). With few exceptions, most early work assumed that the parameters of the relationship were known with certainty. The control problem in a multivariate regression context with unknown parameters was considered by Zellner and Chetty

(1965). Their model is extended in this chapter to include costs of control and to permit some input variables to be uncontrollable.

In the next section optimal levels are established for the control variables for the problem of predicting one period into the future. The last section contains some discussion of the problem of controlling several periods in the future.

14.2 SINGLE PERIOD CONTROL

14.2.1 Model

Let $\mathbf{y}_k : p \times 1$ denote the kth observation of p correlated system outputs (dependent variables), and $\mathbf{x}_k : q \times 1$ denote the kth observation of q system inputs (independent variables). Suppose they are related by the linear model (with the usual linear model assumptions as discussed in Chapter 8)

$$\underset{(p\times 1)}{\mathbf{y}_k} = \underset{(p\times q)(q\times 1)}{\mathbf{B}\mathbf{x}_k} + \underset{(p\times 1)}{\mathbf{u}_k}, \tag{14.2.1}$$

where $k = 1,\ldots,T$. Assume the $\mathbf{x}_k$'s are nonstochastic (the extension of the problem to stochastic $\mathbf{x}_k$ follows directly, as in the regression model), and the errors, $\mathbf{u}_k$, are mutually independent and identically distributed as

$$\mathcal{L}(\mathbf{u}_k) = N(\mathbf{0},\mathbf{\Sigma}), \tag{14.2.2}$$

$k = 1,\ldots,T$. The first future observation is given by

$$\mathbf{y}_{T+1} = \mathbf{B}\mathbf{x}_{T+1} + \mathbf{u}_{T+1}. \tag{14.2.3}$$

For convenience, define

$$\mathbf{z} \equiv \mathbf{y}_{T+1} \qquad \text{and} \qquad \mathbf{w} \equiv \mathbf{x}_{T+1}, \tag{14.2.4}$$

so that

$$\underset{(p\times 1)}{\mathbf{z}} = \underset{(p\times q)(q\times 1)}{\mathbf{B}\mathbf{w}} + \underset{(p\times 1)}{\mathbf{u}_{T+1}}. \tag{14.2.5}$$

Now suppose that the first q_1 components of $\mathbf{w}$ are controllable and the remaining $q_2 \equiv q - q_1$ components, while still independent variables of the regression, are uncontrollable. Partition $\mathbf{w}$ and $\mathbf{B}$ to correspond to this classification of components

$$\underset{(q\times 1)}{\mathbf{w}} = \begin{pmatrix}\mathbf{w}_1\\ \mathbf{w}_2\end{pmatrix}, \qquad \underset{(p\times q)}{\mathbf{B}} = (\mathbf{B}_1 \quad \mathbf{B}_2),$$

where $\mathbf{w}_1 : q_1 \times 1$, $\mathbf{w}_2 : q_2 \times 1$, $\mathbf{B}_1 : p \times q_1$, $\mathbf{B}_2 : p \times q_2$. Then (14.2.5) becomes

$$\mathbf{z} = \mathbf{B}_1\mathbf{w}_1 + \mathbf{B}_2\mathbf{w}_2 + \mathbf{u}_{T+1}. \tag{14.2.6}$$

The system output at time $(T+1)$ will be controlled by minimizing the loss function with respect to the predictive distribution for $\mathbf{z}$. The latter is developed in the next section.

14.2.2 Predictive Distribution

If prior information about $(\mathbf{B},\mathbf{\Sigma})$ is available it should be used to formulate a prior density function, which might then be used to derive the predictive distribution for $\mathbf{z}$ given the previous observations. In the absence of such information it is not unreasonable to use a diffuse prior. Accordingly, let $\mathbf{\Lambda} = \mathbf{\Sigma}^{-1}$, and assume the joint prior density of the parameters is given by [see (3.6.5)]

$$p(\mathbf{B},\mathbf{\Lambda}) \propto \frac{1}{|\mathbf{\Lambda}|^{(p+1)/2}}. \tag{14.2.7}$$

The likelihood function given the observations is

$$p(\mathbf{y}_1,\ldots,\mathbf{y}_T|\mathbf{B},\mathbf{\Lambda}) \propto |\mathbf{\Lambda}|^{T/2} \exp\left\{-\frac{1}{2}\sum_{k=1}^{T}(\mathbf{y}_k - \mathbf{B}\mathbf{x}_k)'\mathbf{\Lambda}(\mathbf{y}_k - \mathbf{B}\mathbf{x}_k)\right\}. \tag{14.2.8}$$

Combining (14.2.7) and (14.2.8) gives, for the posterior density of the parameters,

$$p(\mathbf{B},\mathbf{\Lambda}|\mathbf{y}_1,\ldots,\mathbf{y}_T) \propto |\mathbf{\Lambda}|^{(T-p-1)/2} \exp\{-\tfrac{1}{2}\operatorname{tr}\mathbf{\Lambda}\mathbf{V}\}, \tag{14.2.9}$$

where

$$\mathbf{V} \equiv \sum_{k=1}^{T}(\mathbf{y}_k - \mathbf{B}\mathbf{x}_k)(\mathbf{y}_k - \mathbf{B}\mathbf{x}_k)'.$$

Since

$$\mathcal{L}(\mathbf{z}|\mathbf{B},\mathbf{\Lambda}) = N(\mathbf{B}\mathbf{w},\mathbf{\Lambda}^{-1}),$$

so that

$$p(\mathbf{z}|\mathbf{B},\mathbf{\Lambda}) \propto |\mathbf{\Lambda}|^{1/2} \exp\{-\tfrac{1}{2}(\mathbf{z} - \mathbf{B}\mathbf{w})'\mathbf{\Lambda}(\mathbf{z} - \mathbf{B}\mathbf{w})\}, \tag{14.2.10}$$

and since, by definition, the predictive distribution for $(\mathbf{z}|\mathbf{y}_1,\ldots,\mathbf{y}_T)$ is given by

$$p(\mathbf{z}|\mathbf{y}_1,\ldots,\mathbf{y}_T) = \iint p(\mathbf{z}|\mathbf{B},\mathbf{\Lambda})p(\mathbf{B},\mathbf{\Lambda}|\mathbf{y}_1,\ldots,\mathbf{y}_T)d\mathbf{B}d\mathbf{\Lambda},$$

substitution of (14.2.9) and (14.2.10) gives for the joint predictive density

$$p(\mathbf{z}|\mathbf{y}_1,\ldots,\mathbf{y}_T) \propto \iint|\mathbf{\Lambda}|^{(T-p)/2} \exp\{-\tfrac{1}{2}\operatorname{tr}\mathbf{\Lambda}\mathbf{A}\}d\mathbf{B}d\mathbf{\Lambda}, \tag{14.2.11}$$

where

$$\mathbf{A} \equiv \mathbf{V} + (\mathbf{z} - \mathbf{B}\mathbf{w})(\mathbf{z} - \mathbf{B}\mathbf{w})'.$$

Using the normalizing constant of the Wishart density [see Equation (5.1.1)] to integrate over $\mathbf{A}$ gives

$$p(\mathbf{z}|\mathbf{y}_1, \ldots, \mathbf{y}_T) \propto \int \frac{d\mathbf{B}}{|\mathbf{A}|^{(T+1)/2}},$$

or

$$p(\mathbf{z}|\mathbf{y}_1, \ldots, \mathbf{y}_T) \propto \int \frac{d\mathbf{B}}{|\mathbf{V} + (\mathbf{Bw} - \mathbf{z})(\mathbf{Bw} - \mathbf{z})'|^{(T+1)/2}}. \qquad (14.2.12)$$

Now note that since $\mathbf{V}$ may be written as

$$\mathbf{V} = \sum_{k=1}^{T} [(\mathbf{y}_k - \hat{\mathbf{B}}\mathbf{x}_k) + (\hat{\mathbf{B}}\mathbf{x}_k - \mathbf{B}\mathbf{x}_k)][(\mathbf{y}_k - \hat{\mathbf{B}}\mathbf{x}_k) + (\hat{\mathbf{B}}\mathbf{x}_k - \mathbf{B}\mathbf{x}_k)]',$$

where $\hat{\mathbf{B}}$ denotes the least squares estimator

$$\hat{\mathbf{B}} = \left(\sum_{k=1}^{T} \mathbf{y}_k\mathbf{x}_k'\right)\left(\sum_{k=1}^{T} \mathbf{x}_k\mathbf{x}_k'\right)^{-1},$$

by expanding the binomial in $\mathbf{V}$, it is seen that

$$\mathbf{V} = \mathbf{S} + (\mathbf{B} - \hat{\mathbf{B}})\mathbf{D}(\mathbf{B} - \hat{\mathbf{B}})',$$

where $\mathbf{D} \equiv \Sigma_1^T \mathbf{x}_k\mathbf{x}_k'$, $\mathbf{S} \equiv \Sigma_1^T(\mathbf{y}_k - \hat{\mathbf{B}}\mathbf{x}_k)(\mathbf{y}_k - \hat{\mathbf{B}}\mathbf{x}_k)'$, and the cross-product terms vanish since they correspond to the normal equations. Thus, the matrix in the denominator of (14.2.12) becomes

$$\mathbf{V} + (\mathbf{Bw} - \mathbf{z})(\mathbf{Bw} - \mathbf{z})' = \mathbf{S} + (\mathbf{B} - \hat{\mathbf{B}})\mathbf{D}(\mathbf{B} - \hat{\mathbf{B}})' + (\mathbf{Bw} - \mathbf{z})(\mathbf{Bw} - \mathbf{z})',$$

where $\mathbf{S}$, $\hat{\mathbf{B}}$, $\mathbf{D}$, $\mathbf{w}$, $\mathbf{z}$ do not depend upon $\mathbf{B}$.

Completing the square on $\mathbf{B}$ gives

$$\mathbf{V} + (\mathbf{Bw} - \mathbf{z})(\mathbf{Bw} - \mathbf{z})' = (\mathbf{B} - \mathbf{C}_1)\mathbf{C}_2(\mathbf{B} - \mathbf{C}_1)' + \hat{\mathbf{B}}\mathbf{D}\hat{\mathbf{B}}' + \mathbf{zz}' - (\hat{\mathbf{B}}\mathbf{D} + \mathbf{zw}')(\mathbf{D} + \mathbf{ww}')^{-1}(\hat{\mathbf{B}}\mathbf{D} + \mathbf{zw}')',$$

where

$$\mathbf{C}_1 = (\hat{\mathbf{B}}\mathbf{D} + \mathbf{zw}')(\mathbf{D} + \mathbf{ww}')^{-1},$$

and

$$\mathbf{C}_2 = \mathbf{D} + \mathbf{ww}'.$$

Application of (2.5.6) to $(\mathbf{D} + \mathbf{ww}')^{-1}$ and combining terms shows that (14.2.12) becomes

$$p(\mathbf{z}|\mathbf{y}_1, \ldots, \mathbf{y}_T) \propto \int \frac{d\mathbf{B}}{|\mathbf{C}_3 + (\mathbf{B} - \mathbf{C}_1)\mathbf{C}_2(\mathbf{B} - \mathbf{C}_1)'|^{(T+1)/2}},$$

where

$$\mathbf{C}_3 = \mathbf{S} + \frac{(\mathbf{z} - \hat{\mathbf{B}}\mathbf{w})(\mathbf{z} - \hat{\mathbf{B}}\mathbf{w})'}{1 + \mathbf{w}'\mathbf{D}^{-1}\mathbf{w}}.$$

Recall that if $\hat{\mathbf{\Sigma}}$ is an unbiased estimator of $\mathbf{\Sigma}$, $\mathbf{S} = (T - q)\hat{\mathbf{\Sigma}}$.

Now note that the integrand has the form of a matrix **T**-distribution [see Equation (6.2.6)]. Hence, the integration is carried out easily to give the required predictive density

$$p(\mathbf{z}|\mathbf{y}_1,\ldots,\mathbf{y}_T) \propto \frac{1}{|\mathbf{C}_3|^{(T+1-q)/2}},$$

or

$$p(\mathbf{z}|\mathbf{y}_1,\ldots,\mathbf{y}_T) \propto \frac{1}{[\nu + (\mathbf{z} - \hat{\mathbf{B}}\mathbf{w})'\mathbf{H}(\mathbf{z} - \hat{\mathbf{B}}\mathbf{w})]^{(T+1-q)/2}}, \quad (14.2.13)$$

where

$$\nu \equiv T - q - (p - 1), \qquad \mathbf{H} \equiv \frac{\nu\mathbf{S}^{-1}}{1 + \mathbf{w}'\mathbf{D}^{-1}\mathbf{w}}.$$

That is, the predictive distribution of a future observation given the previous observations is multivariate Student **t** and is centered at $\hat{\mathbf{B}}\mathbf{w}$.

[*Remark:* It was shown by Geisser (1965) that the joint predictive distribution of several future vector observations is matrix **T**. The result in (14.2.13) is a special case.]

14.2.3 Optimal Control

Suppose the system output at time $(T + 1)$, $\mathbf{y}_{T+1}$, is to be kept as close as possible to some target output, $\mathbf{a}: p \times 1$. Let $\mathbf{G}: p \times p$ denote a known, preassigned matrix of constants. Assume the loss attributable to system output deviations from the target levels is representable as

$$(\mathbf{y}_{T+1} - \mathbf{a})'\mathbf{G}(\mathbf{y}_{T+1} - \mathbf{a}),$$

so that there are some losses that are proportional to the squared "error," or deviation from the goal, and some proportional to the sum of error products of different kinds. The elements of **G** should be adjusted to constraints of particular situations, assuming a quadratic form is a close approximation to reality.

Assume there is a cost of controlling the system given by

$$(\mathbf{x}_{T+1} - \mathbf{x}_T)'\mathbf{J}(\mathbf{x}_{T+1} - \mathbf{x}_T),$$

where **J** is a known, preassigned $q \times q$ matrix of constants. Thus, the cost is a quadratic function of the amount by which present values must be changed to effect control.

The total loss is obtained by adding the above two partial losses to give

$$L(\mathbf{z}) = (\mathbf{z} - \mathbf{a})'\mathbf{G}(\mathbf{z} - \mathbf{a}) + (\mathbf{w} - \mathbf{x}_T)'\mathbf{J}(\mathbf{w} - \mathbf{x}_T). \quad (14.2.14)$$

The *future risk*, ρ, is the loss averaged with respect to the predictive distribution. Thus,

$$\rho = E(L(\mathbf{z})|\mathbf{y}_1,\ldots,\mathbf{y}_T) = \int L(\mathbf{z})p(\mathbf{z}|\mathbf{y}_1,\ldots,\mathbf{y}_T)d\mathbf{z}. \quad (14.2.15)$$

Substituting (14.2.13) and (14.2.14) into (14.2.15), applying (3.2.11), and using the moment expressions following (6.2.4) gives

$$\rho = \left(\frac{\text{tr } \mathbf{SG}}{\nu - 2}\right)(1 + \mathbf{w}'\mathbf{D}^{-1}\mathbf{w}) + (\hat{\mathbf{B}}\mathbf{w} - \mathbf{a})'\mathbf{G}(\hat{\mathbf{B}}\mathbf{w} - \mathbf{a}) + (\mathbf{w} - \mathbf{x}_T)'\mathbf{J}(\mathbf{w} - \mathbf{x}_T). \quad (14.2.16)$$

Now partition into controllable and uncontrollable elements and associated matrices.

$$\mathbf{w} = \begin{pmatrix} \mathbf{w}_1 \\ \mathbf{w}_2 \end{pmatrix}, \quad \mathbf{x}_T = \begin{pmatrix} \mathbf{x}_T(1) \\ \mathbf{x}_T(2) \end{pmatrix}, \quad \hat{\mathbf{B}} = (\hat{\mathbf{B}}_1, \hat{\mathbf{B}}_2),$$

$$\mathbf{D}^{-1} = \begin{pmatrix} \mathbf{D}^{11} & \mathbf{D}^{12} \\ \mathbf{D}^{21} & \mathbf{D}^{22} \end{pmatrix}, \quad \mathbf{J} = \begin{pmatrix} \mathbf{J}_{11} & \mathbf{J}_{12} \\ \mathbf{J}_{21} & \mathbf{J}_{22} \end{pmatrix},$$

where

$$\mathbf{w}_1\colon q_1 \times 1,\ \mathbf{x}_T(1)\ q_1\colon \times 1,\ \hat{\mathbf{B}}_1\colon p \times q_1,\ \mathbf{D}^{11}\colon q_1 \times q_1,$$

and $\mathbf{J}_{11}$: $q_1 \times q_1$. The risk may be minimized in the usual way. Differentiating ρ with respect to $\mathbf{w}_1$, the controllable subvector of the future independent variables, setting the result equal to zero, and solving, gives the optimal control vector

$$\underset{(q_1 \times 1)}{\mathbf{w}_1^*} = \left[\hat{\mathbf{B}}_1'\mathbf{G}\hat{\mathbf{B}}_1 + \left(\frac{\text{tr } \mathbf{SG}}{\nu - 2}\right)\mathbf{D}^{11} + \mathbf{J}_{11}\right]^{-1}\left[\hat{\mathbf{B}}_1'\mathbf{G}(\mathbf{a} - \hat{\mathbf{B}}_2\mathbf{w}_2) - \left(\frac{\text{tr } \mathbf{SG}}{\nu - 2}\right)\mathbf{D}^{12}\mathbf{w}_2 + \mathbf{J}_{11}\mathbf{x}_T(1) + \mathbf{J}_{12}(\mathbf{x}_T(2) - \mathbf{w}_2)\right]. \quad (14.2.17)$$

[*Remark:* Note that $\mathbf{w}_1^*$ depends upon $\mathbf{w}_2$, the future values of the independent but uncontrollable variables. Typically, $\mathbf{w}_2$ is assigned by estimating it from previous data (forecasting).]

Example (14.2.1): Take G and J to be identity matrices. $T = 20$, $p = 2$, $q = 3$. Thus, there are 20 observations, with two system outputs and three system inputs (one might correspond to a constant term). It follows that

$$\nu = T - q - (p - 1) = 16,$$

so that if $q_1 = 1$ (one controllable variable), and $q_2 = 2$, from (14.2.17) the optimal control vector is a scalar given by

$$w_1^* = \frac{\hat{\mathbf{B}}_1'(\mathbf{a} - \hat{\mathbf{B}}_2\mathbf{w}_2) - \left(\frac{\text{tr } \mathbf{S}}{14}\right)\mathbf{D}^{12}\mathbf{w}_2 + \mathbf{x}_T(1)}{1 + \hat{\mathbf{B}}_1'\hat{\mathbf{B}}_1 + \left(\frac{\text{tr } \mathbf{S}}{14}\right)\mathbf{D}^{11}}.$$

Suppose the regression data are given by

$$\hat{\mathbf{B}} = \begin{pmatrix} 2 & 0 & -1 \\ 1 & 3 & 2 \end{pmatrix},$$

so that

$$\hat{\mathbf{B}}_1 = \begin{pmatrix} 2 \\ 1 \end{pmatrix}, \quad \hat{\mathbf{B}}_2 = \begin{pmatrix} 0 & -1 \\ 3 & 2 \end{pmatrix};$$

$$\mathbf{D}^{-1} = \begin{pmatrix} 7 & 0 & 0 \\ 0 & 3 & 2 \\ 0 & 2 & 4 \end{pmatrix},$$

so that $\mathbf{D}^{11} = 7$, and $\mathbf{D}^{12} = (0, 0)$; and

$$\mathbf{S} = \begin{pmatrix} 2 & -1 \\ -1 & 3 \end{pmatrix},$$

so that $\text{tr}(\mathbf{S}) = 5$. Suppose that desired target values are given by $\mathbf{a}' = (1,1)$, and that the most recent control setting was $\mathbf{x}_T(1) = 1$. The only remaining variable to be assigned is $\mathbf{w}_2 : 2 \times 1$, where $\mathbf{w}_2 \equiv \mathbf{x}_{T+1}(2)$. That is, $\mathbf{w}_2$ is the vector whose two components are the values of the uncontrollable independent variables at time $(T + 1)$. Suitable values may be obtained by standard forecasting (prediction) methods and it is assumed, for illustrative purposes, that $\mathbf{w}_2' = (2,1)$. Then substitution gives

$$w_1^* = -.24.$$

That is, the vector of independent variables for the time period $(T + 1)$ should be

$$\mathbf{x}_{T+1} = \begin{pmatrix} -.24 \\ 2.00 \\ 1.00 \end{pmatrix}.$$

14.3 MULTIPLE PERIOD CONTROL

If it is desired to control the system at a point in time several periods in the future, the problem is more complex, since it is necessary to take account of the facts that

(1) after controlling the system for one period ahead, a new optimal control setting will be developed for the second period, taking into account the error of the system after one period; and then third period control must take second period control into account, and so on; multiple period control must anticipate each of these optimal controls taking place sequentially;

(2) after one period the observation obtained may be added to the

existing bank of data to improve the estimates of the regression coefficients in the underlying relationship (adaptive control).

The first point may be handled by standard dynamic programming methods (using backward induction). The second point involves updating of the regression estimates.

The joint predictive distribution of observations several periods in the future is matrix **T** (see Section 14.2.2). Using that predictive distribution for the $(T + K)$th period given the previous observations, $k = 1,\ldots,T, T + 1,\ldots,T + K - 1$, an optimal control vector can be found as in the last section. Then the process must be repeated for the $(T + K - 1)$ period given all earlier periods, and so on. It is not clear that a closed form solution to this problem exists for $K > 1$. In fact, the two period prediction case ($K = 2$) considered for just the univariate regression problem by Aoki (1967), and by Zellner (1966), does not appear to have a closed form solution. Hence, it is not likely to have one in the multivariate case.

Study of the multivariate control problem with unknown parameters has been very limited to date in spite of the many and diverse applications to which results in this subject could be put. However, because of the widespread interest that has been expressed in this problem by research workers in many disciplines, and because of the availability of high speed computers, operationally usable solutions to many multivariate control problems should become available in the near future.

REFERENCES

Aoki (1967).
Bather and Chernoff (1967).
Bellman (1961).
Box and Jenkins (1962).
Fisher (1962).
Florentin (1962).
Freund (1956).
Geisser (1965a).
Kihlstrom (1967).
Simon (1956).
Sworder (1966).
Theil (1957).
Tou (1965).
Zellner (1966).
Zellner and Chetty (1965).

EXERCISES

14.1 A governmental program to assist educationally disadvantaged teenagers in improving their position in the job market provides remedial education at specially provided centers. Achievement tests are given to trainees on entering the program, and once per month thereafter.

Let y_k denote the change in score after k weeks in the program, and let x_k denote the number of hours of instruction given to the trainee during his k weeks in the program. Adopt the linear regression model

$$y_k = \beta_0 + \beta_1 x_k + u_k,$$
$$E(u_k) = 0, \qquad \text{var}\,(u_k) = \sigma^2, \qquad k = 1, \ldots, T,$$

where u_k denotes the disturbance term, and assume the disturbances are mutually uncorrelated and Normally distributed.

Suppose that on the basis of observing many individuals, the least squares estimates of the coefficient parameters are found to be $\hat{\beta}_0 = -.43$, $\hat{\beta}_1 = +.43$, and an unbiased estimate of σ^2 is $\hat{\sigma}^2 = 25$.

It is desired to effect a change in score of ten points after 27 weeks of training. Suppose the effective cost for the program attributable to low achievement scores is given by $3(y_{T+1} - 10)^2$ and the cost attributable to control changes is given by $2(x_{T+1} - 130)^2$.

Find the optimal number of hours of instruction to provide by time $T + 1 = 27$, assuming a diffuse prior on the parameters as given in (14.2.7). [*Note:* Assume

$$\bar{x} = \frac{1}{26}\sum_1^{26} x_k \equiv 60, \quad \text{and} \quad \sum_1^{26} x_k^2 = 100{,}000.\Big]$$

14.2 Determine the relationship for the optimal control vector, corresponding to (14.2.17), for the case in which the cost of control is zero.

14.3 In Exercise 14.1, compute the future risk corresponding to the optimal control setting.

14.4 How would the optimal control vector setting corresponding to (14.2.17) change if a generalized natural conjugate prior were used instead of the diffuse prior used in (14.2.7)? [*Hint:* See Section 8.6.2.]

14.5 Evaluate the optional control vector for a problem in which $T = 20, p = 3$,

$$\mathbf{G} = \begin{pmatrix} 3 & 1 & 0 \\ 1 & 1 & 2 \\ 0 & 2 & 1 \end{pmatrix}, \qquad \mathbf{J} = \begin{pmatrix} 1 & 0 & 0 \\ 0 & 0 & 0 \\ 0 & 0 & 0 \end{pmatrix},$$

$q = 2, q_1 = 1$,

$$\mathbf{B} = \begin{pmatrix} 3 & 4 \\ 2 & 1 \\ 1 & 2 \end{pmatrix}, \qquad \mathbf{D} = \begin{pmatrix} 3 & 1 \\ 1 & 4 \end{pmatrix}, \qquad \mathbf{S} = \begin{pmatrix} 4 & 1 & 1 \\ 1 & 2 & 1 \\ 1 & 1 & 4 \end{pmatrix},$$

$w_2 = 2$.

15

STRUCTURING MULTIVARIATE POPULATIONS (MULTIDIMENSIONAL SCALING AND CLUSTERING)

15.1 INTRODUCTION

Other chapters of this book, devoted to models, have been concerned with methods of analyzing collections of data vectors drawn from multivariate populations. The methods of analysis differed according to the objectives of the investigator. However, it was always assumed that the underlying multivariate populations were well established (and often, they were assumed to be Normal populations), even if the distribution parameters were unknown. But multivariate data do not always arrive in this highly structured form. It is therefore fitting in this last chapter to consider some methods which have been proposed for dealing with relatively unstructured multivariate data. Some of the sources of difficulty with structuring multivariate populations are the following.

(1) Data may be qualitative (expressed in nominal scales). For example, one characteristic of object j may be x_j, where x_j has three possible values: red, green, and blue; or x_j may denote one of several categories of race, or automobile model types.
(2) Some data, though quantitative, may be continuous, while others may be discrete (say, yes or no answers to a survey question).
(3) Data may arrive in ordinal form. For example, although the values of x_j may not be available, it may be known that $x_1 \leq x_2 \leq x_3 \leq \cdots \leq x_N$. That is, the data may arrive in ordered or ranked form, and sometimes the ordering relation may only denote preferences

rather than mathematical inequality of real numbers. Moreover, the data may only be partially ordered. For example, perhaps it is known that $x_1 \leq x_2$, and $x_3 \leq x_4$, and some other similar relationships are known, but all x_j's are not ranked relative to one another.

(4) Data groups may be *similar* to one another (such as disease types, or groups of symptoms of diseases), but it is not clear how to structure multivariate populations whose *distances* are directly related to the similarities. Once distances can be established in a Euclidean coordinate system, conventional multivariate analyses can proceed. This is the problem of multidimensional scaling (see Sect. 15.2).

(5) Data characteristics or attributes may not arrive in grouped, or vector form, so that it may be unclear as to how to define the underlying populations. That is, which characteristics should be grouped together so that vectors from a population with those characteristics are most similar or alike, whereas vectors from other populations are dissimilar? This is the problem of *clustering analysis* (see Section 15.3).

(6) In some problems, several or all of the above sources of difficulty will be present in the same problem.

The structuring problems discussed above are difficult to solve. Qualitative data is often treated nonparametrically (no distributional assumptions) using large sample results. This approach is not discussed further here although there is a growing literature in nonparametric multivariate analysis (see, for instance, Puri and Sen, 1971).

Distributions with some continuous and some discrete variables are called *mixed*. Since statistical inference in many models often relies heavily on multivariate Normality, underlying mixed distributions generate inference problems of great magnitude.

Ordinal data are often handled by replacing the data points by their ranks. The usual multivariate models are then applied to the ranked data and asymptotic theory is invoked (if sufficiently large samples are available). That is, the summary statistics, such as the sample mean vector (even when based upon the ranks), are generally Normally distributed in large samples (under a wide variety of assumptions).

Attention is focused in this chapter on the multivariate structuring problems of multidimensional scaling and clustering. Even in these two areas, the results to date are still embryonic. However, in spite of the fact that mathematical rigor is still minimal and statistical inference is almost nonexistent, the methodology that has developed has now reached the point where highly predictable and useful results are being obtained. For these reasons, it may not be too sanguine to expect that firm statistical foundations for these and related models will be established in the near future. In the meantime, it will often prove fruitful to apply existing computer oriented techniques to real world problems, while recognizing the exploratory nature of the results, and relying only upon results from large samples for inferences.

15.2 MULTIDIMENSIONAL SCALING

15.2.1 Background

Multidimensional scaling refers to a collection of techniques which have been developed for inferring multidimensional structure from one-dimensional similarity or proximity data. The similarity data could be correlations or rankings among objects. If a specific functional form is used to relate the similarities to Euclidean distances in some multidimensional coordinate system, the procedures are referred to as *metric* multidimensional scaling. If no functional form is specified, but merely a monotonicity is required between similarities and distances in the sense that

$$d_{ij} < d_{kl},$$

whenever

$$s_{ij} > s_{kl},$$

(d_{ij} = distance between objects i and j, s_{ij} = similarity), the procedures are referred to as *non-metric* multidimensional scaling.

Suppose, for example, in a marketing study of American cars, Buick, Chevrolet, . . . , types of cars are paired off according to their degree of similarity. Suppose it is felt that Chevrolet and Ford are most similar, Chevrolet and Lincoln Continental are least similar, and all other pairs are ranked somewhere in between. Let δ_{ij} denote the rank of the dissimilarity for the pair of model types i and j. Now the δ_{ij} may be ordered. But nothing has been said of the characteristics that have really dictated this ordering of similarities. That is, there is in fact an underlying attribute space of some dimension in which each automobile model type may be represented as a point. In this space each model type may be studied relative to the other model types in terms of its distance from the other models. That is, there is a space and associated metric (distance in this space) which underlies the ordinal data presented as perceived rankings of similarity of automobile model types. In like fashion, cultures may be ranked according to their degrees of similarity (or dissimilarity), securities may be so ranked, and so may a large variety of constructs such as people, economies, job types, educational systems, and perceptions of reality. Multidimensional scaling methods (both metric and non-metric) may be used to convert these similarities into numerical distances. Moreover, the methods do not require the linearity assumptions of regression analysis (Chapter 8) and Factor Analysis (Chapter 10).

Historically, metric scaling techniques were summarized and extended by Torgerson, 1958, and by Coombs, 1958. Shepard (1962a and 1962b) set forth the guiding principles of non-metric multidimensional scaling by requiring the distances in the multidimensional space to follow the same ordering as proximities, i.e., similarities.[1] The objectives of the model were thus established and Shepard indicated a method of solution. A more formal solution was given by Kruskal (1964a and 1964b) who also provided a computer algorithm for obtaining a numerical result. His methodology is summarized below. All of the procedures mentioned are similar in the sense that they provide a structure for the collection of perceptions of a single individual. An extension of these techniques called "individual differences scaling" provides a method for grouping the perceptions of many individuals to generate a composite picture of relative similarities. This extension is described in Carroll and Chang (1970). A useful summary of the methodology and applications of non-metric multidimensional scaling is given in Shepard, Romney, and Nerlove, 1972, Volumes I and II.

15.2.2 Methodology

Suppose there are n objects (products, people, and so on) to be scaled. Define the symmetric input data matrix

$$\underset{(n \times n)}{\boldsymbol{\Delta}} = \begin{pmatrix} \delta_{11} & \cdots & \delta_{1n} \\ \cdot & & \cdot \\ \cdot & & \cdot \\ \cdot & & \cdot \\ \delta_{n1} & \cdots & \delta_{nn} \end{pmatrix}, \qquad (15.2.1)$$

where for $i \neq j$, δ_{ij} denotes the observed dissimilarity between objects i and j (δ_{ii} will never be used and therefore is not defined). That is, in any problem, $\boldsymbol{\Delta}$ is taken as known. Moreover, suppose δ_{ij} is not subject to error or stochastic variation but is specified exactly. (If $\boldsymbol{\Delta}$ is stochastic, the problem is more complicated.)

The δ_{ij} may be obtained by asking a subject to rank order all pairs of the n objects according to his perception of their degree of dissimilarity [there are $M \equiv n(n-1)/2$ such pairs]. In such a case, the δ_{ij}'s could be taken as the M ranks of the ordered pairs. Alternatively, the subject could be asked to place each of the M pairs on a scale (ranging from say,

[1] A technique that does not depend upon the monotonicity property was proposed by Coleman (1969). It is one of several metric approaches which have been proposed.

0 to 10, or any other convenient range) according to his perception of their degree of dissimilarity, with 10 denoting most dissimilar, and 0 denoting least dissimilar.[2] Then the δ_{ij}'s would correspond to the values on the scale.

It has also been suggested that N subjects might each be used to rank the M pairs of objects, and then an average rank could be used for the δ_{ij}'s. This approach was studied with a simulation technique by Klahr (1967) who found that the multidimensional scaling configuration which resulted by averaging the rankings of the subjects was highly correlated with the configurations that resulted from the rankings of the subjects individually.

A third alternative is to take $\boldsymbol{\Delta}$ as the rank correlation matrix of the n objects, based upon N subjects. The off-diagonal elements are then measures of degree of closeness of the objects.

Given the data matrix $\boldsymbol{\Delta}$, the off-diagonal elements must now be ordered (assuming there are no ties) as

$$\delta_{i_1j_1} < \delta_{i_2j_2} \cdots < \delta_{i_Mj_M}, \tag{15.2.2}$$

where $(i_1,j_1),\ldots,(i_M,j_M)$ denote all pairs of unequal subscripts of δ_{ij}. Ties will be treated separately later.

Let $\mathbf{x}_j$: $p \times 1$ denote the location vector of the jth object in a space of dimension p, $j = 1,\ldots,n$. Of course at the start of any scaling problem the $\mathbf{x}_j$'s are unknown. To facilitate understanding of the approach being used, assume a particular set of p-vectors $(\mathbf{x}_1,\ldots,\mathbf{x}_n)$ is selected arbitrarily, just to see how well this set fits the observed data (the dissimilarities, δ_{ij}). The dimension p is also generally unknown at the start of a problem. For the time being, however, assume it too has been selected at some fixed value, arbitrarily.

Define the distance[3] between the ith and jth objects as

$$d_{ij} = [(\mathbf{x}_i - \mathbf{x}_j)'(\mathbf{x}_i - \mathbf{x}_j)]^{1/2}. \tag{15.2.3}$$

A scattergram of the points (d_{ij},δ_{ij}) may now be plotted to study how well the trial distances match the dissimilarities. Figure 15.2.1 illustrates a hypothetical case in which monotonicity is not preserved. Thus, as dissimilarities between objects increase, their distances do not necessarily increase; that is, they sometimes decrease. A case in which there is a perfect match (monotonicity is preserved) is shown in Figure 15.2.2.

Goodness of Fit

Let $\mathbf{x}_1,\ldots,\mathbf{x}_n$ denote an arbitrary and known configuration of n points in p-space; the associated distances are defined in (15.2.3). Define the

[2] There is a growing belief that a different set of ranks is obtained when people are asked to rank according to similarity rather than dissimilarity. For this reason, the scaling could be carried out using both approaches, if the subjects will cooperate.

[3] Non-Euclidean distances such as L_p-norms might be used as alternatives (see Kruskal, 1964a, p. 22).

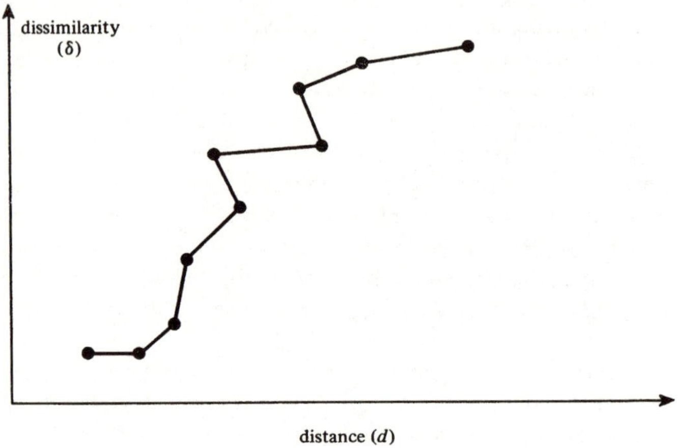

Figure 15.2.1. Scattergram without Monotonicity

$M \times 1$ vector of distances between pairs of points

$$\underset{(M\times 1)}{\mathbf{D}} = (d_{i_1j_1}, \ldots, d_{i_Mj_M})'.$$

Let $\hat{d}_{ij}$ denote a "fitted" value of d_{ij}. Kruskal (1964a) defined the goodness-of-fit measure

$$\text{stress} = S = \left[\frac{(\mathbf{D} - \hat{\mathbf{D}})'(\mathbf{D} - \hat{\mathbf{D}})}{\mathbf{D}'\mathbf{D}}\right]^{1/2}, \tag{15.2.4}$$

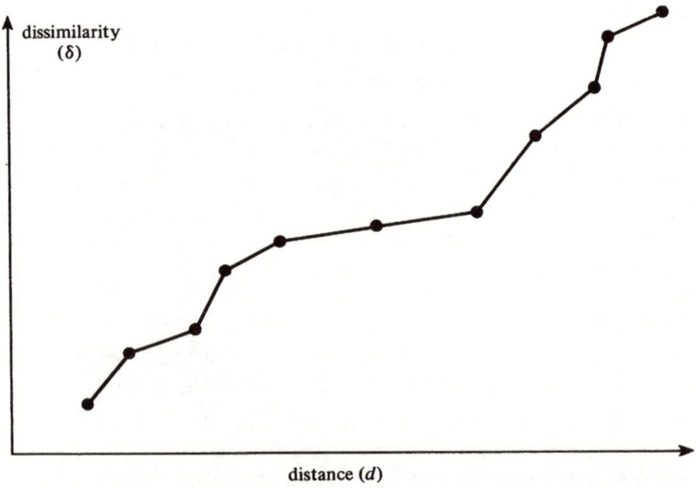

Figure 15.2.2. Scattergram Exhibiting Monotonicity

where $\hat{\mathbf{D}} \equiv (\hat{d}_{ij})$. Thus, the *stress* of the configuration of points $(\mathbf{x}_1,\ldots,\mathbf{x}_n)$ is defined as

$$S(\mathbf{x}_1,\ldots,\mathbf{x}_n) = \min_{\hat{\mathbf{D}}} \left[\frac{(\mathbf{D} - \hat{\mathbf{D}})'(\mathbf{D} - \hat{\mathbf{D}})}{\mathbf{D}'\mathbf{D}}\right]^{1/2}, \tag{15.2.5}$$

subject to the monotonicity constraint

$$\hat{d}_{i_1j_1} \leq \hat{d}_{i_2j_2} \leq \cdots \leq \hat{d}_{i_Mj_M}. \tag{15.2.6}$$

That is, $S(\mathbf{x}_1,\ldots,\mathbf{x}_n)$ is a measure[4] of how good a fit $(\mathbf{x}_1,\ldots,\mathbf{x}_n)$ is to the given input data, $\mathbf{\Delta}$. Since the $\hat{d}_{ij}$'s are constrained to an ordering consistent with that of the δ_{ij}'s, $\hat{\mathbf{D}}$ provides a set of numbers close to $\mathbf{D}$ and correctly ordered. Hence, $S(\mathbf{x}_1,\ldots,\mathbf{x}_n)$ reflects the degree to which monotonicity is preserved in the configuration corresponding to $\mathbf{D}$; the smaller the stress, the closer the match in terms of ordering. As long as $\hat{d}_{ij}$ is within $2d_{ij}$ for each i and j, the stress will be less than unity.

It is easy to check that $S(\mathbf{x}_1,\ldots,\mathbf{x}_n)$ is invariant with respect to changes of location, scale, rotation, and reflection. For example, if for some $a \neq 0$, and b, $\mathbf{x}_j = a\mathbf{y}_j + b$, $j = 1,\ldots,n$, the d_{ij}'s are each divided by a, $\mathbf{D}$ is divided by a, $\hat{\mathbf{D}}$ is divided by a, and $S(\mathbf{x}_1,\ldots,\mathbf{x}_n)$ does not change. As a result, the final configuration of points can be oriented around *any* center and around *any* set of orthogonal axes.

Define the *p-stress* as

$$S_p = \min_{\left\{\substack{\text{all } (\mathbf{x}_1,\ldots,\mathbf{x}_n) \text{ in} \\ p\text{-dimensions}}\right\}} S(\mathbf{x}_1,\ldots,\mathbf{x}_n).$$

Thus, to obtain the p-stress, the points $(\mathbf{x}_1,\ldots,\mathbf{x}_n)$ in p-space are varied and the stress is computed for each variation until the resulting stress is as small as possible. The configuration of points corresponding to the p-stress is the best fitting configuration in p-space.

The best fitting configuration in p-space may be found by means of a computer program developed by Kruskal (1964b). Its characteristics are summarized in Appendix A. In essence, the algorithm involves choosing an arbitrary initial configuration $(\mathbf{x}_1,\ldots,\mathbf{x}_n)$ in p-space and then reducing the associated stress by moving to a new improved configuration. The location of the new configuration is found by the method of steepest descent (gradients).

Choice of Dimension

A p-stress of zero may always be found by choosing $p = n - 1$. That is, there will always be a configuration of n points in $(n - 1)$-space whose interpoint distances can be arbitrarily ordered. In some problems, theo-

[4] For discussion of other goodness-of-fit measures, see Kruskal and Carroll (1969).

retical considerations dictate an appropriate value of p. For example, there may be five fundamental factors which seem to characterize the essence of an object (in terms of its similarity to other objects), and so a solution would be sought in a space of dimension $p = 5$. However, it is more often the case that the investigator is unable to preselect p with any degree of confidence. In such cases, the scaling problem should be solved for each of several values of p. It is generally the case that the p-stress is a decreasing function of p (it will always be theoretically, but sometimes imperfect computer solutions will cause deviations). If the p-stress drops below some tolerably low value such as 5 percent for some value of p, there is no need to go to a space of higher dimension since the fit is already quite good. When the stress is positive but less than 5 percent there will be several points that do not conform to the monotonicity criterion. However, most points will conform and a final scattergram plot will generally provide a visual check of the goodness of fit of the result.

Once a multidimensional scaling analysis has been carried out, other models may be applied to the metric data. For example, the resulting points are often immediately rotated to principal component axes (Chapter 9).

15.2.3 Special Considerations

Ties

When two of the dissimilarities are equal, it does not really matter which of the associated distances is greater. Hence, the computer program does not constrain the associated fitted values, $\hat{d}_{ij}$, $\hat{d}_{kl}$.

Uniqueness of Solution

As stated earlier, n points whose interpoint distances are ordered can always be arranged in $(n - 1)$-space to conform to the ordering. However, if the dimension of the space is held fixed (say the dimension is p) while the number of points increases, it becomes increasingly difficult to arrange the points in p-space. That is, the number of feasible solutions steadily decreases as n increases, with the solution becoming more and more constrained. In fact, Shepard (1966) showed that if the number of points in a p-dimensional convex region increases to infinity and completely fills in the region, the given rank order of the interpoint distances uniquely determines the set of points in p-space to within a similarity transformation [see Equation (2.9.5)] and the distances themselves to within a multiplication by an arbitrary scale factor. For example, if $p = 2$, the ordered interpoint distances of 10 or more points will essentially fix the points in the plane.

Dissimilarities with Error

It is of considerable interest to know how errors in measuring the dissimilarities affect the final multidimensional scaling solution. Young (1968) carried out a simulation to study this question and performed a multivariate analysis of variance on the results. He found that "when the ratio of the number of similarities, M, to the number of coordinates, p, is sufficiently large, nonmetric multidimensional scaling is able to recover the metric information of a data structure even when the structure contains error." This conclusion is based upon errors in the dissimilarities following a Normal distribution with variance of up to 25 percent of the variance of the underlying variance in the configuration.

Interpretation of the Results

The final result of a multidimensional scaling analysis is a configuration of n points in p-space. Although an axis system could be located anywhere and could have any arbitrary orientation relative to the configuration of points, "natural" axes are sometimes suggested by theory, or by the way the points fall. In the latter case, the axes may not have any "natural" physical interpretation, however. This is the same problem that was encountered in principal components analysis and factor analysis (Chapters 9 and 10, respectively).

Example (15.2.1): Greenberg (1969) reported the results of a study of packaged soap and detergents carried out for Proctor and Gamble. In this study, 8 detergent brands were compared by 206 women in a metropolitan area of upstate New York during July 1967. Each woman was asked to rate each of the 28 pairs of brands on a 3 point scale ranging from high to low similarity. Thus, in the notation of Section 15.2.2, $n = 8$, $M = n(n - 1)/2 = 28$. The $\mathbf{\Delta}$ matrix of perceived similarities is 8×8. With minor modifications, its entries were assigned by aggregating the scores of all women for each pair of brands.

The two-dimensional solution shown in Figure 15.2.3 was generated by multidimensional scaling followed by a judiciously selected rotation of axes. The particular rotation used was selected on the basis of complementary data collected from the women during the experiment. Note that the location of the origin was selected on the basis of external factors, and was not determined by the analysis.

Since the investigators felt they might be on somewhat shaky ground in their conclusions, they decided to replicate the experiment 3 months later in a metropolitan area of the Southwest. The same 8 brands were used with the same procedure for scoring similarity of brands. However,

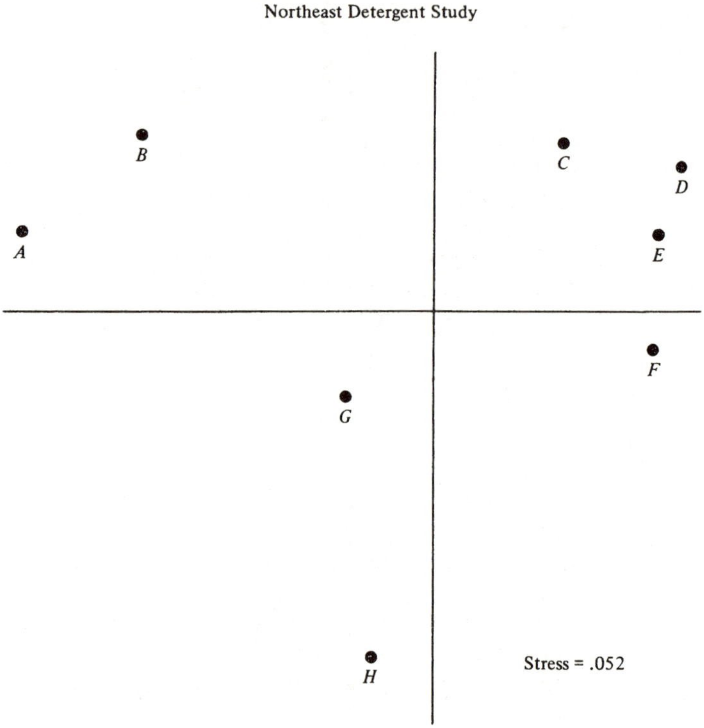

Figure 15.2.3. Multidimensional Scaling

this time, 354 women were used. The two-dimensional solution obtained is shown in Figure 15.2.4. It is seen by comparison with Figure 15.2.3 that the basic configuration was barely altered by changing the geographic location of the experiment. At this time it is difficult to establish whether the differences are statistically significant. Since perceptions are constantly shifting over time and across regions, the relative stability of the resulting map was interpreted as an indication of inherent similarities and differences among brands. As a result, it was felt that maps of this type could be helpful in the positioning of new brands within the existing structure, that they could help in revealing the need to reposition existing brands, and that they could help to determine the direction of change along each axis toward a desired brand image. (Of course the inference question of how much variation in the final configuration could be expected merely because there were only 8 points is still open.)

The multidimensional scaling model was applied to the study of 10 common stocks by Green and Maheshwari (1968). For an application of the model to the study of proximities of perceptions of an art product such as a drawing or a painting, see Klein (1968).

Southwest Detergent Study

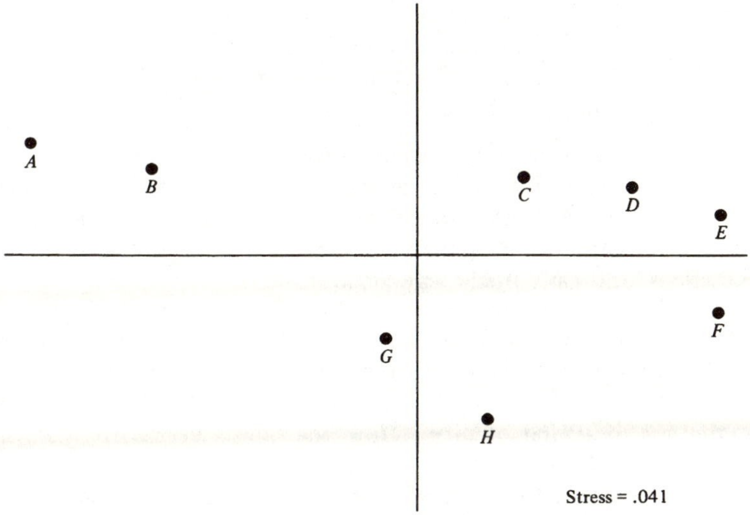

Figure 15.2.4. Multidimensional Scaling

15.3 CLUSTERING ANALYSIS

15.3.1. Introduction

The term "clustering" includes a collection of techniques that are used to group multidimensional entities according to various criteria of their degrees of homogeneity and heterogeneity. For example, the grouping might be on the basis of people's scores on p different types of tests. People with high verbal ability might tend to cluster into one group while people with high abstract reasoning ability or with high artistic ability might tend to cluster into other groups. How close should test scores be before people are grouped into the same cluster is the question of degree of homogeneity desired, and how many clusters there ought to be is the question of the degree of heterogeneity desired. This type of grouping is called *clustering by subject.*

Another type of grouping is called *clustering by attribute.* For example, for security j, let the $(N \times 1)$ vectors of price changes during each of N months be denoted by $\mathbf{x}_j$, $j = 1, \ldots, p$. Then attribute clustering corresponds to grouping securities according to their price change behavior; that is, finding subgroups of the p vectors in N-space which are similar. Generally, N is considerably larger than p.

The subject clustering problem can be formulated more succinctly as follows. Suppose $\mathbf{x}_1, \ldots, \mathbf{x}_N$ are independent observation vectors in p-space (observations on N people). How should these N people be partitioned into K groups ($K < N$) so that in some sense, individuals within a group are homogeneous, and individuals in different groups are heterogeneous?

There are many questions which must be addressed in such a loosely structured problem. They include the following.

(1) How many clusters are inherent in the data?

(2) Since attributes may be measured in different units, should attributes be standardized before they are clustered? This problem is no different than the analogous ones in the principal components, factor analysis, and canonical correlation models.

(3) How should cluster boundaries be established?

(4) How large should the errors be before they are considered intolerable? There will be one type of error made by not grouping similar quantities into the same cluster, and another type of error made by grouping dissimilar quantities into the same cluster. Alternatively, there might be a loss function to minimize.

(5) Should correlated attributes be handled differently than uncorrelated attributes?

(6) Should all possible pairs of N points in p-space (or p points in N-space) be scrutinized for similarities? This can be a problem of tremendous magnitude. Both enumerative and nonenumerative techniques will be discussed.

Problems in many of the sciences have been approached by means of clustering and the reader should refer to Table 15.3.1 for sources of applications to his specific area of interest (the table is by no means comprehensive). The approaches used have been quite diverse in terms of how they handle each of the problems raised in the last paragraph (see Ball (1965), and Wallace (1968) for surveys of many of these methods). See also Anderberg, 1973; Chernoff, 1973; Everitt, 1978; Gnanedesikan, 1977, Chap. 4; Hartigan, 1975. The problem of clustering is an old one. Computers have only recently permitted a wide variety of attempts at solution. The major techniques of clustering analysis are described below.

As pointed out in Section 15.1, in other models discussed in this book the variables possess the minimal structure of belonging to particular populations, a priori. As a consequence, it is usually possible to assume particular distributions for the populations and then make associated inferences. By contrast, in clustering problems, the principal concern is how to establish appropriate populations. Therefore, clustering analysis logically precedes the application of most other multivariate models when the data arrive in this unstructured form. However, the configuration obtained from a multidimensional scaling analysis could reasonably be

Table 15.3.1

Discipline	*Reference to use of clustering*
Accounting	Jensen (1968)
Anthropology	Rao (1952)
Biology	Edwards, Cavalli-Sforza (1965)
	Rogers and Fleming (1964)
	Rogers and Tanimoto (1960)
	Rohlf and Sokal (1962)
	Sokal and Sneath (1963)
Economics	Fisher (1958, 1965)
Education	Fortier and Solomon (1966)
Finance	King (1967)
Linguistics	Dyen (1962)
	Pierce (1962)
Marketing	Green, Frank, and Robinson (1967)
Medicine	Overall (1963)
Personnel Management	Thomas (1952)
Political Science	Harris (1964)
	Rice (1928)
Psychology	Bass (1957)
	McQuitty (1957)
	Tryon (1939)
	Young (1939)
Sociology	Alexander (1963)
	Coleman and MacRae (1960)
	Coleman (1969)

subjected to a clustering analysis if the scaling solution is based upon a sufficiently large number of points so that the resulting configuration is stable or robust.

In the next section enumerative methods of clustering are considered. Section 15.3.3 presents several nonenumerative approaches that have been proposed; and Section 15.3.4 discusses methods for addressing the general questions raised above.

15.3.2 Enumerative Clustering

Some approaches to clustering have involved an enumeration of all ways of partitioning the N points in p-space (or p points in N-space) into clusters, and then selecting the "best" configuration. Typically, there is a four step procedure for enumerative clustering.

(1) First, a measure of distance (called a metric) such as Euclidean distance is computed for all pairs of N points. Call this distance ρ.

(2) A list is prepared of all possible partitions of the N points into clusters.

(3) For each partition a measure of homogeneity of clusters is calculated.
(4) The partition with the "best" homogeneity pattern is selected.

The procedure outlined above seems, on the surface, very reasonable. Unfortunately, however, because of step (2), this approach is usually not feasible with currently available computers. In general, the number of ways to distribute N distinguishable objects into K nonempty subsets is called a Stirling Number of the Second Kind (see, for instance, Abromovitz and Stegun, 1964). For most situations of interest, such numbers are extremely large. For example, King (1967) points out that the number of ways of partitioning 25 distinguishable objects into 5 nonempty subsets is between 2 and 3 quadrillion! Because of this drawback of sheer size of the enumeration problem, a variety of alternatives have been proposed, and several of them are discussed in the next section.

15.3.3 Nonenumerative Clustering

This section discusses several approaches to the systematic clustering of data that do not involve examining every possible partitioning of the data. The approaches in greatest use are based upon the criterion of minimization of the variance (or possibly the distances) within clusters. Moreover, clustering procedures are often "hierarchical" in that points are added sequentially, and at each stage, the existing partition is, in some sense, optimal. These terms are discussed below.

Minimum Variance/Distance Clustering

The minimum variance/distance clustering approach has been considered from many points of view. Fisher (1958) assigned weights w_i to N one-dimensional objects having associated numerical measures a_i, $i = 1,\ldots,N$, and sought a procedure for clustering the N objects into K groups (K is preassigned) so that the total within group variance

$$\sigma^2 = \sum_{i=1}^{N} w_i(a_i - \bar{a}_i)^2$$

is minimized; $\bar{a}_i$ denotes the weighted arithmetic mean of those a's assigned to the cluster to which element i is assigned. In this form such problems arise often in stratified sampling (see, for instance, Hansen, Hurwitz, and Madow, 1953, p. 235).

Cox (1957) was also concerned with this approach but assumed the variables were Normally distributed. Moreover, his vectors were both one- and two-dimensional.

Friedman and Rubin (1967) suggested that subject clustering might often be carried out by minimizing the determinant of the within-groups sum-of-squares matrix, which is also a minimum variance approach.

In a marketing problem, Green, Frank, and Robinson (1967) used a very simple multidimensional procedure based upon Euclidean distance between points. The procedure is not optimal, but in a suboptimal way, sums of squares of distances are minimized so that there is an implied minimum variance interpretation. Their procedure is:

(1) Standardize the N-vectors of attributes relative to the sample so that each element has sample mean zero and sample variance one.
(2) Compute the Euclidean distances between all pairs of points in p-space.
(3) Choose the pair that are closest and find their center of gravity (centroid).
(4) Examine the distances of all remaining points to this centroid. Merge the closest point by computing a new centroid.
(5) Continue merging the closest points into the cluster until either (a) a prespecified number of points is in the cluster, or (b) the next point to be added has a distance to the centroid which exceeds some prespecified "closeness" level.
(6) Of the unclustered points, take the closest two, find their centroid, and repeat steps (4) and (5). The process is continued until all points are clustered.

Computer programs appropriate for this type of clustering are discussed in Bonner (1964), Edwards and Cavalli-Sforza (1965), and Tryon (1958).

Example (15.3.1) Clustering by Subject: Suppose various measured characteristics of cities and metropolitan areas are provided. How should the cities be clustered into homogeneous groups so that for test marketing purposes, a city within a cluster can be selected with assurance that results will project to other cities within the same cluster (where the ultimate marketing will be carried out)?

This was the problem which motivated the work of Green, Frank, and Robinson. They had available 14 characteristics of standard metropolitan areas including population, number of households, retail sales, effective buying income, median age, proportions of males, nonwhites, and unemployed, median education level, number of retail outlets and wholesale outlets, newspaper circulation, television coverage, and transit ad circulation figures; and they secured these data from 88 cities. Thus, $N = 88$ and $p = 14$.

By applying the clustering procedures outlined above and prefixing the number of clusters at 18, they concluded that the cities cluster into the 18 groups delineated in Table 15.3.2.

Table 15.3.2 Clustering of "Cities" for Test Marketing (Results Taken from Green, Frank, and Robinson, 1967)

1	Columbus Dayton Fort Worth Oklahoma City Omaha	7	Phoenix Sacramento San Bernadino San Jose Tucson	13	Allentown Jersey City Louisville Providence York
2	Binghamton Davenport Harrisburg Peoria Worcester	8	Gary Jacksonville Knoxville Nashville San Antonio	14	Cincinnati Miami Milwaukee Patterson Seattle
3	Albany Canton Springfield Toledo Youngstown	9	Atlanta Dallas Houston Indianapolis Kansas City	15	Charleston Ft. Lauderdale Norfolk San Diego Tacoma
4	Bridgeport Hartford New Haven Rochester Syracuse	10	Birmingham Chattanooga Memphis Mobile Shreveport	16	Lancaster Minneapolis New Orleans Richmond Tampa
5	Grand Rapids Orlando Tulsa Wichita Wilmington	11	Baltimore Buffalo Cleveland Newark Pittsburgh	17	Boston Detroit Philadelphia San Francisco
6	Bakersfield Beaumont El Paso Flint Fresno	12	Albuquerque Charlotte Denver Portland Salt Lake City	18	St. Louis Washington

In another minimum variance/distance clustering approach, MacQueen defined a "coarsening parameter," C, which he used to reduce the number of clusters (the partition is made more coarse), and a "refinement parameter," $\tilde{R}$, which he used to increase the number of clusters (the partition is made finer). These terms are taken from topology. The following procedure was suggested.

(1) Preassign C, $\tilde{R}$, and K (the number of clusters).
(2) The first K points (in p-space) available are used as the centroids of K clusters.
(3) Each subsequent sample point is assigned to the nearest centroid on the basis of Euclidean distance; after each assignment, a new centroid is computed for that cluster.
(4) After each new sample point is assigned (and for the initial means as well) the pair of centroids which are closest together is found. This distance is called ρ^*.
(5) If $\rho^* < C$, the two clusters are merged and a new centroid is found (appropriately weighted).
(6) Steps (4) and (5) are repeated until all centroids are separated by at least C. Thus, the existing K may be reduced and the partition thereby made coarser.
(7) For each new sample point considered, its distance to the nearest centroid, Δ, is computed. If $\Delta \leq \tilde{R}$, the point is merged into the cluster. However, if $\Delta > \tilde{R}$, the point is left by itself to form the seed for a new cluster. Thus, the existing number of clusters may increase and the partition may thereby be refined.
(8) All points in the sample are reclustered according to nearness to the final centroids.

MacQueen showed that for the partition that results from this approach, the within cluster variance tends to be small. There is no guarantee it will be small in all cases although it is actually the optimal partition in many cases.

The computer running time for MacQueen's program varies with C, $\tilde{R}$, p, and N. For example, for $N = 200$, $p = 20$, and C and $\tilde{R}$ set so that K remained in the vicinity of about 20, MacQueen reported a running time of about one minute. Program characteristics are given in Appendix A.

Hierarchical Clustering

Hierarchical clustering is another approach for grouping data without completely enumerating all of the possible partitions. If this approach is applied to the problem of clustering by subject, the N points are first considered as N clusters of one point each. Then the number of clusters is reduced to $(N - 1)$ by merging two of the points, according to some criterion of optimality, or using some objective function. The process is continued until as many groups as are desired remain. However, since at each step, the partitioning is optimal, the clustering is nested or hierarchical. This is the approach often used in taxonomic applications where objects are classified into successively more detailed subgroups.

The "tree-like" patterns of classification which result are sometimes called dendrograms.

The hierarchical clustering technique was outlined in general form by Ward (1963) who also provided flow diagrams and computer suggestions. His motivation was an application in personnel training research.

Example (15.3.2) Clustering by Attribute: Two hierarchical clustering procedures were proposed for the attribute clustering problem by King (1967). In his first approach the $p \times p$ sample correlation matrix corresponding to the N data vectors in p-space is formed. Then the pair of attribute variables having maximum correlation coefficient are merged to form a two-point cluster, so that at that point, the p single-point clusters have been reduced to $p - 1$ clusters. A new correlation matrix is then computed and the hierarchical procedure is continued.

In a second approach, King suggests that in some situations, the Λ statistic for testing the significance of canonical correlations (see Chapter 11) might be more appropriate as a clustering criterion than the pairwise correlation coefficients. Let $\mathbf{R}$ denote the correlation matrix of the p attributes, based upon the p-dimensional data vectors $\mathbf{x}_1,\ldots,\mathbf{x}_N$. Let $\mathbf{x}_j' = (\mathbf{y}_j', \mathbf{z}_j')$, $j = 1,\ldots,N$ denote a partitioning of the variables into two groups, and let $\mathbf{R}$ be partitioned accordingly, so that

$$\mathbf{R} = \begin{pmatrix} \mathbf{R}_{11} & \mathbf{R}_{12} \\ \mathbf{R}_{21} & \mathbf{R}_{22} \end{pmatrix}. \tag{15.3.1}$$

It was shown in Chapter 11 that the Λ statistic [the $2/N$ power of the likelihood ratio statistic for testing that $\mathbf{R}_{12} = \mathbf{0}$, see (11.3.3), and (11.3.4)] is given by

$$\Lambda = \frac{|\mathbf{R}|}{|\mathbf{R}_{11}|\,|\mathbf{R}_{22}|} = \prod_{j=1}^{p_1} (1 - \lambda_j^2), \tag{15.3.2}$$

where $\mathbf{R}_{11}: p_1 \times p_1$, $\mathbf{R}_{22}: p_2 \times p_2$, $p_1 + p_2 = p$, and $+\sqrt{\lambda_j^2}$ is the jth canonical correlation for the attribute vectors. At the first stage of hierarchical clustering y_j' would be a scalar so that $R_{11} = 1$. That is, start off with p single point clusters and at the first stage compute Λ for all two-group configurations of $\mathbf{R}$ involving 1 variable and $(p - 1)$ variables in each group. The minimum Λ configuration corresponds to the greatest measure of association between that variable and the group of $(p - 1)$ variables remaining. Therefore, the variable singled out is merged into one of the remaining variables and the hierarchical process of clustering is continued. In later stages R_{11} need not be one since a cluster group might contain several points. The two procedures are analogous at this step and although there is no rigorous proof available, King presents some evidence, based upon two complete enumeration examples, that the Λ criterion approach may be superior to the simple correlation coefficient criterion approach. Examples used involved the clustering of financial securities, and the clustering of traits of school children.

In some hierarchical situations it may be desirable to use a clustering criterion which involves minimizing the sample variance within clusters. Such an objective function is encompassed within the model outlined by Ward (1963), so that in such a case the clustering would be hierarchical with a minimum variance objective function.

15.3.4 Special Problems in Clustering

In Section 15.3.1 certain general questions about clustering were raised. This section proposes some answers to those questions.

In many problems, the context almost dictates the appropriate number of clusters to use. In some cases, K can safely be assumed known a priori. In other problems, fixing the degrees of homogeneity within clusters and heterogeneity among clusters will approximately fix K.

It should be recognized by now that standardizing variables (relative to the sample) before applying a particular multivariate model usually has the drawback of providing answers to a problem different from the one originally posed. However, if the analyst can operate within the framework of the standardized form of the problem, he can take advantage of a potentially tremendous gain in simplicity of computations, a gain in simplicity of interpretation of results, and a gain in meaningfulness of mathematical manipulations of the data.

Several methods for establishing cluster "boundaries" were given. They all involved choosing an objective function for clustering; minimum within cluster variance was one, and maximum correlation coefficient and minimum Λ statistic were others.

The question of how large clustering errors should be may be answered quantitatively by the careful assignment of a loss or objective function.

Because some of the attributes under consideration are often correlated, and therefore, they are really measuring the same underlying factors, the data are sometimes subjected to a preclustering principal components analysis (see, for instance, Green, Frank, and Robinson, 1967). The clustering is then carried out on the principal component variables that are retained. This approach leaves something to be desired since part of the information in the data is lost by deleting some of the principal components, and when samples are small, this effect is magnified. That is, the problem of clustering the original variables is first changed by standardizing the variables; then it is modified by transforming to a few principal components; and finally, there are errors of clustering. So how valid are the results? It is suggested that for many problems, if standardization is followed directly by clustering, without any intermediate analysis, results obtained will be closer to the results originally sought.

15.3.5 Recommended Approach to Clustering

The problem of clustering N data vectors in p-space (or p vectors in N-space) into K clusters can be handled in many ways and the most appropriate technique to use depends upon the problem. If the investigator can reasonably justify doing so, it is recommended that the data vectors be standardized first. Then a decision must be made about how to structure the problem: Is a given number of points in each cluster desired, or is a given number of clusters desired. For subject clustering, if the maximum diameter of a cluster and the distance between clusters can be preset, the procedures of Green, Frank, and Robinson or of MacQueen can be used (computer routines are available). If attribute clustering is desired and if there is uncertainty about the number of clusters to use, the investigator might use the Λ criterion of King. Since the procedure is hierarchical, the output can be studied, and if it is unsatisfactory more clusters can be added sequentially.

If possible, a graphical display technique such as "Promenade" (see Hall *et al.*, 1968) should be used (for problems in no more than three dimensions at a time) with a cathode ray tube display to study the proximity of the elements of each cluster pictorially. In such systems, the investigator can interact with the computer directly and immediately see the results of a change in the clustering pattern displayed on his scope. In this way, human judgment can be used to correct the output of poorly selected clustering criteria and techniques.

Finally, just as it was recommended in the factor analysis model (see *Remark* at the end of Section 10.5), if large samples are available, the sample should be split into several subsamples to see if clusters with the same characteristics emerge when the clustering technique is applied to each subsample separately. This procedure will increase credibility of the results.

REFERENCES

Abromovitz and Stegun (1964).
Alexander (1963).
Anderberg (1973).
Ball (1965).
Bass (1957).
Bonner (1964).
Carroll and Chang (1970).
Chernoff (1973).
Coleman (1969).
Coleman and MacRae (1960).
Coombs (1958).
Cox (1957).
Dyen (1962).
Edwards and Cavalli-Sforza (1965).
Everitt (1978).

Fisher (1958, 1965).
Fortier and Solomon (1966).
Green and Maheshwan (1968).
Green, Frank, and Robinson (1967).
Greenberg (1969).
Hall, Ball, Wolf, and Eusebio (1968).
Hansen, Hurwitz, and Madow, Vol. 1 (1953).
Harris (1964).
Hartigan (1975).
Jensen (1968).
King (1967).
Klahr (1967).
Klein (1968).
Kruskal (1964a, 1964b).
Kruskal and Carroll (1969).
MacQueen (1965).
McQuitty (1957).
Overall (1963).
Pierce (1962).
Rao (1952).
Rice (1928).
Rogers and Fleming (1964).
Rogers and Tanimoto (1960).
Rohlf and Sokal (1962).
Sen and Puri (1969).
Shepard (1962a, 1962b, 1966).
Shepard, Romney, and Nerlove (1972).
Sokal and Sneath (1963).
Torgerson (1958).
Thomas (1952).
Tryon (1939, 1958).
Wallace (1968).
Ward (1963).
Young (1939, 1968).

EXERCISES

15.1 Show that in the multidimensional scaling model if the fitted distances $\hat{d}_{ij}$ are all within twice the preassigned distances d_{ij}, respectively, the stress is a numerical quantity confined to the unit interval.

15.2 In the multidimensional scaling model suppose the M dissimilarities are not known but there are N judges available to provide N sets of ordered dissimilarities. Two procedures are suggested: (a) the ranks assigned by the N judges could be averaged, and (b) the rank correlation matrix for the M dissimilarities could be determined. Compare the merits of carrying out the scaling using method (a) as opposed to method (b).

15.3 Explain how the final configuration of a multidimensional scaling problem would be altered if all dissimilarities were given as points on an interval scale and were then standardized (mean zero and variance unity) relative to the sample before the scaling algorithm was applied.

15.4 Is it possible to use minimum variance clustering and hierarchical clustering in the same problem? Explain your answer.

15.5 Suppose all data vectors in a clustering problem are known a priori to follow multivariate Normal distributions with unknown parameters and equal covariance matrices. Devise a clustering technique for this problem.

15.6 Distinguish between clustering by subject and clustering by attribute.

15.7 In what situations is hierarchical clustering likely to be most useful?

15.8 Give an example of a *metric* multidimensional scaling problem. [Hint: see, e.g., Torgerson, 1958.]

15.9 What is the *conjoint measurement* model? When should it be used? [Hint: see, e.g., Shepard, et al., 1972.]

15.10 Suppose we wish to cluster data from multivariate normal-populations with identical covariance matrices. How might the Mahalanobis distances be used?

appendix A

COMPUTER PROGRAMS FOR MULTIVARIATE MODELS

The compilation of computer program listings presented in this appendix is not intended to be exhaustive. In most instances several alternative program listings are presented for possible use with a particular multivariate model. It was believed at the time of preparation of this appendix that the programs selected were the best available. Many other program listings which, in some sense, may have been superior, were either overlooked, or were not included because of some particular bias of the author. Moreover, the development of software appropriate for use in multivariate analysis has been so rapid in recent years, it will be surprising if this appendix is not greatly out of date in a very short time subsequent to publication. However, it is hoped that use of the program listings included will serve to provide users with an adequate, although possibly not optimal, library of routines that will take most of the pain out of applying the models and techniques of multivariate analysis.

CONTENTS

Descriptions of Multivariate Programs for:

A.1 PROGRAMS FOR MULTIVARIATE ANALYSIS OF COVARIANCE AND VARIANCE

1. **Name of Program:** BC ANV—Generalized Complete Factorial Analysis of Variance

   Code Number of Program and a Location Where It Might Be Found: SDA 3337, Share Libraries

   Developer of Program and Affiliation: L. R. Grosenbaugh
University of California at Berkeley
Berkeley, California

   Machine for Which the Program Was Written: 7094

   Machine Language: Fortran IV

   Limitations: up to 8192 observations
up to 12 factors

   Form of Output: Sums of squares
Degrees of freedom
Bartlett's chi-square test of homogeneity within subclasses

   Reference: Grosenbaugh

   General Description and Remarks: This program handles all completely balanced factorial combinations. Split plots, randomized blocks, and Graeco-Latin squares are handled by manipulation of the input or the output.

2. **Name of Program:** BYU ANOVAR—Generalized Analysis of Variance/Covariance Processor

   Code Number of Program and a Location Where It Might Be Found: SDA 3365, Share Libraries

   Developer of Program and Affiliation: G. Brian Bone
Brigham Young University
Computer Research Center

   Machine for Which the Program Was Written: 7040

   Machine Language: Fortran IV

   Limitations: up to $8(n_v)$ variates
up to $8(nc)$ covariates

up to 5-way factorial analysis
up to 12 effects in hierarchical classification
$\prod_{i=1}^{N} L_i(n_v + nc) \leq 8000$ where L_i is the number of levels of effect i

Form of Output: Sums of squares
Degrees of freedom
Mean squares
Level means F statistics

Reference: Bone

General Description and Remarks: This program performs an exact analysis for any model in the balanced complete design. It also performs an exact analysis for a nested classification with unequal cell frequencies and an approximate analysis on cell means for any other case which would reduce to the balanced complete case when analysis is performed on cell means. This routine will perform multivariate analysis of variance and covariance computations.

3. **Name of Program:** Mesa 97(NYBMUL) (Mesa 98 extends Mesa 97 for use on 360 machines)

Code Number of Program and a Location Where It Might Be Found: UCSL 705
University of Chicago Computation Center
Chicago, Illinois

Developer of Program and Affiliation: Jeremy D. Finn
State University of New York at Buffalo;
see also, D. Bock, Department of Education,
University of Chicago, Chicago, Illinois

Machine for Which the Program Was Written: 7094

Machine Language: Fortran IV

Limitations: up to about 100 variables and about 4000 cells
if more storage is needed parts of the program may be removed

Form of Output: Principal Components of error correlation matrix
"Latent roots"
F-ratios
Degrees of freedom
Sums of squares

Reference: Finn

General Description and Remarks: The program handles any crossed and/or nested design with up to 10 factors. Discriminant analysis, canonical correlations, and regression are additional features of the program. This program will perform multivariate analysis of variance and covariance computations.

4. **Name of Program:** Multivariate Analysis of Variance with Principal Components of All Dispersion (Variance-Covariance) Matrices

Code Number of Program and a
Location Where It Might Be Found: UCSM 339, University of Chicago
Computation Center
Chicago, Illinois

Developer of Program and Affiliation: Cooley and Lohnes (see the Bibliography)

Machine for Which the Program Was Written: 7094

Machine Language: Fortran II

Limitations: up to 99 populations
up to 50 dimensions

Form of Output: Raw sums of squares and cross products matrix
Covariance matrix
Correlation matrix
Principal components of each within group matrix
Determinants and latent roots

Reference: Herzog

General Description and Remarks: This program computes a test of the within variance matrices for equality and a test of equality for the vector of means for one-way designs. The program then executes a principal components analysis of the within group matrices.

5. **Name of Program:** Multivariate Statistical Programs

Code Number of Program and a Location Where It Might Be Found: MSP
Biometric Laboratory
University of Miami
Coral Gables, Florida

Developer of Program and Affiliation: Dean J. Clyde, Eliot M. Cramer, Richard J. Sherin
University of Miami
Coral Gables, Florida

Machine for Which the Program Was Written: 7040

Machine Language: Fortran IV

Limitations: up to 40 variables including covariates
up to 8 factors
up to 20 levels each

Form of Output: Reduced model matrix
Correlations of effects
Error correlations of variables
Estimates of effects tested against the error term
Analysis of regression for covariates
Likelihood ratio test

Reference: Clyde, Cramer, and Sherin

General Description and Remarks: This routine provides an exact solution in either orthogonal or nonorthogonal cases and can handle single or multiple degrees of freedom contrasts in the main effects or interactions. The system also has factor analysis, discriminant analysis, and canonical correlations. Both analysis of variance and covariance can be carried out.

6. **Name of Program:** NO ANAV—General Purpose Analysis of Variance Program

Code Number of Program and a Location Where It Might Be Found: SDA 3027 01
Share Libraries

Developer of Program and Affiliation: R. S. Gardner
U. S. Naval Ordnance Test Station
China Lake, California

Machine for Which the Program Was Written: 7090

Machine Language: Fortran IV

Limitations: up to 8 factors and 2000 observations
sum of main effects and interaction terms ≤ 1500
degrees of freedom for main effects and interactions ≤ 665

Form of Output: Sums of squares
Degrees of freedom
Mean squares
Lists of main effects and interactions

Reference: Gardner

General Description and Remarks: This program can handle any design. It requires NO LSQ4 (SDA 3440) as a companion package.

7. **Name of Program**: A Generalized Univariate and Multivariate Analysis of Variance, Covariance, and Regression Program (MULTIVARIANCE)

Code Number of Program and a Location Where It Might Be Found: MULTIVARIANCE
International Educational Services
P.O. Box A3650
Chicago, Illinois 60690

Developer of Program and Affiliation: Jeremy D. Finn
Department of Educational Psychology
State University of New York At Buffalo
Buffalo, New York 14222

Machine for Which the Program Was Written: IBM 360/370; CDC 6400

Machine Language: FORTRAN IV

Reference: MULTIVARIANCE: Univariate and Multivariate Analysis of Variance, Covariance, Regression and Repeated Measures
USER'S GUIDE; Version VI, Release 1, March 1978

General Description and Remarks: Multivariance is a computer program which will perform a variety of linear multivariate analyses. Multivariance will perform univariate and multivariate linear estimation, and tests of hypotheses for any crossed and/or nested design, with or without concomitant variables. The number of observations in the subclasses may be equal, proportional, or disproportionate. The latter includes the extreme case of unequal group size involving null subclasses, such as might arise if the application is to incomplete experimental designs. The program performs an exact least-squares analysis by the method described by Bock (1963). Sample problem lists include:

Problem 1: Univariate and Multivariate Repression Analysis

Problem 2: One-group Multivariate Regression, Canonical Correlation

Problem 3: One-way Univariate Analysis of Variance and Covariance

Problem 4: Two-way Multivariate Analysis of Variance, Discriminant Analysis

Problem 5: Two-way Fixed Effects Multivariate Analysis of Variance with Null Subclasses

Problem 6: Four-way Fixed Effects Bivariate Analysis of Variance—Unequal Subclass Sizes

Problem 7: Nested Design Mixed Model

Problem 8A: Latin Square Three-way Anova. First Run to Detect Compounding in the Incomplete Anova Model

Problem 8B: Latin Square Three-way Anova. Second Run to Complete Analysis

Problem 9: One-way Four-group Repeated Measures Analysis Pre and Post Tests

Problem 10: Repeated Measures Analysis of Longitudinal Data

For more information about this program see the reference.

8. **Name of Program:** NESTED; Analysis of Variance and Covariance for Data from an Experiment with a Nested (Hierarchical) Structure

Code Number of Program and A Location Where It Might Be Found: NESTED: SAS
SAS Institute Inc.
P.O. Box 10066
Raleigh, N.C. 27605

Developer of Program and Affiliation: James H. Goodnight
SAS Institute Inc.

Machine for Which The Program Was Written: IBM 360/370 (and plug-compatible machine such as Amdahl, Itel, CDC Omega, Magnuson, Ryad, etc.) under OS or OS/DS

Machine Language: Mixture of Assembler, FORTRAN IV and PL/I

Limitations: The data set must first be sorted by classification variables defining the effects (no other explicit limitations).

Form of Output: For each effect in the model, NESTED prints:

1. the coefficients of the variance components making up the expected mean square.

For each dependent variable, NESTED prints an ANOVA table containing

2. sources of variation;
3. degrees of freedom;
4. sums of squares;
5. mean squares;
6. estimates of variance components;
7. percentages;
8. overall mean;
9. the standard deviations
10. the coefficient of variation of the response variable, based on the MSE; for each pair of dependent variables: analysis of covariance table including;
11. the degrees of freedom;
12. the sum of products;
13. the mean product;
14. the estimate of covariance component;
15. the covariance component correlation;
16. the mean square correlation.

Reference: SAS User's Guide (1979 edition)
SAS Institute
Statistical Analysis System

General Description and Remarks: This program performs analysis of variance and covariance for data from a nested design. Each effect is assumed to be a random effect. Although both GLM and VARCOMP in SAS provide similar analysis, NESTED is more efficient, especially for designs involving large numbers of levels and observations.

9. **Name of Program:** Multivariate Analysis of Variance and Covariance

Code Number of Program and a Location Where It Might Be Found: (see GLM; A.10; program no. 8).

10. **Name of Program**: ANOVA

Code Number of Program and A Location Where It Might Be Found: ANOVA; SAS
SAS Institute Inc.
(see also A.2, no. 8)

Developer of Program and Affiliation: James H. Goodnight
SAS Institute Inc.

Machine for whcih the Program was Written: (see A.1, no. 8)

Machine Language: (see A.1, no. 8)

Limitations: Applicable for balanced data only (no other explicit limitations)

Form of Output: The degrees of freedom; the sum of quares; F values for testing the hypothesis that the group means for the effect are equal; the probability value associated with the F value; for each H matrix, the characteristic roots and vector of $E^{-1}H$ (For H and E matrix see the SAS User's Guide); the Hotelling-Lawley Trace; Pillai's Trace; Wilk's Criterion, Roy's maximum root criterion and more.

Reference: SAS User's Guide
(see also A.1, no. 8)

General Description and Remarks: The ANOVA procedure performs analysis of variance for *balanced data* from a wide variety of experimental designs (balanced data only) with the exception of Latin square designs, certain balanced incomplete block designs, completely nested designs, and designs whose cell frequencies are proportional to each other and in addition are proportional to the background population. Note that multivariate analysis is performed by the use of the MANOVA statements.

11. **Name of Program**: Analysis of Multivariate Change Score

Code Number of Program and a Location Where It Might Be Found: 30; BMDP3D. The BMD and BMDP manuals can be ordered from:

University of California Press
2223 Fulton Street
Berkeley, CA 94720
(415) 642-4243

Developer of Program and Affiliation: Sandra Fu, Jerry Douglas, Lanaii Kline and James W. Frome
Department of Biomathematics
University of California
Los Angeles, CA 90024

Machine for Which The Program Was Written: CDC 6000/CYBER; Honeywell; Univac 1100 series; Univac 70/90 series; PDP-11; HP-3000; Riad-20; ICL system 4 and 2900 series; Hitachi Hitac series; Fujitsu Facom series; Xerox sigma 7; Telefunken TR 440

Machine Language: FORTRAN IV

Limitations: The maximum number of variable V depends on the number of group G. The following table can be used as a guide:

G	2	3	4	5	10
V without Hotel and CORR	240	160	120	96	48
V with Hotel and CORR	53	42	36	31	21

it is not limited by the number of cases in the data.

Form of Output: For each variable: group mean, standard deviation errors, min., max.; pooled and separated t values; an F value for comparison of variance; 20-interval histogram for each group. Mahalanobis' D^2, Hotelling T^2, F value with degrees of freedom and p value (option).

Reference: 1. BMDP-79, Biomedical Computer Programs
P-series 1979
Health Science Computing Facility
Department of Biomathematics
School of Medicine
UCLA
University of California Press

2. BMDP Users' Digest
BMDP Program Software
Department of Biomathematics
UCLA
Los Angeles, CA 90024

General Description and Remarks: P3D perform one-group t test, with or without assumption of equality of variance. For more than two groups, all possible pairs of grousp are compared. The means of several variables can be simultaneously tested for equality between two groups by Hotelling T^2 and Mahalanobis' D^2. Correlations between the variables in each group can be printed.

12. **Name of Program:** Multivariate Analysis of Variance and Covariance.

Code Number of Program and a Location Where It Might Be Found: (see P7M, A.3, no. 8)

Reference: BMDP-79
Biomedical Computer Program P-series, 1979

General Description and Remarks: Multivariate Analysis of variance and covariance can be performed in BMDP7M using the Stepwise Discriminant Analysis program. When P7M runs with its option, a one-way multivariate analysis is computed. Multiway analysis of variance and covariance can be performed at the cost of running the program several times.

13. **Name of Program:** Multivariate Analysis of Variance and Covariance

Code Number of Program and a Location Where It Might Be Found: BMD12V
Health Science Computing Facility
CHS Bldg., AV-111
University of California
Los Angeles, CA 90024
(see also A.1, no. 11)

Developer of Program and Affiliation: Paul Sampson, a member of the staff of Health Sciences Computing Facility, UCLA.

Machine for Which the Program Was Written: (see A.1, no. 12)

Machine Language: FORTRAN IV

Limitations: a. $n \leq 10$
b. $cp(p+1)/2 + m \leq 15000$
c. $c \leq 100$

where:
n = number of analysis of variance indices
c = number of components in the analysis of variance part of the model
P = total part of variables
m = number of means which must be stored. This may be as large as half the amount of data,

depending on the design, and includes at least the means to be printed.

Form of Output:
1. Covariance matrix for each analysis of variance component (optional)
2. Cell means for each variable
3. For each univariate analysis of covariance:
   a) Regression coefficients under each hypothesis.
   b) Analysis of variance table including source, sum of squares, mean square, degrees of freedom, and *F*-statistics for each analysis of variance component of the model and each covariate.
   c) Adjust cell means.
4. For each multivariate analysis of variance or covariance the following is tabulated for each analysis of variance component and each covariate:
   a) Generalized Variance.
   b) U-statistic and degrees of freedom.
   c) Approximate *F*-statistic and degrees of freedom.

Reference: BMD, Biomedical Computer Program
Health Science Computing Facility
Department of Biomathematics
School of Medicine
UCLA
University of California Press
January 1, 1978; (W. J. Dixon, editor)

General Description and Remarks: This program will perform univariate or multivariate analysis of variance or covariance for any hierarchical design with equal cell sizes. This includes nested and partially crossed, and fully crossed designs. Several analyses may be performed for each problem.

A.2 PROGRAMS FOR CANONICAL CORRELATIONS

1. **Name of Program**: Canonical Correlation Analysis

Code Number of Program and a Location Where It Might Be Found: P6M; BMDP6M

Health Sciences Computing Facility
UCLA
Los Angeles, CA 90024
(see also A.1, no. 11)

Developer of Program and Affiliation: James Frane
Health Sciences Computing Facility
UCLA
Los Angeles, CA 90024

Machine for Which the Program Was Written: (see A.1, no. 11)

Machine Language: FORTRAN IV

Limitations: A maximum of 100 variables can be analyzed in double precision and 150 in single precision.

Form of Output: Univariate summary statistics and data listing for the first five cases. Correlation matrix, squared multiple correlations for each variable with the others, in its set, canonical correlations with their associated eigenvalues, optional canonical variables scores and coefficients, covariance matrix, and bivariate plots of any variable or canonical variable vs. any other variables or canonical variables.

Reference: BMDP-79, Biomedical Computer Programs, P-series. BMDP User's Digest
(see also A.1, no. 11)

General Description and Remarks: P6M computes canonical correlation analysis for two sets of variables and Bartlett's test for the significance of the remaining eigenvalues. Input can be data, a covariance matrix or a correlation matrix.

2. **Name of Program:** Harvard Canonical Correlations Program

Code Number of Program and a Location Where It Might Be Found: UCSM 304
University of Chicago Computation Center
Chicago, Illionois

Developer of Program and Affiliation: Cooley and Lohnes (see Bibliography)

Machine for Which the Program Was Written: 7090

Machine Language: Fortran II

Limitations: up to 50 variables

Form of Output: Correlation matrix
Likelihood ratio
Canonical correlation

Reference: Herzog

General Description and Remarks: This program computes a full set of canonical correlations.

3. **Name of Program:** Mesa 97 (NYBMUL)

Code Number of Program and a Location Where It Might Be Found: UCSL 705
University of Chicago Computation Center
Chicago, Illionois

Developer of Program and Affiliation: Jeremy D. Finn
State University of New York at Buffalo
Buffalo, New York

Machine for Which the Program Was Written: 7094

Machine Language: Fortran IV

Limitations: up to about 100 variables and 4000 cells
if more storage is needed parts of the program may be removed

Form of Output: Canonical correlations
Squared correlations
Percent of variation in dependent variables accounted for
Coefficient for dependent and independent variables

Reference: Finn

General Description and Remarks: Discriminant analysis regression and multivariate analysis of variance are additional features of this program.

4. **Name of Program:** Multivariate Statistical Programs

Code Number of Program and a Location Where It Might Be Found: MSP

Biometric Laboratory
University of Miami
Coral Gables, Florida

Developer of Program and Affiliation: Dean J. Clyde, Eliot M. Cramer, Richard J. Sherin
University of Miami
Coral Gables, Florida

Machine for Which the Program Was Written: 7040

Machine Language: Fortran IV

Limitations: up to 50 variables per group
up to 2500 observations

Form of Output: Canonical correlations
Scores for each observation

Reference: Clyde, Cramer, and Sherin

General Description and Remarks: This program also has factor analysis, multivariate analysis of variance, and discriminant analysis.

5. **Name of Program:** P-STAT Princeton Statistical Package

Code Number of Program and a Location Where It Might Be Found: P-STAT
Princeton University Computation Center
Princeton, New Jersey

Developer of Program and Affiliation: Roald Buhler
Princeton University
Princeton, New Jersey

Machine for Which the Program Was Written: 7094

Machine Language: Fortran II, Fap

Limitations: Canonical correlations: up to 60 variables, 40 per group
Biserial correlation up to 50 rows (r) or columns (c); $rc \leq 4500$
Asymmetric Pearson product moment correlations up to 150 rows (r) or columns (c); $rc \leq 3600$
Symmetric Pearson product moment correlations: up to 75 rows and columns
Tetrachoric correlations: up to 120 rows and columns

Form of Output: Correlations

Reference: Buhler

General Description and Remarks: This system can handle canonical, biserial, Pearson product moment, and tetrachoric correlations. In addition the system can handle factor analysis.

6. **Name of Program:** SSL PAC—Social Science Program Library Statistical Package

Code Number of Program and a Location Where It Might Be Found: UCSL 509
University of Chicago Computation Center
Chicago, Illinois

Developer of Program and Affiliation: Michael Black
University of Chicago

Machine for Which the Program Was Written: 7094

Machine Language: Fortran II

Limitations: up to 80 variables per group

Form of Output: Latent roots
Correlation coefficients
Common variance represented by each canonical variate
Likelihood ratio

Reference: Black

General Description and Remarks: This system also does factor analysis and discriminant analysis.

7. **Name of Program:** Canonical Correlation Analysis

Code Number of Program and a Location Where It Might Be Bound: SUBPROGRAM CANCORR; SPSS
Statistical Package for the Social Sciences

Developer of Program and Affiliation: James Tuccy
Northwestern University
with assistance of Paul Vincent Warwick
(University of Washington, Seattle)

Machine for Which the Program Was Written: IBM 360/370 series; the CDC 6000 series, the Univac 1108-1110 series and Xerox Sigma series; Burroughs B6000-B7000 and PDP-11

Machine Language: FORTRAN IV

Limitations: No more than 250 variables on the VARIABLES = list of CANCORR. No more than 200 variables on the RELATE = list of the CANCORR workspace: required core storage space depends on the computer, and can be calculated; for details please see the SPSS, p. 525.

Form of Output: A summary table including, the magnitude of the eigenvalues, the canonical correlations, Wilk's Lambda statistics; means and standard deviations of each variable, correlation matrix, canonical variate scores.

Reference: SPSS
Statistical Package for Social Sciences
2nd edition

General Description and Remarks: This subprogram performs canonical correlation analysis on the data. Input may be either raw data or a correlation matrix. It automatically ouputs the canonical correlations; tests of statistical significance for the correlations; the canonical variate coefficients.

8. **Name of Program:** Canonical Correlation.

Code Number of Program and a Location Where It Might Be Found: CANCORR, SAS
SAS Institute Inc.
(see also A.1, no. 8)

Developer of Program and Affiliation: Warren S. Sarle
SAS Institute Inc.
P.O. Box 10066
Raleigh, N.C. 27605

Machine for Which the Program Was Written: (see A.1, no. 8)

Machine Language: (see A.1, no. 8)

Limitations: No explicit limitations.

Form of Output: Corrected sums-of-square-and-product matrix for variables in group 1, group 2 and between groups; the means of canonical variables generated in group 1 and group 2; the canonical correlations; Bartlett's chi-square statistics; the error degrees of freedom; the probability values for chi-square statistics; normalized eigenvector, the correlations between groups 1's canonical variables and group 1's variable; the correlations between group 2's canonical variables and group 2's variables.

Reference: SAS User's Guide (1979 edition)
SAS Institute
Statistical Analysis System

General Description and Remarks: This program computes canonical correlation between 2 sets of variables. Typically, one group consists of predictor or independent variables, and the second group contains criterion or dependent variables.

9. **Name of Program:** Canonical Correlation

Code Number of Program and a Location Where It Might Be Found: (see MULTIVARIANCE, A.1, no. 7)

A.3 PROGRAMS FOR CLASSIFICATION AND DISCRIMINATION

1. **Name of Program:** Classificatory and Regional Analysis by Discriminant Iterations

Code Number of Program and a Location Where It Might Be Found: UCSM 503
University of Chicago Computation Center
Chicago, Illinois

Developer of Program and Affiliation: Emilio Casetti
Northwestern University
Evanston, Illinois

Machine for Which the Program Was Written: 7094

Machine Language: Fortran II

Limitations: up to 150 observations
up to 20 populations

up to 10 dimensions per observation
the data must be orthonormalized

Form of Output: Loading of the dimensions in each discrimination function
Scores of each observation with respect to each discriminant function
Sum of the latent roots at each step

Reference: Casetti

General Description and Remarks: This program identifies and evaluates classifications. The discriminant iteration applies a discriminant procedure to all objects of a classification with which the procedure is functionally related.

2. **Name of Program:** Discriminant Analysis for Several Groups

Code Number of Program and a Location Where It Might Be Found: BMD 05 M
Health Sciences Computing Facility
UCLA
Los Angeles, California

Developer of Program and Affiliation: UCLA Health Sciences Computing Facility

Machine for Which the Program Was Written: 7094

Machine Language: Fortran IV

Limitations: up to 5 groups (g)
g to 25 dimensions
up to 175 observations per group

Form of Output: Cross products matrix
Covariance matrix and its inverse
Mahalanobis' D^2
Coefficients and constants
Scores for each observation

Reference: Health Sciences Computing Facility

General Description and Remarks: This program computes a set of linear functions so that an individual may be classified into one of several groups. The group's assignment procedure is derived from a model of a multivariate Normal distribution of observations within groups such that the covariance matrix is the same for all groups. The individual is classified into the group for which the estimated probability density is largest.

3. **Name of Program:** Discriminant Analysis for Two Groups

Code Number of Program and a Location Where It Might Be Found: BMD 04 M
Health Sciences Computing Facility
UCLA
Los Angeles, California

Developer of Program and Affiliation: UCLA Health Sciences Computing Facility

Machine for Which the Program Was Written: 7094

Machine Language: Fortran IV

Limitations: up to 25 (p) dimensions
from p to 300 observations for each group

Form of Output: Deviation sums of squares and cross products matrix and its inverse
Discriminant function coefficients
Mahalanobis' D^2 and associated F statistic

Reference: Health Sciences Computing Facility

General Description and Remarks: This program provides the discriminant function between two groups for the discrimination between the mean indices of each group. The difference between the means divided by the standard deviation of the indices is maximized.

4. **Name of Program:** Mesa 97 (NYBMUL)

Code Number of Program and a Location Where It Might Be Found: UCSL 705
University of Chicago Computation Center
Chicago, Illinois

Developer of Program and Affiliation: Jeremy D. Finn
State University of New York at Buffalo
Buffalo, New York

Machine for Which the Program Was Written: 7094

Machine Language: Fortran IV

Limitations: up to about 100 variables and about 4000 cells
if additional storage is needed parts of the program may be removed

Form of Output: Variance of the canonical variate
Roy's criterion
Discriminant function coefficients
Hotelling's trace criterion
Bartlett's chi-square test for significance of successive canonical variates
Canonical form of least squares estimates

Reference: Finn

General Description and Remarks: This program does a discriminant function analysis for each between cell hypothesis. Canonical correlations, multivariate analysis of variance, and regression are preceding portions of the program.

5. **Name of Program:** Multivariate Statistical Analyzer

Code Number of Program and a Location Where It Might Be Found: Harvard University Computation Center
Cambridge, Massachusetts

Developer of Program and Affiliation: Kenneth Jones
Harvard University
Cambridge, Massachusetts

Machine for Which the Program Was Written: 7090

Machine Language: Fortran II

Limitations: up to 80 dimensions
up to 50 groups
(Mahalanobis' D^2 for up to 30 dimensions and 20 groups)

Form of Output: Coefficients and constants
Scores for each observation
Mahalanobis' D^2

Reference: Jones

General Description and Remarks: This system can also handle factor analysis, analysis of covariance, and discrimination problems.

6. **Name of Program:** Multivariate Statistical Programs

Code Number of Program and a
Location Where It Might Be Found: MSP
Biometric Laboratory
University of Miami
Coral Gables, Florida

Developer of Program and Affiliation: Dean J. Clyde, Eliot M. Cramer, Richard J. Sherin
University of Miami
Coral Gables, Florida

Machine for Which the Program Was Written: 7040

Machine Language: Fortran IV

Limitations: 2 populations
up to 100 dimensions
up to 99,999 observations per population
no missing data

Form of Output: Discriminant function
Scores for each observation

Reference: Clyde, Cramer, and Sherin

General Description and Remarks: This routine provides discriminant analysis for 2 groups. Additional system routines include multivariate analysis of variance, canonical correlations, and factor analysis.

7. **Name of Program:** SSLPAC—Social Science Program Library Statistical Package

Code Number of Program and a
Location Where It Might Be Found: UCSL 509
University of Chicago Computation Center
Chicago, Illinois

Developer of Program and Affiliation: Michael Black
University of Chicago
Chicago, Illinois

Machine for Which the Program Was Written: 7094

Machine Language: Fortran II

Limitations: up to 80 dimensions (*d*)

number of discriminants $\leq d$ or $\leq$ number of populations-1

Form of Output: Latent roots
Eta coefficient
Percent of extracted variance represented by each discriminant
Likelihood ratio

Reference: Black

General Description and Remarks: This system also does factor analysis and canonical correlations.

3. **Name of Program:** Stepwise Discriminant Analysis

Code Number of Program and a Location Where It Might Be Found: BMD 07 M
Health Sciences Computing Facility
UCLA
Los Angeles, California

Developer of Program and Affiliation: UCLA Health Sciences Computing Facility

Machine for Which the Program Was Written: 7094

Machine Language: Fortran IV

Limitations: up to 80 dimensions
2 to 80 groups

Form of Output: Within groups correlation and covariance matrices
U and F statistics
Discriminant functions
Classification matrix
Latent roots
Canonical correlations and coefficients for canonical variables
Posterior probability of being in each group for each observation
Mahalanobis' D^2

Reference: Health Sciences Computing Facility

General Description and Remarks: This program performs multiple discriminant analysis in a stepwise fashion. At each step the variable with the highest F value enters the set of discriminating variables. A variable is deleted if its F value falls too low. The program also computes canonical correlations and coefficients for canonical variables.

9. **Name of Program:** Stepwise Discriminant Analysis

Code Name of Program and a Location Where It Might Be Found: P7M; BMDP7M
Health Sciences Computing Facility
UCLS
Los Angeles, CA 90024
(see also A.1, no. 11)

Developer of Program and Affiliation: Robert Jennrich, HSCF
Paul Sampson, HSCF
James Frane, HSCF

Machine for Which the Program Was Written: (see A.1, no. 11)

Machine Language: FORTRAN IV

Limitations: The maximum number of:
2 groups, 10 variables, and 1800 cases; or
2 groups, 50 variables, and 1350 cases; or
3 groups, 50 variables, and 1150 cases; or
5 groups, 50 variables, and 850 cases; or
10 groups, 50 variables, and 450 cases.

Form of Output: Means, standard deviations and coefficient of variations Wilk's D or U Statistics, degrees of freedom; Hotelling's T^2, summary table; F-to-enter value; Mahalanobis's D, plots, and more.

Reference: 1. BMDP-79, Biomedical Computer Program P-series, 1979
2. BMDP User's Digest
(see also A.1, no. 11)

General Description and Remarks: P7M performs a discriminant analysis between two or more groups. The variables used in computing the linear classification function are chosen in a stepwise manner. Both forward or backward selection is possible. (For only two groups stepwise logistic regression (PLR) and all possible subsets regression (P9R) can also be used if your grouping variable is coded as zero or one.)

10. **Name of Program:** Discriminant Analysis

Code Number of Program and a Location Where It Might Be Found: DISCRIM, SAS
SAS Institute Inc.
(see also A.1, no. 8)

Developer of Program and Affiliation: Warren S. Sarle
SAS Institute Inc.

Machine for which the Program Was Written: (see A.1, no. 8)

Machine Language: (see A.1, no. 8)

Limitations: No explicit limitations.

Form of Output: Values of the classification variables, frequencies, prior probabilities. If the pooled covariance matrix is used, the linearized discriminant function; and (optionally: simple descriptive statistics, within-group covariance matrix, within-group correlation matrix, pooled covariance matrix, the classification results for each observations, the chi-square test of homogeneity of the within-group covariance matrices), the generalized squared distance between groups, a summary of the performance of the classification criterion.

Reference: SAS User's Guide, 1979 edition
SAS Institute
Statistical Analysis System

General Description and Remarks: DISCRIM Program develops a discriminant model that it uses to classify each observation into one of the groups and then summarizes the performance of this discriminant model.

11. **Name of Program:** NEIGHBOR; Nearest Neighbor Discriminant Analysis

Code Number of Program and a Location Where It Might Be Found: NEIGHBOR; SAS
SAS Institute Inc.
(see also A.1, No. 8)

Developer of Program and Affiliation: James H. Goodnight
SAS Institute Inc.

Machine for Which the Program Was Written: (see A.1, no. 8)

Machine Language: (see A.1, no. 8)

Limitations: No explicit limitations.

Form of Output: Values of classification variable; frequencies; prior probabilities; optionally, classification result for each observation; the actual group for each observation; the group into which the developed criterion would classify it; the posterior probability and a summary of the performance of the classification criterion.

Reference: SAS User's Guide, 1979 edition
SAS Institute
Statistical Analysis System.

General Description and Remarks: The NEIGHBOR procedure performs a nearest neighbor discriminant analysis, classifying observations into groups according to either the nearest neighbor rule or the K-nearest neighbor rule. The default K value is 1.

12. **Name of Program:** Discriminant Analysis

Code Number of Program and a Location Where It Might Be Written: DISCRIMINANT; SPSS Statistical Package for the Social Sciences

Developer of Program and Affiliation: James Tuccy
Vogelback Computing Center
Northwestern University
William R. Klecka
University of California

Machine for Which the Program Was Written: (see A.2, no. 7)

Machine Language: FORTRAN IV

Limitations: Pairwise deletion of missing data is not available. The amount of core storage space in bytes can be calculated (depending on the computer being used). For this see the SPSS manual p. 461.

When the default amount of workspace is available (70,000 bytes), the typical usage will allow the following maximum combinations:

53 variables, 2 groups, and 8 subanalysis; or
53 variables, 4 groups, and 3 subanalysis; or
50 variables, 10 groups, and 6 subanalysis; or
46 variables, 20 groups, and 2 subanalysis.

Form of Output: Classification result table; discriminant scores; plot for each group; territorial map, discriminant function coefficients, classification function; membership probability for all groups; pooled within-group covariance and correlation matrices; univariate F ratios; Box's M and its associated F test; group covariance matrices, total covariance matrix; and some other options (also see general description and remarks).

Reference: SPSS
Statistical Package for Social Sciences
2nd edition

General Description and Remarks: This subprogram performs discriminant analysis either by entering all discriminant variables directly or through a variety of stepwise methods selection the "best" set of discriminant variables. The available criteria for the stepwise selection are: minimum Wilks' Lambda, minimum Mahalanobis distance, largest minimum between-groups F, largest increase in average multiple correlation, and largest increase in Rao's V.

13. **Name of Program:** Discriminant Analysis

Code Number of Program and a Location Where It Might Be Found: (see MULTIVARIANCE, A.1. no. 7)

A.4 PROGRAMS FOR CLUSTERING

1. **Name of Program:** BC TRY

Code Number of Program and a Location Where It Might Be Found: UCSM 602
University of California Computation Center
Berkeley, California

Developer of Program and Affiliation: Robert Tryon, University of California at Berkeley
Daniel Bailey, University of Colorado

Machine for Which the Program Was Written: 7090

Machine Language: Fortran II

Limitations: up to 20 dimensions as definers of each cluster
up to 120 dimensions

Form of Output: Diagonal values
Unrotated factor coefficients
Cluster indicators

Reflection indicators
Cluster scores

Reference: Tryon and Bailey

General Description and Remarks: This cluster analysis includes empirical key cluster analysis where the program selects the definers of the clusters, preset key cluster analysis where the user inputs the definers, preset dimension analysis where the number of clusters is set by the user, noncommunality key cluster analysis where factor coefficients are calculated with the use of diagonal elements, oblique noncommunality cluster analysis, spherical analysis and an approximation to least squares key cluster analysis.

The system also includes *V*-analysis and *O*-analysis to compress studies into computer size problems. In addition this system contains factor analysis routines.

2. **Name of Program:** *k*-Means

Code Number of Program and a Location Where It Might Be Found: (see developer of program)

Developer of Program and Affiliation: James MacQueen
University of California at Los Angeles
Graduate School of Business
Los Angeles, California

Machine for Which the Program Was Written: 7094

Limitations: at least 20 dimensions are possible
at least 560 sample points are possible, at least 20 means are possible

Form of Output: Points in each cluster with 18 associated attributes
Distance of each point to nearest mean
Distances between means
Average for each cluster
Within cluster variance

Reference: MacQueen

General Description and Remarks: Program is available only from the author. User must preset the number of clusters desired, and two parameters representing the degree of homogeneity (closeness of points) within a cluster desired, and the degree of heterogeneity between clusters desired.

3. **Name of Program:** Stepwise Cluster I

Code Number of Program and a Location Where It Might Be Found: Write to developer of program.

Developer of Program and Affiliation: B. King, University of Chicago
Graduate School of Business
Chicago, Illinois

Machine for Which the Program Was Written: 7094

Machine Language: Fortran IV

Limitations: up to 100 variables

Form of Output: Clusters of points; at each stage the correlation coefficient for merging clusters (by centroid) is provided.

Reference: King (1967)

General Description and Remarks: To cluster 60 variables takes well under 2 minutes including compilation.

4. **Name of Program:**

Code Number of Program and a Location Where It Might be Found: CLUSTAN
Program Library Unit
Edinburgh University
18 Buccleuch Place
Edinburgh EH8 9LM
SCOTLAND

Developer of Program and Affiliation: Dr. D. Wishart, et al.
c/o Department of Computational Science
University of St. Andrews
North Haugh, St. Andrews KY16 9SX
SCOTLAND

Machine Language: FORTRAN IV

Limitations: a) General

N (population size)	999
MN (number of continuous variables)	200
MB (number of binary attributes)	400

b) Principal components analysis and correlations MN must not exceed 80

Form of Output: The form of output depends on the procedure used. Some outlines are as follows: Principle component and cluster diagnostics for the RELOCATE classifications; cluster diagrams for principal components; distance matrix; standard mode analysis; estimates of the modes in the sample; cluster diagnostics for the *K*-linkage list; Ward's part-optimum result; alphabetic case labels; plot of a dendogram for Ward's method; cluster diagram including cluster circles and/or convex hulls; plot of significant distances for a threshold value of the smallest mode cluster fusion coefficient; scatter diagram and the minimum spanning tree; error sum of squares; correlation matrix.

Reference: CLUSTAN USER MANUAL
(3rd edition) by D. Wishart
Inter-University/Research Councils Series
Report No. 47
January, 1978

General Description and Remarks: CLUSTAN is an integrated package of Fortran IV programs for the collective study and use of various cluster analysis and other multivariate methods. The analysis involving *k*-means (RELOCATE) or Ward's method (HIERARCHY) will find tight "minimum variance" clusters. Those involving MODE, single linkage or KDEND, will seek "natural" clusters which do not necessarily have to be tight. The input data may be continuous measurements, multi-state attributes or binary presence/absence observations on a sample population of *n* objects. Alternatively, any matrix of similarities or distances may be supplied, for which the underlying variables need not have been observed. The SPSS procedure may also be used to read a data matrix from an SPSS system file (for the details, see the reference).

5. **Name of Program**: Cluster Analysis

Code Number of Program and a Location Where It Might Be Found: CLUSTER; SAS
SAS Institute Inc.
(see also A.1, no. 8)

Developer of Program and Affiliation: Warren S. Sarle
SAS Institute Inc.

Machine for Which the Program Was Written: (see A.2, no. 8)

Machine Language: (see A.1, no. 8)

Limitations: Due to the size of the distance matrix and computations involved, cluster should not be applied to data sets of more than about 250 observations.

Form of Output: 1. The number of cluster; 2. The maximum diameter of a cluster; 3. The number of distances within clusters; 4. The total number of distances less than the maximum diameter; 5. The ratio of 3 and 4; 6. Optionally it prints a cluster map; 7. The minimum, average, and maximum distances within and between clusters; 8. List of observation within each cluster; 9. The means of the variables within and between clusters.

Reference: SAS User's Guide, 1979 edition
SAS Institute
Statistical Analysis System

General Description and Remarks: The CLUSTER procedure, designed to help identify clusters of observations that have similar attributes, performs a hierarchical cluster analysis; it is used when there is no prior or theoretical classification information about the data. The technique is based on an algorithm outlined by Johnson.

6. **Name of Program**: Cluster Analysis of Variables

Code Number of Program and a Location Where It Might Be Found: P1M; BMDP1M
Health Sciences Computing Facility
Los Angeles, CA 90024
(see also A.1, no. 11)

Developer of Program and Affiliation: Designed by John Hartigan, Yale University. Programmed by Howard Gilbert and Steve Chasen; later revisions were made by Steve Chasen and James Frane.

Machine for Whcih the Program Was Written: (see A.1, no. 11)

Machine Language: FORTRAN IV

Limitations: P1M can analyze up to 140 variables. (The capacity of programs can be increased, see Appendix B in BMDP book).

Form of Output: A summary table of the clusters formed; a tree showing the clusters formed at each step, with an explanation of the clustering process; (Optional) Correlation matrix and shaded correlation matrix display.

Reference: 1. BMDP-79, Biomedical Computer Programs P-series, 1979
2. BMDP User's Digest (see also A.1, no. 11)

General Description and Remarks: P1M provides a choice of four measures of similarity (association) for clustering variables, and three criteria for linking or combining clusters. The amalgamating process continues in a stepwise fashion while a single cluster is formed that contains all the variables. Optionally a similarity or distance matrix can be used as input to P1M.

7. **Name of Program**: Cluster Analysis of Cases

Code Name of Program and a Location Where It Might Be Found: P2M; BMDP2M
Health Sciences Computing Facility
UCLA
Los Angeles, CA 90024
(see also A.1, no. 11)

Developer of Program and Affiliation: Designed by Lazlo Engelman, HSCF. Programmed by Lazlo Engelman and Sandra Fu. Later additions were made by James Frane.

Machine for Which the Program Was Written: (see A.1, no. 11)

Machine Language: FORTRAN IV

Limitations:

Number of variables:	5	10	20
Maximum number of cases when distance matrix is printed:	500	400	300
Maximum number of cases when distance matrix is not printed:	630	520	370

(Program capacity can be increased.)

Form of Output: Vertical or horizontal tree diagram describing the sequence of a table that lists the amalgamation distance and the mean of each variable as each new cluster is formed.

Optional data listing in original units or standardized form.
Optional matrix of the distances between the cases.
Optional shaded distance matrix graphical display.

Reference: 1. BMDP 79, Biomedical Computer Programs, P-series, 1979.
2. BMDP User's Digest (see also A.1, no. 11)

General Description and Remarks: P2M forms clusters of cases (observations) based on one of four distance measures. The distance measures are the Euclidean distance (L_2); the L_p distance, the chi-square statistics or phi-square. Initially each case is considered to be in a cluster of its own. The problem joins cases and/or clusters of cases in a stepwise process until all cases are combined into one cluster. The algorithm uses the average distance (average linkage) as a criterion for joining (amalgamating) clusters.

8. **Name of Program**: Block Clustering

Code Number of Program and a Location Where It Might Be Found: P3M; BMDP3M
Health Sciences Computing Facility
UCLA
Los Angeles, CA 92521
(see also A.1, no. 11)

Developer of Program and Affiliation: P3M was designed by John Hartigan, Yale University and programmed by John Hartigan and Jerry Douglas. Recent revisions were made by Lanaii Kline.

Machine for Whcih the Program was Written: (see A.1, no. 11)

Machine Language: FORTRAN IV

Limitations: The following table provides a guide to the maximum number of cases that can be analyzed.

Number of variables:	5	10	20	50
Maximum number of cases:	720	440	240	90

(Appendix B in BMDP book describes how to increase the capacity of the program).

Form of Output: A block diagram to describe the identified submatrices. Two tree diagrams (one for cases, the other for variables) to describe the clustering sequence, a prediction table to recover the coded data matrix from the block diagram, codes used for each variable and their frequencies.

Reference: 1. BMDP-79, Biomedical Computer Programs, P-series, 1979
2. BMDP User's Digest
(see also A.1, no. 11)

General Description and Remarks: P3M simultaneously forms clusters of both the cases and variables in a data matrix. The method is primarily appropriate when each variable has a few distinct rules. Each variable is treated as categorical or nominal (not ordered). An iterative technique is used to identify blocks that have a similar pattern.

9. **Name of Program**: *K*-Means Clustering of Cases

Code Number of Program and a Location Where It Might Be Found: PKM; BMDPKM
Health Sciences Computing Facility
UCLA
Los Angeles, CA 90024
(see also A.1, no. 11)

Developer of Program and Affiliation: PKM was designed by John Hartigan and Laszlo Engelman (HSCF), and programmed by William Eddy and Laszlo Engelman.

Machine for Which the Program was Written: (see A.1, no. 11)

Machine Language: FORTRAN IV

Limitations: With 10 variables the program can process approximately 450 cases into 10 clusters in 15,000 storage words ($M = 15{,}000$). (For exact formula see appendix B in BMDP manual.)

Form of Output: A description of each variable in each cluster: the mean, min., max., and variance. For each cluster, the distance of the cluster center to each case is printed and histograms display these distances for: a) cases in the cluster;

b) cases not in the cluster; a scatter plot of the orthogonal projection of cases into the plane defined by the centers of the three most populous clusters; a summary of the cluster means and standard deviations for each variable; for each variable an analysis of variance with descriptive F-ratio that compares the between-cluster mean square to the within-cluster mean square; cluster profile; pooled within-cluster covariance and correlation matrices; optional distance between cluster centers, optional cross tabulation of clusters with user-specified variables.

Reference: 1. BMDP-79, Biomedical Computer Programs, P-series, 1979.
2. BMDP User's Digest (see also A.1, no. 11)

General Description and Remarks: PKM partitions a set of cases (observations) into clusters based on the Euclidean distance measure between the cases and the centers of the clusters. The program begins with user-specified clusters or with all the data in one cluster and splits one cluster into two clusters at each step. More than one number can be specified for K (the number of clusters) so that results are obtained for several values of K. See also Program No. 2 (MacQueen).

A.5 PROGRAMS FOR FACTOR ANALYSIS

1. **Name of Program:** Analysis of Three Mode Matrices

Code Number of Program and a Location Where It Might Be Found: UCSM 407
University of Chicago Computation Center
Chicago, Illinois

Developer of Program and Affiliation: James Walsh
University of Washington
Seattle, Washington

Machine for Which the Program Was Written: 7090

Machine Language: Fortran II

Limitations: up to 50 mode a variables
up to 25 mode b variables
up to 5 mode c variables

Form of Output: Factor loading vectors
Latent roots
Orthonormals

Reference: Walsh

General Description and Remarks: This program does a principal axis factor analysis on 3 mode matrices.

2. **Name of Program:** BC TRY

Code Number of Program and a Location Where It Might Be Found: UCSM 602
University of California Computation Center
Berkeley, California

Developer of Program and Affiliation: Robert Tryon—University of California at Berkeley
Berkeley, California
Daniel Bailey, University of Colorado
Boulder City, Colorado

Machine for Which the Program Was Written: 7090

Machine Language: Fortran II

Limitations: up to 120 variables

Form of Output: Factor coefficients
Augmented factor coefficients
Cumulative partial communalities
Latent roots
Latent vectors
Factor scores

Reference: Tryon and Bailey

General Description and Remarks: Factor analysis routines include principal axis, canonical and augmenting factoring procedures. Varimax and quartimax rotations are available.

This program also can do cluster analysis and can sample to reduce large problems.

3. **Name of Program:** Direct Factor Analysis Program

Code Number of Program and a
Location Where It Might Be Found: UCSL 601
University of Chicago Computation Center
Chicago, Illinois

Developer of Program and Affiliation: B. Wright, D. MacRae
University of Chicago
Chicago, Illinois

Machine for Which the Program Was Written: 7090

Machine Language: Fortran II

Limitations: up to a 160 by 160 data matrix
up to 25 factors

Form of Output: Latent vectors
Residuals

Reference: Wright and MacRae

General Description and Remarks: This program performs direct factor analysis of a data matrix. After finding latent roots and residuals from the data the process is repeated.

4. **Name of Program:** Essoteric

Code Number of Program and a
Location Where It Might Be Found: UCSM 331
University of Chicago Computation Center
Chicago, Illinois

Developer of Program and Affiliation: Frank Steidler
Esso Research and Engineering Company
Linden, New Jersey

Machine for Which the Program Was Written: 7090

Machine Language: Fortran II

Limitations: up to 100 variables and factors
input formats are restricted

Form of Output: Correlation matrix
Latent roots
Normalized latent vectors
Varimax factor matrix

Varimax factor variance
Factor scores

Reference: Steidler

General Description and Remarks: This program does a factor analysis and includes varimax rotation. The program can also handle a principal components problem.

5. **Name of Program:** General Factor Analysis

Code Number of Program and a Location Where It Might Be Found: BMD 03 M
Health Sciences Computing Facility
UCLA
Los Angeles, California

Developer of Program and Affiliation: UCLA Health Science Computing Facility

Machine for Which the Program Was Written: 7094

Machine Language: Fortran IV

Limitations: 2 to 80 variables (p)
2 to 9999 cases
2 to p factors

Form of Output: Correlation matrix
Latent roots and latent vectors
Factor matrix
Check factor matrix
Original and successive variances
Factor scores

Reference: Health Sciences Computing Facility

General Description and Remarks: This program provides a principal components solution and an orthogonal rotation of the factor matrix. Communalities are estimated from the squared multiple correlation coefficient or the maximum absolute row values or they may be specified by the user.

6. **Name of Program:** Mesa 85

Code Number of Program and a Location Where It Might Be Found: UCSL 510
University of Chicago Computation Center
Chicago, Illinois

Developer of Program and Affiliation: B. Wright, C. Bradford, R. Strecker, F. Bamberger
University of Chicago
Chicago, Illinois

Machine for Which the Program Was Written: 7090

Machine Language: Fortran II

Limitations: up to 100 variables
up to 20 factors

Form of Output: Latent roots
Factor matrix before and after rotation
Factor scores

Reference: Wright, Bradford, Strecker, and Bamberger

General Description and Remarks: Factor analysis routines include procedures for maximum row r on the diagonal, multiple R^2 on the diagonal, image covariance analysis, and varimax rotations.

The program also can do principal components analysis and univariate regression analysis.

7. **Name of Program:** Multivariate Statistical Analyzer

Code Number of Program and a Location Where It Might Be Found: Harvard University Computation Center
Cambridge, Massachusetts

Developer of Program and Affiliation: Kenneth Jones
Harvard University
Cambridge, Massachusetts

Machine for Which the Program Was Written: 7090

Machine Language: Fortran II

Limitations: up to 28 factors
maximum likelihood and alpha factor analysis: up to 80 variables
principal factor analysis and image covariance analysis: up to 100 variables

Form of Output: Factor loading matrix
Factor scores
Latent roots

Reference: Jones

General Description and Remarks: This system contains principal factor analysis, maximum likelihood factor analysis, alpha factor analysis, and image covariance routines. In addition, varimax, quartimin, biquartimin, and covarimin rotations are available. The system also can do disciminant, principal components, and covariance analysis.

8. **Name of Program:** Multivariate Statistical Programs

Code Number of Program and a Location Where It Might Be Found: MSP
Biometric Laboratory
University of Miami
Coral Gables, Florida

Developer of Program and Affiliation: Dean J. Clyde, Eliot M. Cramer, Richard J. Sherin
University of Miami
Coral Gables, Florida

Machine for Which the Program Was Written: 7040

Machine Language: Fortran IV

Limitations: up to 190 dimensions and 99,999 observations

Form of Output: Intercorrelations
Principal components
Factor loadings
Varimax rotations

Reference: Clyde, Cramer, and Sherin

General Description and Remarks: The system also has multivariate analysis of variance, discriminant analysis, and canonical correlations.

9. **Name of Program:** P—STAT Princeton Statistical Package

Code Number of Program and a Location Where It Might Be Found: P—STAT
Princeton University Computation Center
Princeton, New Jersey

Developer of Program and Affiliation: Roald Buhler
Princeton University
Princeton, New Jersey

Machine for Which the Program Was Written: 7094

Machine Language: Fortran II, Fap

Limitations: iterative routine: up to 100 variables and factors
noniterative routine: up to 150 variables and factors
rotations: up to 150 rows and 40 columns
congruence coefficients: up to 100 rows and 50 columns

Form of Output: Congruence coefficients
Factor coefficients
Factor scores
Rotated matrix

Reference: Buhler

General Description and Remarks: This routine calls subprograms for each operation. Rotations include varimax, quartimax, and equimax operations. The system can also do several types of correlation analysis.

10. Name of Program: RMLFA—Restricted Maximum Likelihood Factor Analysis

Code Number of Program and a Location Where It Might Be Found: UCSM 722
University of Chicago Computation Center
Chicago, Illinois

Developer of Program and Affiliation: K. G. Jöreskog, G. Gruvaeus
Education Testing Service
Princeton, New Jersey

Machine for Which the Program Was Written: 7044

Machine Language: Fortran IV

Limitations: up to 30 variables
up to 10 factors
up to 120 free parameters to be estimated
positive definite correlation matrix

Form of Output: Factor loading matrix
Factor correlations and uniqueness
Maximum likelihood solution
Goodness-of-fit test

Reference: Jöreskog and Gruvaeus

General Description and Remarks: Any values may be specified in advance for any number of factor loadings, factor correlations, and unique variances. The remaining free parameters, if any, are estimated by the maximum likelihood method.

11. **Name of Program:** SSL PAC—Social Science Program Library Statistical Package

Code Number of Program and a Location Where It Might Be Found: UCSL 509
University of Chicago Computation Center
Chicago, Illinois

Developer of Program and Affiliation: Michael Black
University of Chicago
Chicago, Illinois

Machine for Which the Program Was Written: 7094

Machine Language: Fortran II

Limitations: up to 110 variables and factors
Varimax rotation: up to 150 variables and factors
Binovmamin rotation: up to 150 variables and 30 factors
Procrustes rotation: up to 75 variables and factors
Rotoplot rotation: up to 90 variables and factors

Form of Output: Variance table
Factor matrix
Rotated matrices

Reference: Black

General Description and Remarks: The system uses principal axis factor analysis procedures. The system can also do discriminant analysis and canonical correlations.

12. **Name of Program:** UMLFA—Unrestricted Maximum Likelihood Factor Analysis

Code Number of Program and a Location Where It Might Be Found: UCSM 718
University of Chicago Computation Center
Chicago, Illinois

Developer of Program and Affiliation: K. G. Jöreskog
Educational Testing Service
Princeton, New Jersey

Machine for Which the Program Was Written: 7044

Machine Language: Fortran IV

Limitations: up to 75 variables
up to 30 factors
positive definite correlation matrix

Form of Output: Factor matrix
Unique variances
Varimax rotated matrix
Degrees of freedom
Wilson-Hilferty transformation
Lawley's chi-square

Reference: Jöreskog

General Description and Remarks: This program performs an unrestricted maximum likelihood factor analysis on a given correlation matrix. Goodness of fit is tested by Lawley's chi-square test based on the likelihood ratio technique.

13. **Name of Program:** Varimax

Code Number of Program and a Location Where It Might Be Found: UCSM 704
University of Chicago Computation Center
Chicago, Illinois

Developer of Program and Affiliation: Cooley and Lohnes (see Bibliography)

Machine for Which the Program Was Written: 7090

Machine Language: Fortran II

Limitations: up to 75 factors
up to 100 variables

Form of Output: Rotated matrix

Reference: Lankford

General Description and Remarks: This program performs either a varimax rotation or iterates for a number of times specified by the user.

14. **Name of Program:** Factor Analysis

Code Number of Program and a Location Where It Might Be Found: FACTOR; SAS
SAS Institute Inc.
(see also A.1, no. 8)

Developer of Program and Affiliation: Warren S. Sarle
SAS Institute Inc.

Machine for Which the Program Was Written: (see A.1, no. 8)

Machine Language: (see A.1, no. 8)

Limitations: A maximum of 250 variables may be used with FACTOR

Form of Output: 1. Means, standard deviations, number of observations, and variable labels (if any); 2. Correlation matrix; 3. Prior estimates of communalities; 4. Eigenvalues; 5. Optionally, eigenvectors; 6. If image analysis is performed, the anti-image correlation matrix, the image-scaled correlation matrix; plus eigenvalues of the image; 7. Initial factor loadings; 8. Communality estimates; 9. Optionally, the rotated factor matrix and the transformation matrix; 10. Optionally, the target matrix and the factor structure matrix; 11. Optionally, plots of the factor patterns; 12. Optionally, the scoring coefficient matrix; 13. Optionally, the inter-factor correlations.

Reference: SAS User's Guide, 1979 edition
SAS Institute
Statistical Analysis System

General Description and Remarks: The FACTOR procedure performs a factor analysis for variables in a SAS data set. Principle axis factoring, image analysis, alpha factor analysis, and iterated principle axis factoring are all available, along with varimax equimax, quartimax, and promax (oblique) rotation techniques. Prior estimates of communalities may be specified in several different ways.

15. Name of Program: Factor Analysis

Code Number of Program and a Location Where It Might Be Found: FACTOR: SPSS
Statistical Package for the Social Sciences

Developer of Program and Affiliation: Jae-On Kim
University of Iowa, et al.

Machine for Which the Program Was Written: (see A.2, no. 7)

Machine Language: FORTRAN IV

Limitations: No more than 100 variables. Work space required depends on the computer. For IBM 360/370, the default value of 70,000 bytes of work space is available, a maximum of 62 variables may be entered on the VARIABLES = list. At least 180,228 bytes of work space is required to perform a factor analysis on 100 variables. Factor scores for each case may not be output when matrix input is used.

Form of Output: Correlation matrix for input variables, initial factor loadings, factor-*pattern* matrix, factor-*estimate* or factor-score coefficient matrix, factor-*structure* matrix, correlation matrix for terminal factors.

General Description and Remarks: Subprogram FACTOR performs a factor-analytic technique. Input may be raw data, a correlation matrix, or a factor matrix. Five different factoring are available: 1. principle factoring without iteration, 2. principle factoring with iterations, 3. Rao's canonical factoring, 4. alpha factoring, and 5. image factoring. Four alternative rotational methods may be applied, three orthogonal solutions are varimax, quartimax, and equimax. In the oblique rotation, the user can control the degree of correlation between factors. Graphical plotting is also available.

16. **Name of Program:** Factor Analysis

Code Number of Program and a Location Where It Might Be Found: P4M; BMDP4M
Health Sciences Computing Facility
UCLA
Los Angeles, CA 90024
(see also A.1, no. 11)

Developer of Program and Affiliation: P4M was designed by James Frane (HSCF), with major contributions from Robert Jennrich (HSCF), and Paul Sampson. It was programmed by James Frane with major contributions from Paul Sampson. It supersedes BMDO8M, which was developed by Jennrich and Paul Sampson.

Machine for Which the Program Was Written: (see A.1, no. 11)

Machine Language: FORTRAN IV

Limitations: If maximum likelihood analysis is requested, up to 10 factors and 60 variables can be analyzed; other methods will handle 100 variables.

Form of Output: Univariate summary statistics; rotated and unrotated factor loadings and their plots; display of sorted rotated factor loading; factor score coefficient, scores for each case, and factor score plots; Mahalanobis distances from each case to the centroid of all cases for original data; factor scores and their differences; correlation matrix, squared multiple correlation of each variable with all others, eigenvalues; optional display of the correlation in sorted and shaded form; optional listing of data or standard scores, covariance matrix, inverse of correlation or covariance matrix, partial correlations, residual correlations.

Reference: 1. BMDP-79, Biomedical Computer Programs P-series, 1979

2. BMDP User's Digest (see also A.1, no. 11)

General Description and Remarks: P4M performs a factor analysis of a correlation or covariance matrix. It provides four methods of initial factor extraction, principal components, maximum likelihood, Kaiser's second Generation Little Jeffy, or iterated principal factor analysis. Input can be data or correlation or covariance matrix, factor loadings or factor score coefficients.

17. **Name of Program:** Factor Analysis

Code Number of Program and a Location Where It Might Be Found: (see LISREL IV, A.10, no. 12)

A.6 PROGRAMS FOR GENERAL MATRIX OPERATIONS

Developer of Programs: Cooley and Lohnes (see Bibliography)

General Description and Remarks: This book has subroutines for basic matrix operations such as finding a matrix inverse and finding latent roots and latent vectors.

2. **Name of Program:** MATRIX

Code Number of Program and a Location Where It Might Be Found: MATRIX; SAS
SAS Institute Inc.
(see also A.1, no. 8)

Developer of Program and Affiliation: John P. Sall
SAS Institute Inc.

Machine for Which the Program Was Written: (see A.1, no. 8)

Machine Language: Mixture of assembler, FORTRAN IV and PL/1

Limitations: No single matrix should contain more than 32,767 elements; each active matrix needs 2 + n row X n column doublewords. From 200K region, about 100K should be available for work space, enough for 125 10 X 10 matrices or one 110 X 110 matrix.

Form of Output: Depends on the kind of calculation (almost all the matrix operations including logical comparison and sorting can be done by MATRIX procedure. For details see the SAS User's Guide.

Reference: SAS User's Guide, 1979 edition
SAS Institute
Statistical Analysis System

General Description and Remarks: The MATRIX procedure is a complete programming language in which operations are performed on entire matrices values. The language is patterned directly after matrix notation.

A.7 PROGRAMS FOR LATENT STRUCTURE ANALYSIS

1. **Name of Program:** LASY: A Computer program for the Latent Class Model.

Code Number of Program and a Location Where It Might Be Found: UCSM 803
University of Chicago Computation Center
Chicago, Illinois

Developer of Program and Affiliation: N. Henry
Bureau of Applied Social

Research
Columbia University

Machine for Which the Program Was Written: 7094

Machine Language: Fortran II

Limitations: Up to 10 latent classes and up to 29 items. First-, second-, and third-order data are required as inputs.

Form of Output: Estimated recruitment probabilities (latent class frequencies), and estimated latent probabilities.

Reference: A. Herzog

General Description and Remarks: Another program called BAN provides maximum likelihood estimates of the latent parameters. However, this approach was not discussed in the text.

2. **Name of Program:** Unrestricted and Restricted Maximum Likelihood Latent Structure Analysis

Code Number of Program and a Location Where It Might Be Found: MLLSA
Population Issues Research Center
The Pennsylvania State University
22 Burrowes Building
University Park, PA 16802

Developer of Program and Affiliation: Clifford C. Clogg
Population Issues Research Center
The Pennsylvania State University

Machine Language: FORTRAN IV G level 21

Limitations: The arrays in the program will handle six manifest variables, twelve latent classes, and at least three (but no more than four) classes per manifest variable. Both dimension statements can be modified to handle other problems. The area required for the arrays when the default is used is less than 28K. The program is dimensioned to handle large arrays, and requires about 280K. This can be changed, however.

Form of Output: Maximum likelihood estimates of recruitment probabilities and latent probabilities, goodness of fit tests.

Reference: Working Paper 1977-09, Population Issues Research Center, The Pennsylvania State University (July 1977).

General Description and Remarks: The MLLSA program is able to estimate a wide variety of unrestricted and restricted latent structure models for a contingency table involving *m* polytomous manifest variables.

A.8 PROGRAMS FOR MULTIDIMENSIONAL SCALING

1. **Name of Program:** MDSCAL; A multidimensional scaling program.

Code Number of Program and a Location Where It Might Be Found: UCSM 402
University of Chicago Computation Center
Chicago, Illinois

Developer of Program and Affiliation: J. B. Kruskal
Bell Telephone Labs, Inc.
Murray Hill, New Jersey

Machine for Which the Program Was Written: 709, 7090, 7094, 7040

Machine Language: Fortran II

Limitations: Up to 80 objects may be scaled in up to 10 dimensions. The data may include up to 3200 entries.

Form of Output: Coordinates of all objects are provided.

Reference: Kruskal (1964b).

General Description and Remarks: The scaling computation stops if the stress reaches a sufficiently small value or if it has not been declining rapidly enough.

2. **Name of Program:** TORSCA

Code Number of Program and a Location Where It Might Be Found: L. L. Thurstone Psychometric Lab.
University of North Carolina
Chapel Hill, North Carolina

Developer of Program and Affiliation: F. W. Young, W. S. Torgerson

Machine for Which the Program Was Written: 360/75

Machine Language: Fortran IV

Form of Output: Coordinates of all objects are provided

Reference: F. W. Young and W. S. Torgerson, 1967.

A.9 PROGRAMS FOR PRINCIPAL COMPONENTS

1. **Name of Program:** Essoteric

Code Number of Program and a Location Where It Might Be Found: UCSM 331
University of Chicago Computation Center
Chicago, Illinois

Developer of Program and Affiliation: Frank Steidler
Esso Research and Engineering Company
Linden, New Jersey

Machine for Which the Program Was Written: 7090

Machine Language: Fortran II

Limitations: up to 100 variables and factors
input formats are restricted

Form of Output: Correlation matrix and its inverse
Principal components analysis of correlation matrix

Reference: Steidler

General Description and Remarks: This program also will do factor analysis.

2. **Name of Program:** Mesa 85

Code Number of Program and a Location Where It Might Be Found: UCSL 510
University of Chicago Computation Center
Chicago, Illinois

Developer of Program and Affiliation: B. Wright, C. Bradford, R. Strecker, F. Bamberger
University of Chicago

Machine for Which the Program Was Written: 7090

Machine Language: Fortran II

Limitations: up to 100 variables
up to 20 factors

Form of Output: Latent roots
Factor matrix before and after rotation
Factor cores

Reference: Wright, Bradford, Strecker, and Bamberger

General Description and Remarks: This program can also do factor analysis. In addition after this analysis it can run a regression on the results.

3. **Name of Program:** Multivariate Analysis of Variance with Principal Components of All Dispersion (Variance-Covariance) Matrices

Code Number of Program and a Location Where It Might Be Found: UCSM 339
University of Chicago Computation Center
Chicago, Illinois

General Description and Remarks: This program does a principal components analysis of within group matrices as part of its multivariate analysis of variance. See the multivariate analysis of variance program, UCSM 339.

4. **Name of Program:** Multivariate Statistical Analyzer

Code Number of Program and a Location Where It Might Be Found: Harvard University Computation Center
Cambridge, Massachusetts

Developer of Program and Affiliation: Kenneth Jones
Harvard University
Cambridge, Massachusetts

Machine for Which the Program Was Written: 7090

Machine Language: Fortran II

Limitations: up to 100 variables
up to 28 factors

Form of Output: Correlation coefficients
Latent roots
Latent vectors

References: Jones

General Description and Remarks: This system can also handle factor analysis, discrimination, and analysis of covariance.

5. **Name of Program:** Principal Components Analysis

Code Number of Program and a Location Where It Might Be Found: BMD 01 M
Health Sciences Computing Facility
UCLA
Los Angeles, California

Developer of Program and Affiliation: UCLA Health Sciences Computing Facility

Machine for Which the Program Was Written: 7094

Machine Language: Fortran IV

Limitations: 2 to 25 variables
3 to 400 cases

Form of Output: Correlation coefficients
Latent roots
Latent vectors
Rank order of each standardized case

Reference: Health Sciences Computing Facility

General Description and Remarks: This program computes the principal components of standardized data and rank orders each standardized case by the size of each principal component separately.

6. **Name of Program:** Regression on Principal Components

Code Number of Program and a Location Where It Might Be Found: BMD 02 M
Health Sciences Computing Facility
UCLA
Los Angeles, California

Developer of Program and Affiliation: UCLA Health Sciences Computing Facility

Machine for Which the Program Was Written: 7094

Machine Language: Fortran IV

Limitations: up to 25 independent variables
up to 20 dependent variables
3 to 200 cases

Form of Output: Correlation coefficients
Latent roots
Latent vectors
Rank order of each standardized case
Regression coefficients
Reduction in sums of squares of residuals

Reference: Health Sciences Computing Facility

General Description and Remarks: This program first computes the principal components of standardized data and rank orders each standardized case by the size of each principal component separately. Each dependent variable is then regressed on the first one, first two, and first three and all the principal components.

7. **Name of Program:** Principal Components Analysis and Correlations

Code Number of Program and a Location Where It Might Be Found: (see CLUSTAN, A.4, no. 4)

8. **Name of Program:** Principal Components

Code Number of Program and a Location Where It Might Be Found: (see P4M, BMDP4M, A.5, no. 16)

9. **Name of Program:** Principal Components

Code Number of Program and a Location Where It Might Be Found: (see FACTOR, SPSS, A.5, no. 15)

10. **Name of Program:** Factor Analysis (Principal Axis Factoring)

Code Number of Program and a Location Where It Might Be Found: (see FACTOR, SAS, A.5, no. 14)

11. **Name of Program:** The PRINCOMP Procedure

Code Number of Program and a Location Where It Might Be Found: PROC PRINCOMP, SAS
SAS Institute Inc.
(see also A.1, no. 8)

Developer of Program and Affiliation: John P. Sall
SAS Institute Inc.

Limitations: No explicit limitations.

Form of Output: The eigenvalues and eigenvectors, along with either the covariance or correlation matrix.

Reference: SAS Technical Report P-110 (June 29, 1979)

General Description and Remarks: The PRINCOMP Procedure computes principal components of variables in a SAS data set and outputs them to a new data set. The principal components may be computed with respect to either the correlation matrix or the covariance matrix. The eigenvalues and eigenvectors are extracted using a *QL* routine from Wilkinson and Reinsch. Although the factors and score procedure (in SAS) can also be used to compute principal components, they are not as convenient as PRINCOMP.

A.10 PROGRAMS FOR MULTIVARIATE REGRESSION AND DISCRETE DATA ANALYSIS

1. **Name of Program:** Mesa 85

   Code Number of Program and a Location Where It Might Be Found: UCSL 510
   University of Chicago Computation Center
   Chicago, Illinois

   General Description and Remarks: This program will run a regression on the principal components it has generated. See the principal components write-up.

2. **Name of Program:** Mesa 97 (NYBMUL)

   Code Number of Program and a Location Where It Might Be Found: UCSL 705
   University of Chicago Computation Center
   Chicago, Illinois

   Developer of Program and Affiliation: Jeremy D. Finn
   State University of New York at Buffalo
   Buffalo, New York

   Machine for Which the Program Was Written: 7094

   Machine Language: Fortran IV

Limitations: up to about 100 variables and about 4000 cells
if more storage is needed parts of the program may be removed

Form of Output: Raw and standardized regression coefficients
Standard errors
Partial correlations
Variance-covariance factors
Estimated cell means
Residuals

Reference: Finn

General Description and Remarks: Discriminant analysis, canonical correlations, and multivariate analysis of variance are additional features of the program.

3. **Name of Program:** Program for Computing Two and Three Stage Least Squares Estimates and Associated Statistics

Code Number of Program and a Location Where It Might Be Found: 2—3SLS
Center for Mathematical Studies
in Business and Economics
University of Chicago
Chicago, Illinois

Developer of Program and Affiliation: A. Stroud, A. Zellner, and L. C. Chau
Revised by H. Thornber and A. Zellner
University of Chicago
Chicago, Illinois

Machine for Which the Program Was Written: 7094

Machine Language: Fortran II

Limitations: up to 70 variables and 20 equations
up to 30 independent variables in any equation

Form of Output: Unrestricted reduced form estimates
2 stage estimates
3 stage estimates
Generalized regression estimates
Covariance and correlation matrices
Residual statistics

Reference: Stroud, Zellner, and Chau

General Description and Remarks: This program calculates two and three stage least squares estimates for the multivariate regression model. In addition the program can be used for the estimation of generalized regression equations. Trans-generations are permitted.

4. **Name of Program:** Regression on Principal Components

Code Number of Program and a Location Where It Might Be Found: BMD 02 M
Health Sciences Computing Facility
UCLA
Los Angeles, California

General Description and Remarks: This program will run a regression on the principal components it has generated. See the principal components program BMD 02 M.

5. **Name of Program:** Multivariate General Linear Hypothesis

Code Number of Program and a Location Where It Might Be Found: BMD11V
Health Sciences Computing Facility
UCLA
Los Angeles, CA 90024
(see also A.1, no. 13)

Developer of Program and Affiliation: Paul Sampson
HSCF
UCLA

Machine for Which the Program Was Written: (see A.1, no. 11)

Machine Language: FORTRAN IV

Limitations: With p independent variables and q dependent variables, the following restriction must be satisfied for each hypothesis being tested.

$$(p + q)^2 + [r,q]p + [r,q]r + [r,s]q + qs < 9000$$

where r is the number of rows in A, s is the number of rows in C, and $[X, Y]$ denotes the larger of X and Y. In any case, if $(p + q) < 55$, the inequality is satisfied. No transgenerations are available.

Form of Output:
1. Gross-product matrix $(X, Y)'(X, Y)$;
2. Regression coefficients, $B = (X'X)^{-1}X'Y$ and residual cross-product matrix $E = Y'Y - Y'XB$;

3. For each hypothesis, $A, C, D, ABC' - D$, $A(X'X)^{-1}A'$ and CEC′ matrices are printed.
4. For each hypothesis, the hypothesis sum of products matrix, U-statistic, approximate F-statistic, and degrees of freedom are printed.

Reference: BMD, Biomedical Computer Programs (1973 edition) (see also A.1, no. 13)

General Description and Remarks: The program performs a multiple regression where the dependent variable is a vector. It computes U-statistics and approximate F-statistics to test hypotheses of the form $A\beta C' = D$ where β is a matrix of regression coefficients and where $A, C,$ and D are matrices specified by the user. Estimates of $\Gamma = A\beta C' - D$ and the covariance matrix of its estimator are also obtained. With proper specification it can be used to carry out balanced or unbalanced multivariate analyses of variance and covariance.

6. **Name of Program**: Multiway Frequency Tables–The Log-Linear Model

Code Number of Program and a Location Where It Might Be Found: P3F; BMDP3F
Health Sciences Computing Facility
UCLA
Los Angeles, CA 90024
(see also A.1, no. 11)

Developer of Program and Affiliation: P3F was designed by Morton Brown, (HSCF) and programmed by Morton Brown and Koji Yamasaki, it is maintained by Lawrence Young.

Machine for Which the Program Was Written: (see A.1, no. 11)

Machine Language: FORTRAN IV

Limitations: Total number of cells for all tables must be less than 12,000. A single table of 12,000 cells can be formed but not analyzed; to be analyzed a table must have less than 3,000 cells. If neither CUTP nor CODES is specified the program assigns space for 10 levels for each factor.

Form of Output: Up to seven-factor multiway tables and all marginal tables. χ^2 and likelihood ratio χ^2

statistics for testing (1) that all $k + 1$ and higher factor interactions are zero, and (2) all k factor interactions are zero. Likelihood ratio χ^2 tests of marginal and partial association; optional expected values and two types of residuals (standardized and Freeman-Tukey); optional parameter estimates, and estimates divided by their standard error; optional orthogonal decomposition of the log-linear parameters.

Reference: 1. BMDP, Biomedical Computer Programs P-series, 1979
2. BMDP User's Digest
(see alo A.1, no. 11)

General Description and Remarks: P3F forms multiway contingency tables and analyzes them by fitting a log-linear model to the cell frequencies. Relationships between the factors of the table described either by forming a model for the data or by testing and ordering the importance of the interactions between the factors. Input may be a single observation per case or frequencies already in Table form.

7. **Name of Program**: Partial Correlation and Multivariate Regression

Code Number of Program and a Location Where It Might Be Found: P6R; BMDP6R
Health Sciences Computing Facility
UCLA
Los Angeles, CA 90024
(see also A.1, no. 11)

Developer of Program and Affiliation: James Frane
HSCF
UCLA

Machine for Which the Program Was Written: (see A.1, no. 11)

Machine Language: FORTRAN IV

Limitations: A maximum of 100 variables can be analyzed in double precision.

Form of Output: Univariate summary statistics; correlation matrix; R^2 of each independent variable with all other independent variables; R^2 of each dependent variable with the independent variables; partial correlation between the dependent variables after removing the effects of the independent variables; optional covariance matrix and partial covariances; optional regression coefficients for each dependent variable, with tests of significance and the covariances and correlations of the regression coefficients; optional scatter plots and normal probability plots.

Reference: 1. BMDP, Biomedical Computer Programs P-series, 1979
2. BMDP User's Digest
(see also A.1, no. 11)

General Description and Remarks: P6R computes the partial correlations of a set of variables after removing the linear effects of a second set of variables. The computations of the partial correlations includes the computations of the regression coefficients for predicting one set of variables from another set of variables (multivariate regression).

8. **Name of Program:** GLM—General Linear Model Procedure

Code Number of Program and a Location Where It Might Be Found: GLM; SAS
SAS Institute Inc.
(see also A.1, no. 8)

Developer of Program and Affiliation: James H. Goodnight, et al.
SAS Institute Inc.

Machine For Which the Program Was Written: (see A.1, no. 8)

Limitations: (No explicit limitations)

Form of Output: Sums of squares, degree of freedom, mean square, F-value. The Hotelling-Lawley value, Pillai's value, Wilks criterion, Roy's maximum root criterion statistics, partial correlation of dependent variables given the independent variable, residual matrix and more (see the GLM/reference).

Reference: SAS User's Guide, 1979 edition
SAS Institute
Statistical Analysis Aystem

General Description and Remarks: The GLM procedure uses the principle of least squares to fit a linear model. GLM performs both univariate and multivariate analysis, including sample linear regression, multiple linear regression, analysis of variance and covariance (univariate or multivariate), and partial correlation analysis.

9. **Name of Program**: Log Linear Models

Code Number of Program and a Location Where It Might Be Found: FUNCAT; SAS
SAS Institute Inc.
(see also A.1, no. 8)

Developer of Program and Affiliation: John P. Sall
SAS Institute Inc.

Machine for Which the Program Was Written: (see A.1, no. 8)

Machine Language: (see A.1, no. 8)

Limitations: No explicit limitations.

Form of Output: Table of frequencies for the design by the response classification; table of probability estimates; design matrix in linear model; estimated correlation matrix; estimated covariance matrix; response function values; crossed effects (interactions); nested effects; ANOVA table.

Reference: SAS User's Guide, 1979 edition
SAS Institute
Statistical Analysis System

General Description and Remarks: FUNCAT models a function of categorical responses as a linear model. It uses generalized least squares to produce minimum chi-square estimates according to the methods proposed by Grizzle, Starmer, and Koch. The estimates are asymptotically efficient. This method assumes a dense sample, with all cells filled with observations.

10. **Name of Program**: CATDAP (a categorical data analysis program package)

Code Number of Program and a Location Where It Might Be Found: CATDAP
Mr. Y. Sakamoto
The Institute of Statistics Mathematics
4-6-7 Minami-Azabu,

Minato-Ku
Tokyo, 106, JAPAN

Developer of Program and Affiliation: (Designed by Sakamoto programmed by K. Katsura and Y. Sakamoto)
Mr. Y. Sakamoto
The Institute of Statistical Mathematics
4-6-7 Minami-Azabu, Minato-Ku
Tokyo, 106, JAPAN

Machine Language: FORTRAN IV

Limitations: It can process survey data consisting of questionnaire items up to one hundred. (It is applicable even in the cases where we know very little about the shape of the population distribution from which the samples are drawn.)

Form of Output: List of explanatory variables arranged in ascending order of AIC, two-way tables arranged in ascending order of AIC, summary of AIC's for the two-way tables, gray shading display of all the AIC's response variables, contingency table with the optimal combination of explanatory variables, summary of subsets of explanatory variables arranged in ascending order of AIC and more.

Reference: CATDAP, A categorical data analysis program package by K. Katsura and Y. Sakamoto, Computer Science Monographs no. 14 (A publication of the Institute of Statistical Mathematics. The Institute of Statistical Mathematics, 6-7, 4-chome, Minami-Azaba, Minato-Ku, Tokyo, 106, Japan.

General Description and Remarks: CATDAP is a package of programs for the analysis of cross-classified data. The package consists of the following two main programs, CATDAP-01 and CAPDAP-02. The following table characterizes the problems dealt with by the programs CATDAP-01 and CATDAP-02.

	Explanatory		
Response variable	Categorical	Continuous	Mixed
Categorical	CATDAP-01/02	CATDAP-02	CATDAP-02

The program uses a stepwise selection procedure with respect to *F*. CATDAP-01 can handle many variables simultaneously as response variables. The basic statistics adopted in this package is obtained by the applications of the statistics AIC to the model developed by Sakamoto and Sakamoto and Akaike.

11. **Name of Program:** Everyman's Contingency Table Analysis (ECTA)

Code Number of Program and a Location Where It Might Be Found: ECTA
Department of Statistics
University of Chicago
1118 East 58th Street
Chicago, IL 60637

Developer of Program and Affiliation: Professor Leo Goodman and Robert Fay
Department of Statistics
University of Chicago
1118 East 58th Street
Chicago, IL 60637

Machine for Which the Program Was Written: IBM 360/370, CDC, Univac (can be used with other computers, see the reference)

Machine Language: FORTRAN IV

Limitations: With 4000 words all functions can be performed on any table of size 500, including the special case of $512 = 2^9$. Fitting can be performed on a table of about 1200, if no estimation is requested.

Form of Output: Table of observed and fitted values, estimates of parameters of model, effects, χ^2 and degrees of freedom and more.

Reference: ECTA Program
Description for Users
(see also developer of program and affiliation)

General Description and Remarks: The ECTA program is able to calculate log-linear fits for hierarchical models for contingency tables. It is also able to estimate the parameters of the models. The program is able to fit quasi-independent models and other related models for tables with missing cells.

12. **Name of Program:** Analysis of Linear Structural Relationship by the Method of Maximum Likelihood

Code Number of Program and a Location Where It Might Be Found: LISREL IV
International Educational Services
1525E E. 53rd Street,

Rm. 829
Chicago, IL 60615

Developer of Program and Affiliation: Karl G. Jöreskog and Dag Sörbom
University of Uppsala

Machine for Which the Program Was Written: IBM, CDC, etc.

Machine Language: FORTRAN IV

Limitations: The IBM version of LISREL IV has two system dependent features required to implement dynamic core allocation and overtime protections. For detail see the reference.

Form of Output: The printed output consists of:

1. Standard output
2. The matrix being analyzed and starting values
3. Miscellaneous results computed from the LISREL ESTIMATES
4. Factor scores regression
5. Technical output
6. Standard errors, *t*-values and correlation matrix of the LISREL estimates
7. First derivatives
8. Standardized solution

The standard output is always obtained; all other parts are optional. The standard output consists of the title with parameter listing, the parameter specifications, the LISREL (maximum likelihood) estimates and the result of the test of goodness of fit.

Reference: LISREL IV (user's Guide)
Analysis of Linear Structural Relationships by the Method of Maximum Likelihood by Karl G. Jöreskog and Dag Sörbom
University of Uppsala
Copyright, 1978, by National Educational Resources, Inc.
Distributed by
International Educational Services
1525 E. 53rd Street, Rm. 829
Chicago, IL 60615

General Description and Remarks: LISREL is a general computer program for estimating the unknown coefficients in a set of linear structural equations. The variables in the equation system may be either directly observed variables or unmeasured hypothetical construct variables or latent variables which are not observed but related to other observed variables. The model allows for both errors in equations (disturbances) and errors in the observed variables (errors of measurement) and yields estimates of the residual covariance matrix and the measurement error covariance matrix as well as estimates of the casual effects in the structural equations provided that all these parameters are identified. The programs cover a wide range of models useful in the social and behavioral sciences, for example, path analysis models, econometric models, recursive or interdependent factor analysis and covariance structure models. (The features necessary to do a simultaneous analysis in several groups are described in Chapter III of the reference.)

13. **Name of Program**: Log-linear probability model

Code Number of Program and a Location Where It Might Be Found: LOGLIN
Professor Marc Nerlove
Department of Economics
Northwestern University
Evanston, IL

Developer of Program and Affiliation: K. M. Mauer of the Rand Corporation and R. Olsen of Northwestern University, Economics Dept.

Machine for Which the Program Was Written: IBM 360/65

Machine Language: FORTRAN

Limitations: A maximum of four jointly dependent dichotomous variables; the main effects may depend upon up to 16 exogenous explanatory variables (and a constant term); interaction terms are assumed constant.

Form of Output: Maximum likelihood estimates of coefficients of all explanatory variables, main effects, and interaction terms in both saturated and unsaturated models.

Reference: See Nerlove and Press, 1973, in Appendix C.

General Description and Remarks: This program implements computation for the multivariate log-linear/logistic regression model, with polytomous response variables, and where the main effects may depend upon exogenous explanatory variables.

A.11 BAYESIAN COMPUTER PROGRAMS

I. Normal Linear Regression Model Programs

Program Number

1. Linear Bayesian regression analysis (Allum)–BAYREG
2. Bayesian inference in normal regression (Leamer and Leonard)–SEARCH
3. Bayesian regression analysis (Abowd/Zellner)–BRAP
4. Interactive elicitation of opinion for a normal linear model (Kadane/Dickey)
5. Bayesian regression program (Drèze)–BRP
6. Multiple regression analysis with incomplete observations (Dagenais)–IOGQ3P
7. Stepwise regression (Thornber/Zellner)–B-34T, May 1, 1967.
8. Multiple regression, ANOVA, utility functions (Isaacs/Novick)–CADA

II. Bayesian Programs In Sampling Problems

9. Estimating category probabilities with undecided respondents (Press and Yang)–BECP
10. Sequential analysis in decision problems (Leventhal)–SEQUAN
11. Sampling problems in auditing (Ilderton)–BAYVAR
12. Attributes sampling (Martz)–EBSPE
13. Empirical Bayes estimates for parameters in Lot Acceptance Quality Control (Martz)–EBSSP

III. Confidence Interval Programs

14. Confidence intervals for cdfs., for hazard rate functions, and reliability problems (Stewart)
15. Confidence intervals for the Fieller-Creasy Problem (Kappenman/Geisser)

Program #1

Name of Program: BAYREG

Function: Linear Bayesian Regression Analysis

1. Means and variance-covariance matrix for the

prior distribution of the regression coefficients (assumed multivariate normal).
2. Estimated standard deviation of the response variable.

Output: 1. Listing of prior information.
2. Bayesian regression coefficients.

Computer Language and Machine: FORTRAN IV with IBM 370.

Documentation: Available on request.

Availability: Program is part of Central Electricity Generating Board, England, Statistical System STATSYS. This system would not function as a "stand alone," without modification.

Remarks: 1. Program developed was based on work of D. V. Lindley and A. F. M. Smith by Applied Statistics Section of CEGB's Computing Services Department.
2. Sent by W. J. Allum, Section Head, CEGB, London, England.

Program #2

Name of Program: SEARCH

Function: User-oriented Bayesian inference package to carry out formal pooling of user's prior beliefs and data evidence as long as they can be described completely by a location and a precision or variance covariance matrix. In particular, the package can be used on normal linear regression models.

Input: Input on cards, formatted or free-formatted.

Output: 1. Diagnostic messages to aid in correcting syntax errors.
2. Analytic outputs:
Summary of prior and data information received
Contract Curve Analyses
Extreme Bounds Analyses (computes extreme bounds for any linear function specified).

Computer Language and Machine: IBM 370/168 at Harvard/MIT; FORTRAN.

Documentation: "SEARCH–Manual for Bayesian Inference," by Herman B. Leonard, Technical Paper No. 14, Sept. 1977 (SEARCH Version 3.3HU April 18, 1977), Harvard Institute of Economic Research, Harvard University; Cambridge, Massachusetts. The manual describes the Harvard University variant of Version 3.3, released on April 18, 1977. This variant differs from the standard version only in that it exploits special features of the line printer control channels Provided on the Harvard/MIT 370/168.

Availability: Available from

Prof. Edward E. Leamer
Dept. of Economics
UCLA
405 Hilgard Ave.
Los Angeles, CA 90024
(213) 825-3925

Prof. Herman B. Leonard
64 Linnaean St.
Cambridge, MA 02138
(617) 495-3290

Remarks:

1. Developed by Edward E. Leamer and Herman B. Leonard.
2. Programmed by Herman Leonard, Kathy Burgoyne, and Robert Topel.
3. Research concerning development was supported by NSF Grants GS-31929 and SOC 76-08863.

Program #3

Name of Program: BRAP: Bayesian Regression Analysis Package, Version 1.0

Function: Perform Bayesian analysis of the normal linear multiple regression model with multivariate normal errors under diffuse or natural conjugate prior assumptions. Both the prior and posterior distributions for the regression coefficients can be analyzed. The posterior distribution for linear combinations of coefficients, the realized error terms, the predictive distribution of the dependent variable and several parameterizations of the error variance can also be analyzed.

Input: Formatted or unformatted input from cards, disk, or tape. Allows an addition of a FORTRAN transformation subroutine.

Output: Computes and plots prior and posterior distributions for parameters, moments of these distributions,

	and other measures of interest to investigators. Also, provides standard posterior information under both diffuse and natural priors.
Computer Language and Machine:	FORTRAN with IBM 370/168.
Documentation:	Abowd, John M., "BRAP User's Manual, Version 1.0 of 9/8/77," H. G. B. Alexander Research Foundation, Graduate School of Business, University of Chicago, September 1977, 65 pp.
Availability:	Available from Professor A. Zellner, University of Chicago, Graduate School of Business.
Remarks:	1. Contributors include F. Finnegan, S. Grossman, C. Plosser, A. Siow, J. Stafford, W. Vandaele, and A. Zellner. 2. Support for development of this program package was provided by NSF Grant GF-40033 and SOC 73-05547.

Program #4

Name of Program:	Interactive Elicitation of Opinion for a Normal Linear Model.
Function:	To interactively elicit subjective prior distributions for normal linear regression models using the conjugate prior family. Questions are asked about the quantiles of the predictive distribution of the dependent variable conditional on the independent variable.
Input:	By means of computer terminal.
Output:	When all the parameters have been elicited, the program yields a fitted prior distribution. If data is available, a posterior and posterior predictive distribution are also returned.
Computer Language and Machine:	Macro-language for TROLL Econometric Modeling 370/168 at Cornell.
Documentation:	"TROLL Experimental Programs: Bayesian Regression," Document D0070, Center for Computational Research in Economics and Management Science at MIT.
Availability:	Available from Professor J. Kadane, Department of

Statistics, Carnegie-Mellon University.

Remarks:

1. Implemented on the MIT system TROLL.
2. Research supported in part by the National Science Foundation and in part by the National Highway Research Program.
3. Reference: Kadane, Joseph B., Dickey, James M., Winkler, Robert L., Smith, Wayne S., and Peters, Stephen C., "Interactive Elicitation of Opinion for a Normal Linear Model."
4. Program expected to be implemented in CADA Monitor (1979 version). (Availabe from Professor Novick. See Program 8).
5. The program may be used nationwide on the Telenet system using a telephone terminal.

Program #5

Name of Program: BRP: Bayesian Regression Program

Function: To execute the necessary computations for Bayesian inference in various standard econometric models.

Input: Raw data by cards for batch jobs. A future version may allow for input conversationally from terminals.

Output: Posterior parameters, marginal posteriors of $\tilde{\beta}$, of precision and standard deviations; results of classical regression; conditional posterior (with given precision), posterior residuals and predictive density function; and conditional posterior distribution of some coefficients, given some others, marginally the precision. Input data are echoed.

Computer Language and Machine: FORTRAN IV with IBM 370/158 at the University of Louvain, Belgium.

Documentation: Bauwens, L. and Tompa, H., "Bayesian Regression Programme (BRP)," CORE User's Manual, Set No. A-5, May 1977.

Tompa, H., "Poly-*t* Distributions," CORE User's Manual, Set No. C-9, May 1977.

Availability: Available from Professor Jacques Drèze, CORE, 34 Voie du Roman Pays, 1348 Louvain-la-Neuve, Belgium, 010/41.81.81.

Remarks:

1. Development of BRP by Hans Tompa is taking place at CORE under the "Programme National d'Impulsion à la Recherche en Informatique" of the Belgian Government (Contract I/14 bis 6), with advice provided by Jacques Drèze and Jean-Francois Richard, and assistance from Luc Bauwens, Jean-Paul Bulteau and Philippe Gille.
2. BRP calls upon another program, PTD, to evaluate poly-t densities.

Program #6

Name of Program: IOGQ3P Incomplete observations–Gaussian quadrature–3 parameters.

Function: To carry out posterior inferences in a normal linear regression when some of the observations are missing.

Input: Raw data by cards, tape or disk.

Output: Marginal posterior distributions of the regression parameters.

Computer Language and Machine: CDC CYBER 74 (in FORTRAN IV).

Documentation: Dagenais, M. G., "Multiple Regression Analysis with incomplete observations, from a Bayesian Viewpoint," in *Studies in Bayesian Econometrics and Statistics,* ed. by A. Zellner and S. Fienberg, North-Holland Publishing Co., 1974, à paraître.

Availability: Available from Professor Marcel G. Dagenais, Université de Montréal, Montréal, Canada

Remarks:

1. The developer of the program was Tran Cong Liem, of the Université de Montréal.
2. Sent by Jacques Beaudry, Université de Montréal.

Program #7

Name of Program: (B34T, 1st May, 1967) A Stepwise Regression Program (CDC version).

Function: Stepwise Regression, transformation of raw data corrects for first order autocorrelation, also computes the marginal posterior of the regression coefficient, the autocorrelation coefficient, and the population R^2.

Input: By card, tape, or read off a logical unit in binary or BCD. Input can be raw data read in variable format or cross-product matrix.

Output: Can punch transformed data, regression coefficient and cross-product matrix. Plots marginal posteriors and residuals.

Computer Language and Machine: CDC CYBER 74 version in FORTRAN IV and assembly language (COMPASS). (The original version, available from the University of Chicago, is in IBM 7094 version in FORTRAN II).

Documentation: Dr. H. Thornber, "Manual for (B34T, 8 March, 1966) a Stepwise Regression Program," Technical Report 6603, Center for Mathematical Studies in Business and Economics, University of Chicago, also (BAYES addendum to Technical Report 6603).

Availability: IBM version and documentation available from Center for Mathematical Studies in Business and Economics, University of Chicago. CDC version available from Professor Marcel G. Dagenais, Université de Montréal, Montréal, Canada.

Remarks:

1. Sent by Jacques Beaudry, Université de Montréal, Montréal, Canada.
2. This program was developed by Dr. H. Thornber under the supervision of Professor Arnold Zellner; the development was financed by Professor Zellner's NSF grant, "Bayesian Inference in Econometrics and Related Topics."

Program #8

Name of Program: Computer-Assisted Data Analysis (CADA), 1977.

Function: Multipurpose: General linear model, studies of distributions arising in Bayesian inference, assessment of utility functions.

Input: Raw data typed in, in real-time, on on-line console.

Output: Analysis of Beta-Binomial model, two parameter normal model, multinomial model, various utility functions, multiple factor ANOVA, comparison of means and *m*-group and Beta-Distribution proportions. Evaluations of the following distributions:

Student t, Beta, Inverse-Chi, Normal, Behrens-Fisher, Inverse Chi-Square, Chi-Square, Snedecor's F, Binomial, Pascal, Beta Binomial, Beta Pascal, Poisson, Gamma, Calculation of statistics for univariate and bivariate data. Multiple linear regression analysis, as well as simultaneous estimation of regressions in m groups, can be performed.

Computer Language and Machine: BASIC with HP 2000 ACCESS, DEC-PDP-11-RSTS, DEC-PDP-10, CEC CYBER-NOS, IBM 370 VS, and Univac 1100.

Documentation: "The Bayesian Computer-Assisted Data Analysis (CADA) Monitor—1977" by Melvin R. Novick, Gerald L. Isaacs, and Dennis DeKeyrel, the University of Iowa, Department of Education.

Availability: Available from Professor Melvin R. Novick, Department of Education and Statistics, University of Iowa.

Remarks:

1. The first BASIC version of CADA was written by Isaacs and Christ in the Basic dialect for the CDC 3600 at the University of Massachusetts. This was then translated into versions for the Hewlett-Packard 2000C and the Digital Equipment Corporation PDP-11.
2. This new and expanded version of CADA offers additional statistical analysis not available in the previous version and is better organized so data is passed automatically from one module to another.
3. The System of Computer-Assisted Data Analysis (CADA) was developed at the University of Iowa (Novick 1971, 1973).
4. Sent by Melvin R. Novick, University of Iowa.
5. Development: Melvin R. Novick, Principal Investigator; Gerald L. Isaacs, Project Leader; Dennis F. Dekeyrel, Project Analyst.
6. The work was partially supported by National Science Foundation Grants #EPP73-00164 and #SED 77-18432.
7. Available on the following media:
   HP 200 ACCESS Dump Tape (9 trk, 800 BPI)
   CDC CYBER-NOS Dump Tape (9 trk)
   Magnetic Tape (9 trk, 800 BPI, ASCH)
   Magnetic Tape (9 trk, 800 BPI, EBCDIC).

Program #9

Name of Program: Bayesian Estimation of Category Probabilities in Sample Surveys Involving "Undecided" Respondents (BECP).

Function: Estimating category probabilities when some of the respondents attempt to conceal their true beliefs about sensitive questions by responding in the "don't know," "no opinion," or "undecided" category. The program utilizes information obtained from "decided" respondents on the *main question* of interest, and on some *subsidiary questions* posed to the same respondents. The respondents in the "undecided" category on the main question are classified into one of the unambiguous categories based on their responses to the same set of subsidiary questions answered by the respondents who were unambiguous on the main question.

Input:

1. Prior probability for the category (if non-vague prior is used;
2. data from "undecided" respondents on subsidiary questions;
3. data from "decided" respondents on main and subsidiary questions.

Output: The estimated category probabilities. In addition, standard errors are provided.

Computer Language and Machine: FORTRAN IV, IBM 360

Documentation: A manual for the Computer Program for "Bayesian Estimation of Category Probabilities in Sample Surveys Involving Second-Guessing 'Undecided' Respondents," by C. E. Yang, September 1978.

Availability: Available from Professor C. E. Yang, Department of Marketing, School of Business, University of Southern California, Los Angeles, CA.

Remarks:

1. Reference: S. J. Press and C. E. Yang, "A Bayesian Approach to Second-Guessing 'Undecided' Respondents," *JASA*, 69, (March 1974), 58–67.
2. This program was written by W. Y. Lau.

Program #10

Name of Program:	SEQUAN—Programs for Sequential Analysis, consisting of BASSM—Bayesian Sequential Sampling Plan for a many-decision problem. BASS2—Bayesian Sequential Sampling Plan for a 2 decision problem. GASP—General Analysis of Sequential Plans.
Function:	Design and analysis of sequential sampling plans. Calculation of the Bayes sequential strategy for a general class of hypothesis tests of an unknown parameter.
Input:	FORTRAN function sub-programs to specify functional ingredients of problem and data cards, punched with fixed formats to provide all parameter values.
Output:	Bayes sequential strategies.
Computer Language and Machine:	FORTRAN
Documentation:	Available on request.
Availability:	Available from Barry Leventhal, University College London Computer Centre, London, England.
Remarks:	1. Written and sent by Barry Leventhal, University College London Computer Centre. 2. Program assumes exponential family distribution of each observation.

Program #11

Name of Program:	"Use of Bayesian Methods in Applying Statistical Sampling to Auditing," consisting of four time-sharing programs: BAYVAR, KRAFT 1, BAYSI 1, BAYSI 2.
Function:	BAYVAR is used to apply the Bayesian Method to sampling for variables. KRAFT 1, BAYSI 1, BAYSI 2 apply the Bayes Formula to sampling for attributes.
Input:	Raw data on paper tape. Conversational.

Output: *BAYVAR* If current year's audit are consistent with prior audit, confidence level for each confidence interval, on the basis of the prior probability distribution and the current results are performed. Otherwise, a message for determination of cause of large difference. Unit and ratio costs questioned and related standard errors.

KRAFT 1 Required sample size and total sample reliability for the error rates selected in the prior probability distribution.

BAYSI 1 Minimum sample sizes required.

BAYSI 2 Minimum sample sizes required.

Computer Language and Machine: BASIC.

Documentation: "Use of Bayesian Methods in Applying Statistical Sampling to Auditing," currently used by the Defence Contract Audit Agency.

Availability: Available from Robert Ilderton, Chief, Techniques Research and Special Audits Branch, H.Q. Defense Contract Audit Agency, Alexandria, Virginia 22314.

Remarks: Sent by Robert Ilderton, Techniques Research and Special Audits Branch.

Program #12

Name of Program: EBSPE—Empirical Bayes Posterior Performance Evaluation of any Single-Sampling-by-Attributes Sampling Plan.

Function: Used to conduct an empirical Bayes posterior performance evaluation of any specified single-sampling-by-attributes sampling plan.

Input: Raw data punched on cards.

Output: Empirical Bayes posterior quantities; operating characteristics curves, and optional printout of lot-by-lot input data.

Computer Language and Machine: FORTRAN IV with IBM 370/145.

Documentation: Martz, Harry F., Jr., "EBSPE—A FORTRAN Computer Program for Conducting an Empirical Bayes

Posterior Performance Evaluation of any Single-Sampling-by-Attributes Sampling Plan," Technical Report No. 4, Department of Industrial Engineering, Texas Tech. University, December 1974.

Availability: Approved for public release by Office of Naval Research, Statistics and Probability Program; CODE-436, Arlington, Virginia 22217, U.S.A. Available from Professor Harry F. Martz, Jr., Department of Industrial Engineering, Texas Tech. University, Texas.

Remarks:
1. Written and sent by Harry F. Martz, Jr., Texas Tech. University.
2. Binomial sampling is assumed throughout.

Program #13

Name of Program: *EBSSP–Determines Empirical Bayes Single–Sampl*ing Plans for Specified Consumer and Producer Risks.

Function: To find the empirical Bayes single-sample-by-attributes inspection sampling plan that achieves a specified empirical Bayes posterior producer's risk and a specified empirical Bayes posterior consumer's risk.

Input: Raw data punched on cards.

Output: Optional printout at each iteration of the empirical Bayes estimates of producer's risk, consumer's risk and unconditional probability of lot acceptance and expected lot fraction defective in lots accepted; optimum required sampling plan and associated quantities; and optional printout of the lot-by-lot input data.

Computer Language and Machine: FORTRAN IV with IBM 370/145.

Documentation: "EBSSP–A FORTRAN Computer Program for Determining Empirical Bayes Single-Sampling Plans for Specified Consumer and Producer Risks," by Harry F. Martz, Jr., Technical Report No. 3, Department of Industrial Engineering, Texas Tech. University, December 1974.

Availability: Approved for public release by Office of Naval

Research, Statistics and Probability Program; CODE 436; Arlington, Virginia 22217, U.S.A. Available from Professor Harry F. Martz, Jr., Department of Industrial Engineering, Texas Tech. University, Texas.

Remarks:
1. Binomial sampling is assumed throughout.
2. Written and sent by Harry F. Martz, Jr., Texas Tech. University.

Program #14

Name of Program: Multiparameter Univariate Bayesian Analysis Using Monte Carlo Integration.

Function: To handle problems in univariate inference with Bayesian analysis and Monte Carlo integration. The program can handle usual random sampling data, censored data, interval data, success-failure data at different stresses, and success-failure data versus time.

Input: On cards, files, or tapes.

Output: Plots Bayesian confidence limits for cumulative distribution functions, failure (hazard) rate functions, probability of failure versus stress functions, and reliability versus time curves. Optimal decisions can be computed if a loss function is specified.

Computer Language and Machine: FORTRAN; UNIVAC 1110 and SC-4020 plotter.

Documentation: A user's manual will be available in January, 1979. Reference: Stewart, Leland, "Multiparameter Univariate Bayesian Analysis Using Monte Carlo Integration."

Availability: Dr. Leland Stewart, Lockheed Palo Alto Research Laboratory, Palo Alto, California 94304 U.S.A.

Remarks:
1. The project was coordinated by Dr. Leland Stewart.
2. The program allows the families of distributions used in the program to have up to ten parameters.
3. The program can incorporate random sampling variation and uncertainty about the assumptions

on which a required extrapolation beyond the range of the data is based.

4. The program has been used extensively over the past ten years at Lockheed.

Program #15

Name of Program:	Bayesian and Fiducial Solutions for the Fieller-Creasy Problem.
Function:	To provide Bayesian confidence intervals for the Fieller-Creasy Problem.
Input:	Raw data on cards.
Output:	The plot of the density function of η, the ratio of the means, a closed interval for η, and alternate posterior limits for η.
Computer Language and Machine:	FORTRAN IV.
Documentation:	"Bayesian and Fiducial Solutions for the Fieller-Creasy Problem" by Russell F. Kappenman, Seymour Geisser, and Charles E. Antle, in *Sankhyā, Indian Journal of Statistics,* Series B, Vol. 32, Parts 3 and 4, 1970.
Availability:	FORTRAN program given in the reference.
Remarks:	Written by Russell F. Kappenman, Pennsylvania State University, Seymour Geisser, University of Minnesota, and Charles E. Antle, Pennsylvania State University. Sent by Seymour Geisser, University of Minnesota.

appendix B

TABLES

Table B.1 Cumulative Normal Distribution

$$F(z) = \int_{-\infty}^{z} \frac{1}{\sqrt{2\pi}} \exp(-t^2/2)dt$$

z	.00	.01	.02	.03	.04	.05	.06	.07	.08	.09
.0	.5000	.5040	.5080	.5120	.5160	.5199	.5239	.5279	.5319	.5359
.1	.5398	.5438	.5478	.5517	.5557	.5596	.5636	.5675	.5714	.5753
.2	.5793	.5832	.5871	.5910	.5948	.5987	.6026	.6064	.6103	.6141
.3	.6179	.6217	.6255	.6293	.6331	.6368	.6406	.6443	.6480	.6517
.4	.6554	.6591	.6628	.6664	.6700	.6736	.6772	.6808	.6844	.6879
.5	.6915	.6950	.6985	.7019	.7054	.7088	.7123	.7157	.7190	.7224
.6	.7257	.7291	.7324	.7357	.7389	.7422	.7454	.7486	.7517	.7549
.7	.7580	.7611	.7642	.7673	.7704	.7734	.7764	.7794	.7823	.7852
.8	.7881	.7910	.7939	.7967	.7995	.8023	.8051	.8078	.8106	.8133
.9	.8159	.8186	.8212	.8238	.8264	.8289	.8315	.8340	.8365	.8389
1.0	.8413	.8438	.8461	.8485	.8508	.8531	.8554	.8577	.8599	.8621
1.1	.8643	.8665	.8686	.8708	.8729	.8749	.8770	.8790	.8810	.8830
1.2	.8849	.8869	.8888	.8907	.8925	.8944	.8962	.8980	.8997	.9015
1.3	.9032	.9049	.9066	.9082	.9099	.9115	.9131	.9147	.9162	.9777
1.4	.9192	.9207	.9222	.9236	.9251	.9265	.9279	.9292	.9306	.9319
1.5	.9332	.9345	.9357	.9370	.9382	.9394	.9406	.9418	.9429	.9441
1.6	.9452	.9463	.9474	.9484	.9495	.9505	.9515	.9525	.9535	.9545
1.7	.9554	.9564	.9573	.9582	.9591	.9599	.9608	.9616	.9625	.9633
1.8	.9641	.9649	.9656	.9664	.9671	.9678	.9686	.9693	.9699	.9706
1.9	.9713	.9719	.9726	.9732	.9738	.9744	.9750	.9756	.9761	.9767
2.0	.9772	.9778	.9783	.9788	.9793	.9798	.9803	.9808	.9812	.9817
2.1	.9821	.9826	.9830	.9834	.9838	.9842	.9846	.9850	.9854	.9857
2.2	.9861	.9864	.9868	.9871	.9875	.9878	.9881	.9884	.9887	.9890
2.3	.9893	.9896	.9898	.9901	.9904	.9906	.9909	.9911	.9913	.9916
2.4	.9918	.9920	.9922	.9925	.9927	.9929	.9931	.9932	.9934	.9936
2.5	.9938	.9940	.9941	.9943	.9945	.9946	.9948	.9949	.9951	.9952
2.6	.9953	.9955	.9956	.9957	.9959	.9960	.9961	.9962	.9963	.9964
2.7	.9965	.9966	.9967	.9968	.9969	.9970	.9971	.9972	.9973	.9974
2.8	.9974	.9975	.9976	.9977	.9977	.9978	.9979	.9979	.9980	.9981
2.9	.9981	.9982	.9982	.9983	.9984	.9984	.9985	.9985	.9986	.9986
3.0	.9987	.9987	.9987	.9988	.9988	.9989	.9989	.9989	.9990	.9990
3.1	.9990	.9991	.9991	.9991	.9992	.9992	.9992	.9992	.9993	.9993
3.2	.9993	.9993	.9994	.9994	.9994	.9994	.9994	.9995	.9995	.9995
3.3	.9995	.9995	.9995	.9996	.9996	.9996	.9996	.9996	.9996	.9997
3.4	.9997	.9997	.9997	.9991	.9997	.9997	.9997	.9997	.9997	.9998

Table B.2 Cumulative Chi-Square Distribution

$$F(u) = \int_0^u \frac{x^{(n-2)/2}e^{-x/2}\,dx}{2^{n/2}[(n-2)/2]!}$$

n \ $F(u)$	.005	.010	.025	.050	.100	.250	.500	.750	.900	.95	.975	.990	.995
1	.000	.000	.001	.004	.016	.102	.455	1.32	2.71	3.84	5.02	6.63	7.88
2	.010	.020	.051	.103	.211	.575	1.39	2.77	4.61	5.99	7.38	9.21	10.6
3	.072	.115	.216	.352	.584	1.21	2.37	4.11	6.25	7.81	9.35	11.3	12.8
4	.207	.297	.484	.711	1.06	1.92	3.36	5.39	7.78	9.49	11.1	13.3	14.9
5	.412	.554	.831	1.15	1.61	2.67	4.35	6.63	9.24	11.1	12.8	15.1	16.7
6	.676	.872	1.24	1.64	2.20	3.45	5.35	7.84	10.6	12.6	14.4	16.8	18.5
7	.989	1.24	1.69	2.17	2.83	4.25	6.35	9.04	12.0	14.1	16.0	18.5	20.3
8	1.34	1.65	2.18	2.73	3.48	5.07	7.34	10.2	13.4	15.5	17.5	20.1	22.0
9	1.73	2.09	2.70	3.33	4.17	5.90	8.34	11.4	14.7	16.9	19.0	21.7	23.6
10	2.16	2.56	3.25	3.94	4.87	6.74	9.34	12.5	16.0	18.3	20.5	23.2	25.2
11	2.60	3.05	3.82	4.57	5.58	7.58	10.3	13.7	17.3	19.7	21.9	24.7	26.8
12	3.07	3.57	4.40	5.23	6.30	8.44	11.3	14.8	18.5	21.0	23.3	26.2	28.3
13	3.57	4.11	5.01	5.89	7.04	9.30	12.3	16.0	19.8	22.4	24.7	27.7	29.8
14	4.07	4.66	5.63	6.57	7.79	10.2	13.3	17.1	21.1	23.7	26.1	29.1	31.3
15	4.60	5.23	6.26	7.26	8.55	11.0	14.3	18.2	22.3	25.0	27.5	30.6	32.8

Table B.2 *(continued)*

n \ $F(u)$	.005	.010	.025	.050	.100	.250	.500	.750	.900	.95	.975	.990	.995
16	5.14	5.81	6.91	7.96	9.31	11.9	15.3	19.4	23.5	26.3	28.8	32.0	34.3
17	7.50	6.41	7.56	8.67	10.1	12.8	16.3	20.5	24.8	27.6	30.2	33.4	35.7
18	6.26	7.01	8.23	9.39	10.9	13.7	17.3	21.6	26.0	28.9	31.5	34.8	37.2
19	6.84	7.63	8.91	10.1	11.7	14.6	18.3	22.7	27.2	30.1	32.9	36.2	38.6
20	7.43	8.26	9.59	10.9	12.4	15.5	19.3	23.8	28.4	31.4	34.2	37.6	40.0
21	8.03	8.90	10.3	11.6	13.2	16.3	20.3	24.9	29.6	32.7	35.5	38.9	41.4
22	8.64	9.54	11.0	12.3	14.0	17.2	21.3	26.0	30.8	33.9	36.8	40.3	42.8
23	9.26	10.2	11.7	13.1	14.8	18.1	22.3	27.1	32.0	35.2	38.1	41.6	44.2
24	9.89	10.9	12.4	13.8	15.7	19.0	23.3	28.2	33.2	36.4	39.4	43.0	45.6
25	10.5	11.5	13.1	14.6	16.5	19.9	24.3	29.3	34.4	37.7	40.6	44.3	46.9
26	11.2	12.2	13.8	15.4	17.3	20.8	25.3	30.4	35.6	38.9	41.9	45.6	48.3
27	11.8	12.9	14.6	16.2	18.1	21.7	26.3	31.5	36.7	40.1	43.2	47.0	49.6
28	12.5	13.6	15.3	16.9	18.9	22.7	27.3	32.6	37.8	41.3	44.5	48.3	51.0
29	13.1	14.3	16.0	17.7	19.8	23.6	28.3	33.7	39.1	42.6	45.7	48.6	52.3
30	13.8	15.0	16.8	18.5	20.6	24.5	29.3	34.8	40.3	43.8	47.0	50.9	53.7

SOURCE: The entries in this table appeared in "Tables of percentage points of the incomplete beta function and of the chi-square distribution," *Biometrika*, Vol. 32 (1941). They are reproduced here with the kind permission of the author, C. M. Thompson, and the editor of *Biometrika*.

Table B.3 Cumulative "Student's" Distribution

$$F(t) = \int_{-\infty}^{t} \frac{[(n-1)/2]!\; dx}{[(n-2/2]!\ \sqrt{\pi n}\ (1 + x^2/n)^{(n+1)/2}}$$

n \ F	.55	.60	.65	.70	.75	.80	.85
1	.158	.325	.510	.727	1.000	1.376	1.963
2	.142	.289	.445	.617	.816	1.061	1.386
3	.137	.277	.424	.584	.765	.978	1.250
4	.134	.271	.414	.569	.741	.941	1.190
5	.132	.267	.408	.559	.727	.920	1.156
6	.131	.265	.404	.553	.718	.906	1.134
7	.130	.263	.402	.549	.711	.896	1.119
8	.130	.262	.399	.546	.706	.889	1.108
9	.129	.261	.398	.543	.703	.883	1.100
10	.129	.260	.397	.542	.700	.879	1.093
11	.129	.260	.396	.540	.697	.876	1.088
12	.128	.259	.395	.539	.695	.873	1.083
13	.128	.259	.394	.538	.694	.870	1.079
14	.128	.258	.393	.537	.692	.868	1.076
15	.128	.258	.393	.536	.691	.866	1.074
16	.128	.258	.392	.535	.690	.865	1.071
17	.128	.257	.392	.534	.689	.863	1.069
18	.127	.257	.382	.534	.688	.862	1.067
19	.127	.257	.391	.533	.688	.861	1.066
20	.127	.257	.391	.533	.687	.860	1.064
21	.127	.257	.391	.532	.686	.859	1.063
22	.127	.256	.390	.532	.686	.858	1.061
23	.127	.256	.390	.532	.685	.858	1.060
24	.127	.256	.390	.531	.685	.857	1.059
25	.127	.256	.390	.531	.684	.856	1.058
26	.127	.256	.390	.531	.684	.856	1.058
27	.127	.256	.389	.531	.684	.855	1.057
28	.127	.256	.389	.530	.683	.855	1.056
29	.127	.256	.389	.530	.683	.854	1.055
30	.127	.256	.389	.530	.683	.854	1.055
40	.126	.255	.388	.529	.681	.851	1.050
60	.126	.254	.387	.527	.679	.848	1.046
120	.126	.254	.386	.526	.677	.845	1.041
∞	.126	.253	.385	.524	.674	.842	1.036

Table B.3 *(continued)*

n \ F	.90	.95	.975	.99	.995	.9995
1	3.078	6.314	12.706	31.821	63.657	636.619
2	1.886	2.910	4.303	6.965	9.925	31.598
3	1.638	2.353	3.182	4.541	5.841	12.941
4	1.533	2.132	2.776	3.747	4.604	8.610
5	1.476	2.015	2.571	3.365	4.032	6.859
6	1.440	1.943	2.447	3.143	3.707	5.959
7	1.415	1.895	2.365	2.998	3.499	5.405
8	1.397	1.860	2.306	2.896	3.355	5.041
9	1.383	1.833	2.262	2.821	3.250	4.781
10	1.372	1.812	2.228	2.764	3.169	4.587
11	1.363	1.796	2.201	2.718	3.106	4.437
12	1.356	1.782	2.179	2.681	3.055	4.318
13	1.350	1.771	2.160	2.650	3.012	4.221
14	1.345	1.761	2.145	2.624	2.977	4.140
15	1.341	1.753	2.131	2.602	2.947	4.073
16	1.337	1.746	2.120	2.583	2.921	4.015
17	1.333	1.740	2.110	2.567	2.898	3.965
18	1.330	1.734	2.101	2.552	2.878	3.922
19	1.328	1.729	2.093	2.539	2.861	3.883
20	1.325	1.725	2.086	2.528	2.845	3.850
21	1.323	1.721	2.080	2.518	2.831	3.819
22	1.321	1.717	2.074	2.508	2.819	3.792
23	1.319	1.714	2.069	2.500	2.807	3.767
24	1.318	1.711	2.064	2.492	2.797	3.745
25	1.316	1.708	2.060	2.485	2.787	3.725
26	1.315	1.706	2.056	2.479	2.779	3.707
27	1.314	1.703	2.052	2.473	2.771	3.690
28	1.313	1.701	2.048	2.467	2.763	3.674
29	1.311	1.699	2.045	2.462	2.756	3.659
30	1.310	1.697	2.042	2.457	2.750	3.646
40	1.303	1.684	2.021	2.423	2.704	3.551
60	1.296	1.671	2.000	2.390	2.660	3.460
120	1.289	1.658	1.980	2.358	2.617	3.373
∞	1.282	1.645	1.960	2.326	2.576	3.291

SOURCE: This table has been reproduced from Table III of Fisher and Yates: *Statistical Tables for Biological, Agricultural and Medical Research*, published by Oliver and Boyd, Edinburgh, and by permission of the authors and publishers.

Table B.4 Cumulative F Distribution

$$G(F) = \int_0^F \frac{[(m+n-2)/2]!m^{m/2}n^{n/2}x^{(m-2)/2}(n+mx)^{-(m+n)/2}\,dx}{[(m-2)/2]![(n-2)/2]!}$$

m degrees of freedom in numerator; n in denominator

G	n	$m \longrightarrow$ 1	2	3	4	5	6	7	8
.90	↓	39.9	49.5	53.6	55.8	57.2	58.2	58.9	59.4
.95		161	200	216	225	230	234	237	239
.975	1	648	800	864	900	922	937	948	957
.99		4.05*	5.00*	5.40*	5.62*	5.76*	5.86*	5.93*	5.98*
.995		16.2*	20.0*	21.6*	22.5*	23.1*	23.4*	23.7*	23.9*
.90		8.53	9.00	9.16	9.24	9.29	9.33	9.35	9.37
.95		18.5	19.0	19.2	19.2	19.3	19.3	19.4	19.4
.975	2	38.5	39.0	39.2	39.2	39.3	39.3	39.4	39.4
.99		98.5	99.0	99.2	99.2	99.3	99.3	99.4	99.4
.995		199	199	199	199	199	199	199	199
.90		5.54	5.46	5.39	5.34	5.31	5.28	5.27	5.25
.95		10.1	9.55	9.28	9.12	9.01	8.94	8.89	8.85
.975	3	17.4	16.0	15.4	15.1	14.9	14.7	14.6	14.5
.99		34.1	30.8	29.5	28.7	28.2	27.9	27.7	27.5
.995		55.6	49.8	47.5	46.2	45.4	44.8	44.4	44.1
.90		4.54	4.32	4.19	4.11	4.05	4.01	3.98	3.95
.95		7.71	6.94	6.59	6.39	6.26	6.16	6.09	6.04
.975	4	12.2	10.6	9.98	9.60	9.36	9.20	9.07	8.98
.99		21.2	18.0	16.7	16.0	15.5	15.2	15.0	14.8
.995		31.3	26.3	24.3	23.2	22.5	22.0	21.6	21.4
.90		4.06	3.78	3.62	3.52	3.45	3.40	3.37	3.34
.95		6.61	5.79	5.41	5.19	5.05	4.95	4.88	4.82
.975	5	10.0	8.43	7.76	7.39	7.15	6.98	6.85	6.76
.99		16.3	13.3	12.1	11.4	11.0	10.7	10.5	10.3
.995		22.8	18.3	16.5	15.6	14.9	14.5	14.2	14.0
.90		3.78	3.46	3.29	3.18	3.11	3.05	3.01	2.98
.95		5.99	5.14	4.76	4.53	4.39	4.28	4.21	4.15
.975	6	8.81	7.26	6.60	6.23	5.99	5.82	5.70	5.60
.99		13.7	10.9	9.78	9.15	8.75	8.47	8.26	8.10
.995		18.6	14.5	12.9	12.0	11.5	11.1	10.8	10.6
.90		3.59	3.26	3.07	2.96	2.88	2.83	2.78	2.75
.95		5.59	4.47	4.35	4.12	3.97	3.87	3.79	3.73
.975	7	8.07	6.54	5.89	5.52	5.29	5.12	4.99	4.90
.99		12.2	9.55	8.45	7.85	7.46	7.19	6.99	6.84
.995		16.2	12.4	10.9	10.1	9.52	9.16	8.89	8.68

* Denotes (entry)·(1000)

Table B.4 (*continued*)

G	n	m → 9	10	12	15	20	30	60	120	∞
.90	↓	59.	60.2	60.7	61.2	61.7	62.3	62.8	63.1	63.3
.95		241	242	244	246	248	250	252	253	254
.975	1	963	969	977	985	993	1000	1010	1010	1020
.99		6.02*	6.06*	6.11*	6.16*	6.21*	6.26*	6.31*	6.34*	6.37*
.995		24.1*	24.2*	24.4*	24.6*	24.8*	25.0*	25.2*	25.4*	25.5*
.90		9.38	9.39	9.41	9.42	9.44	9.46	9.47	9.48	9.49
.95		19.4	19.4	19.4	19.4	19.5	19.5	19.5	19.5	19.5
.975	2	39.4	39.4	39.4	39.4	39.4	39.5	39.5	39.5	39.5
.99		99.4	99.4	99.4	99.4	99.4	99.5	99.5	99.5	99.5
.995		199	199	199	199	199	199	199	199	199
.90		5.24	5.23	5.22	5.20	5.18	5.17	5.15	5.14	5.13
.95		8.81	8.79	8.74	8.70	8.66	8.62	8.57	8.55	8.53
.975	3	14.5	14.4	14.3	14.3	14.2	14.1	14.0	13.9	13.9
.99		27.3	27.2	27.1	26.9	26.7	26.5	26.3	26.2	26.1
.995		43.9	43.7	43.4	43.1	42.8	42.5	42.1	42.0	41.8
.90		3.93	3.92	3.90	3.87	3.84	3.82	3.79	3.78	3.76
.95		6.00	5.96	5.91	5.86	5.80	5.75	5.69	5.66	5.63
.975	4	8.90	8.84	8.75	8.66	8.56	8.46	8.36	8.31	8.26
.99		14.7	14.5	14.4	14.2	14.0	13.8	13.7	13.6	13.5
.995		21.1	21.0	20.7	20.4	20.2	19.9	19.6	19.5	19.3
.90		3.32	3.30	3.27	3.24	3.21	3.17	3.14	3.12	3.11
.95		4.77	4.74	4.68	4.62	4.56	4.50	4.43	4.40	4.37
.975	5	6.68	6.62	6.52	6.43	6.33	6.23	6.12	6.07	6.02
.99		10.2	10.1	9.89	9.72	9.55	9.38	9.20	9.11	9.02
.995		13.8	13.6	13.4	13.1	12.9	12.7	12.4	12.3	12.1
.90		2.96	2.94	2.90	2.87	2.84	2.80	2.76	2.74	2.72
.95		4.10	4.06	4.00	3.94	3.87	3.81	3.74	3.70	3.67
.975	6	5.52	5.46	5.37	5.27	5.17	5.07	4.96	4.90	4.85
.99		7.98	7.87	7.72	7.56	7.40	7.23	7.06	6.97	6.88
.995		10.4	10.2	10.0	9.81	9.59	9.36	9.12	9.00	8.88
.90		2.72	2.70	2.67	2.63	2.59	2.56	2.51	2.49	2.47
.95		3.68	3.64	3.57	3.51	3.44	3.38	3.30	3.27	3.23
.975	7	4.82	4.76	4.67	4.57	4.47	4.36	4.25	4.20	4.14
.99		6.72	6.62	6.47	6.31	6.16	5.99	5.82	5.74	5.65
.995		8.51	8.38	8.18	7.97	7.75	7.53	7.31	7.19	7.08

*Denotes (entry)·(1000)

Table B.4 (*continued*)

G	n $m \longrightarrow$ 1	2	3	4	5	6	7	8
.90	↓ 3.46	3.11	2.92	2.81	2.73	2.67	2.62	2.59
.95	5.32	4.46	4.07	3.84	3.69	3.58	3.50	3.44
.975	8 7.57	6.06	5.42	5.05	4.83	4.65	4.53	4.43
.99	11.3	8.65	7.59	7.01	6.63	6.37	6.18	6.03
.995	14.7	11.0	9.60	8.81	8.30	7.95	7.69	7.50
.90	3.36	3.01	2.81	2.69	2.61	2.55	2.51	2.47
.95	5.12	4.26	3.86	3.63	3.48	3.37	3.29	3.23
.975	9 7.21	5.71	5.08	4.72	4.48	4.32	4.20	4.10
.99	10.6	8.02	6.99	6.42	6.06	5.80	5.61	5.47
.995	13.6	10.1	8.72	7.96	7.47	7.13	6.88	6.69
.90	3.29	2.92	2.73	2.61	2.52	2.46	2.41	2.38
.95	4.96	4.10	3.71	3.48	3.33	3.22	3.14	3.07
.975	10 6.94	5.46	4.83	4.47	4.24	4.07	3.95	3.85
.99	10.0	7.56	6.55	5.99	5.64	5.39	5.20	5.06
.995	12.8	9.43	8.08	7.34	6.87	6.54	6.30	6.12
.90	3.18	2.81	2.61	2.48	2.39	2.33	2.28	2.24
.95	4.75	3.89	3.49	3.26	3.11	3.00	2.91	2.85
.975	12 6.55	5.10	4.47	4.12	3.89	3.73	3.61	3.51
.99	9.33	6.93	5.95	5.41	5.06	4.82	4.64	4.50
.995	11.8	8.51	7.23	6.52	6.07	5.76	5.52	5.35
.90	3.07	2.70	2.49	2.36	2.27	2.21	2.16	2.12
.95	4.54	3.68	3.29	3.06	2.90	2.79	2.71	2.64
.975	15 6.20	4.77	4.15	3.80	3.58	3.41	3.29	3.20
.99	8.68	6.36	5.42	4.89	4.56	4.32	4.14	4.00
.995	10.8	7.70	6.48	5.80	5.37	5.07	4.85	4.67
.90	2.97	2.59	2.38	2.25	2.16	2.09	2.04	2.00
.95	4.35	3.49	3.10	2.87	2.71	2.60	2.51	2.45
.975	20 5.87	4.46	3.86	3.51	3.29	3.13	3.01	2.91
.99	8.10	5.85	4.94	4.43	4.10	3.87	3.70	3.56
.995	9.94	6.99	5.82	5.17	4.76	4.47	4.26	4.09
.90	2.88	2.49	2.28	2.14	2.05	1.98	1.93	1.88
.95	4.17	3.32	2.92	2.69	2.53	2.42	2.33	2.27
.975	30 5.57	4.18	3.59	3.25	3.03	2.87	2.75	2.65
.99	7.56	5.39	4.51	4.02	3.70	3.47	3.30	3.17
.995	9.18	6.35	5.24	4.62	4.23	3.95	3.74	3.58
.90	2.79	2.39	2.18	2.04	1.95	1.87	1.82	1.77
.95	4.00	3.15	2.76	2.53	2.37	2.25	2.17	2.10
.975	60 5.29	3.93	3.34	3.01	2.79	2.63	2.51	2.41
.99	7.08	4.98	4.13	3.65	3.34	3.12	2.95	2.82
.995	8.49	5.80	4.73	4.14	3.76	3.49	3.29	3.13

Table B.4 *(continued)*

G	n ↓ m →	9	10	12	15	20	30	60	120	∞
.90		2.56	2.54	2.50	2.46	2.42	2.38	2.34	2.31	2.29
.95		3.39	3.35	3.28	3.22	3.15	3.08	3.01	2.97	2.93
.975	8	4.36	4.30	4.20	4.10	4.00	3.89	3.78	3.73	3.67
.99		5.91	5.81	5.67	5.52	5.36	5.20	5.03	4.95	4.86
.995		7.34	7.21	7.01	6.81	6.61	6.40	6.18	6.06	5.95
.90		2.44	2.42	2.38	2.34	2.30	2.25	2.21	2.18	2.16
.95		3.18	3.14	3.07	3.01	2.94	2.86	2.79	2.75	2.71
.975	9	4.03	3.96	3.87	3.77	3.67	3.56	3.45	3.39	3.33
.99		5.35	5.26	5.11	4.96	4.81	4.65	4.48	4.40	4.31
.995		6.54	6.42	6.23	6.03	5.83	5.62	5.41	5.30	5.19
.90		2.35	2.32	2.28	2.24	2.20	2.15	2.11	2.08	2.06
.95		3.02	2.98	2.91	2.84	2.77	2.70	2.62	2.58	2.54
.975	10	3.78	3.72	3.62	3.52	3.42	3.31	3.20	3.14	3.08
.99		4.94	4.85	4.71	4.56	4.41	4.25	4.08	4.00	3.91
.995		5.97	5.85	5.66	5.47	5.27	5.07	4.86	4.75	4.64
.90		2.21	2.19	2.15	2.10	2.06	2.01	1.96	1.93	1.90
.95		2.80	2.75	2.69	2.62	2.54	2.47	2.38	2.34	2.30
.975	12	3.44	3.37	3.28	3.18	3.07	2.96	2.85	2.79	2.72
.99		4.39	4.30	4.16	4.01	3.86	3.70	3.54	3.45	3.36
.995		5.20	5.09	4.91	4.72	4.53	4.33	4.12	4.01	3.90
.90		2.09	2.06	2.02	1.97	1.92	1.87	1.82	1.79	1.76
.95		2.59	2.54	2.48	2.40	2.33	2.25	2.16	2.11	2.07
.975	15	3.12	3.06	2.96	2.86	2.76	2.64	2.52	2.46	2.40
.99		3.89	3.80	3.67	3.52	3.37	3.21	3.05	2.96	2.87
.995		4.54	4.42	4.25	4.07	3.88	3.69	3.48	3.37	3.26
.90		1.96	1.94	1.89	1.84	1.79	1.74	1.68	1.64	1.61
.95		2.39	2.35	2.28	2.20	2.12	2.04	1.95	1.90	1.84
.975	20	2.84	2.77	2.68	2.57	2.46	2.35	2.22	2.16	2.09
.99		3.46	3.37	3.23	3.09	2.94	2.78	2.61	2.52	2.42
.995		3.96	3.85	3.68	3.50	3.32	3.12	2.92	2.81	2.69
.90		1.85	1.82	1.77	1.72	1.67	1.61	1.54	1.50	1.46
.95		2.21	2.16	2.09	2.01	1.93	1.84	1.74	1.68	1.62
.975	30	2.57	2.51	2.41	2.31	2.20	2.07	1.94	1.87	1.79
.99		3.07	2.98	2.84	2.70	2.55	2.39	2.21	2.11	2.01
.995		3.45	3.34	3.18	3.01	2.82	2.63	2.42	2.30	2.18
.90		1.74	1.71	1.66	1.60	1.54	1.48	1.40	1.35	1.29
.95		2.04	1.99	1.92	1.84	1.75	1.65	1.53	1.47	1.39
.975	60	2.33	2.27	2.17	2.06	1.94	1.82	1.67	1.58	1.48
.99		2.72	2.63	2.50	2.35	2.20	2.03	1.84	1.73	1.60
.995		3.01	2.90	2.74	2.57	2.39	2.19	1.96	1.83	1.69

Table B.4 *(continued)*

G	n	$m \rightarrow 1$	2	3	4	5	6	7	8
.90	↓	2.75	2.35	2.13	1.99	1.90	1.82	1.77	1.72
.95		3.92	3.07	2.68	2.45	2.29	2.18	2.09	2.02
.975	120	5.15	3.80	3.23	2.89	2.67	2.52	2.39	2.30
.99		6.85	4.79	3.95	3.48	3.17	2.96	2.79	2.66
.995		8.18	5.54	4.50	3.92	3.55	3.28	3.09	2.93
.90		2.71	2.30	2.08	1.94	1.85	1.77	1.72	1.67
.95		3.84	3.00	2.60	2.37	2.21	2.10	2.01	1.94
.975	∞	5.02	3.69	3.12	2.79	2.57	2.41	2.29	2.19
.99		6.63	4.61	3.78	3.32	3.02	2.80	2.64	2.51
.995		7.88	5.30	4.28	3.72	3.35	3.09	2.90	2.74

G	n	$m \rightarrow 9$	12	10	15	20	30	60	120	∞
.90	↓	1.68	1.65	1.60	1.54	1.48	1.41	1.32	1.26	1.19
.95		1.96	1.91	1.83	1.75	1.66	1.55	1.43	1.35	1.25
.975		2.22	2.16	2.05	1.94	1.82	1.69	1.53	1.43	1.31
.99	120	2.56	2.47	2.34	2.19	2.03	1.86	1.66	1.53	1.38
.995		2.81	2.71	2.54	2.37	2.19	1.98	1.75	1.61	1.43
.90		1.63	1.60	1.55	1.49	1.42	1.34	1.24	1.17	1.00
.95		1.88	1.83	1.75	1.67	1.57	1.46	1.32	1.22	1.00
.975	∞	2.11	2.05	1.94	1.83	1.71	1.57	1.39	1.27	1.00
.99		2.41	2.32	2.18	2.04	1.88	1.70	1.47	1.32	1.00
.995		2.62	2.52	2.36	2.19	2.00	1.79	1.53	1.36	1.00

SOURCE: The entries in this table have appeared in "Tables of percentage points of the inverted beta distribution," *Biometrika*, Vol. 33 (1943), with the kind permission of the authors M. Merrington and C. M. Thompson, and the editor of *Biometrika*.

Table B.5 Heck Charts for MANOVA

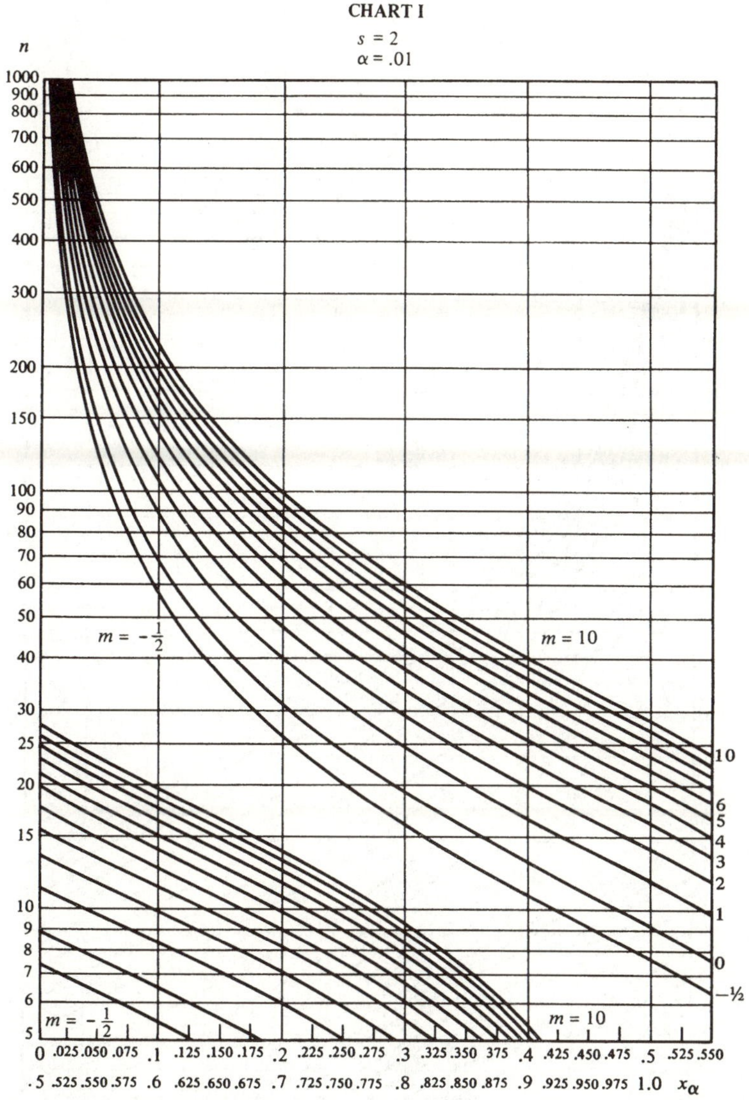

SOURCE: The charts in Table B.5 appeared in "Charts of some upper percentage points of the distribution of the largest characteristic root," by D. L. Heck, *Annals of Mathematical Statistics*, Vol. 31 (1960), pp. 625–642. They are reproduced here with the kind permission of the authors and the publishers of the *Annals*. The two sets of twelve curves are plotted on a folded scale. The curves in each set are for the range $m = -\frac{1}{2}$, 0, 1, 2, . . ., 10.

Table B.5 (*continued*)

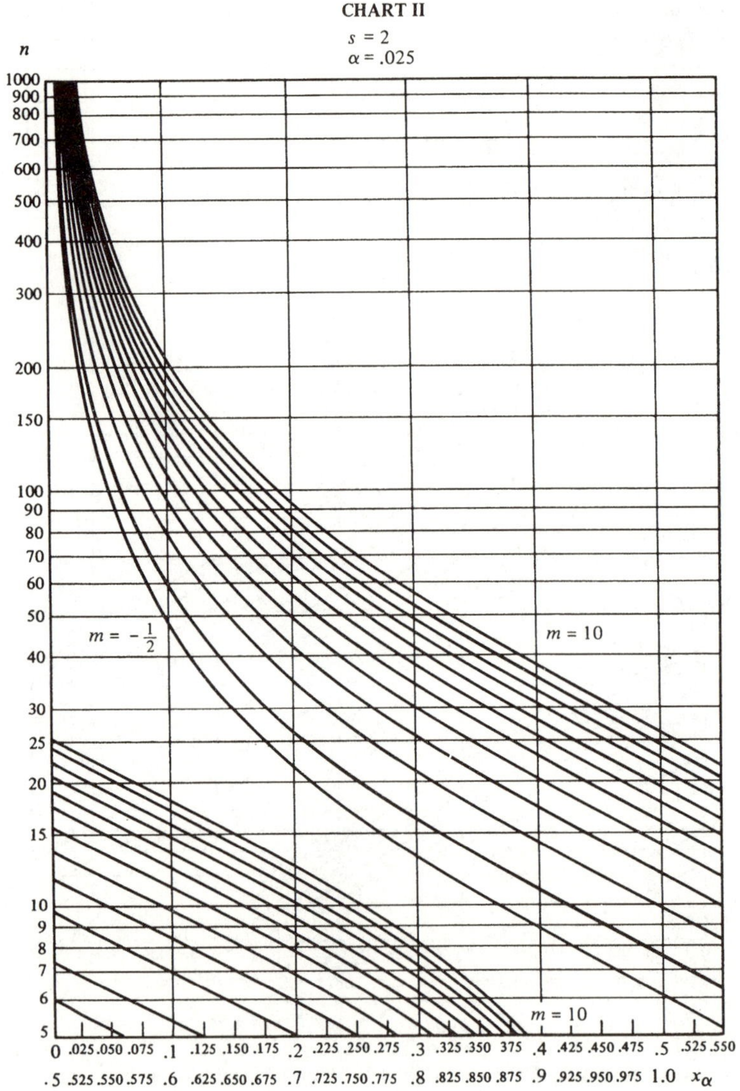
CHART II
s = 2
α = .025
n
1000
900
800
700
600
500
400
300
200
150
100
90
80
70
60
50
40
30
25
20
15
10
9
8
7
6
5
m = − 1/2
m = 10
m = 10
0 .025 .050 .075 .1 .125 .150 .175 .2 .225 .250 .275 .3 .325 .350 375 .4 .425 .450 .475 .5 .525 .550
.5 .525 .550 .575 .6 .625 .650 .675 .7 .725 .750 .775 .8 .825 .850 .875 .9 .925 .950 .975 1.0 x_α

Table B.5 *(continued)*

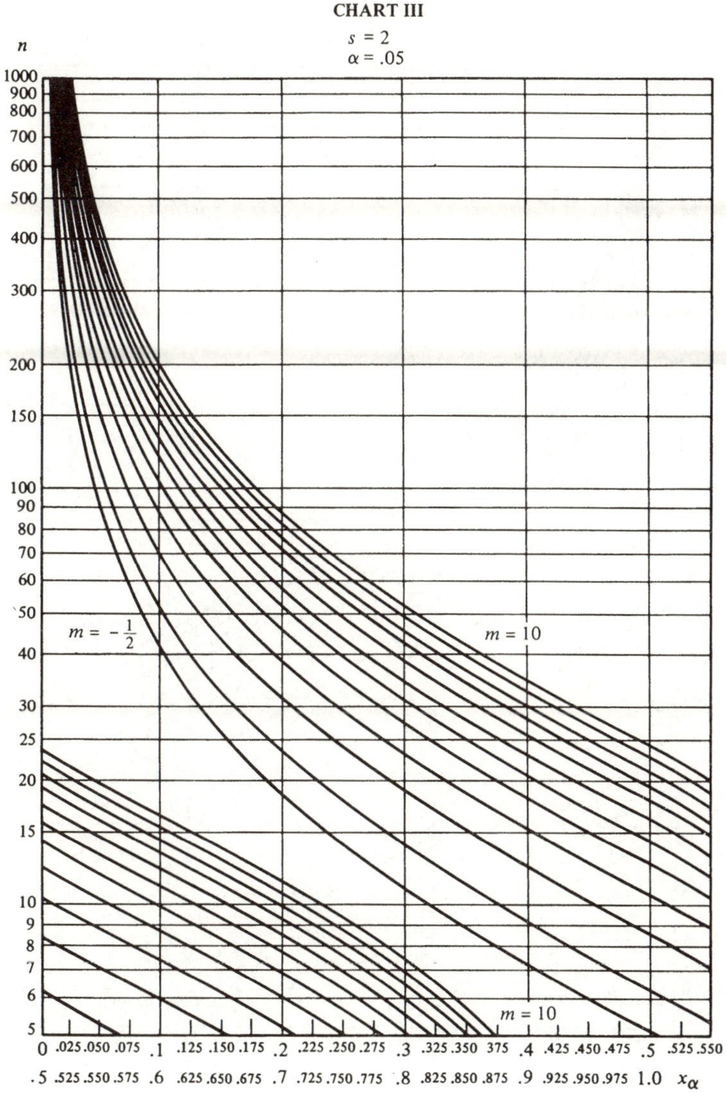

Table B.5 (*continued*)

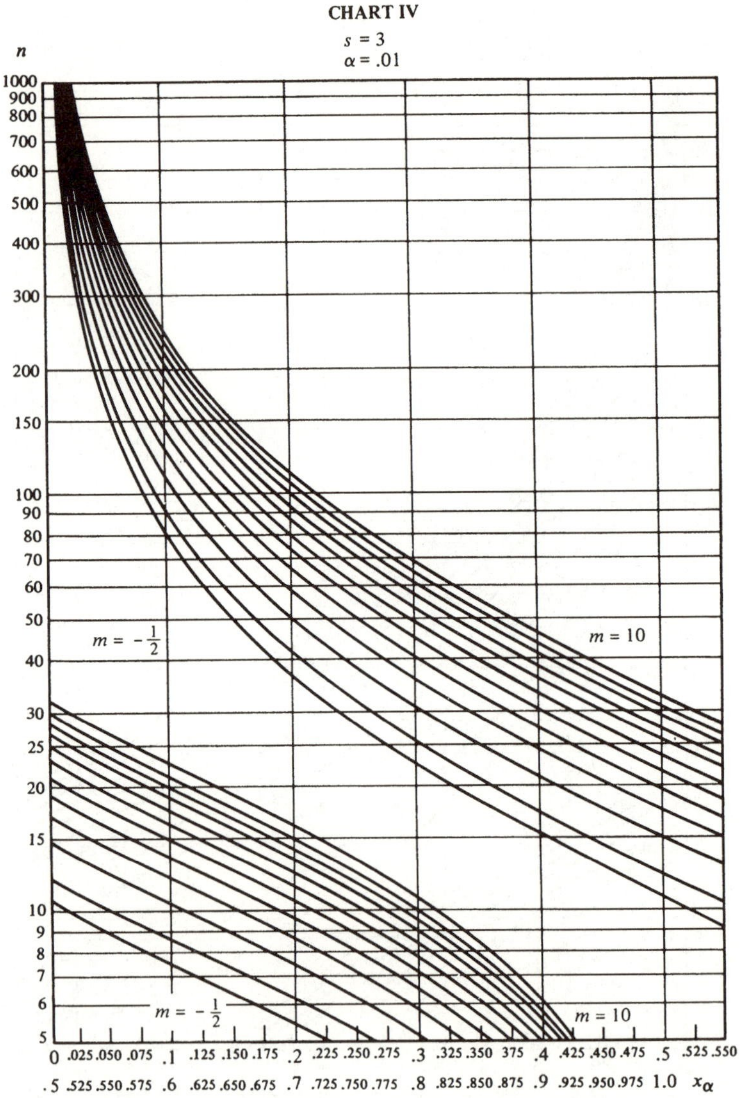

Table B.5 (*continued*)

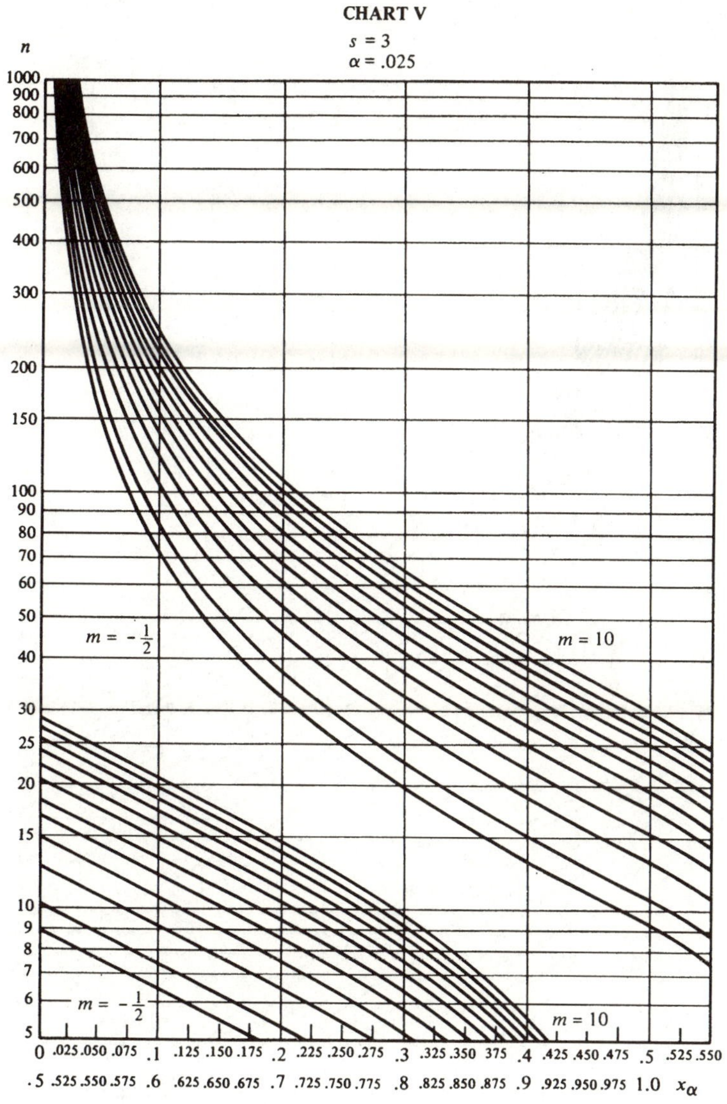

Table B.5 (*continued*)

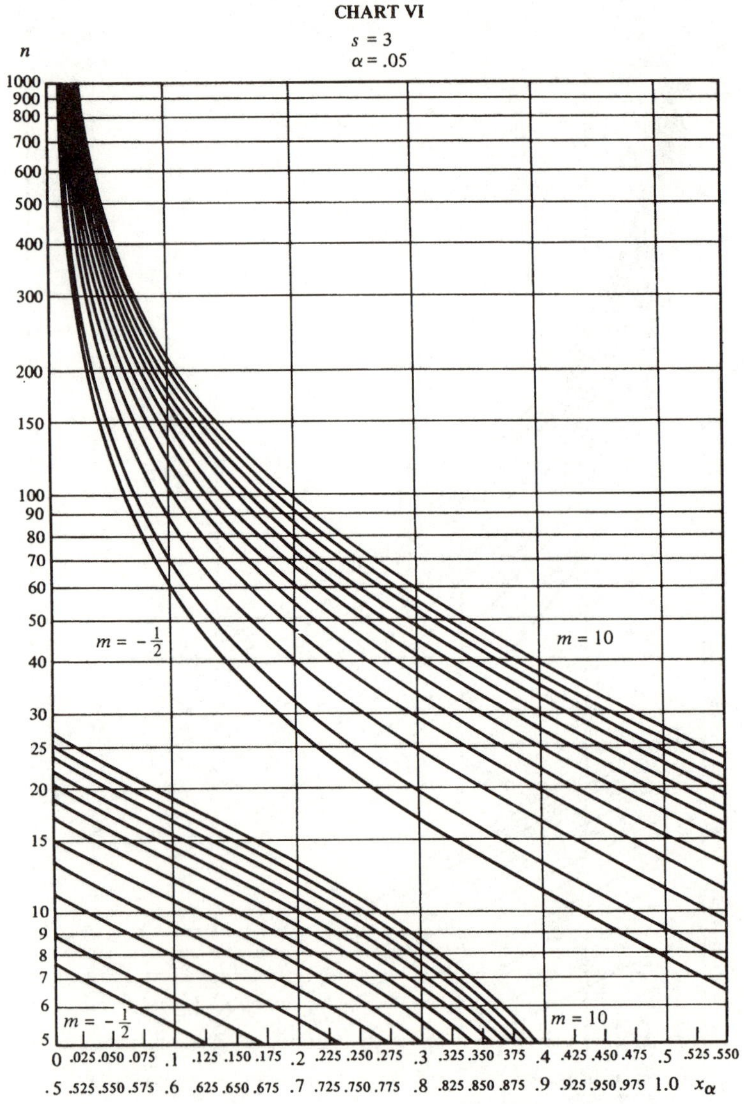

Table B.5 *(continued)*

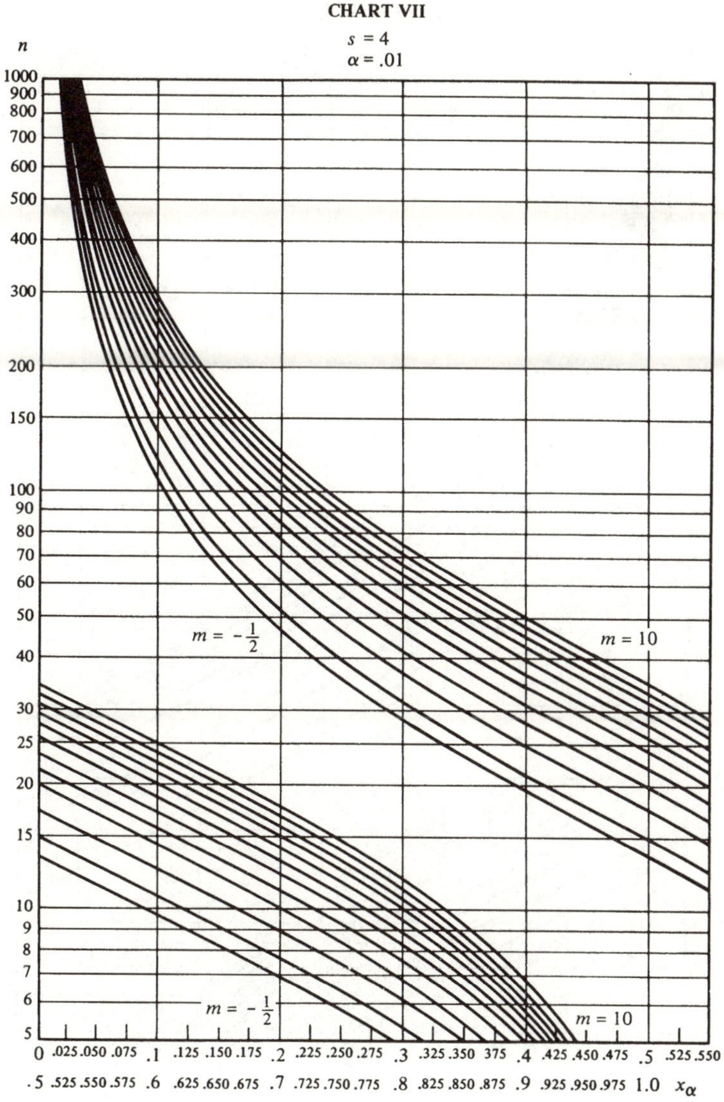

Table B.5 (*continued*)

CHART VIII

$s = 4$
$\alpha = .025$

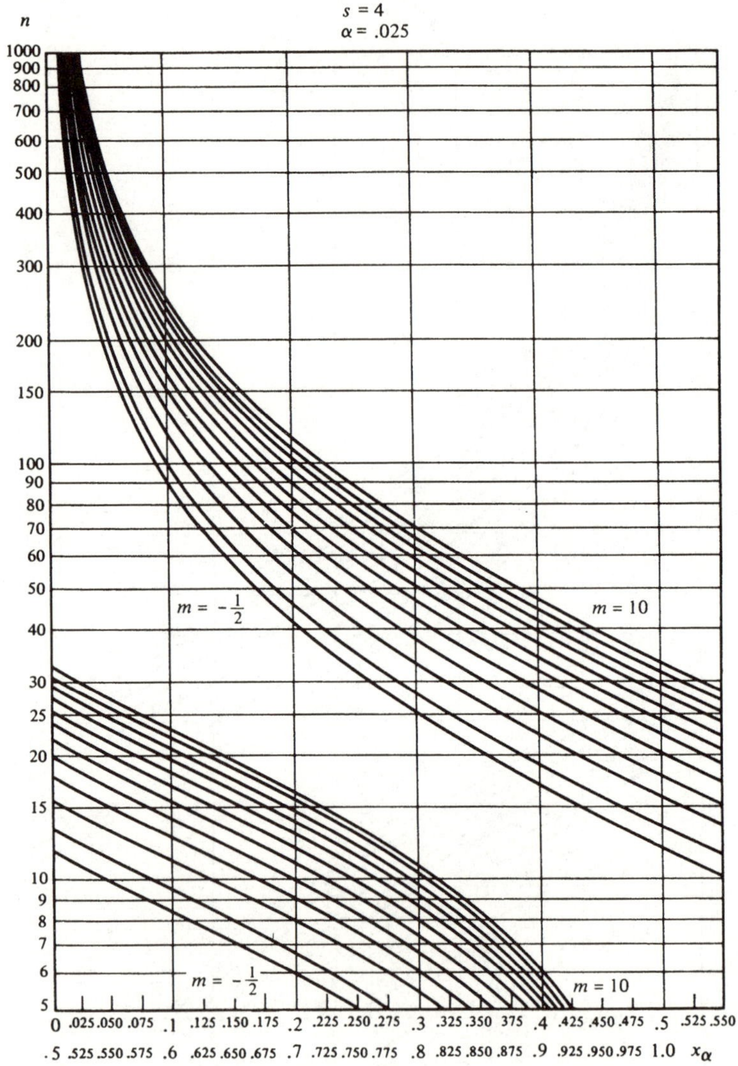

Table B.5 *(continued)*

CHART IX

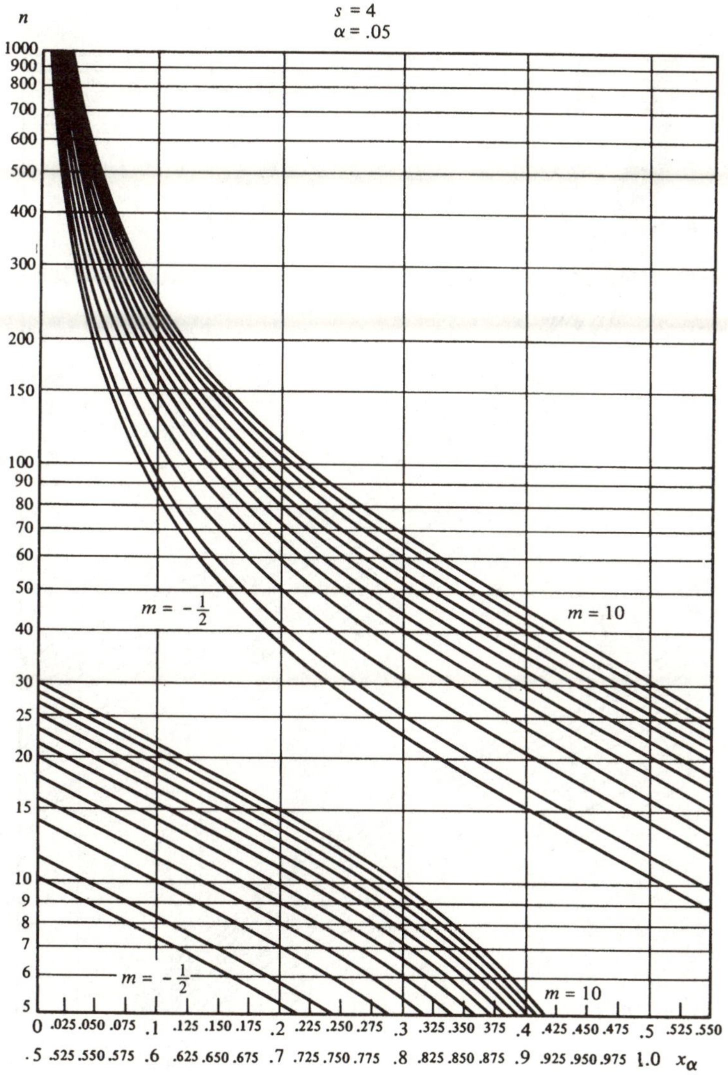

Table B.5 *(continued)*

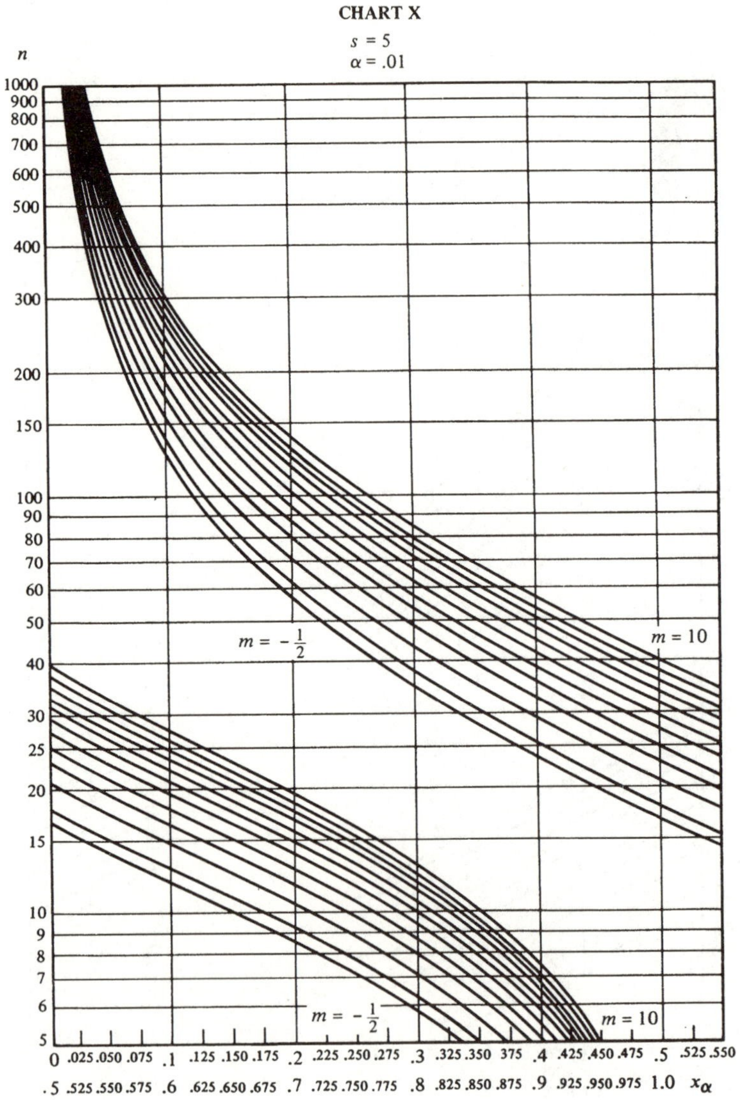

Table B.5 *(continued)*

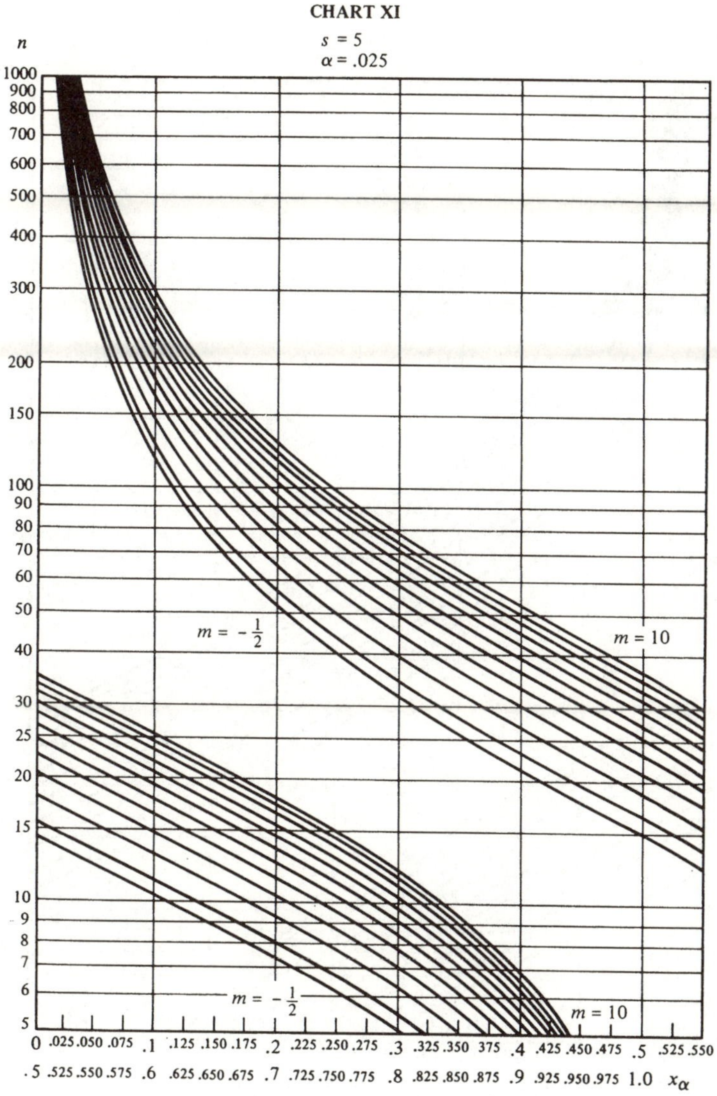

Table B.5 (*continued*)

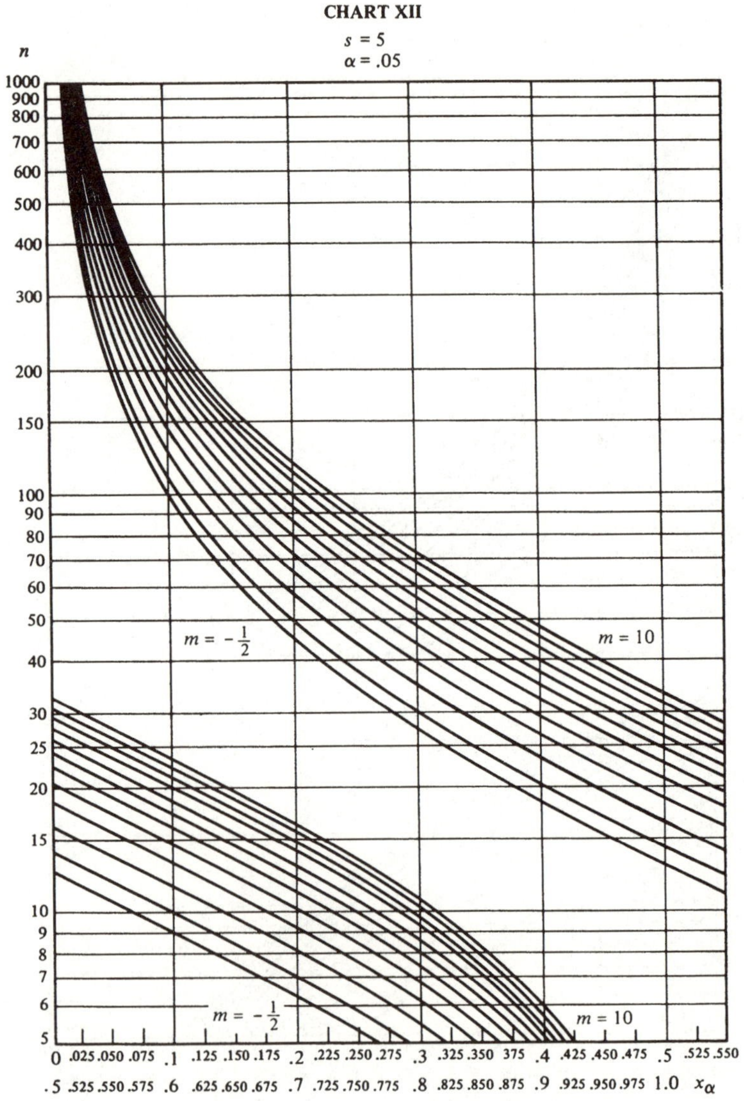

appendix C

BIBLIOGRAPHY

Abromovitz, M. and Stegun, I. A., *Handbook of Mathematical Functions with Formulas, Graphs, and Mathematical Tables.* U. S. Dept. of Commerce, National Bureau of Standards, App. Math. Series 55, 1964.

Adelman, I. and Adelman, F. "The Dynamic Properties of the Klein-Goldberger Model," *Econometrica*, Vol. 27 (1959), pp. 596–625.

Afifi, A. A. and Elashoff, R. M., "Missing Observations in Multivariate Statistics I. Review of the Literature," *Journal of the American Statistical Association*, Vol. 61 (1966), pp. 595–605.

——— and ———, "Missing Observations in Multivariate Statistics II. Point Estimation in Simple Linear Regression," *Journal of the American Statistical Association*, Vol. 62 (1967), pp. 10–29.

——— and ———, "Missing Observations in Multivariate Statistics III. Large Sample Analysis of Simple Linear Regression," *Journal of the American Statistical Association*, Vol. 64 (1969a), pp. 337–358.

——— and ———, "Missing Observations in Multivariate Statistics IV. A Note on Simple Linear Regression," *Journal of the American Statistical Association*, Vol. 64 (1969b), pp. 359–365.

Ahamad, B., "An Analysis of Crimes by the Method of Principal Components," *Journal of Applied Statistics*, Vol. 16 (1967), pp. 17–35.

Ahmad, Q. *Indian Cities: Characteristics and Correlates*, Department of Geography Research Paper No. 102. Chicago: University of Chicago, 1965.

Aitken, A. C., "On Least Squares and Linear Combination of Observations," *Proceedings of the Royal Society of Edinburgh*, Vol. 55 (1935), pp. 42–48.

Akaike, H., "Information Theory and an Extension of the Maximum Likelihood Principle," *2nd International Symposium on Information Theory*, edited by B. N. Petrov and F. Csaki, Budapest: Akademiai Kiado (1973), pp. 267–281.

———, "On Entropy Maximization Principle," in *Applications of Statistics,* edited by P. R. Krishnaiah, Amsterdam: North Holland Publishing Co., 1977.

Alessi, S. L., see Lawshe, C. H., Jr.

Alexander, C. N., "A Method for Processing Sociometric Data," *Sociometry,* Vol. 26 (1963), pp. 268-279.

Amemiya, T. and Fuller, W. A., "A Comparative Study of Alternative Estimations in a Distributed Lag Model," *Econometrica,* Vol. 35 (1967), pp. 509-529.

Anderberg, M. R., *Cluster Analysis for Applications,* Academic Press, Inc., 1973.

Amemiya, T. and Fuller, W. A., "A Comparative Study of Alternative Estimations in a Distributed Lag Model," *Econometrica,* Vol. 35 (1967), pp. 509–529.

Anderson, T. W., "The Noncentral Wishart Distribution and Its Application to Problems in Multivariate Statistics." Doctoral dissertation, Princeton University Library, 1945.

———, "The Noncentral Wishart Distribution and Certain Problems of Multivariate Statistics," *Annals of Mathematical Statistics,* Vol. 17 (1946), p. 409.

———, "Classification by Multivariate Analysis," *Psychometrika,* Vol. 16 (1951), pp. 31–50.

———, "On Estimation of Parameters in Latent Structure Analysis," *Psychometrika,* Vol. 19 (1954), pp. 1–10.

———, *An Introduction to Multivariate Statistical Analysis.* New York: John Wiley and Sons, 1958.

———, "Asymptotic Theory for Principal Component Analysis," *Annals o Mathematical Statistics,* Vol. 34 (1963), pp. 122–148.

———, "A Test for Equality of Means when Covariance Matrices Are Unequal," *Annals of Mathematical Statistics,* Vol. 34, No. 2 (1963), pp. 671–672.

———, Das Gupta, S., and Styan, G. P. H., *A Bibliography of Multivariate Statistical Analysis,* Huntington, N.Y.: Krieger Publishing Co., Inc., 1977.

——— and Girshick, M. A., "Some Extensions of the Wishart Distribution," *Annals of Mathematical Statistics,* Vol. 15 (1944), pp. 345–357.

——— and Rubin, H., "Statistical Inference in Factor Analysis," *Proceedings of the Third Berkeley Symposium on Mathematical Statistics and Probability,* Vol. 5. Berkeley: University of California Press, 1956, pp. 111–150.

Ando, A. and Kaufman, G. M., "Bayesian Analysis of the Independent Multinormal Process—Neither Mean nor Precision Known," *Journal of the American Statistical Association,* Vol. 60 (1965), pp. 347-358.

Andrews, D., Gnanedesikan, R., and Warner, J., "Methods for Assessing Multivariate Normality," *Multivariate Analysis III,* edited by P. R. Krishnaiah, 1973.

Aoki, M., *Optimization of Stochastic Systems.* New York: Academic Press, 1967.

Armstrong, J. S., "Derivation of Theory by Means of Factor Analysis or Tom

Swift and His Electric Factor Analysis Machine," *The American Statistician* (December, 1967), pp. 17–21.

Arnold, S. F., "A Coordinate-Free Approach to Finding Optimal Procedures for Repeated Measures Designs," *Annals of Mathematical Statistics,* Vol. 7, (1979b), pp. 812–822.

———, "Linear Models with Exchangeably Distributed Errors," *Journal of the American Statistical Association,* Vol. 74, (1979a), pp. 194–199.

———, "Applications of Products to the Generalized Compound Symmetry Problem," *Annals of Mathematical Statistics,* Vol. 4, (1976), pp. 227–233.

———, "Application of the Theory of Products of Problems to Certain Patterned Covariance Matrices," *Annals of Mathematical Statistics,* Vol. 1, (1973), pp. 682–699.

Aykac, A. and Brumat, C., *New Developments in the Applications of Bayesian Methods,* North Holland Publishing Co., 1977.

Baehr, M. E. and Renck, R., "The Definition and Measurement of Employee Morale," *Administrative Science Quarterly,* Vol. 3 (1958), pp. 157–184.

Baggaley, A. R., *Intermediate Correlational Methods.* New York: John Wiley and Sons, 1964.

Bailey, D., see Tryon, Robert C.

Balestra, P., "Le Derivation Matricelle," Collection de L'IME, No. 12, Sirey, Paris.

Ball, G. H., "Data Analysis in the Social Sciences: What About the Details?" *Proceedings—Fall Joint Computer Conference* (1965), pp. 533–559.

———, see Hall *et al.*

Bamberger, F., see Wright, B. D.

Banks, S., "The Relationship between Preference and Purchase of Brands," *Journal of Marketing,* Vol. 15 (October, 1950), pp. 145–157.

———, see Kirsch, A. D.

Baranchik, A. J., "Multiple Regression and Estimation of the Mean of a Multivariate Normal Distribution," Technical Report No. 51, Department of Statistics, Stanford University, 1964.

Barnard, G. A., "Sampling Inspection and Statistical Decisions," *Journal of the Royal Statistical Society,* Series B, Vol. 16 (1954), pp. 151–174.

Barnett, V. D., "Order Statistics Estimators of the Cauchy Distribution," *Journal of the American Statistical Assn.,* Vol. 61 (1966), pp. 1205–1218.

Baron, M., see Harper, R.

Barrett, R. S., "The Process of Predicting Job Performance," *Personnel Psychology,* Vol. 11 (1958), pp. 39–57.

Bartlett, M. S., "Further Aspects of the Theory of Multiple Regression," *Proceedings of the Cambridge Philosophical Society,* Vol. 34 (1938), pp. 33–40.

———, "A Note on Tests of Significance in Multivariate Analysis," *Proceedings of the Cambridge Philosophical Society*, Vol. 35 (1939), pp. 180–185.

———, "The Statistical Significance of Canonical Correlation," *Biometrika*, Vol. 32 (1941), pp. 29–38.

———, "Factor Analysis in Psychology as a Statistician Sees It," *Uppsala Symposium on Psychological Factor Analysis*. Uppsala, Sweden: Almqvist and Wiksell, 1953, pp. 23–34.

———, "A Note on the Multiplying Factors for Various Chi-Squared Approximations," *Journal of the Royal Statistical Society*, Series B, Vol. 16 (1954), pp. 296–298.

Basmann, R. L., "A Note on the Exact Finite Sample Frequency Functions of Generalized Classical Linear Estimators in a Leading Three-Equation Case," *Journal of the American Statistical Association*, Vol. 58 (1963), pp. 161–171.

Bass, B. M., "Iterative Inverse Factor Analysis—A Rapid Method for Clustering Persons," *Psychometrika*, Vol. 22 (1957), pp. 105–107.

Bather, J. and Chernoff, H., "Sequential Decisions in the Control of a Spaceship," *Proceedings of the 5th Berkeley Symposium on Mathematical Statistics and Probability*, Vol. III. Berkeley: University of California Press, 1967.

Beale, E. M. L., Kendall, M. G. and Mann, D. W., "The Discarding of Variables in Multivariate Analysis," *Biometrika*, Vol. 54 (December, 1967), pp. 357–366.

Beckenbach, E. F. and Bellman, R., *Inequalities*. New York: Springer-Verlag, 1961.

Bellman, R., *Introduction to Matrix Analysis*. New York: McGraw-Hill, 1960.

———, *Adaptive Control Processes*. Princeton: Princeton University Press, 1961.

Bennett, B. M., "Note on a Solution of the Generalized Behrens-Fisher Problem," *Annals of the Institute of Statistical Mathematics*, Vol. 2 (1951), pp. 87–90.

Berkson, J., "Application of the Logistic Function to Bio-assay," *Journal of the American Statistical Association*, Vol. 39 (1944), pp. 357–365.

Berry, B. J. L., "A Method for Deriving Multifactor Uniform Regions," *Przeglad Geografrczny*, Vol. 33 (1961), pp. 263–279.

Birch, M. W., "Maximum Likelihood in Three-way Contingency Tables," *Journal of the Royal Statistical Society*, Series B, Vol. 25, (1963), pp. 220–233.

Bishop, Y. M. M., Fienberg, S. E., and Holland, P. W., *Discrete Multivariate Analysis: Theory and Practice*, Cambridge, Mass.: The MIT Press, 1975.

Black, M., "SSL PAC." Chicago: Computation Center, University of Chicago, 1965.

Blanchard, R. E., see Drewes, D. W.

Bloch, D., "A Note on the Estimation of the Location Parameter of the Cauchy Distribution," *Journal of the American Statistical Association*, Vol. 61 (1966), pp. 852–855.

Bock, R. D., "Programming Univariate and Multivariate Analysis of Variance," *Technometrics*, Vol. 5 (1963), pp. 95–117.

Bolanovich, D. J., "Statistical Analysis of an Industrial Rating Chart," *Journal of Applied Psychology*, Vol. 30 (1946), pp. 23–31.

Bone, G., "BYU ANOVAR," SHARE, 1966 (See SHARE).

Bonner, R. E., *Clustering Program.* Yorktown Heights, New York: International Business Machines Corporation, 1964.

Boot, J. C. G. and de Witt, G. M., "Investment Demand: An Empirical Contribution to the Aggregation Problem," *International Economic Review*, Vol. 1 (1960), pp. 3–30.

Bose, R. C. and Roy, S. N., "The Exact Distribution of the Studentized D^2-Statistic," *Sankhyā*, Vol. 4 (1938), pp. 10–38.

———, see Roy, S. N.

Bowker, A. H., "A Representation of Hotelling's T^2 and Anderson's Classification Statistic W in Terms of Simple Statistics," *Contributions to Probability and Statistics: Essays in Honor of Harold Hotelling*, ed. by I. Olkin *et al.* Stanford, Calif.: Stanford University Press, 1960, pp. 142–149.

Box, G. E. P., "A General Distribution Theory for a Class of Likelihood Ratio Criteria," *Biometrika*, Vol. 36 (1949), pp. 317–346.

——— and Cox, D. R., "An Analysis of Transformations," *Journal of the Royal Statistical Society*, Series B, Vol. 26 (1964), pp. 211–252.

——— and Jenkins, G. M., "Some Statistical Aspects of Adaptive Optimization and Control," *Journal of the Royal Statistical Society*, Series B, Vol. 24 (1962), pp. 297–343.

———, and Tiao, G. C., *Bayesian Inference in Statistical Analysis,* Addison-Wesley Publishing Co., 1973.

Bradford, C., see Wright, B. D.

Brown, S. E., see Henderson, P. L.

Buhler, R., *Princeton Statistical Package.* Princeton: Princeton University Press, 1966.

Carrol, J. B. and Schweiker, R. F., "Factor Analysis in Educational Research," *Review of Educational Research,* Vol. 21 (1951), pp. 368–388.

Carroll, J. D., see Kruskal, J. B.

——— and J. J. Chang, "Analysis of Individual Differences in Multidimensional Scaling via an N-way Generalization of Eckart-Young Decomposition," *Psychometrika*, Vol. 35 (1970), pp. 283–319.

Casetti, E., *Classifactory and Regional Analysis by Discriminant Iterations.* Evanston, Illinois: Northwestern University, 1964.

Cattell, R. B. (ed.), *Handbook of Multivariate Experimental Psychology.* Chicago: Rand McNally and Co., 1966.

———, *The Scientific Use of Factor Analysis in Behavioral and Life Sciences,* New York: Plenum Press, 1978.

Cavalli-Sforza, L. L., see Edwards, A. W. F.

Chambers, J. M., "On Methods of Asymptotic Approximation for Multivariate Distributions," *Biometrika,* Vol. 54 (1967), pp. 367–383.

Chang, J. J., see Carroll, J. D.

Chau, L. C., see Stroud, A.

Chernin, K. E., see Ibragimov, I. A.

Chernoff, H., see Bather, J.

——— and Zacks, S., "Estimating the Current Mean of a Normal Distribution which is Subjected to Changes in Time," Technical Report No. 91, Department of Statistics, Stanford University, 1963.

———, "Using Faces to Represent Points in k-Dimensional Space Graphically," *Journal of the American Statistical Association*, Vol. 68, (1973), pp. 361–368.

Chetty, V. C., see Zellner, A.

Chow, G. C., "A Comparison of Alternative Estimators for Simultaneous Equations," *Econometrica*, Vol. 32, No. 2 (October, 1964), pp. 532–553.

——— and Ray-Chaudhuri, D. K., "An Alternative Proof of Hannan's Theorem on Canonical Correlation and Multiple Equation Systems," *Econometrica*, Vol. 35 (1967), pp. 139–142.

Christ, C. F., "Aggregate Econometric Models," *American Economic Review*, Vol. 46 (1956), pp. 385–408.

Claycamp, H. J., "Characteristics of Owners of Thrift Deposits in Commercial Banks and Savings and Loan Associations," *Journal of Marketing Research*, Vol. 2 (1965), pp. 163–170.

Clyde, D. J., Cramer, E. M. and Sherin, R. J., *Multivariate Statistical Programs*. Coral Gables, Florida: Biometric Laboratory, University of Florida, 1966.

Coleman, J. S., "Clustering in N Dimensions by Use of a System of Forces." Manuscript, Dept. of Social Relations, The Johns Hopkins University, 1969.

——— and MacRae, D., Jr., "Electronic Processing of Sociometric Data for Groups up to 1000 in Size," *American Sociological Review*, Vol. 25 (1960), pp. 722–726.

Constantine, A. G., "Some Non-Central Distribution Problems in Multivariate Analysis," *Annals of Mathematical Statistics*, Vol. 34 (1963), pp. 1270–1285.

———, "The Distribution of Hotelling's Generalized T_0^2," *Annals of Mathematical Statistics*, Vol. 37 (February, 1966), pp. 215-225.

Cooley, W. W. and Lohnes, P. R., *Multivariate Procedures for the Behavioral Sciences*. New York: John Wiley and Sons, 1962.

Coombs, C. H., "An Application of a Nonmetric Model for Multidimensional Analysis of Similarities," *Psychological Reports*, Vol. 4 (1958), pp. 511–518.

Cornfield, J., see Geisser, S.

Cornish, E. A., "The Multivariate **t**-Distribution Associated with a Set of Normal Sample Deviates," *Australian Journal of Physics*, Vol. 7 (1954), pp. 531–542.

Courant, A. and Hilbert, D., *Methods of Mathematical Physics*, Vols. I and II. New York: Interscience Publishers, 1953.

Cox, D. R., "Note on Grouping," *Journal of the American Statistical Association*, Vol. 52 (1957), pp. 543–547.

———, *Analysis of Binary Data*. London: Methuen and Co. Ltd., 1970.

Cramer, E. M., see Clyde, D. J.

Cramér, H., *Mathematical Methods of Statistics*. Princeton: Princeton University Press, 1946.

Davis, P. J., *Circulant Matrices*, John Wiley & Sons, Inc., 1979.

Deemer, W. L. and Olkin, I., "The Jacobians of Certain Matrix Transformations Useful in Multivariate Analysis, Based on Lectures of P. L. Hsu at the University of North Carolina," *Biometrika*, Vol. 38 (1951), pp. 345–367.

deFinetti, B., *Theory of Probability*, Vols. 1 and 2, John Wiley and Sons, Inc., 1974, 1975.

DeGroot, M., *Optimal Statistical Decisions*, McGraw-Hill Publishing Co., 1970.

Dempster, A. P., "On a Paradox Concerning Inference about a Covariance Matrix," *Annals of Mathematical Statistics*, Vol. 34 (1963), pp. 1414–1418.

———, "Estimation in Multivariate Analysis." *Multivariate Analysis*, ed. by P. R. Krishnaiah, New York: Academic Press, 1966.

———, *Elements of Continuous Multivariate Analysis*. Reading, Mass.: Addison-Wesley Publishing Co., Inc., 1969.

deSilva, B. M., "A Class of Multivariate Symmetric Stable Distributions," *Journal of Multivariate Analysis*, Vol. 8, No. 3, (1978), p. 335.

deWitt, G. M., see Boot, J. C. G.

Dickey, J. M., "Multivariate Generalizations of the Multivariate *t*-Distribution and the Inverted Multivariate *t*-Distribution," *Annals of Mathematical Statistics*, Vol. 38 (April, 1967), pp. 511–518.

Drewes, D. W. and Blanchard, R. E., "A Factorial Study of Labor Arbitration Cases," *Personnel Psychology*, Vol. 12 (1959), pp. 303–310.

Dudek, E. E., see Lawshe, C. H., Jr.

Dunnett, C. W., "A Multiple Comparison Procedure for Comparing Several Treatments with a Control," *Journal of the American Statistical Association*, Vol. 50 (1955), pp. 1096–1121.

——— and Sobel, "A Bivariate Generalization of Student's *t*-Distribution with Tables for Certain Special Cases," *Biometrika*, Vol. 41 (1954), pp. 153–169.

Dunsmore, I. R., "A Bayesian Approach to Classification," *Journal of the Royal Statistical Society*, Series B, Vol. 28 (1966), pp. 568–577.

Dvoretzky, A., "Asymptotic Normality of Sums of Dependent Random Vectors," in *Multivariate Analysis IV*, edited by P. R. Krishnaiah, Amsterdam: North Holland Publishing Co., (1977), pp. 23–34.

Dwyer, P. S., "Some Applications of Matrix Derivatives in Multivariate Analysis," *Journal of the American Statistical Association*, Vol. 62, No. 318 (1967), pp. 607–625.

——— and MacPhail, M. S., "Symbolic Matrix Derivatives," *Annals of Mathematical Statistics*, Vol. 19 (1948), pp. 517–534.

Dyen, I., "The Lexicostatistical Classification of the Malayopolynesian Languages," *Language*, Vol. 38 (1962), pp. 34–46.

Edwards, A. W. F. and Cavalli-Sforza, L. L., "A Method for Cluster Analysis," *Biometrics*, Vol. 21 (1965), pp. 362–375.

Effron, B. and Morris, C., "Stein's Paradox in Statistics," *Scientific American* magazine, (May, 1977), 119–127.

Eisenhart, C., Hastay, M. W., and W. A. Wallis (eds.), *Techniques of Statistical Analysis.* New York: McGraw-Hill, 1947.

Ekeblad, F. A. and Stasch, S. F., "Criteria in Factor Analysis," *Journal of Advertising Research*, Vol. 7, No. 3 (1967), pp. 48–57.

Erdélyi, A., *Higher Transcendental Functions*, McGraw Hill, 1953.

Ericson, W. A., "Optimum Stratified Sampling Using Prior Information," *Journal of the American Statistical Association*, Vol. 60 (1965), pp. 750–771.

Eusebio, J. W., see Hall, *et al.*

Evans, F. B., "Psychological and Objective Factors in the Prediction of Brand Choice, Ford versus Chevrolet," *Journal of Business*, Vol. 32 (October, 1959), pp. 340–369.

Everitt, B., *Graphical Techniques for Multivariate Data*, London: Heinemann Educational Books, Ltd., 1978.

Ewart, E., Seashore, S. E. and Tiffin, J., "A Factor Analysis of an Industrial Merit Rating Scale," *Journal of Applied Psychology*, Vol. 25 (1941), pp. 481–486.

Exton, H., *Multiple Hypergeometric Functions and Applications*, Chichester: Ellis Howard, Ltd., 1976.

———, *Handbook of Hypergeometric Integrals*, Chichester: Ellis Howard, Ltd., 1978.

Fama, E. F., "Portfolio Analysis in a Stable Paretian Market," *Management Science*, Vol. 11 (1965a), pp. 404–419.

———, "The Behavior of Stock Market Prices," *Journal of Business*, Vol. 38 (1965b), pp. 34–105.

———, "Risk Return and Equilibrium," Report No. 6831, Center for Mathematical Studies in Business and Economics, Graduate School of Business, University of Chicago, June, 1968.

——— and Miller, M. H., *The Theory of Finance*, Holt, Rinehart and Winston, Inc. 1972.

——— and Roll, R., "Some Properties of Symmetric Stable Distributions," *Journal of the American Statistical Association*, Vol. 63 (1968), pp. 817–836.

——— and ———, "Parameter Estimates for Symmetric Stable Distributions," manuscript, University of Chicago, 1969.

Farley, J. U., "Testing a Theory of Brand Loyalty," *Proceedings of the American*

Marketing Association (December, 1963), pp. 308–315.

———, "Why Does 'Brand Loyalty' Vary Over Products?" *Journal of Marketing Research*, Vol. 1 (November, 1964), pp. 9–14.

Feeney, G. J. and Hester, D. D., "Stock Market Indices: A Principal Components Analysis." *Risk Aversion and Portfolio Choice*, ed. by D. D. Hester and J. Tobin. New York: John Wiley and Sons, 1967.

Ferber, R. and Wales, H. G., *Motivation and Market Behavior*. Homewood, Illinois: Richard D. Irwin, Inc., 1958.

Ferguson, T. S., "On the Existence of Linear Regression in Linear Structural Relations," *University of California Publications in Statistics*, Vol. 2 (1955), pp. 143–166.

———, "A Representation of the Symmetric Bivariate Cauchy Distribution," *Annals of Mathematical Statistics*, Vol. 33 (December, 1962), pp. 1256–1266.

Fienberg, S. E., *The Analysis of Cross-Classified Categorical Data*, Cambridge, Mass.: The MIT Press, 2nd ed., 1980.

——— and Zellner, A., *Studies in Bayesian Econometrics and Statistics in Honor of Leonard J. Savage*, North Holland Publishing Co., 1975.

Finn, J., *Multivariate Analysis of Variance*. Buffalo, New York: New York State University, 1968.

Finney, D. J., *Probit Analysis* (2nd ed.). Cambridge, England: Cambridge University Press, 1952.

Fisher, F. M., see Rothenberg, T. J.

Fisher, R. A., "On the Probable Error of a Coefficient of Correlation Deduced from a Small Sample, *Metron*, Vol. 1, No. 4 (1921), p. 1.

———, *Statistical Methods for Research Workers*, Edinburgh, Scotland: Oliver and Boyd, 1925.

———, "The General Sampling Distribution of the Multiple Correlation Coefficient," *Proceedings of the Royal Society of London*, Series A, Vol. 121 (1928), pp. 654–673.

———, "The Use of Multiple Measurements in Taxonomic Problems" *Annals of Eugenics*, Vol. 7 (1936), pp. 179–188.

Fisher, W. D., "On Grouping for Maximum Homogeneity," *Journal of the American Statistical Association*, Vol. 53 (1958), pp. 789–798.

———, "Estimation in the Linear Decision Model," *International Economic Review*, Vol. 3 (1962), pp. 1–29.

———, "Two Way Cluster Analysis." Talk delivered at the University of Chicago, 1965.

———, "Simplification of Economic Models," *Econometrica*, Vol. 34, No. 3 (1966), pp. 563–584.

———, *Clustering and Aggregation in Economics*. Baltimore, Md.: The Johns Hopkins Press, 1969.

Fix, E., "Distributions which Lead to Linear Regressions," *Proceedings of the*

Berkeley Symposium on Mathematical Statistics and Probability. Berkeley: University of California Press, 1949.

Fleming, H., see Rogers, D. J.

Florentin, J. J., "Adaptive Control of a Simple Bayesian System," *Journal of Electronics and Control*, Vol. 13 (1962), pp. 473–488.

Forrester, J. W., *Industrial Dynamics*. Cambridge: The M.I.T. Press, 1961.

Fortier, J. J. and Solomon, H., "Clustering Procedures." *Multivariate Analysis*., ed. by P. R. Krishnaiah. New York: Academic Press, 1966.

Foster, F. G., "Upper Percentage Points of the Generalized Beta Distribution II," *Biometrika*, Vol. 44 (1957), pp. 441–453.

———, "Upper Percentage Points of the Generalized Beta Distribution III," *Biometrika*, Vol. 45 (1958), pp. 492–503.

——— and Rees, D. D., "Upper Percentage Points of the Generalized Beta Distribution I," *Biometrika*, Vol. 44 (1957), pp. 237–247.

Frank, R. E., Kuehn, A. A. and Massy, W. F., *Quantitative Techniques in Marketing Analysis*. Homewood, Illinois: Richard D. Irwin, Inc., 1962.

———, Massy, W. F. and Morrison, D. G., "Bias in Multiple Discriminant Analysis," *Journal of Marketing Research*, Vol. 2 (1965), pp. 250–258.

———, see Green, P. E.

Fraser, D. A. S., *The Structure of Inference*, New York: John Wiley and Sons, 1968.

Freund, R. J., "The Introduction of Risk into a Programming Model," *Econometrica*, Vol. 24 (1956), pp. 253–263.

Friedman, H. P. and Rubin, J., "On Some Invariant Criteria for Grouping Data," *Journal of the American Statistical Association*, Vol. 62 (1967), pp. 1159–1178.

Gabriel, K. R., "A Comparison of Some Methods of Simultaneous Inference in MANOVA." *Multivariate Analysis-II*, ed. by P. R. Krishnaiah. Academic Press, 1969, pp. 67–86.

Gantmacher, F. R., *The Theory of Matrices*, Vols. I and II. New York: Chelsea Publishing Company, 1959.

Gardner, R. S., "NO ANAV," SHARE, 1964 (see SHARE).

Geisel, M. S., see Zellner, A.

Geisser, S., "Multivariate Analysis of Variance for a Special Covariance Case," *Journal of the American Statistical Association*, Vol. 58 (1963), pp. 660–669.

———, "Posterior Odds for Multivariate Normal Classifications," *Journal of the Royal Statistical Society*, Series B, Vol. 26 (1964), pp. 69–76.

———, "Bayesian Estimation in Multivariate Analysis," *Annals of Mathematical Statistics*, Vol. 36 (1965a), pp. 150–159.

———, "A Bayes Approach for Combining Correlated Estimates," *Journal of the American Statistical Association*, Vol. 60 (1965b), pp. 605–607.

———, "Predictive Discrimination." *Multivariate Analysis*, ed. by P. R. Krishnaiah. New York: Academic Press, 1966, pp. 149–163.

———, "Estimation Associated with Linear Discriminants," *Annals of Mathematical Statistics*, Vol. 38 (1967), pp. 807–817.

——— and Cornfield, J., "Posterior Distributions for Multivariate Normal Parameters," *Journal of the Royal Statistical Society*, Series B, Vol. 25 (1963), pp. 368–376.

Girshick, M. A., "Principal Components," *Journal of the American Statistical Association*, Vol. 31 (1936), pp. 519–528.

———, "On the Sampling Theory of Roots of Determinantal Equations," *Annals of Mathematical Statistics*, Vol. 10 (1939), pp. 203–224.

———, see Anderson, T. A.

Glickman, A. S., "Factor Analysis of Personnel Components of Ship Performance," *Operations Research*, Vol. 6 (1958), pp. 106–115.

Gnanadesikan, R., see Roy, S. N.

———, see Smith, H.

———, *Methods for Statistical Data Analysis of Multivariate Observations*, John Wiley & Sons, Inc., 1977.

Gnedenko, B. V. and Kolmogorov, A. N., *Limit Distributions for Sums of Independent Random Variables.* Cambridge, Mass.: Addison-Wesley Publishing Co., 1954.

Gokhale, D. V. and Press, S. J., "Assessment of a Prior Distribution for the Correlation Coefficient in a Bivariate Normal Distribution," Tech. Report No. 58, Department of Statistics, University of California, Riverside, (Aug. 1979).

Goldberger, A. S., *Econometric Theory.* New York: John Wiley and Sons, 1964.

Goldstein, M. and Dillon, W. R., *Discrete Discriminant Analysis*, New York: John Wiley and Sons, Inc., 1978.

Goodman, L. A., *Analyzing Qualitative/Categorical Data*, Cambridge, Mass.: Abt Books, 1978.

Graybill, F. A., *An Introduction to Linear Statistical Models.* New York: McGraw-Hill, 1961.

———, *Introduction to Matrices with Applications in Statistics*, Wadsworth Publishing Co., 1969.

Green, B. F., "Latent Structure Analysis and Its Relation to Factor Analysis," *Journal of the American Statistical Association*, Vol. 47 (1952), pp. 71–76.

Green, P. E., Frank, R. E. and Robinson, P. J., "Cluster Analysis in Test Market Selection," *Management Science*, Vol. 13, No. 8 (April, 1967), pp. 387–400.

———, Halbert, M. H. and Robinson, P. J., "Canonical Analysis: An Exposition and Illustrative Application," *Journal of Marketing Research*, Vol. 3 (February, 1966), pp. 32–39.

——— and Maheshwari, A., "Common Stock Perception and Preference: An

Application of Multidimensional Scaling," *Journal of Business*, Vol. 42 (1969), pp. 439-457.

Greenberg, M. G., "Some Applications of Nonmetric Multidimensional Scaling." Presented at the Annual Meeting of the American Statistical Association, August, 1969.

Grenander, U. and Szegö, G., *Toeplitz Forms and Their Applications*. Berkeley: University of California Press, 1958.

Griliches, Z., "Distributed Lags: A Survey," *Econometrica*, Vol. 35 (1967), pp 16–49.

Grosenbaugh, L. R., "BC ANV," SHARE, 1965 (see SHARE).

Grubbs, F. E., "Tables of 1 Per Cent and 5 Per Cent Probability Levels for Hotelling's Generalized T^2 Statistics (Bivariate Case)." Aberdeen Proving Grounds, Ballistic Research Laboratories Technical Note No. 926, 1954.

Guttman, I., see Tiao, G. C.

Haberman, S. J., *Analysis of Frequency Data*, Chicago: University of Chicago Press, 1974.

———, *Analysis of Qualitative Data*, Vol. 1, Introductory Topics, New York: Academic Press, 1978.

———, *Analysis of Qualitative Data*, Vol. 2, New Developments, New York: Academic Press, 1979.

Hadamard, J. "Résolution d'une Question Relative aux Determinants," *Bulletin of Scientific Mathematics*, Vol. 2 (1893), pp. 240–248.

Halbert, M. H., see Green, P. E.

Hall, D. J., Ball G. H., Wolf, D. E. and Eusebio, J. W., "Promenade—An Interactive Graphics Pattern-Recognition System," Stanford Research Institute Report, 14 June, 1968, Menlo Park, California.

Hannan, E. J., "Canonical Correlation and Multiple Equation Systems in Economics," *Econometrica*, Vol. 35 (1967), pp. 123–138.

Hansen, M. H., Hurwitz, W. N. and Madow, W. G., *Sample Survey Methods and Theory*, Vol. 1. New York: John Wiley and Sons, 1953.

Harman, H. H., *Modern Factor Analysis*. Chicago: University of Chicago Press, 1967.

———, *Modern Factor Analysis*, 3rd Edition, Revised, Chicago: The University of Chicago Press, 1976.

Harper, R., "Factor Analysis as a Technique for Examining Complex Data on Foodstuffs," *Applied Statistics*, Vol. 5, No. 1 (1956), pp. 32–48.

——— and Baron, M., "Factorial Analysis of Rheological Measurements on Cheese," *Nature*, Vol. 182 (1948), p. 821.

——— and ———, "The Application of Factor Analysis to Tests on Cheese," *British Journal of Applied Physics*, Vol. 2 (1951), pp. 35–41.

Harrell, T. W., see Schreiber, R. J.

Harris, C. C., Jr., "A Scientific Method of Districting," *Behavioral Science*, Vol.

9, No. 3 (July, 1964), pp. 219–225.

Harrison, R., "Sources of Variation in Managers' Job Attitudes," *Personnel Psychology*, Vol. 12 (1960), pp. 425–434.

Hartigan, J. A., *Clustering Algorithms*, New York: John Wiley & Sons, Inc., 1975.

Hayes, H. D., see Immer, F. R.

Haynsworth, E. V., "Special Types of Partitioned Matrices," *Journal of Research of the National Bureau of Standards*, Vol. 65, No. 1 (1961), pp. 7-12.

Health Sciences Computing Facility, *BMD—Biomedical Computer Programs*. Los Angeles: University of California, 1965.

Heck, D. L., "Charts of Some Upper Percentage Points of the Distribution of the Largest Characteristic Root," *Annals of Mathematical Statistics*, Vol. 31 (1960), pp. 625–642.

Helmert, F. R., "Die Genauigkeit der Formel von Peters zer Berechnung des wahrscheinlichen Beobachtungsfehlers Direkter Beobachtungen gleicher Genauigkeit," *Astronomische Nachrichten*, Vol. 88, No. 2096 (1876).

Henderson, P. L., Hind, J. F. and Brown, S. E., "Sales Effects of Two Campaign Themes," *Journal of Advertising Research*, Vol. 1, No. 6 (1961), pp. 2–11.

Henry, N. W., see Lazarsfeld, P. E.

Hester, D. D., see Feeney, G. J.

Herz, C. S., "Bessel Functions of Matrix Argument," *Annals of Mathematics*, Vol. 61 (1955), pp. 474–523.

Herzog, A., *Harvard Canonical Correlation Program*. Chicago: University of Chicago Computation Center, 1963.

———, *Multivariate Analysis of Variance with Principal Components of All Dispersion (Variance-Covariance) Matrices*. Chicago: University of Chicago Computation Center, 1963.

Hilbert, D., see Courant, A.

Hind, J. F., see Henderson, P. L.

Hooper, J. W., "Simultaneous Equations and Canonical Correlation Theory," *Econometrica*, Vol. 27 (1959), pp. 245–256.

Hotelling, H., "The Generalization of Student's Ratio," *Annals of Mathematical Statistics*, Vol. 2 (1931), p. 360.

———, "Analysis of a Complex of Statistical Variables into Principal Components," *Journal of Educational Psychology*, Vol. 24 (1933), pp. 417–441, 498–520.

———, "The Most Predictable Criterion," *Journal of Educational Psychology*, Vol. 26 (1935), pp. 139-142.

———, "Relations between Two Sets of Variates," *Biometrika*, Vol. 28 (1936), pp. 321–377.

———, "Air Testing Sample Bombsights," prepared for the Applied Mathematics Panel, National Defense Research Committee, by the Statistical Research

Group, Columbia University. SRG Report No. 328, AMP Report No. 5.1 R (November, 1944).

———, "Multivariate Quality Control, Illustrated by the Air Testing of Sample Bombsights." *Techniques of Statistical Analysis*, ed. by Eisenhart, C., Hastay, M. W., and W. A. Wallis. New York: McGraw-Hill, 1947, pp. 113–184.

———, "A Generalized *T* Test and Measure of Multivariate Dispersion," *Proceedings of the 2nd Berkeley Symposium on Mathematical Statistics and Probability*. Berkeley: University of California Press, 1951, pp. 23–41.

Houle, A., Biography: Bayesian Statistics, Faculté des sciences de l'administration, Université Lavel, 1973.

Houthakker, H. S., see Prais, S. J.

Howard, A. H. and Schutz, H. G., "A Factor Analysis of a Salary Job Evaluation Plan," *Journal of Applied Psychology*, Vol. 36 (1952), pp. 243–246.

Hsu, P. L., "Notes on Hotelling's Generalized T^2," *Annals of Mathematical Statistics*, Vol. 9 (1938), pp. 231–243.

Hua, Loo-Keng, *Harmonic Analysis of Functions of Several Complex Variables in the Classical Domains*, Moscow (in Russian), 1959.

Huang, D. S., see Zellner, A.

Hughes, J. B., see Smith, H.

Hurwitz, W. N., see Hansen, M. H.

Ibragimov, I. A. and Chernin, K. E., "On the Unimodality of Stable Laws," English translation in *Theory of Probability and Its Applications*, in *SIAM Society Journal*, Vol. 4 (1959), pp. 417–419.

Ito, K., "A Comparison of the Powers of Two Multivariate Analyses of Variance Tests," *Biometrika*, Vol. 49 (1962), pp. 455–462.

———, "On the Effect of Heteroscedasticity and Non-Normality upon Some Multivariate Test Procedures." *Multivariate Analysis-II*, ed. by P. R. Krishnaiah. New York: Academic Press, 1969, pp. 87–120.

Jackson, J. E. and Morris, R. H., "An Application of Multivariate Quality Control to Photographic Processing," *Journal of the American Statistical Association*, Vol. 52 (1957), pp. 186–199.

James, A. T., "The Distribution of the Latent Roots of the Covariance Matrix," *Annals of Mathematical Statistics*, Vol. 31 (1960), pp. 151–158.

———, "Zonal Polynominals of the Real Positive Definite Symmetric Matrices," *Annals of Mathematics*, Vol. 74 (1961), pp. 456–469.

———, "Distributions of Matrix Variates and Latent Roots Derived from Normal Samples," *Annals of Mathematical Statistics*, Vol. 35 (1964), pp. 475–501.

———, "Normal Multivariate Analysis and the Orthogonal Group," *Annals of Mathematical Statistics*, Vol. 25, (1954), pp. 40–75.

James, W. and Stein, C., "Estimation with Quadratic Loss," *Fourth Berkeley Symposium on Mathematical Statistics and Probability*. Berkeley: University of California Press, 1960, pp. 361–379.

Janosi, P. E. de, "Factors Influencing the Demand for New Automobiles," *Journal of Marketing*, Vol. 23 (1959), pp. 412-418.

Jeffreys, H., *Theory of Probability*, 3rd ed. Oxford: Clarendon Press, 1961.

Jenkins, G. M., see Box, G. E. P.

Jensen, M. C., "Risk, the Pricing of Capital Assets and the Evaluation of Investment Portfolios," *Journal of Business*, Vol. 42 (1969), pp. 167–247.

Jensen, R. E., "Contiguous Performance Groupings of Companies—In Multiple Characteristic Dimensions Using Cluster Analysis Methods of Numerical Classification," Working Paper No. 17, Department of Accounting and Financial Administration, Michigan State University, 1968.

Johnson, N. L. and Kotz, S., *Distributions in Statistics: Continuous Multivariate Distributions,* New York: John Wiley & Sons, Inc., 1972.

Johnston, J., *Econometric Methods.* New York: McGraw-Hill, 1963.

Jones, K., *Multivariate Statistical Analyzer*, Cambridge, Mass.: Harvard Coop., 1964.

Jones, L. V., "Analysis of Variance in Its Multivariate Developments." *Handbook of Multivariate Experimental Psychology,* ed. by R. B. Cattel. Chicago: Rand McNally and Co., pp. 244–266.

Jöreskog, K. G., *Statistical Estimation in Factor Analysis.* Stockholm, Sweden: Almqvist and Wiksell, 1963.

———, "UMFLA—A Computer Program for Unrestricted Maximum Likelihood Factor Analysis," Research Memorandum 66-20, Revised edition. Princeton: Educational Testing Service, 1967.

———, "Some Contributions to Maximum Likelihood Factor Analysis," *Psychometrika,* Vol. 32 (1967), pp. 443–482.

———, "A General Method for Analysis of Covariance Structures," *Biometrika,* Vol. 57 (1970), pp. 239–251.

———, "Analysis of Covariance Structures, "in Multivariate Analysis III, edited by P. R. Krishnaiah, Academic Press, (1973), pp. 263–285.

——— and Sörbom, D., *Advances in Factor Analysis and Structural Equation Models,* Cambridge, Massachusetts: Abt Books, 1979.

Jorgenson, D., "Rational Distributed Lag Functions," *Econometrica,* Vol. 34 (1966), pp. 135–149.

Judge, G. G., and Bock, M. E., *The Statistical Implications of Pre-Test and Stein-Rule Estimators in Econometrics,* Amsterdam: North Holland Publishing Co., 1978, Chapter 8.

Kabe, D. G., "A Note on the Exact Distributions of the GCL Estimators in Two Leading Overidentified Cases," *Journal of the American Statistical Association,* Vol. 58 (1963), pp. 535–537.

———, "On the Exact Distributions of the GCL Estimators in a Leading Three-Equation Case," *Journal of the American Statistical Association,* Vol. 59 (1964), pp. 881–894.

Kadane, J. B., Dickey, J. M., Winkler, R. L., Smith, W. S., and Peters, S. C. "Interactive Elicitation of Opinion for a Normal Linear Model," *Jour. Amer. Stat. Assn.*, 75 (1980), 845–854.

Kadushin, C., "The Friends and Supporters of Psychotherapy," *American Sociological Review*, Vol. 31 (1966), pp. 786–802.

Kaiser, H. F., "The Varimax Criterion for Analytic Rotation in Factor Analysis," *Psychometrika*, Vol. 23 (1958), pp. 187–200.

Kamen, J. M., see Pilgrim, F. J.

Karlin, S., *Total Positivity*, Vol. I. Stanford, California: Stanford University Press, 1968.

——— and Trauax, D. "Slippage Problems," *Annals of Mathematical Statistics*, Vol. 31 (1960), pp. 296–324.

Kaufman, G. M., "Some Bayesian Moment Formulae," Report No. 6710, Center for Operations Research and Econometrics, Catholic University of Louvain, Heverlee, Belgium, 1967.

———, see Ando, Albert.

———, "Conditional Prediction and Unbiasedness in Structural Equations," *Econometrica*, 37 (1969), No. 1, 44–49.

——— and Press, S. J., "Bayesian Factor Analysis," Abstract No. 73t-25, *Bulletin of the Institute of Mathematical Statistics*, Vol. 2, No. 1, Issue No. 7, (March 1973) (submitted Dec. 15, 1972).

——— and ———, "Bayesian Factor Analysis," Working Paper No. 413, Faculty of Commerce and Business Administration, University of British Columbia, (Sept. 1976).

Kempthorne, O. (ed.), *Statistics and Mathematics in Biology*. Ames, Iowa: Iowa State College Press, 1954.

Kendall, M. C., *A Course in Multivariate Analysis*. Griffin Statistical Monographs. New York: Hafner Publishing Company, 1961.

———, see Beale, E. M. L.

——— and Stuart, A., *The Advanced Theory of Statistics*, Vol. 2. New York: Hafner Publishing Company, 1961.

——— and ———, *The Advanced Theory of Statistics*, Vol. 3. New York: Hafner Publishing Company, 1961.

Kettenring, J. R., "Canonical Analysis of Several Sets of Variables," *Biometrika*, 58 (1971), 433–451.

Kihlstrom, R., "Economic Applications of a Bayesian Decision Model," Working Paper No. 2; Department of Economics, University of Minnesota, 1967.

King, B., "Market and Industry Factors in Stock Price Behavior," *Journal of Business*, Vol. 39 (1966), pp. 139–190.

———, "Step-Wise Clustering Procedures," *Journal of the American Statistical Association*, Vol. 62 (March, 1967), pp. 86–101.

King, W. R., "Marketing Expansion—A Statistical Analysis," *Management Science*, Vol. 9 (July, 1963), pp. 563–573.

———, "Performance Evaluation in Marketing Systems," *Management Science*, Vol. 10 (1964), pp. 659–666.

———, "Toward a Methodology of Market Analysis," *Journal of Marketing Research*, Vol. 2 (1965), pp. 236–243.

———, "Structural Analysis and Descriptive Discriminant Functions," *Journal of Advertising Research*, Vol. 7, No. 2 (1967), pp. 39–43.

Kirsch, A. D. and Banks, S., "Program Types Defined by Factor Analysis," *Journal of Advertising Research*, Vol. 2, No. 3 (1962), pp. 29–31.

Klahr, D., "Decision Making in a Complex Environment: The Use of Similarity Judgments to Predict Preferences," working paper, Graduate School of Business, University of Chicago, August, 1967.

Klein, S. P., "Using Points of View and Multidimensional Scaling Analysis to Describe Esthetic Judgments," *Proceedings 76th Annual Convention of American Psychological Association*, 1968.

Kleinman, D. C., "Estimating the Parameters of Stable Probability Laws." Preliminary thesis proposal, Graduate School of Business, University of Chicago, 1965.

Kloek, T. and Mennes, L. B. M., "Simultaneous Equations Estimation Based on Principal Components of Predetermined Variables," *Econometrica*, Vol. 28 (January, 1960), pp. 45–61.

Kolmogorov, A. N., see Gnedenko, B. V.

Koyck, L. M., *Distributed Lags and Investment Analysis*. Amsterdam: North Holland Publishing Company, 1954.

Krishnaiah, P. R. (ed.), *Multivariate Analysis*. New York: Academic Press, 1966.

——— (ed.), *Multivariate Analysis-II*. New York: Academic Press, 1969.

———, (ed.), *Multivariate Analysis III*, New York: Academic Press, 1973.

———, (ed.), *Multivariate Analysis IV*, Amsterdam: North Holland Publishing Co., 1977.

———, (ed.), *Multivariate Analysis V*, Amsterdam: North Holland Publishing Co., 1980.

Kruskal, J. B., "Multidimensional Scaling by Optimizing Goodness of Fit to a Non-Metric Hypothesis," *Psychometrika*, Vol. 29 (March, 1964), pp. 1–27.

———, "Non-Metric Multidimensional Scaling—A Numerical Method," *Psychometrika*, Vol. 29 (June, 1964), pp. 115–129.

———, "Factor Analysis and Principal Components," *International Encyclopedia of Statistics*, edited by W. Kruskal, and J. Tanur, Vol. I, (1978), 307–330.

——— and Carroll, J. D., "Geometric Models and Badness-of-Fit Functions." *Multivariate Analysis-II*, ed. by P. R. Krishnaiah. New York: Academic Press, 1969.

Kshirsagar, A. M., "Some Extensions of the Multivariate **t**-Distribution and the Multivariate Generalization of the Distribution of the Regression Coefficient," *Proceedings of the Cambridge Philosophical Society*, Vol. 57 (1960), pp. 80–85.

Kuehn, A. A., see Frank, R. E.

Kullback, S., *Information Theory and Statistics*. New York: John Wiley and Sons, 1959.

Laha, R. G., see Lukacs, E.

Langmuir, C. R., see Rulon, P. J.

Lankford, P., *Varmax*. Chicago: University of Chicago Computation Center.

Lawley, D. N., "The Estimation of Factor Loadings by the Method of Maximum Likelihood," *Proceedings of the Royal Society of Edinburgh*, Vol. 60 (1940), pp. 64–82.

——— and Maxwell, A. E., *Factor Analysis as a Statistical Method*. London: Butterworth and Company, 1963.

Lawshe, C. H., Jr. and Alessi, S. L., "Studies in Job Evaluation: IV. Analysis of Another Point Rating Scale for Hourly Paid Jobs and the Adequacy of an Abbreviated Scale," *Journal of Applied Psychology*, Vol. 30 (1946), pp. 310–319.

——— Dudek, E. E. and Wilson, R. F., "Studies in Job Evaluation: VII. A Factor Analysis of Two Point Rating Methods of Job Evaluation," *Journal of Applied Psychology*, Vol. 32 (1948), pp. 118–129.

——— and Maleski, A. A., "Studies in Job Evaluation: III. An Analysis of Point Ratings for Salary Paid Jobs in an Industrial Plant," *Journal of Applied Psychology*, Vol. 30 (1946), pp. 117–128.

——— and Satter, G. A., "Studies in Job Evaluation: I. Factor Analysis of Point Ratings for Hourly-Paid Jobs in Three Industrial Plants," *Journal of Applied Psychology*, Vol. 28 (1944), pp. 189–202.

——— and Wilson, R. F., "Studies in Job Evaluation: V. An Analysis of the Factor Comparison System as It Functions in a Paper Mill," *Journal of Applied Psychology*, Vol. 30 (1946), pp. 426–434.

Lazarsfeld, P. F., "The Logical and Mathematical Foundation of Latent Structure Analysis," *Measurement and Prediction*, ed. by S. A. Stouffer *et al*. Princeton: Princeton University Press, 1950, Chapter 10.

——— and Henry, N. W., *Latent Structure Analysis*. Boston: Houghton Mifflin Company, 1968.

LeCam, L., "On the Asymptotic Theory of Estimation and Testing Hypotheses," *Proceedings of the Third Berkeley Symposium on Mathematical Statistics and Probability*, Vol. 1. Berkeley: University of California Press, 1956, pp. 129–156.

Leitch, R. A., and Paulson, A. S., "Estimation of Stable Law Parameters, Stock Price Behavior Application," *Journal of the American Statistical Association*, Vol. 70 (1975), pp. 690–698.

Lévy, P., "Theorie des Erreurs. La Loi de Gauss et Les Lois Exceptionnelles," *Bull. Soc. Math.*, Vol. 52 (1924), pp. 49–85.

———, *Calcul des Probabilités*, Paris: Gauthier-Villars, 1925.

———, *Théorie de l'Addition des Variables Aléatoires*. Paris: Gauthier-Villars, 1937.

———, *Théorie de l'Addition des Variables Aléatoires*, 2nd ed. Paris: Gauthier-Villars, 1954.

Lindley, D. V., "The Use of Prior Probability Distributions in Statistical Inference and Decisions," *Proceedings of the Fourth Berkeley Symposium on Mathematical Statistics and Probability*, Vol. 1. Berkeley: University of California Press, 1961, pp. 453–468.

———, *Introduction to Probability and Statistics*, Vols. I and II. Cambridge: The University Press, 1965.

———, *Bayesian Statistics: A Review*, Philadelphia: SIAM, 1972.

———, and Novick, M. R., "The Role of Exchangeability in Inference," *Ann. Statist.*, 9, No. 1, 1982, 45–58.

Loève, M., *Probability Theory*, 2nd ed. Princeton: D. Van Nostrand Company, 1960.

Lohnes, P. R., see Cooley, W. W.

Lukacs, E., *Characteristic Functions*. London: Griffin, 1960.

——— and Laha, R. G., *Applications of Characteristic Functions*. London: Griffin, 1954.

Luther, N. Y., "Decomposition of Symmetric Matrices and Distributions of Quadratic Forms," *Annals of Mathematical Statistics*, Vol. 36 (April, 1965), pp. 683–690.

MacPhail, M. S., see Dwyer, P. S.

MacQueen, J., "Some Methods for Classification and Analysis of Multivariate Observations," *Proceedings of the Fifth Berkeley Symposium on Mathematical Statistics and Probability*, Vol. 1. Berkeley: University of California Press, 1967, pp. 218–297.

MacRae, D., Jr., see Coleman, J. S.

———, see Wright, B. D.

Madansky, A., "The Fitting of Straight Lines when Both Variables Are Subject to Error," *Journal of the American Statistical Association*, Vol. 54 (1959), pp. 173–205.

Madow, W. G., see Hansen, M. H.

Maheshwari, A., see Green, P. E.

Maleski, A. A., see Lawshe, C. H., Jr.

Malinvaud, E., *Statistical Methods of Econometrics*. Chicago: Rand McNally and Company, 1966.

Mandelbrot, B., "The Variation of Certain Speculative Prices." *The Random Character of Stock Market Prices*, ed. by P. H. Cootner. Cambridge: The M.I.T. Press, 1964.

Mann, D. W., see Beale, E. M. L.

Mantel, N., "Models for Complex Contingency Tables and Polychotomous Response Curves," *Biometrics*, Vol. 22 (1966), pp. 83–110.

Markowitz, H. M., *Portfolio Selection*. New York: John Wiley and Sons, 1959.

Martin, J. J., *Bayesian Decision Problems and Markov Chains*. New York: John Wiley and Sons, 1967.

Massy, W. F., "Applying Factor Analysis to a Specific Marketing Problem," *Proceedings of the American Marketing Association* (December, 1963), pp. 291–307.

———, "Principal Components Regression in Exploratory Statistical Research," *Journal of the American Statistical Association*, Vol. 60 (March, 1965), pp. 234–256.

———, "On Methods: Discriminant Analysis of Audience Characteristics," *Journal of Advertising Research*, Vol. 5 (1965), pp. 39–48.

———, see Frank, R. E.

Matusita, K., "Classification Based on Distance in Multivariate Gaussian Cases," *Proceedings of the Fifth Berkeley Symposium on Mathematical Statistics and Probability*, Vol. 1. Berkeley: University of California Press, 1967, pp. 299–304.

Maxwell, A. E., see Lawley, D. N.

McDonald, R. P., "A General Approach to Non-Linear Factor Analysis," *Psychometrika*, Vol. 27 (1962), pp. 397–415.

McElroy, F. W., "A Necessary and Sufficient Condition that Ordinary Least Squares Estimators Be Best Linear Unbiased," *Journal of the American Statistical Association*, Vol. 62 (1967), pp. 1302–1304.

McQuitty, L. L., "Elementary Linkage Analysis for Isolating Orthogonal and Oblique Types and Typal Relevancies," *Educational and Psychological Measurement*, Vol. 17 (1957), pp. 207–229.

Megee, M., "On Economic Growth and the Factor Analysis Method," *Southern Economic Journal*, Vol. 31 (1965), pp. 215–228.

Mennes, L. B. M., see Kloek, T.

Midler, J., *Optimal Control of a Discrete-Time Stochastic System Linear in the State*. Santa Monica, California: RAND Corporation, 1967.

Miller, R., *Simultaneous Statistical Inference*. New York: McGraw-Hill, 1966.

Millican, R. D., "Application of Factor Analysis to Consumption Expenditures," *Review of Economics and Statistics*, Vol. 41 (1959), pp. 74-76.

Mirsky, L., *Introduction to Linear Algebra*. New York: Oxford University Press, 1955.

Moore, E. H., *General Analysis*, Part I. Philadelphia: Memoirs of the American Philosophical Society, Vol. I, 1935.

Morris, R. H., see Jackson, J. E.

Morrison, D. F., *Multivariate Statistical Methods*. New York: McGraw-Hill, 1967.

Morrison, D. G., see Frank, R. E.

Mosteller, F. and Wallace, D. L., *Inference and Disputed Authorship: The Federalist.* Reading, Mass.: Addison-Wesley Publishing Co., 1964.

Mukherjee, B. N., "A Factor Analysis of Some Qualitative Attributes of Coffee," *Journal of Advertising Research*, Vol. 5, No. 1 (March, 1965).

Mulaik, S. A., *The Foundations of Factor Analysis*, New York: McGraw Hill, Inc., 1972.

Muthen, B., "Contributions to Factor Analysis of Dichotomous Variables," *Psychometrika*, Vol. 43, (1978), pp. 551–560.

———, "A Structural Probit Model With Latent Variables," *Journal of the American Statistical Association*, Vol. 74, No. 368, (1979), pp. 807–811.

Nagar, A. L., "The Bias and Moment Matrix of the General k-Class Estimators of the Parameters in Simultaneous Equations," *Econometrica*, Vol. 27 (October, 1959), pp. 575–595.

Narain, R. D., "On the Completely Unbiased Character of Tests of Independence in Multivariate Normal Systems," *Annals of Mathematical Statistics*, Vol. 21 (1950), pp. 293–298.

Nel, D. G., "On Matrix Differentiation," *South African Stat. Jour.*, (1980), Vol. 14, 137–193

Nerlove, M., *Distributed Lags and Demand Analysis for Agriculture and Other Commodities.* Washington, D. C.: United States Department of Agriculture, Handbook No. 141, 1958.

———, *Estimation and Identification of Cobb-Douglas Production Functions.* Chicago: Rand McNally and Company, 1965.

——— and Press, S. J., "Univariate and Multivariate Log-Linear and Logistic Models," R-1306-EDA-NIH, The Rand Corporation, Santa Monica, Ca, 1973.

———, "Review of Bishop et al. 1975 Discrete Multivariate Analysis: Theory and Practice," Cambridge: MIT Press, *Bulletin of the American Mathematical Society*, Vol. 84, (March 1978), pp. 470–480.

———, "Multivariate Log-Linear Probability Models for the Analysis of Qualitative Data," Discussion Paper No. 1, Center for Statistics and Probability, Northwestern University, (1976).

———, "Multivariate Log-Linear Probability Models for the Analysis of Qualitative Data," *Bulletin of the International Statistical Institute*, in press, (1980a).

———, "Multivariate Log-Linear Probability Models in Econometrics," Invited Paper for *Journal of the American Statistical Association* and Invited General Methodology Lecture, American Statistical Association meeting, Houston, TX, August 1980. Draft June 2, 1980, (1980b).

——— and Wage, S., "On the Optimality of Adaptive Forecasting," *Management Science*, Vol. 10 (1964), pp. 207–224.

Neyman, J. and Scott, E. L., "Consistent Estimates Based on Partially Consistent Observations," *Econometrica*, Vol. 16. (1948), p. 1.

Okamoto, M., "Optimality of Principal Components." *Multivariate Analysis-II*, ed. by P. R. Krishnaiah. New York: Academic Press, 1969, pp. 673–685.

Olkin, I., "Note on 'The Jacobians of Certain Matrix Transformations Useful in Multivariate Analysis,' " *Biometrika*, Vol. 40 (1953), pp. 43–46.

——— and Press, S. J., "Testing and Estimation for a Circular Stationary Model," *Annals of Mathematical Statistics*, Vol. 40 (1969), pp. 1358–1373.

——— and Rubin, H., "Multivariate Beta Distributions and Independence Properties of the Wishart Distribution," *Annals of Mathematical Statistics*, Vol. 35 (March, 1964), pp. 261–269.

———, see Deemer, W. L.

——— and Siotani, M., "Asymptotic Distribution of Functions of a Correlation Matrix," Tech. Report No. 6, Laboratory for Quantitative Research in Education, Department of Statistics, Stanford University, March 16, 1964.

Overall, J. E., "A Configural Analysis of Psychiatric Diagnostic Stereotypes," *Behavioral Science*, Vol. 8 (1963), pp. 221-219.

Owen, D. B., *Handbook of Statistical Tables*, Reading, Mass.: Addison-Wesley Publishing Co., 1962.

Parkhurst, A. M. and James, A. T., "Zonal Polynomials of Order 1 through 12," in *Selected Tables in Mathematical Statistics*, Vol. II, ed. by H. L. Harter and D. B. Owen, Providence, R. I.: American Mathematical Society, 1974, pp. 199-388.

Parks, R. W., "Efficient Estimation of a System of Regression Equations when Disturbances Are Both Serially and Contemporaneously Correlated," *Journal of the American Statistical Association*, Vol. 62 (1967), pp. 500-509.

Paulauskas, V. J., "Some Remarks on Multivariate Stable Distributions," *Journal of Multivariate Analysis*, Vol. 6, (1976), pp. 356-368.

Paulson, A. S., Holcomb, E. W., and Leitch, R. A. "The Estimation of the Parameters of the Stable Laws," *Biometrika*, Vol. 67, (1975), pp. 163-169.

Pearson, K., "On Lines and Planes of Closest Fit to Systems of Points in Space," *Philosophical Magazine*, Series 6, Vol. 2 (1901), pp. 559–572.

———, *Tables of the Incomplete Beta-Function*. Cambridge: The University Press, 1934.

Penrose, R., "A Generalized Inverse for Matrices," *Proceedings of the Cambridge Philosophical Society*, Vol. 51 (1955), pp. 406–413.

Pierce, J. E., "Possible Electronic Computation of Typological Indices for Linguistic Structures," *International Journal of American Linguistics*, Vol. 28 (1962), pp. 215–226.

Pilgrim, F. J. and Kamen, J. M., "Patterns of Food Preferences through Factor Analysis," *Journal of Marketing*, Vol. 24 (1959), pp. 68–72.

Pillai, K. C. S., *On Some Distribution Problems in Multivariate Analysis*. Chapel

Hill: Institute of Statistics, University of North Carolina, 1954 (Mimeograph Series No. 88).

Plackett, R. L., *The Analysis of Categorical Data,* New York: Hafner Press, 1974.

———, *Statistical Tables for Tests of Multivariate Hypotheses.* Manila: Statistical Center, University of the Philippines, 1960.

Port, S. C., *Hitting Times and Potentials for Recurrent Stable Processes.* Santa Monica, California: The RAND Corporation, Memorandum RM-5104-PR, August, 1966.

Powers, L., see Immer, F. R.

Prais, S. J. and Houthakker, H. S., *The Analysis of Family Budgets.* Cambridge: Cambridge University Press, 1955.

Pratt, J. W., "Risk Aversion in the Small and in the Large," *Econometrica,* Vol. 32 (1964), pp. 122–136.

Prescott, E. C., "Adaptive Decision Rules for Macro Economic Planning." Doctoral dissertation, Graduate School of Industrial Administration, Carnegie Institute of Technology, 1967.

Press, S. J., "Some Hypothesis Testing Problems Involving Multivariate Normal Distributions with Unequal and Intraclass Structured Covariance Matrices," Technical Report No. 12, Department of Statistics, Stanford University, 1964, pp. 127–131.

———, "Structured Multivariate Behrens-Fisher Problems," *Sankhyā,* Series A, Vol. 29, Part 1 (1967a), pp. 41–48.

———, "On the Sample Covariance from a Bivariate Normal Distribution," *Annals of the Institute of Statistical Mathematics,* Vol. 19 (1967b), pp. 355–361.

———, "A Modified Compound Poisson Process with Normal Compounding," *Journal of the American Statistical Association,* Vol. 63 (1968), pp. 607–613.

———, "The *t*-Ratio Distribution," *Journal of the American Statistical Association,* Vol. 64 (1969a), pp. 242–252.

———, "On Serial Correlation," *Annals of Mathematical Statistics,* Vol. 40 (1969b), pp. 188–196.

———, "A Compound Poisson Process Model for Multiple Security Analysis," Random Counts in Scientific Work, Vol. 3, ed. by G. P. Patil. University Park: Pennsylvania State University Press, 1970.

———, see Olkin, I.

———, "Multivariate Stable Distributions," *Journal of Multivariate Analysis,* Vol. 2, (1972), pp. 444-462.

———, "Stable Distributions: Probability, Inference, and Applications in Finance–A survey and a Review of Recent Results," in *Statistical Distributions in Scientific Work,* Vol. 1, ed. by G. P. Patil, S. Kotz, and J. K. Ord, Boston, Massachusetts: D. Reidel Publishing Co., (1975), pp. 87–102.

———, "Qualitative Controlled Feedback for Forming Group Judgments and

Making Decisions," *Journal of the American Statistical Association,* Vol. 73, (Sept. 1978).

———, "Matrix Intraclass Covariance Matrices with Applications in Agriculture," Tech. Report #49, Department of Statistics, University of California, Riverside, (Feb. 1979).

———, "Bayesian Inference in MANOVA," *Handbook of Statistics,* Vol. 1, ed. by P. R. Krishaiah, North Holland Publishing Co., (1980a).

———, "Bayesian Computer Programs," Chapter 27 in *Bayesian Analysis in Econometrics and Statistics,* ed. by A. Zellner, Vol. 1, North Holland Publishing Co., (1980b) pp. 429–442.

———, "Multivariate Group Judgments by Qualitative Controlled Feedback," in *Multivariate Analysis V,* ed. by P. R. Krishnaiah, North Holland Publishing Co., (1980c), 581–591.

——— et al., "Authentication by Keystroke Timing: Some Preliminary Results," R-2526-NSF, The Rand Corporation, Santa Monica, CA, May (1980d).

———, "Bayesian Inference in Group Judgment Formulation and Decision Making Using Qualitative Controlled Feedback," *Trabajos de Estadistica,* 1980, Vol. 31, 397–415. (*Bayesian Statistics,* edited by Bernardo et al.)

——— and Scott, A., "Missing Variables in Bayesian Regression," in *Studies in Bayesian Econometrics and Statistics,* ed. by A. Zellner and S. Fienberg, North Holland Publishing Co., 1975, pp. 259–272.

——— and Scott, A., "Missing Variables in Bayesian Regression II," *Journal of the American Statistical Association,* Vol. 71, (June 1976), pp. 366–369.

——— and Wilson, S., "Choosing Between Logistic Regression and Discriminant Analysis," *Journal of the American Statistical Association,* Vol. 73, (Dec. 1978), pp. 699–705.

——— and Yang, C., "A Bayesian Approach to Second Guessing Undecided Respondents," *Journal of the American Statistical Association,* Vol. 69, (1974), pp. 58–67.

——— and Zellner, A., "Posterior Distribution for the Multiple Correlation Coefficient with Fixed Regressors, *Journal of Econometrics,* Vol. 8, (Dec. 1978), pp. 307–321.

Puri, M. L., see Sen, P. D.

Quenouille, M. H., *The Analysis of Multiple Time Series,* Griffin's Statistical Monograph, New York: Hafner Publishing Company, 1957.

Raiffa, H. and Schlaifer, R., *Applied Statistical Decision Theory.* Boston: Harvard University Press, 1961.

Rainville, E. D., *Special Functions,* New York: MacMillan Co., 1960.

Rao, C. R., *Advanced Statistical Methods in Biometric Research.* New York: John Wiley and Sons, 1952.

———, "Estimation and Tests of Significance in Factor Analysis," *Psychometrika,* Vol. 20 (1955), pp. 93–111.

———, "A Note on a Generalized Inverse of a Matrix with Applications to Problems in Mathematical Statistics," *Journal of the Royal Statistical Society*, Series B, Vol. 24 (1962), pp. 152–158.

———, "The Use and Interpretation of Principal Components Analysis in Applied Research," *Sankhyā*, Series A, Vol. 26 (1964), pp. 329–358.

———, *Linear Statistical Inference and Its Applications*. New York: John Wiley and Sons, 1965.

———, "Least Squares Theory Using an Estimated Dispersion Matrix and Its Application to Measurement of Signals," *Proceedings of the 5th Berkeley Symposium on Mathematical Statistics and Probability*, Vol. 1. Berkeley: University of California Press, 1967, pp. 355–372.

———, "A Note on a Previous Lemma in the Theory of Least Squares and Some Further Results," *Sankhyā*, Series A, Vol. 30 (1968), pp. 245–252.

———, "Matrix Approximations and Reduction of Dimensionality in Multivariate Statistical Analysis," in *Multivariate Analysis V*, ed. by P. R. Krishnaiah, Amsterdam: North Holland Publishing Co., 1980.

Ray-Chaudhuri, D. K., see Chow, G. C.

Rees, D. H., see Foster, F. G.

Renck, R., see Baehr, M. E.

Rice, S. A., *Quantitative Methods in Politics*. New York: Knopf, 1928.

Richards, A. B., "Input-Output Accounting for Business," *Accounting Review*, Vol. 35 (July, 1960), pp. 429–436.

Roberts, H. V., *Statistical Inference and Decision*. (To be published.) New York: John Wiley and Sons.

Robinson, P. J., see Green, P. E.

Rogers, D. J. and Fleming, H., "A Computer Program for Classifying Plants: II. A Numerical Handling of Non-Numerical Data," *Bioscience*, Vol. 14 (1964), pp. 15–28.

——— and Tanimoto, T. T., "A Computer Program for Classifying Plants," *Science*, Vol. 132 (1960), pp. 1115–1118.

———, "Analysis of Two Point-Rating Job Evaluation Plans," *Journal of Applied Psychology*, Vol. 30 (1946), pp. 579–585.

Rohlf, F. and Sokal, R., "The Description of Taxonomic Relations by Factor Analysis," *Systematic Zoology*, Vol. 11 (1962).

Roll, R., see Fama, E. F.

———, "The Efficient Market Model Applied to U. S. Treasury Bill Rates." Unpublished doctoral dissertation, Graduate School of Business, University of Chicago, 1968.

Rothenberg, T. J., "A Bayesian Analysis of Simultaneous Equation Systems," Report 6315, Netherlands School of Economics, Econometric Institute, 1963.

———, Fisher, F. M. and Tilanus, C. B., "A Note on Estimation from a Cauchy

Sample," *Journal of the American Statistical Association*, Vol. 59 (1964), pp. 460–463.

Roux, J. J. J., "On Generalized Multivariate Distributions," *South African Statistical Journal*, Vol. 5, (1971), pp. 91–100.

Roy, S. N., *Some Aspects of Multivariate Analysis*. New York: John Wiley and Sons, 1957.

——— and Bose, R. C., "Simultaneous Confidence Interval Estimation," *Annals of Mathematical Statistics*, Vol. 24 (1953), pp. 513–536.

——— and Gnanadesikan, R., "Some Contributions to ANOVA in One or More Dimensions, II," *Annals of Mathematical Statistics*, Vol. 30 (1959), pp. 318–340.

——— and Sarhan, A. E., "On Inverting a Class of Patterned Matrices," *Biometrika*, Vol. 43 (1956), pp. 227–231.

———, see Bose, R. C.

Rubin, H., see Anderson, T. W.

———, see Olkin, I.

Rulon, P. J., Tiedeman, D. V., Tatsuoka, M. and Langmuir, C. R., *Multivariate Statistics for Personnel Classification*. New York: John Wiley and Sons, 1967.

Rutherford, D. E., "Some Continuant Determinants Arising in Physics and Chemistry-II," *Proceedings of the Royal Society of Edinburgh*, Vol. 63 (1952), p. 232.

Sarhan, A. E., see Roy, S. N.

Satter, G. A., see Lawshe, C. H., Jr.

Savage, L. J., *The Foundations of Statistics*. New York: John Wiley and Sons, 1954.

——— *et al.*, *The Foundations of Statistical Inference*. London: Methuen and Company, Ltd., 1962.

Scheffé, H., "On Solutions of the Behrens-Fisher Problem Based on the *t*-Distribution," *Annals of Mathematical Statistics*, Vol. 14 (1943), pp. 35–44.

———, *The Analysis of Variance*. New York: John Wiley and Sons, 1959.

Schlaifer, R., see Raiffa, H.

———, *Analysis of Decisions Under Uncertainty*, McGraw-Hill Book Co., 1969.

Schnore, L. F., "The Statistical Measurement of Urbanization and Economic Development," *Land Economics*, Vol. 37 (1961), pp. 229–245.

Schreiber, R. J., Smith, R. G., Jr. and Harrell, T. W., "A Factor Analysis of Employee Attitudes," *Journal of Applied Psychology*, Vol. 36 (1952), pp. 247–250.

Schutz, H. G., see Howard, A. H.

Schweiker, R. F., see Carroll, J. B.

Scott, E. L., see Neyman, J.

Scott, J. T., "Factor Analysis and Regression," *Econometrica*, Vol. 34, No. 3 (1966), pp. 552–562.

Seal, H., *Multivariate Statistical Analysis for Biologists*. New York: John Wiley and Sons, 1964.

Seashore, S. E., see Ewart, E.

Sen, P. K. and Puri, M. L., *Nonparametric Methods in Multivariate Analysis.* New York: John Wiley and Sons, 1971.

Sen Gupta, A., "On the Problems of Construction and Statistical Inference Associated with a Generalization of Canonical Variables," Tech. Rept. under ONR Contract N00014-75-C-0442, Dept. of Stat., Stanford Univ., 1981.

Shannon, C. E., "The Mathematical Theory of Communication," *Bell System Technical Journal*, Vol. 27 (1948), pp. 379–423, 623–656.

SHARE. Head, DP Program Information Department, International Business Machines Corporation, 40 Saw Mill River Road, Hawthorne, New York.

Sharpe, W. F., "A Simplified Model for Portfolio Analysis," *Management Science* (January, 1963), pp. 277–293.

Shepard, R. N., "The Analysis of Proximities: Multidimensional Scaling with an Unknown Distance Function," Part One, *Psychometrika*, Vol. 27 (June, 1962), pp. 125–139.

———, "The Analysis of Proximities," Part Two, *Psychometrika*, Vol. 27 (September, 1962), pp. 219–246.

———, "Metric Structures in Ordinal Data," *Journal of Mathematical Psychology*, Vol. 3 (1966), pp. 287–315.

———, Romney, A. K., and Nerlove, S. B., editors, *Multidimensional Scaling: Theory and Applications in the Behavioral Sciences*, New York: Seminar Press, 1972.

Sherin, R. J., see Clyde, Dean J.

Simon, H. A., "Dynamic Programming Under Uncertainty with a Quadratic Criterion Function," *Econometrica*, Vol. 24 (1956), pp. 74–81.

Sitgreaves, R., "On the Distribution of Two Random Matrices Used in Classification Procedures," *Annals of Mathematical Statistics*, Vol. 23 (1952), pp. 263–270.

Slater, L. J., *Confluent Hypergeometric Functions*, Cambridge University Press, 1960.

Smith, H., Gnanadesikan, R., and Hughes, J. B., "Multivariate Analysis of Variance (MANOVA)," *Biometrics*, Vol. 18 (1962), pp. 22–41.

Smith, R. G. Jr., see Schreiber, R. J.

Sneath, P. H. A., see Sokal, R. R.

Sobel, M., "Multivariate Hermite Polynomials, Gram-Charlier Expansions and Edgeworth Expansions," Technical Report No. 18, Department of Statistics, University of Minnesota, Minneapolis, Minnesota, 1963.

———, see Dunnett, C. W.

Sokal, R. R. and Sneath, P. H. A., *Principles of Numerical Taxonomy.* San Francisco: W. H. Freeman Company, 1963.

———, see Rohlf, F.

Solomon, H., *Studies in Item Analysis and Prediction*. Stanford, California: Stanford University Press, 1961.

———, see Fortier, J. J.

Srivastava, M. S., "Some Tests for the Intraclass Correlation Model," *Annals of Mathematical Statistics*, Vol. 36 (1965), pp. 1802–1806.

Stasch, S. F., see Ekeblad, F. A.

Steidler, F., *Essoteric*. Linden, New Jersey: Esso Research and Engineering. Revised, University of Chicago Computation Center.

Stein, C., "Inadmissibility of the Usual Estimator for the Mean of a Multivariate Normal Distribution," *Proceedings of the Third Berkeley Symposium on Mathematical Statistics and Probability*. Berkeley: University of California Press, 1956, pp. 197–206.

———, "Confidence Sets for the Mean of a Multivariate Normal Distribution," *Journal of the Royal Statistical Society*, Series B, Vol. 24 (1962), pp. 265–285.

———, "Approximation of Improper Prior Measures by Prior Probabieliy Measures," *Proceedings of an International Research Seminar*, ed. by Jrzyt Neyman and Lucien M. LeCam. New York: Springer-Verlag, 1965, pp. 215–240.

———, see James, W.

Stoetzel, J., "A Factor Analysis of the Liquor Preferences of French Consumers," *Journal of Advertising Research*, Vol. 1, No. 2 (December, 1960), pp. 7–11.

Stone, R., "The Analysis of Market Demand," *Journal of the Royal Statistical Society*, New Series, Vol. 108 (1945), pp. 286–391.

———, "On the Interdependence of Blocks of Transactions," *Supplement to the Journal of the Royal Statistical Society*, Vol. 9 (1947), pp. 1–32.

Stracker, R., see Wright, B. D.

Stroud, A., Zellner, A. and Chau, L. C., "Program for Computing Two- and Three-Stage Least Squares Estimates and Associated Statistics," Workshop Paper No. 6308, revised by H. Thornber and A. Zellner, July 4, 1965, Social Systems Research Institute, University of Wisconsin.

Stuart, A., see Kendall, M. G.

Student (W. S. Gossett), "On the Probable Error of a Mean," *Biometrics*, Vol. 6 (1908), p. 1.

Swamy, P. A. V. B., "Statistical Inference in Random Coefficient Models and Its Application in Economic Analysis," *Annals of Mathematical Statistics*, Abstract No. 6, Vol. 38 (1967), p. 1940.

———, "Statistical Inference in Random Coefficient Models and Its Application in Economic Analysis." Ph.D. dissertation, Department of Economics, University of Wisconsin (1968).

Sworder, D., *Optimal Adaptive Control Systems*. New York: Academic Press, 1966.

Szegö, G., see Grenander, U.

Tanimoto, T. T., see Rogers, D. J.

Tatsuoka, M., see Rulon, P. J.

Teigen, R. L., "Demand and Supply Functions for Money in the United States: Some Structural Estimates," *Econometrica*, Vol. 32 (October, 1964), pp. 476–509.

Theil, H., "A Note on Certainty Equivalence in Dynamic Planning," *Econometrica*, Vol. 25 (1957), pp. 346–349.

———, *Principles of Econometrics*. New York: John Wiley and Sons, 1970.

Thomas, L. L., "A Cluster Analysis of Office Operations," *Journal of Applied Psychology*, Vol. 36 (1952), pp. 238–242.

Thornber, H., "Manual for (B34T, 8 March 66) a Stepwise Regression Program," Report No. 6603, Center for Mathematical Studies in Business and Economics, Graduate School of Business, University of Chicago.

———, H., see Zellner, A.

Thurstone, L. L., *Multiple Factor Analysis*. Chicago: The University of Chicago Press, 1947.

Tiao, G. C. and Guttman, I., "The Inverted Dirichlet Distribution with Applications," *Journal of the American Statistical Association*, Vol. 60 (1965), pp. 793–805.

——— and Zellner, A., "Bayes' Theorem and Use of Prior Knowledge in Regression Analysis," *Biometrika*, Vol. 51 (1964a), pp. 219–230.

——— and ———, "On the Bayesian Estimation of Multivariate Regression," *Journal of the Royal Statistical Society*, Series B, Vol. 26 (1964b), pp. 217–285.

Tiedeman, D. V., see Rulon, P. J.

Tiffin, J., see Ewart, E.

Tilanus, C. B., see Rothenberg, T. J.

Tintner, G., "Some Applications of Multivariate Analysis to Economic Data," *Journal of the American Statistical Association*, Vol. 41 (1946), pp. 472–500.

———, "Some Formal Relations in Multivariate Analysis," *Journal of the Royal Statistical Society*, Series B, Vol. 12 (1950), pp. 95–101.

———, *Econometrics*. New York: John Wiley and Sons, 1952.

Tobin, J., "The Application of Multivariate Probit Analysis to Economic Survey Data," Cowles Foundation Discussion Paper No. 1, Yale University, 1955.

Torgerson, W. S., *Theory and Methods of Scaling*, New York: John Wiley and Sons, 1958.

——— see Young, F. W.

Tou, J. T., *Modern Control Theory*. New York: McGraw-Hill, 1965.

Tracy, D. S., and P. S. Dwyer, "Multivariate Maxima and Minima with Matrix Derivatives," *Jour. Amer. Stat. Assn.*, 64 (1969), 1576–1594.

Truax, D., see Karlin, S.

Tryon, R. C., *Cluster Analysis*. Ann Arbor, Mich.: Edwards Brothers, 1939.

———, "Cumulative Communality Cluster Analysis," *Educational and Psychological Measurement*, Vol. 18 (1958), pp. 3–35.

——— and Bailey, D., "BC TRY," University of California at Berkeley and University of Colorado, 1965.

Tsui, K-W. and Press, S. J., "Simultaneous Bayesian Estimation of the Parameters of Independent Poisson Distributions," Tech. Report #33, Department of Statistics, University of California, Riverside, (Nov. 1977a).

——— and ———, "Simultaneous Estimation of Several Poisson Parameters Under *k*-Normalized Squared Error Loss," Working Paper #456, Faculty of Commerce, University of British Columbia, (April 1977b).

Tukey, J. W., "Dyadic ANOVA, and Analysis of Variance for Vectors," *Human Biology*, Vol. 21 (1949), pp. 65–110.

Twedt, D. W., "A Multiple Factor Analysis of Advertising Readership," *Journal of Applied Psychology*, Vol. 36 (1952), pp. 207–215.

Twery, R. J., "Detecting Patterns of Magazine Reading," *Journal of Marketing*, Vol. 22 (1958), pp. 290–294.

Varga, R. S., *Matrix Iterative Analysis*. Englewood Cliffs, N.J.: Prentice-Hall, 1962.

Villegas, C., "On the A Priori Distribution of the Covariance Matrix," *Annals of Mathematical Statistics*, Vol. 40 (1969), pp. 1098–1099.

Vincent, N. L., "A Note on Stoetzel's Factor Analysis of Liquor Preferences," *Journal of Advertising Research*, Vol. 2 (1962), pp. 24–27.

Vinograd, B., "Canonical Positive Definite Matrices Under Internal Linear Transformation," *Proceedings of the American Mathematical Society*, Vol. 1 (1950), pp. 159–161.

Votaw, D. F., Jr., "Testing Compound Symmetry in a Normal Multivariate Distribution," *Annals of Mathematical Statistics*, Vol. 19 (1948), pp. 447–473.

Wage, S., see Nerlove, M.

Wald, A., "On a Statistical Problem Arising in the Classification of an Individual into One of Two Groups," *Annals of Mathematical Statistics*, Vol. 15 (1944), pp. 145–163.

———, "Note on the Consistency of the Maximum Likelihood Estimate," *Annals of Mathematical Statistics*, Vol. 20 (1949), p. 595.

Wales, H. G., see Ferber, R.

Walker, A. M., "On the Asymptotic Behavior of Posterior Distribution," *Journal of the Royal Statistical Society*, Series B, Vol. 31, No. 1 (1969), pp. 80–88.

Walker, M. A., "Some Critical Comments on 'An Analysis of Crimes by the Method of Principal Components' by B. Ahamad," *Journal of Applied Statistics*, Vol. 16 (1967), pp. 36–39.

Wallace, D. L., "Clustering," *International Encyclopedia of the Social Sciences.* New York: Crowell Collier, 1968.

———, see Mosteller.

Wallace, T. D. and Hussain, A., "The Use of Error Components Models in Combining Cross Section with Time Series Data," *Econometrica*, Vol. 37 (1969), pp. 55–72.

Walsh, J. A., "Three Mode Factor Analysis," *Educational and Psychological Measurement*, Vol. 24 (1964), p. 669.

Ward, J. H., Jr., "Hierarchical Grouping to Optimize an Objective Function," *Journal of the American Statistical Association*, Vol. 58 (1963), pp. 236–244.

Watson, G. N., *A Treatise on the Theory of Bessel Functions* , 2nd ed. Cambridge: Cambridge University Press, 1944.

Waugh, F. V., "Regressions between Sets of Variables," *Econometrica*, Vol. 10 (1942), pp. 290–310.

Wegner, P., "Relations between Multivariate Statistics and Mathematical Programming," *Applied Statistics*, Vol. 12 (1963), pp. 146–150.

Wherry, R. J., "Factor Analysis of Morale Data: Reliability and Validity," *Personnel Psychology*, Vol. 11 (1958), pp. 78–89.

Whittle, P., *Hypothesis Testing in Time Series Analysis.* Uppsala, Sweden: Almqvist and Wiksells, 1951.

Wilks, S. S., "Certain Generalizations in the Analysis of Variance," *Biometrika*, Vol. 24 (1932), p. 471.

———, "The Sampling Theory of Systems of Variances, Covariances, and Intraclass Covariances," *American Journal of Mathematics*, Vol. 58 (1936), pp. 426–432.

———, "The Large-Sample Distribution of Likelihood Ratio for Testing Composite Hypotheses," *Annals of Mathematical Statistics*, Vol. 9 (1938), p. 60.

———, "Sample Criteria for Testing Equality of Means, Equality of Variances, and Equality of Covariances in a Normal Multivariate Distribution," *Annals of Mathematical Statistics*, Vol. 17 (1946), pp. 257–281.

———, *Mathematical Statistics.* New York: John Wiley and Sons, 1962.

Wilson, R. F., see Lawshe, C. H., Jr.

Wise, J., "The Autocorrelation Function and the Spectral Density Function," *Biometrika*, Vol. 42 (1955), pp. 151–159.

Wishart, J., "The Generalized Product Moment Distribution in Samples Drawn from a Normal Multivariate Population," *Biometrika*, Vol. 20 (1928), pp. 38–40.

Wolf, G. H., see Hall *et al.*

Woodbury, M., Technical Report No. 42, Princeton University, Princeton, N. J.

Wright, B. D., Bradford, C., Stracker, R., and Bamberger, F., *Mesa 85*. Chicago: University of Chicago Computation Center, 1965.

——— and MacRae, D., *Direct Factor Analysis Program*. Chicago: University of Chicago Computation Center, 1966.

Young, F. W., "Nonmetric Multidimensional Scaling: Development of an Index of Metric Determinacy," The L. L. Thurstone Psychometric Laboratory, University of North Carolina Report No. 68, August, 1968.

——— and Torgerson, W. S., "TORSCA, a Fortran IV Program for Shepard-Kruskal Multidimensional Scaling Analysis," *Behavioral Science*, Vol. 12 (1967), p. 498.

Young, G., "Factor Analysis and the Index of Clustering," *Psychometrika*, Vol. 4 (1939), pp. 201–208.

Zacks, S., see Chernoff, H.

Zellner, A., "An Efficient Method of Estimating Seemingly Unrelated Regressions and Test for Aggregation Bias," *Journal of the American Statistical Association*, Vol. 57 (1962), pp. 348–368.

———, "Estimators for Seemingly Unrelated Regressions: Some Exact Finite Sample Results," *Journal of the American Statistical Association*, Vol. 58 (1963), pp. 977–992.

———, "On Controlling and Learning about a Normal Regression Model." Preliminary paper presented at Graduate School of Business, University of Chicago Workshop, 1966.

———, *An Introduction to Bayesian Inference in Econometrics*, New York: John Wiley and Sons, 1971.

———, *Bayesian Analysis in Econometrics and Statistics, Essays in Honor of Harold Jeffreys*, editor, North Holland Pub. Co., 1980.

——— and Chetty, V. K., "Prediction and Decision Problems in Regression Models from the Bayesian Point of View," *Journal of the American Statistical Association*, Vol. 60 (1965), pp. 608–616.

——— and Geisel, M. S., "Sensitivity of Control to Uncertainty and Form of the Criterion Function," Report No. 6717, Center of Mathematical Studies in Business and Economics, Department of Economics and Graduate School of Business, University of Chicago, Chicago, Illinois, May, 1967.

——— and Huang, D. S., "Further Properties of Efficient Estimators for Seemingly Unrelated Regression Equations," *International Economic Review*, Vol. 3 (1962), pp. 300–313.

——— and Lee, T. H., "Joint Estimation of Relationships Involving Discrete Random Variables," *Econometrica*, Vol. 33 (1965), pp. 382–394.

——— and Thornber, H., "Computational Accuracy and Estimation of Simultaneous Equation Econometric Models," *Econometrica*, Vol. 34, No. 3 (1966), pp. 727–729.

——— see Stroud, A.

——— see Tiao, G. C.

INDEX

* Page numbers in italics denote principal sources.

Discriminatory power, 408–410

SOLUTIONS MANUAL
for
Applied Multivariate Analysis: Using Bayesian and Frequentist Methods of Inference

By

S. James Press

Technical Report No. 132

with special assistance from
Dr. Kazuo Shigemasu
Tokyo Institute of Technology

Department of Statistics
University of California
Riverside, California 92521

March, 1985

Preface

This solutions manual is intended to serve as a guide to solutions to the exercises in the book, Applied Multivariate Analysis: Using Bayesian and Frequentist Methods of Inference. In some cases alternative solutions are possible and in some cases, alternative proofs may be shorter.

There are a total of 171 exercises and solutions. The number of exercises per chapter is roughly proportional to the length of the chapter, ranging from 5, in Chapter 14, to 21, in Chapter 8.

Dr. Kazuo Shigemasu of the Tokyo Institute of Technology was very helpful in the preparation of this manual. We both worked all of the exercises and cross checked each other on the solutions.

We strongly believe that all solutions presented are correct, and we have verified our numerical solutions by computer, wherever possible. Nevertheless, errors have a way of cropping up, in spite of how much care has been taken in preparation. We hope there aren't any. We also hope that this manual will aid students, teachers, and research workers in improving their understanding of the material presented in the text.

I would like to express my gratitude to my many students over the years who, by their stimulating questions, helped me to formulate exercises and solutions that would hopefully be challenging and pedagogical, and at the same time interesting.

Chapter 2

2.1 From eqn. (2.4.9),

$$|\Sigma| = |A-B|^{p-1}|A+B(p-1)|$$

$$A-B = \begin{pmatrix} 5 & & 19/6 \\ & \ddots & \\ & & 5 \end{pmatrix}, \qquad A+B(p-1) = \begin{pmatrix} 17 & & 43/6 \\ & \ddots & \\ & & 17 \end{pmatrix}.$$

Use eqn. (2.4.7) to get

$$|\Sigma| = \left(\frac{59}{6}\right)^{k-1} \left(\frac{11}{6}\right)^{2(k-1)} \left(\frac{59+43k}{6}\right)\left(\frac{11+19k}{6}\right)^2 .$$

(1) The pattern of scores is the same in each of the three cities; variances are the same across tests, and covariances are the same for each pair of tests.

(2) Covariance between test scores on the same test for a pair of cities is the same for all pairs of cities.

2.2 (a) $$\Sigma^{-1} = \frac{1}{2}\begin{pmatrix} 1.0794 & -.1905 & -3.1746\text{x}10^{-2} & -.1905 \\ & 1.0794 & -.1905 & -3.1746\text{x}10^{-2} \\ & & -.1905 & 1.0794 \\ & & & 1.0794 \end{pmatrix}.$$

(b) $$\Sigma^2 = \frac{1}{4}\begin{pmatrix} 1.09 & .44 & .28 & .44 \\ & 1.09 & .44 & .28 \\ & & 1.09 & .44 \\ & & & 1.09 \end{pmatrix}.$$

Easy to predict sales with high accuracy.

2.3 A is idempotent, so latent roots are 0, 1.

2.4 Latent roots of A are:

$\lambda_1 = 6.93202,\ \lambda_2 = 3.10862,\ \lambda_3 = 1.95936;$

$$\Gamma = \begin{pmatrix} .598880 & -.473031 & .646207 \\ .632113 & -.216225 & -.744097 \\ .491707 & .854101 & .169516 \end{pmatrix}.$$

A is a covariance matrix and non-singular. Latent roots of A^2 are:

$\lambda_1 = 48.0530,\ \lambda_2 = 9.6635,\ \lambda_3 = 3.8391.$

2.5 (a) $$\Sigma^{-1} = \frac{1}{2}\begin{pmatrix} 3 & 3 & -1 \\ & 5 & -1 \\ & & 1 \end{pmatrix}.$$

(b) $\lambda_1 = 3.73205,\ \lambda_2 = .50000,\ \lambda_3 = .26795$

all latent roots are positive.

(c) as the incidence of crime of type (1) increases it tends to decrease for crime of type (2): and conversely.

2.6 $\Sigma^+ = \text{pseudoinverse} = \Delta\left(D^*_\lambda, 0\right)\Gamma'$,

$$\Sigma^+ = \begin{pmatrix} .011465 & .064805 & .04187 \\ .022599 & .048022 & .002825 \\ -.041708 & .010635 & .094052 \end{pmatrix}.$$

Σ is not positive definite (it is not even symmetric).

The pseudoinverse of Σ yields the smallest possible norm of solutions of $\Sigma x = y$ in the class of all generalized inverses of Σ; i.e., $(x'x)$ is minimized. (See Remark in Example (2.12.4)).

2.7 $\min_i (\lambda_i) \le \frac{x'Ax}{x'x} \le \max_i (\lambda_i)$.

Use eqn. (2.8.5) to find

$2 - \sqrt{3} \le C(x) \le 2 + \sqrt{3}$, or (.268, 3.732).

2.8 (a) $x' \overline{\log x} = \sum_i x_i \log x_i$; $x'e = \sum_i x_i$.

Differentiate.

(b) $\frac{d}{dx} f(x) = \overline{\log}\, x = 0$ implies $x = e$ for stationary points.

(c) $H = D_{1/x} = \begin{pmatrix} \frac{1}{\bar{x}_1} & & 0 \\ & \ddots & \\ 0 & & \frac{1}{\bar{x}_p} \end{pmatrix}$.

(d) $x = e$ yields a global minimum because $H > 0$.

2.9 (a) $\begin{pmatrix} 9B & (9/2)B, \ldots, (9/2)B \\ & 9B \ddots \quad \vdots \\ & \quad 9B \end{pmatrix}$.

(b) $|A \otimes B| = 12^{k(k-1)}[9 + \frac{9}{2}(k-1)]^k \; [4 + \frac{4}{3}(k-1)]^k$.

(c) $\mathrm{tr}\,(A \otimes B) = 36k^2$.

(d) $(A \otimes B)^{-1} = A^{-1} \otimes B^{-1}$, $A^{-1} = \frac{2}{9}\left(I - \frac{ee'}{k+1}\right)$

$B^{-1} = \frac{3}{8}\left(I - \frac{ee'}{2+k}\right)$.

(e) λ_j and θ_k are latent roots of A, B, respectively. Latent roots of $A \otimes B$ are $\delta_{jk} \equiv \lambda_j \theta_k$, where

$\lambda_1 = 9 + \frac{9}{2}(k-1)$, $\lambda_2 = \ldots = \lambda_k = \frac{9}{2}$,

$\theta_1 = 4 + \frac{4}{3}(k-1)$, $\theta_2 = \ldots = \theta_k = \frac{8}{3}$.

(f) $A \otimes B$ is a covariance matrix with intraclass covariance structure (see Part (a)).

2.10 Use eqn. (2.8.6): $\lambda_1 = \frac{43}{6}k + \frac{59}{6}$,

$\lambda_2 = \ldots = \lambda_k = \frac{59}{6}$,

$\lambda_{k=1} = \ldots = \lambda_{3+k-1} = \frac{11}{6} + \frac{19}{6}k$,

$\lambda_{3+k} = \ldots = \lambda_{3k} = \frac{11}{6}$.

2.11 Find $\frac{\partial f(Z)}{\partial Z} = -2[\Sigma - E(Z)]$ and set it equal to zero.

Find $\frac{\partial^2 f}{\partial Z \partial Z'} = H = 2I > 0$, for a minimum.

2.12 Using eqn. (2.6.7) gives

$$f = |V|^{\frac{n-p-1}{2}} \int |V_{11} - V_{12}V_{22}^{-1}V_{21}|^{\frac{n-p-1}{2}} dV_{12}.$$

Factor out V_{11}, and transform V_{12} into Z by

$Z = V_{11}^{-1/2}V_{12}V_{22}^{-1/2}$. Use eqn. (2.15.4) to obtain the Jacobian.

2.13 $A\beta = a$, where: $a \equiv (1,0,6)'$,

$$\beta = (x,y,z)',\ A = \begin{pmatrix} 3 & 2 & -1 \\ 7 & -1 & 2 \\ -2 & 3 & -5 \end{pmatrix}.$$

2.14 (a) $b'b = 66$; (b) $a'b = 12$;

(c) $aa' = \begin{pmatrix} 9 & 3 & 6 \\ 3 & 1 & 2 \\ 6 & 2 & 4 \end{pmatrix}$; (d) $5a + 2b = (13,19,18)'$.

(e) $a'Hb = 43$.

2.15 (a) 2; (b) 3; (c) 4.

2.16 (a) $\lambda_1 = 17.657 = 4(3+\sqrt{2})$; $\lambda_2 = 6.343 = 4(3-\sqrt{2})$;

$$\Gamma = \begin{pmatrix} .923880 & -.382683 \\ .382683 & .923880 \end{pmatrix}.$$

(b) $\lambda_1 = 9.41421 = 8 + \sqrt{2}$; $\lambda_2 = 6.58579 = 8 - \sqrt{2}$;

Γ is the same as in part (a).

(c) $\lambda_1 = .6306 = (3+\sqrt{2})/7$; $\lambda_2 = .22654 = (3-\sqrt{2})/7$;

Γ is the same as in part (a).

(d) $\lambda_1 = 19.4853 = 11+6\sqrt{2}$; $\lambda_2 = 2.5147 = 11-6\sqrt{2}$;

Γ is same as in part (a).

(e) $\lambda_1 = 2.1010$; $\lambda_2 = 1.25928$;

Γ is same as in part (a).

2.17 (a) $x'Ax$, where $x \equiv (x_1, x_2)'$, and

$$A \equiv \begin{pmatrix} 2 & 2 \\ 2 & 1 \end{pmatrix}.$$

A is symmetric, but not positive definite ($|A| = -2$).

(b) $x'Bx$, where $x \equiv (x_1, x_2)'$, and

$$B \equiv \begin{pmatrix} 3 & -1 \\ -1 & 4 \end{pmatrix}. \quad B > 0.$$

Chapter 3

3.1

$$R = \begin{pmatrix} 1 & .1 & .6 & .1 \\ & 1 & .01 & .5 \\ & & 1 & .01 \\ & & & 1 \end{pmatrix}.$$

3.2 $p(\theta,\Sigma|\bar{X},V) \propto |\Sigma|^{-11} \exp(-1/2)\{\text{tr}\Sigma^{-1}H+12(\theta-\alpha)'\Sigma^{-1}(\theta-\alpha)\},$

where: $\alpha \equiv (2.09,2.92,1.54,1.08,2.00)'$,

and $$H \equiv \begin{pmatrix} 14.009 & 2.292 & 1.854 & 5.108 & 8.2 \\ & 9.917 & .742 & .483 & 3.4 \\ & & 6.229 & 2.458 & 4.6 \\ & & & 9.917 & 5.8 \\ & & & & 21 \end{pmatrix}.$$

Hint: $(X|\theta,\Sigma) \sim N(\theta,\Sigma/N)$, $(V|\Sigma) \sim W(\Sigma,5,10)$.

$(\bar{X}|\theta,\Sigma)$ and $(V|\Sigma)$ are independent,

$(\theta|\Sigma) \sim N(2e,\Sigma)$, $\Sigma \sim W^{-1}(H,5,10)$.

3.3 (a) $E(Y|X) = (X+1)/3$.

(b) $\text{var}(Y|X) = 3$; $\text{var}(X|Y) = 6.75$.

(c) $P\{X \leq 6\} = P\{Z \leq 1/3\} = .6293$, $Z \sim N(0,1)$.

(d) $P\{4 \leq X \leq 6,\ 1 \leq Y \leq 3\} = .11331$

Hint: See, for example, p. 184 of Owen.

3.4 $(\bar{z}|\theta,\Sigma) \sim N(\theta,\Sigma/N)$, $N = 100$, $p = 3$.

$(V|\Sigma) \sim W(\Sigma,p,N-1)$, $\bar{z}$ and V are independent.

$$p(\theta,\Sigma) \propto \frac{1}{|\Sigma|^{(p+1)/2}} .$$

$(y|\theta,\Sigma) \sim N(\theta,\Sigma)$.

(a) $p(y|\bar{z},V) = \int\int g(y|\theta,\Sigma)h(\theta,\Sigma|\bar{z},V)d\theta d\Sigma$.

$$p(y|\bar{z},V) \propto \{(n-p+1) + (y-\bar{x})'H^{-1}(y-\bar{x})\}^{-[n-p+1)+p]/2}.$$

That is, $(y|\bar{z},V)$ is multivariate Student t, centered at $\bar{x}$ $(E[y|\bar{z},V] = \bar{x})$, with $(n-p+1)$ degrees of freedom (see eqn. (6.2.4)).

(b) From eqn. (6.2.4)',

$$p(y_1|\bar{z},V) \propto \{(n-p+1)+(y_1-\bar{x}_1)^2 H_{11}{}^{-1}\}^{-[(n-p+1)+1]/2},$$

where: $H_{11} = \frac{(N+1)}{(N)} \frac{v_{11}}{(n-p+1)} = \left(\frac{101}{100}\right) \frac{v_{11}}{n-2}$,

$V \equiv (v_{ij})$. Then, if

$$\tau \equiv \frac{(y_1 - \bar{x}_1)\sqrt{(n-2)}}{\sqrt{v_{11}}} \sim t_{n-p+1} = t_{n-2}.$$

So: with $n=99$, $N=n-1$, $\bar{z}_1=62$, $v_{11}=13$,

$$P \equiv P\{65 \le y_1 \le 70|z_1,\ldots,z_{100}\}$$

$$= P\{8.19 \le \tau \le 21.85|\bar{x},V\},$$

and $\tau \sim t_{97}$. Therefore, $P \approx 0$.

3.5 $W \equiv AY+BZ$. $W \sim N(\phi, \Sigma^*)$,

where: $\phi = A\theta_1 + B\theta_2$,

$\Sigma^* = A\Sigma_{11}A'+B\Sigma_{21}A'+A\Sigma_{12}B'+B\Sigma_{22}B'$.

3.6 $z = (x-\theta_0)'\Sigma_0^{-1}(x-\theta_0) \sim \chi_2^2(.05) \equiv 5.99$.

$z = .704$ which is not significant. So cannot reject H_0.

3.7 (a) $(X_1 | X_2, X_3) \sim N(\frac{X_2+X_3}{6}, \frac{1}{3})$;

(b) $(X_1, X_2 | X_3) \sim N(\phi, \Omega)$, $\phi = (\frac{X_3}{4}+1, \frac{X_3}{2}+6)'$,

$$\Omega = \begin{pmatrix} \frac{3}{4} & \frac{5}{2} \\ & 15 \end{pmatrix};$$

(c) $(X_3 | X_1, X_2) \sim N(\frac{10}{7} X_1 - \frac{X_2}{7} + \frac{36}{7}, \frac{20}{7})$.

3.8 $X \sim N(\theta, \Sigma)$, $\theta \equiv (2,1)'$,

$$\Sigma = \begin{pmatrix} 25 & 12 \\ 12 & 36 \end{pmatrix}, \ \rho = .4.$$

3.9 (a) $h(p | r_0, n_0) \propto p^{r_0+\alpha-1} (1-p)^{n_0-r_0+\beta-1}$,

$\alpha > 0$, $\beta > 0$. (Note: $g(p) \propto p^{\alpha-1}(1-p)^{\beta-1}$.)

(b) $f(r | n, r_0, n_0) = \int_0^1 f^*(r | n, p) h(p | r_0, n_0) dp$

$\propto \{\Gamma(r)\Gamma(n-r)\Gamma(r+r_0+\alpha)\Gamma(n+n_0+\beta-r-r_0)\}^{-1}$.

3.10 (a) Assume $EX = EY = 0$.

$E(XY) = \int xy dF(x,y) \underset{\text{indep.}}{=} \int xy dF_x(x) dF_y(y)$

$= E(X)E(Y) = 0$.

(b) Write the density of a bivariate normal distribution and set $\rho = 0$. The density then factors into the marginal densities.

(c)

X \ Y	0	1	
0	p_{00}	p_{01}	p_{0+}
1	p_{10}	p_{11}	p_{1+}
	p_{+0}	p_{+1}	1

$p_{00} + p_{01} + p_{10} + p_{11} = 1.$

$$XY = \begin{cases} 0, & \text{if } X = 0 \text{ or } Y = 0 \\ 1, & \text{if } X = 1 \text{ and } Y = 1. \end{cases}$$

$E(XY) = p_{11}$;

$E(X) = p_{1+}$; $E(Y) = p_{+1}$.

$\rho \equiv \text{corr}(X,Y) = p_{11} - p_{1+}p_{+1} = 0.$

It must be shown that (1) $p_{0+}p_{+0} = p_{00}$;

(2) $p_{0+}p_{+1} = p_{01}$;

(3) $p_{+0}p_{1+} = p_{10}$.

To show (1), use the fact that because $p_{1+}p_{+1} = p_{11}$,

$$p_{0+}p_{+0} = (1-p_{+1})(1-p_{1+}) = 1-p_{+1}-p_{1+}+p_{+1}p_{1+} = 1-p_{+1}-p_{1+}+p_{11}$$

$$p_{0+}p_{+0} = (p_{00}+p_{01}+p_{10}+p_{11}) - (p_{01}+p_{11}) - (p_{10}+p_{11}) + (p_{11})$$

$$p_{0+}p_{+0} = p_{00}.$$

Similarly for (2) and (3).

(d) Suppose $Y|X$ follows a normal distribution, and suppose

$$X = \begin{cases} 1, & \text{with probability } p \\ 0, & \text{with probability } 1-p. \end{cases}$$

Assume: $(Y|X=1) \sim N(\theta_1, \sigma^2)$

$(Y|X=0) \sim N(\theta_0, \sigma^2)$.

Then, the density of (X,Y) is

$$g(x,y) = \frac{p}{\sigma\sqrt{2\pi}} e^{(-1/2\sigma^2)(y-\theta_1)^2} + \frac{(1-p)}{\sigma\sqrt{2\pi}} e^{(-1/2\sigma^2)(y-\theta_0)^2}.$$

3.11 The prior in (3.6.5) is the Jeffreys invariant prior. Dempster's linear invariant distributions are a special case of the Jeffreys prior. So when using the Jeffreys priors we can expect posterior inferences to be sensitive to scale changes.

3.12 For $x > 0$,

$$\int_x^\infty x e^{-t^2/2} dt \le \int_x^\infty t e^{-t^2/2} dt = e^{-x^2/2}.$$

3.13 (a) $\phi_x(t) = \cos t = \frac{e^{it}}{2} + \frac{e^{-it}}{2}$.

By inspection, X must be a discrete random variable with atoms at +1 and −1. That is,

$$P\{X=+1\} = P\{X=-1\} = \frac{1}{2}.$$

(b) $\phi_x(t) = \cos^2 t = \left(\frac{e^{it}+e^{-it}}{2}\right)^2$

$$= \frac{1}{4} e^{2it} + \frac{1}{4} e^{-2it} + \frac{1}{2}.$$

By inspection this corresponds to a discrete random variable:

$$X = \begin{cases} 2, & \text{probability} = 1/4 \\ -2, & 1/4 \\ 0, & 1/2. \end{cases}$$

(c) $\phi(t) = \frac{1}{4} + \frac{1}{8} e^{it} + \frac{5}{8} e^{-2it}$.

By inspection this corresponds to the discrete random variable

$$X = \begin{cases} 0, & 1/4 \\ 1, & 1/8 \\ -2, & 5/8 \end{cases} .$$

3.14 Assume we have observed

$$\underset{(p\times 1)}{y_1}, \ldots, \underset{(p\times 1)}{y_n}, \quad y_j \equiv (y_{ij}), \; y_{ij} > 0.$$

Now use the multivariate Box-Cox transformation of Andrews et al., 1973. Let

$$\underset{(p\times 1)}{z_j^{(\lambda)}} = \left(z_{ij}^{(\lambda_i)}\right), \quad z_{ij}^{(\lambda_i)} = \frac{y_{ij}^{\lambda_i} - 1}{\lambda_i},$$

and assume: $z_j^{(\lambda)} \sim N(\theta,\Sigma)$, for $j = 1,\ldots,N$, $\lambda \equiv (\lambda_i)$, for some (θ,Σ). Estimate (θ,Σ,λ) from the data by MLE. Then test the hypothesis

H: y_j's are normal (i.e., $\lambda = (1,\ldots,1)' \equiv e'$), vs.

A: A_j's are not normal.

Use the asymptotic Wilks result that under H,

$2 \log \left[\text{L.F.}(\lambda=\hat{\lambda}) - \text{L.F.}(\lambda=e)\right] \sim \chi_p^2$, where L.F. denotes the value of the maximized likelihood function (e denotes the vector of ones).

Chapter 4

4.1 $\bar{X}|\theta \sim N(\theta,\Sigma_0/N)$, $p(\theta) \propto$ constant. Then $\theta|\bar{X} \sim N(\bar{X},\Sigma_0/N)$.

Define $A^{-1} = -\dfrac{\partial^2 \log p(\bar{X}|\theta)}{\partial\theta\partial\theta'}$, and find $A^{-1} = N\Sigma_0^{-1}$. Since

$\hat{\theta}$(MLE) $= \bar{X}$, result is the same as in Theorem (4.6.1) in that

$\sqrt{N}\,\Sigma_0^{-1/2}(\theta-\bar{X})|\bar{X} \sim N(0,I)$.

4.2 $\sqrt{N}[f(\bar{x}) - f(\theta)] \longrightarrow N(0,\phi'\Sigma\phi)$, and $\bar{x} \xrightarrow{P} \theta$.

But $f(\bar{x}) - f(\theta) = a'[\overline{\log}\ \bar{x} - \overline{\log}\ \theta]$,

$$\phi = \frac{\partial f}{\partial \bar{x}}\bigg|_{\bar{x}=\theta} = \left(\frac{a_1}{\theta_1},\ldots,\frac{a_p}{\theta_p}\right)',$$

$$t^2 \equiv \phi'\Sigma\phi = \sum_i \sum_j \left(\frac{a_i}{\theta_i}\frac{a_j}{\theta_j}\sigma_{ij}\right).$$

So $\sqrt{N}[a'(\overline{\log}(\bar{x}) - \overline{\log}(\theta)] \longrightarrow N(0,t^2)$.

4.3 (a) $\theta_6 = 1 - \sum_1^5 \theta_j = 1 - e'\theta^*$, $\theta^* \sim N(\theta_0,\Sigma_0)$,

$e'\theta^* \sim N(e'\theta_0, e'\Sigma_0 e)$. So

$\theta_6 \sim N(1-e'\theta_0,\ e'\Sigma_0 e) = N(\frac{1}{6}\ ,\ 15)$.

(b) $P\{\theta_6 \leq 4\} = .8389$;

(c) $(\theta_6|\bar{x}) \sim N(1-e'\bar{x}, \frac{1}{n}\ e'\Phi e)$.

4.4 $(\bar{x}|\theta) \sim N(\theta,\frac{1}{n})$.

$\lim_{N\to\infty} \mathcal{L}\{A^{-1/2}(\theta-\bar{x})|\bar{x}\} = N(0,1)$.

$$A^{-1} = -\frac{\partial^2 \log p\ (\bar{x}|\theta)}{\partial\theta^2} = N.$$

So $\lim_{N\to\infty} \mathcal{L}\{\sqrt{N}(\theta-\bar{x})|\bar{x}\} = N(0,1)$.

4.5 $f(\bar{x}|\theta) \propto \exp\{(-\frac{N}{2})(\bar{x}-\theta)'(\bar{x}-\theta)\}$

$$A^{-1} = -\frac{\partial^2 \log f(\bar{x}|\theta)}{\partial\theta\partial\theta'} = NI$$

$$\lim_{N\to\infty} \{\sqrt{N}(\theta-\bar{x})|\bar{x}\} = N(0,I).$$

4.6 $|I+W|^\alpha = e^{\alpha \log|I+W|} = e^{\alpha \log|I+D_\lambda|}$,

where: $D_\lambda = \begin{pmatrix} \lambda_1 & & 0 \\ & \ddots & \\ 0 & & \lambda_p \end{pmatrix}$, and $|W-\lambda I|=0$.

Then, $|I+W|^\alpha = e^{\alpha \log \prod_1^p (1+\lambda_i)} = e^{\alpha \sum_1^p \log(1+\lambda_i)}$

$$= \exp\{\alpha \sum_1^p [\lambda_i - \frac{\lambda_i^2}{2} + \dots]\}, \lambda_i < 1,$$

$$= \exp\{\alpha[\mathrm{tr}(W) - \frac{1}{2}\mathrm{tr}\, W^2 + \dots]\}.$$

For $\max_i (\lambda_i) < < 1$,

$$|I + W|^\alpha \cong e^{\alpha \mathrm{tr}(W)}.$$

4.7 Let $H(\theta) = \left.\frac{\partial^2 f(X)}{\partial X \partial X'}\right|_{X=\theta}$. Then,

$$f(X) = f(\theta) + (X-\theta)'q + \frac{1}{2}(X-\theta)'H(\theta)(X-\theta) + \dots$$

4.8 $V \sim W(\Sigma, p, n)$, $p \leq n$, $\Sigma > 0$.

$U \equiv \frac{V}{n}$, $f(U) = |U|$, $E(U) = \Sigma$, $f[E(U)] = |\Sigma|$.

$$\lim_{n\to\infty} \mathcal{L}\left\{\sqrt{n}\left(\frac{|U|}{|\Sigma|} - 1\right)\right\} = N(0,2p).$$

$|U| = \frac{|V|}{n^p}$. So

$$\lim_{n\to\infty} \mathcal{L}\left\{ \frac{1}{n^{p-1/2}} \left(\frac{|V|}{|\Sigma|} - n^p \right) \right\} = N(0,2p).$$

See Anderson, 1958, p. 173.

4.9 (a) $\operatorname{plim}_{N\to\infty} (s_{xy}) = E(XY) = \sigma_{12}$.

(b) $\operatorname{plim}_{N\to\infty} (\frac{1}{N}\Sigma X_i^3) = E(X^3) = 0$.

(c) $\operatorname{plim}_{N\to\infty} (\frac{1}{N_i}\Sigma Y_i^4) = E(Y^4) = 3\sigma_{22}^2$.

Note: $E(Y^4) = \int_{\infty}^{\infty} \frac{y^4}{\sigma_{22}\sqrt{2\pi}} \exp\{-y^2/2\sigma_{22}\}dy$.

4.10 (a) $L(y|\theta) = \prod_i^p \theta_i^{y_i}$.

Let $\hat{\theta}$ denote the MLE of θ.

$\hat{\theta} = \frac{y}{n}$, $E\hat{\theta} = \theta$.

Let $f(\theta) \equiv \gamma = \theta'\theta$. So $\hat{\gamma} = \hat{\theta}'\hat{\theta}$.

(b) $\hat{\gamma} = \gamma + (\hat{\theta}-\theta)'(2\theta) + (\hat{\theta}-\theta)'(\hat{\theta}-\theta)+\ldots$

$$\begin{aligned} E(\hat{\gamma}) &\approx \gamma + E(\hat{\theta}-\theta)'(\hat{\theta}-\theta) \\ &\approx \gamma + E\operatorname{tr}(\hat{\theta}-\theta)'(\hat{\theta}-\theta) \\ &\approx \gamma + E\operatorname{tr}(\hat{\theta}-\theta)(\hat{\theta}-\theta)' \\ &\approx \gamma + \operatorname{tr}[E(\hat{\theta}-\theta)(\hat{\theta}-\theta)'] \\ &\approx \gamma + \operatorname{tr}\Sigma. \end{aligned}$$

But $\Sigma = \frac{1}{n}\begin{pmatrix} \theta_1(1-\theta_1) & & \\ & \ddots & -\theta_i\theta_j \\ & & \theta_p(1-\theta_p) \end{pmatrix}$.

So $\operatorname{tr}(\Sigma) = \frac{1}{n}(1-\theta'\theta) = \frac{1}{n}(1-\gamma)$.

(c) $\lim \mathcal{L}\{\sqrt{n}(\hat{\gamma}-\gamma)\} = N(0,\sigma^2)$

where: $\sigma^2 = \phi'\Sigma^*\phi$, $\Sigma^* = n\Sigma$,

$$\phi = \frac{\partial f}{\partial \theta} = 2\theta, \quad \sigma^2 = 4\theta'\Sigma^*\theta,$$

$\Sigma^* = n\Sigma = D_\theta - \theta\theta'$. (See Σ in Part (b)).

$\sigma^2 = 4\theta'(D_\theta - \theta\theta')\theta = 4\theta' D_\theta \theta - 4(\theta'\theta)^2$.

But $\theta' D_\theta \theta = \sum_i \theta_i^3$, and $\theta'\theta = \gamma$.

Chapter 5

5.1 $(V|\Sigma) \sim W(\Sigma,p,n)$, $p \le n = N-1$;

$$\Sigma = 4\begin{pmatrix} 1 & & \rho \\ & \ddots & \\ \rho & & 1 \end{pmatrix} = 4[(1-\rho)I+\rho ee'].$$

$|\Sigma| = \lambda_1\lambda_2^{q+1}$. Use of (2.5.6) gives Σ^{-1}.

$$p(V|\Sigma) \propto \frac{|V|^{(n-p-1)/2}\exp\ (-1/8)\mathrm{tr}V\ \frac{I}{1-p} - \frac{\rho ee'}{(1-\rho)[1+(q-1)\rho]}}{\{4q(1-\rho)^{q-1}\ [1+(q-1)\rho]\}^{n/2}}.$$

5.2 $V = \begin{pmatrix} V_{11} & V_{12} \\ V_{21} & V_{22} \end{pmatrix}$. $V_{11} \sim W(\Sigma_{11},q,n)$.

$$\Sigma_{11} = 4\begin{pmatrix} 1 & \rho \\ \rho & 1 \end{pmatrix},\ \Sigma_{11}^{-1} = \frac{1}{4(1-\rho^2)}\begin{pmatrix} 1 & -\rho \\ -\rho & 1 \end{pmatrix}.$$

$|\Sigma_{11}| = 4(1-\rho^2)$.

$$p(V_{11}|\Sigma_{11}) \propto \frac{(V_{11}V_{22}-V_{12}^2)^{(n-3)/2}}{[4(1-\rho^2)]^{n/2}}\ \exp\left\{-\frac{1}{8(1-\rho^2)}[V_{11}+V_{22}-2\rho V_{12}]\right\}.$$

5.3 $$p(V_{11\cdot 2}) = \frac{C_{11\cdot 2}|V_{11\cdot 2}|^{(n-q-1)/2}}{|\Sigma_{11\cdot 2}|^{(n-q+2)/2}}\ \exp\left\{(-\frac{1}{2})\mathrm{tr}\Sigma_{11\cdot 2}^{-1}V_{11\cdot 2}\right\},$$

where: $|\Sigma_{11\cdot 2}| = a^2-b^2$, $C_{11\cdot 2}^{-1} = 2^{n-q+2}\pi^{1/2}\Gamma(\frac{n-q+2}{2})\ \Gamma(\frac{n-q}{2})$,

$\Sigma_{11\cdot 2} = \begin{pmatrix} a & b \\ b & a \end{pmatrix}$, and

$$a = \frac{4(q^2-5q+6)\rho^3 + (5q-11)\rho^2 - (q^2+2q-9)\rho-4}{(q-3)\rho^2 - (q-4)\rho-1},$$

$$b = \frac{4(q^2-4q+3)\rho^3 + 8\rho^2 - (q^2-4q+8)\rho}{(q-3)\rho^2 - (q-4)\rho - 1},$$

and $\Sigma_{11\cdot 2}^{-1} = \frac{I}{a-b} - \frac{bee'}{(a-b)[a+(q-3)b]}$.

5.4 $V^* = AVA'$, $W^* = BWB'$, $U = V^*+W^*$.

$V^* \sim W(\Sigma_1, p, n)$, $W^* \sim W(\Sigma_2, p, m)$,

$\Sigma_1 = A\Sigma A'$, $\Sigma_2 = B\Sigma B'$.

The density of U is the convolution of V* and W*. [see, e.g., Parzen, E. (1960). Modern Probability Theory and its Applications, New York: John Wiley and Sons, Inc., p. 317.]

Thus,

$$f(U) = \int g_{V^*}(X) h_{W^*}(U-X)\,dX,$$

where: $g_{V^*}(X)$ and $h_{W^*}(Y)$ denote the densities of V* and W*.

Therefore,

$$f(U) \propto \int \frac{|X|^{(n-p-1)/2}|U-X|^{(m-p-1)/2}}{|\Sigma_1|^{n/2}|\Sigma_2|^{m/2}}$$

$$\cdot \exp\{(-\tfrac{1}{2})[\operatorname{tr}(\Sigma_1^{-1}X + \Sigma_2^{-1}(U-X))]\}dX.$$

Alternatively, in terms of characteristics functions:

$$f(U) = \frac{1}{(2\pi)^p}\int_{-\infty}^{-\infty} \cdots \int_{-\infty}^{-\infty} \frac{\exp\{-i\operatorname{tr}(TU)\}\ dT}{|I-2i\Sigma_1 T|^{n/2}|I-2i\Sigma_2 T|^{m/2}}.$$

Note: $\Sigma_1 = \begin{pmatrix} 4 & 8\rho & \cdots & 8\rho & 4\rho \\ & 16 & 16\rho \cdots & 16\rho & 8\rho \\ & & \ddots & \vdots & \vdots \\ & & & 16 & 8\rho \\ & & & & 4 \end{pmatrix},$

and $$\Sigma_2 = \begin{pmatrix} 4 & 12\rho & \cdots & 12\rho & 4\rho \\ & 36 & 36\rho \cdots & 36\rho & 12\rho \\ & & \ddots & \vdots & \vdots \\ & & & 36 & 12\rho \\ & & & & 4 \end{pmatrix}.$$

5.5 (a) $E(v_{11}) = 27$;

(b) $\text{var}(v_{11}) = 162$;

(c) $P\{v_{11} \leq 40\} \cong .8515.$

Hint: $v_{11} \sim W(\sigma_{11}, 1, 9)$,

or $\frac{v_{11}}{\sigma_{11}} \sim \chi^2_n$. Since $\sigma_{11} = 3$, $n = 9$,

$$P\{v_{11} \leq 40\} = P\{\chi_9^2 \leq \frac{40}{3}\}.$$

(d) Using eqn. (2.8.5),

$$|\Sigma| = \prod_{k=1}^{p} \left(3+2 \cos \frac{k\pi}{p+1}\right), \quad k = 1,\ldots,p.$$

5.6 $(X_1,\ldots,X_n|\Sigma) \underset{\text{i.i.d.}}{\sim} N(0,\Sigma)$, $\Sigma \sim W^{-1}(G,p,\nu)$,

for $\nu > 2p$. $x \equiv (X_1,\ldots,X_n)$.

$$p(\Sigma|x) \propto |\Sigma|^{-(n+\nu)/2} \exp\{(-1/2)\text{tr}\Sigma^{-1}(V+G)\}.$$

For a new observation Y,

$$p(Y|x) = \int p(y|\Sigma)p(\Sigma|x)d\Sigma$$

$$p(Y|x) \propto \{(\nu+n-2p) + (\nu+n-2p)Y'(V+G)^{-1}Y\}^{-(\nu+n-p)/2}$$

5.7 $$p(V) = \frac{1}{|\Sigma|^{(n/2)}} e^{-1/2\text{tr}\Sigma^{-1}V}.$$

Let $\mathcal{L} \equiv \log p$, and define $\Lambda \equiv \Sigma^{-1}$.

$$\mathcal{L} = \frac{n}{2} \log |\Lambda| - \frac{1}{2} \operatorname{tr} \Lambda V.$$

Differentiate with respect to $\hat{\Sigma}$ using eqns. (2.14.2) and (2.14.6)'. (See also Section 7.1.2).

$$\frac{\partial \mathcal{L}}{\partial \Lambda} = \frac{n}{2} [2\Lambda^{-1} - \operatorname{diag} \Lambda^{-1}] - \frac{1}{2} [2V - \operatorname{diag} V].$$

Setting $\frac{\partial \mathcal{L}}{\partial \Lambda} = 0$ and solving gives $\hat{\Sigma} = V/n$.

To show this result gives a maximum of $\mathcal{L}$, see p. 182.

5.8 From eqn. (6.2.2), the density of $\hat{\beta}_{13}$ may be written as

$$f(x) = \sum_{\alpha_1} \sum_{\alpha_2} a_{\alpha_1 \alpha_2} g(x | \alpha_1, \alpha_2),$$

where: $g(x) \propto \dfrac{1}{(1+x^2)^{\alpha_1 + 3/2}}$; $\int_{\infty}^{\infty} g(x)dx = 1$;

a Student t-density. So if μ_r denotes the rth moment of a Student t-variate with $(2\alpha_1 + 2)$ degrees of freedom,

$$E(\hat{\beta}_{13}^r) = \sum_{\alpha_1} \sum_{\alpha_2} a_{\alpha_1 \alpha_2} \mu_r.$$

Note that $\mu_1 = 0$, $\mu_2 = \dfrac{2\alpha_1 + 2}{2\alpha_1}$.

Let s denote the standard error.

$$s = \sqrt{\operatorname{var} \hat{\beta}_{13}} = [E(\hat{\beta}_{13}^2) - E^2(\hat{\beta}_{13})]^{1/2}.$$

But $E(\hat{\beta}_{13}) = 0$, $E(\hat{\beta}_{13}^2) = \sum_{\alpha_1} \sum_{\alpha_2} a_{\alpha_1 \alpha_2} \left(\dfrac{2\alpha_1 + 2}{2\alpha_1}\right)$.

So $s = [\sum_{\alpha_1}\sum_{\alpha_2} a_{\alpha_1\alpha_2}(1+\frac{1}{\alpha_1})]^{1/2}$.

But $\sum_{\alpha_1}\sum_{\alpha_2} a_{\alpha_1\alpha_2} = 1$. So $s = [1 + \sum_{\alpha_1}\sum_{\alpha_2} (a_{\alpha_1\alpha_2}/\alpha_1)]^{1/2}$.

5.9 $\phi_V(T) = E\exp\{itr(TV)\}$

$$= C\int_{V>0} \frac{|V|^{(n-p-1)/2}}{|\Sigma|^{n/2}} e^{-1/2tr\Sigma^{-1}V+itrTV}dV,$$

where: C is defined by $\int_{V>0} p(V)dV = 1$.

Combining terms

$$\phi_V(T) = \frac{C}{|\Sigma|^{n/2}}\int|V|^{(n-p-1)/2}e^{-1/2trV\Omega^{-1}}dV,$$

where: $\Omega^{-1} \equiv \Sigma^{-1}-2iT$. Rewriting,

$$\phi_V(T) = \frac{|\Omega|^{n/2}}{|\Sigma|^{n/2}}\int\frac{C|V|^{(n-p-1)/2}}{|\Omega|^{n/2}}e^{-trV\Omega^{-1}}dV$$

$$= \frac{|\Omega|^{n/2}}{|\Sigma|^{n/2}}.$$

5.10 $J(V\to U) = |U|^{-(p+1)}$, so $g(U) \propto p[V(U)]\cdot\frac{1}{|U|^{(p+1)}}$,

which is the density of an inverted Wishart distribution (see Section 5.2.1).

5.11 Use result in Exercise 5.4, or directly, that if $U = v_1 + v_2$, the density of U is the convolution

$f(U) = \int g_{v_1}(X)g_{v_2}(U-X)dX$, or if $\Sigma_1 = \Sigma_2$,

$$f(U) \propto \frac{|U|^{(n-p-1)/2}}{|\Sigma|^n}e^{-1/2tr\Sigma^{-1}U}F(U),$$

where: $F(U) = \int |X|^{(n-p-1)/2} |I-U^{-1/2}XU^{-1/2}|^{(n-p-1)/2} dX.$

Transforming from X to $Y = U^{-1/2}XU^{-1/2}$,

$J(X \to Y) = |U^{-1/2}|^{-(p+1)}$, and

$F(U) \propto |U|^{n/2}$. So $U \sim W(\Sigma, p, 2n)$.

For $\Sigma_1 \neq \Sigma_2$ the result is not Wishart and is much more complicated.

5.12 (a) $$E\left(\frac{|V|^k}{|\Sigma|^k}\right) = 2^{kp} \prod_{j=1}^{p} \left[\frac{\Gamma\left(\frac{n+2k-j+1}{2}\right)}{\Gamma\left(\frac{n-j+1}{2}\right)}\right],$$

(b) $$E\left(\frac{|V|^k}{|\Sigma|^k}\right) = \frac{1}{2^{kp}} \prod_{j=1}^{p} \left[\frac{\Gamma\left(\frac{n-2k-p-j}{2}\right)}{\Gamma\left(\frac{n-p-j}{2}\right)}\right].$$

Chapter 6

6.1 $T^2 = 3200/7, \ \dfrac{T^2}{N-1} \cdot \dfrac{N-P}{p} = 226.26$

$F_{2,98}(.01) = 2.37$; so reject H.

6.2 $(\bar{x}|\theta,\Sigma) \sim N(\theta,\Sigma/N), \ (V|\Sigma) \sim W(\Sigma,p,n).$

$$p(\theta,\Sigma) \propto \frac{1}{|\Sigma|^{(p+1)/2}} .$$

$$p(\theta,\Sigma|\bar{x},V) \propto |\Sigma|^{-(n+p+2)/2} \exp\{(-1/2)[V+N(\theta-\bar{x})(\theta-\bar{x})']\}.$$

$$p(\theta|\bar{x},V) \propto \{(N-p) + N(N-p)(\theta-\bar{x})'V^{-1}(\theta-\bar{x})\}^{-[\frac{(N-p)+p}{2}]}$$

6.3 Let x denote a new observation.

$$p(x|\bar{x},V) = \int p(x|\theta,\Sigma)p(\theta,\Sigma|\bar{x},V)d\theta d\Sigma$$

$$p(x|\bar{x},V) \propto \{(N-p) + \frac{N(N-p)}{(N+1)}(x-\bar{x})'V^{-1}(x-\bar{x})\}^{-1/2[(N-p)+p]}$$

6.4 $f(r|\theta) = \binom{N}{r_1,\ldots,r_p} \prod_{i=1}^{p} \theta_i^{r_i}, \ \sum_{i=1}^{p} r_i = N.$

Let $g(\theta) \propto \prod_{i=1}^{p} \theta_i^{(\nu_i - 1)}, \ \sum_{1}^{p} \theta_i = 1.$

$h(\theta|r) \propto \prod_{1}^{p} \theta_i^{\nu_i + r_i - 1}$, or

$(\theta|r) \sim D(\nu_1+r_1,\ldots,\nu_{p-1}+r_{p-1};\nu_p+r_p).$

6.5 $Z = \binom{X}{Y}$, $Z \sim$ Log Normal (θ,Σ). From Theorem (6.4.1),

$E(X) = \exp\{\theta_1+\sigma_1^2/2\} = e$

$E(Y) = \exp\{\theta_2+\sigma_2^2/2\} = e^{9/2}$.

(a) $E(3X+2Y) = 3e + 2e^{9/2}$.

(b) $\mathrm{var}(3X+2Y) = 9\,\mathrm{var}X + 4\,\mathrm{var}Y + 12\,\mathrm{cov}(X,Y)$.

From Corollary (6.4.1),

$\mathrm{var}(X) = e^8 - e^2$, $\mathrm{var}(Y) = e^{12} - e^9$

$\mathrm{cov}(X,Y) = e^{15/2} - e^{11/2}$.

$\mathrm{var}(3X+2Y) = 9(e^8-e^2) + 4(e^{12}-e^9) + 12\,(e^{15/2}-e^{11/2})$.

6.6 X and Y follow the same stable law, so a linear combination can only have a different scale parameter. The density of z is

(a) $f(z) = \dfrac{(h^*/2\pi)^{1/2}}{z^{3/2}}\, e^{-h^*/2z}$; $h^* = h\left(C_1^{1/2}+C_2^{1/2}\right)^2$.

(b) $\phi_z(t) = \exp\left\{-(h^*)^{1/2}|t|^{1/2}\left(1+\dfrac{it}{|t|}\right)\right\}$.

6.7 $X \equiv C_1'Y + C_2'Z$. Using Theorem (6.5.3),

$\mathcal{L}(Y) = S_4(1,a;\Omega;3/2)$,

$\mathcal{L}(Z) = S_4(1,b;\psi;3/2)$,

and Y and Z are independent. So

$\mathcal{L}(X) = S_1\left(2,-1/2;\ \dfrac{21}{64};\ \dfrac{369}{128}\right)$.

6.8 The result of Exercise 6.4 might be used for the posterior distribution. Alternatively:

likelihood: $f(r|p) \propto \prod_1^k p_i^{r_i}, \quad \sum_1^k r_i = n.$

$p = (p_i), \quad r = (r_i).$

prior : $g(p) \propto \prod_1^k p_i^{\nu_i - 1}, \quad \sum_1^k p_i = 1.$

posterior: $h(p|r) \propto \prod_1^k p_i^{\nu_i + r_i - 1}.$

Let r^* = vector of new observations.

$f(r^*|r) = \int f(r^*|p)\, h(p|r)dp$

$$f(r^*|r) = \binom{N}{r_i^*,\ldots,r_k^*} \frac{\Gamma\left[\sum_1^k(\nu_i+r_i)\right]\prod_1^k \Gamma(\nu_i+r_i+r_i^*)}{\prod_1^k \Gamma(\nu_i+r_i)\Gamma\left[\sum_1^k(\nu_i+r_i+r_i^*)\right]}$$

6.9 Use Theorem (6.5.1). For $\alpha \neq 1$, $\phi_y(t)$ is a characteristic function of a multivariate stable distribution.

For $\alpha = 1$, $\phi_y(t)$ is not a multivariate stable characteristic function.

6.10 Let $z \equiv -(1/2) \sum_{j=1}^m (t'\Omega_j t)^{\alpha/2}$. If $x_j \equiv \Omega_j^{1/2} t$,

$$z = -(1/2) \sum_{j=1}^m (x_j' x_j)^{\alpha/2}.$$

Let $y_j = (x_j'x_j)^{\alpha/2}$; so $z = -(1/2) \sum_{j=1}^{m} y_j$.

$$\frac{\partial y_j}{\partial x_j} = \alpha\ (x_j'x_j)^{\alpha/2-1}x$$

$$\theta_j \equiv \frac{\partial^2 y_j}{\partial x_j \partial x_j'} = \alpha\ (x_j'x_j)^{(\alpha/2)-2}\{[(\alpha/2)-1]2x_j x_j' + \alpha(x_j'x_j)I\}.$$

If y_j is convex, $\sum_j y_j$ is convex, and z is concave. But y_j is convex if $\theta_j > 0$. To show $\theta_j > 0$, examine the quadratic form

$$w'\theta_j w = \{(x_j'x_j)(w'w) - (\tfrac{2-\alpha}{\alpha})(w'x_j)^2\}\{\alpha^2(x_j'x_j)^{(\alpha/2)-2}\}.$$

The Cauchy inequality says

$$(x_j'x_j)(w'w) \geq (w'x_j)^2.$$

So, $w'\theta_j w \geq 0$, if $1 \leq \alpha \leq 2$, which implies $\theta_j > 0$. If $0 < \alpha < 1$, the argument doesn't apply. So z is concave if and only if $1 \leq \alpha \leq 2$.

To examine the case of $\alpha = 1$, examine the latent roots of θ_j. Use

$$\theta_j^* = \frac{\theta_j}{\alpha(x_j'x_j)^{(\alpha/2)-2}}.$$

$$|\theta_j^* - \lambda I| = |[(\alpha/2)-1](2x_j x_j') + \alpha x_j'x_j I - \lambda I| = 0$$

or $(\alpha x_j'x_j - \lambda)^p \left|I + \frac{(\alpha-2)}{(\alpha x_j'x_j - \lambda)} x_j x_j'\right| = 0.$

So $\lambda = \alpha x_j' x_j$, or $\lambda^* = 2(\alpha-1)x_j' x_j$. If $\alpha = 1$, $\lambda^* = 0$. If $\alpha > 1$, $\lambda^* > 0$. Thus, latent roots of θ_j are > 0 if $1 < \alpha \leq 2$, but the case of $\alpha = 1$ must be evaluated separately because it yields a zero latent root. Note that for $\alpha = 1$, $w'\theta_j w \geq 0$ by the Cauchy inequality.

6.11 If $\phi_y(t)$ is in the same family we must be able to write it as

$$\log \phi_y(t) = -\frac{1}{2}\sum_1^2 (t'\Omega_j t)^{1/2}.$$

If $\Omega_1 = 4\begin{pmatrix} 4 & 6 \\ 6 & 9 \end{pmatrix}$, $\Omega_2 = 4\begin{pmatrix} 36 & -30 \\ -30 & 25 \end{pmatrix}$,

$$\frac{1}{2}(t'\Omega_1 t)^{1/2} = |2t_1 + 3t_2|, \quad \frac{1}{2}(t'\Omega_2 t)^{1/2} = |6t_1 - 5t_2|.$$

So $\phi_y(t)$ is in fact included in the family defined by eqn. (6.5.9), but note that $|\Omega_1| = 0$, $|\Omega_2| = 0$, but $\sum_1^2 \Omega_j > 0$, and Ω_1 and Ω_2 are not proportional.

6.12 Note that $0 < U < 1$. From Theorem (6.2.6), and Anderson, 1958, p. 194,

$$\mathcal{L}(U) = \mathcal{L}(U_{p,q,m-q}) = \left(\prod_1^p X_i\right),$$

where $X_1, \ldots, X_p$ are independent, and

$$\mathcal{L}(X_i) = \beta\left(\frac{m-q-i+1}{2}, \frac{q}{2}\right).$$

$$\text{So } E(U) = \prod_1^p EX_i = \prod_1^p \frac{(m-q-i+1)}{(m-i+1)}.$$

6.13 Using eqn. (6.6.11),

$$E|Z|^k = \int C_n |Z|^k {}_pF_q(a;b;BRBZ)\,|Z|^{n-(\frac{m+1}{2})} e^{-trBZ} dZ,$$

where $C_n = |B|^n \{\Gamma_m(n)\, {}_{p+1}F_q(a,n;b;BR)\}^{-1}$.

$$E|Z|^k = \frac{C_n}{C_{n+k}}\ .$$

Chapter 7

7.1 Define $\overline{\log}\, x_j \equiv (\log x_{1j},\ldots,\log x_{pj})'$. Then, sufficient statistics for (θ,Σ) are

$$\left\{ \sum_{j=1}^{n} \overline{\log}\, x_j,\ \sum_{j=1}^{n} \left(\overline{\log}\, x_j\right)\left(\overline{\log}\, x_j\right)' \right\}.$$

7.2 Define $\Gamma = p^{-1/2}\begin{pmatrix} e' \\ \equiv \end{pmatrix}$, where the rows can be anything, as long as they are mutually orthogonal, and the elements of the first row are equal.

Define $D_\lambda = \text{diag}\,(\lambda_1,\lambda_2,\ldots,\lambda_2)$, and note that we can write $\Sigma = \Gamma D_\lambda \Gamma'$.

Let $y_j = \Gamma x_j$, $\phi = \Gamma\theta$. So we have transformed the parameter space from (θ,ρ,σ^2) to $(\phi,\lambda_1,\lambda_2)$. The transformed log-likelihood becomes

$$\ell = -\frac{N}{2}\log\left|D_\lambda\right| - \frac{1}{2}\Sigma(y_j - \phi)' D_\lambda^{-1}(y_j - \phi).$$

$\frac{\partial\ell}{\partial\phi} = 0$ yields $\hat{\phi} = \bar{y}$, or $\hat{\theta} = \bar{x}$.

$\frac{\partial\ell}{\partial\lambda_1} = 0$ yields $\hat{\lambda}_1 = \frac{w_{11}}{N}$, where

$W = \Sigma\,(y_j - \bar{y})(y_j - \bar{y})'$; $W = (w_{ij})$.

$\frac{\partial\ell}{\partial\lambda_2} = 0$ yields $\lambda_2 = \frac{\sum_2^p w_{ii}}{N(p-1)}$.

Note that $W = \Gamma V \Gamma'$ so that

$$w_{11} = \frac{1}{p}\, e'Ve,\quad \sum_2^p w_{ii} = \text{tr}V - \frac{1}{p}\, e'Ve.$$

$\lambda_1 = \sigma^2[1+(p-1)\rho]$, $\lambda_2 = \sigma^2(1-\rho)$.

Solving the resulting equations in λ_1, λ_2 for $\hat{\sigma}^2$ and $\hat{\rho}$ gives $\hat{\sigma}^2 = \frac{1}{NP}\operatorname{tr}V$, $\hat{\rho} = \frac{e'Ve-\operatorname{tr}V}{(p-1)\operatorname{tr}V}$.

7.3 (a) $\hat{\theta}_{\text{Bayes}} = E(\theta|\bar{x},s) = (70.9, 82.7, 57.3)'$.

(b) See eqn. (7.1.11), or differentiate with respect to Σ.

$\hat{\Sigma}_{\text{Bayes}} = E(\Sigma|\bar{x},s)$.

$$p(\Sigma|\bar{x},S) \propto \frac{1}{|\Sigma|^{14/2}} e^{-1/2\operatorname{tr}\Sigma^{-1}[V+G+(1+N)\Delta]},$$

where: $\Delta = \frac{(\phi\phi'+N\overline{xx}')}{(1+N)} - \frac{(\phi+N\bar{x})(\phi+N\bar{x})'}{(1+N)^2}$.

Since $N = 10$,

$$E(\Sigma|x,s) = \frac{V+G+11\Delta}{6} = \begin{pmatrix} 6.82 & 3.95 & -.29 \\ & 34.86 & -26.53 \\ & & 33.20 \end{pmatrix}.$$

(c) $\hat{\theta}_{\text{Bayes}} = \hat{\theta}_{\text{MLE}} = \bar{x} = (71,84,56)'$;

$$\hat{\Sigma}_{\text{Bayes}} = \hat{\Sigma}_{\text{MLE}} = S = \begin{pmatrix} 4 & 1 & 1 \\ & 3 & 2 \\ & & 2 \end{pmatrix}.$$

7.4 $\hat{\theta}_{\text{Stein}} = [\max(o,a)]\bar{x}$,

where: $a = 1-\frac{c}{\bar{x}'s^{-1}\bar{x}}$, $c = \frac{p-2}{N-(p-2)}$.

$a = 1 - .00003858$

$\hat{\theta}_{\text{Stein}} = (.9999614)\bar{x}$,

$\hat{\theta}_{\text{MLE}} = \bar{x}$.

The Stein estimator and the MLE are so close because p is small.

7.5 (a) $Z = \sqrt{N}\,\Sigma_0^{-1/2}(\bar{x}-\theta) \sim N(0,I)$.

Under H, $Z'Z \sim \chi_3^2(.95) = 7.81$

$Z'Z = 8.93$, so reject H.

(b) $(\phi \mid \bar{x}) \sim N(\phi_0, I)$.

(c) $(\phi_\perp \mid \bar{x}) \sim N(\phi_{0\perp}, 1)$.

H: $\theta_\perp = 5$, A: $\theta_\perp \neq 5$.

A 95% credibility interval is $\left(-2 + \frac{\sqrt{10}}{2}, 2 + \frac{\sqrt{10}}{2}\right)$ or (-.38, 3.54). It includes zero, so we cannot reject H.

7.6 Use two sample Hotelling's T^2 test and find

$T^2 = 93.89$, against $F_{2,27} = 3.35$. So reject H.

7.7 Asymptotically, the effect of the prior disappears and results are analogous to those of frequentist theory, except we condition on the data instead of the parameters. The result therefore is the same as that obtained by using the frequentist procedures in Section 7.3.3. We could also use the large sample normal approximations and use the testing method in the Hint.

7.8 Under H, $Z \equiv \frac{r}{\sqrt{1-r^2}}\sqrt{N-2} \sim t_{N-2} = t_9(.005) = 3.25$,

$$r = \frac{\sqrt{6}}{4} = .612.$$

$Z = 2.32$

So we cannot reject H: $\rho = 0$.

7.9 Let $\hat{R}$ denote the sample correlation matrix.

$$|\hat{R}| = \frac{7}{24} = .292.$$

Reject H if $(\frac{7}{24})^5 = .00212 < C$, where for large N,

$$Z \equiv -(N-1-\frac{2p+5}{6}) \log |\hat{R}| \sim \chi^2_{(p/2)(p-1)^2}.$$

Since χ^2_6 (.10) = 10.6, and Z = 8.82, we cannot reject H.

7.10 $\theta|\bar{x} \sim N(\bar{x}, \Sigma_0/N)$.

Let $y = N(\theta-\bar{x})'\Sigma_0^{-1}(\theta-\bar{x})$. Then, $(y|\bar{x}) \sim \chi^2_2$.

Since $\chi^2_2(.05) = 5.99$, and y = 25, reject H.

7.11 Use Hotelling's two sample T^2 with p = 19, or use the Bayesian approach of Section 7.1.6 for two samples, with p = 19.

The implied methodology is 19 simultaneous Student t-tests with equal variances assumed. We might assume that Radday assumed the tests were independent with p-values p_i, and he computed

$$p = \prod_1^{19} p_i,$$

and found $p = 10^{-5}$. If the p_i were equal, this would imply a commom $p_i = .545$.

It would be better to use multivariate procedures since they are more sensitive to departure from the null hypothesis of commom authorship. Also, tests are not independent, so some other method such as the use of the Bonferoni inequality should be used.

7.12 $f_2(\theta|x) = \int g(\theta,\phi|x)d\phi = \int f_1(\theta|\phi,x)p(\phi|x)d\phi$

$f_1(\theta|x,\phi) \propto f_0(x|\theta,\phi)\ h(\theta|\phi)$

$f_2(\theta|x) \propto \int f_0(x|\theta,\phi)\ h(\theta|\phi)\ p(\phi|x)d\phi.$

After integrating,

$(\theta|x) \sim N(Gx,\ H^{-1}),$

where: $Gx = \dfrac{h_1 x}{h_1+h_2} + \dfrac{h_2 \overline{x}e}{h_1+h_2}$,

$$H^{-1} = \frac{I}{h_1+h_2} + \frac{h_2 ee'}{(h_1+h_2)(h_1 p)}.$$

$E(\theta_1|x)$ = first component of Gx

$$= \left(\frac{h_1}{h_1+h_2}\right) x_1 + \left(\frac{h_2}{h_1+h_2}\right) \overline{x}.$$

7.13 $r = .672$

h: $\rho = 0$, vs. A: $\rho \neq 0$.

$$z = \frac{r\sqrt{N-2}}{(1-r^2)^{1/2}} \sim t_{N-2} = t_8 = 2.306$$

$z = 2.5666$, so reject H.

7.14 Use Hotellings T^2 - statistic for two samples.

$T^2 = 2187.5$

$$U = \frac{N_1+N_2-p-1}{N_1+N_2-2)p}\, T^2 \sim F_{p,N_1+N_2-p-1}$$

$U = 1082.59$, $F_{2,97}(.05) = 3.12$

Reject H.

7.15 Use Section 7.3.3 and eqn. (7.3.6).

$G = -7.5772.$

Reject H: $\Sigma_1 = \Sigma_2$, if $G < C$.

From eqn. (7.3.8), under H it is approximately true that

$\mathcal{L}\{-aG\} = \chi_3^2(.05) = 7.81,$

where: $a = .7779$. We can find $-aG = 7.41$. So we cannot reject H. We therefore conclude that $\Sigma_1 = \Sigma_2$.

CHAPTER 8

Erratum: Note that on p. 236, the (1,3) element of $\hat{\Sigma}$ should be 3.012.

8.1 (a) $Y = XB + U$, $Y:(N\times p)$, $X:(N\times q)$

$p + q \leq N$, $r(x) = q$, $E(U) = 0$, $U' = (v_1,\dots,v_N)$,

$$Ev_iv_j' = \begin{cases} \Sigma > 0, & i = j \\ 0, & i \neq j \end{cases}$$

$v_j \sim N(0,\Sigma)$,

X is random, uncontrolled, and observable, and independent of U, and functionally independent of B, which is deterministic.

(b) yes

$$\hat{B} = (X'X)^{-1}X'Y = (X'X)^{-1}X'(XB+U) = B+E(X'X)^{-1}X'U$$
$$= B+E\left[(X'X)^{-1}X'\right] E(U) = B.$$

8.2 (a) From eqn. (8.3.6), $\hat{\beta} = (X'\Sigma^{-1}X)^{-1}X'\Sigma^{-1}y$, $\hat{\beta} = (-2,3)'$.

(b) $\sigma^2 = 0$.

8.3 $\tilde{\beta} = (-2,3)'$.

$\tilde{\beta} = (X'\Sigma^{-1}X)^{-1}X'\Sigma^{-1}y$, $\Sigma = (1-\rho)I+\rho ee'$.

Use eqn. (2.5.9).

Note that $\hat{\beta}_{OLS} = (X'X)^{-1}X'Y = \tilde{\beta}$.

8.4 $$\lambda^{2/N} = \lambda^{2/5} = \frac{\left|\hat{\Sigma}_{H_0 U H_1}\right|}{\left|\hat{\Sigma}_{H_0}\right|}, \quad \hat{\Sigma}_{H_0 U H_1} = \begin{pmatrix} 3.84 & 3.59 & 3.21 \\ & 5.17 & 3.88 \\ & & 3.12 \end{pmatrix}$$

$\left|\hat{\Sigma}_{H_1}\right| = .07$.

$$\hat{\Sigma}_{H_0} = \begin{pmatrix} 3.15 & 2.65 & 2.21 \\ & 3.97 & 2.54 \\ & & 2.26 \end{pmatrix}; \quad \left|\hat{\Sigma}_{H_0}\right| = 2.435.$$

$\lambda^{2/5} = .029$. Since $q_1 < p$, use

$\mathcal{L}(U_{p,q_1,n}) = \mathcal{L}(U_{q_1,p,n+q_1-p})$, or

$\mathcal{L}(U_{3,1,3}) = \mathcal{L}(U_{1,3,1})$.

$P\{-\frac{3}{2}\log U_{1,3,1} \leq z\} \approx P\{\chi_3^2 \leq z\} = 7.81.$

$-\frac{3}{2}\log U_{1,3,1} = 12.96.$

So reject H_0.

8.5 (1) regressor matrices should be mutually othogonal, so

$X_i'X_j = 0$, for all i, j=1,...,p.

(2) Want correlations across equations to be non-zero.

8.6 $Y = XB+U$, $Y:(N\times p)$, $X:(N\times q)$.

$v_j \sim N(0,\Sigma)$. From eqn. (8.6.7)

$$p(B|Y,X) \propto \frac{1}{|V+(B-\hat{B})'(X'X)(B-\hat{B})|^{N/2}}.$$

For $\beta_j:(q\times 1)$, and Theorem (8.6.1),

$$p(\beta_j|Y,X) \propto \frac{1}{\{v_{jj}+(\beta_j-\hat{\beta}_j)'(X'X)(\beta_j-\hat{\beta}_j)\}^{(N-p+1)/2}},$$

$$p(\beta_1|Y,X) \propto \frac{1}{\{v_{11}+(\beta_1-\hat{\beta}_1)'(X'X)(\beta_1-\hat{\beta}_1)\}^{(N-p+1)/2}}.$$

$$\left(\frac{\beta_{11}-\hat{\beta}_{11}}{w_{11}^{1/2}}\Big|X,Y\right) \sim t_n = 63.66,$$

$$w_{11} = .939,$$

where: $W^{-1} = \left(\frac{X'X}{v_{11}/n}\right)$, $W = (w_{ij})$.

Cannot reject H.

8.7 To test H_1 vs. A_1, use Hotelling's T^2-test, or a Bayesian two sample t-test.

To test H_2 vs. A_2, need methods of MANOVA, and so get U-functions with Box approximations.

8.8 Use bivariate, one way layout, fixed effects model with three covariates, MANOCOVA.

$$\underset{(2\times1)}{y_{ij}} = \mu+\alpha_i + \underset{(2\times3)}{\Lambda} \cdot \underset{(3\times1)}{x_{ij}} + e_{ij}, \quad i=1,2,3; \; j=1,\ldots,J_i.$$

y_{ij} = target miss distance, in feet (bivariate) for aircraft i, under the jth set of test conditions of altitude, speed, and dive angle.

$x_{ij}(1)$ = altitude of aircraft i in jth test;

$x_{ij}(2)$ = speed of aircraft i in jth test;

$x_{ij}(3)$ = dive angle of aircraft i in jth test;

$$x_{ij} = (x_{ij}(1), x_{ij}(2), x_{ij}(3))'.$$

$\underset{(2\times1)}{e_{ij}}$ = error vector, $Ee_{ij} = 0$,

$E(e_{ij}e'_{ij}) = \sigma^2 I$, $\text{cov}(e_{ij}, e_{kl}) = 0$,

$i \neq k$, $j \neq l$; e_{ij} is normally distributed.

Test H: $\alpha_1 = \alpha_2 = \alpha_3$, vs. A: α_i not all equal. Study contrasts. Could also do Bayesian inference and find distribution of $\theta_i = \mu + \alpha_i$, given all of the data. Then examine the distribution of $\theta_i - \theta_j = \alpha_i - \alpha_j$.

8.9 $r(HG^{-1}) = 1$. See Footnote 9, p. 273.

$r(HG^{-1}) = r\,(G^{-1/2}HG^{-1/2}) = r(H)$

= number of latent roots of H, λ_j, different from zero.

If $r(HG^{-1}) = 1$, there is only one $\lambda_j \neq 0$.

$r(H) = \min(p,q)$, where $q = q_1$, q_1^*, or q_1^{**}, depending upon whether H corresponds to rows, columns, or interactions. So $q = 1$, regardless.

Let λ denote the non-zero latent root. (See Section 8.7.3).

The LRS becomes

$$\frac{1}{\prod_1^p (1+\lambda_j)} = \frac{1}{\lambda}.$$

The "trace criterion" becomes:

$$T_0^2 = \sum_{j=1}^{p} \lambda_j = \lambda.$$

The "largest root criterion" becomes:

$$\frac{\lambda}{1+\lambda} = \frac{1}{1+\frac{1}{\lambda}}.$$

All three are monotone functions of one another, so all are equivalent.

8.10 $p(Y|X,B,\Lambda) \propto |\Lambda|^{N/2} \exp\{(-1/2)[\mathrm{tr}\Lambda V+(\beta-\hat{\beta})'(\Lambda \otimes X'X)(\beta-\hat{\beta})]\}$.

Let $\beta \sim N(\emptyset,F)$, $F > 0$; $\Lambda \sim W(H,p,m)$, $p \leq m$.

$p(B,\Lambda) \propto |\Lambda|^{(m-p-1)/2} \exp\{(-1/2)[\mathrm{tr}\ H^{-1}\Lambda+(\beta-\emptyset)'F^{-1}(\beta-\emptyset)]\}$

$p(B,\Lambda|Y,X) \propto p\,(Y|X,B,\Lambda)p(B,\Lambda)$.

8.11 $$R = \frac{(y-\bar{y})'(\hat{y}-\bar{y}e)}{\{(y-\bar{y})'(y-\bar{y})(\hat{y}-\bar{y})'(\hat{y}-\bar{y})\}^{1/2}} \equiv \frac{N}{D}.$$

$(y-\hat{y}'(\hat{y}-\bar{y}e) = 0$, see Section 8.2.4.

$$N = [(y-\hat{y}) + (\hat{y}-\bar{y}e)]'(\hat{y}-\bar{y}e)$$

$$= (y-\hat{y})'(\hat{y}-\bar{y}e) + (\hat{y}-\bar{y}e)'(\hat{y}-\bar{y}e)$$

$$= (\hat{y} - \bar{y}e)'(\hat{y}-\bar{y}e).$$

$$R = \frac{\{(\hat{y}-\bar{y})'(\hat{y}-\bar{y})\}^{1/2}}{\{(y-\bar{y}e)'(y-\bar{y}e)\}^{1/2}};$$

this is the definition of R^2 (see eqn. (8.2.6)).

8.12 Let $t_1 = X$, $t_2 = (Y,Z);$, where $t \sim N(\theta,\Sigma)$.

From p. 73, Section 3.5,

$$t_2 | t_1 \sim N(\theta_2 + \Sigma_{21}\Sigma_{11}^{-1}(t_1-\theta_1, \Sigma_{22.1})).$$

$$\Sigma_{22.1} = \text{var}(Y,Z|X)$$
$$= \left(\begin{array}{c|c} \sigma_Y^2-\sigma_{XY}^2/\sigma_X^2 & \sigma_{YZ}-\sigma_{XY}\sigma_{XZ}/\sigma_X^2 \\ \hline & \sigma_Z^2- \sigma_{XZ}^2/\sigma_X^2 \end{array}\right).$$

$$\rho_{YZ|X} = \frac{\sigma_{YZ}\sigma_X^2-\sigma_{XY}\sigma_{XZ}}{\{(\sigma_X^2\sigma_Y^2-\sigma_{XY}^2)(\sigma_X^2\sigma_Z^2-\sigma_{XZ}^2)\}^{1/2}}.$$

For relation to regression residuals, see e.g., Johnston, Second Edition, pp. 61-64.

If all σ's are replaced by $\hat{\sigma}$'s, which are MLE's in the associated regressions, then $\hat{\rho}$ that corresponds is the ρ given in the problem.

8.13 From pp. 270, 271, use the harmonic mean of cell frequencies as a good approximation when cell frequencies are unequal, but fairly close. If the sample size is small and we are doing only a one or two way layout in low dimensions, the approxima- is reasonable. The approximation will be poor if the problem is large (large sample size, high dimension, many-way layout) or if the differences in the cell frequencies are caused by the experiment. If the approximation will not be good use a computer and calculate the exact solution.

8.14 $(\theta | \phi, F) \sim N(\phi, F)$, $F > 0$.

$(\Sigma | G, n) \sim W^{-1}(G, p, n)$, $2p \le n$, $p = 2$.

Assess: $\phi = \theta_0 = (25, 30)'$.

Assess: $E(\Sigma) = [G/(n-2p-2)] = \Sigma_0$

$$\Sigma_0 = \begin{pmatrix} 100 & 50 \\ 50 & 100 \end{pmatrix}. \quad \text{Take } n = 300,$$

$$\text{so } G = 294\, \Sigma_0 = 29{,}400 \begin{pmatrix} 1 & 1/2 \\ 1/2 & 1 \end{pmatrix}.$$

$$\text{Assess: } F = \begin{pmatrix} \sigma_1^2 & \rho\sigma_1\sigma_2 \\ & \sigma_2^2 \end{pmatrix}, \text{ with}$$

$6\sigma_1 = 50$, $6\sigma_2 = 20$, $\rho = 1/2$; or

$$F = \begin{pmatrix} 69.44 & 13.89 \\ & 11.11 \end{pmatrix}, \ \phi + (25,30)'.$$

$(X | \theta, \Sigma) \sim N(\theta, \Sigma)$.

$$p(\theta,\Sigma) \propto |\Sigma|^{-n/2}\exp\{(-1/2)[\mathrm{tr}\Sigma^{-1}G+(\theta-\phi)'F^{-1}(\theta-\phi)]\}$$

Posterior:

$$p(\theta,\Sigma|X,V) \propto |\Sigma|^{-(N+n)/2}\exp\{(-1/2)[(\theta-\phi)'F^{-1}(\theta-\phi)+\mathrm{tr}\Sigma^{-1}W]\}$$

where: $W \equiv (G+V) + N(\bar{X}-\theta)(\bar{X}-\theta)'$.

Integrating,

$$p(\theta|\bar{X},V) \propto \frac{\exp\{(-1/2)(\theta-\phi)'F^{-1}(\theta-\phi)\}}{\{1+N(\theta-\bar{X})'(G+V)^{-1}(\theta-\bar{X})\}^{(N+n-p-1)/2}}.$$

This kernal is the product of a normal density kernal and a multivariate Student t-density kernal.

All of the results given are just one way to assess these densities. The assessments, therefore, are not unique. Other reasonable assumptions would also be valid.

8.15 (a) $\hat{y} = 2.5t-.5$

(b) $\hat{\rho} = [\sum_{1}^{n-1} \hat{e}_\alpha \hat{e}_{\alpha+1} / \sum_{1}^{n-1} \hat{e}_\alpha^2] = -.80.$

(c) $\hat{e}_1 = 0;\ \hat{e}_2 = .5;\ \hat{e}_3 = 1.0;\ \hat{e}_4 = -.5.$

Yes. We expect $\rho < 0$ because the $\hat{e}_\alpha$'s alternate in sign.

8.16 Define R_p^2 as the squared sample multiple correlation coefficient based upon p explanatory variables in the regression.

$$R_p^2 = 1 - \frac{\hat{u}'\hat{u}}{\sum_1^n (y_j-\bar{y})^2}\ ;\quad \underset{(N\times1)}{y} = \underset{(N\times p)}{X_1}\underset{(p\times1)}{\beta} + u;\ \mathrm{var}(u) = \sigma_1^2 I.$$

Let $\underset{(N\times1)}{z} = (z_i)$, and consider a new model with additional independent observations z_i. The new model becomes

$$\underset{(N\times1)}{y} = (\underset{(N\times p)}{X_1}\ \underset{(N\times1)}{z})\binom{\alpha}{\gamma} + \underset{(N\times1)}{v} \equiv X_2\,\delta + v,$$

$\mathrm{var}(v) = \sigma_2^2 I,\ \delta = (\alpha',\gamma)',X_2 = (X_1,z).$

$$R^2_{p+1} = 1 - \frac{\hat{v}'\hat{v}}{\sum_1^n (y_j-\bar{y})^2}.$$

$$\frac{\hat{v}'\hat{v}}{\Sigma(y_j-\bar{y})^2} = \frac{\hat{v}'\hat{v}}{\hat{u}'\hat{u}} \cdot \frac{\hat{u}'\hat{u}}{\Sigma(y_j-\bar{y})^2} .$$

$$1 - R^2_{p+1} = \frac{\hat{v}'\hat{v}}{\hat{u}'\hat{u}} \left(1-R^2_p\right).$$

Define:

r = the partial sample correlation coefficient between y and z

holding the first p-variables fixed.

$$\frac{\hat{v}'\hat{v}}{\hat{u}'\hat{u}} = 1 - r^2.$$

So $(1-R^2_{p+1}) = (1-r^2)(1-R^2_p) \leq 1-R^2_p$,

or $R^2_p \leq R^2_{p+1}$.

An alternative proof can be carried out using projection operators.

8.17 $\text{tr}(R) = p = \sum_1^p \lambda_j$, so $\bar{\lambda} = 1$. If the average $\lambda_j = 1$, and all $\lambda_j \neq 1$, there must be at least one $\lambda_j > 1$.

8.18 Rewrite the model as:

$Y = XB\Gamma^{-1} + U\Gamma^{-1} \equiv X\pi + V$,

where: $\pi = B\Gamma^{-1}$, $V = U\Gamma^{-1}$.

π can be estimated in the usual way, but unless there is prior information, or special structure on B and Γ, they cannot be estimated separately.

B and Γ can be estimated by three stage least squares, once there are enough constraints for the system to be identified.

8.19 $(y|X,\beta) \sim N(X\beta, \sigma^2 I_N)$; $(\underset{(1\times1)}{y^*}|z,\beta) \sim N(z'\beta, \sigma^2)$.

Let $\tilde{y} = E(y^*) = z'\beta$.

From the proof of Theorem (8.6.6),

$$p(\beta, h|\text{sample}) \propto h^{(\frac{N}{2}-1)} e^{-\frac{gh}{2}},$$

where $g \equiv (N-q)\sigma^2 + (\beta-\hat{\beta})'(X'X)(\beta-\beta)$.

Integrating,

$$p(\beta|\text{sample}) \propto \{(N-q)\hat{\sigma}^2 + (\beta-\hat{\beta})'(X'X)(\beta-\hat{\beta})\}^{-(\frac{N-q}{2})-\frac{q}{2}}.$$

From p. 137, Theorem (6.2.1),

$$p(\tilde{y}|\text{sample}) \propto \{(N-q)\hat{\sigma}^2[z'(X'X)^{-1}z] + (\tilde{y}-z'\hat{\beta})^2\}^{-\nu/2},$$

where: $\nu \equiv (N-q+1)$.

So: $$\left.\frac{\tilde{y}-z'\hat{\beta}}{\hat{\sigma}[z'(X'X)^{-1}z]^{1/2}}\right| \text{sample} \sim t_{N-q}.$$

8.20 $$\underset{(p\times1)}{y_{ijkl}} = \mu + \alpha_i^{(1)} + \alpha_j^{(2)} + \alpha_k^{(3)} + \beta_{ij}^{(1,2)} + \beta_{ik}^{(1,3)} + \beta_{jk}^{(2,3)}$$

$$+ \gamma_{ijk}^{(1,2,3)} + e_{ijkl},$$

where: $i=1,\ldots,I$; $j=1,\ldots,J$; $k=1,\ldots,K$; $\ell=1,\ldots,L$; $L > 1$;

$\alpha_.^{(1)} = \alpha_.^{(2)} = \alpha_.^{(3)} = 0$; $\beta_{i.}^{(1,2)} = \beta_{.j}^{(1,2)} = \beta_{.k}^{(1,3)} = \beta_{i.}^{(1,3)} = 0$;

$\beta_{j.}^{(2,3)} = \beta_{.k}^{(2,3)} = 0$; $\gamma_{.jk}^{(1,2,3)} = \gamma_{i.k}^{(1,2,3)} = \gamma_{ij.}^{(1,2,3)} = 0$.

$e_{ijkl} \sim N(0,\Sigma)$, $\Sigma > 0$, all e_{ijkl} independent.

Define: $G = \underset{ijkl}{\Sigma\Sigma\Sigma\Sigma}[y_{ijkl}-y_{ijk.}][\ldots]'$;

$H_I = JKL\,\Sigma[y_{i\ldots}-y_{\ldots.}][\ldots]'$; etc.

H_0: all $\alpha_i^{(1)} = 0$.

Reject H_0 if

$$z = \frac{|G|}{|G+H_I|} < \text{constant.}$$

Under H_0, $\mathcal{L}(z) = \mathcal{L}(U_{p,q_I,n})$,

where: $q_I = I-1$, $n = IJK(L-1)$.

Other tests are analogous.

8.21 $$L(v|p) = \prod_{i=1}^{n} \prod_{j=1}^{p} p_{ij}^{v_{ij}}$$

Take $g(p) \propto \prod_{j=1}^{p} p_{ij}^{\alpha_j - 1}$.

$$h(p_i|v) \propto \prod_{i=1}^{n} \prod_{j=1}^{p} p_{ij}^{v_{ij} + \alpha_j - 1}.$$

From Section 6.3.2,

$$(p_i|v) \sim D(v_{i1}+\alpha_1, \ldots, v_{i,p-1}+\alpha_{p-1}; v_{ip}+\alpha_p);$$

$$(p_{i1}|v) \sim D(v_{i1}+\alpha_1; \sum_{j=2}^{p} (v_{ij}+\alpha_j)).$$

So: $$E(p_1|v) = \frac{v_1+\alpha_1}{\sum_{j=1}^{p} (v_j+\alpha_j)}.$$

CHAPTER 9

9.1 (a) Variances are: $\hat{\lambda}_1 = 8.13861$, $\hat{\lambda}_2 = 2.29726$, $\hat{\lambda}_3 = 1.86413$.

Principal components are

$z_1 = \alpha_1'x$; $z_2 = \alpha_2'x$; $z_3 = \alpha_3'x$,

where: $\Gamma \equiv (\alpha_1, \alpha_2, \alpha_3)$

$$= \begin{pmatrix} -.5786 & .5592 & .5937 \\ -.6165 & -.7765 & .1306 \\ -.5340 & .2905 & -.7940 \end{pmatrix}.$$

(b) Assume N = 101 is large enough for Theorem (9.3.2) to be applicable.

$10(\hat{\lambda}_1 - \lambda_1) \sim N(0, 2\lambda_1^2)$; so under H, $\hat{\lambda}_1 \sim N(8, 1.28)$.

The largest latent root of $\hat{\Sigma}$ is $\hat{\lambda}_1 = 8.13861$. So we cannot reject H: $\lambda_1 = 8$.

9.2 $$R = \begin{pmatrix} 1.0000 & .4767 & .5165 \\ & 1.0000 & .4824 \\ & & 1.0000 \end{pmatrix}.$$

$\hat{\lambda}_1 = 1.984$; $\hat{\lambda}_2 = .533$; $\hat{\lambda}_3 = .483$.

$$\Gamma \equiv (\alpha_1, \alpha_2, \alpha_3) = \begin{pmatrix} -.581 & .448 & .679 \\ -.568 & -.821 & .056 \\ -.583 & .353 & -.732 \end{pmatrix}.$$

The interpretation is difficult in small samples because we don't have exact distribution theory.

9.3 (a) One appproach would be to use the result on the bottom of p. 315.

Another approach is the following. The latent roots of Σ are: $\hat{\lambda}_1 = \hat{\sigma}^2[1+(p-1)\rho]$, $\hat{\lambda}_2 = \hat{\lambda}_3 = \hat{\sigma}^2(1-\rho)$.

$V \sim W(\Sigma, p, n)$, and if $W = \Gamma V \Gamma$; where $\Gamma = \frac{1}{\sqrt{3}}\begin{pmatrix} e' \\ \hline \\ \hline \end{pmatrix}$, $\Gamma\Gamma' = I$, and the rows of Γ can be anything as long as they are mutually orthogonal, of length one, and the first has equal elements,

$$W \sim W(D,p,n),\ D = \Gamma\Sigma\Gamma' = \begin{pmatrix} \alpha & & \\ & \beta & 0 \\ 0 & & \beta \end{pmatrix},$$

where $\alpha = \lambda_1$, $\beta = \lambda_2 = \lambda_3$.

By MLE, $\hat{\alpha} = \frac{w_{11}}{n}$, $\hat{\beta} = \frac{w_{22}+w_{33}}{2n}$,

where $W = (w_{ij})$. Since $w_{11} = \frac{e'Ve}{3}$,

and $tr(W) = tr(V) = \sum_1^3 w_{jj}$

$\hat{\alpha} = 8.037$, $\hat{\beta} = 2.07$.

(b) Use p. 316, eqn. (9.3.5):

$\frac{10(\hat{\beta}-\beta)}{\beta} \sim N(0,1)$, or β belongs to the confidence interval: (1.73, 2.57).

9.4 Use Theorem (9.3.2)

λ_1: (6.3732, 11.25672)

λ_2: (1.79895, 3.17740)

λ_3: (1.45977, 2.57833).

9.5 $\hat{\lambda}_1 = 10.4236$, $\hat{\lambda}_2 = .8503$, $\hat{\lambda}_3 = .0001$.

$$\Gamma \equiv (\alpha_1, \alpha_2, \alpha_3) = \begin{pmatrix} -.5466 & .7730 & .3221 \\ -.6473 & -.6340 & .4232 \\ -.5313 & -.0228 & -.8469 \end{pmatrix}.$$

$z_i = \alpha_i' x_i, i = 1,2,3.$

9.6 The density contours of a multivariate Student t-distribution are the same as those of a normal distribution with the same first two moments. So the interpretation must be the same.

9.7 For $x_j = (y_j, z_j)'$, $j=1,\ldots,N$, assume

$y_t = \rho y_{t-1} + u_t$, $|\rho| < 1$,

$z_t = \tau z_{t-1} + v_t$, $|\tau| < 1$,

for all $t = 2,\ldots,N$,

$\text{var}(u_t) = \sigma_1^2$, $\text{var}(v_t) = \sigma_2^2$

for all $t = 1,2,3,\ldots,N$. See p. 224 and find

$$\Sigma = \sigma_{ij}, \quad \sigma_{11} = \frac{\sigma_1^2}{1-\rho^2}, \quad \sigma_{22} = \frac{\sigma_2^2}{1-\tau^2},$$

$$\sigma_{12} = \frac{\sigma_1 \sigma_2 \theta}{\{(1-\rho^2)(1-\tau^2)\}^{1/2}}.$$

Step (1) - estimate σ_1^2, σ_2^2, ρ^2, τ^2, θ;

Step (2) - estimate Σ;

Step (3) - find latent roots and vectors of $\hat{\Sigma}$; estimators of ρ, τ are obtained using eqn. (8.3.17).

9.8 g measures dispersion or spread of the λ_j's. Small g implies the latent roots are close to one another, so many roots may be required to account for most of the variance. Large g implies the roots are spread, so a few λ_j's may account for most of the variance.

CHAPTER 10

10.1 Define $Y_i = 1$ (0) if a congressman votes yes (no) on bill i. Assume

$$P\{Y_i = 1, Y_j = 1 | z\} = P\{Y_i = 1 | z\} P\{Y_j = 1 | z\},$$

for $i \neq j$, and adopt the latent class model of Latent Structure Analysis with z a specific latent class, such as an agricultural or liberal block.

10.2 The correlation matrix is readily interpretable; fator loadings become correlations. The disadvantage is that because in small samples we don't have appropriate distribution theory, we must use a scale free estimation procedure, such as MLE, or MINRES.

10.3 MLE factor anlysis enjoys the scale invariance property; it has a sound statistical basis, but gives non-unique solutions, unless we impose artificial constraints. We can test hypotheses and do confirmatory factor analysis.

Principal-Factor Factor Analysis is easy to carry out computationally since we just use the correlation matrix, substitute communalities for diagonal elements, and then do principal components analysis (which has a unique solution for those communalities, but the solution changes if we change the estimted communalities). We don't get Heywood cases of negative specific variance estimates. Finally, we can carry out interval estimation.

10.4 $p_{i_1 \ldots i_K} \equiv P\{Y_1 = i_1, \ldots, Y_K = i_K\}$, $i_1, \ldots, i_K = 0,1$.

Reorder lexicographically:

$q_1 = p_{1\ldots 1}$, $q_2 = p_{1,\ldots,1,0}$, $\ldots q_J = p_{0\ldots 0}$, where $J = 2^K$.

The likelihood is $L(q) = \prod_{j=1}^{J} q_j^{n_j}$, $q \equiv (q_j)$; n_j denotes the number of subjects in category j. $p_{i_1 \ldots i_K} = \sum_{\alpha=1}^{m} \nu_\alpha \lambda_{\alpha i_1} \ldots \lambda_{\alpha i_K}$, where ν_α and $\lambda_{\alpha i_j}$ are the probabilities defined for these classes in the text.

Define: $\phi_{i_1 \ldots i_K;\alpha} \equiv \lambda_{i_1 \alpha} \ldots \lambda_{i_K \alpha}$

$$\underset{(J\times 1)}{q} = \underset{(J\times m)}{\phi} \cdot \underset{(m\times 1)}{\nu}, \quad \nu = (\nu_\alpha), \text{ where: } \Phi = \left(\phi_{i_1,\ldots,i_k;\alpha}\right).$$

Note: the rows of ϕ are lexicographically ordered with the columns obtained by the α-ordering.

Let f(q) denote the priori density of q. If ϕ and ν are independent, a priori, $f(q) = g(\phi)\, h(\nu)$.

Take $h(\nu)$ to be Dirichlet, and the elements of ϕ independent,

$$g(\phi) = \prod_{\alpha=1}^{m} \prod_{i_1=0}^{1} \cdots \prod_{i_K=0}^{1} g^*\left(\phi_{i_1 \ldots i_k;\alpha}\right)$$

$$= \prod_{\alpha} \prod_{i_1} \cdots \prod_{i_k} g_1\left(\lambda_{i_1 \alpha}\right) \ldots g\left(\lambda_{i_k \alpha}\right).$$

As a prior we could take

$$f(q) \propto \prod_{j=1}^{J} q_j^{a_j - 1}, \quad \sum_{j=1}^{J} q_j = 1,\ a_j > 0, \text{ a Dirichlet distribution.}$$

Alternatively, depending upon which parameters were of interest, we could take each $\lambda_{i_j \alpha}$ to be beta distributed,

$$f^*(\nu,\Lambda) = \prod_{k=1}^{m} \nu_k^{b_k-1} \prod_{\alpha=1}^{m} \prod_{i_1=0}^{1} \cdots \prod_{i_K=0}^{1} \lambda_{i_1\alpha}^{\gamma_{i_1\alpha}-1} \cdots \lambda_{i_k\alpha}^{\gamma_{i_k}-1}$$

$$(1-\lambda_{i_1\alpha})^{\delta_{i_1}-1} \cdots (1-\lambda_{i_k\alpha})^{\delta_{i_k}-1}, \text{ for } b_k > 0,\ \gamma_{i_j\alpha} > 0,\ \delta_{i_j} > 0.$$

10.5 An oblique model will be more appropriate when we believe the factors are all interrelated, and each is dependent upon some simpler, second order, factors.

10.6 (a) Use of multiple R^2 as a lower bound;

(b) Use the maximum value of $|r_{ij}|$ in a row or column of the correlation matrix $R = (r_{ij})$.

10.7 There would be no change in the model since

$$E\{(Y_iY_j - EY_iEY_j)|\alpha\} = 0$$

implies, and is implied by

$$P\{Y_i = 1, Y_j = 1|\alpha\} = P\{Y_i = 1|\alpha\}\ P\{Y_j = 1|\alpha\}$$

i.e., uncorrelated implies independent for binary random variables. (See Exercise 3.10(c).)

10.8 Some questions that might be asked are

(1) Would you be willing to live next door to a family of another race (yes or no)?

(2) Would you be willing to marry a person of another race (yes or no)?

(3) Would you be willing to eat with a person of another race on a regular basis (yes or no)?

(4) Would you hire a person of another race to work for you(yes or no)?

(5) Would you work for a person of another race (yes or no)?

(6) Do you like to understand the cultures of other races (yes or no)?

(7) Do you think people of other races are inferior to yours (yes or no)?

<u>Model</u>

Assume $q_j = \sum_{\alpha=1}^{3} \nu_\alpha \lambda_{j\alpha}$, $q_{ij} = \sum_{\alpha=1}^{3} \nu_\alpha \lambda_{i\alpha}\lambda_{j\alpha}$.

Use a sample to estimate the q's, and the q_{ij}'s, and then the ν_α's and the $\lambda_{i\alpha}$'s

10.9 $p_k(\alpha) = P\{$Subject k belongs to latent class α, given his response $\eta_k\}$,

where $\eta_k \equiv (i_1,\ldots,i_k)$, $i_j = 0,1$. By Bayes Theorem,

$p_k(\alpha) \propto P_1\{\eta_k|$Subject k belongs to latent class $\alpha\}$

$\times\, P_2\{$Subject k belongs to latent class $\alpha\}$.

Take $P_2 = \frac{1}{m}$, $P_1 = \lambda_{\alpha i_1}\cdots\lambda_{\alpha i_K}$.

So $\hat{p}_k(\alpha) \propto \frac{1}{m}\lambda_{\alpha i_1}\cdots\lambda_{\alpha i_K}$.

Classify Subject k into the category α with the highest posterior probability $\hat{p}_k(\alpha)$.

$$\hat{p}_k(\alpha) = \frac{\frac{1}{m}\hat{\lambda}_{\alpha i_1}\cdots\hat{\lambda}_{\alpha i_k}}{\hat{\pi}_{i_1\ldots i_k}},$$

where the λ's are estimated by Latent Structure Analysis, the π's are estimated as sample portions. We use (1/m) instead of $(\hat{\nu}_\alpha)$ because we have been advised to do it that way.

10.10 The estimated recruitment probabilities are $\hat{\nu}_1$ = .4750572, and $\hat{\nu}_2$ = .5249428. The estimated latent (conditional) probabilities are $\hat{\lambda}_{ij}$:

Item/ Response	Class 1	Class 2
1 - 1	.9119	.1844
1 - 2	.0881	.8156
2 - 1	.9683	0.0000
2 - 2	.0317	1.0000
3 - 1	0.0000	.9144
3 - 2	1.0000	.0856
4 - 1	.8647	.1509
4 - 2	.1353	.8491

p = degrees of freedom for k = 4, m = 2; so p = 5.

The chi-square statistic is 6.237. Since χ_5^2 (.95) = 11.1, we cannot reject H that a two class model fits the data.

10.11 The Factor Analysis model has continuous factors and continuous observable variables. The latent Structure Analysis has categorical observable variables and categorical factors (classes).

10.12 There are some difficulties with this exercise.

10.13 The advantages of Principal Components Analysis over Factor Analysis are:

(1) as long as the latent roots are distinct, we get unique answers;

(2) there are no issues of post-analysis rotations;

(3) we need not decide whether to select an oblique or an orthogonal model;

(4) we need not assume an underlying linear structure.

10.14 We start with the correlation matrix R. Let

$R = \Lambda\Lambda' + D_\psi$.

Form $\varepsilon \equiv (R - \Lambda\Lambda') - \text{diag}\,(R - \Lambda\Lambda')$.

In the MINRES procedure, choose Λ to minimize the residuals; that is, the sum of squares of the off-diagonal elements of ε. Thus, the factor laoding matrix is estimated by minimizing the residual matrix elements between the observables and the model structure.

10.15 No. Because chances are that the results are an artifact of the model rather than a real effect. We could still use the results to cluster variables in an exploratory analysis, however.

CHAPTER 11

11.1 $U_1 = \alpha_1'X$, $V_1 = \gamma_1'Y$

$U_2 = \alpha_2'X$, $V_2 = \gamma_2'Y$,

where: $\alpha_1 = (.992,.227)'$, $\gamma_1 = (.966,.266)'$,

$\alpha_2 = (-.128,.974)'$, $\gamma_2 = (-.257,.964)'$,

$\lambda_1 \equiv \text{Corr}(U_1,V_1) = .60904$,

$\lambda_2 \equiv \text{Corr}(U_2,V_2) = .49344$.

11.2 The result is the same as in Exercise 11.1, and it is predictable from Lemma (11.2.1). See the Remark on the bottom of page 361.

11.3 (a) $\lambda_1 \neq 0$ (reject H); $-\nu \log \Lambda_1 = 72.399$, $\nu = 97.5$
$\chi_4^2\ (.95) = 9.49$.

(b) $\lambda_2 \neq 0$ (reject H); $-\nu \log \Lambda_2 = 27.205$, $\chi_1^2\ (.95) = 3.84$.

11.4 Adopt the model: $\underset{(p\times1)}{y_{ij}} = \mu + \alpha_i + e_{ij}$; $i = 1,\ldots,q$. Combine the q-groups pairwise and do cononical correlations analysis on each pair. For example, suppose $q = 2$, and let $\hat{\Sigma} \equiv (\hat{\Sigma}_{ij})$, $i,j = 1,2$. Test whether the canonical correlations are zero. If zero, then the two groups have the same α_i; otherwise, not. In the case of general q, repeat the test for all pairs.

11.5 First remove the correlation by modeling it using an autoregressive scheme. For example, if $Y_i \equiv Y(t_i)$, and a first order lag will suffice,

$$\hat{Y}_i = \hat{\rho} Y_{i-1},$$

and we can use adjusted data

$$(Y_1^*, Y_2^*, \ldots, Y_N^*) \equiv (Y_1, \hat{\rho} Y_1, \hat{\rho}^2 Y_1, \ldots, \hat{\rho}^{N-1} Y_1).$$

Form the covariance matrix

$$S^* = \frac{1}{N} \sum_1^N (Y_i^* - \overline{Y}^*)(Y_i^* - \overline{Y}^*)',$$

and carry out a canonical correlations analysis. Do the analogous thing for higher order lagged structures.

11.6 A multiple correlation coefficient is the special case of a canonical correlation coefficient in which one of the two sets of variates has dimension unity.

11.7 Define $Y(t): p \times 1$ as the p-vector of the p-economic variables that comprise the index, at time t. Define

$$X(t) = [Y(t)', Z(t)']',$$

where: $t = 1, \ldots N$, and $Z(t) \equiv Y(t+N)$ so that there are a total of 2N observation points. Compute the covariance matrix

$$\hat{\Sigma} = \frac{1}{N} \sum_{t=1}^N [X(t) - \overline{X}][X(t) - \overline{X}]',$$

for $\overline{X} = \frac{1}{N} \sum_{t=1}^N X(t)$. Partition $\hat{\Sigma}$ as

$$\Sigma = \begin{pmatrix} \hat{\Sigma}_X & \hat{\Sigma}_{XY} \\ & \hat{\Sigma}_Y \end{pmatrix}.$$

Compute the largest canonical correlation ρ, and test it for significance, i.e., test $H: \rho = 0$.

11.8 They are complementary to one another in that both are expressible in zonal polynomial form (see Section 6.6). Define $\hat{\Lambda} \equiv (\hat{\lambda}_1^2,\ldots,\hat{\lambda}_{p_1}^2)$, and $\Lambda \equiv (\lambda_1^2,\ldots,\lambda_{p_1}^2)$. Then, $(\hat{\Lambda}|\Lambda)$ follows the same distribution as $(\Lambda|\hat{\Lambda})$, with all of the variables interchanged, where $(\hat{\lambda}_j,\lambda_j)$ are sample and population canonical correlation coefficients, respectively. (See Geisser, 1965a.)

11.9 Canonical correlations are generalizations of Pearson product-moment correlations and are therefore intended to be applied to continuous, normally distributed data. When the data are discrete, the interpretation as a measure of departure from linearity becomes less tenable and highly questionable. When the data are categorical it is difficult to know how to interpret linear combinations of variates.

11.10 In generalized (seemingly unrelated) regression the generalized variance of the estimated coefficient vector in any equation is inversely proportional to the squared canonical correlations between the design matrices in pairs of equations (see eqn. (8.5.15)). Thus, if X_j denotes the design matrix in equation j, j = 1, 2, if $X_1'X_2 = 0$, the canonical correlation is zero.

This is the case of greatest estimation efficiency (by the Zellner method) since then, the generalized variance is at a minimum. X_1 and X_2 become closer, so that $X_1'X_2 \approx X_1'X_1$, the canonical correlation becomes approximately unity, and the generalized variance becomes very large (creating a decreasingly useful coefficient estimator.

Chapter 12

12.1 $C = \Sigma^{-1}e/e'\Sigma^{-1}e = (4/7,2/7,1/7)$.

12.2 $C = (p^{-1},p^{-1},p^{-1})' = (1/3,1/3,1/3)'$.

See example (12.4.2).

12.3 $C = (2/11,3/11,6/11)'$.

12.4 Use equation (12.5.6). The non-negativity constraints are not satisfied, so convex programming must be used for the tabacco securities in this portfolio.

For the oils $C_2 = (0,.1887)'$, and for the utilities,

$$C_3 = (.2264,.1132)'.$$

12.5 An "unrestricted" efficient portfolio is one in which assets are selected so as to maximize expected return while minimizing risk, subject to various constraints. A "restricted" efficient portfolio is one in which we have already selected the assets for the portfolio; we ignore expected return and minimize risk subject to constraints.

12.6 Because $0 < \alpha < 1$, the risk function is neither convex nor concave. So to maximize median return while minimizing risk is a problem of finding a local extreme value instead of a global extreme value. This is difficult to accomplish in a high dimensional problem and the distributional behavior is difficult to interpret.

12.7 Model I: all assets jointly follow a symmetric stable distribution with the same characteristic exponent.

Model II: there are K groups of assets, and each group is like Model I, but the characteristic exponents across groups are not generally equal.

12.8 We ignore expected return and selection of assets, but we merely decide how to allocate our resources among the predesignated assets so as to minimize risk.

12.9 Evaluate the sample cdf, $F_n(x)$, and use it to approximate

$$P\{X > x\} \simeq 1 - F_n(x) \simeq \frac{\text{constant}}{x^{\alpha}} .$$

Then plot log P{X > x) against log x. The slope of the line is an estimate of α. (See p. 387, top; and also Mandelbrot, 1964.)

Chapter 13

13.1 Assume the parameters are known. See Example (13.2.1), eqn. (13.2.3). Classify z into π_1 if:

$$Q(z) \equiv (z-\phi)'\Sigma_2^{-1}(z-\phi) - (z-\theta)\Sigma_1^{-1}(z-\theta) \geq K,$$

where: $K \equiv 2 \log\left\{\frac{|\Sigma_1|^{1/2}}{|\Sigma_2|^{1/2}} \cdot \frac{p_2}{p_1} \cdot \frac{c_{12}}{c_{21}}\right\}$.

13.2 Take $p_1 = p_2$, $c_{12} = c_{21}$. $K = 2.6532$.

$Q(z_1) = 24.7324$; $Q(z_2) = 80.6197$

$Q(z_3) = 47.2394$; $Q(z_4) = -6.8451$.

Classify z_1 into π_1; classify z_2 into π_1;

classify z_3 into π_1; classify z_4 into π_2.

13.3 Use Theorem (13.3.1); p. 402. $L_{12} = .2252$.

Define:

$$R(z) \equiv \frac{p(z|\text{data},\pi_1)}{p(z|\text{data},\pi_2)}.$$

Classify z into π_1 if $R > 1$; otherwise into π_2.

$$R(z) = \frac{(.2252)\left\{1+\frac{19}{399}(z-(6,6)')'\begin{pmatrix}4 & -2\\ -2 & 6\end{pmatrix}(z-(6,6)')\right\}^{10}}{\left\{1+\frac{(10)(9)}{(99)(71)}(z-(4,7)')'\begin{pmatrix}8 & -3\\ -3 & 10\end{pmatrix}(z-(4,7)')\right\}^{5}}.$$

13.4 $R(z_1) = 143.80$ so classify z_1 into π_1.

$R(z_2) = 575{,}959.73$, so classify z_2 into π_1.

$R(z_3) = 6928.33$, so classify z_3 into π_1.

$R(z_4) = .1829$, so classify z_4 into π_2.

13.5 Results are the same. The Bayesian procedure is preferable because

(a) the sample size is small and the classical procedure is really only applicable for large samples; the Bayesian procedure is generally applicable;

(b) the Bayesian approach requires fewer assumptions;

(c) the Bayesian approach provides a predictive odds ratio, while the classical procedure provides only conclusions.

13.6 For K = 5, N = 239, and eqn. (13.4.3), the discriminatory power Q becomes $Q = \frac{(nK-N)^2}{N(K-1)}$.

The number of correct classifications, n, should increase (so Q should increase). The Confusion matrix would become more diagnonal looking.

13.7 Select a set of p-discriminator words. Let x_{ij}: (px1) denote the p-vector of: frequency-of-use-of each of the p-words in sample j, for author-i; i=1,2; j=1,...N_i. Sample j is the j<u>th</u> journal article. We could then assume equal misclassification costs; equal prior probabilities (or we could take)

$p_1 = (N_1)/(N_1+N_2)$, $p_2 = (N_2)/N_1+N_2))$.

Let z denote the vector of word frequencies of use for each of the p-words in the article in question. Use Theorem (13.3.1) to decide authorship, assuming normality of the data (if it is non-normal, it could be transformed to normality, or we could use the actual distributions).

13.8 Let $p(z) \equiv P\{E|z\}$, where E denotes the event that observation z: $p \times 1$ belongs to $\pi_1 \equiv N(\theta,\Sigma_1)$; $\bar{E}$ denotes the event that z belongs to $\pi_2 \equiv N(\phi,\Sigma_2)$. Assume sample sizes are large; prior probabilities are equal; $\Sigma_1 \neq \Sigma_2$. By Bayes theorem,

$$p = \frac{p(z|E)p(E)}{p(z|E)p(E) + p(z|\bar{E})p(\bar{E})} .$$

Then,

$$p(z) = \frac{1}{1 + 3.7683 \exp\{(-1/2)Q(z)\}} ,$$

where: $Q(z) = (z-\phi)'\Sigma_2^{-1}(z-\phi) - (z-\theta)'\Sigma_2^{-1}(z-\theta)$.

$p(z)$ is a monotone increasing function of Q.

So classify z into π_1, if $p(z) > .5$.

$p(z_1) = .99984$; classify z_1 into π_1;

$p(z_2) \simeq 1$; classify z_2 into π_1;

$p(z_3) \simeq 1$; classify z_3 into π_1;

$p(z_4) = .00858$; so classify z_4 into π_2.

The results are identical with those in Exercises (13.2) and (13.4).

13.9 Let $z(t) \equiv [v(t),\theta(t)]'$, $t = 1,\ldots,N$, where: $v(t)$ denotes speed at time t, and $\theta(t)$ azimuth angle at time t. Assume $z(t) \sim N(\phi,\Sigma)$; (ϕ,Σ) unknown. If the aircraft is friendly, $(\phi,\Sigma) = (\phi_1,\Sigma_1)$: if the aircraft is unfriendly, we could take $(\phi,\Sigma) = (\phi_2,\Sigma_2)$, or $(\phi,\Sigma) \neq (\phi_1,\Sigma_1)$.

We could classify the unknown aircraft at each time point, and we could augment the data base using the results of each classification. Use a sequential classification rule (see Geisser, 1966).

13.10 The Bayesian procedure permits $\Sigma_1 \neq \Sigma_2$, it does not require large samples, and a complete predictive distribution is found, instead of just a classification decision. See also Exercise 13.5.

13.11 If the continuous components are normal and the other components are binary, it can be shown that the implied classification procedure follows from logistic regression, but classical procedures are inappropriate. Moreover, logistic regression provides danger signals to warn of pathological data. See also Press and Wilson, 1978.

13.12 Use a different facial characteristic to represent each component of the symptom vector. Try to determine by inspection the face the one in question most closely resembles (each face corresponds to a possible disease the patient might have).

Chapter 14

14.1 Use eqn. (14.2.17), take $x_T \equiv (x_T^*, 1)'$; $w \equiv (x_{T+1}^*, 1)' \equiv (w_1, w_2)'$.

$w_1^* = 106.8$.

14.2 Take J = 0. Then,

$$w_1^* = [\hat{B}_1' G \hat{B}_1 + (\frac{tr(SG)}{\nu-2}) D^{11}]^{-1}$$

$$\cdot [\hat{B}_1' G(a - \hat{B}_2 w_2) - (\frac{(tr(SG)}{\nu-2}) D^{12} w_2].$$

14.3 Use eqn. (14.2.16). Note errors on pages 420, 421,, Section 14.2.3, that J is (qxq), rather than (pxp); also in Example (14.2.1), G and J are identity matrices, but G ≠ J. In this exercise, J: (2x2), $\rho = 4968.86$.

14.4 While this exercise can be started, it is not straightforward to carry it to completion.

14.5 In this exercise, $\hat{B} = \begin{pmatrix} 3 & 4 \\ 2 & 1 \\ 1 & 2 \end{pmatrix}$, and we are to evaluate the optimal control vector.

$$w_1^* = \frac{1}{53.6234} \{11a_1 + 7a_2 + 5a_3 + x_T(1) - 121.688\}.$$

Chapter 15

15.1 Take $\hat{d}_{ij} \leq 2d_{ij}$, all i,j. Let

$$z = \frac{(D-\hat{D})'(D-\hat{D})}{D'D}, \quad \underset{(M\times 1)}{D} = (d_{1,1},\ldots,d_{1,M})'.$$

Let $D = (a_i)$, so that $z = \sum_i (a_i - \hat{a}_i)^2 / \sum_i a_i^2$.

But $\hat{a}_i \leq 2a_i$, $\hat{a}_i - a_i \leq a_i$, so $(a_i - \hat{a}_i)^2 \leq a_i^2$.

So $z \leq (\Sigma a_i^2 / \Sigma a_i^2) = 1$.

15.2 We should get different results since rank correlations and average ranks are generally not linearly related. Using average ranks should approximate results obtained by using individual judges (see p. 429 and the reference to Klahr). Correlational information ignores certain relationships present in the original preferences. Rank correlation is less affected by idiosyncrasies of judges.

15.3 It would be unaffected since stress is unaffected by location and scale changes.

15.4 Yes. Use minimum variance within groups as the clustering criterion at each hierarchical stage. See p. 443; also Ward, 1963.

15.5 Same as Exercise 15.10.

15.6 Let $X \equiv (x_1, \ldots x_N)$, x_j: (px1): the N x_j's are assumed to be independent.

Clustering by attribute is to cluster the N points in p-space; i.e., cluster the columns of X.

Clustering by subject is to cluster the p-points in N-space, which is to cluster the rows of X. These vectors are correlated.

15.7 In situations where there is a natural nesting we are seeking, such as in taxonomical problems. It is also useful when we don't have a reasonable measure of homogeneity within groups.

15.8 See Torgerson's "Complete Method of Triads" for obtaining comparative distances between stimuli (Torgerson, 1958, pp. 263-268).

15.9 See Shepard et al. 1972. Also, P. E. Green and V. R. Rao, 1971, Jour. of Marketing Research, 355-363, "Conjoint Measurement for Quantifying Judgemental Data".

15.10 (See also Exercise 15.5.) If the parameters are all known we can use discrimination procedures on each data vector i.e., assume there are K known populations $\pi_j \equiv N(\theta_j, \Sigma)$, $j = 1,\ldots,K$. Then use eqn. (13.2.7).

If the parameters are unknown, evaluate, $d_{ij} = (z_i - z_j)'(z_i - z_j)$ for all $N(N-1)/2$ pairs (z_i, z_j), $i,j = 1,\ldots N$. Next, rank the d_{ij}'s and assume the smallest d_{ij} has z_i and z_j in the same population. Let that be π_1.

Compute (θ_1, Σ) for these two z_i's (as the sample mean and covariance matrix). Now examine the next ranking d_{ij}. If it includes one of the first two z_i's test the new z_i to see if it belongs to π_1. Let z^* denote the new z_i. Compute

$\delta = (z^*-\hat{\theta}_1)'\hat{\Sigma}^{-1}(z^*-\hat{\theta}_1)$.

If z^* belongs to π_1, $\delta \sim \chi_p^2$. We can test the hypothesis that z^* belongs to π_1. If we accept the hypothesis, we merge z^* into π_1 and repeat the procedure for the next ranking d_{ij}. If we reject, assume $z^* \sim N(\theta_2,\hat{\Sigma})$. Now examine another z_i and see if it belongs to π_1 or π_2, or neither, where

$\pi_2 \equiv N(\hat{\theta}_2,\hat{\Sigma})$, $\hat{\theta}_2 = z^*$.

If neither, this defines $\pi_3 \equiv N(\hat{\theta}_3,\hat{\Sigma})$. If it belongs to π_1 or π_2, merge it into the population and recompute the parameter estimates; etc.

Errata

Page 256 Eqn. (8.6.23) should be:

$$p(\boldsymbol{\beta},\boldsymbol{\Sigma}|\boldsymbol{Y},\boldsymbol{X}) \propto \frac{\exp\left\{\left(-\frac{1}{2}\right) tr\left[\boldsymbol{\Sigma}^{-1}(\boldsymbol{V}+\boldsymbol{G})\right]\right\}}{|\boldsymbol{\Sigma}|^{(N+m)/2}} \bullet \exp\left\{\left(-\frac{1}{2}\right)(\boldsymbol{\beta}-\overline{\boldsymbol{\beta}})'\boldsymbol{M}(\boldsymbol{\beta}-\overline{\boldsymbol{\beta}})\right\}$$

$$\bullet \exp\left\{\left(-\frac{1}{2}\right)(\hat{\boldsymbol{\beta}}-\boldsymbol{\varphi})'(\boldsymbol{\Sigma}^{-1}\otimes \boldsymbol{X'X})\left[\boldsymbol{\Sigma}^{-1}\,\text{Ä}\,\boldsymbol{X'X}+\boldsymbol{F}^{-1}\right]^{-1}\boldsymbol{F}^{-1}(\hat{\boldsymbol{\beta}}-\boldsymbol{\varphi})\right\}, \quad (8.6.23)$$

Page 256 Eqn. (8.6.24) should be:

$$p(\boldsymbol{\Sigma}|\boldsymbol{Y},\boldsymbol{X}) \propto \frac{\exp\left\{\left(-\frac{1}{2}\right) tr\left[\boldsymbol{\Sigma}^{-1}(\boldsymbol{V}+\boldsymbol{G})\right]-\boldsymbol{H}\right\}}{|\boldsymbol{\Sigma}|^{(N+m)/2}\left|\boldsymbol{F}^{-1}+\boldsymbol{\Sigma}^{-1}\otimes\boldsymbol{X'X}\right|^{1/2}}, \quad (8.6.24))$$

where

$$\boldsymbol{H} \equiv \left(\frac{1}{2}\right)(\hat{\boldsymbol{\beta}}-\boldsymbol{\varphi})'(\boldsymbol{\Sigma}^{-1}\text{Ä}\boldsymbol{X'X})\left[\boldsymbol{\Sigma}^{-1}\otimes\boldsymbol{X'X}+\boldsymbol{F}^{-1}\right]^{-1}\boldsymbol{F}^{-1}(\hat{\boldsymbol{\beta}}-\boldsymbol{\varphi}).$$

A CATALOG OF SELECTED
DOVER BOOKS
IN SCIENCE AND MATHEMATICS

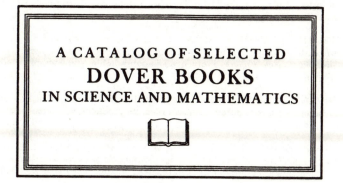

Mathematics

FUNCTIONAL ANALYSIS (Second Corrected Edition), George Bachman and Lawrence Narici. Excellent treatment of subject geared toward students with background in linear algebra, advanced calculus, physics, and engineering. Text covers introduction to inner-product spaces, normed, metric spaces, and topological spaces; complete orthonormal sets, the Hahn-Banach Theorem and its consequences, and many other related subjects. 1966 ed. 544pp. 6⅛ x 9¼. 40251-7

ASYMPTOTIC EXPANSIONS OF INTEGRALS, Norman Bleistein & Richard A. Handelsman. Best introduction to important field with applications in a variety of scientific disciplines. New preface. Problems. Diagrams. Tables. Bibliography. Index. 448pp. 5⅜ x 8½. 65082-0

VECTOR AND TENSOR ANALYSIS WITH APPLICATIONS, A. I. Borisenko and I. E. Tarapov. Concise introduction. Worked-out problems, solutions, exercises. 257pp. 5⅝ x 8¼. 63833-2

THE ABSOLUTE DIFFERENTIAL CALCULUS (CALCULUS OF TENSORS), Tullio Levi-Civita. Great 20th-century mathematician's classic work on material necessary for mathematical grasp of theory of relativity. 452pp. 5⅝ x 8¼. 63401-9

AN INTRODUCTION TO ORDINARY DIFFERENTIAL EQUATIONS, Earl A. Coddington. A thorough and systematic first course in elementary differential equations for undergraduates in mathematics and science, with many exercises and problems (with answers). Index. 304pp. 5⅜ x 8½. 65942-9

FOURIER SERIES AND ORTHOGONAL FUNCTIONS, Harry F. Davis. An incisive text combining theory and practical example to introduce Fourier series, orthogonal functions and applications of the Fourier method to boundary-value problems. 570 exercises. Answers and notes. 416pp. 5⅜ x 8½. 65973-9

COMPUTABILITY AND UNSOLVABILITY, Martin Davis. Classic graduate-level introduction to theory of computability, usually referred to as theory of recurrent functions. New preface and appendix. 288pp. 5⅜ x 8½. 61471-9

ASYMPTOTIC METHODS IN ANALYSIS, N. G. de Bruijn. An inexpensive, comprehensive guide to asymptotic methods–the pioneering work that teaches by explaining worked examples in detail. Index. 224pp. 5⅜ x 8½ 64221-6

APPLIED COMPLEX VARIABLES, John W. Dettman. Step-by-step coverage of fundamentals of analytic function theory–plus lucid exposition of five important applications: Potential Theory; Ordinary Differential Equations; Fourier Transforms; Laplace Transforms; Asymptotic Expansions. 66 figures. Exercises at chapter ends. 512pp. 5⅜ x 8½. 64670-X

INTRODUCTION TO LINEAR ALGEBRA AND DIFFERENTIAL EQUATIONS, John W. Dettman. Excellent text covers complex numbers, determinants, orthonormal bases, Laplace transforms, much more. Exercises with solutions. Undergraduate level. 416pp. 5⅜ x 8½. 65191-6

Physics